Applied SEISMOLOGY

Applied SEISMOLOGY

A Comprehensive Guide to Seismic Theory and Application

By
Mamdouh R. Gadallah
and
Ray L. Fisher

PennWell®

by PennWell Corporation
1421 South Sheridan/P.O. Box 1260
Tulsa, Oklahoma 74112/74101

800.752.9764
+1918.831.9242
sales@pennwell.com
www.pennwellbooks.com
www.pennwell.com

Managing Editor: Marla Patterson
Production Editor: Sue Rhodes Dodd
Book and Cover Designer: Clark Bell

Library of Congress Catataloging in Publication Data

Gadallah, Mamdouh R.
Applied seismology : a comprehensive guide to seismic theory and application / by Mamdouh R. Gadallah, Ray L. Fisher.--1st American ed.
p. cm.
Includes bibliographical references.
ISBN 1-59370-022-9
1. Seismic prospecting. 2. Seismic reflection method--Data processing 3. Seismic refraction method--Data processing. 4. Seismology. 5. Geophysics--Mathematics. I. Fisher, Ray L. II. Title.

TN269.8.G32 2004
622'.1592--dc22

2004009843

Printed in the United States of America

1 2 3 4 5 09 08 07 06 05

We wish to dedicate this book to our wives—Jean Gadallah and Ileaine Fisher. It was because of their understanding that we were able to spend the necessary hours writing this book. Further, their encouragement in this endeavor was a giant factor in our completing it. So, we dedicate this book, with love and gratitude, to our wives.

Contents

Caption and Courtesy List

Preface

After spending many years working in the earth sciences, we became aware that there was no single book that covered all aspects of petroleum exploration, development, and exploitation. It is the aim of this book to fill that perceived need.

The downturn of the industry over the past two decades, the reduction of the work force, and the drop in earth science--student enrollments in colleges and universities have produced a gap between experienced and inexperienced professionals. We know from the history of the petroleum industry that there can be a sudden increase in the need for professionals in our industry. In such an event, there will be a need to rapidly train and upgrade the professional level of geophysicists, geologists, engineers, managers, and others in our industry.

We also see a need for a book that is practical as well as theoretical, easily understood, and comprehensive. We see the inclusion of workshops as another positive of this book. These allow the students to review and reinforce the concepts, theory and practical applications provided in the book.

Mamdouh R. Gadallah
Ray L. Fisher
Houston, Texas

Acknowledgments

We are grateful for all who so kindly allowed us to use some of their illustrations in our book. Specifically, we thank:

Dr. Luis Canales, Vice-President of WesternGeco
American Association of Petroleum Geologists (AAPG)
Society of Economic Paleontologists and Mineralogists (SEPM)
Seismograph Services Corporation
CGG of America

Special thanks to the Exxon explorationists and geoscientists who developed the new theories of seismic sequence analysis. Their efforts allowed us to reach an understanding of the valuable theories and ideas found in this book. We regret it if any of those people were not duly credited.

We particularly want to acknowledge the help and encouragement we received from our former manager, and still good friend, Dr. Quentin Spradling.

We also wish to thank many others, too numerous to name, who through the years have shared their knowledge and expertise with us. It is only because of these people that we are able to make this book so comprehensive.

Mamdouh R. Gadallah
Ray L. Fisher
Houston, Texas

1 Overview and Summary

Introduction

Until 1859, petroleum exploration was a rather simple and straightforward procedure. One simply looked for oil seepage at the surface—particularly near streams and from oil springs. Petroleum was used principally for medicinal purposes at that time, so the approach yielded a sufficient supply to meet demands.

In 1859, Colonel E. L. Drake completed the first successful well, drilled specifically for oil (although wells drilled earlier for other purposes had yielded oil). Actually, Drake used the early method of petroleum exploration since his well was located near a known oil seep along Oil Creek in western Pennsylvania. Soon, there were many wells being drilled up and down Oil Creek.

These early successes led to an exploration method often called *creekology* in which accumulations of oil were associated with low spots along and near streams. Hills and plateaus were not considered suitable drilling sites.

Trendology was another early exploration method arising from the observation that oil pools and fields frequently occurred along almost straight lines for many miles. In other words, after early discovery of two or more fields, lines connecting these were extended in both directions and wells located along the line. Actually, this is a relatively sound procedure, which is still used under certain conditions. Locating wells near oil seeps is also a good method.

As more data accumulated, geologists were able to develop scientific theories concerning oil and gas accumulation. These theories led to practical application and discovery of petroleum deposits from observations of surface geology. Observations of surface rocks and landforms enabled trained geologists to infer whether proper conditions for petroleum accumulation existed at depths of hundreds or thousands of feet.

The invention of the internal combustion engine near the beginning of the 20th century greatly increased the demand for petroleum. Likewise, drilling activity increased and was extended to a wider geographic area and to greater depths. As a result, surface geology in many cases was no longer a reliable guide to the location of drilling sites.

A number of scientifically based petroleum exploration methods have been developed over the years. Where drilled wells are available, various kinds of well surveys are made by lowering instruments into the borehole to obtain directly measured data for analysis to determine the kinds of subsurface rocks present and to identify their various properties.

Other methods yield data useful in assessing the probability of petroleum accumulation without the necessity of drilling to very great depths. Among these are observations of magnetic anomalies caused by local variations in magnetic properties of subsurface rocks, observations of anomalies in gravitational attraction due to variations in rock densities, and observations of seismic waves (earth vibrations or earthquake waves) that propagate through subsurface rocks and return to the surface.

The seismic method, particularly the form called *seismic reflection*, usually yields the most precise and accurate results. It is by far the most popular geophysical exploration method and the one this book will investigate.

Seismic Refraction Exploration Overview

By the end of the 19th century, earthquake seismologists brought seismic theory to a high degree of development. Studies of earthquake waves led to a very good understanding of the internal structure of the earth as a whole, and a rather detailed model of the crust in Europe resulted from study of refracted arrivals from earthquakes. Interest in finding new deposits of petroleum resulted in application of seismic refraction techniques to the problem.

The refraction method was quite successful in locating salt domes, first along the Baltic coast of Germany and later along the Gulf of Mexico coastal regions in the United States. Very large charges of dynamite were used to generate the seismic waves over rather long distances with adequate amplitudes to be recorded by the primitive instruments of the day.

Although refraction data can be analyzed and interpreted rather easily, it has a fatal flaw—it cannot give accurate results unless the seismic velocity in each rock layer is higher than the one above it. An exception to this situation, a *velocity inversion*, causes incorrect layer velocities and thicknesses to be calculated.

Seismic Reflection Exploration Overview

During World War I, the Germans had a huge cannon (nicknamed Big Bertha) capable of firing shells from Germany to Paris, France. When the cannon was fired, its recoil was so intense that it produced seismic waves that could be recorded in Paris. Since the seismic waves traveled faster than the cannon shells, the French were able to warn Paris residents before the shell arrived (Fig. 1–1).

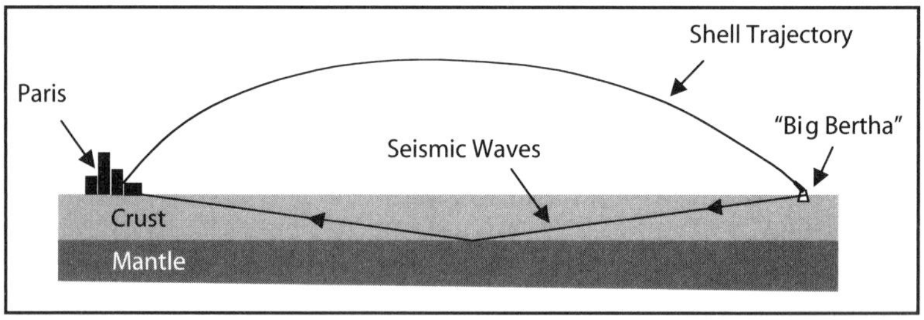

Fig. 1–1 *Seismic Waves from Big Bertha's Firing*

Dr. Clarence Karcher was stationed in France during this time and was intrigued by this application of seismic wave recording. After the war, Dr. Karcher began experimenting with reflected seismic waves when he was a professor of physics at the University of Oklahoma. His experiments demonstrated that

- seismic waves reflected from interfaces between rock layers could be recorded
- the times it took for these seismic waves to return to the surface after reflection could be used to determine their depths and attitudes

Dr. Karcher later formed the first seismic reflection exploration company, Geophysical Research Corporation, which became a training ground for the industry.

The seismic reflection method is quite simple in concept. It requires

- a source to generate seismic waves (explosives only until around 1950)
- seismic receivers to detect seismic waves and convert them to electrical signals (seismometers or geophones)
- instruments to record amplitude and elapsed time between firing the source and arrival of the reflection

Many things have changed over the years since Dr. Karcher's pioneering work.

- Explosives are no longer the only choice for the energy sources. Vibrators are most often used on land and airguns for marine surveys.
- Detectors (geophones on land, hydrophones for marine) have become smaller and more powerful.
- Recording has progressed from analog recording on photographic paper to digital recording on magnetic tape.
- Processing has evolved from a few manual operations to sophisticated software applications on computers.

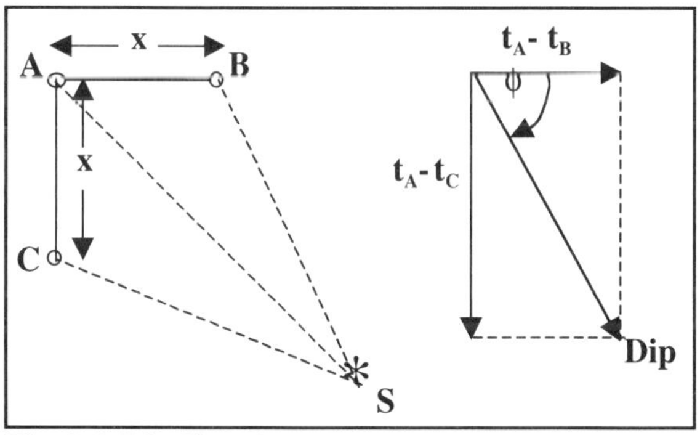

Fig. 1–2 *Dip Shooting*

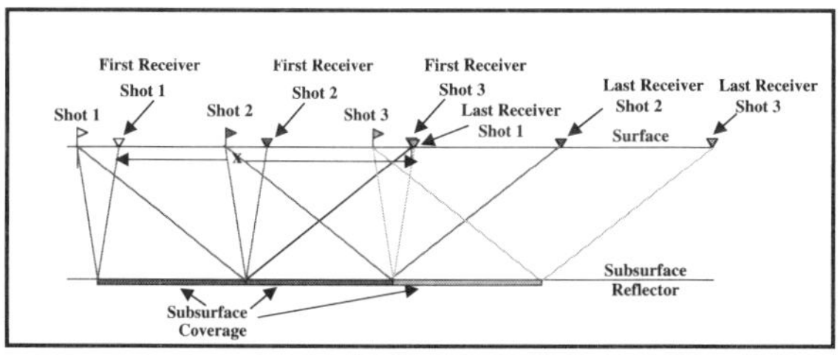

Fig. 1–4 *Continuous or Single-fold Subsurface Coverage*

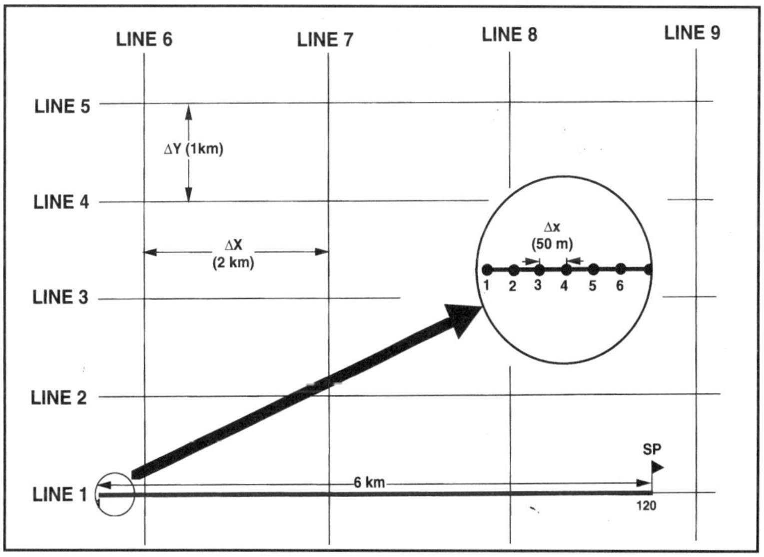

Fig. 1–3 *Typical 2-D Seismic Geometry*

Seismic acquisition techniques have also undergone many changes. The earliest technique was called *dip shooting.* Detectors were laid out in an L-shaped pattern with receivers at *A*, *B*, and *C* and a source at *S* (Fig. 1–2). Reflection times at each location; $t_A < t_B$, and t_C, were measured and time differences t_A- t_B and t_A- t_C were then plotted. Their vector sum gave dip, in seconds (sec) or milliseconds (ms), and dip direction, ϕ.

Dip shooting gave way to continuous (or single-fold) coverage in which receivers were laid out along intersecting lines (Fig. 1–3). For a given shot, receivers were laid out over a distance *x* that gave subsurface coverage of *x*/2. For the next shot, source point and receivers were moved a distance *x*/2 along the line (Fig. 1–4).

Moving the *spread* (source plus receivers) one-half spread length between shots produced continuous subsurface coverage whereas dip shooting left gaps in subsurface coverage. Later, a method called multi-fold or *common mid-point* (CMP) *shooting* was developed. In this method, the spread is moved less than one-half spread length resulting in overlap of subsurface coverage. Moving one-fourth spread length means that the same reflections are recorded by two different shots at two different receivers at two different shot-to-receiver distances but the midpoint between shots and receivers is the same! This is called *2-fold shooting*. Increasing the overlap increases the fold as shown in the following examples:

- move-up of 1/6 spread gives 3-fold
- move-up of 1/8 spread gives 4-fold
- move-up of 1/12 spread gives 6-fold

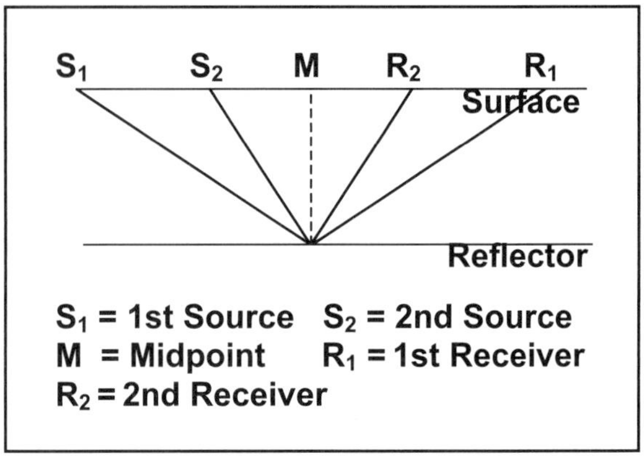

Fig. 1–5 *CMP Shooting*

CMP shooting was not really practical at the time it was introduced because analog recording was still being used. Figures 1–5 and 1–6 illustrate CMP shooting.

The method described to this point is now called 2-D shooting. In 2-D shooting, one is forced to assume that all reflections are recorded in a vertical plane below the surface spread, which often is not true. The rather coarse sampling of the subsurface limits the ability to detect and correctly map subsurface features. Figure 1–7 gives an example of a subsurface feature that is very difficult to accurately map with 2-D data but can be with 3-D data.

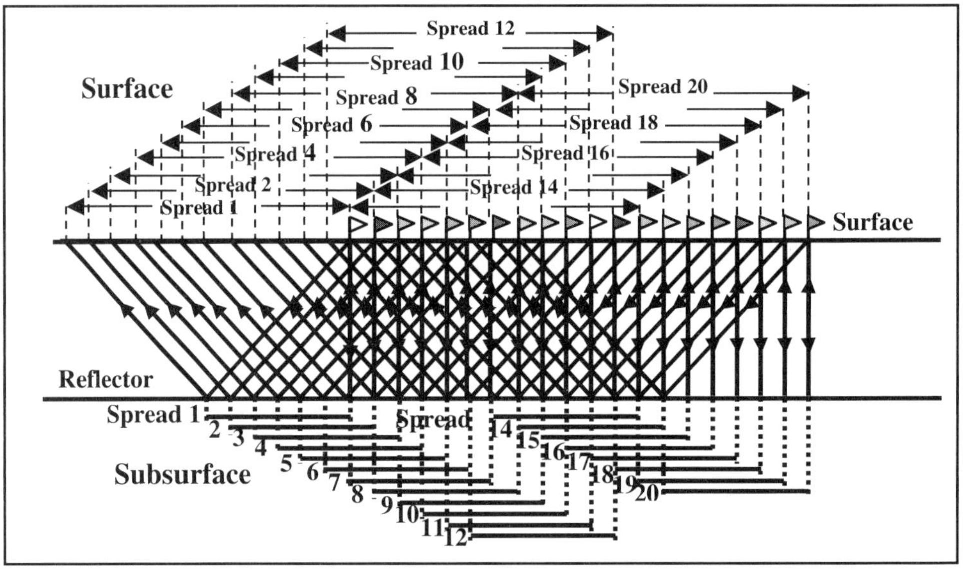

Fig. 1–6 *Multi-fold Shooting*

3-D shooting gives much finer sampling of the subsurface by placing receiver lines much closer together and by having sources that are not generally in line with the receivers. (Fig. 1–8).

Similar advances occurred in marine shooting where, instead of a single streamer being towed, as many as 12 have been used.

Project coordination. On large jobs, it is advisable to have one overall project manager/supervisor who is responsible for the following:

- technical/geophysical coordination
- operational coordination
- client contacts/communications
- planning

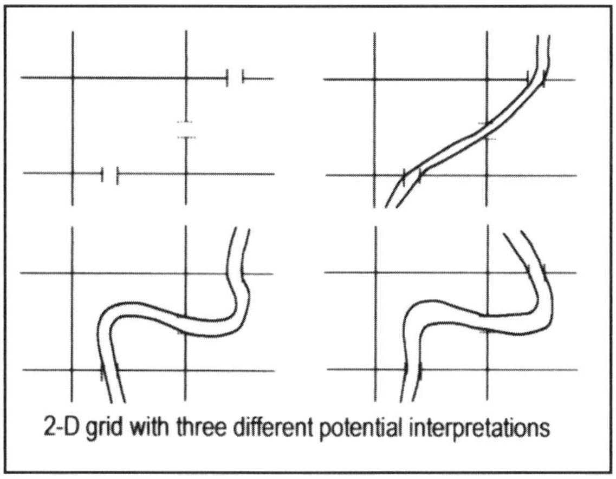

Fig. 1–7 *Meandering Stream Channel*

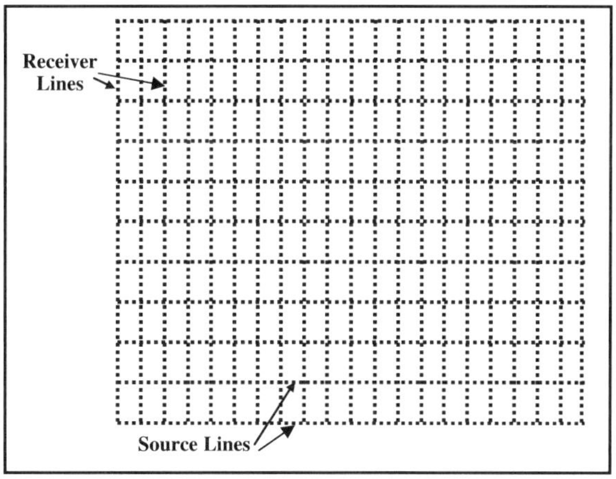

Fig. 1–8 *3-D Prospect Layout*

This person should be involved from the start of the job. Survey designs are normally completely specified long before recording starts, but operational factors (weather and other considerations) may change. Technical and/or cost improvements can frequently be made by intelligent personnel scheduling, equipment use, helicopter flight and load scheduling, and recovery shot planning. Major steps in seismic reflection exploration include

- pre-planning
- data acquisition
- data processing
- interpretation

Pre-planning. The objective here is to determine the economic viability of the proposed project and, if viable, design the best survey possible with available funds to satisfy the geologic and geophysical requirements. Another purpose is to anticipate problems and develop solutions for them, before work has started.

Steps in the preplanning stage include the following:

- select primary and secondary targets
- estimate their potential return
- determine survey main objectives
- establish priorities for all objectives
- document both objectives and priorities.
- allocate acquisition and processing budgets based on
    - data quantity and quality specifications
    - reasonable schedules and deadlines
- locate and modify lines of survey as required by
    - natural and man-made obstructions
    - environmental and permit restrictions
- specify acquisition methods and equipment types
- determine acquisition parameters

Geologic objective. The geologic objective must be described with regard to trap type (structural, stratigraphic, or combination); depth, thickness, and areal extent; and maximum dip expected. In addition, regional dip needs to be specified.

Acquisition parameters. Parameters that must be defined include the following:

- spacing of receiver groups
- number of receiver groups
- line spacing
- number of lines
- maximum and minimum source-to-receiver distances (offsets)
- source spacing
- source type
- recording geometry

Data acquisition functions include the following:

- permitting
    - permission must be obtained from all owners or authorities prior to entering and working in the area
- positioning
    - precise locations of source and receiver positions must be known
- signal generation
    - seismic waves having appropriate amplitudes and frequency spectra must be generated
- recording
    - seismic waves must be detected and converted to electrical signals
    - electrical signals must be transmitted to the recording system with minimum attenuation and distortion
    - signals must be recorded in a form that provides easy retrieval while preserving as much of the information contained in the original signal as possible

Data processing. The objective of seismic data processing is to produce an as accurate as possible image of the subsurface target. Inputs are field records on magnetic tape, survey/navigation data on magnetic tape, reports, maps, and other items on a variety of media. Outputs are seismic (cross) sections, horizontal sections (3-D), and various special displays.

The same basic processes are applied to both 2-D and 3-D data but the nature of the 3-D data volume is such that some different processes are applied and others are applied quite differently.

Data interpretation. The objective of seismic data interpretation is to convert the *geophysical image* of the subsurface (output of seismic data processing) into a *geologic image* or model of the subsurface. Approaches to the problem include: *modeling, inversion,* and *amplitude versus offset studies* (AVO). *Vertical seismic profiling* (VSP) and *seismic stratigraphy* have also been found to be useful tools.

2 Geological Background

Introduction

Geology as a science can really be traced to one man, James Hutton (1726–1797), who promulgated the three basic laws of geology.

Uniformitarianism

- The processes that formed rocks throughout geological time are the same that form rocks today—the present is the key to the past.
- Sandstone is formed from sand deposited as beaches, in rivers, as dunes, etc.
- Limestone is formed in reefs as carbonate banks resulting from calcite coming out of solution.
- Present environments of deposition infer conditions under which material comprising ancient rocks was deposited.

Law of Superposition

- The oldest rocks are usually at the bottom of a sequence of rocks.
- Youngest rocks are at the top of such a sequence.
- Structural deformation may alter this relationship and the resultant change can be used to interpret deformation events.

Law of Horizontality

- Sedimentary rocks are formed from material usually deposited in layers that are nearly flat or have very small slope.
- Subsequent structural deformations may result in steep slopes in rock layers.
- Stratification, bedding or layering is usually observed in sedimentary rocks.
- Variations in sediment deposition may be seasonal or caused by climatic changes, change in sediment source, or structural deformation.

Geology has many subdivisions or specialties. This text discusses a few of these.

- **Historical Geology** – Studies dealing with changes in geology with time. This owes much to the theory of evolution that was originated by Charles Darwin.
- **Structural Geology** – The study of rock deformations and shapes of geologic features.
- **Petroleum Geology** – Studies dealing with the origin, composition, and accumulation of oil and natural gas.

Geologic Time Scale

The Earth is approximately 4.6 billion (4.6×10^9) years old. The geologic record, however, does not extend back that far. The oldest rocks found on continents are about 3.3 billion years old. Rocks along continental margins tend to be much younger than near centers of continents. Most current land surfaces have been alternately covered by shallow seas then uplifted above sea level.

Deposition of rock-forming sediments has occurred along shorelines and ocean bottoms for more than 3.3 billion years. This deposition has covered much of the Earth's surface. Evidence of ancient life forms is included with the sediments, as fossils. Fossils in older rocks are of simpler, more primitive life forms. This suggests that physical changes in the Earth can be correlated with evolution of plants and animals.

Relative time dates rocks and rock groups as older or younger than others, primarily based on fossil evidence. Radioactive dating provides absolute times. Geologic time divisions (see Table 2–1) are not uniform, as they are based on large-scale geologic events, called *orogenies*. The largest time divisions, called *eras*, are based on worldwide orogenies.

Table 2–1 *Divisions of Geologic Time*

Era	Period	Epoch	Absolute Age (millions of years)
Cenozoic	Quaternary	Holocene	0.01
		Pleistocene	2
			5.
	Tertiary	Pliocene	38
		Miocene	55
		Eocene	65
		Paleocene	140
Mesozoic	Cretaceous		200
	Jurassic		250
	Triassic		290
Paleozoic	Permian		320
	Pennsylvanian		365
	Mississippian		405
	Devonian		445
	Silurian		500
	Ordovician		570
Cambrian			4500
Precambrian			

Eras are subdivided into *periods*, whose boundaries are marked by less intense but worldwide (or nearly so) orogenies. Periods are subdivided into *epochs* and epochs into *ages*. Boundaries of the latter two subdivisions are not usually found to agree worldwide.

Rock groups are also identified by their ages. *Time rock groups* are subdivided into eras, systems, series, and stages in decreasing order. Charts later in this chapter provide details on the geologic time scale and time rock groups.

Paleontology—paleo for short—is the branch of geology that deals with the age of rocks. It uses the preserved remains of plants and animals in sedimentary rocks, called *fossils*. The age of fossils is based on the theory of evolution that assumes that there was a continuous change and extinction of plants and animals through geologic time.

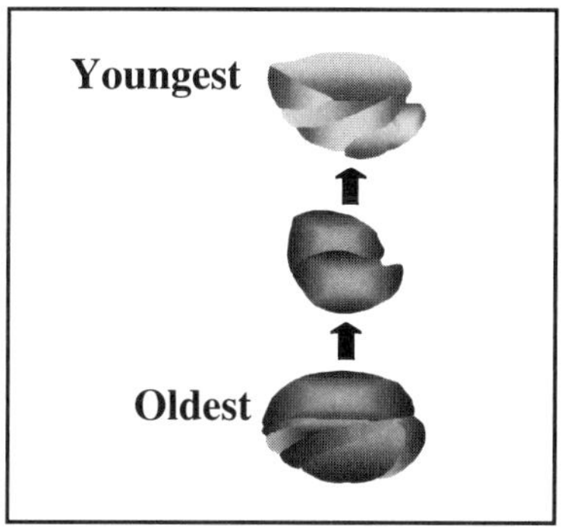

Fig. 2–1 *A Fossil Sequence*

Fossils are used to date rock layers by invoking the geologic Law of Superposition. Figure 2–1 shows a fossil sequence and from the Law of Superposition. We infer that the fossil at the top of the sequence is the youngest (most recent) and the one at the bottom is the oldest.

The age of fossils in rock can be used to fix location in the subsurface (Fig. 2–2). The wellsite geologist looks at fossils in the rock material, called *cuttings*, that are flushed to the surface while drilling wells. Since the drilling process destroys large fossils, small microfossils are the most useful in wells because the drilling does not destroy them. These fossils are the remains of small plants and animals that lived primarily in the ocean.

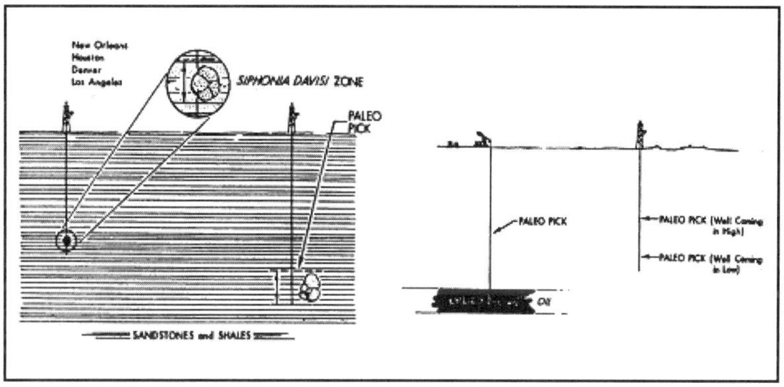

Fig. 2–2 *Dating Rocks from Fossils*

Internal Structure of the Earth

Structurally, the Earth has three main parts (Fig. 2–3). At the center is the *core*, composed mostly of nickel and iron. The inner core, with a radius of about 1230 kilometers (km) appears to be a solid. The outer core, with a thickness of 2220 km, behaves as a liquid. Because of its core, the Earth has electric and magnetic fields. The interaction of the Earth's magnetic field with the solar wind gives rise to the northern lights and also helps protect us from harmful radiation.

Outside the core is the *mantle*, rich in silicates and other rock-forming minerals. The mantle also has two parts. The inner mantle has a thickness of about 1900 km. The outer mantle's thickness is about 970 km. It is thought that the difference between the inner and outer mantle is physical rather than chemical—caused by the increase in temperature and pressure.

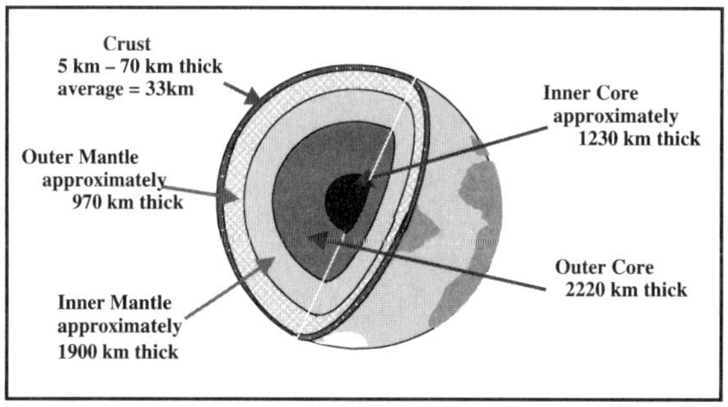

Fig. 2–3 *Internal Structure of the Earth*

The outermost part of the Earth is called the *crust*. The thickness of the crust is highly variable, ranging from as little as 5 km at a few points in the oceans to about 70 km beneath the highest mountain ranges.

There are two types of crust—*oceanic* and *continental*. The oceanic crust has a single layer composed of basalts. The continental crust has two layers—a granitic layer and a basaltic layer. There are also transition zones at continental margins. There is a very distinct boundary between the crust and the mantle where seismic wave propagation velocities become appreciably and abruptly faster in the mantle. This boundary is called the *Mohorovicic* discontinuity or more informally as the *Moho*.

The upper mantle, just below the Moho, is quite complex. There is a concentration of radioactive elements in this zone, and their decay causes heating of the upper mantle, resulting in a narrow zone of variable thickness that behaves as a viscous liquid. This is called the *asthenosphere*. The *lithosphere* includes the uppermost part of the mantle down to just below the asthenosphere and the crust (Fig. 2–4).

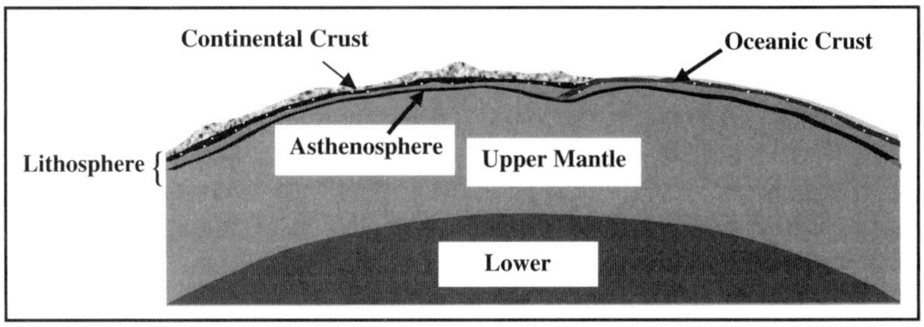

Fig. 2–4 *The Lithosphere*

Numerous theories to explain the distribution of continents, mountain chains, and oceanic ridges and trenches have been proposed (Fig. 2–5). Some early geologists attempted to explain these by assuming the Earth was shrinking. It was shown that this could not account for all the distributions. Others proposed the opposite—an expanding Earth! This could not explain the distribution satisfactorily either.

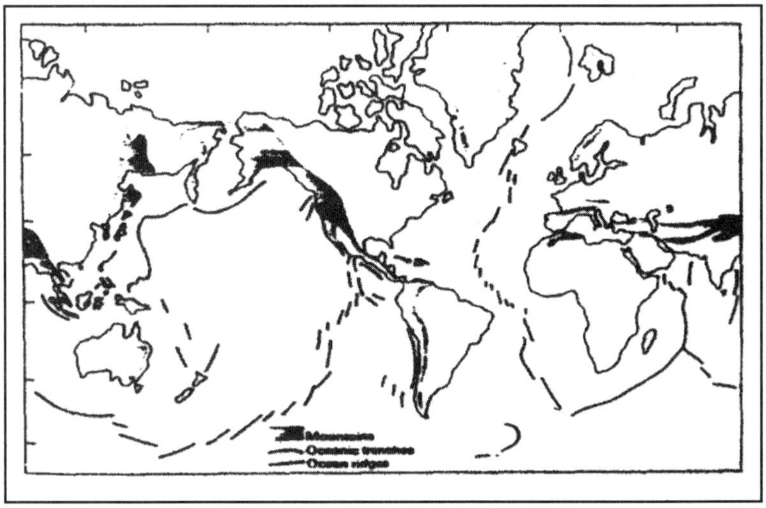

Fig. 2–5 *Mountains, Ridges, and Rises of the World*

In the mid-1950s, a startling dicovery was made. Rocks on the seafloor near the Mid-Atlantic Ridge were very young. There were no signs of older rocks there but progressively older rocks were found on the seafloors in both directions away from the ridge. The Atlantic Ocean floor was getting larger!

Sampling the seafloors all over the world found similar patterns. The conclusion reached was that new oceanic crust was being formed at the oceanic ridges. This led to the *Seafloor Spreading Theory* illustrated in Figure 2–6. Magma, molten mantle material, is forced upward through the top of the ridge system. It then flows downward, pushing the older crustal material away from the ridge and creating new crust.

Studies of magnetic polarization in seafloor rocks corroborated this conclusion. It is known that the Earth's magnetic poles have switched between north and south many times in the past. Magnetic paticles in magma are polarized in the direction of the Earth's magnetic field as they cool, and they retain this polarity even though the Earth's field reverses. It was found that all across the floor of the Atlantic Ocean, paralleling the ridge, there are strips of alternating magnetic polarity. This can be explained by the seafloor spreading theory.

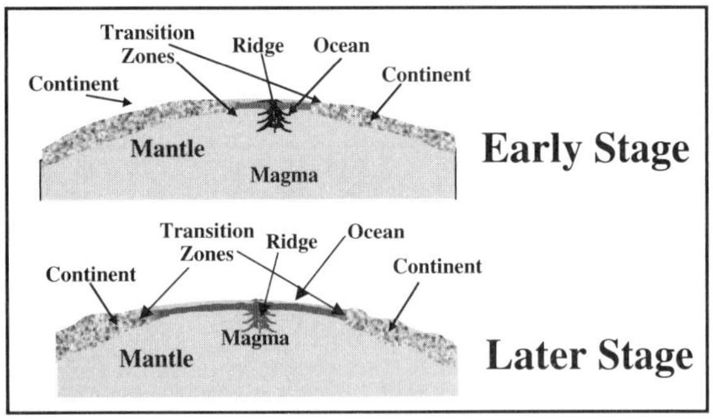

Fig. 2–6 *Seafloor Spreading*

These findings led to the ressurection of an old idea—*continental drift*. Many people had noted that the east coast of South America and the west coast of Africa could be fitted together quite nicely. North America and Europe also could be fitted together with a few shifts. A comprehensive theory of continental drift was proposed by Alfred Wegener in papers published in 1912 and 1924. Unfortunately, the theory lacked a convincing mechanism to drive this drift. Seafloor spreading coupled with the idea of upper mantle convection currents provided the mechanism (Fig. 2–7).

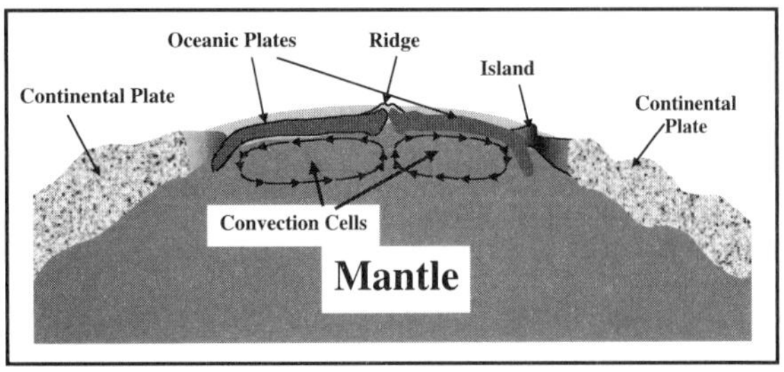

Fig. 2–7 *Mantle Convection Current*

Continental drift was reborn and expanded by the *Tectonic Plate Theory*. According to this theory, the lithosphere is not continuous but instead composed of separate pieces called *tectonic plates* that move over the Earth's surface relative to one another in a variety of ways. The upper mantle, as previously noted, behaves as a viscous liquid. Because of differential heating, convection cells develop with hotter material rising toward the surface, cooling, and sinking back down. Because of its viscosity, the mantle material drags the oceanic plates with it away from the oceanic ridges toward the continental plates. The oceanic plates, being heavier than continental plates, dive down under the continental plates and are consumed into the mantle.

The lithosphere is divided into many tectonic plates. Figure 2–8 shows the plates and their boundaries. Note that there are three different types of plate boundaries—consuming (where one plate dives below another and is consumed in the mantle), extensional (plates pull apart as along the Mid-Atlantic Ridge), and transform (where plates move horizontally past one another, as along the San Andreas Fault of California). Figure 2–9 shows the plate boundary types in more detail with a tectonic plate model.

Zones of intense earthquake activity exist along the west coast of South America and along the island arcs of the western Pacific ocean. In these regions, eathquake foci show a pattern of increasing depth as one moves from the ocean toward the continents or across the island arcs. These patterns of earthquake occurrence are explained by the Pacific and Nazca plates plunging under the American, Eurasian, and Indian plates.

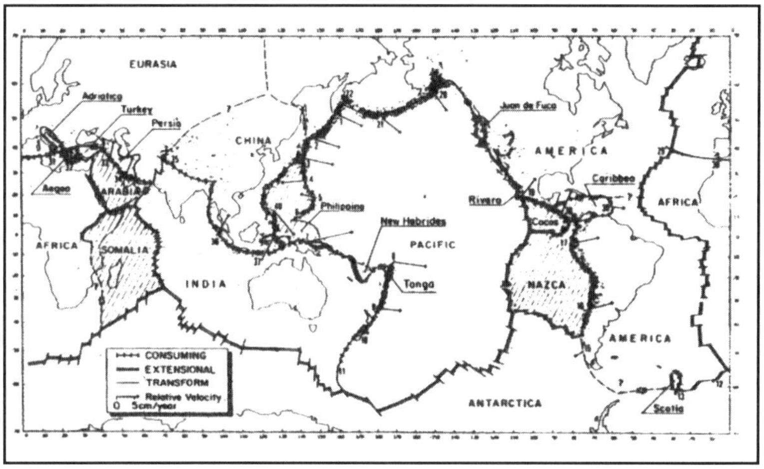

Fig. 2–8 *Tectonic Plate Boundaries*

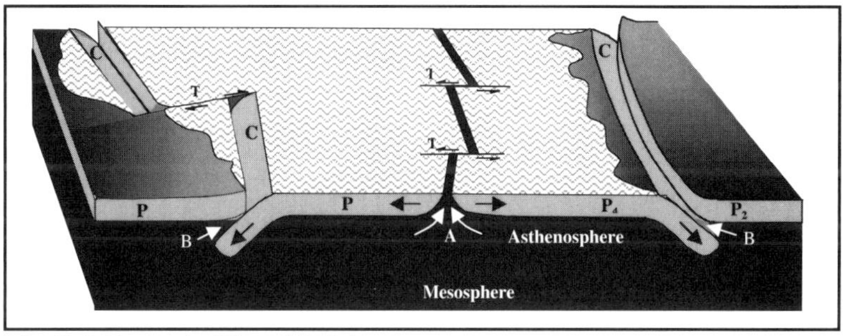

Fig. 2–9 *Tectonic Plate Model*

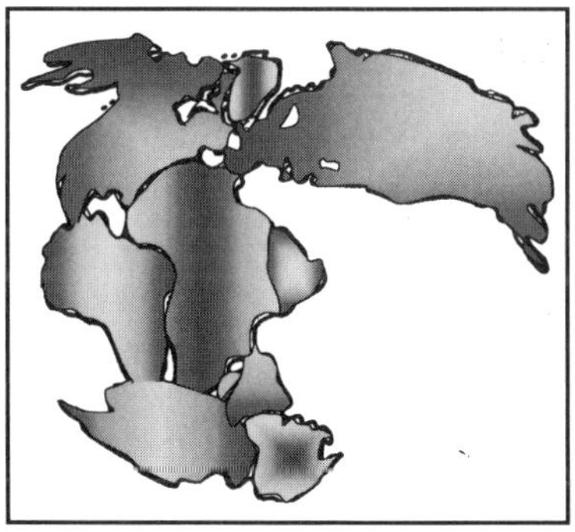

Fig. 2–10 *Pangea, the Universal Continent, about 200 Million Years Ago*

Another part of the tectonic plate theory is that about 200 million years ago, there was only one continent (given the name *Pangea*) as shown in Figure 2–10. The fit of the continents is based not on just shapes of continental margins but also on similarity of fossils, rocks, and geological features.

The continents are still moving. Figure 2–11 shows the continents in their present position and their projected positions 50 million years from now.

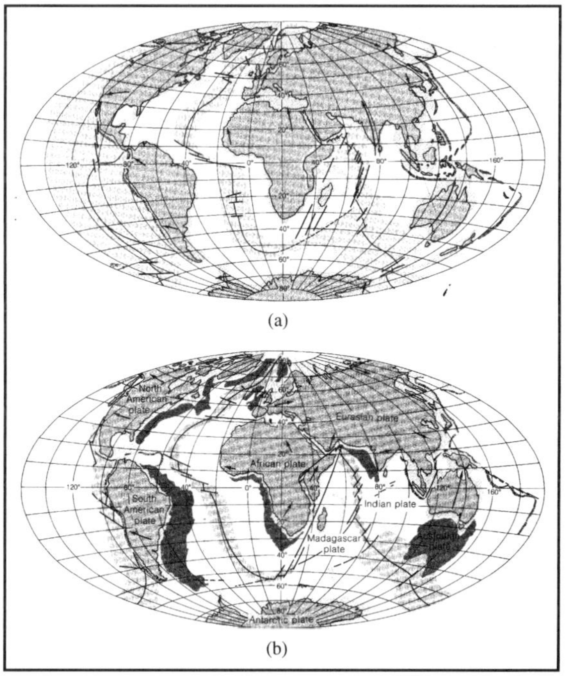

Fig. 2–11 *The Continents in Their Present Positions and Their Projected Positions 50 Million Years from Now*

Rocks in the Earth's Crust

To this point, we have discussed only the large-scale features and structures of the Earth. Our real interest, however, is centered on the uppermost parts of the Earth's crust. This is shown in Figure 2–12.

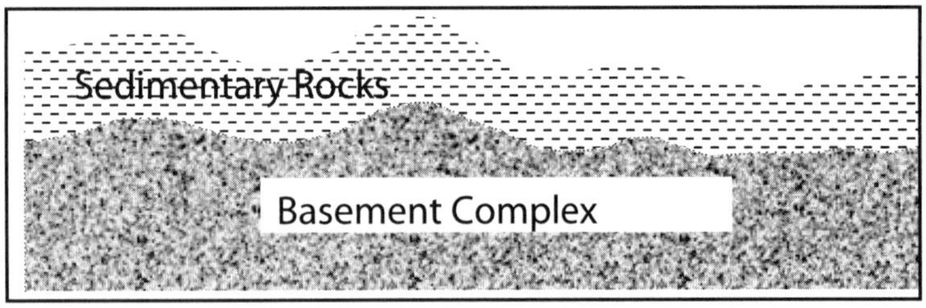

Fig. 2–12 *Top Part of the Earth's Crust*

Rocks and soils of the Earth's crust are aggregates of mineral grains. Minerals are naturally occurring chemical compounds. Two of the more common minerals are quartz (SiO_2) and calcite ($CaCO_3$). Some rocks are composed of many different minerals. For example, granite contains black mica, feldspar, and quartz. Other rocks, such as limestone, are composed of only one mineral—calcite in the case of limestone.

There are three classes of crustal rocks—*igneous, sedimentary,* and *metamorphic.*

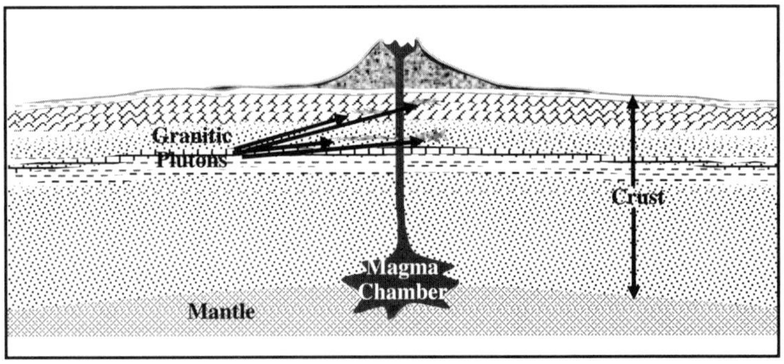

Fig. 2–13 *Formation of Igneous Rocks*

Igneous rocks are created when cooling from a molten state, and are subdivided into two types. *Volcanic* (basaltic or extrusive) igneous rocks form through quick cooling of lava flows. They are glassy or fine-grained crystalline rocks. *Plutonic* (granitic or intrusive) igneous rocks result from the slow cooling of magma beneath the surface. They are coarse-grained crystalline rocks.

Igneous rocks are associated with zones of weakness in the Earth's crust, such as an island arc or crest of an oceanic ridge (Fig. 2–13).

Sedimentary rocks are formed in layers from sediments transported from one place and deposited in another. There are three types of sedimentary rocks.

- *Clastic* rocks are compacted and cemented rock fragments eroded from pre-existing rocks. Examples include sand stone and shale.
- *Chemical* rocks are precipitated from solution. They include carbonates, salts, and sulfates.
- *Organic* rocks are composed of the remains of plants and animals. Coal is an example of an organic rock.

Metamorphic rocks are formed from rocks altered by high temperature and/or pressure but without melting. Limestone, a sedimentary rock, can be changed into marble, a metamorphic rock. Shale, also a sedimentary rock, can be changed into slate, a metamorphic rock. Metamorphic rocks are associated with mountains and the basement complex. As shown in Figure 2–14, any of the three rock classes can be changed into either of the other two.

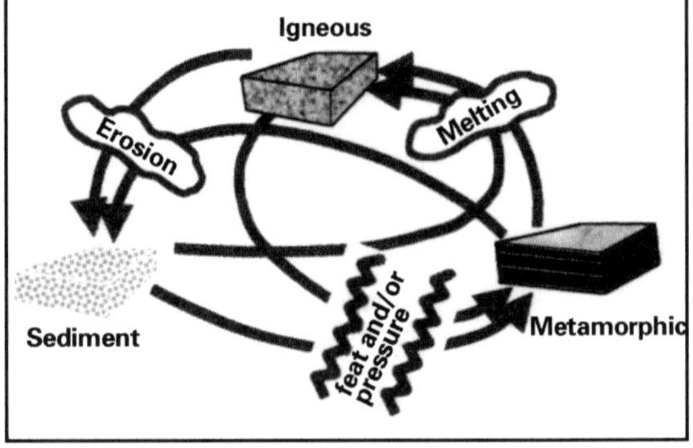

Fig. 2–14 *The Rock Cycle*

The processes of *weathering, erosion* and *deposition* form clastic sedimentary rocks. Weathering is the process in which physical, chemical, and biologic agents that break the rocks down into smaller pieces or chemical components act upon exposed rocks. In erosion, products of weathering are transported from their place of origin to another site. Deposition is when eroded rock fragments settle to the bottom of a stream, ocean, etc.

Clastic sedimentary rocks are formed when grains resulting from erosion of existing rocks are transported by gravity, wind, and water to an area of deposition. The area of deposition may be a topographic low or depression such as a lake, sea, or ocean. The rocks tend to form as relatively flat, parallel layers or strata in a marine environment. The greatest thicknesses of sedimentary rocks are found in *sedimentary basins*. These were once oceans or seas that received sediments from rivers that drained large land areas.

Sediments tend to have a high (15 to 45%) *porosity* immediately after deposition. Porosity is the percent of rock volume made up of pores or open spaces. Porosity may be subsequently decreased by *cementation* and *compaction*. Cementation is the precipitation of salts and minerals such as calcite, sulfates, quartz, and salts between loose grains. Compaction is the compression of rock material by weight of sediments deposited on it.

Permeability, the rate of fluid flow through a rock, tends to vary with uniformity in the size in grains deposited. Compaction and cementation tend to decrease permeability. Rocks in the subsurface are usually full of fluid, and subsurface water or other fluids can flow through permeable rocks.

Water flowing through rock can cause cementation or *leaching*—dissolving of rock material. Leaching may be selective, dissolving only certain grain types, but it can also improve porosity and permeability of rocks. Primary porosity results from original holes between fragments. Secondary porosity results from leaching of rock material.

Grain size is important to porosity and permeability. Consequently, geologists have created a nomenclature to describe grain size as shown in Table 2–2.

Table 2–2 *Clastic Grain Size Classification*

Class	Diameter (mm)
Boulders	>256
Cobbles	64 to 256
Pebbles	4 to 64
Granules	2 to 4
Sand	1/16 to 2
Silt	1/256 to 1/16
Clay	< 1/256

Sorting describes the size distribution of rock grains. Well-sorted means that all grains are approximately the same size, e.g. clean sands. Poorly sorted means that there is a wide range of intermixed grain sizes, e.g. dirty sands that contain sand plus finer material. In graded sorting, grain sizes vary linearly with depth in a rock segment (Fig. 2–15).

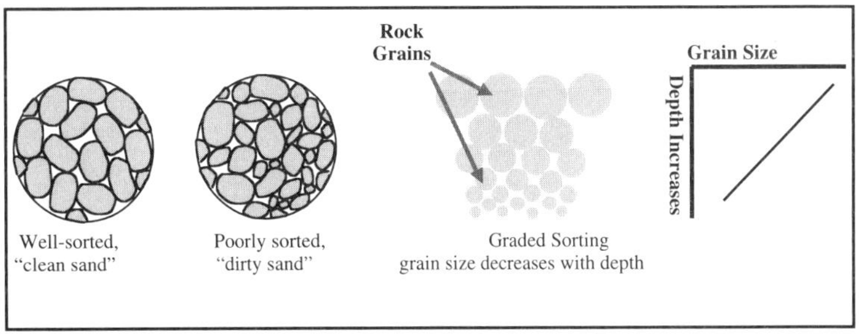

Fig. 2–15 *Sorting*

The depositional environment can be inferred from the interpretation of sedimentary rocks based on preserved clues such as ripple marks, mud cracks, fossils (evidence of ancient animals and plants), and on environmental types as follows:

- alluvial—seasonal water flow or landslides
- fluvial—river deposits
- glacial
- marine

Interpretation of sedimentary rocks is also based on rock type—clastic or carbonate—and fluid content—fresh water, brine, oil, or gas.

The most prevalent sedimentary rock types are shale, sandstone, and limestone. They comprise 99% of the Earth's sedimentary rocks. Shale has a small grain size and a high organic content. This is important to petroleum formation. Sandstone has a larger grain size than shale, and its grains are predominately silica fragments. Sandstone has high porosity and permeability in many cases. Limestone is either chemically precipitated calcite or built up from skeletal remains of animals. It is of oceanic (marine) origin. Limestone may be chemically altered to dolomite. More petroleum has been found in limestone and dolomite than any other rock type.

Limestone, sandstone, and shale are not always found in pure forms but frequently grade from one to another as shown in Figure 2–16.

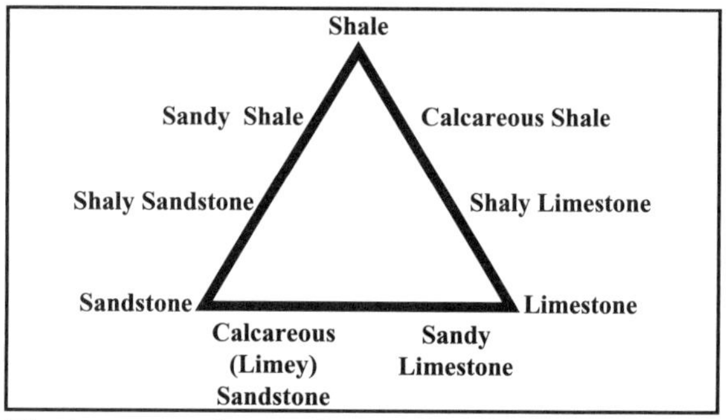

Fig. 2–16 *Prevalent Sedimentary Rock Types*

Petroleum is found within porous and permeable rocks. Pores are needed to store the petroleum. Permeability is necessary for petroleum to enter rocks. Porosity is the percentage of rock volume made up of pores or openings between fragments. Porosity in rocks ranges from 1 to 45%. Porosity is classified as follows:

- poor, 1–8%
- fair, 9–15%
- good, 16–25%
- excellent, more than 25%

Permeability is a measure of fluid flow and depends on interconnection of pores. The *darcy* is the unit of permeability measurement, but because this is such a large unit, *millidarcies* (md = .001 darcies) are most often used. Permeability can be measured in the laboratory by determining the rate of hydrogen flow through a rock under a standard pressure. Adjustments are made to correct from gas to liquid flow. Relative permeability measurements are

- poor, 0.1–50 md
- fair, 50–250 md
- good, 250–1000 md
- excellent, more than 1000 md

Clastic sedimentary rocks are most likely to be porous and permeable since they consist of fragments eroded from other rocks. Carbonate rocks (banks and reefs) may also be porous and permeable. Igneous and metamorphic rocks can be of interest if they have sufficient cracks (fractures) to produce porosity and permeability. Fractured igneous and metamorphic rocks can have high permeability but usually have low porosity, limiting the amount of petroleum they can hold.

Deformation of Crustal Rocks

Weathering breaks down rocks at the Earth's surface into smaller pieces or fragments called *sediments*. The breakdown can be by mechanical means (a physical break in the rock) or chemical (dissolution of parts of the rock by groundwater or other liquids such as dilute acids).

Erosion is the transportation of sediments from their site of origin. Gravity, in the form of landslides or rock fragments simply rolling down slope, can be a cause of erosion. Wind picks up smaller sediments and drops them at various distances from their original position. Glaciers slowly move downhill, and, as they move, pick up sediments beneath them. Later, melting allows these sediments to be deposited elsewhere. Moving water—waves or currents in rivers and oceans—is perhaps the most active agent of erosion. Most sediments are deposited at the bottom of oceans or other bodies of water.

Unconformities are surfaces of erosion or non-deposition. Of particular interest to petroleum exploration are *angular unconformities*. In angular unconformities, rock layers below the erosional surface are tilted at an angle while layers above the erosional surface are more nearly horizontal. An unconformity represents a period of major mountain building and subsequent erosion. Angular unconformities can form giant hydrocarbon traps as seen in the East Texas and Prudhoe Bay Fields. The East Texas field is shown in Figure 2–17.

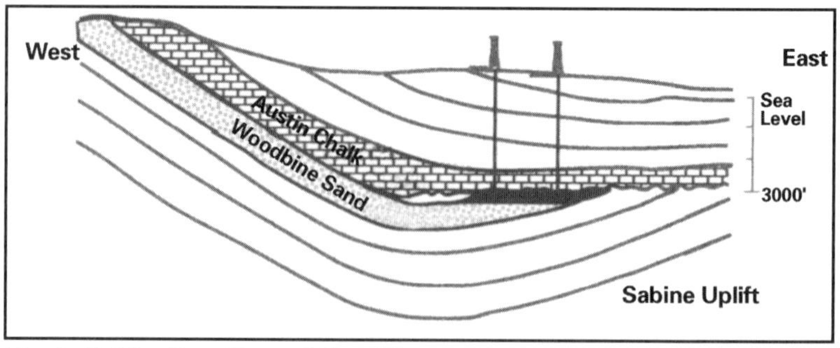

Fig. 2–17 *The East Texas Oil Field*

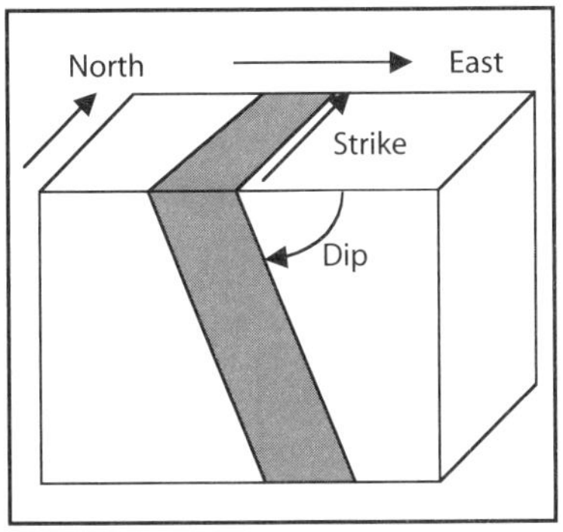

Fig. 2–18 *Strike and Dip*

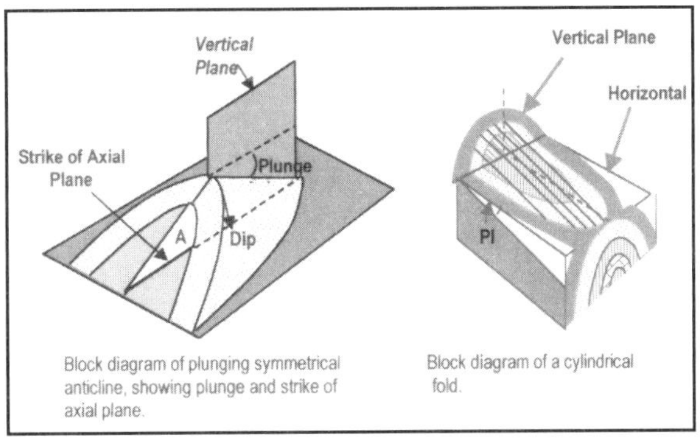

Fig. 2–19 *Strike, Dip, and Plunge*

Structural position of tilted, sedimentary layers is described using *strike* and *dip*. Strike is the direction relative to north of the line formed by the intersection of the top (or bottom) of a layer with a horizontal plane. Dip is the angle that a plane surface (layer or bedding) makes with a horizontal plane. Both the dip and strike are usually specified for a dipping layer. In Figure 2–18 the strike is 0°, which is due north. Some structures require another parameter, *plunge*, to describe their positions. This is shown in Figure 2–19.

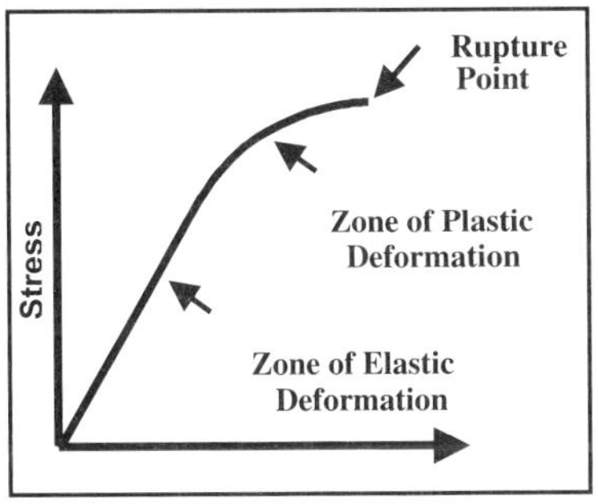

Fig. 2–20 *Stress-Strain Relationship*

Crustal layers are deformed by stresses (force per unit area) caused by

- collision of tectonic plates at plate boundaries
- weight of overlying rock layers (overburden)
- *isostasy*, or buoyancy of lithospheric units in underlying mantle

Layer deformation is a function of the size of the applied stress and the physical properties of the layer. Figure 2–20 is a plot of stress and strain. If the stress is within elastic limits of a given layer, then an applied stress results in a *strain* (the relative change in dimensions or shape) directly proportional to the stress. When stress exceeds the elastic limit, the strain is no longer proportional to the stress and there is a zone of plastic deformation, called *folding*. At the upper limit of plastic deformation the layer ruptures, fracturing into separate layers with relative movement between them. This is called *faulting*.

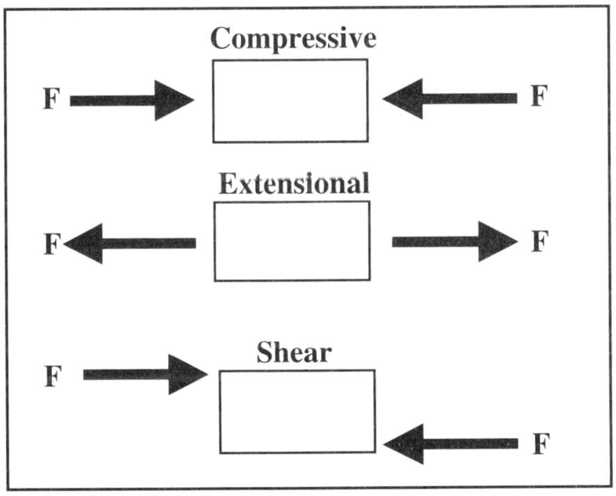

Fig. 2 21 *Types of Stress*

There are three general types of deforming stresses: *compressive*, collinear forces (F) that push rocks together; *extensional*, which are also collinear but pull rocks apart; and *shear*, in which forces are offset from one another causing one face of a rock to slide past the opposite one. These stresses are illustrated in Figure 2–21.

Folding of crustal rocks. Folds (Fig. 2–22) occur when horizontal sedimentary layers are subjected to a compressive stress but the plastic limit is not exceeded. *Deformation* is the bending of the rock layers. Bends in rock layers are called folds. Ruptures of the rock layers occur when the plastic limit is exceeded. Ruptures are called faults.

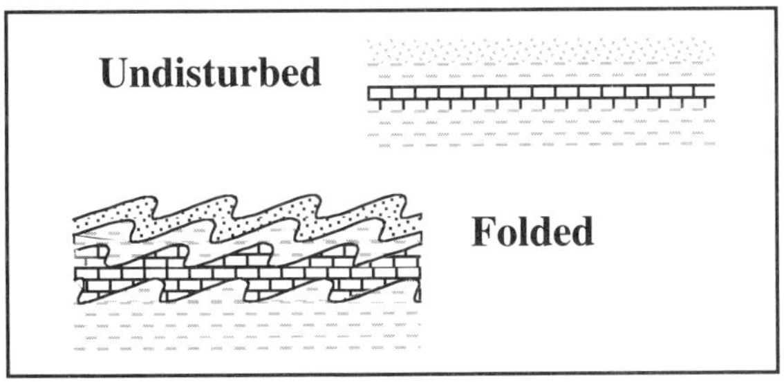

Fig. 2–22 *Folding of Rocks*

Mountain building involves intense, large-scale folding and it is very common to have both folding and faulting. Folds tend to remain even after faulting occurs. Folds in the Earth's crustal layers are excellent places for hydrocarbons to be found. Early geologists and geophysicists worked to find folds in rocks.

There are many kinds or classes of folds. A *monocline* is when rock layers are tilted (dip) in one direction. Note that the dip can vary in a monocline, but the dip direction stays the same. A *syncline* is when rock layers are folded down into a depression. An anticline is when rock layers are folded up into an arch.

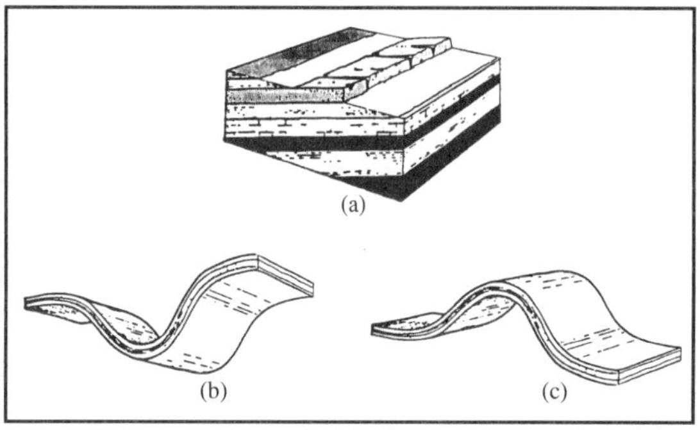

Fig. 2–23 *Monoclines, Synclines, and Anticlines*

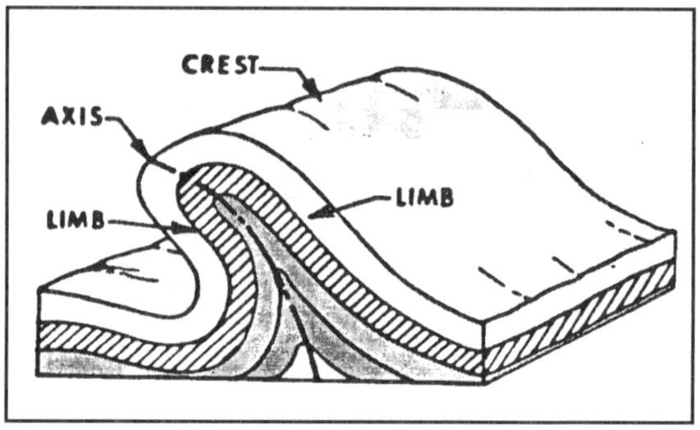

Fig. 2–24 *Terms Describing Anticlines*

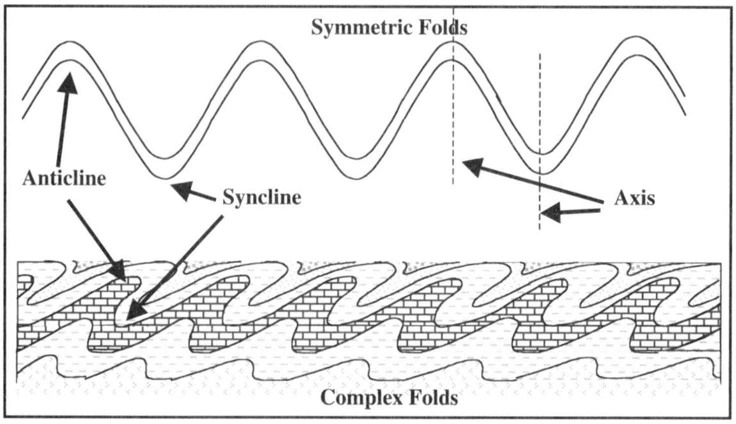

Figure 2–25 *Symmetric and Asymmetric Folds*

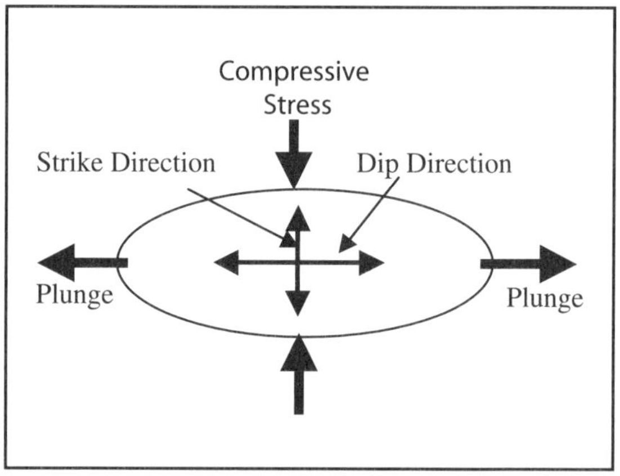

Fig. 2–26 *Anticlinal Dome, Overhead View*

Anticlines make excellent hydrocarbon traps. Figure 2–23 shows monoclines, synclines, and anticlines.

Figure 2–24 illustrates some terms used to describe anticlines. The *crest* is simply the highest point on the anticline. The *axis* is an imaginary line or surface about which the folding takes place. The *limbs* are the parts of the fold that slope down on each side of the crest. It is not unusual to have a series of folds with anticlines and synclines in succession.

A symmetric anticline is one with limbs dipping down from the crest at the same angle but in opposite directions. Symmetric synclines have limbs that dip up from the trough (low) at the same angle but in opposite directions. Limbs of asymmetric anticlines and synclines have different dips on each limb. Usually one has a much steeper dip than the other. See Figure 2–25.

A dome is a circular or elliptical anticline. It makes a perfect hydrocarbon trap. Figure 2–26 illustrates the features of a dome.

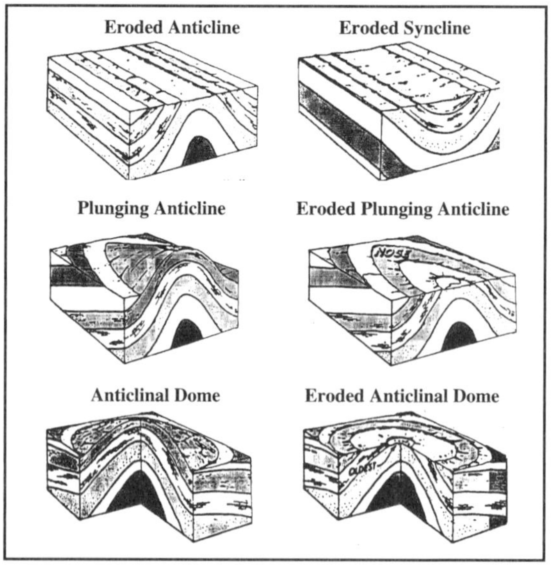

Fig. 2–27 *Eroded Folds*

Young folds are ridges or valleys that have not been extensively eroded. *Old folds* have been eroded flat (*pleneplaned*). The surface pattern of an eroded cylindrical fold is symmetrical stripes. An anticline has the oldest rocks in the center. A syncline has the youngest rocks in center as seen in Figure 2–27. The surface erosion pattern of any anticline is for the oldest rocks to be in the center.

In a *plunging fold*, the axis is not level. The surface pattern of an eroded, plunging fold is called a *nose*. Plunging anticlines and plunging synclines are most common. A plunging anticline has the oldest rocks in the center. A plunging syncline has the youngest rocks in the center.

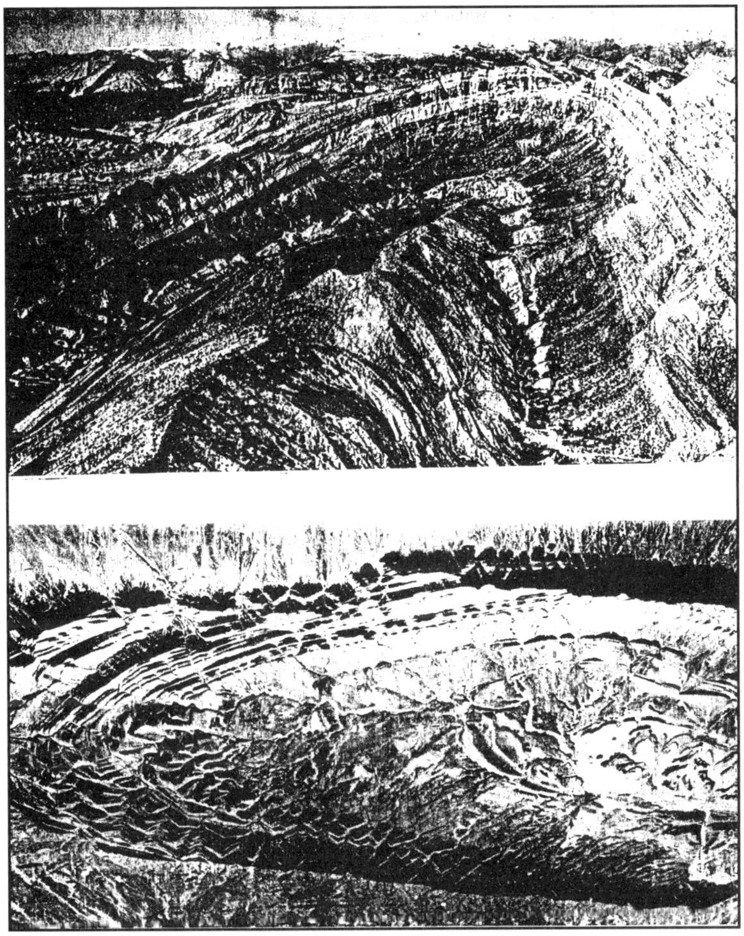

Fig. 2–28 *An Eroded, Plunging Syncline in Northwest Africa and an Eroded, Plunging Anticline in the Zagros Mountains of Iran*

Figure 2–28 shows photographs of a real eroded syncline and eroded anticline.Without knowledge of rock ages, it would be impossible to tell the anticline from the syncline.

Joints are natural fractures in rocks. They are caused by compressive or tensional stress. The applied stress exceeds the plastic limit of a rock, causing rupturing or fracturing. However, there is no relative movement of one side of fracture relative to the other. Joints usually occur as relatively evenly spaced sets with the same general orientation throughout an area. Homogeneous rocks exhibit fairly uniform fracture patterns (Fig. 2–29).

Joints enhance porosity with more space to store fluids and permeability by providing fluid flow paths in subsurface rocks. Joints also increase the surface area exposed to water and ice, thus weathering and erosion of rocks with joint patterns is enhanced.

Faults are fractures in rock layers accompanied by relative motion between the resulting faults. Movement on one side of a fracture relative to the other side can be of three different types. In *dip slip*, movement is up and down along the fault plane. Such faults are classified as *normal* and *reverse*, depending on which fault block appears to have moved up. In *strike slip*, movement is strictly horizontal. *Oblique slip* is a combination of dip slip and strike slip motion. Fault motion is illustrated in Figure 2–30.

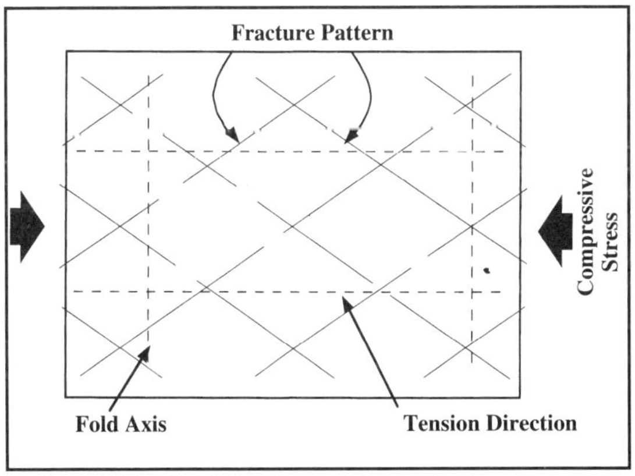

Fig. 2–29 *Joints*

Figure 2–31(a) shows a type of dip slip fault called a normal or *gravity* fault. Also shown in the figure are some terms used to describe faults. The *fault plane* is the surface along which the two blocks created by the fault move. The *footwall* is the block below the fault plane. The *hanging wall* is the block above the fault plane. *Heave* is the amount of horizontal movement between the two fault blocks, and *throw* is the amount of vertical movement between the two fault blocks. A normal fault is one where the hanging wall has moved down relative to the footwall. Note that this is relative motion. The footwall could have moved up or the hanging wall down or a combination of the two.

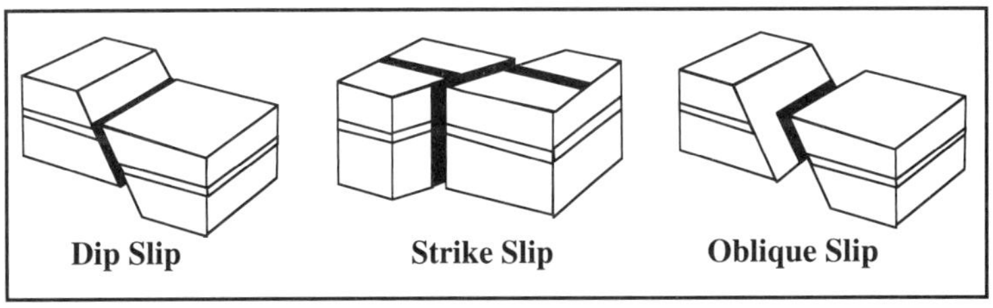

Fig. 2–30 *Fault Motion*

Figure 2–31(b) shows a reverse fault. In this case, the hanging wall has moved up relative to the footwall. If the fault motion is the same but the dip of the fault plane is less than 45o, then it is called a *thrust* fault.

The stress causing a normal fault is tension that pulls the two blocks apart. The weight of the hanging wall causes it to fall. In a well cutting a normal fault there will be a missing section—layers that are present in Earth's section are not penetrated by the well bore The stress causing a reverse or thrust fault is compression. A well cutting a reverse fault will have a repeat section—a particular layer is present twice in a well bore (Fig. 2–32).

In strike slip faults, the stress applied to the beds is shear. Motion of the fault blocks is horizontal along the strike of the fault plane. Displacements can be very large like those along the San Andreas Fault of California or the Great Glen Fault of Scotland (Fig. 2–33). Dip slip and strike slip faults are both important to exploration for hydrocarbons (oil and gas). Many anticlinal traps worldwide are related to normal and reverse faults. California's oil production along the Pacific Ocean coast is directly related to the strike slip motion of the San Andreas Fault.

Horsts and *grabens* (Fig. 2–34) are formed by parallel normal faults. The graben is the down dropped block and the horst is the ridge between grabens.

Growth faults are a type of normal fault formed in river deltas and coastal plains (depositional basins). Large volumes of loose sediments are rapidly deposited. Movement along the fault occurs at the same time as sediment deposition, providing accommodation space for more sediment. Movement is caused by the weight of the water and saturated sediments. The fault plane is arcuate, concave toward the basin. It is steep at the ground surface and becomes less steep at depth. It *soles* out at depth—becomes parallel to layers at depth.

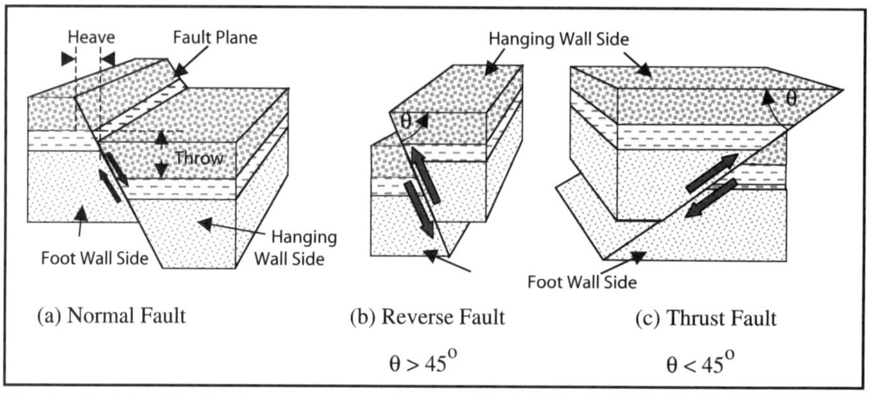

Fig. 2–31 *Dip Slip Faults*

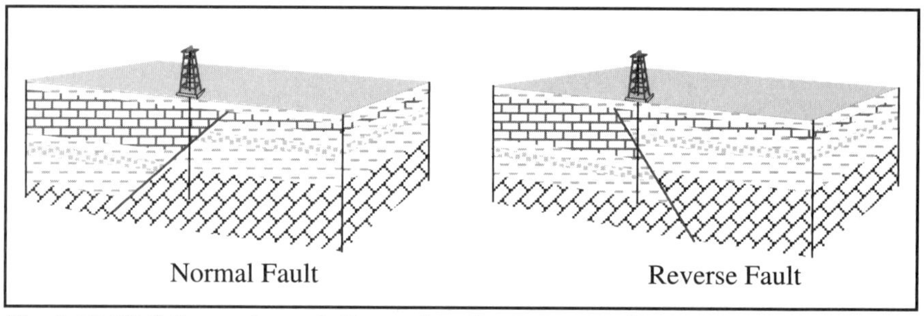

Fig. 2–32 *Well Bores through Normal and Reverse Faults*

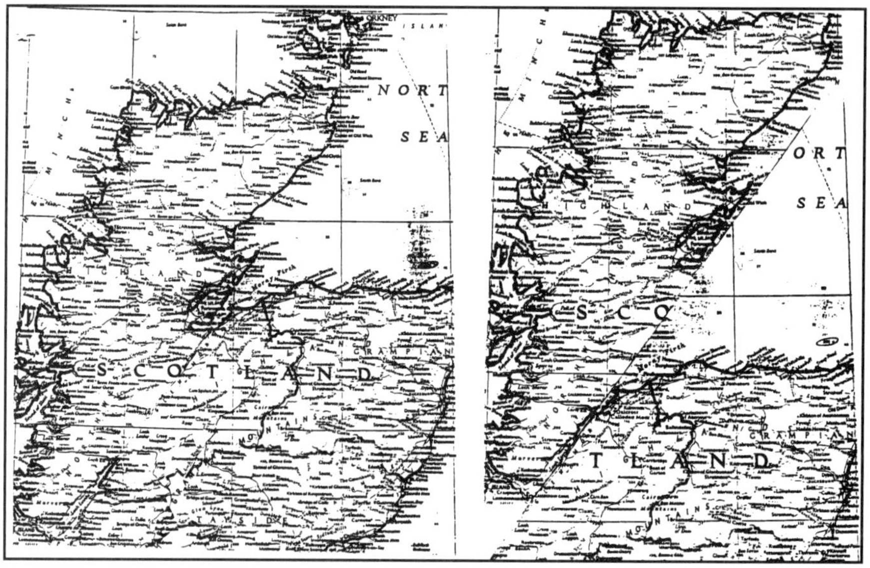

Fig. 2–33 *Fault Displacement along the Great Glen Fault*

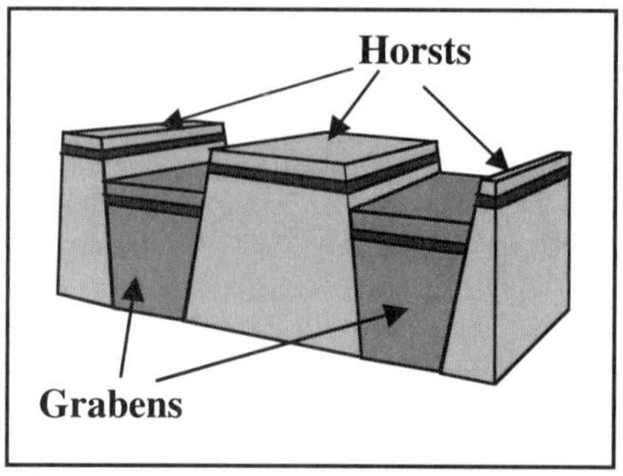

Fig. 2–34 *Horsts and Grabens*

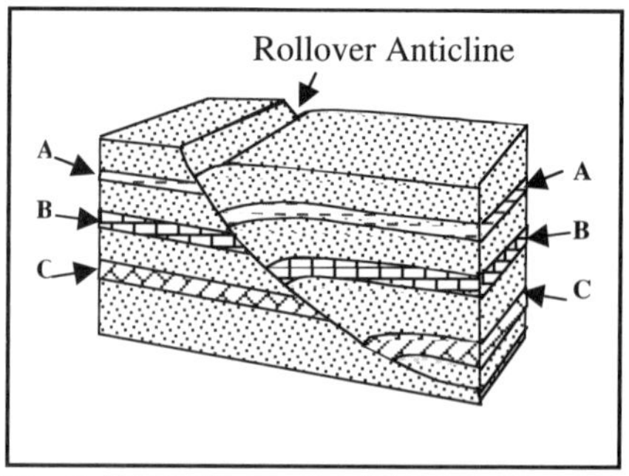

Fig. 2–35 *Rollover Anticline*

Rollover anticlines are broad anticlines formed on the basin sides of growth faults. They are very important oil and gas traps in the Gulf of Mexico and Nigeria. Figure 2–35 shows a growth fault with time markers *A, B,* and *C*. Movement on the fault causes accommodation space on the downside of the fault. Between *A* and *B* and between *B* and *C*, beds thicken due to higher sedimentation rates on the down side of fault. The result is the formation of rollover anticlines.

The Nature of Petroleum

Materials grouped within the broad category of petroleum are mixtures of various *hydrocarbons*—compounds of hydrogen and carbon. The chemistry of hydrocarbons can be very complex but methane (CH_4), the simplest of all hydrocarbons, is the basic building block for all of them. Petroleum may be a gas, a liquid, or a solid. Simple hydrocarbons such as methane and acetylene are found in gasses emitted by volcanoes.

The similarity of the array of complex hydrocarbons found in petroleum and in organisms points to an organic origin for petroleum. Figure 2–36 illustrates the formation and combustion of hydrocarbons in what is called the *fuel cycle*. Sunlight combined with the process of photosynthesis promotes plant growth. Animals then eat the plants.

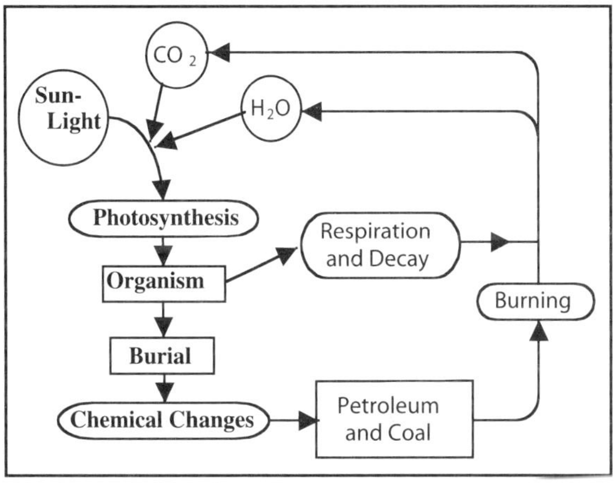

Fig. 2–36 *The Fuel Cycle*

Small marine organisms are assumed to be the raw organic ingredients for petroleum. Sediments must bury this organic material quickly. Decay of the organisms must be in an *anaerobic* environment, which means there is little or no oxygen present. The anaerobic condition is necessary for petroleum to be a fuel, since burning petroleum is simply very rapid oxidation.

Oceans are an abundant source of the types of the small organisms required and are also repositories of vast quantities of sediment. Most sources of hydrocarbons are found in rocks deposited in marine environments.

The chemical composition of petroleum is summarized in Table 2-3. Hydrocarbons are, by definition, compounds of hydrogen and carbon. The other elements in Table 2–3 are, thus, impurities. The worst of these is sulfur, which must be removed in the refining process. *Sweet crude* oil has less than 1% sulfur content and is more valuable than *sour crude* oil that has more than 1% sulfur content.

Table 2–3 *Chemical Composition of Petroleum*

Chemical Elements	Crude Oil	Natural Gas
Carbon	82–87%	65–80%
Hydrogen	12–15%	1–25%
Sulfur	0.1–5.5%	0–0.2%
Nitrogen	0.1–4.5%	1–15%
Oxygen	0.1–1.5%	-------

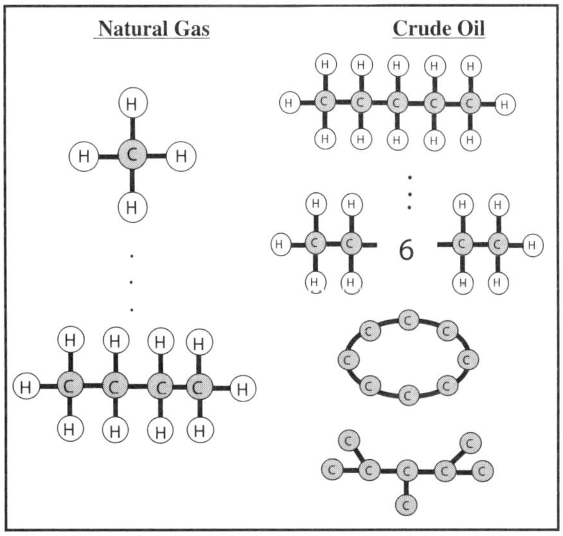

Fig. 2–37 *Molecular Forms of SB Hydrocarbons*

Hydrocarbon molecules take many forms. Some of these are shown in Figure 2–37. Natural gas is a mixture of the smaller and lighter molecules that take the form of atomic chains. The molecules form by sharing electrons in the valence bands. Hydrogen has one electron in its valence band and needs one more to fill its valence band. Carbon has four electrons in its valence band and needs four more to fill it.

The simplest hydrocarbon, methane, does this with one carbon atom and four hydrogen atoms. The chain is built by adding one carbon atom and two hydrogen atoms a time. Ethane (C_2H_6) is the next in the series, and the two carbon atoms share one electron with each other and one with the three hydrogen atoms around it. Butane (C_3H_8) and propane (C_4H_{10}) are the next in the series. Each carbon atom shares one electron with its four neighboring atoms—either hydrogen or carbon.

The situation with crude oil is more complex. The hydrocarbon molecules are larger and heavier and occur in more than one form. *Paraffins* have long chains of up to 60 carbon atoms. Physically, paraffins are very waxy. This can cause production problems, as the paraffins tend to clog up production equipment. *Naphthenes* (cycloparaffins) have closed carbon rings. Crude oils with high naphthene levels tend to also have high asphalt content. *Aromatics* (benzenes) also have carbon rings but some carbon atoms share two electrons with a neighboring carbon atom. Aromatics yield high-octane gasoline. They have a strong fruity odor. *Asphaltics* are brown to black and solid to semi-solid. They have a high boiling point. A typical crude oil is 30% paraffins, 49% naphthenes, 15% aromatics and 6% asphaltics.

The constituents of crude oil strongly affect refining. Asphalt-based crude oils have large hydrocarbon molecules and little or no paraffin wax. They yield high quality gasoline and asphalt for road paving. Asphalt-based crude oils are black in color. Paraffin-based crude oils have little or no asphalt. They yield paraffin wax, high quality lubricating oils and kerosene and are greenish in color. Sulfur must be removed from crude oils in the refining process. As previously mentioned, high sulfur content crude oils are less valuable.

Crude oil density or gravity is a measure of viscosity. It is measured in degrees API.

API° = (141.5/sp. gr.) - 131.5, where sp.gr. = specific gravity of the fluid in grams/cubic centimeter (gm/cc).

Here are some typical API values.

- Water—10
- Heavy Crude—5 to 20 (viscous, high sulfur content)
- Average Crude—30 to 40
- Light Crude—45 to 55 (very fluid, most valuable)

Typically, natural gas is composed of the following hydrocarbons: methane, 88%, ethane, 5%; butane 2%; and propane, 1%. There are also usually some gas impurities that do not burn. These include carbon dioxide (CO_2), nitrogen, and helium. Hydrogen sulfide (H_2S) is also found in some natural gas deposits.

Carbon dioxide and nitrogen are colorless and odorless. They are often injected into reservoirs to recover additional oil, a process called *enhanced recovery*. Helium, which is truly inert because it does not enter into any chemical reactions, may also be found in natural gas. Hydrogen sulfide is extremely poisonous and corrosive. Its presence requires use of stainless steel well equipment and also gives the name of *sour* gas to natural gas that contains it because of its rotten egg smell.

In many oil fields, natural gas is dissolved in the oil. This is called *formation gas*. The amount of dissolved gas is given by the gas-to-oil ratio (GOR). This is defined as the number of cubic feet of gas dissolved in one barrel of oil in the subsurface reservoir rock. It is pressure dependent (and depth dependent, since pressure increases with depth). The presence of dissolved gas causes a shrinkage factor with values ranging from 1.0 to 0.6.) *Shrinkage factor* is the decrease in the volume of oil as it is brought to the surface and the dissolved gas bubbles out of it due to the decrease in pressure. *Formation volume factor* (FVF) is the number of barrels of oil in the reservoir required to produce one stock tank barrel of oil. *Oil in* place refers to the oil in the subsurface reservoir. *Stock tank barrel* refers to oil at the surface (at atmospheric pressure).

Condensate is a term for the extremely light, liquid hydrocarbons that exist as gas under subsurface reservoir conditions of high temperature and pressure. They condense as a liquid when the gas is produced. For this reason, condensate is also called drip gas, casing head gas, or natural gasoline (pure gasoline). Condensate has very high API gravity. It is transparent or yellowish in color. *Wet gas* contains condensate and *dry gas* contains no condensate.

Crude oil production is measured in barrels of oil per day (BOPD) and one barrel equals 42 gallons (US). Natural gas production is usually measured in cubic feet under standard conditions (atmospheric pressure). Standard abbreviations include Mcf (1000 cubic feet) and MMcf (1 million cubic feet). Natural gas production is sometimes measured in British Thermal Units (BTUs). This is a measurement of heat content. One cubic foot of natural gas yields 900 to 1200 BTUs. A *BTU* is the amount of heat required to raise the temperature of one cubic foot of water one degree Fahrenheit.

Oil and Gas Accumulation

Requirements for the accumulation of oil and gas in an oil trap or gas trap follow.

- A *source rock* that contains organic material from which hydrocarbon is formed
- A *reservoir rock* that stores the accumulated hydrocarbons
- A *trap* or *seal*—a structure or rock material change that blocks the flow of hydrocarbons out of the reservoir rock and allows them to accumulate
- A *migration path* or fluid flow conduit (Fig. 2–38) that allows hydrocarbon-forming material to flow from the source rock to the reservoir rock

The vast majority of reservoir-quality rocks are not capable of being a source rock. Reservoir rocks or faults might be the migration path for an accumulation of hydrocarbons.

Clastic sedimentary rocks are most conducive to accumulation of petroleum. To see this, look at the basic components of clastic sedimentary rock.

- The *matrix* comprises the rock grains or fragments.
- *Cements* may be absent in unconsolidated rocks, but they bind fragments or grains together in the lithification process.
- *Pores* are openings filled with a fluid (oil, gas, or water).

The sandstone matrix in Figure 2–39a consists of rounded fragments that have large pores between grains. The pores tend to be connected, allowing fluid to flow easily through a rock. The shale matrix consists of *platy*, clay particles that overlap each other.

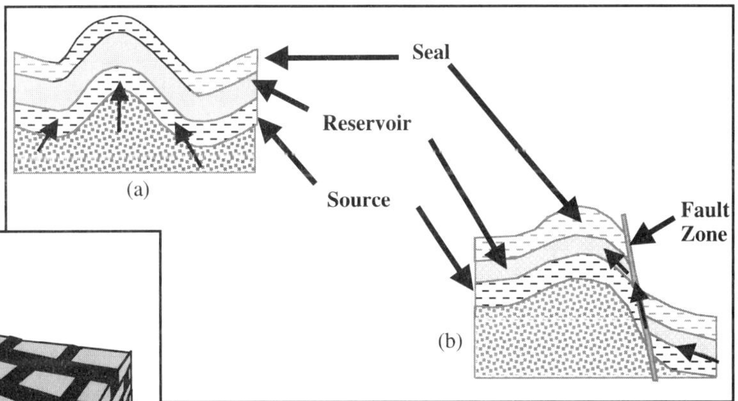

Fig. 2–38 *Migration of Hydrocarbons*

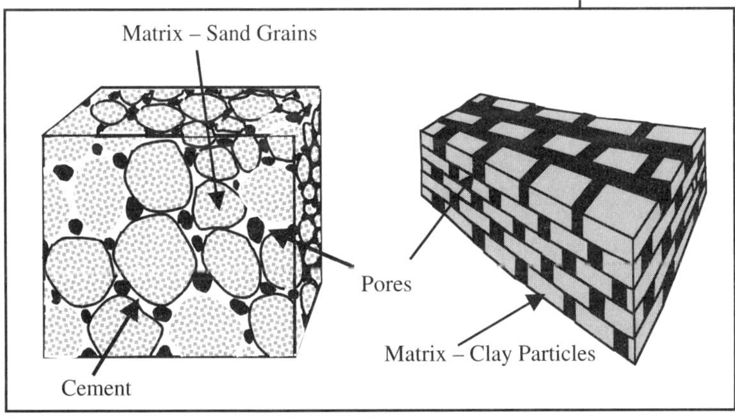

Fig. 2–39 *Components of Clastic Sedimentary Rocks*

Shale, illustrated in Figure 2–39b, may have high porosity but lacks interconnection of pores. Shale is usually impermeable and makes a good seal. Igneous and metamorphic rocks are neither porous nor permeable. However, fracturing can make any rock porous and permeable. Since oil and gas are less dense than water they will tend to migrate up in rocks due to buoyancy. This buoyancy migration will tend to be in reservoir rock or open fractures.

The holes or voids that give reservoir rocks porosity and permeability are of two types:

- *Intergranular*—openings between sedimentary grains
- *Intragranular*—openings within leached sedimentary grains

Intergranular porosity can result from original spaces between grains at the time of deposition remaining after lithification or from fractures after lithification. Solution cavities in limestones can produce interconnected pores or voids that allow fluid flow through the reservoir rock. Table 2–4 lists classifications of rock porosity and Table 2–5 gives classifications of permeability.

Table 2–4 *Reservoir Rock Porosity*

Ranges of Values	Description
0 to 5%	Negligible
5 to 10%	Poor
10 to 15%	Fair
15 to 20%	Good
20 to 25%	Very Good
greater than 25%	Excellent

Table 2–5 *Reservoir Rock Permeability*

Ranges of Values	Description
less than 0.1 to 1 md	Poor
1 to 10 md	Fair
10 to 100 md	Good
100 to 1000 md	Very Good
greater than 1000 md	Excellent

Reservoir rock porosity is either primary, with the remainder of the original spaces between sand grains and particles or fossil pieces, or secondary, meaning it was formed after deposition and burial. Secondary porosity results from solution in *fossil molds* (or vugs) and from fractures, such as between crystals in limestone and dolomite. Porosity may be measured by visual examination of well cuttings taken from core samples, a standard laboratory procedure, or from wire-line well logs.

In deep reservoirs, as cementation and compaction increase, porosity and permeability decrease. Gas reservoirs occur at greater depths but require less porosity and permeability to be productive.

Reservoir rock permeability indicates how easily fluids can flow through a rock and thus, depends on interconnection of the pores. Permeability is measured using a *perm plug*, a cylindrical piece of rock drilled from a core. Tight sands and limestone have permeabilities of less than 5 md.

Source rocks are sedimentary rocks in which organic matter is preserved. Black sediments have high organic content. Coals are preserved woody material. Black shales have 1 to 5% organic matter. They are the most common source rock.

Generation, migration, and accumulation of petroleum. Methane (swamp gas) is generated at shallow depths via biogenic or bacterial activity. However, this methane is generated at too shallow a depth for large quantities to be trapped. (Some efforts are being made to capture such gas produced at municipal landfills.) Bacterial action decreases with increasing depth and temperature.

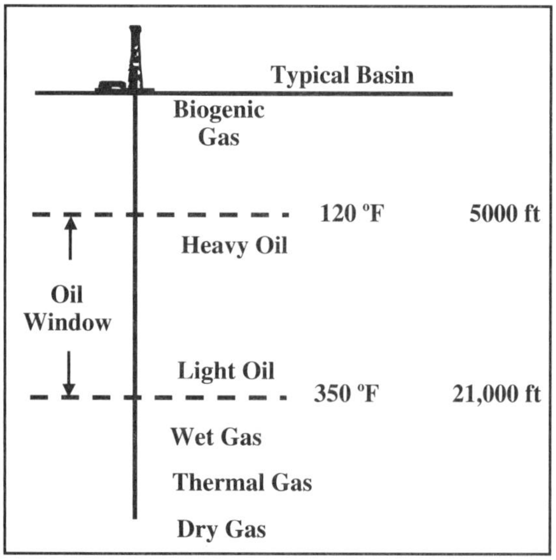

Fig. 2–40 *The Oil Window*

Crude oil is generated very slowly, taking millions of years at temperatures of from 120°F to 350°F. Heavy oils with low API gravity are generated at the lower temperatures in this range. (Heavy oil may be biodegraded lighter oil.) Light oils are generated at the higher temperatures in the range. Thermal gas is generated at temperatures above 350° F. At these temperatures crude oil breaks down into graphite (C) and gas.

Organic matter and coal generate gas. Wet gas with associated condensate is generated at shallower depths. Dry gas with no liquids is generated at deeper depths.

The oil window (Fig. 2–40) is the subsurface region of oil generation. The temperature range of 120°F to 350°F corresponds to a depth range of about 5000 ft (1524 m) to 21,000 ft (6400 m). Heavy oil is generated at the top of the oil window and light oil at the bottom. Similarly, wet gas is generated just below the oil window and dry gas at deeper depths.

The reason for the differences in type of petroleum generated at different depths and temperature ranges becomes clear when the nature of heat is investigated. Heat is actually molecular motion. The higher the temperature, the faster the molecules move. The larger the hydrocarbon molecule, the less stable it is at high temperatures. Thus temperature establishes a ceiling on molecular size.

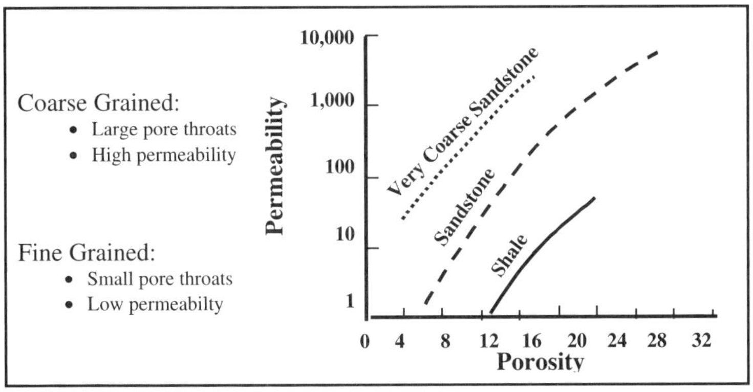

Fig. 2–41 *Effect of Grain Size on Porosity and Permeability*

Note that the preceding deals only with the depths of petroleum generation and not the depths of its occurrence. Depth indicates the heaviest hydrocarbon molecules that can be found, not the lightest. Dry gas can be found at even shallow depths.

Barren basins result when there are no source rocks, the source rocks are not buried deeply enough, or the organic material available is the wrong type of organic source for oil.

In migration, hydrocarbons are expulsed from their shale source rock. Shales are compacted with burial, decreasing pore space and expelling fluids (hydrocarbons and water). Fracturing caused by overburden load also releases hydrocarbon from shales, as does volume increase resulting from generation of oil and gas from solid organic matter.

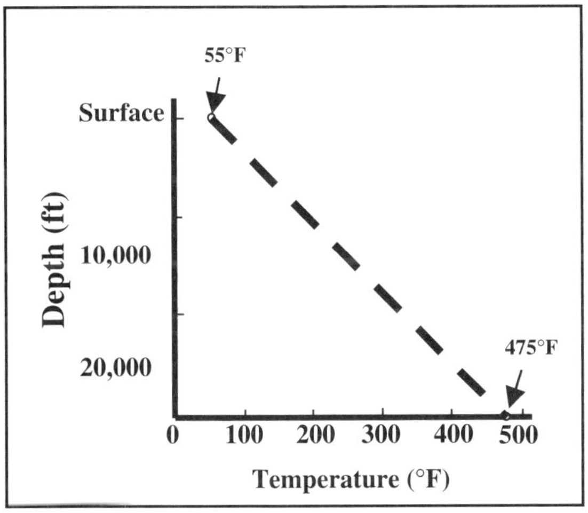

Fig. 2–42 *Geothermal Gradient in Sedimentary Basins*

Horizontal migration can be up to 200 miles. Oil begins its migration from within the oil window. Gas can start its migration from below the oil window.

Reservoir rock fluid saturations refer to relative percentage of fluids in pores. Water saturation plus oil saturation equals 100%. *Wetting fluid* (also called *water wet*) is the fluid that covers the pore wall surfaces. Most reservoirs are water wet because water was present first. Oil forced its way into pore space but could not push water off of pore walls due to surface tension. Increasing water saturation decreases the volume of oil in the reservoir and the permeability of the oil-fluid phase, making it harder for oil to push past water coating pore throats.

Water saturation is calculated from wire-line well logs—induction or electrical (resistivity)—and core samples. *Irreducible* or *residual water saturation* is water that cannot be driven from pore space by the intrusion of oil. It is water held by surface tension to rock surfaces in pores and is on the order of 20 to 30%.

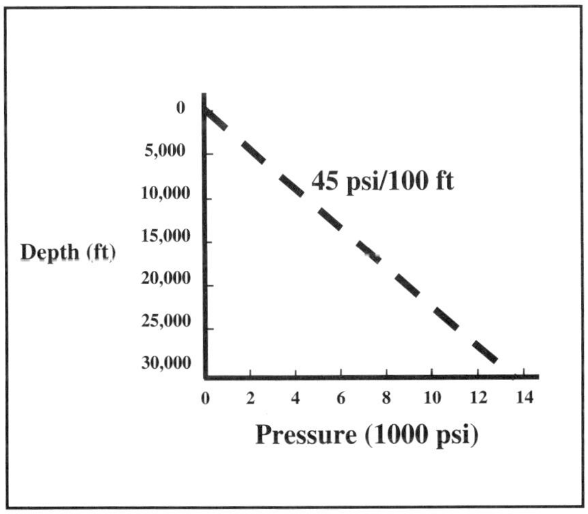

Fig. 2–43 *Hydrostatic Pressure Gradient*

Fine-grained rocks have smaller pores, meaning they have little or no permeability. Shales are impermeable, thus, shales make excellent seals for hydrocarbon traps. Rocks consisting of interwoven crystals also have low permeability (Fig. 2–41).

Ground water tends to become more saline with depth. Salinity is measured in parts per million (ppm) of salt. Fresh water has 0 ppm, brackish water has more than 5000 ppm, and brine more than 300,000 ppm. Water found in oil and gas wells is usually brine.

Temperatures increase with depth in the Earth. The *geothermal gradient* (Fig. 2–42) measures the rate of increase. The average for the Earth as a whole is 2°F/100 ft. However, in sedimentary basins it is only about 1.4°F/100 ft. This difference is due to rocks with a lower thermal conductivity as well as the increased thickness above the mantle.

Reservoir, formation, fluid, and bottom-hole pressure also increase with depth. The *hydrostatic pressure gradient* (Fig. 2–43) measures the rate of pressure increase with depth. Its average value is 45 pounds per square inch (psi) /100 ft. Note that atmospheric pressure under standard conditions is only 14.7 psi. Pressure increases because of the weight of the overlying water column.

Hydrocarbon Traps

Since most sedimentary rocks are formed in water, they tend to be saturated—their pores are completely filled with water. Hydrocarbons migrate up through overlying permeable sediments and rocks since hydrocarbons are less dense than water. If there is no low permeability layer (permeability barrier), then the hydrocarbons will be vented to the atmosphere. Thus, a hydrocarbon trap requires a seal—an impermeable rock or fault to prevent upward flow of hydrocarbons and restrict lateral flow out of the reservoir rock.

Hydrocarbons are usually trapped at a high point in a reservoir rock. That can be a fold in a rock, as in an anticline, or at the termination of reservoir rock, as in a *pinch-out*—rock changes from reservoir quality to tight, not permeable or porous.

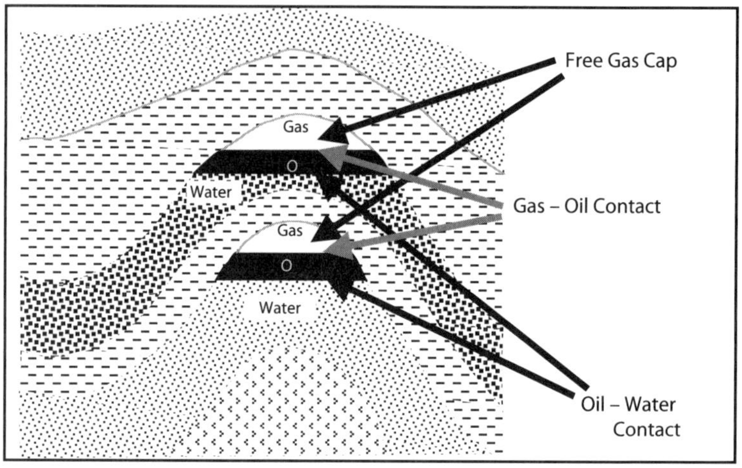

Fig. 2–44 *Anticline Trap*

There is a separation of gas, oil, and water in a trap. Gas is above the oil and oil floats on water. Gas may be above the water if there is no oil in the trap. Some important levels in a trap are the oil-water contact and the gas-oil contact. In gas reservoirs there is a gas-water contact.

Hydrocarbon traps are located by seismic reflection prospecting, rock outcrops, oil and gas surface seeps, and by drilling.

There are three general classes of hydrocarbon traps:

- *Structural traps* are formed by changes in shape and attitude of rock layers caused by folding and faulting.
- *Stratigraphic traps* are formed by changes in rock character, such as a facies change or erosional up-dip truncation.
- *Strati-structural* or *combination* traps combine elements of structural and stratigraphic traps.

An anticlinal dome is, perhaps, the perfect oil trap (Fig. 2–44). They are similar to inverted bowls so that migrating hydrocarbons cannot escape the trap unless they completely fill the anticline's closure. *Closure* is the difference in elevation between the top of the reservoir rock structure and the spill point. Tilting of a structure (regional dip) causes the resulting accumulation to be smaller (Fig. 2–45).

Traps are not necessarily always full. A hydrocarbon trap is caused by the combination of the buoyancy of oil and gas in water and the dead-end migration flow path. A dome with overlying shale provides a tight seal. A trap may have oil only, oil and gas, or gas only in it. The amount of petroleum in a trap is a function of the porosity and the height of the hydrocarbon column, usually up to a maximum of the amount of structural closure. The areal extent of closure is used to calculate the volume of petroleum in the trap. This volume determines whether the trap is valuable. Closure is determined from structure maps of the reservoir rock.

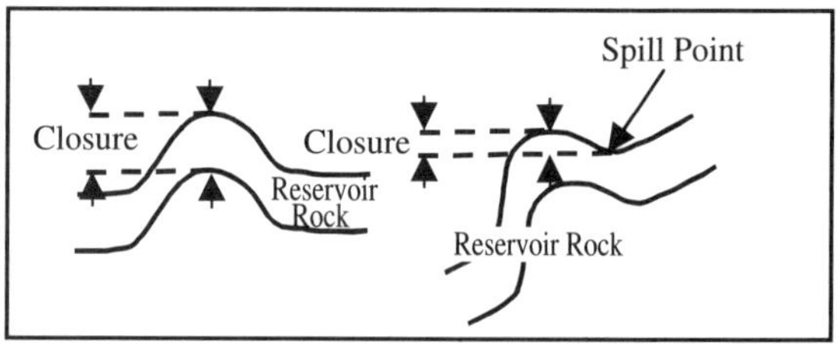

Fig. 2–45 *Reservoir Spill Point*

The relationship between gas and oil in a trap is a function of pressure and temperature.

A well drilled into a *free gas cap* would produce only gas. A free gas cap occurs because oil cannot dissolve any more gas, it is saturated. When gas is dissolved in oil with no gas cap, the oil is undersaturated. A trap with gas only means that the petroleum was formed at high temperature. It is said to be a *mature* hydrocarbon. See Figure 2–46.

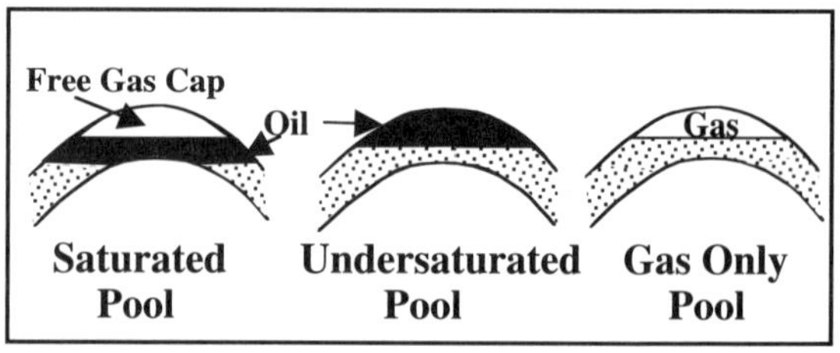

Fig. 2–46 *Reservoir Fluid Types*

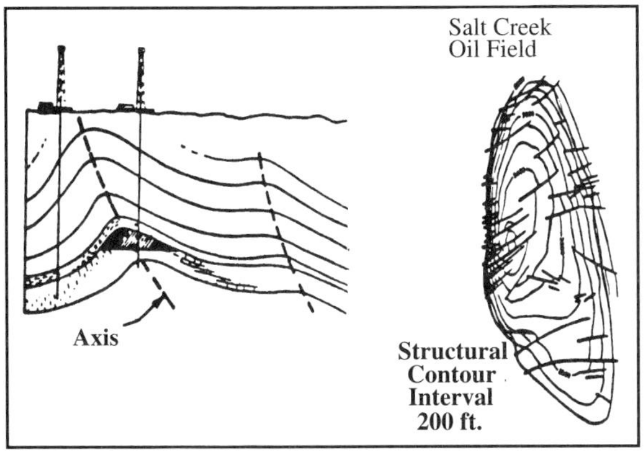

Fig. 2–47 *Asymmetrical Anticline Trap*

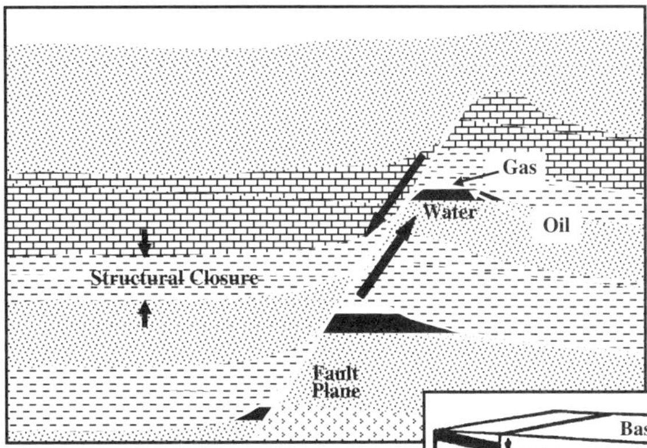

Fig. 2–48 *Fault Trap*

Figure 2–47 shows an asymmetrical anticline trap. The top of the structure is offset with depth. The structure top at the surface or at shallower layers is offset from the top at deeper layers.

Pay zone refers to the length of the producing reservoir rock in an oil or gas well. The thickness of the pay zone is a function of the position of the well on the structure. A well on top of a structure will have a thicker pay zone than a well with less closure or closer to the edge of the trap.

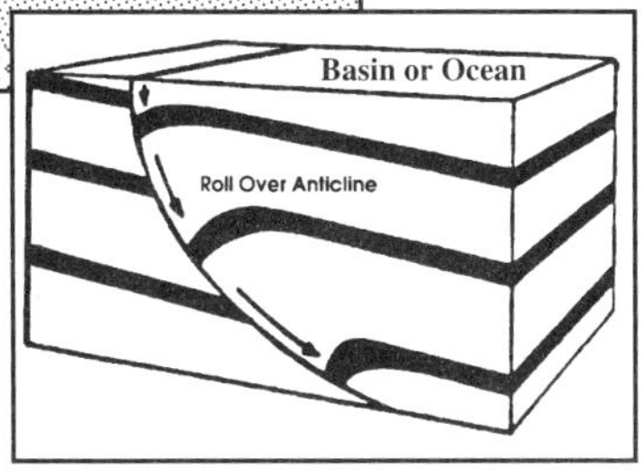

Fig. 2–49 *Growth Fault or Down-to-the-basin Fault*

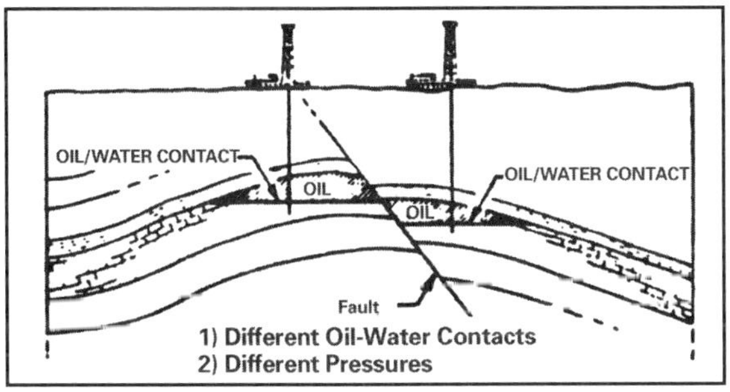

Fig. 2–50 *Fault Cutting Reservoir into Separate Accumulations*

If movement of fault blocks places a porous and permeable rock under an impermeable rock and the fault plane is closed to fluid flow, then a trap is formed. Traps can form on either side of a fault as seen in Figure 2–48.

If the oil-water contacts are at different levels across a fault plane, then the fault surface must be a seal as seen in Figure 2–48.

In growth or down-to-the-basin faults that have rollover anticlines, high quality sandstones tend to be concentrated on the down side of fault. They are called localized *depocenters* or centers of deposition. See Figure 2–49.

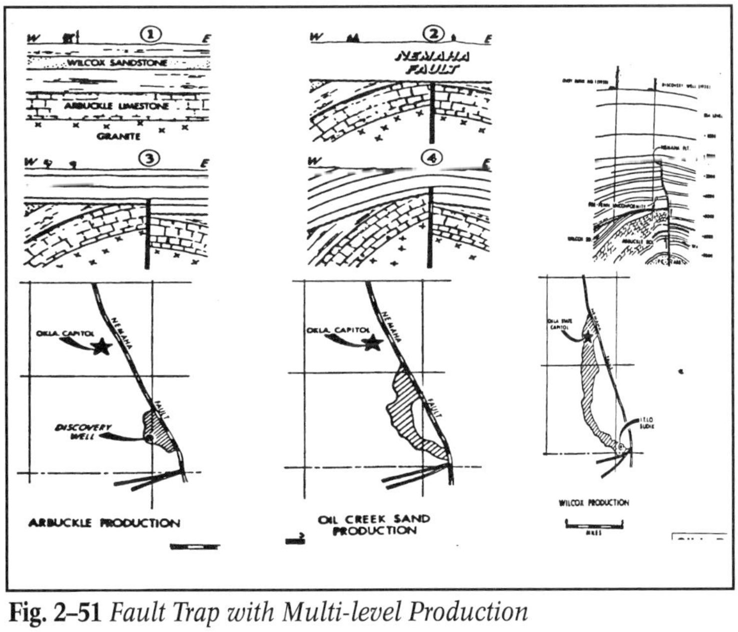

Fig. 2–51 *Fault Trap with Multi-level Production*

Faults can divide an existing reservoir in two, as shown in Figure 2–50. A fault trap, as well as an anticline, can have production at multiple levels. This is shown in Figure 2–51.

In a thrust fault there is the potential for *drag fold* on either side of fault surface. The drag fold is formed by the friction generated by one side of the fault pushing against the other (compressional stress). Overthrust belts are regions of low angle faults. They are usually associated with mountain ranges on land. Examples include the Rocky Mountains or the western overthrust belt of the United States.

The Painter Reservoir Field in Wyoming, discovered in 1977, is an example of traps found in overthrust regions. A drag fold on the thrust fault (left side of Fig. 2–52) forms the trap. The nugget sandstone (eolian dunes) is the reservoir rock.

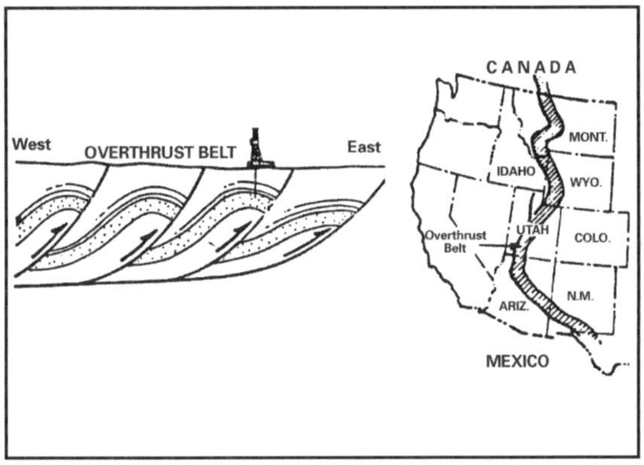

Fig. 2–52 *Western Overthrust Belt*

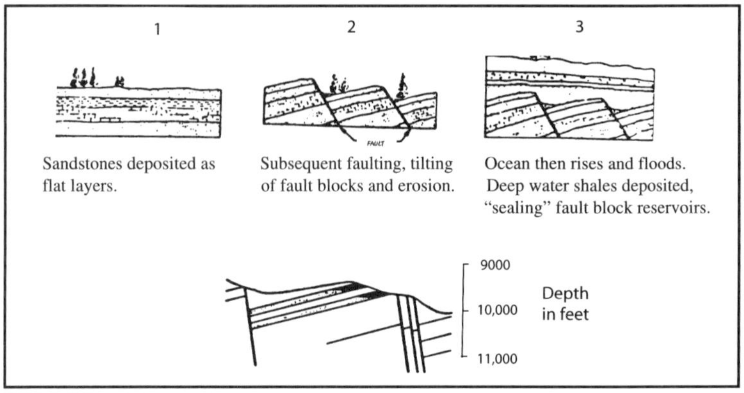

Fig. 2–53 *The Stratfjord Field*

A trap can also be formed by an up-dip facies change. That is, the reservoir rock changes from porous and permeable to tight or impermeable. Up-dip erosion of a porous layer—an erosional truncation—can also form a trap. The Stratfjord Field in the North Sea is a good example. It is an eroded, tilted fault block. Ultimate recovery is estimated at three billion barrels of oil (Fig. 2–53).

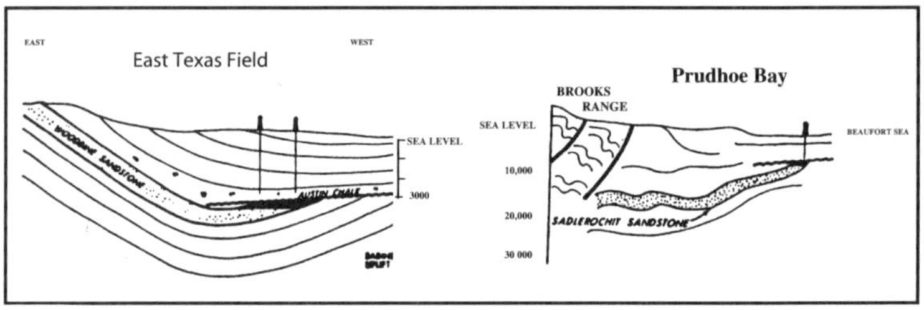

Fig. 2–54 *Oil Fields Formed by Angular Unconformities*

Angular unconformities, resulting from termination of the reservoir rock by erosion, form excellent traps. Two giant oil fields—the East Texas and Prudhoe Bay (Alaska) fields—were formed by angular unconformities (Fig. 2–54).

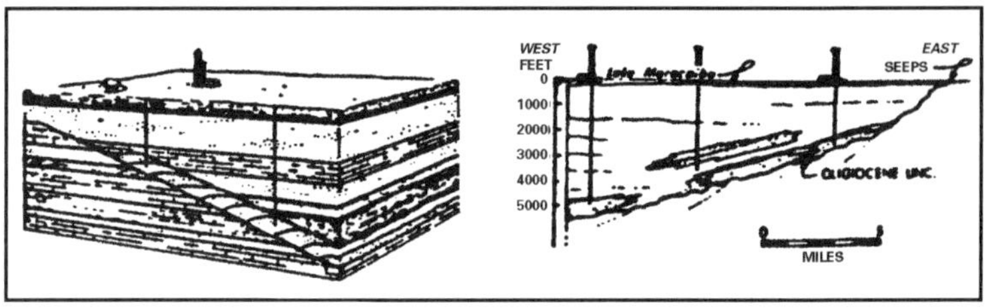

Fig. 2–55 *Production from Sands above Angular Unconformity, and Cross Section through the Bolivar Coastal Fields in Lake Maracaibo, Venezuela*

Sandstone reservoirs deposited on an angular unconformity take the form of buttress sands or onlap sands. They are beach sands deposited by rising sea level. An example is the Bolivar Coastal Fields in Lake Maracaibo, Venezuela. There are seeps along oligocene angular unconformity, seen at the right side of Figure 2–55. Ultimate recovery from this field is estimated at much more than 30 billion barrels of oil.

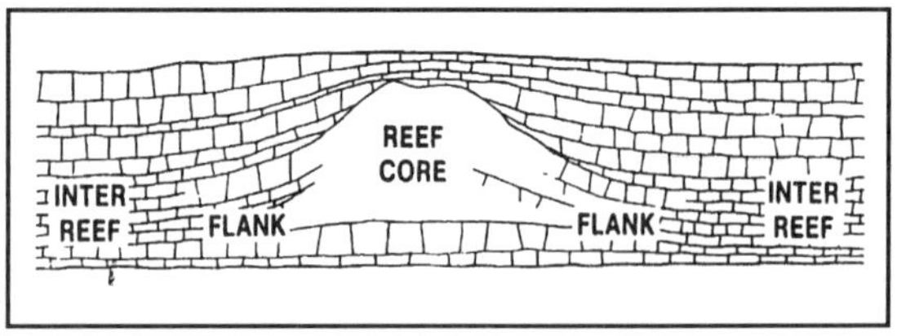

Fig. 2–56 *Reef Facies*

Reefs, organic debris in the form of marine animal skeletons, also form traps. Porosity is usually nearly all secondary, that is, solution porosity that forms vugs or openings in the reef body. Most of the primary porosity is cemented up at the time of deposition due to contemporaneous calcite cementation. The trap is a porous reef facies surrounded by a tight marine shale that acts as both seal and source. Figure 2–56 shows reef facies.

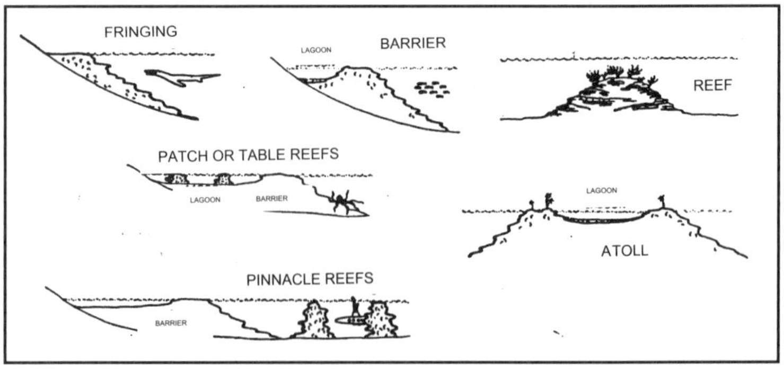

Fig. 2–57 *Reef Configurations*

Reefs come in a variety of configurations, as seen in Figure 2–57. The term reef refers to limestone deposits built up by a community of microscopic animals, not just one type

Although reefs existing today are associated with tropical or subtropical waters, oil fields with reef traps are found in parts of the world that are now far from tropical. Since reefs are built up close to a continent or island arc, they tend to occur as trends—along straight lines or arcs. Figures 2–58 and 2–59 show two such areas in Alberta, Canada and Michigan.

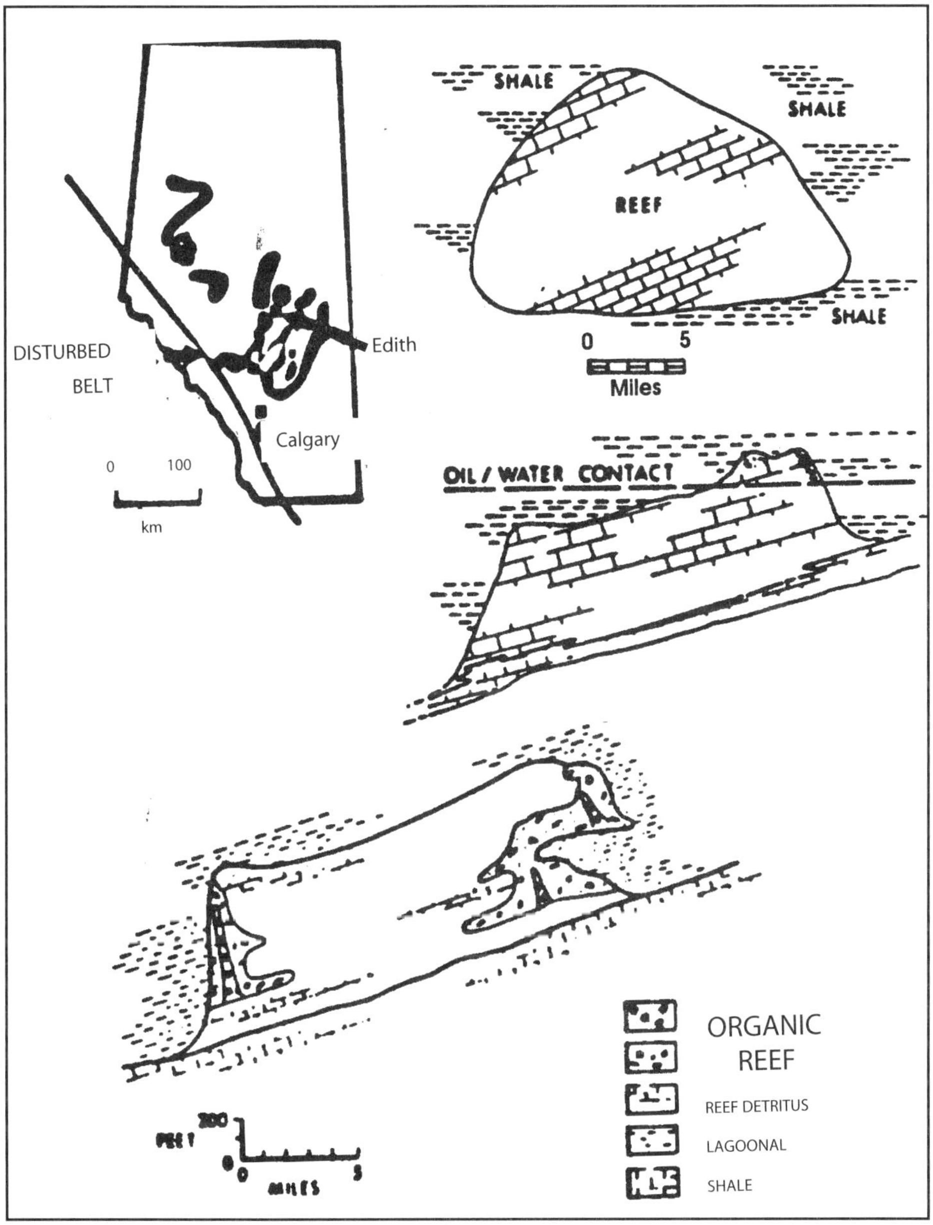

Fig. 2–58 *The Alberta Reef Trend and the Redwater Oil Field*

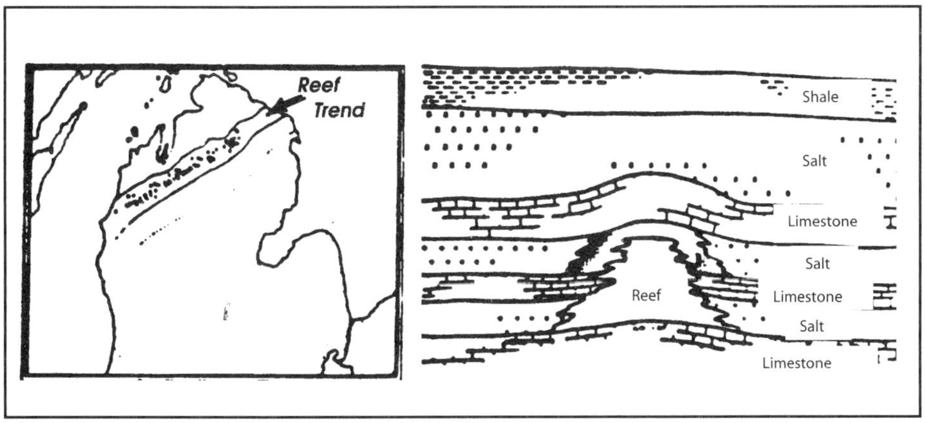

Fig. 2–59 *Michigan Basin Silurian Reef Trend*

Figure 2–58 also shows configuration of reefs and reservoirs in the Redwater oil field, discovered in 1949. The field is a carbonate platform covering 200 square miles and is up to 800 feet thick. Reefs rim the platform. Production is controlled by the oil-water contact (OWC), as seen in the lower center drawing of Figure 2–58. Ultimate recovery from the Redwater field is an estimated 850 million barrels of oil.

The left side of Figure 2–59 shows the Michigan Basin Silurian *pinnacle reef* trend and the right side shows a typical pinnacle reef in the basin.

Lenticular sandstone reservoirs are usually encased in shale. They may be completely filled with petroleum. Lenticular sandstones may be formed by sites like beaches, islands, river channels, or point bars. The shape of the sand body, in profile, indicates its original nature. See Figure 2–60.

Figure 2–61 shows the Bush City Trend in Kansas. It is a river channel fill, 800 ft below the surface, 13 miles long, 1000 to 3000 ft wide, and up to 35 ft thick.

Up-dip pinch-outs of coastal plain sands frequently occur. The Texas coast is a good example. The traps formed by these pinch-outs are usually small in size. See Figure 2–62.

The Hugoton gas field is the largest gas field in North America. It extends from southwestern Kansas across the Oklahoma panhandle into the Texas panhandle in the South central part of the United States. Estimated ultimate production is 70 trillion (one million million) cubic feet of gas. The field is formed by a permeable carbonate reservoir rock that becomes impermeable going from east to west. The seal for the trap is the overlying Wichita salt section. Subsequent uplift by the Stratford Arch gives the trap its structural component with an up-dip seal. Source rocks are from the Anadarko basin to the east. The field is, thus formed by a combination trap.

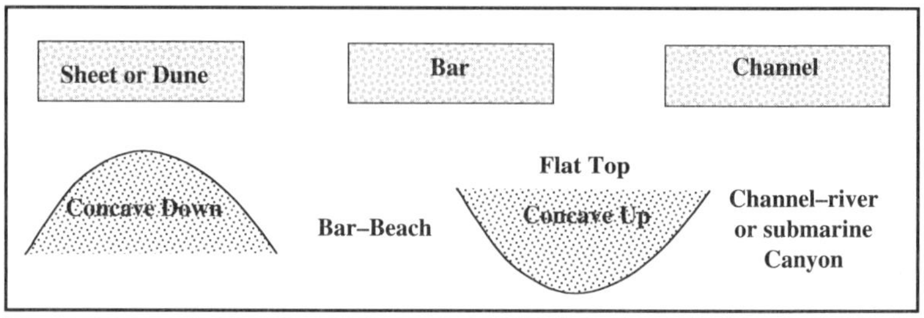

Fig. 2–60 *Lenticular Sand Shapes and Origins*

Another type of combination trap is a salt dome, pillow, or plug. These are salt masses that rise by buoyancy through overlying, denser sediments. They move upward in the form of a plug, usually termed a *salt diapir*. The salt was originally deposited in flat layers, from evaporation in salt flats or precipitation from highly saline waters. Salt is crystalline and nonporous. It is also not compactable. It stays at a constant density of 2.2 gm/cm^3. The overlying sands and shales originally had high porosities, but porosities decrease with burial and densities increase. Salt, sand, and shale have the same density at a depth of 2000 ft. Below 2000 ft, sand and shale are more dense than salt, and the salt begins to deform or flow. It moves laterally, then upward at zones of weakness due to buoyancy caused by the density gradient.

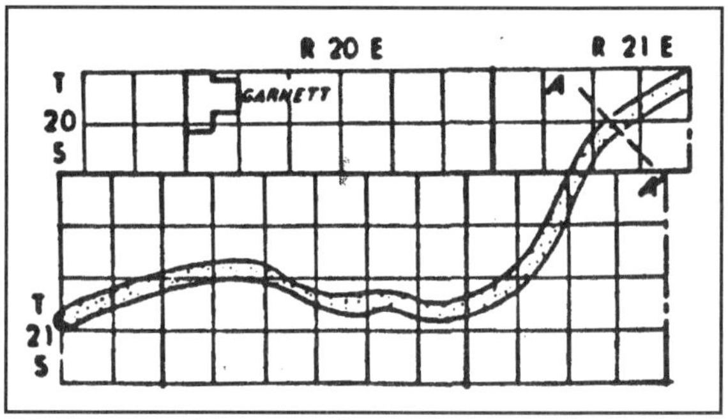

Fig. 2–61 *Bush City Pool*

The overburden, consisting of domed and faulted sediments, is shallow on top of a salt dome. Some salt domes have a *cap rock*—material that was just above the salt when it began rising. The first U.S. production was from uplifted and pierced sediments along flanks of salt bodies. The famous Spindletop oil field, near the eastern gulf coast of Texas, is of this type. The Ekofisk Oil Field, in the North Sea, produces from Ekofisk chalk that was fractured by an uplifting salt dome. Its ultimate production is estimated at 6 billion (6 thousand million) barrels of oil. Figure 2–63 shows sketches of both of these fields.

Fractured reservoirs are rocks that would otherwise be tight or impermeable. Shales, cherts, chalks, and granitic or metamorphic basement can be made porous and permeable by fracturing. For example, the basement in the center of an anticline is often highly fractured. See Figure 2–64.

Weathering of impermeable rocks can also form reservoir rocks. For example, *granite wash* (Fig. 2–65) is a sandstone soil zone on flank of eroded, granite basement high. It can be up to 100 feet thick and usually occurs as a rim around the basement high.

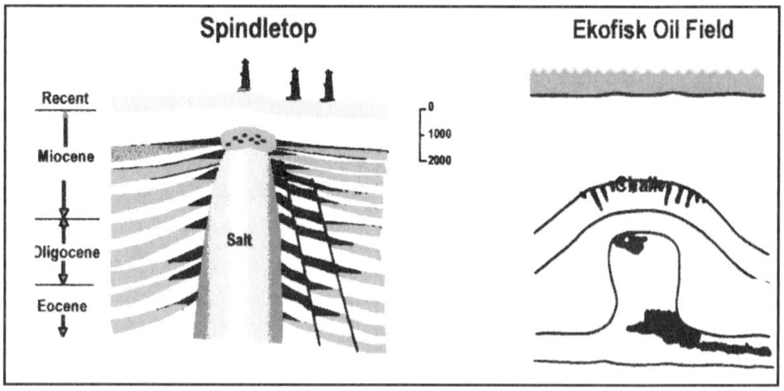

Fig. 2–63 *Giant Oil Fields Formed by Salt Domes*

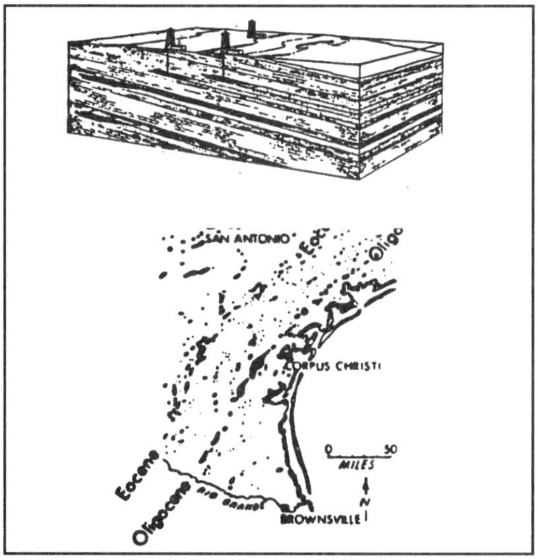

Fig. 2–62 *Coastal Sand Pinch-outs*

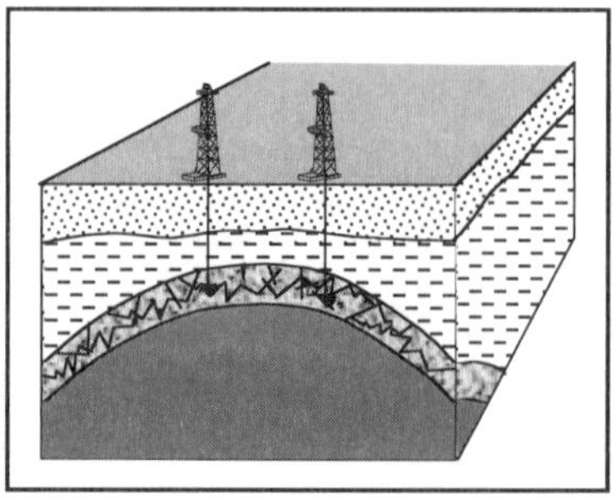

Fig. 2–64 *Fractured Reservoir Rock*

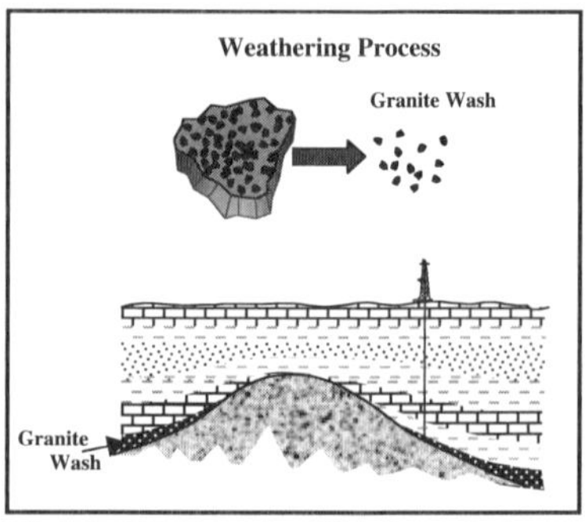

Fig. 2–65 *Granite Wash*

Workshop

1. Prior to the breakup of Pangea, there was a small sea between what is now New Brunswick, Canada, and Morocco. Since oil has been found off the coast of New Brunswick, do you think this makes it more or less likely that oil can be found off the Atlantic coast of Morocco?

2. Some tectonic plates are plunging beneath other tectonic plates, resulting in parts of the plunging plates being absorbed into the mantle. What relevance does this have to the amount of oil to be found in the vicinity of the plunging plates?

3. What processes are involved in the formation of clastic sedimentary rocks?

4. The amount of calcium carbonate that can be dissolved in water decreases as water temperature decreases. The temperature of water in seas and oceans gets lower as water depth increases. What do these two statements imply about the depths at which limestones form from precipitation?

5. Shale is formed from clay particles and sandstone is formed from sand. Which type of rock would you expect to form farther from shore in a sea or ocean?

6. Two sandstone reservoirs have equal volumes. One reservoir has a porosity of 20% and a permeability of 100 md. The other reservoir has a porosity of 16% and a permeability of 400 md. Which is likely to contain more petroleum?

7. How are the geologic laws used, along with fossil evidence, to determine the geologic time scale?

8. Do you think that the findings of historical geology—the branch of geology related to geologic time and time sequence of geologic events—support the theory of evolution? Why or why not?

9. In North America, the Pennsylvanian and Mississippian are considered to be Periods of the Paleozoic Era. In Europe, the Carboniferous Period covers essentially the same time span, and the Pennsylvanian and Mississippian are not recognized as Periods. What does this imply about the orogeny that separated the Pennsylvanian and Mississippian Periods?

10. Two fossils, designated *X* and *Y*, are in found in two wells, *A* and *B*. Fossil *X* is found in formation 1 at a depth of 359 m below sea level in Well A and at 423 m below sea level in Well B. Fossil *Y* is found in formation 2 at a depth of 571 m below sea level in Well A and at 807 m below sea level in Well B. What does this imply about formations between 1 and 2 between the two wells?

11. Based on outcrops, a sandstone bed is found to have a strike of N 23° W. When looking along the strike, the bed dips down to the right. In what direction should the dip be measured?

12. Assume that Figures 2–66 through 2–69 show surface expressions of eroded geologic structures. What types of structure do they represent?

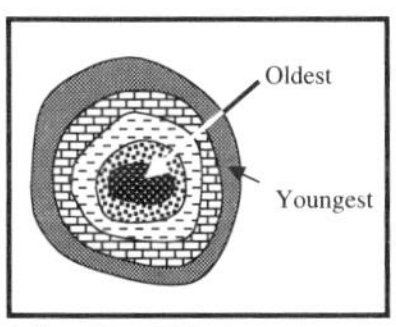

Fig. 2–66

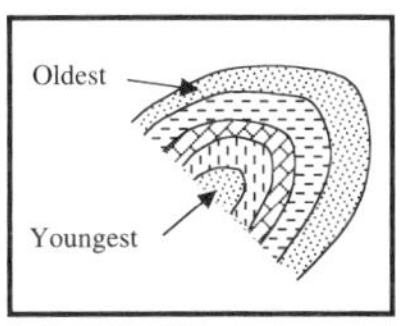

Fig. 2–67

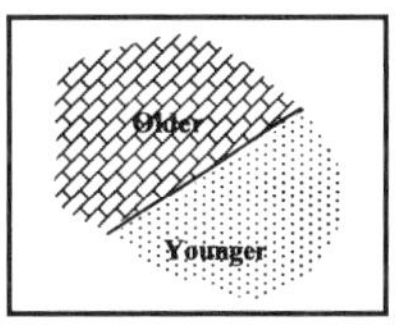

Fig. 2–68

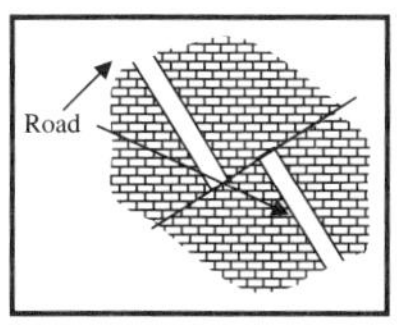

Fig. 2–69

a.

b.

c.

d.

13. Sketch a normal fault that has a heave of 20 m and a throw of 50 m. What is the dip angle?

14. What kind of fault does Figure 2–70 represent?

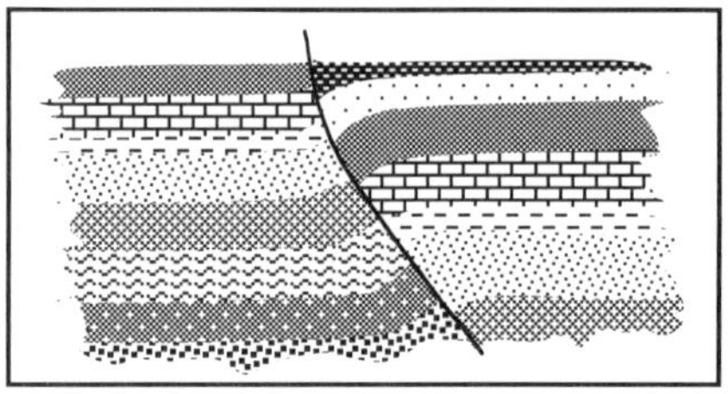

Fig. 2–70

15. Why are sulfur and oxygen undesirable elements in crude oil?

16. Ranges of crude oil densities in Degrees API for heavy crude, average crude and light crude are as follows:

    - Water—10
    - Heavy Crude—5 to 20 (viscous, high sulfur content)
    - Average Crude—30 to 40
    - Light Crude—45 to 55 (very fluid, most valuable)

    Calculate the corresponding ranges of specific gravity.

17. We know that temperature is a measure of heat content, and temperature increases with depth in the earth. Further, heat is actually the amount of molecular motion in a solid or fluid. How might these facts account for the absence of the larger hydrocarbon molecules when the source rocks are below the oil window?

18. The average thermal gradient is about 2° F, but in sedimentary basins the thermal gradient is about 1.4 °F. How do you account for the difference?

19. Determine the trap type in Figures 2–71 through 2–74.

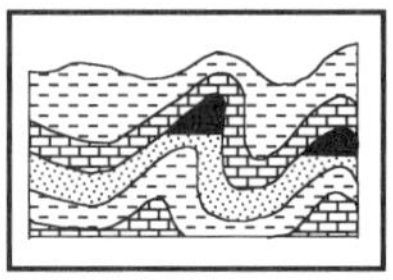

Fig. 2–71

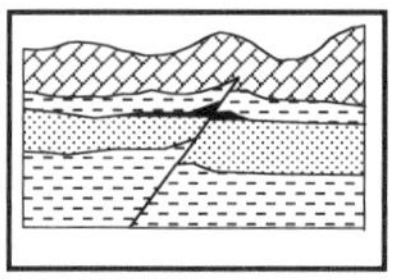

Fig. 2–72

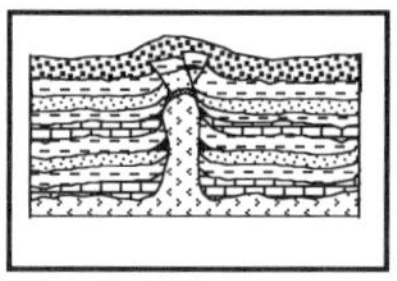

Fig. 2–73

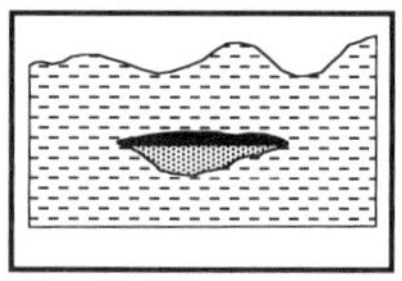

Fig. 2–74

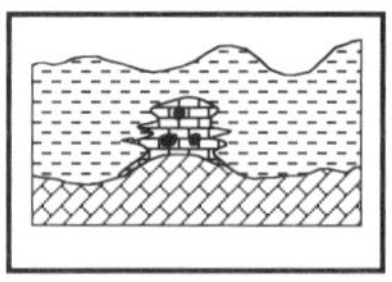

Fig. 2–75

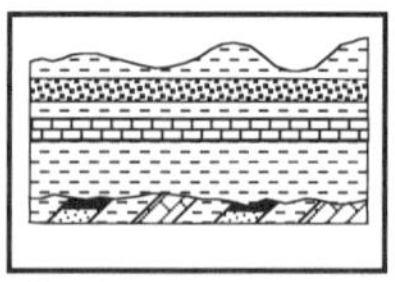

Fig. 2–76

a.

b.

c.

d.

e.

f.

20. Indicate for the rocks in Table 2–6 whether they are porous or nonporous and permeable or impermeable. Could all these rocks be porous and permeable in some situations? Explain your answer.

Table 2–6 Rock Grid

Rock	Porous	Nonporous	Permeable	Impermeable
Granite				
Limestone				
Salt				
Sandstone				
Shale				

3 Geophysical and Mathematical Background

Basic Geophysical Theory

The theory upon which seismic exploration for oil is based comes from a branch of physics called *classical mechanics* that was developed by the late 17th to the early 18th century. Of particular interest to us is the development of the *wave equation*. The wave equation is applicable to virtually any kind of wave motion—mechanical or electromagnetic. The wave equation can be written as:

$$\nabla^2 u = \frac{1}{V^2}\frac{\partial^2 u}{\partial t^2} \quad \text{or} \quad \frac{\partial^2 u}{\partial x^2} + \frac{\partial^2 u}{\partial y^2} + \frac{\partial^2 u}{\partial z^2} = \frac{1}{V^2}\frac{\partial^2 u}{\partial t^2}$$

Where u is the wavefield; x, y, and z are the three coordinates of a three-dimensional coordinate system (for example, x may be distance along some particular direction, y is distance at right angles to x, and z is vertical depth or altitude); t is time; and V is propagation velocity or rate at which the wave progresses.

Seismic waves are mechanical waves in the solid earth. Solutions of the wave equation are facilitated if certain simplifying assumptions are made about the medium through which the seismic waves are propagating. These assumptions usually are that the medium is perfectly *elastic*, *homogeneous*, and *isotropic*.

An elastic solid is one in which *stress* is proportional to *strain*. Stress is a force applied perpendicular or parallel to a surface of a body divided by the area of that surface. Strain is the ratio of change in dimension(s) caused by an applied stress to the original dimension(s). Different constants of proportionality are used depending on the type of stress applied.

Young's modulus (E) is the constant for linear stress as seen in Figure 3–1a. It is defined as the force per unit area divided by the ratio of change in length to original length. Young's modulus ignores the effect of applying a linear stress to a body. The applied stress does cause a strain in the direction of the applied stress, but it also causes a strain in the orthogonal directions. Figure 3–1b shows the *bulk modulus* (k) that is used to relate stress to volumetric strain. Bulk modulus is defined as the force per unit area acting on a surface divided by the ratio of change in volume to the original volume. In Figure 3–1b, it can be seen that a stress applied vertically changes the vertical dimension of a cube from *d* to ($d - \Delta h$), but the two horizontal dimensions are changed from *d* to ($d + \Delta d$). Thus, a stress applied in one direction results in strains not only in the direction of the stress but in the orthogonal direction as well. It can be seen that the compression in the vertical direction decreases the vertical dimension while increasing the horizontal dimensions.

The ratio of the strain in an orthogonal direction to the strain collinear with the stress is called *Poisson's ratio* (σ) and can be seen in Figure 3–1c. Poisson's ratio varies from 0 to 0.5. The drawing in Figure 3–1d illustrates the *shear modulus* (μ).The stress in this case is across, rather than perpendicular to, the surface on which it acts. The deformation is in shape rather than volume or linear dimension. Shear modulus is defined as the force per unit area acting on a surface divided by the angle through which one dimension is rotated.

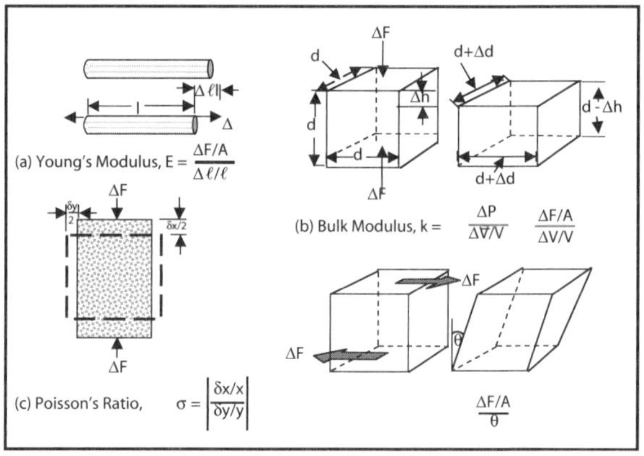

Fig. 3–1 *Basic Elastic Moduli and Poisson's Ratio*

A homogeneous body is identical throughout. If a homogeneous body were divided into pieces, then every piece would be chemically and physically identical, no matter how small the pieces. This, of course, cannot be realized in the rocks through which seismic waves propagate. Rocks are usually made of fragments or interlocking crystals. If such rocks were cut into very small pieces, differences in pieces would be observed. Thus, rocks are *inhomogeneous*—not homogenous.

An isotropic medium has the same physical properties in all directions. Of particular concern are the propagation velocities of the medium—the time rate at which seismic waves travel through the medium. In an *anisotropic* (not isotropic) medium, propagation velocity does depend on direction of propagation. There are many types of anisotropy. One type, called *transverse isotropy*, behaves as though it were composed of thin isotropic layers. As a result, there is a difference between horizontal and vertical propagation velocities. Another type of anisotropy is found in rocks that have cracks running through them. In such cases, the velocity parallel to the cracks differs from velocity orthogonal to the cracks.

Wave propagation

Solutions of the wave equation for perfectly elastic, homogenous and isotropic media describe four different types of waves. Two of these propagate through the body of the medium and are called *body waves*. One is called a *P-wave* and the other is called an *S-wave*. Both are shown in Figure 3–2.

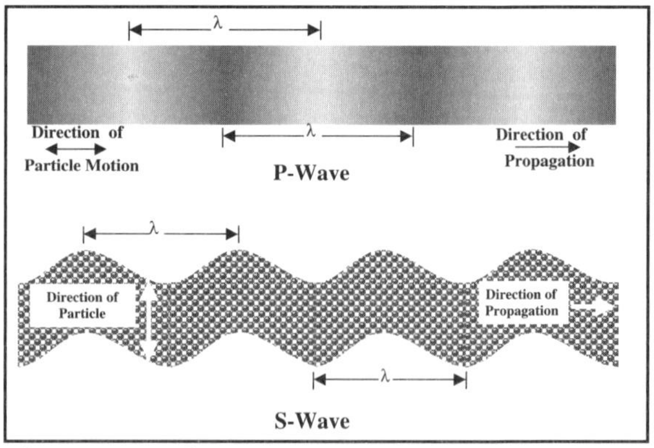

Fig. 3–2 *P- and S-waves*

A P-wave is, basically, a sound wave or pressure wave. The name, however, comes from *primus* or *first* because the P-wave is always the first-arriving seismic wave. Rock particles are alternately compressed and pulled apart as the wave passes by. Also shown in Figure 3–2 is *wavelength* (λ). The wavelength is the distance over which all possible amplitudes are reached. It can be measured from peak (maximum amplitude) to peak or from trough (minimum or maximum negative amplitude) to trough. P-wave particle motion is in the direction of propagation. P-wave velocity in a particular medium is the fastest of all seismic waves. If the elastic moduli and *bulk density* (ρ) are known, P-wave velocity can be calculated using the following equations.

$$V_P = \sqrt{\frac{k + \frac{4}{3}\mu}{\rho}} \tag{3.1}$$

or

$$V_P = \sqrt{\frac{E(1+\sigma)(1-2\sigma)}{\rho(1-\sigma)}} \tag{3.2}$$

S-waves, also called *shear* waves, are the second type of body wave. On earthquake seismograms, they are the second-arriving wave packets, hence the designation S-wave. While P-waves can propagate in solids or fluids, S-waves can propagate in solids only. S-wave velocity can be calculated from the following equations.

$$V_S = \sqrt{\frac{\mu}{\rho}} \tag{3.3}$$

or

$$V_S = \sqrt{\frac{E}{2\rho(1+\sigma)}} \tag{3.4}$$

P-wave velocity is always faster than S-wave velocity. Dividing Equation 3.2 by Equation 3.4 gives the P- to S-wave velocity ratio as:

$$\frac{V_P}{V_S} = \sqrt{\frac{2(1-\sigma)}{1-2\sigma}} \tag{3.5}$$

$$0 \le \sigma \le 0.5, \frac{V_P}{V_S} \ge \sqrt{2}$$

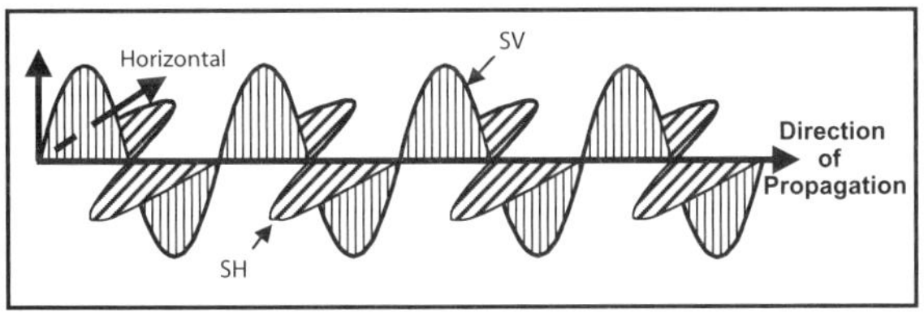

Fig. 3–3 *SV- and SH-waves*

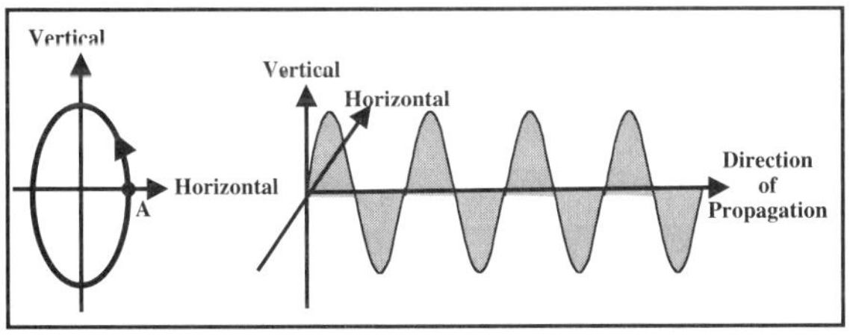

Fig. 3–4 *Rayleigh Waves*

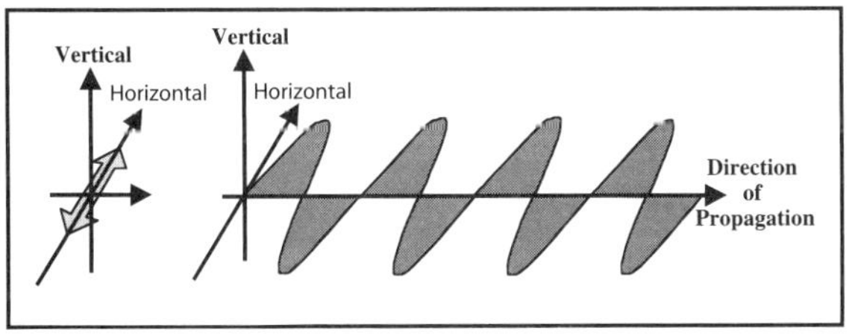

Fig. 3–5 *Love Waves*

S-wave motion can be resolved onto vertical and horizontal components, called *SV* and *SH*. In some anisotropic media, the S-wave train becomes separated into separate SV and SH wave trains because the two have different propagation velocities. Figure 3–3 illustrates SV- and SH-wave motion.

The other two solutions of the wave equation are called *surface waves*. These are restricted to the surface of solid media. There are two kinds of surface waves. One is called a *Rayleigh wave,* in honor of Lord Rayleigh, a 19th century British natural scientist who first developed the theory. Rayleigh wave motion is described as retrograde elliptical. That is, as the wave passes, a ground particle describes an ellipse, starting at point *A* in Figure 3–4. This is actually a combination of vertical and radial motion. Radial is horizontal motion in the direction of propagation.

Figure 3–4 shows Rayleigh wave motion. Rayleigh wave propagation velocity is about 92% of S-wave velocity.

Ground roll, a frequently encountered type of noise observed on exploration seismic records, is a Rayleigh wave. Ground roll can be the largest amplitude event on a seismic record and mask the reflected events that are desired.

The second type of surface wave is called a *Love wave*. Love waves are also named for a British scientist who developed the theory for them. As shown in Figure 3–5, Love wave motion is essentially SH motion but is restricted to the surface of the medium. Love wave propagation velocity is slightly slower than Rayleigh wave velocity.

Love waves are not usually a problem on exploration seismic records. However, if special energy sources are used to generate SH-waves, Love waves can be a severe problem because of their high amplitudes.

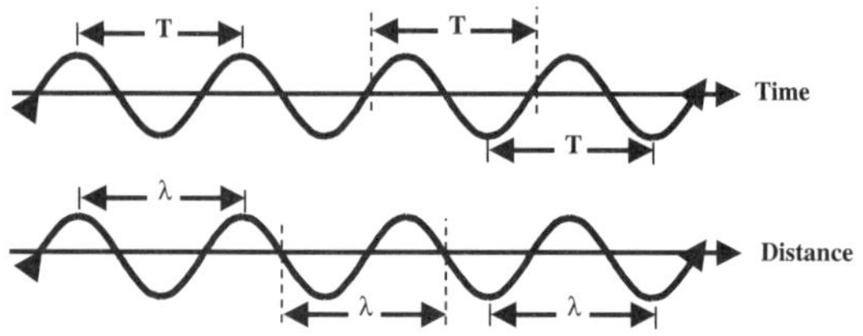

Parameter	Symbol	Units of Measurement
Period	T	seconds (s)
Frequency	f	Hertz (Hz)
Wavelength	λ	meters (m) or feet (ft)
Wave Number	k	cycles /m (c/m) or cycles/foot (c/ft)
Velocity	V	meters/second (m/s) or feet/second (ft/s)

$f = 1/T$ $k = 1/\lambda$ $V = f\lambda$

Fig. 3–6 *Basic Wave Parameters*

Figure 3–6 illustrates and defines some basic wave parameters. Two sinusoids are shown in Figure 3–6. The top one is a function of time and the bottom one is a function of distance. A *cycle* is defined as the interval over which all possible amplitudes occur. A *period* is the time, in seconds, required to complete one cycle. *Frequency*, the reciprocal of period, is the number of cycles per unit of time. Frequency is measured in Hertz (Hz), where 1 Hz equals one cycle/second.

A *wavelength* is the distance over which one cycle is completed. The reciprocal of wavelength is *wave number* (sometimes called spatial frequency). A sinusoidal wave is described by the mathematical expression

$$\psi(t,x) = A \cos[2\pi(ft-kx)+\phi] \quad (3.6)$$

where

$\psi(t,x)$ = the wave field

A = amplitude

t = time

x = distance

ϕ = phase

Amplitude is usually defined as the maximum deflection from the rest position (zero to peak) but is sometimes measured from peak to trough (peak-to-trough amplitude). Phase is an angular quantity that is usually measured in radians or degrees. Phase indicates the initial position of the wave in a cycle.

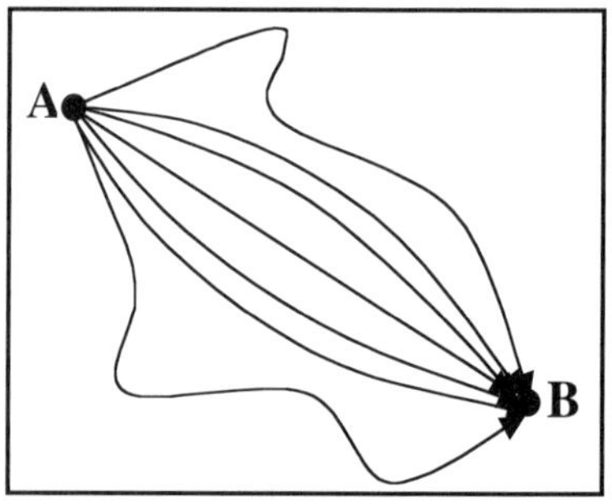

Fig. 3–7 *Fermat's Principle*

Much theoretical work in, or applicable to, seismic exploration is based on ray theory. Rays are simply lines (not necessarily straight) that show the paths waves take between two points. A very important concept in ray theory is *Fermat's Principle* (Fig. 3–7). This principle can be stated many ways. In Sheriff's *Encyclopedic Dictionary of Exploration Geophysics*, the first statement is, "The seismic raypath between two points is that for which the first-order variation of travel time with respect to all neighboring paths is zero." Further on, Sheriff states, "If the intervening media have different speeds, the path will not be straight but will be such that the overall travel-time is minimized (usually)." The resulting raypath is also called the *least time path* or *brachistochrone.*

Seismic exploration is normally conducted in areas called *sedimentary basins*. There, rocks exist as thick sequences called strata. Strata differ not only in rock type but also in geophysical parameters such as propagation velocity and bulk density. When a seismic wave is incident on the boundary between two strata or layers, the normal circumstance is that some of the incident energy is reflected in the first layer and some is transmitted into the second layer. Whether a P-wave or an SV-wave is incident on the boundary, there will usually be reflected P- and S-waves and transmitted P- and S-waves. *Snell's Law* (Fig. 3–8) predicts the resulting ray paths for the waves generated at the interface. In each of the four cases, a P-wave is incident on the interface at an angle θ_0—the angle between the ray and perpendicular to the boundary. P- and S-wave velocities in layer 1 are V_{P1} and V_{S1}, respectively. P- and S-wave velocities in layer 2 are V_{P2} and V_{S2}, respectively.

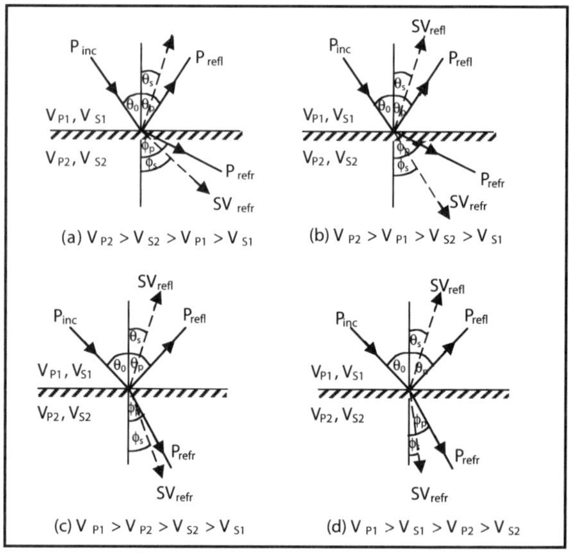

Fig. 3–8 *Snell's Law*

In Figure 3–8a, the P-wave and S-wave velocities in the second layer are faster than the P-wave velocity in the first layer. Consequently, the transmitted P- and SV-waves are bent, or refracted, at angles greater than the incident angle. That is, they are bent toward the horizontal. In the first layer, the angle of reflection for the P-wave equals the incident angle, but the angle of reflection for the reflected SV-wave is smaller. This is because the S-wave velocity is slower than the P-wave velocity. The refraction angle of the SV-wave in the second layer is also smaller than that of the refracted P-wave.

In Figure 3–8b, the P-wave velocity in the second layer is still the fastest but the S-wave velocity in the second layer is slower than the P-wave velocity in the first layer. Thus, the angle of refraction for the SV-wave is now smaller than the angle of incidence.

In Figure 3–8c, the P-wave velocity in the first layer is faster than the P-wave velocity in the second layer so the angle of refraction for the P-wave is less than the angle of incidence—bent toward the vertical). Since the S-wave velocity in layer 2 is faster than the S-wave velocity in layer 1, the angle of refraction for the SV-wave is greater than the angle of reflection for the SV-wave.

In Figure 3–8d, the P-wave velocity in the first layer is faster than the P-wave velocity in the second layer and the S-wave velocity in the first layer is also faster than the S-wave velocity in the second layer. As a result, the angle of refraction for the P-wave is less than the angle of incidence, and the angle of refraction for the SV-wave is less than the angle of reflection for the SV-wave.

The following summarizes Snell's law:

$$\frac{\sin\theta_o}{V_{P1}} = \frac{\sin\theta_P}{V_{P1}} = \frac{\sin\theta_S}{V_{S1}} = \frac{\sin\phi_P}{V_{P2}} = \frac{\sin\phi_S}{V_{S2}} = p \qquad (3.7)$$

where

p is the ray constant

The first two terms of Equation 3.7 correspond to reflection in the upper layer and since the velocities are the same, $\sin\theta_p = \sin\theta_0$. This is the *Law of Reflection*. Generalizing from the first, fourth, and fifth terms, the *Law of Refraction* can be written as $\sin\theta_2 = \frac{V_2}{V_1} \sin\theta_1$, where θ_2 is the angle of refraction, V_2 is velocity in the lower layer for the refracted wave, θ_1 is the angle of incidence, and V_1 is velocity in the upper layer for the incident wave.

There are two exceptions to the preceding discussion. The first is the case of an incident SH-wave. In this case, no P-waves or SV-waves are generated so only reflected and refracted SH-waves result.

$$\frac{\sin\Theta_0}{V_{p1}} = \frac{\sin\Theta_S}{V_{p1}} = \frac{\sin\phi_S}{V_{p2}} \qquad (3.8)$$

The second exception is when the angle of incidence exceeds the critical angle. The critical angle is the angle of incidence at which the angle of refraction is 90°. Actually, there are two critical angles, one for P-waves and one for S-waves. In the first case:

$$\frac{\sin\Theta_{cp}}{V_{P1}} = \frac{\sin 90°}{V_{P2}} = \frac{1}{V_{P2}} \text{ so } \sin\Theta_{cp} = \frac{V_{P1}}{V_{P2}} \qquad (3.9)$$

Similarly, for the SV:

$$\frac{\sin\Theta_{cs}}{V_{P1}} = \frac{\sin 90°}{V_{S2}} = \frac{1}{V_{S2}} \text{ so } \sin\Theta_{cs} = \frac{V_{P1}}{V_{S2}} \qquad (3.10)$$

The existence of a critical angle requires that the layer into which transmission occurs have a higher velocity than in the incident layer. That is, $V_{P2} > V_{P1}$ for critical P-waves and $V_{S2} > V_{P1}$, for critical S-waves. If the angle of incidence is larger than the critical angle, then no energy is transmitted into the lower layer.

From the ray theory point of view, a critically refracted wave travels along the interface between layers and is refracted back into the upper layer at the critical angle, as shown in Figure 3–9. The waves refracted back into the upper layer are called *head waves* or sometimes *first-break refractions* because at certain distances from a source they are the first-arriving energy.

The problem with ray theory is that a ray is one dimensional, showing only direction and only an infinitesimal amount of energy can be transmitted along a ray. To explain phenomena involving finite amounts of energy requires use of *wavefronts*.

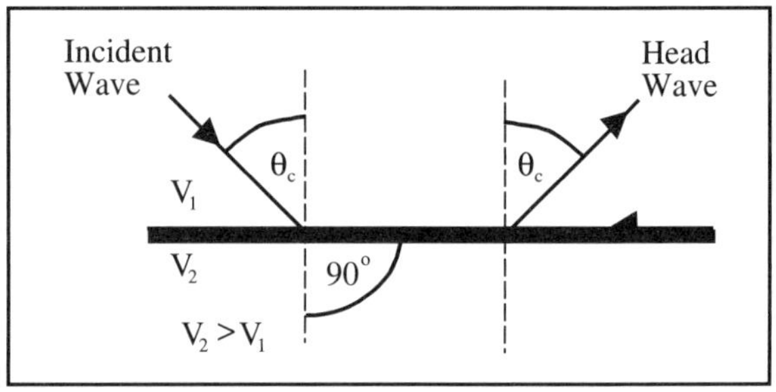

Fig. 3–9 *Critical Refraction and Head Waves*

A wave front can be defined as the locus of all points reached at the same time by a wave. It turns out that rays intersect wavefronts at right angles. Figure 3–10 illustrates the relationship between wavefronts and rays.

Figure 3–11 shows wavefronts corresponding to the time required for energy to be reflected from the bottom of the first layer and return to the surface at the energy source point. Since a constant velocity is assumed, the direct and reflected wavefronts are spherical. The refracted wavefront, however, is not spherical because it does not come from a single point. The figure also shows that the head wave energy does not penetrate the lower layer. Its wavefront, which has the shape of a truncated cone, results from a coupling between the reflected and refracted wavefronts.

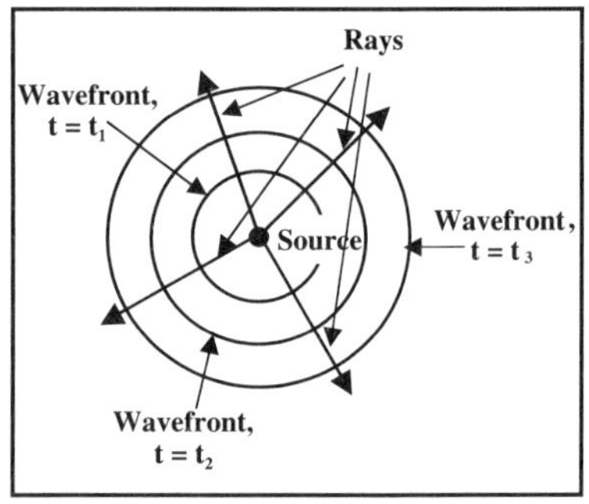

Fig. 3–10 *Wavefronts and Rays*

Huygen's Principle (Fig. 3–12) is very important to studies of seismic wave propagation. Its definition from Sheriff's *Encyclopedic Dictionary of Exploration Geophysics* is, "The concept that every point on an advancing wavefront can be regarded as the source of a secondary wave and the later wavefront is the envelope tangent to all the secondary waves."

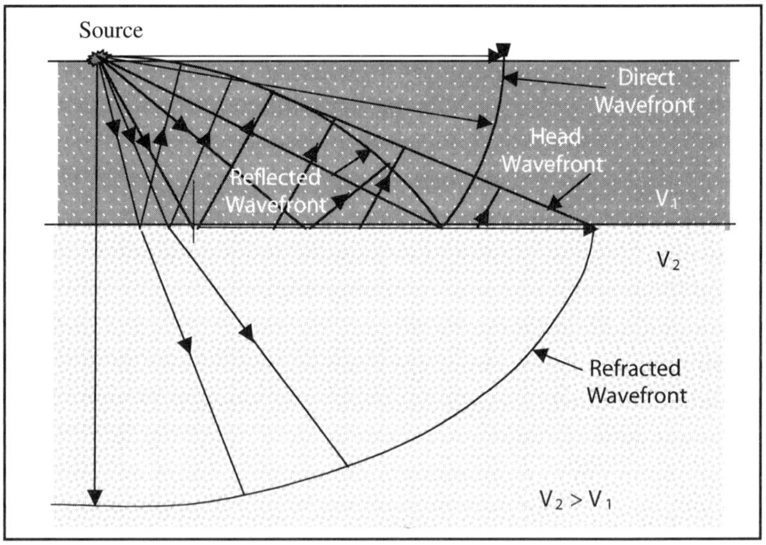

Fig. 3–11 *Wavefronts from a Point Energy Source*

Seismic amplitudes

To this point, no mention has been made of the amplitudes of reflected and refracted seismic waves. Zoeppritz derived a relationship governing the reflection and transmission coefficients for plane waves as a function of angle of incidence and six parameters, three on either side of the reflecting interface. These are V_p, V_s, and density. The equations are complex and their solutions are laborious. Figure 3–13 is representative of P- and SV-wave reflection coefficients derived from the Zoeppritz equations.

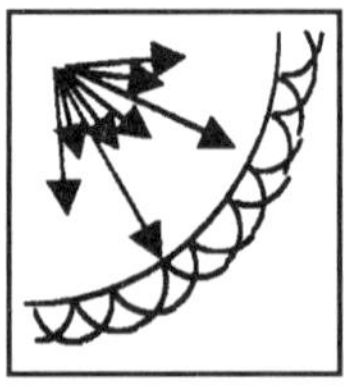
Fig. 3–12 *Huygen's Principle*

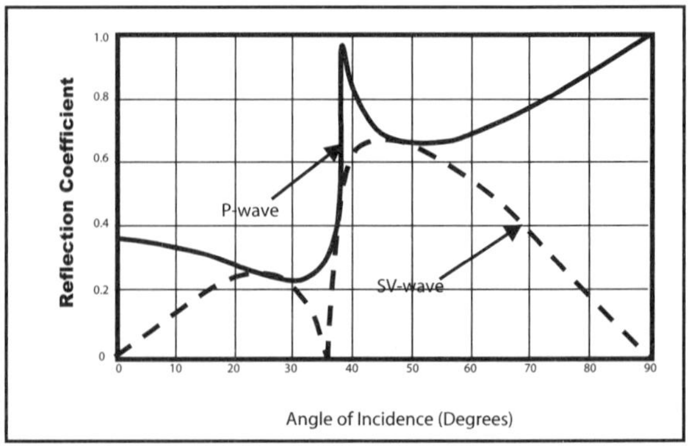

Fig. 3–13 *P- and SV-wave Reflection Coefficients*

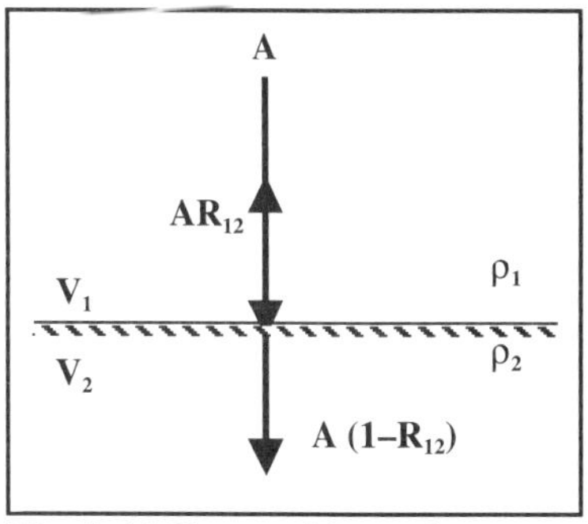

Fig. 3–14 *P-wave Normal Incidence Reflection and Transmission*

At normal incidence (angle of incidence equals 0°) of a P-wave, no reflected or refracted SV-wave is generated. (Fig. 3–14)

As a result, a simplified expression for P-wave *reflection* and *transmission coefficients* at normal incidence can be developed. These are:

$$R_{12} = \frac{\rho_2 V_2 - \rho_1 V_1}{\rho_2 V_2 + \rho_1 V_1} \tag{3.11}$$

and

$$T_{12} = 1 - R_{12} = \frac{2\rho_1 V_1}{\rho_2 V_2 + \rho_1 V_1} \tag{3.12}$$

where

R_{12} = P-wave normal incidence reflection coefficient at the interface separating layers 1 and 2

T_{12} = P-wave normal incidence transmission coefficient at the interface separating layers 1 and 2

ρ_1, ρ_2 = Bulk density of layers 1 and 2, respectively

V_1, V_2 = P-wave velocities of layers 1 and 2, respectively

Introducing the concept of *acoustic* (or *seismic*) *impedance*, $Z = \rho V$, the reflection and transmission coefficients can be written as:

$$R_{12} = \frac{Z_2 - Z_1}{Z_2 + Z_1} \tag{3.13}$$

and

$$T_{12} = \frac{2Z_1}{Z_2 + Z_1} \tag{3.14}$$

where

Z_1, Z_2 = acoustic impedance of layers 1 and 2, respectively

The seismic record

The objective of seismic data acquisition is to obtain data that can be related to subsurface geology. At various points on or near the Earth's surface, an energy source is used to generate seismic waves that travel downward and are reflected at interfaces between geologic strata that differ in acoustic impedance. The reflected energy returns to the surface where it is detected and recorded. This reflected energy is called *signal* or *primary reflections*. From the times at which the primary reflections were recorded and their amplitudes, it is desired to obtain the Earth's *reflectivity function*. This is a set of impulses with amplitudes proportional to the reflection coefficient of each interface and time of occurrence equal to the *two-way reflection time*. The impulse response can then be related to the geologic section from which data were acquired (Fig. 3–15).

Unfortunately, the primary reflection impulse response cannot be recorded directly. It is impossible to record impulses; instead reflections are recorded as *wavelets* whose shapes depend upon the energy source used and transmission losses along the reflection path, as seen in Figure 3–16. For that matter, a seismic record cannot be constrained to record signal only.

Anything on a seismic record other than primary reflections is called *noise*. There are several kinds of noise, but all can be placed in two categories—*source-generated* and *ambient noise*. Table 3–1 summarizes source-generated noise observed on seismic records.

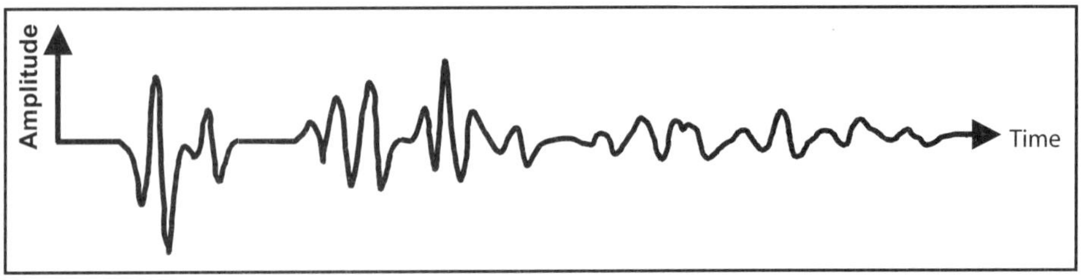

Fig. 3–16 *Noise-free Seismic Trace Derived from Earth Reflectivity Function*

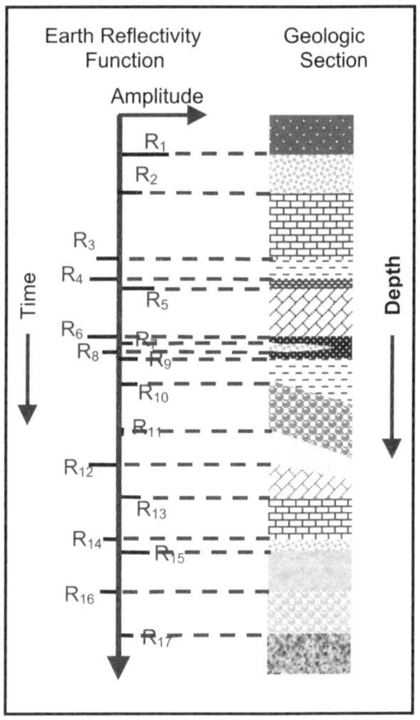

Fig. 3–15 *Earth Reflectivity Function*

Table 3–1 *Source-generated Noise*

Type	Description	Appearance on Seismogram
Direct Waves	Waves propagating directly from shot	High amplitude first events on near traces
Ground roll	Wave propagation along ground surface	Usually low frequency and low apparent velocity
Air waves	Sound waves from shot	Very low apparent velocity (1100 ft/sec)
Shallow refraction	Near-surface refracted wave	High amplitude, often first events on far traces
Reverberations	Ghosts; tails on shot pulse	Tails on reflected events
Multiples (long period)	Repeated reflection	Reflections at even time intervals after the primary reflection
Guided waves	Multiples of direct wave and shallow refractions	Events paralleling first breaks
Diffractions	Wave energy propagated in all directions from a point such as a fault interface	Coherent hyperbolic events

Either the direct arrivals or the shallow refraction (head wave) are the first events seen on seismic records are either direct arrivals or refractions. Direct arrivals travel a direct path from source to the nearest receivers. Refractions, or head waves (Fig. 3–17), travel from source down to the base of near-surface layer, then more or less horizontally, and, finally, back up to the farther receivers.

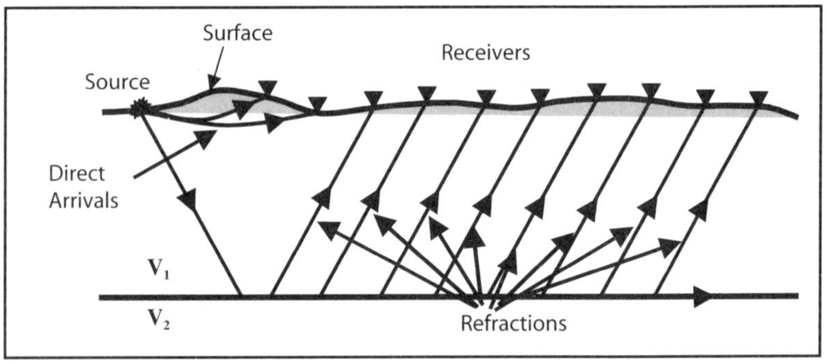

Fig. 3–17 *Direct Waves and Shallow Reflections*

As previously discussed, ground roll is really a Rayleigh wave. The heterogeneity of the surface layer distorts the retrograde elliptical particle motion somewhat. Velocity increases with depth in the surface layer. The longer wavelength components of the ground roll penetrate more deeply in the surface layer than the shorter wavelengths and thus see higher velocities. Longer wavelength means lower frequency since $f = V/\lambda$ (Fig. 3–6). Ground roll occurs as a set of dispersed wave trains with the lower frequency, longer wavelength components arriving first.

Airwaves are simply sound generated by the source traveling through the air.

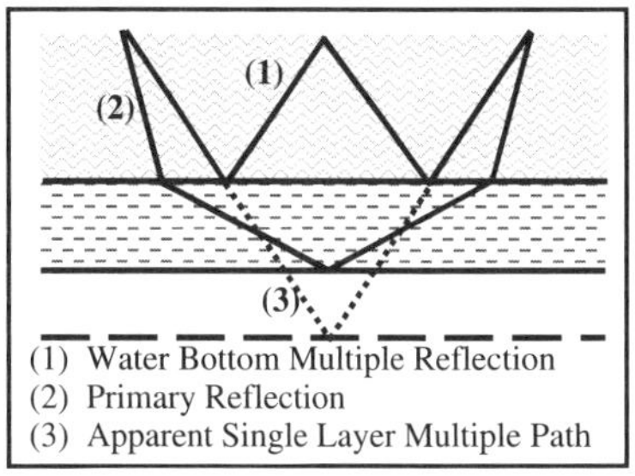

Fig. 3–18 *Reverberations or Water Bottom Multiple Reflections*

Multiple reflections are of three types.

- *Simple multiples* or *reverberations* (multiple reflections in a water layer) simply repeat the primary reflection raypath as seen in Figure 3–18. Simple multiples are not restricted to the water layer.
- *Peg-leg multiples* add a reflection in the surface layer to the primary reflection raypath.
- *Interbed multiples* have a repeat reflection within a layer.

Figure 3–19 illustrates peg leg and interbed multiples.

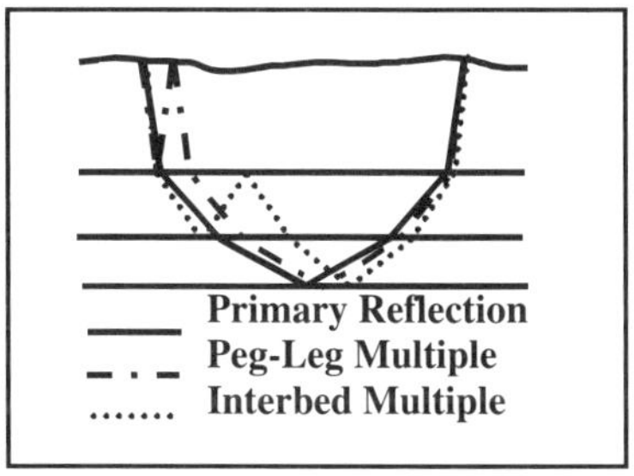

Fig. 3–19 *Multiple Reflections*

Guided waves are trapped energy in surface layers—reflected refractions that appear to be multiples of first break refractions. Figure 3–20 illustrates guided wave raypaths.

Diffractions are scattered energy from a subsurface point, such as the truncation of a reflector caused by a fault. Figure 3–21 illustrates diffractions.

The second type of noise, or unwanted recording, is called ambient noise. Perhaps the best way to describe ambient noise is that it is what would be recorded if a seismic record were made without an energy source being fired. Table 3–2 lists various types of ambient noise.

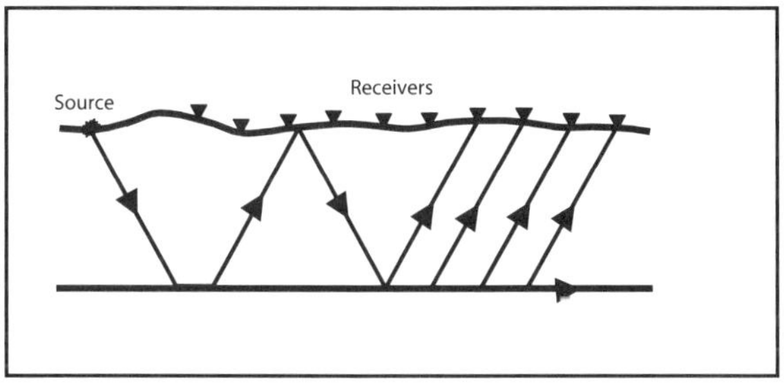

Fig. 3–20 *Guided Waves*

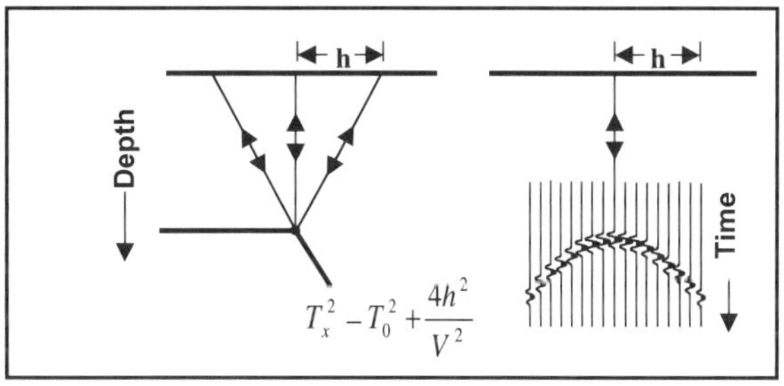

Fig. 3–21 *Diffractions*

Table 3–2 *Ambient Noise*

Type	Description	Appearance on Seismogram
Traffic	Cars, animals, people	Often seen as spikes
Wind	Air currents	High frequency
Earthquakes	Any motion produced from within the earth by other than the seismic source	Usually very low frequency
Highline noise	Induced noise from nearby power lines	50 or 60 Hz
Shot hole	Debris falling from the blown shot hole	Both high and low frequency, fanning from the source point
Instrument	From recording instruments	Usually high frequency, noise should be very low level

Figure 3–22a is the noise-free trace of Figure 3–16 and Figure 3–22b shows the same trace with the addition of ambient noise.

A seismic shot record is a set of seismic traces recorded at one time from a single source. The number of traces on a record varies considerably, sometimes more than a thousand traces per record. Figure 3–23, a schematic land seismic record, shows the different kinds of events recorded on a seismic record. Figure 3–24 shows a real seismic shot record.

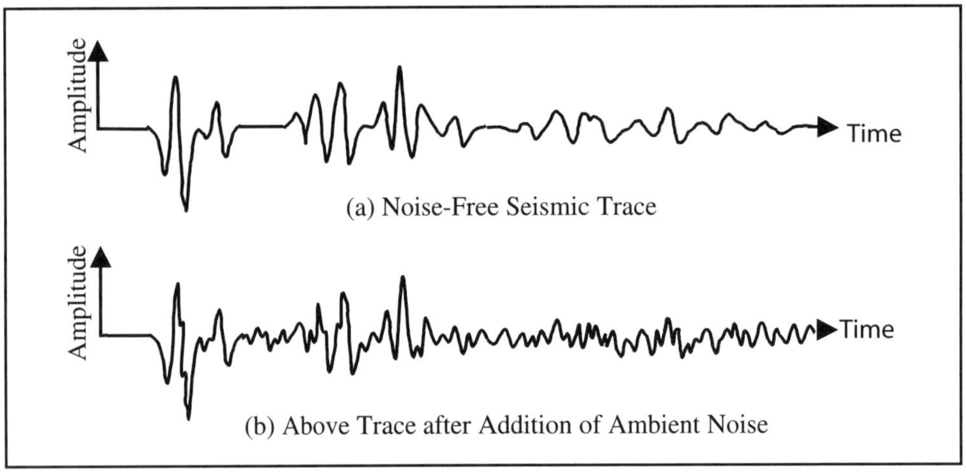

Fig. 3–22 *Effect of Ambient Noise on Seismic Trace*

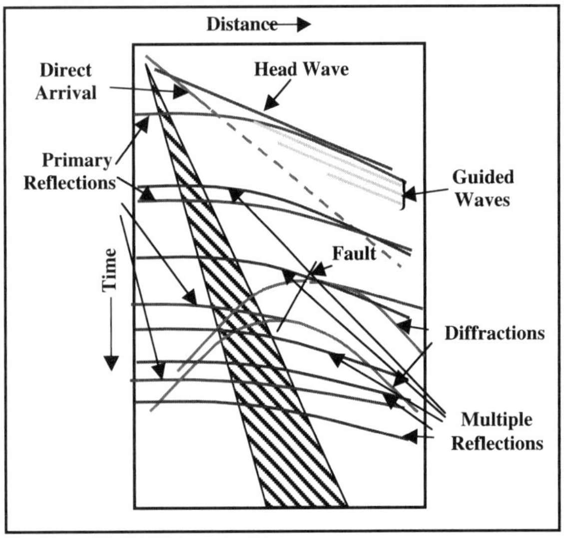

Fig. 3–23 *Schematic Land Seismic Record*

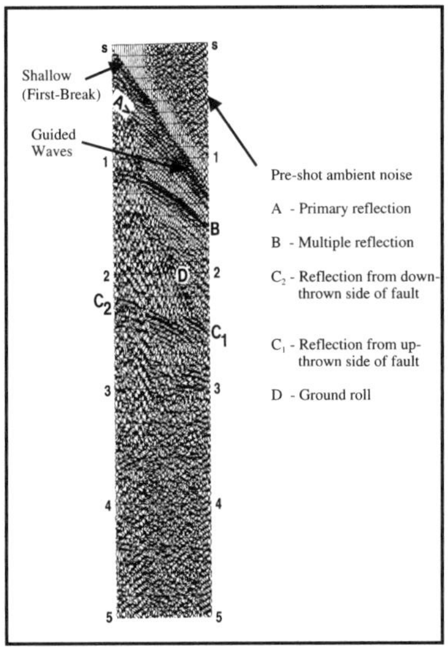

Fig. 3–24 *A Seismic Shot Record*

Mathematical Theory and Concepts

As with any scientific-based discipline, seismic exploration makes considerable use of mathematics. Much of seismic data processing is based on a branch of mathematics called *Statistical Communication Theory*. Details of that theory are beyond the scope of this text but some of the basic concepts and applications of it will be presented. One aspect of modern seismic exploration methods results in a simplification of mathematical tools required. Since seismic data are recorded as *sampled* data, integration reduces to summation and differentiation to subtraction.

Sampled data

In the early days of seismic exploration, data were recorded in analog form. That is, there was a representation of every change in the wavefield. At first, seismic data were recorded optically (light on photographic paper). Later, analog magnetic tape was used. In this case, the amplitude of the seismic data was represented by a continuously varying magnetic intensity. Since the late 1960s, digital recording has been used. Digital recording samples the incoming analog signal at regular time intervals and records the data as a binary number representing the voltages of the samples. A continuous signal is input to the sampler and a set of numbers or time series is output. Figure 3–25 illustrates sampling and reconstruction of the sampled data to analog form.

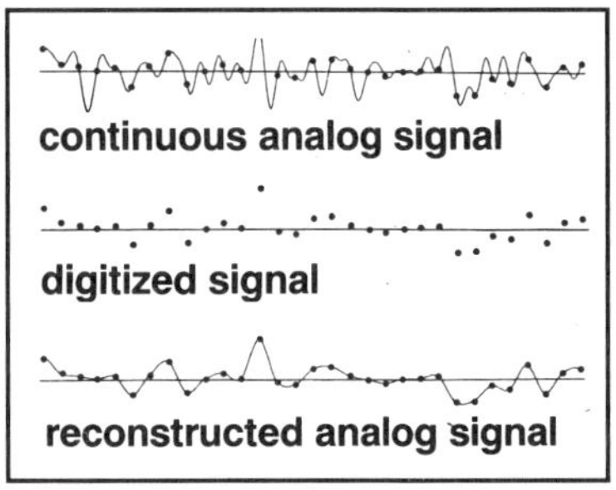

Fig. 3–25 *Digital Recording*

In most cases, the ability to correctly reconstruct a digital signal depends upon the frequency content of the signal and the sampling increment.

Figure 3–26 shows the effect of sampling at different *sample increments* or *sample periods*. In Figure 3–26a, the input is a 25-Hz sinusoid. The reconstructions of the outputs sampled at 2 ms, 4 ms, and 8 ms are the same as the input. In Figure 3–26b, the input is a 75-Hz sinusoid. The reconstructions of the outputs sampled at 2 ms and 4 ms are the same as the input, but the output sampled at 8 ms is a 25-Hz sinusoid! In Figure 3–26c the input is a 150-Hz sinusoid. The reconstruction of the output sampled at 2 ms is the same as the input, but the 4 ms output is a 100-Hz sinusoid and the 8 ms output is a 25-Hz sinusoid!

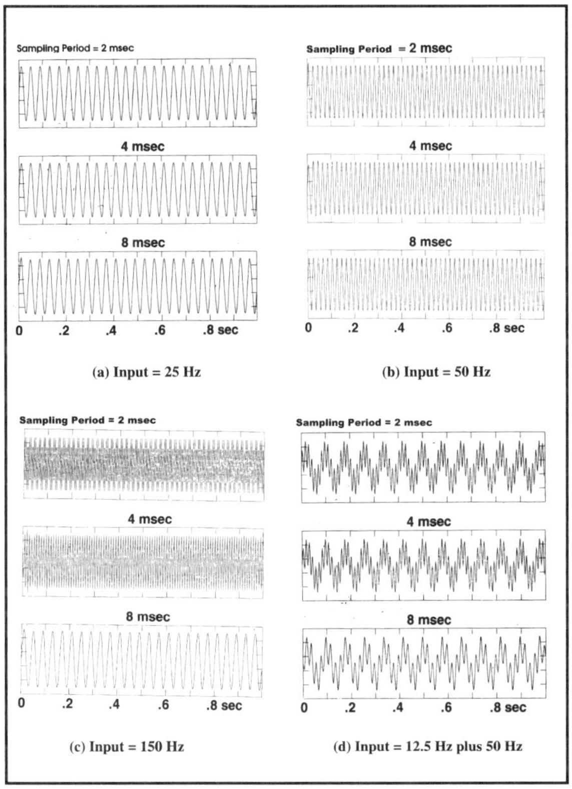

Fig. 3–26 *Effect of Sample Period*

In Figure 3–26d, the input is the sum of 12.5-Hz and 75-Hz sinusoids but the reconstructed output is the sum of 12.5 and 25-Hz sinusoids. What is being demonstrated here is the phenomenon called *aliasing*.

The *Sampling Theorem* can be stated as follows:

An analog signal which is band-limited to frequencies less than f_o is completely described by samples taken at intervals of time Δt, where $\Delta t < 1/2f_o$. Conversely, then, an analog signal band-limited to signals less than f_o can be completely recovered from samples taken at intervals of time Δt, if $f_o < 1/2\Delta t$. If, however, a signal sampled at a sample interval Δt contains frequencies higher than $f_N = 1/2\Delta t$, where f_N is the Nyquist or alias frequency, it cannot be correctly recovered (using conventional processing techniques) because of a distortion called aliasing.

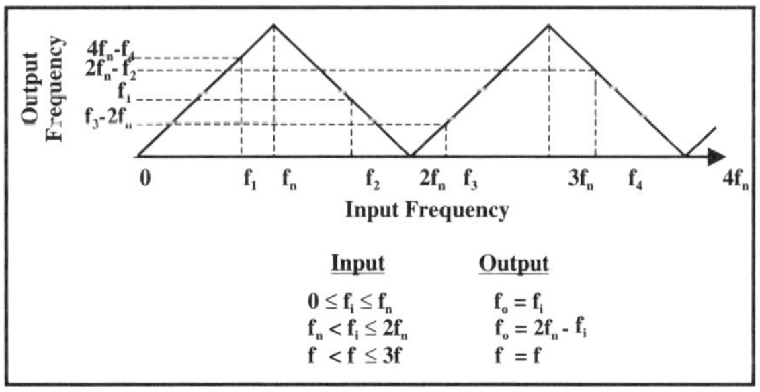

Fig. 3–27 *Aliasing*

Analyzing the data of Figure 3–26, it can be seen that the 25-Hz sinusoid is below Nyquist for all three sample periods. The 75-Hz signal is lower than f_N for 2 and 4 ms sampling but above f_N for 8 ms sampling. From Table 3–3, the Nyquist frequency for 8 ms is 62.5 Hz and 75 Hz is 12.5 Hz more than f_N. The output frequency f_o is 50 Hz or 12.5 Hz less than f_N.

Table 3–3 *Nyquist Frequency*

DT (ms)	f_N (Hz)
0.5	1000
1	500
2	250
4	125
8	62.5

At 4 ms sampling, the 150-Hz sinusoid is 25 Hz above f_N and its output of 100 Hz is 25 Hz below f_N. At 8 ms sampling, the 150-Hz sinusoid is 87.5 Hz above f_N or and 25 Hz above $2f_N$. Its output of 25 Hz is equal to the difference between the input and $2f_N$. Figure 3–27 is a chart for calculating output frequencies relative to input and multiples of f_N.

To prevent aliasing, a filter must be applied before sampling or resampling to a larger sample period, and a filter must be applied to limit frequencies to below Nyquist.

Convolution and correlation

Earlier in this chapter, the concept of the earth reflectivity function was introduced. This is a special kind of *impulse response*. An impulse response is measured by inputting a unit impulse (spike or Dirac delta function) into the system and recording the system output. Figure 3–28 illustrates this by considering the earth to be a system. Each reflecting horizon produces an impulse with amplitude equal to its vertical reflection coefficient and a time delay equal to the two-way reflection time to the horizon.

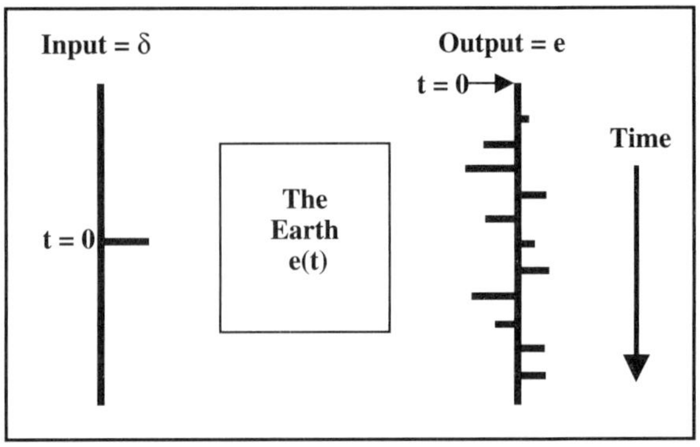

Fig. 3–28 *Measuring the Earth's Impulse Response*

Figure 3–16 presents a noise-free seismic trace derived from an earth impulse response but does not explain how it was derived. The method used is called *convolution*. Given the impulse response of a system, *e(t)* is the output of the system, *o(t)* an arbitrary input, and *w(t)* is given by the convolution of the input and the system impulse response. Expressed mathematically, this is:

$$o(t) = w(t) * e(t)$$

where convolution is symbolized by the asterisk *

For sampled data, convolution of the sampled set $\{x_n\}$ with sampled set $\{y_m\}$ is calculated from:

$$w_n = \sum_{i=0}^{m} x_i y_{i-n} = \sum_{j=0}^{p} x_{j-n} y_j \qquad (3.15)$$

where

$\{x_m\}$ = the *m+1* sample values of the input $x\ (t) = \{x_0, x_1, x_2, \ldots, x_m\}$

$\{y_p\}$ = the *p+1* sample values of the input $y\ (t) = \{y_0, y_1, y_2, \ldots, y_p\}$

The convolution of $\{x_m\}$ and $\{y_p\}$ can be calculated by performing the following steps:

1. Time reverse either $\{x_m\}$ or $\{y_p\}$.
2. Shift the reversed series to the left until only one sample of $\{x_m\}$ and $\{y_{-p}\}$ or $\{x_{-m}\}$ and $\{y_p\}$ are aligned in time.
3. Multiply the aligned *x* and *y* values. This is the first output.
4. Shift the reversed series one sample to the right.
5. Multiply the aligned *x* and *y* values and sum the products. This is the next output.
6. Repeat steps 4 and 5 until no samples are aligned.

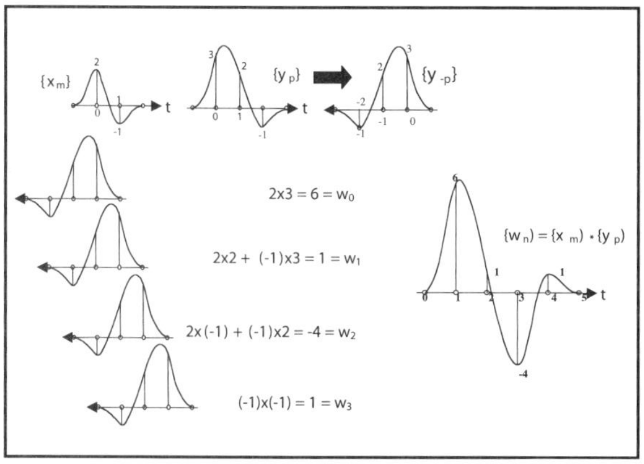

Fig. 3–29 *Convolution*

Figure 3–29 illustrates convolution using $\{x_m\} = \{2, -1\}$ and $\{y_p\} = \{3, 2, -1\}$.

Convolution is an operation describing the interaction between an input and a system of some sort. Computationally, the outcome of the operation is the same whichever time series is time reversed. Also, note that the convolution of any time series with a unit impulse is simply the input time series. Convolution with a time-delayed unit impulse gives a time-delayed version of the input time series.

Cross-correlaton is a process or procedure for measuring the similarity of one time series to another. Computationally, correlation and convolution appear similar. For sampled data, the cross-correlation of $\{x_m\}$ with $\{y_p\}$ is:

$$\varphi_\lambda^{xy} = \sum_{i=0}^{m} x_i y_{i-\lambda} \qquad (3.16)$$

The steps taken to crosscorrelate $\{y_p\}$ onto $\{x_m\}$ follow.

1. Shift $\{y_p\}$ to the left until only one sample of the two time series is aligned.
2. Cross multiply the two aligned values. This is the cross-correlation for the $-n^{th}$ lag, where n is the number of sample periods $\{y_p\}$ was shifted left.
3. Shift $\{y_p\}$ to the right one sample period and cross multiply the two aligned value.
4. Sum the cross products. This is the cross-correlation for the $-(n-1)^{th}$ lag
5. Repeat steps 3 and 4 until only one sample of each time series are aligned. The output lag value increments by +1 for each shift to the right.

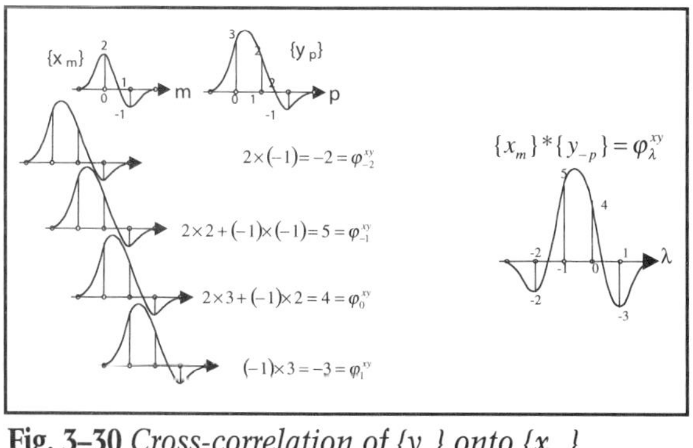

Fig. 3–30 *Cross-correlation of $\{y_p\}$ onto $\{x_m\}$*

Figure 3–30 uses the same two time series that were convolved together to demonstrate cross-correlation. Note how different the cross-correlation is from the convolution of these two wavelets. Also, it can be seen that the two wavelets are most similar at a lag of –1. Since convolution requires that one time series be time reversed but cross-correlation does not, it can be seen that convolution of a time-reversed time series with another is equivalent to cross-correlation — $\{x_m\}*\{y_{-p}\}=\varphi_{-\lambda}^{xy}$.

Another difference between cross-correlation and convolution is that the output *does* depend upon which of the time series is shifted past the other. Figure 3–31 demonstrates this by crosscorrelating $\{x_m\}$ onto $\{y_p\}$, i.e. $-\varphi_\lambda^{yx}$. Comparing the two cross-correlations shows that one is the time reversed version of the other, that is $\varphi_\lambda^{yx} = \varphi_{-\lambda}^{xy}$.

Autocorrelation is simply the cross-correlation of one time series with itself. Mathematically

$$\varphi_\lambda^{xx} = \sum_{i=0}^{m} x_i x_{i-\lambda} \qquad (3.17)$$

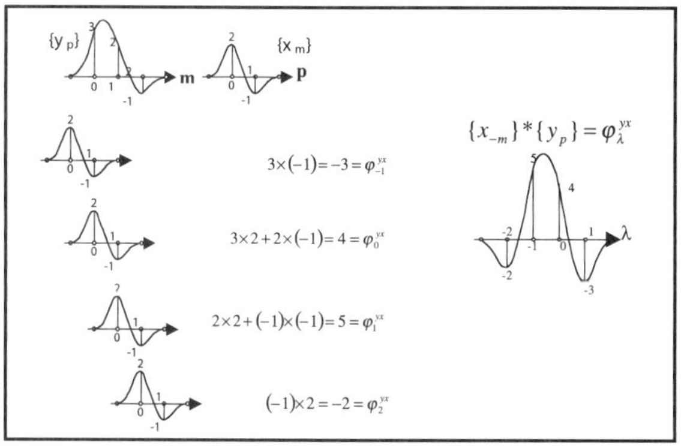

Fig. 3–31 *Cross-correlation of $\{x_m\}$ onto $\{y_p\}$*

Figure 3–32 illustrates the autocorrelation of $\{x_m\}$. Note that the autocorrelation is symmetrical about $\lambda = 0$. This means that the autocorrelation is an even function. Note also that the maximum value of the autocorrelation is at $\lambda = 0$, and this value equal to the sum of the squares of all the amplitudes in the time series. This is the *power* of the series or wavelet.

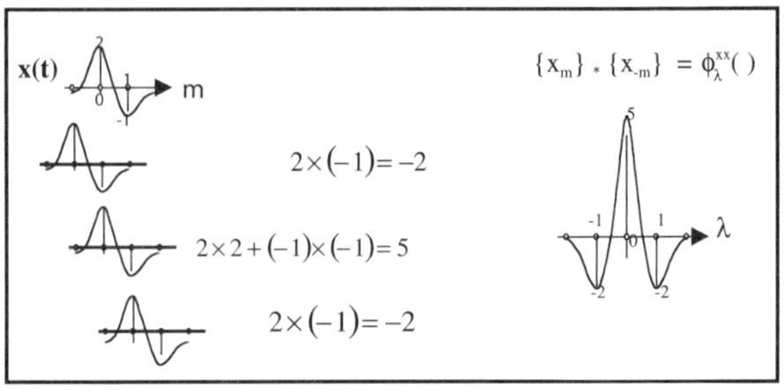

Fig. 3–32 *Autocorrelation of* $\{x_m\}$

Time and frequency domains

To this point, only time series and wavelets defined in terms of amplitudes as functions of time have been discussed. That is, only the *time domain* has been used. In the time domain, a signal is described as having certain amplitudes at certain times. Most signals, traces, etc. can be described in the *frequency domain.*

In the frequency domain, data are described in terms of amplitudes and phases at certain frequencies. The simplest example to illustrate this is a single-frequency sinusoid, illustrated in Figure 3-33. On the right of the figure, a mass is shown suspended from a spring and a light source is attached to the mass. The light shines on photographic film that moves from right to left. When the film is developed, a picture such as that shown at the top or bottom of Figure 3–33 will be seen.

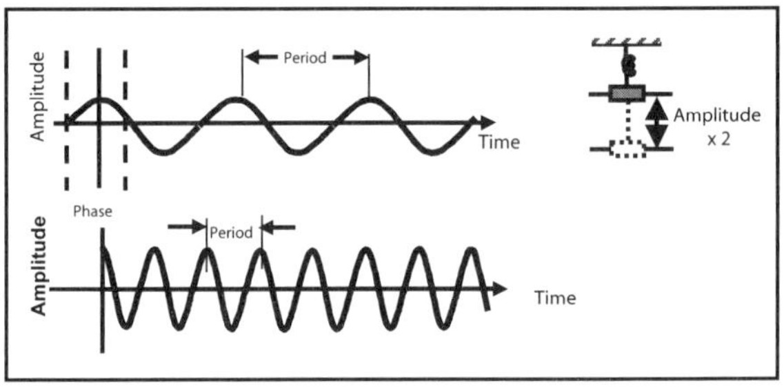

Fig. 3–33 *Single Frequency Sinusoids*

The trace seen on the developed film can be described by

$$y\,(t) = A\cos\,(2\pi t/T+\phi)$$

where

T is the period of time for the suspended mass to complete one cycle of its motion (from its highest to lowest position and back)

ϕ is the phase added to indicate the position of the mass in its cycle relative to the rest position (one cycle corresponds to 2π radians or 360°)

Alternatively, the trace can be described by:

$$y\,(t) = A\cos\,(2\pi ft+\phi)$$

where

f = frequency = $1/T$

The latter of these expressions is the usual time domain description of a single-frequency sinusoid.

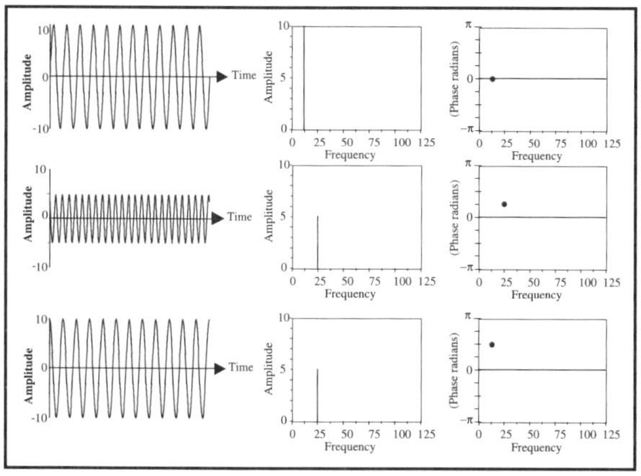

Fig. 3–34 *Amplitude and Phase Spectra for Single-frequency Sinusoids*

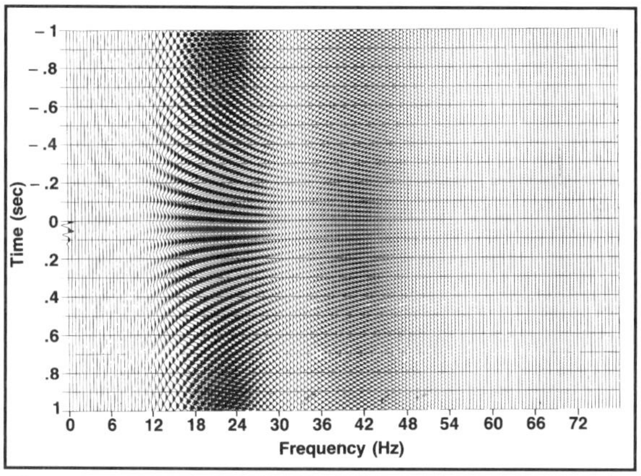

Fig. 3–35 *Signal Synthesis and Decomposition*

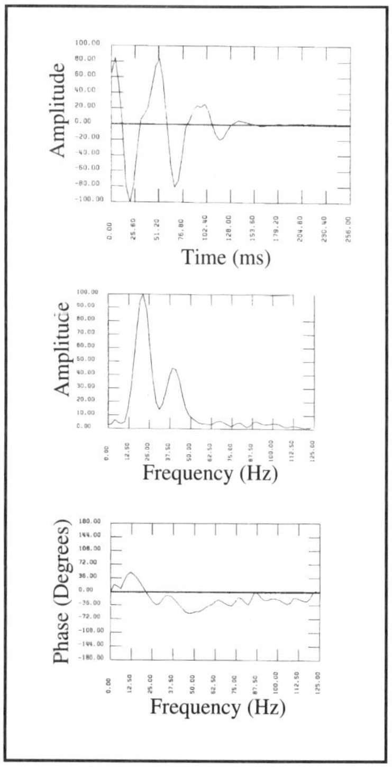

Fig. 3–36 *Time and Frequency Domains*

To describe something in the frequency domain requires both an *amplitude spectrum* and a *phase spectrum*. The amplitude spectrum simply gives an amplitude at each frequency. The phase spectrum similarly gives the phase at each frequency. A single frequency sinusoid has only one amplitude and one phase value, as shown in Figure 3–34. The top and bottom sinusoids have the same frequency and amplitude but different phase. The top and middle sinusoids differ in frequency, amplitude, and phase.

By combining many sinusoids having different amplitudes, phases, and frequencies, complex signals can be synthesized. Conversely, many complex signals can be decomposed into constituent sinusoids having different amplitudes, phases, and frequencies. An example is shown in Figure 3–35.

On the far left of Figure 3–35 is a time domain signal or wavelet similar in appearance to many seismic wavelets. To the right of the wavelet is a set of single frequency sinusoids, with frequencies ranging from 0.5 Hz to 78 Hz. The wavelet is synthesized by summing all these sinusoids. On the other hand, the wavelet can be decomposed into these wavelets.

A plot of each sinusoid's amplitude against its frequency yields an amplitude spectrum. A plot of each sinusoid's phase against its frequency yields a phase spectrum. These are shown in Figure 3–36 for the data of Figure 3–35. Also shown in Figure 3–36 is the time domain wavelet. The amplitude and phase spectra describe the sinusoids that can be summed to synthesize the time domain wavelet. Thus, it can be seen that the time and frequency domain descriptions are equivalent.

The mechanism that allows transformation from the time to the frequency domain is the *Fourier Transform* and the *Inverse Fourier Transform* is used to transform from the frequency domain to the time domain. For sampled data, the Fourier Transform of the time series $\{g_n\}$ is given by:

$$F\{g\,(\Delta T)\} = G\,(m\Delta f)\ \sum_{n=0}^{N} g_n e^{-12\pi mnf\Delta T},\ \mathrm{m}=0, 1, 2, \ldots \mathrm{M} \tag{3.18}$$

where

$F\{g\,(\Delta T)\}$= Fourier Transform of $\{g(DT)\}$

$G\,(m\Delta f)$ = frequency spectrum of the time series $\{g_n\}$ at the frequency $m\Delta f$

g_n = n^{th} amplitude of the time series $\{g_n\}$, n= 0, 1, 2, … N

Δf = frequency increment = $f_n/M\Delta T$ = sample period

i = $\sqrt{-1}$

Taking advantage of the relationship, $e^{-i\theta} = \cos\theta + i\sin\theta$, Equation 3.18 can be written as:

$$G(m\Delta f) = \sum_{n=0}^{N} \{g(\Delta T)\}\,[\cos(2\pi mn\Delta f\Delta T) - i\,\sin(2\pi mn\Delta f\Delta T)] \tag{3.19}$$

It can be seen that, in the general case, the frequency spectrum is complex. Thus, it can be written as:

$$G(m\Delta f) = \mathrm{Re}[G(m\Delta f)] + \mathrm{Im}[G(m\Delta f)]$$

where

$$\mathrm{Re}[G(m\Delta f)] = \sum_{n=0}^{N} g_n \cos(2\pi mn\Delta f\Delta T), \tag{3.20a}$$

$$\mathrm{Im}[G(m\Delta f)] = \sum_{n=0}^{N} g_n \sin(2\pi mn\Delta f\Delta T) \tag{3.20b}$$

The amplitude spectrum $A\ (m\Delta f)$ is given by the absolute value of the frequency spectrum.

$$A(m\Delta f) = |G(m\Delta f)| = \sqrt{\mathrm{Re}[G(m\Delta f])^2 + \mathrm{Im}[G(m\Delta f])^2} \tag{3.21a}$$

The phase spectrum $\phi(m\Delta f)$ is given by:

$$\phi(m\Delta f) = -\tan^{-1}\left(\frac{\mathrm{Im}[G(m\Delta f)]}{\mathrm{Re}[G(m\Delta f)]}\right) \tag{3.21b}$$

From the preceding, an alternate way of writing the frequency spectrum is:

$$G(m\Delta f) = A(f)e^{i\phi(f)} \tag{3.22}$$

To completely describe something in the frequency domain requires both an amplitude and a phase spectrum.

The inverse Fourier Transform is calculated from:

$$F^{-1}[G\ (m\Delta f)] = g(n\Delta T) = \sum_{m=0}^{M} G(m\Delta f)e^{i2pmnf\Delta T},\ n = 0, 1, 2, \ldots N \tag{3.23}$$

where

$F^{-1}[G(m\Delta f)]$ = inverse Fourier Transform of $G(m\Delta f)$

As the length of the time series increases, the computer time required to transform from the time to the frequency domain (or vice versa) increases very rapidly. Consider a 5 s trace, sampled at 2 ms. There are (5/0.002 + 1) = 2501 samples in the trace. A modern seismic record usually has hundreds of traces per record and thousands of records in the total data volume. Transforming this quantity of data into the frequency domain would be prohibitively expensive using Equation 3.18 or 3.19. Fortunately, an algorithm (actually, a family of algorithms) has been developed that speeds up the process sufficiently to allow transformation back and forth between time and frequency domains to be quite economical. The algorithm to transform from the time domain to the frequency domain is called the Fast (or Finite) Fourier Transform (FFT), and the algorithm for transforming from the frequency domain to the time domain is the inverse FFT or IFFT.

The Fourier Transform has many interesting properties. For example, it is linear. This means, for example, that if $F\{x_m\} = X(m\Delta f)$ and $F\ \{y_n\} = Y(m\Delta f)$, then

$$F[ax_m + b\ y_n] = aX(m\Delta f) + bY(m\Delta f) \tag{3.24}$$

Reversing a wavelet in time is equivalent to changing the polarity of its phase spectrum but the amplitude spectrum stays the same (Fig. 3–37).

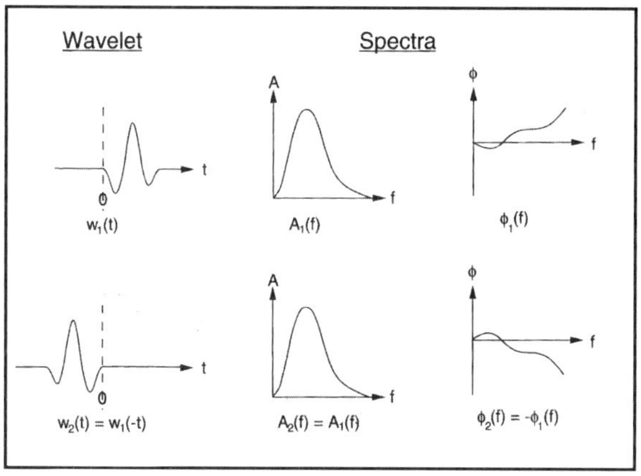

Fig. 3–37 *Effect of Time Reversal*

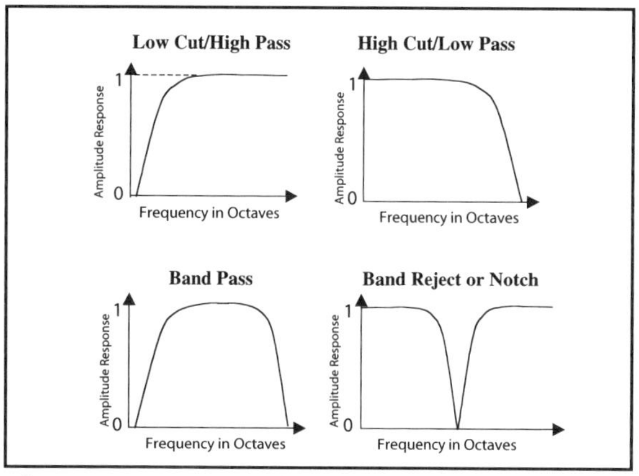

Fig. 3–38 *Frequency Filter Types*

Convolution in the time domain is equivalent to multiplying in the frequency domain and this is equivalent to multiplying amplitude spectra and summing phase spectra.

$$F[\{x_m\} * \{y_n\}] = [A_x(f)e^{i\phi_x(f)}] \times \left[A_y(f)e^{i\phi_y(f)}\right] = A_x(f)A_y(f)e^{\,i[\phi_x(f)+\phi_y(f)]} \tag{3.25}$$

Multiplication in the time domain is equivalent to convolving frequency spectra in the frequency domain. Cross-correlation in the time domain is equivalent to multiplying amplitude spectra and subtracting phase spectra in the frequency domain. To see this, recall that cross-correlation is equivalent to convolving one time series with another time-reversed time series.

$$F[\{\phi_\lambda^{xy}] = F[\{x_m\} * \{y_{-n}\}]$$

$$= [A_x\{f\}_e{}^{i\phi_x(f)}].\left[A_y(f)e^{i\phi_y(f)}\right] = A_y(f)e^{i[\phi_x(f)-\phi_y(f)]} \tag{3.26}$$

Autocorrelation in the time domain is equivalent to squaring the amplitude spectrum in the frequency domain the phase spectrum is always zero. Since autocorrelation is the cross-correlation of a time series with itself, it follows from Equation 3.26 that:

$$F[\{\phi_\lambda^{xy}] = F\,[\{x_m\} * \{x_{-m}\}] = A_x(f)A_x(f)e^{i[\phi_x(f)-\phi_x(f)]} = [A_x(f)]^2 \tag{3.27}$$

When the term filtering is used alone, it usually means frequency filtering. There are four types of filters, based on what they do. These are illustrated by amplitude responses of Figure 3–38.

Filters are defined by parameters called *cutoff frequency* and *attenuation rate.* Before defining these, definitions of two terms, *decibels* (dB) and *octaves* need to be provided. Decibels are a logarithmic measure of relative amplitude. They are calculated from:

$$A_{dB} = 20\log\left(\frac{A}{A_0}\right) \tag{3.28}$$

where A_0 is the normalization amplitude (often the largest amplitude for a time series , trace, etc.)

Octaves are a measure of relative frequency or frequency range. If frequency f_2 equals $2f_1$, then f_2 is one octave above f_1. If frequency f_2 equals $4f_1$, then f_2 is two octaves above f_1. If frequency f_2 equals $1/2\, f_1$, then f_2 is one octave below f_1. In the general case, the number of octaves, n, between frequencies f_2 and f_1 is given by:

$$n = \frac{\log\left(f_2/f_1\right)}{\log 2} \tag{3.29}$$

The cutoff frequency f_{co} is the frequency at which the response is –3dB. The attenuation rate is usually stated as the number of dB/octave the filter amplitude response decreases beyond the cutoff frequency.

When data are filtered, the input amplitude spectrum is multiplied by the filter amplitude response. At frequencies where the filter amplitude response is less than one, attenuation takes place as seen in Figure 3–39. Thus the idea of filtering is to reduce or attenuate frequency components where noise dominates over signal. A low cut (high pass) filter attenuates frequencies less than the cutoff frequency. A high cut (low pass) filter attenuates frequencies above the cutoff. A bandpass filter attenuates frequencies below its low cutoff frequency and above its high cutoff frequency. (The difference between these two cutoff frquencies is the pass band.) A band reject filter attenuates frequencies between its low and high cutoff frequencies.

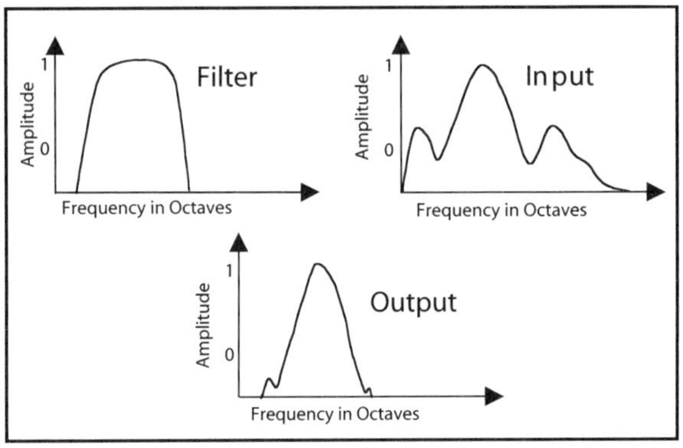

Fig. 3–39 *Bandpass Filtering*

Figure 3–40 illustrates a special kind of high cut filter—an *anti-alias filter*. If the sampled input contains frequency components greater than $1/2\Delta T$ where ΔT is the sample period, then aliasing occurs. To avoid this, an anti-alias filter is usually applied. In Figure 3–40, the input data were filtered with the filters whose amplitude responses are shown on the far right before sampling.

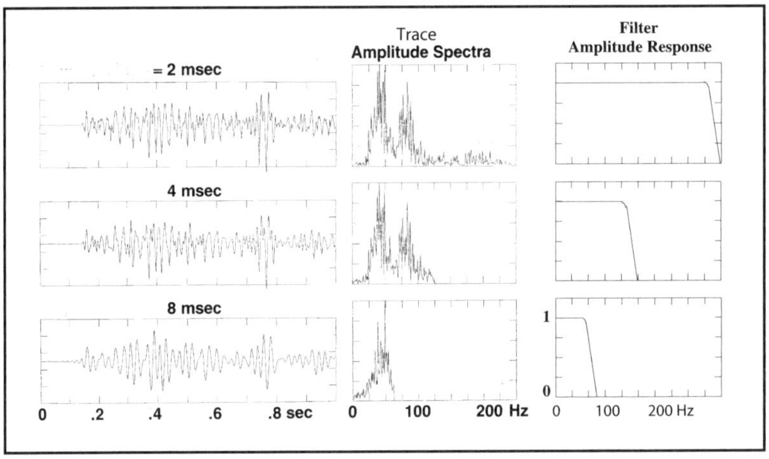

Fig. 3–40 *Anti-alias Filters*

Filters have phase responses as well as amplitude responses. When a filter is applied to input data, the output phase spectrum is the sum of the filter phase response and the input phase spectrum. Thus, filtering is the frequency domain equivalent of convolving an input time series with a filter impulse response.

Phase and its effect on waveforms

As previously stated, phase is an angular measurement of position of a sinusoid at the time observation begins. The terms phase *lead* and *lag* are also used. Figure 3–41 defines these terms and the method of phase measurement. Note that the cosine wave is the sinusoid of reference since it is a maximum at time zero and is an even or symmetrical waveform. It is also zero phase.

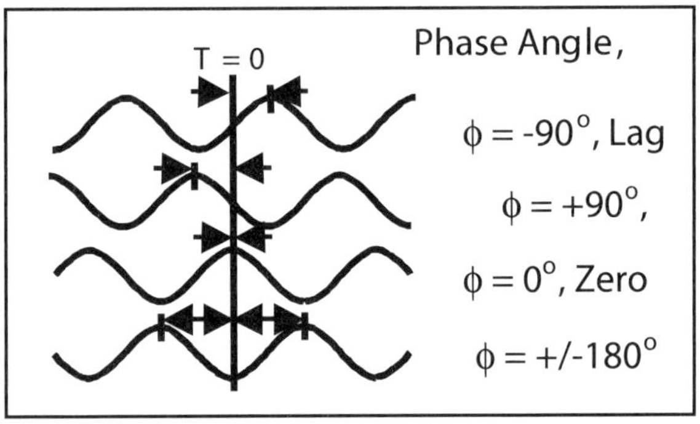

Fig. 3–41 *Phase Definitions*

A wavelet can be zero phase only if all its constituent frequency components are zero phase. Such a wavelet is shown in Figure 3–42a. Figure 3–42b shows the effect of adding 90° to the phase of every frequency component. The amplitude spectrum has not been changed, but simply adding 90° to the phase spectrum makes a drastic change in the waveform. Figure 3–43 shows the changes in waveform as the phase spectrum is modified by successive additions in 90° increments.

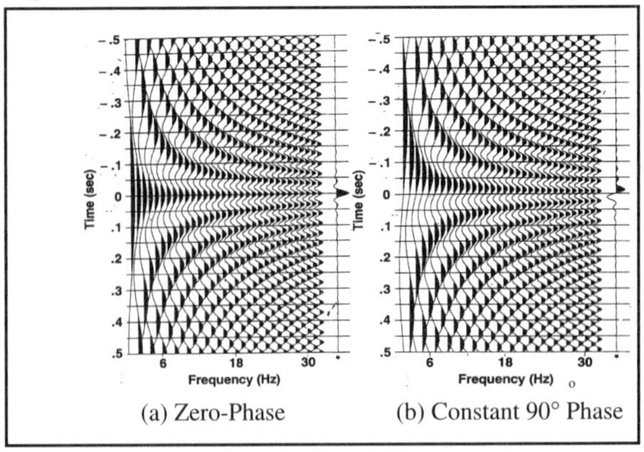

Fig. 3–42 *Effect of Adding 90° to Phase Spectrum*

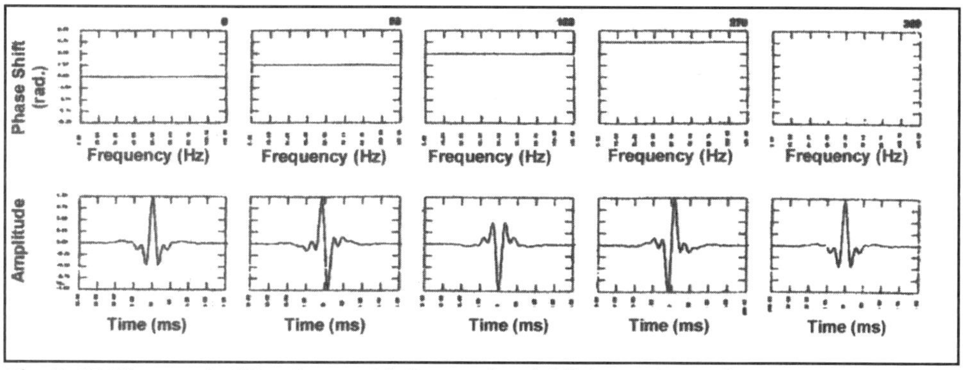

Fig. 3–43 *Changes in Waveform with Successive Additions of 90° Phase*

If a linear phase, i.e. phase proportional to the magnitude of frequency, is added to each frequency component, the effect is quite different. The waveform is unchanged but it is shifted in time. A wavelet can be truly zero phase only if it is centered at time zero. If a zero-phase wavelet is shifted in time, it becomes a linear-phase wavelet (Fig. 3–44).

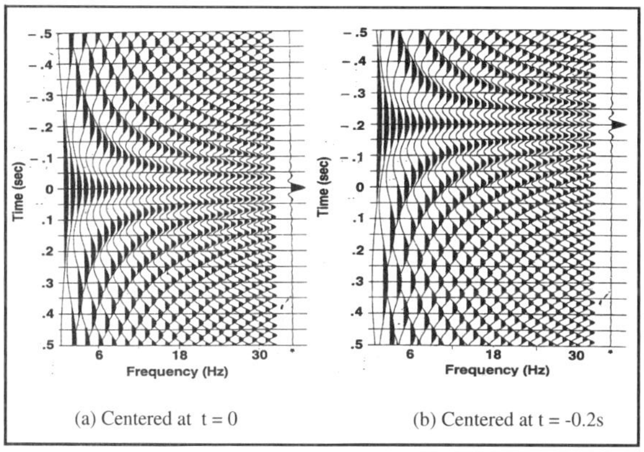

Fig. 3–44 *Effect of Adding Linear Phase to Phase Spectrum*

A linear phase spectrum is a straight line with nonzero slope. A negative slope corresponds to a delay (wavelet is shifted toward a later time). The greater the slope, the larger the time delay (Fig. 3–45). A positive slope corresponds to a negative shift or to an earlier time.

Any phase shift other than linear causes a change in wavelet shape. This is shown in Figure 3–46. From left to right in this figure are a zero-phase wavelet, a constant -90° wavelet advanced 25 ms in time, a constant +90° wavelet delayed 25 ms in time, and a wavelet with an arbitrary phase spectrum. Note that all wavelets shown in this figure have the same amplitude spectrum.

As has been shown, there can be a family of wavelets having the same amplitude spectrum but different phase spectra. One of these wavelets has the smallest phase spectrum. This is the *minimum-phase* or *minimum-delay* wavelet. A minimum-phase wavelet also has the greatest amount of its energy at the beginning of the wavelet as possible. Another wavelet out of this family has the largest phase spectrum. This is called the *maximum-phase* or *maximum-delay* wavelet. All other wavelets in the family are called *mixed-phase* wavelets. See Figure 3–47.

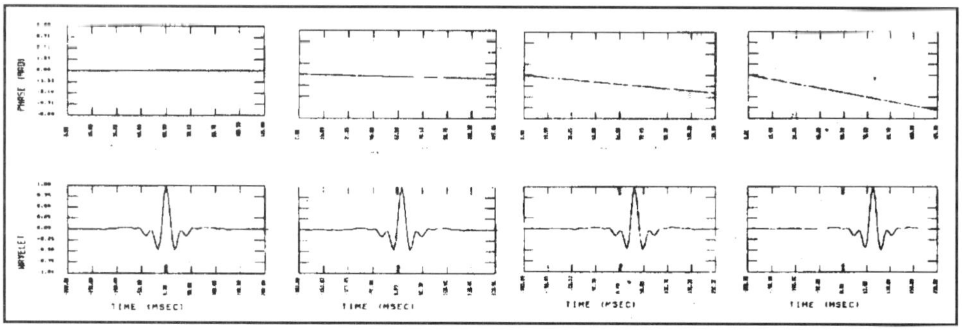

Fig. 3–45 *Linear Phase Slope and Time Shift*

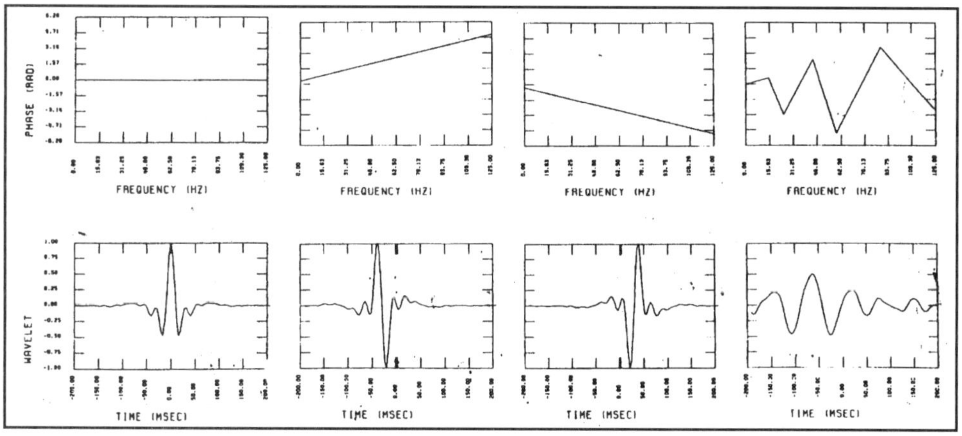

Fig. 3–46 *Phase and Wavelet Shape*

The time series $\{x_n\} = \{x_0, x_1, x_2, ..., x_N\}$ can be expressed as the convolution of $(N+1)/2$, two-point wavelets or doublets:

$$\{\{x_n\} = \{x_0, x_1, x_2, \ldots, x_N\} = \{a_1, b_1)*\{a_2, b_2) *\{a_3, b_3)*\ldots*\{a_M, b_M\}$$

where

$$M = (N+1)/2$$

If $|a_m| > |b_m|$ for m = $1, 2, 3, \ldots, M$, then $\{x_n\}$ is minimum phase. If $|a_m| < |b_m|$ for $m = 1, 2, 3, \ldots, M$, then $\{x_n\}$ is maximum phase.

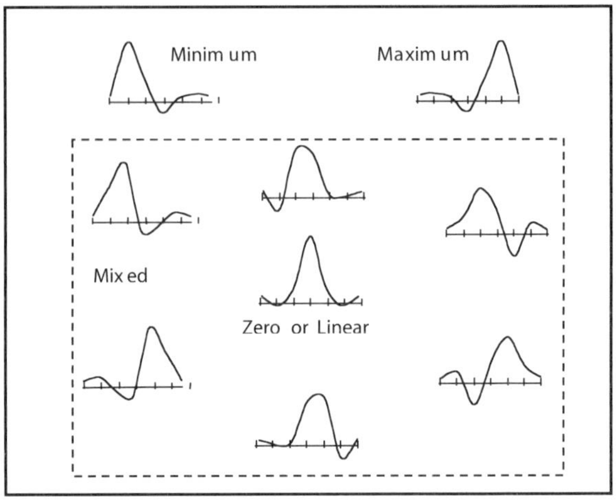

Fig. 3–47 *Minimum- , Maximum- , and Mixed-phase Wavelets*

If $|a_m| > |b_m|$ for some values of m and $|a_m| < |b_m|$ for the other values of m, then $\{x_n\}$ is mixed phase. A zero-phase wavelet is a special kind of mixed-phase wavelet. To see this, let $\{a, b\}$ be a minimum-phase doublet where a is the value at $t = 0$ and b is the value at $t = \Delta T$. The time reversed version of this wavelet is $\{b, a\}$ where b is the value at $t = -\Delta T$ and a is the value at $t = 0$. Convolving these two wavelets gives $\{b, a\}*\{a, b\} = \{ab, a + b, ab\}$. In the resulting wavelet the values at $t = -\Delta T$ and $t = \Delta T$ are equal making the wavelet symmetrical about $t = 0$ and, thus, a zero-phase wavelet. But if $\{a, b\}$ is minimum phase, then $\{b, a\}$ must be maximum phase and $\{ab, a + b, ab\}$ is mixed phase.

Effect of bandwidth

In the discussion of bandpass filters, the passband was defined as the frequency difference between the low and high cutoff frequencies or –3 dB points. Similarly, the bandwidth of an amplitude spectrum is defined as the frequency difference between the –3 dB points. While bandwidth can be expressed in Hz or octaves, the latter is usually more important.

Figure 3–48 illustrates the effect of bandwidth by showing eight zero-phase wavelets and their constituent frequency components. At the top left is the wavelet obtained by summing 1-Hz and 2-Hz cosine waves, giving a bandwidth of one octave. At the bottom is the wavelet obtained by summing 1-Hz through 123-Hz cosine waves (6.94 octaves). There is a steady decrease in wavelet duration or increase in wavelet sharpness. If the bandwidth is increased without limit, then the unit impulse is obtained, as it has an infinite bandwidth.

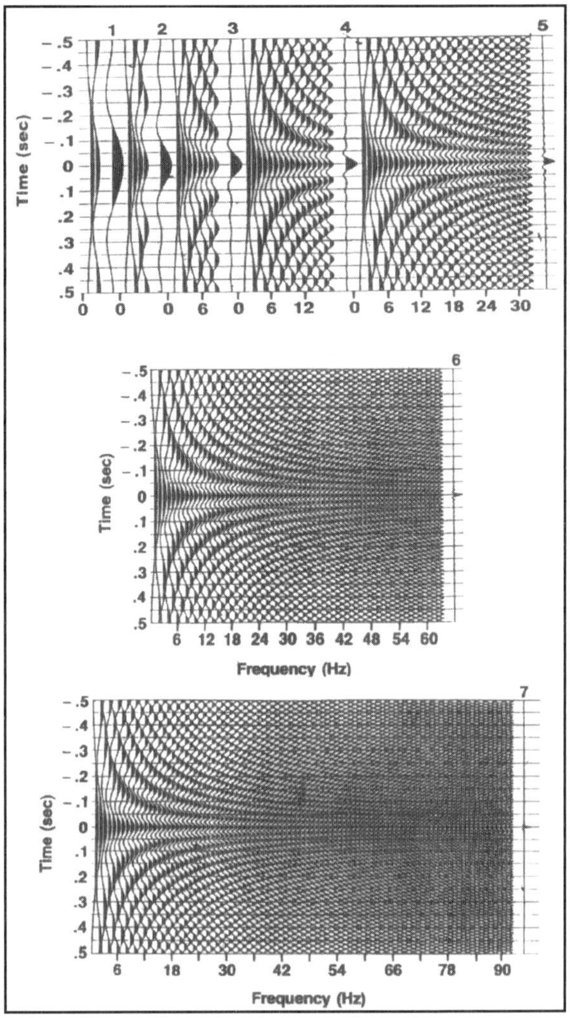

Fig. 3–48 *Effect of Bandwidth*

A wide bandwidth is desirable in many regards. For example, the wider the bandwidth the better one can image the subsurface. Another important aspect of bandwidth is its effect on *resolution*. *Vertical resolution* is defined as the ability to detect two closely spaced reflectors in the subsurface. It has been found that the limit of vertical resolution is ¼ of the dominant wavelength. Since $V = f\lambda$, then $\lambda/4 = V/4f$ where λ = wavelength, V = velocity between the two reflectors, and f is the dominant frequency. Table 3–4 shows how resolution gets worse as velocity increases and frequency decreases. Velocity tends to increase and frequency to decrease with depth. Thus, resolution usually gets worse with depth.

Table 3–4 *Vertical Resolution*

V (m/sec)	f (Hz)	λ/4 (m)
2000	50	10
3000	40	18
4000	30	33
5000	20	62

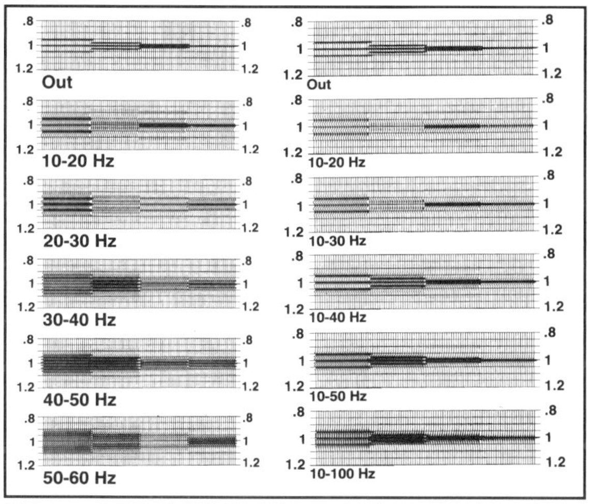

Fig. 3–49 *Bandwidth and Vertical Resolution*

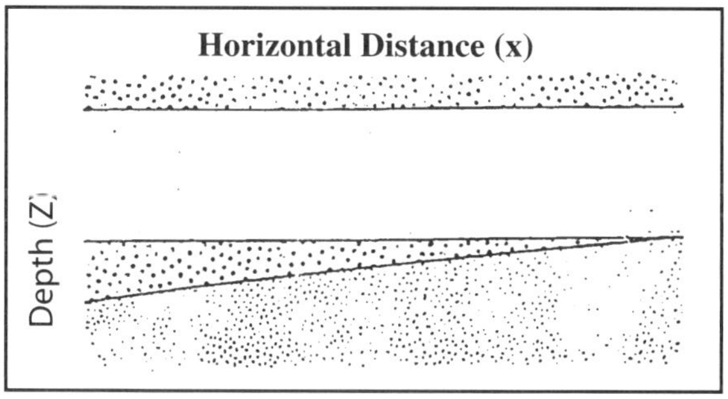

Fig. 3–50 *A Wedge or Pinchout*

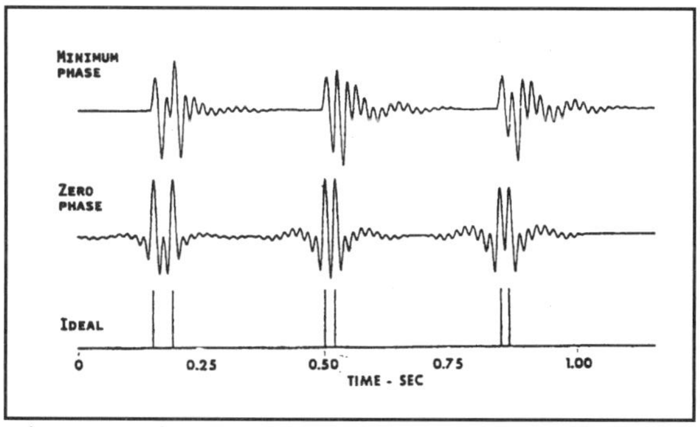

Fig. 3–51 *Phase and Resolution*

Figure 3–49 shows the effect of bandwidth on vertical resolution. There are four sets of three reflectors. The first set is 50 ms apart, the second 25 ms apart, the third 12.5 ms apart, and the fourth 6.25 ms apart. With the maximum bandwidth (filters out), the individual reflections can be seen for the first two sets and just barely seen in the third set. Individual reflections cannot be seen with just 6.25 ms separation. On the left, starting with the second from the top, the bandwidth has been limited to 10 Hz. On the 10 Hz to 20 Hz (one octave), the reflections at 12.5 ms separation seem to have blended into one long event. Below this, as the 10 Hz bandwidth moves to higher frequencies, the ability to detect separate reflections becomes progressively more difficult. Note, also, that the bandwidth in octaves has shrunk from 1.0 to 0.26. So, even though the bandwidth has remained a constant 10 Hz and become higher frequency, vertical resolution has become worse. The right side of Figure 3–49 shows the improvement in vertical resolution as the bandwidth increases from one octave (10 to 20 Hz) to 3.32 octaves (10 to 100 Hz). Thus, good vertical resolution requires a large bandwidth. Figure 3–49 is used courtesy of WesternGeco.

Figure 3–50 shows a geologic situation where good vertical resolution is needed to detect the termination of the pinchout. This could be very important in deciding the location of a well.

Figure 3–51 illustrates another important aspect of resolution—the wavelet type. In this figure, three pairs of reflections are shown with separations of about 40 ms, 20 ms, and 10 ms. At the bottom, the reflection waveform is a spike. At the top, the waveform is minimum phase and in the center the waveform is zero phase. Separate reflections can be seen for all three separations in the case of the spikes and the zero-phase wavelets but it is difficult to identify the time of the second reflection at 20 ms and 10 ms separation for the minimum-phase wavelet.

Reflection time for a minimum-phase wavelet is measured at the start of the wavelet where the trace first breaks down or up. When two minimum-phase wavelets overlap, it makes picking the second reflection difficult unless the wavelet duration is on the order of the reflection separation. Reflection time, for a zero-phase wavelet, is at the central peak, or trough, so picking reflection times is much easier and accurate for zero phase wavelets. This is one reason why zero-phase wavelets are usually output in seismic data processing.

Reflection time for a zero-phase wavelet is at the central peak, or trough, so picking reflection times is much easier and accurate for zero-phase wavelets. This is one reason why zero-phase wavelets are usually output in seismic data processing.

Horizontal resolution. This describes the ability to detect the position of reflector terminations. An example is shown in Figure 3–52. A fault is indicated when a reflection horizon is broken with one side later in time than the other. In Figure 3–53, the actual subsurface displacement ranges from $^1/_{16}$ to $^1/_2$ of a wavelength. For displacements less than $^1/_4$ wavelength, the fault cannot be discerned.

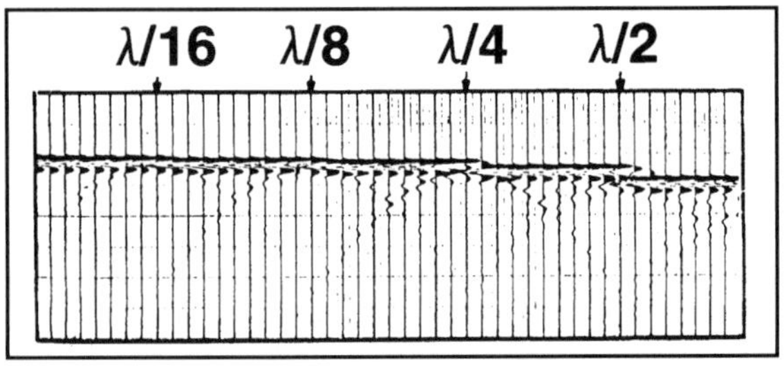

Fig. 3–52 *Horizontal Resolution and Fault Displacement*

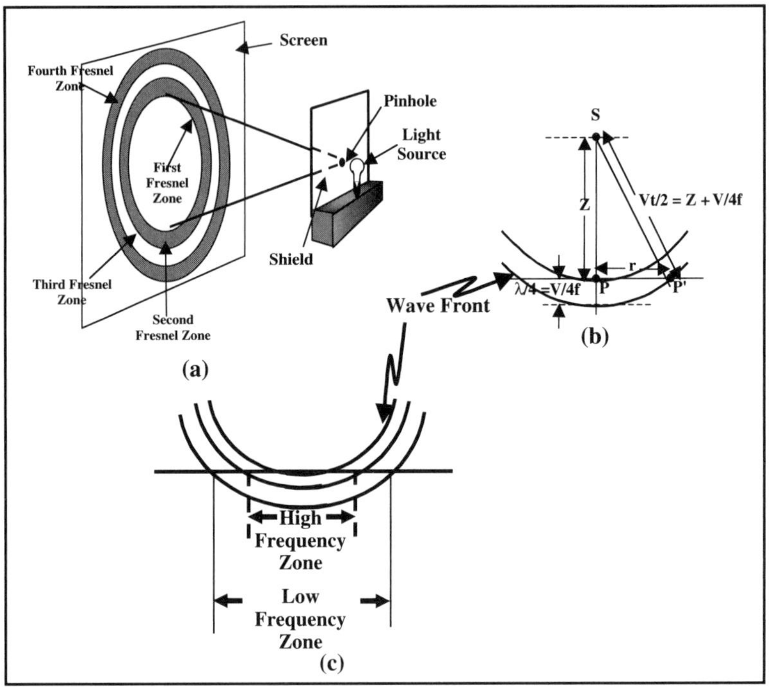

Fig. 3–53 *The Fresnel Zone*

Horizontal resolution is related to the *Fresnel Zone*. The Fresnel Zone is a phenomenon observed for many kinds of wave propagation and was first observed in studies of light. Consider the setup in Figure 3–53a. A light source shines through a pinhole onto a screen in a darkened room. Instead of seeing only a circle of light on the screen, there is a central circle of light surrounded by light and dark bands. This is a result of diffraction of light through the pinhole. The central circle of light is the first Fresnel Zone. It is the only part of the screen that is fully illuminated. Subsequent circular bands of light and dark are the second, third, fourth, etc. Fresnel Zones.

Diffraction also occurs in propagation of seismic waves. In Figure 3–53b, a horizontal reflector is located a depth Z vertically below a seismic energy source at point S. A wavefront reaches the reflector at point P. After traveling $^1/_4$ wavelength farther, the wavefront is at point P^1, a horizontal distance r from P. The distance r is the radius of the first Fresnel Zone. Since SPP^1 is a right triangle, the radius of the first Fresnel Zone in terms of velocity V, reflection time t, and frequency f, can be determined.

$$\left(\frac{Vt}{2}\right)^2 = Z^2 + r^2 = \left(\frac{Vt}{2} - \frac{V}{4f}\right)^2 + r^2$$

$$r^2 = \left(\frac{Vt}{2}\right)^2 - \left(\frac{Vt}{2} - \frac{V}{4f}\right)^2 \tag{3.30}$$

$$r = \sqrt{\left(\frac{Vt}{2}\right)^2 - \left(\frac{Vt}{2} - \frac{V}{4f}\right)^2} = \frac{V}{2}\sqrt{t^2 - \left(t - \frac{1}{2f}\right)^2} \approx \frac{V}{2}\sqrt{\frac{t}{f}}$$

By analogy with optics, it can be seen that the only the first Fresnel Zone is fully illuminated and, thus, its radius defines horizontal resolution. Figure 3–53c shows that higher frequencies have smaller Fresnel zones than lower frequencies. Table 3–5 also illustrates this and the variation with time. Note that frequency usually decreases with time so the lower frequency values in Table 3–5 are more realistic at the later times.

Table 3–5 *Variation of the Fresnel Zone with Time and Frequency*

$r = (z\lambda/2)^{1/2} = (V/2)(t/f)^{1/2}$					
V = 3000 m/s					
Fresnel Zone (m)					
t(s) \ f (Hz)	10	20	30	40	50
1	474	335	273	237	212
2	670	474	387	336	300
3	821	580	474	410	367
4	948	670	547	474	424

The F-K domain

In seismic exploration detectors are placed at discrete locations and the reflected wavefields recorded at those points (Fig. 3–54). So, in seismic exploration, sampling is done in space as well as time.

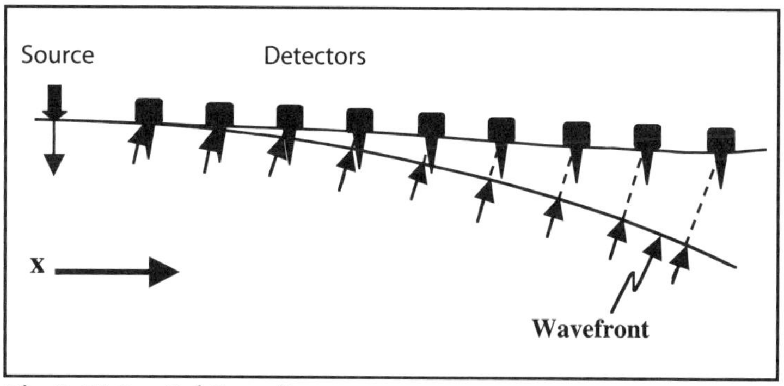

Fig. 3–54 *Spatial Sampling*

Just as the Fourier Transform can be used to transform a function of time to a function of frequency, so can it be used to transform a function of space x to a function of wave-number k and vice versa.

$y(x) \Leftrightarrow Y(k)$

where

Wavenumber is the reciprocal of wavelength λ

$k = 1/\lambda = f/V$

f = frequency

V = velocity

If $\{g(n\Delta x)\}$ is the series of amplitudes obtained by sampling, then the wavefield at discrete spatial locations, spaced Δx apart is as follows:

$$F\{g(n\Delta x)\} = G(m\Delta k) = \sum_{n=0}^{N} g(n\Delta x)\}e^{i2\pi mn\Delta k\Delta x},\ m = 0, 1, 2, \ldots, M \tag{3.31}$$

$$F^{-1}\{G(m\Delta k)\} = \sum_{m=0}^{M} G(m\Delta k)\}e^{i2\pi mn\Delta k\Delta x},\ n = 0, 1, 2, \ldots, N \tag{3.32}$$

Just as in temporal sampling, sampling at too large an interval results in aliasing. The aliasing, or Nyquist, wavenumber is:

$$k_n = \frac{1}{2\Delta x}$$

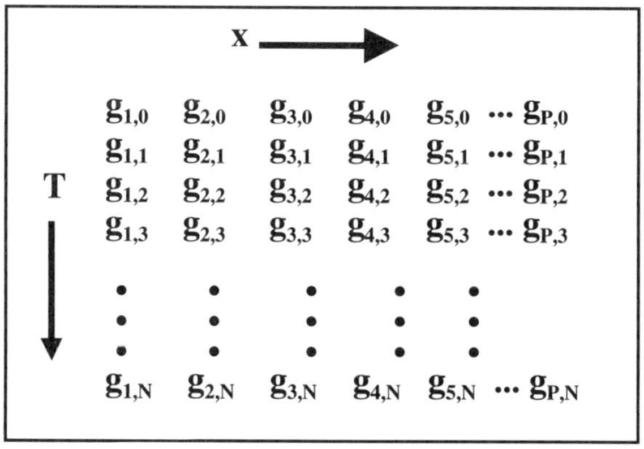

Fig. 3–55 *A Seismic Record as a Two-dimensional Array*

Since the record is sampled in time and space, it can be represented by a two-dimensional array of numbers $\{g(n\Delta T, p\Delta x)\}$. (Fig. 3–55) A two-dimensional Fourier Transform is used to transform the record into $\{G(n\Delta T, m\Delta k)\}$. This transforms the data from the *T-X* domain into the *F-K* domain. A two-dimensional Fourier Transform changes from the *F-K* domain into the *T-X* domain.

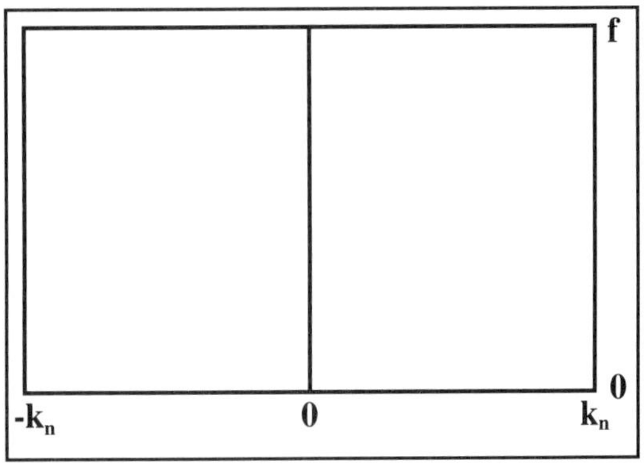

Fig. 3–56 *The F-K Plane*

Because data are sampled in both time and space, data outside a plane that is between $-k_n$ and $+k_n$ in the wavenumber direction and between zero and f_n in the frequency direction is called the *F-K* plane. Thus, the *F-K* plane for data display is as shown in Figure 3–56.

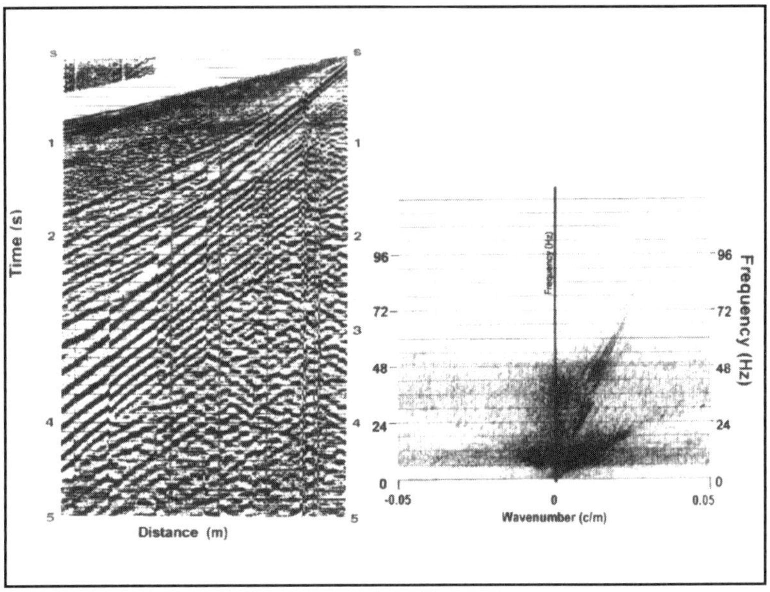

Fig. 3–57 *Noise Record in T-X and F-K Domains*

Figure 3–57 shows a noise record (one shot to identify all source-generated noises) in both the *T-X* and *F-K* domains. Note that linear events in *T-X* are linear events in *F-K*. However, a steeply dipping event in *T-X* appears to have low dip in *F-K*, and vice versa.

Noise such as ground roll and air waves propagate more or less horizontally, as do direct arrivals, first break refractions, and guided waves. These are called linear events because their propagation velocities can be determined from $V=\Delta x\Delta t$, where Δx is a distance increment and t is the corresponding time increment measured on the record. Primary reflections, multiple reflections, and diffractions do not propagate horizontally, but these events appear in the *F-K* plane in terms of an apparent horizontal velocity. This is illustrated in Figure 3–58.

In Figure 3–58, a plane reflected wavefront (line *AB*) is incident on the surface at an angle of emergence of θ. The wavefront arrives at the nth receiver group at an offset x_n at time *t*. It arrives at the $(n+1)^{th}$ group at an offset $x+\Delta x$ at time $t+\Delta t$. Thus, in the time interval Δt, the wavefront travels the distance BB^1. From the right triangle ABB^1, the distance $BB^1 = \Delta x \sin\Theta$. But since the propagation velocity is *V*, then $BB^1 = V\Delta t$ also. So, $V\Delta t = \Delta x \sin\theta$, and with a little algebraic manipulation:

$$\frac{\Delta x}{\Delta t} = \frac{V}{\sin\Theta} = V_{app} \tag{3.33}$$

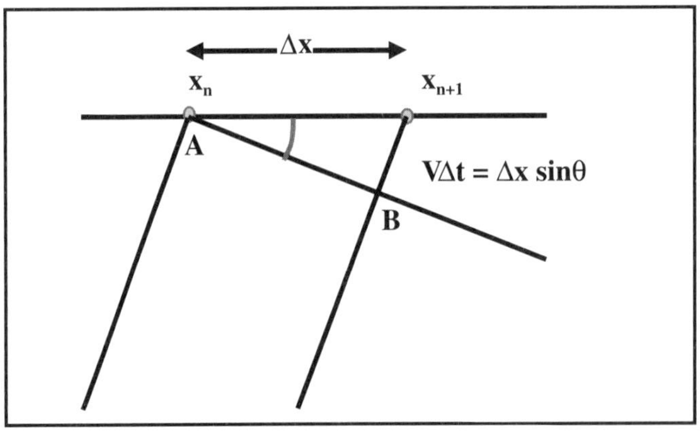

Fig. 3–58 *Plane Wave Incident on Surface*

where

V_{app} is the apparent horizontal velocity

When $\theta = 90°$, (horizontal propagation) $V_{app} = V$

When $\theta = 0°$, $V = \infty$

At any angle of emergence other than 90°, $V_{app} > V$

If a plane wave is reflected from a reflector that is dipping at an angle θ, then the effect will be as described in the previous section. There is, however, another way to look at this. The reciprocal of apparent velocity is dip.

$$dip = \frac{\Delta x}{\Delta t} = \frac{1}{V_{app}} \tag{3.34}$$

When dip is measured from a seismic record or vertical section, it is usually measured in msec/trace. Since two adjacent traces correspond to a distance Δx, and *1s = 1000* msec, the dip in msec/trace, *D*, can be related to apparent velocity by:

$$k = \frac{fk_N}{500} D \tag{3.36}$$

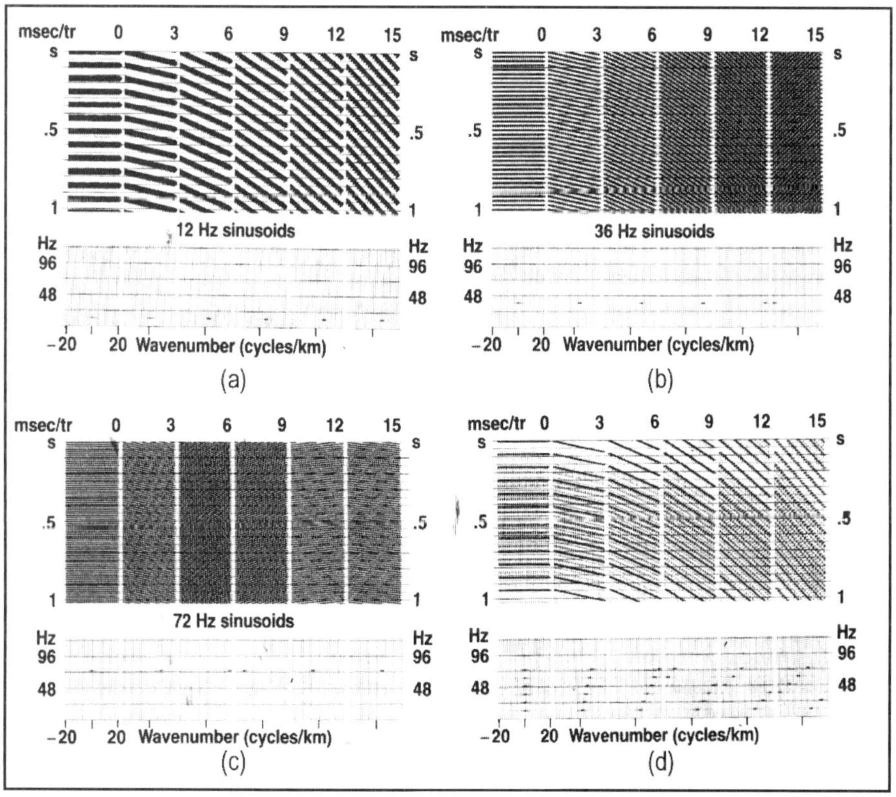

Fig. 3–59 *Spatial Aliasing as a Function of Dip and Frequency*

Spatial aliasing occurs when $k > k_N$, and from Equation 3.36, this depends on both frequency and dip. This is shown in Figure 3–59. Single frequency sinusoids at 12, 36, and 72 Hz are shown in Figure 3-59 sections *a*, *b*, and *c*, respectively. Figure 3 59d shows a combination of 12, 24, 36, 48, and 72 Hz sinusoids. In each case, they are shown at dips of 0 through 15 msec/trace in 3 msec/trace increments. No aliasing occurs at any dip in the 12 Hz case. In Figure 3–59b, the 36 Hz sinusoid aliases at a dip of 15 msec/trace (*15X 36/500 > 1*). In the *T-X* domain, the dip aliases to -15 msec/trace. In the *F-K* domain, the wave number aliases to –18.6 c/km.

In Figure 3–59c, the 72 Hz sinusoid aliases at dips of 9, 12, and 15 msec/trace. In Figure 3–59d, the 72 Hz sinusoid first aliases at a dip of 9 msec/trace. At 12 msec/trace, dip both the 72 and 60 Hz sinusoids alias. At 15 msec/trace dip, only the 12 and 24 Hz sinusoids are not aliased. Thus, spatial aliasing increases as dip and frequency increase.

Figure 3–60a shows, in the *T-X* domain, a single linear event. This is also a linear event in the *F-K* domain, where it can be seen that the signal waveform is band-limited to about 6–48 Hz. Figure 3–60b shows, in the *T-X* domain, a set of parallel linear events with the same signal as in Figure 3–60a. In the *F-K* domain, however, only one event can be seen. Thus, *all events with same dip plot in the same part of the* F-K *plane.*

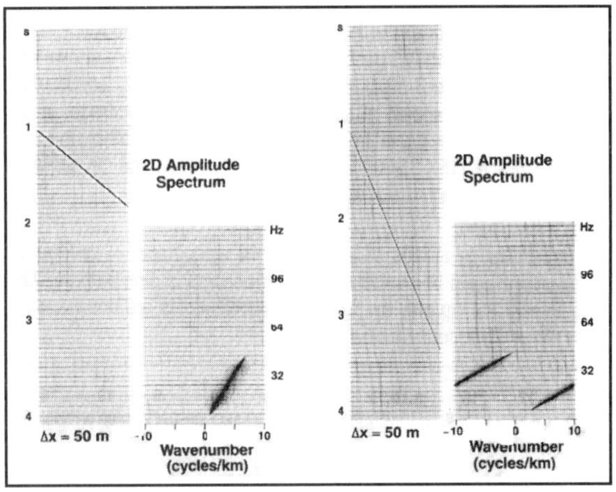

Figure 3–60 *Linear Events in T-X and F-K*

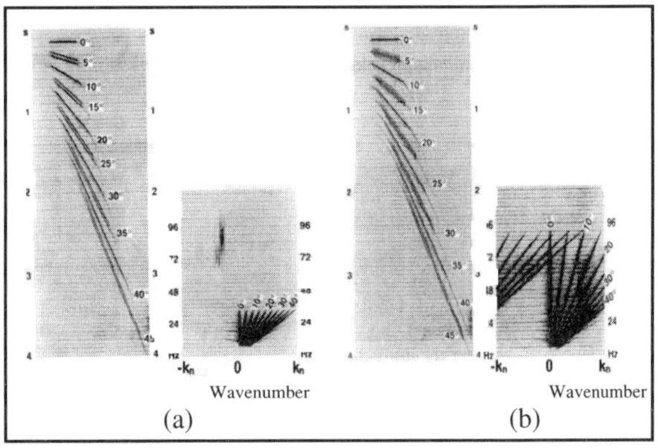

Fig. 3–61 *Linear Dipping Events*

None of the events shown in Figure 3–60 alias because the highest frequency present is 48 Hz. In Figure 3–61a, linear events with dips from 0° to 45° in 5° increments are shown. Signal is band limited from 6 to 36 Hz and no aliasing occurs. Figure 3–61b has linear events with the same dips, but the upper frequency limit has been raised to 90 Hz. As a result, aliasing occurs for dips of 20° and steeper.

It appears that spatial aliasing can be avoided by limiting dips or by limiting bandwidth—neither of which is desirable in most cases. There is, however, an alternative. As shown in Figure 3–62, the smaller the group interval, the higher the frequency that signal can contain without aliasing. That is why the highest frequency required to properly image the exploration objective is an important parameter in designing data acquisition programs.

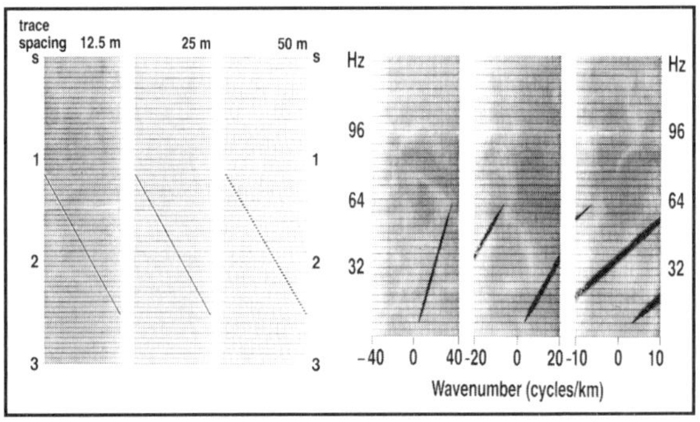

Fig. 3–62 *Effect of Group Interval on Spatial Aliasing*

The great advantage of transformation from the *T-X* domain into the *F-K* domain is that noise, such as ground roll, that crosses through signal in *T-X* can be separated from signal in *F-K*. This is illustrated in Figure 3–63. As shown, signal is usually located near $k = 0$. Direct arrivals, first break refractions, and ground roll are all linear events that plot as straight lines to the right of signal (slower velocity). All of these may alias and cross signal in the higher frequency range. Where, and if, these events cross signal depends on group interval.

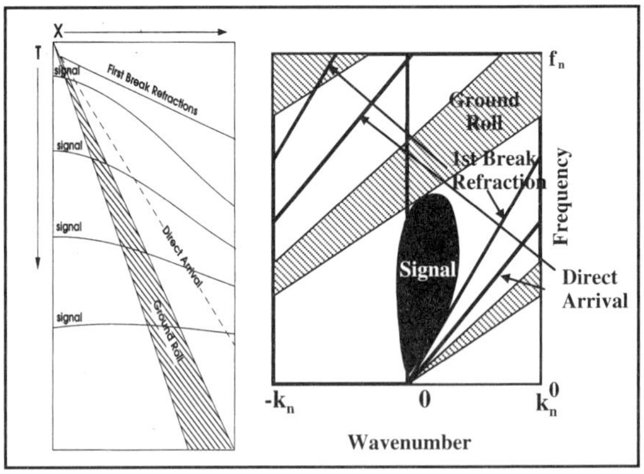

Fig. 3–63 *Separation of Signal and Noise in F-K*

The Z-transform

The *Z-transform* is another mathematical operation used in seismic exploration. It transforms a time series into a power series, allowing simple algebra to be used when working with time series.

Given the time series $\{x_t\} = \{x_0, x_1, x_2, x_3, x_N\}$, its Z-transform is:

$$Z\{x_t\} = x_0 + x_1 z + x_2 z^2 + x_3 z^3 + \ldots + x_N z^N \tag{3.37}$$

The Z-transform is related to the Fourier transform because $z = e^{-i2\pi fT}$.

Given the two time series $\{xt\} = \{2, -1\}$ and $\{yt\} = \{3, 2, -1\}$, we saw earlier that the convolution of $\{xt\}$ and $\{yt\}$ was $w(t) = \{6, 1, -4, 1\}$. Taking the Z-transforms of the two time series gives $X(z) = 2 - z$ and $Y(z) = 3 + 2z - z2$.

Multiplying $X(z)$ and $Y(z)$ gives $X(z)Y(z) = 6 + 4z - 2z^2 - 3z - 2z^2 + z^3 = 6 + z - 4z^2 + z^3$. The inverse Z-transform is $Z^{-1}\{6 + z - 4z^2 + z^3\} = \{6,1,-4,1\}$. That is, the coefficient of each power of z is the corresponding value of the time series, with the power of z determining the delay in sample periods. The product of the two Z-transforms is the Z-transform of the convolution, so it is possible to obtain a convolution with a simple algebraic operation.

Remembering that the cross-correlation of $\{y_t\}$ with $\{x_t\}$ is:

$$\{\phi_{xy}\} = \{x_t\} * \{y_{-t}\} \text{ and } \{y_{-t}\} = \{-1,2,3\}$$

then

$$Z\{y_{-t}\} = -z^{-2} + 2z^{-1} + 3$$

$$Z\{\phi_{xy}\} = (2 - z)(-z^{-2} + 2z^{-1} + 3) = -2z^{-2} + 4z^{-1} + 6 + z^{-1} - 2 - 3z$$

$$= -2z^{-2} + 5z^{-1} + 4 - 3$$

Taking the inverse Z-transform yields: $\{\phi xy\} = \{-2,5,\underline{4}, -3\}$, which agrees with the earlier result.

Another useful application of the Z-transform is the factoring of a power series into doublets.

$$X(z) = x_0 + x_1 z + x_2 z^2 + x_3 z^3 + \ldots + x_N z^N$$

$$= (a_1+b_1 z)(a_2+b_2 z)(a_3+b_3 z)\ldots(a_N+b_N z)$$

If $|a_n| > |b_n|$, then the doublet is minimum phase, and if $|a_n| < |b_n|$, then the doublet is maximum phase. A time series of any length is minimum phase only if all the doublets into which it is factored are minimum phase.

Multiplication by z^n is equivalent to delaying the time series by n sample periods. For example $\{x_t\} = \{4, -2, 1\}$, so $X(z) = 4 - 2z + z^2$. Let $W(z) = zX(z) = 4z - 2z^2 + z^3$. The inverse of $W(z) = \{wt\} = \{0. 4.-2, 1)$. Note that while $\{xt\}$ is minimum phase, $\{w_t\}$ is mixed phase because the delay makes one of the roots of $\{w_t\}$ equal to zero.

It is often desired to find the inverse of a wavelet, etc. Given the wavelet $w(t)$, its inverse $w^{-1}(t)$, is defined by:

$$w(t) * w^{-1}(t) = \delta(t) \tag{3.38}$$

where $\delta(t)$ is the Dirac delta function. What we call a spike is described mathematically as $A\delta(t)$, where A is the amplitude of the spike.

$$Z\{w(t)\} = W(z),\ Z\{w^{-1}(t)\} = W^{-1}(z), \text{ and } Z\{\delta(t)\} = 1.$$

Thus, in the Z-domain, $W(z)\,W^{-1}(z) = 1$, which gives $W^{-1}(z) = 1/W(z)$. Given the time series $\{\underline{2}, -1\}$, its Z-transform is $2 - z$ and the Z-transform of its inverse is $W^{-1}(z) = 1/(2 - z) = 1/2 + z/4 + z^2/8 + z^3/16 + \ldots$. The inverse is infinitely long but it is a converging series because the input is minimum phase $(2 > |-1|)$ If we use a truncated version of the inverse, $H(z) = 1/2 + z/4 + z^2/8$, then $W(z)\,H(z) = 1 - z^3/8$ and the inverse is $(1, 0, 0, -1/8)$ Since the desired output is $\{1, 0, 0, 0\}$, the error is $-1/8$ in the fourth sample. Increasing the length (number of samples) of the inverse will, if it is minimum phase, decrease the error but never eliminate it. A better way to estimate an inverse will be presented in the seismic data processing section.

The radon transform

The *Radon Transform* transforms data from the *T-X* domain into the τ-p domain where p is the ray parameter, $p = \Delta t/\Delta x$, *and* $\tau = t - px$. Figure 3–64 illustrates the ray parameter, which is the constant in Snell's Law.

The Radon transform (also called slant stack) is calculated as follows:

1. Determine the minimum moveout (ΔT_{min}) and maximum moveout (ΔT_{max}) on the trace that will include all desired events. The maximum moveout will probably be for the shallowest reflection and the minimum movement will probably be on the deepest reflection of interest.

2. Calculate the number of p-traces (N_p) from

   $N_p = (\Delta T_{max} - \Delta T_{min})/\Delta T$

   where ΔT is the sample interval.

   Alternately, the number of p-traces can be specified.

3. For the first p-trace, apply a linear moveout correction to every trace of the CMP gather

   $\Delta T_n = (x_n/x_{max})\ \Delta T_{max}$

where:

ΔT_n = moveout correction applied to nth trace

x_n = offset corresponding to nth trace

x_{max} = maximum offset in gather

ΔT_{max} = maximum movement

4. Stack all traces. The output is the first p-trace.

5. Decrement ΔT_{max} by ΔT (sample interval) to obtain the linear moveout correction for the second p-trace.

6. Apply the revised linear moveout and stack to obtain the second p-trace.

7. Repeat steps 5 and 6 until all p-traces are generated.

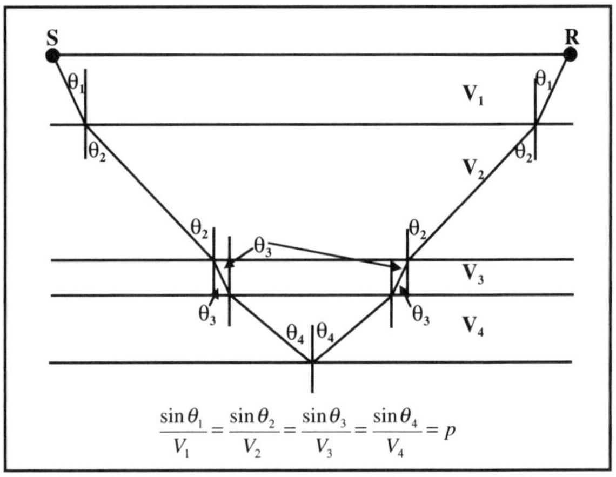

Fig. 3–64 *Definition of the Ray Parameter*

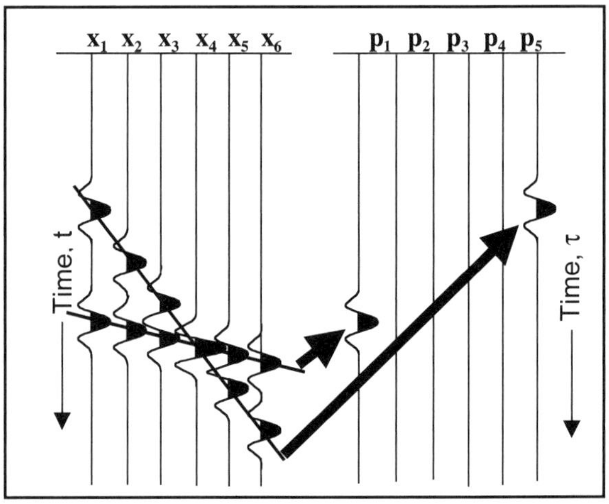

Fig. 3–65 *The Radon Transform*

Figure 3–65 illustrates the Radon Transform with a simple example. The record in the *T-X* domain has two dipping events. The shallower dip is designated ΔT_{min} and the steeper dip is designated ΔT_{max}.

Six values of p are selected, so $\Delta T = (\Delta T_{max} - \Delta T_{min})/5$ with $p_1 = \Delta T_{min}$ and $p_6 = \Delta T_{max}$. Ideally, the two dipping events in *T-X* will map into single wavelets on the p_1 and p_6 traces in τ-p.

Figure 3–66 provides a schematic representation of how events in *T-X* map into τ-p. Table 3–6 summarizes the mapping in terms of event shapes in each domain.

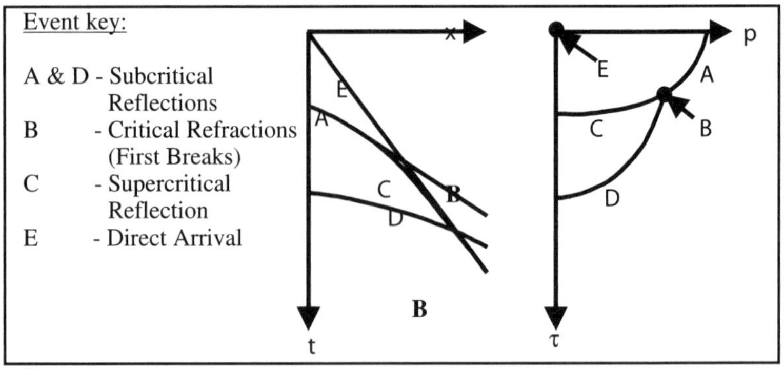

Fig. 3–66 *Event Mapping between T-X and τ-p Domains*

Table 3–6 *T-X and τ-p Relationships*

T-X	τ-p
Hyperbola	Ellipse
$t^2 = t^2 + (x/V)^2$	$\tau^2 = t^2\,(1 - V^2p^2)$
Line	Point
Point	Line

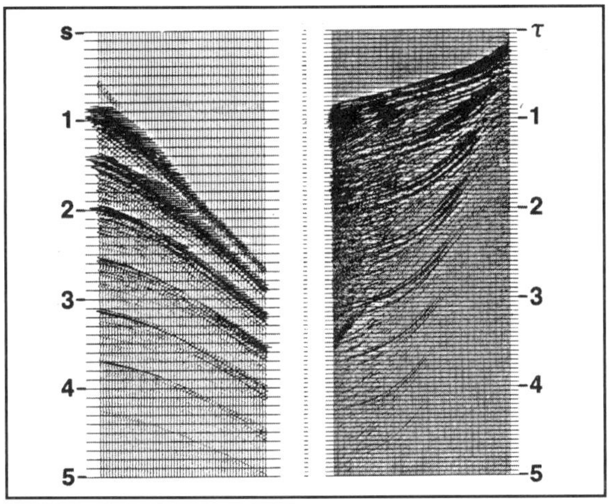

Fig. 3-67 *CMP Record in T-X Domain and τ-p*

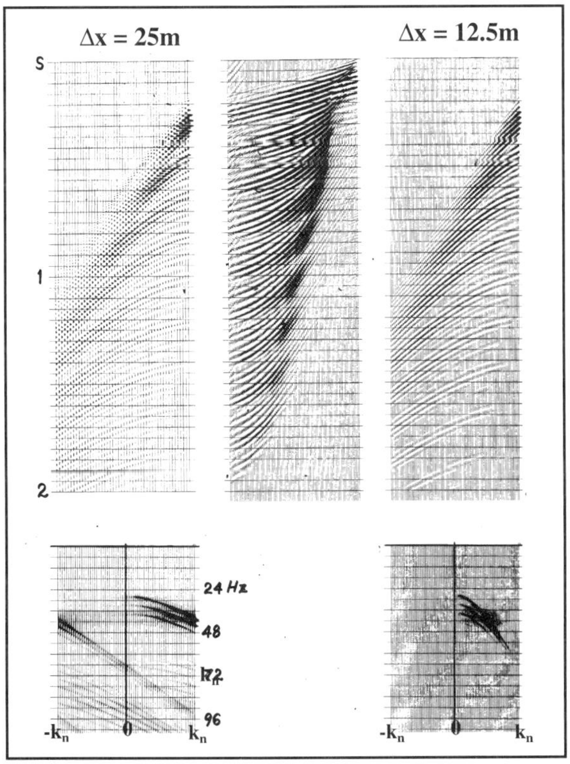

Fig. 3–68 *Trace Interpolation with the Radon Transform*

Figure 3–67 shows a CMP record in the *T-X* and *τ-p* domains. Note that the hyperbolic reflection events in *T-X* become elliptical in *τ-p*.

Radon transform applications include:

- multiple suppression
- time-variant dip filtering
- trace interpolation
- refraction inversion
- velocity analysis
- migration before stack

Figure 3–68 illustrates trace interpolation using the Radon Transform. In this example, the data are transformed into the *τ-p* domain with a trace interval of 25 m and twice as many traces are output with a trace interval of 12.5 m. In this way, spatial aliasing can be avoided without having to acquire the data at 12.5 m trace intervals.

The complex trace

Seismic detectors record motion and, thus, only measure kinetic energy. In a dynamic system total energy is the sum of kinetic and potential energy. For example, as a clock pendulum swings back and forth, it has its maximum velocity at the rest position and at that point all of its energy is kinetic energy ($KE = \frac{1}{2}mv^2$). At its maximum deflection points on each side, its velocity is zero, but it is at a height h above its rest position and all energy is potential energy ($PE = mgh$). See Fig. 3–69. In between these two extremes, the total energy is the sum of kinetic and potential energy.

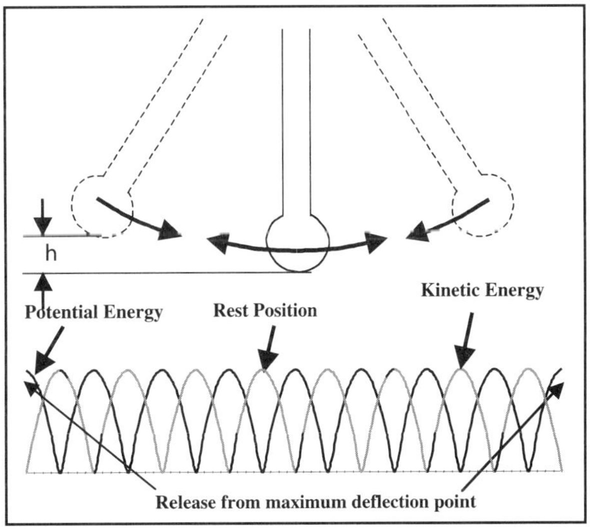

Fig. 3–69 *A Clock Pendulum and Its Energy Cycle*

The *complex* trace, or *analytic signal,* attempts to show total energy in seismic wave propagation. The analytic signal, $u(t)$, is defined as:

$u(t) \quad = x(t) + iy(t)$

where

$x(t) \quad$ = recorded seismic trace and

$y(t) \quad$ = *quadrature trace*

$$= H\{x(t)\} = \frac{1}{\pi t}\, x(t) \tag{3.39}$$

$H(t) \quad$ = *Hilbert Transform*

$$\therefore u(t) = [\delta((t) + \frac{i}{\pi t}].\ x(t) \tag{3.40}$$

The imaginary part of a complex number or variable is usually considered to be orthogonal to the real part. So the complex trace shows the total energy at any one time as the vector sum of the real and imaginary or quadrature trace.

Complex trace analysis makes use of the real trace and the quadrature trace along with three derived quantities—instantaneous amplitude, instantaneous phase, and instantaneous frequency.

Instantaneous Amplitude:

$$R(t) = [x(t)]^2 + [y(t)]^2 \tag{3.41}$$

Instantaneous Frequency:

$$\phi(t) = \tan^{-1}\left[\frac{x(t)}{y(t)}\right] \tag{3.42}$$

Instantaneous Phase:

$$w(t) = \frac{d\phi(t)}{dt} \tag{3.43}$$

Figure 3–70 illustrates these quantities.

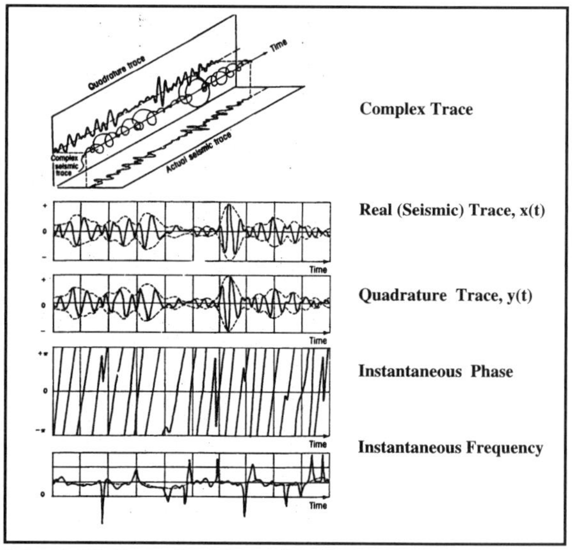

Fig. 3–70 *Complex Trace Analysis*

Workshop

1. Match the list of properties with the wave types by writing the appropriate letter(s) after the wave type.

   A Particle motion in the direction of propagation.

   B. Particle motion orthogonal to direction of propagation.

   C. Particle motion describes retrograde ellipse.

   D. Propagates through body of medium.

   E. Propagates along surface of medium.

   F. Propagates in solids.

   G. Propagates in fluids.

   P-wave: ________________________.

   S-wave: ________________________.

   Rayleigh wave: __________________.

   Love wave: _____________________.

2. P-waves always propagate faster than S-waves in the same medium. True or False?

3. A P-wave that is propagating in a medium having a velocity of 2000 m/s is incident on a medium having a velocity of 2500 m/s at an angle of 15 degrees from the normal to the interface. Determine the angles at which all resulting waves propagate in both media.

4. What are the values of the reflected amplitude, *A1*, and the transmitted amplitude, *A2*, in Figure 3–71?

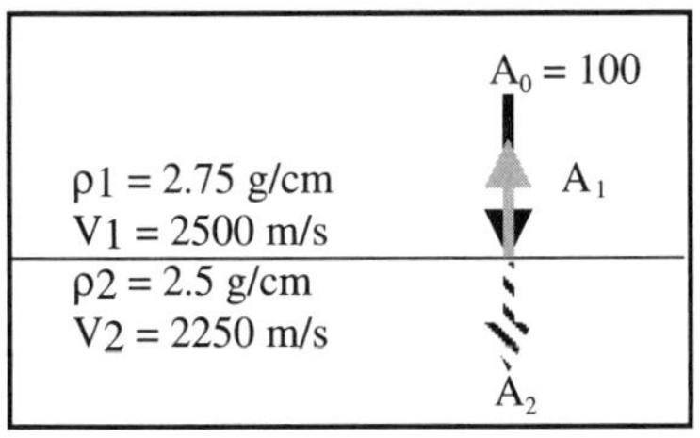

Fig. 3–71

5. Complete Table 3–7 by determining the output frequencies for each input frequency and sample period.

Table 3–7 *Output Frequencies for Various Sample Periods*

Input Frequency (Hz)	Sample period = 1 ms Output Frequency (Hz)	Sample period = 2 ms Output Frequency (Hz)	Sample period = 4 ms Output Frequency (Hz)	Sample period = 8 ms Output Frequency (Hz)
60				
120				
180				
240				
300				

6. Given the impulse response of a system shown in Figure 3–72, what is the output when the input is as shown in the following?

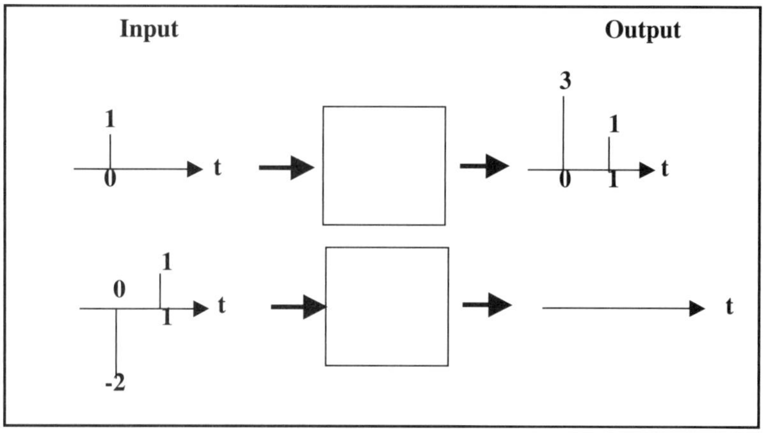

Fig. 3–72

7. Given wavelet $a = \{5, -2\}$ and wavelet $b = \{-3, 1\}$, calculate the cross-correlations, ϕab and ϕba.

8. Calculate the autocorrelation of wavelet a in exercise 7.

9. Which of the following are minimum-phase wavelets? The first value in each case is at time zero.

   a. 6, -1. –2

   b. 3, 4, -4

   c. 0, 12, -1, -6

   d. –2, 5, -2

   e. 28, -27, 5

10. Identify the phase of the wavelets shown in Figure 3–73 as zero phase, linear phase, +90° phase and -90° phase.

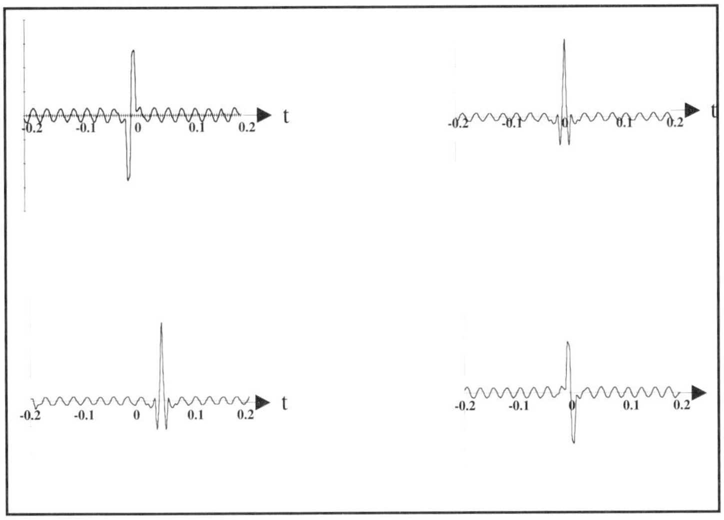

Fig. 3–73

11. Which wavelet has the larger bandwidth, *A* or *B*?

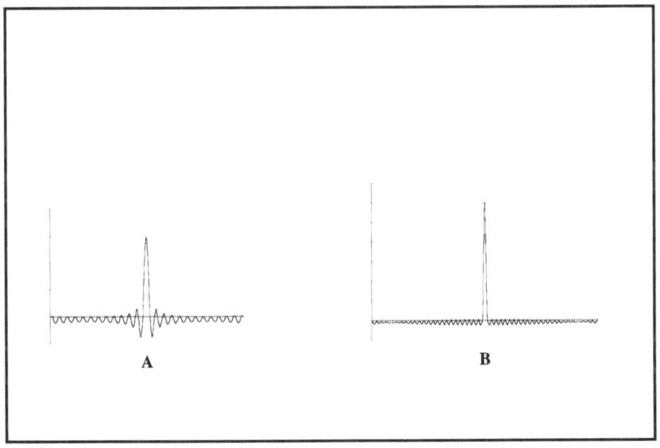

Fig. 3–74

12. The velocity in the target zone is expected to be in the range of 3200 m/s to 3600 m/s at times of 2.7 s to 3.0 s.

    a. What maximum frequency is required to give a vertical resolution of 25 m?

    b. What maximum frequency is required to give a horizontal resolution of 25 m?

13. It is desired to have frequencies up to 50 Hz in the target zone. If the velocity and maximum dip are expected to be 3600 m/s and 35°, respectively, what is the largest group interval that will assure having the desired frequency?

14. Use the Z-transform method to calculate the convolution of wavelets *a* and *b* of exercise 7.

15. Determine the inverse of wavelet *a* in exercise 7.

16. Shown in Figure 3–75 is a schematic representation of a shot record. Sketch in the indicated spaces, the appearance of this record in the *F-K* and *τ-p* domains.

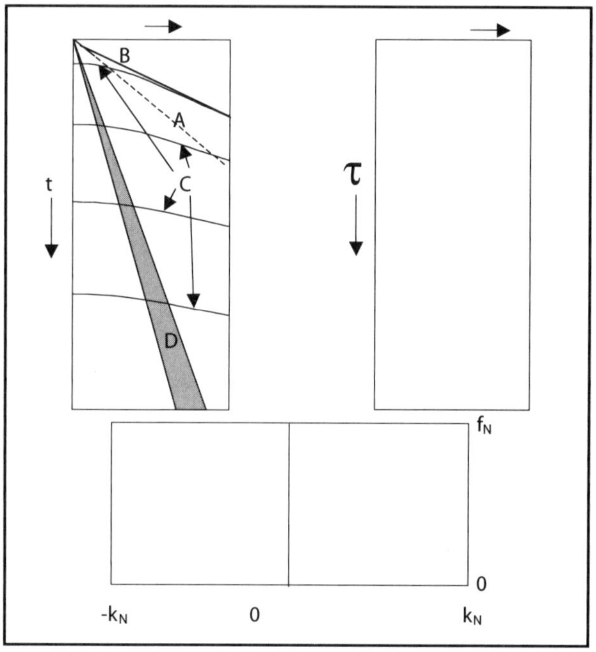

Fig. 3–75

4 Seismic Refraction Exploration

Introduction

A variety of seismic refraction techniques, differing mostly in the geometry of source and receiver configurations, have been used. Inline shooting calls for *shot point* (the position at which the charges are detonated) and detectors laid out along a straight line, arc, or fan shooting involves detectors laid out along an arc of a circle with the shot-point position at the center of the circle, and broadside shooting calls for detectors laid out in a straight line with the shot point offset perpendicular from the center of the line of detectors.

Inline geophone spreads may have as many as 1024 *stations*—location for one or the center of a group of detectors. Spacing between successive detectors or groups ranges from around 50 ft to more than 600 ft in a few cases. Also varying considerably, and in many cases not constant throughout the survey, is the distance from the shot point and the nearest detector position, called *offset distance*. Offset distances of 20 to 30 miles are not uncommon.

Refraction Methods

Inline shooting methods. Figure 4–1 illustrates a layout for *intensive inline refraction* shooting along a 48,000-ft north-south line. The numbers 1 through 6 denote positions of 6 shot points equally spaced along the line. A spread of 24 geophone groups with 400-ft spacing is laid out symmetrically between two successive shot

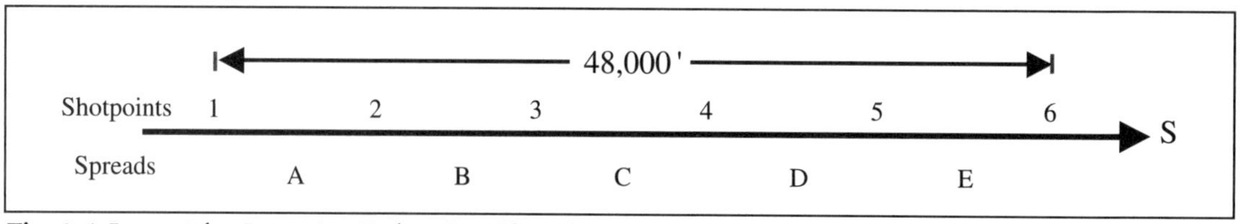

Fig. 4–1 *Layout for Intensive Refraction Shooting*

points. With the spread at *A*, shots are fired at each shot point in succession and records made. Then, the spread is moved to positions *B, C, D,* and *E*, with the same shooting procedure. Thirty records are thereby obtained. This procedure gives complete subsurface coverage but involves a great deal of work and many workers.

At least two shooting and drilling crews should be used to attain good efficiency, and two recording crews might also be used to improve speed of operation further. Thus, considering the number of workers involved, the equipment necessary, and the amount of explosives detonated (perhaps hundreds of pounds at some shot points), this would be an expensive operation—particularly if continued along this or other lines.

A modification of this procedure, called *simple inline forward-and-reverse coverage*, dispenses with shot points two, three, four, and five. The spread is moved as previously described, but shots are fired into the spread from only the two end shot points.

The savings in cost, time, and personnel may be sufficient to offset the loss in completeness, particularly if the geology is rather simple. Timing the first arrivals, or the times at which more or less abrupt trace deflections are first noted, begins interpretation of the 10 records obtained. Times picked for each trace must be corrected for variations in surface elevation and in the near-surface, low-velocity zone. An additional time correction is usually applied to correct to a reference surface such as sea level.

Corrected times are plotted on what is commonly called a *T-X* plot. The vertical scale of the plot is corrected refraction time, and the horizontal scale is the distance between shot and receiver corresponding to the trace from which the time was read. For each such distance, there will be two times corresponding to refraction times from the north and the south ends of the profile.

As shown in Figure 4–2, straight lines are usually fitted to the plotted points. Note that the points normally do not fit exactly to a straight line, but that the best line in a least-square error sense is fitted to appropriate segments of each set of points. It can be seen that if the refractor and surface are both plane and parallel, then refraction time between successive detectors is equal to the separation between the detectors divided by the refraction velocity. Hence, the straight lines shown in Figure 4–2 have slopes that are the reciprocals of refraction velocities.

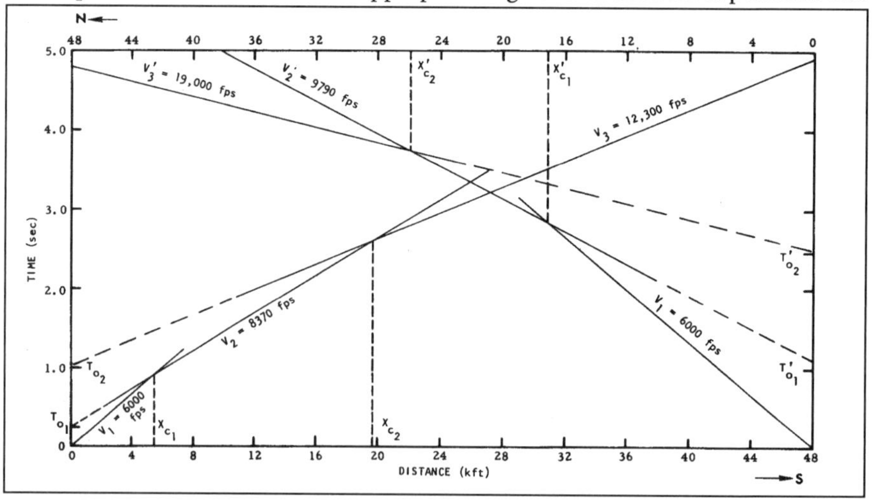

Fig. 4–2 *T-X Plot for Inline Refraction Shooting*

Inspection of Figure 4–2 reveals that the velocities obtained for the second and third refractor differ according to whether refraction paths run from north to south or from south to north. The interpretation of this situation is that the refractors are not horizontal but dipping. Relationships between dip angle, apparent velocities, and actual velocities allow us to compute dip angles, and actual velocities values obtained for the second refractor are a 40° dip to the south and a velocity of 9000 ft per second (fps). For the third refractor, we compute a 70° dip to the south (with respect to the surface) and a velocity of 15,000 fps.

Layer thickness can be computed from either of two sets of data obtained from the *T-X* plot of Figure 4–2—the first set being the intercept times T_{01}, T'_{01}, T_{02} and T'_{02} that are found by projecting the straight line fits back to zero distance axes and the second set being the critical distances X_{01}, X'_{01}, X_{02}, and X'_{02}.

Using these derived data, we construct the model shown in Figure 4–3. Note that this model, which assumes plane refractors, is the simplest one fitting the observations. A more detailed model would likely show more irregular refractors, such as the subsurface relief of each horizon. A refraction profile should also be run at right angles to the one described to obtain true dip of the refractors since the maximum dip may be in a direction other than south.

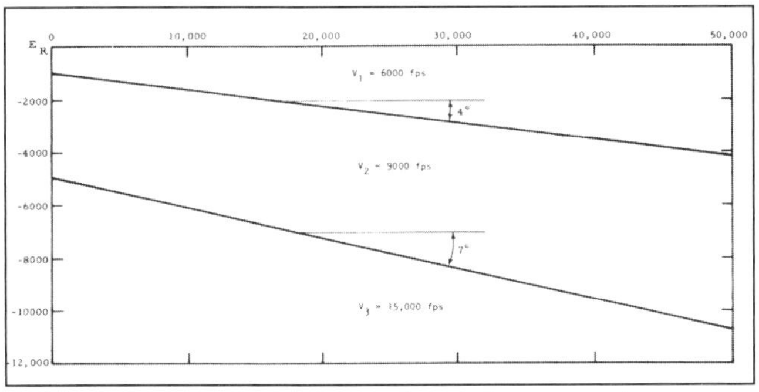

Fig. 4–3 *Simplified Earth Model Based on Data from Figure 4–2*

Another point concerning the interpretation of refraction data is that there are situations in which the method breaks down unless accounted for beforehand from prior information. One situation is the existence of a layer or layers in which the seismic velocity is less than in the overlying layer. Critical refraction cannot occur in this situation, so no refractions are received from this layer for use in estimating layer thickness and velocity. Thickness of other layers and depths to lower ones will be in error due to the *velocity inversion.*

Another situation resulting in incomplete data is the presence of one or more layers, with thickness such that the refraction from the deeper layer arrives at the surface first. If only first arrivals are picked at each distance range, no points corresponding to this layer will be plotted, or the number plotted may be insufficient to show clearly the existence of the layer.

A number of other approaches for inline refraction surveys have been used, most resulting in further reduction of required personnel, materials, and equipment but creating losses in the completeness of the data and the data interpretation. Also, in some cases, there is need for additional assumptions regarding subsurface variations. Consequently, applications of such procedures are limited to geologically simple areas and to reconnaissance of new areas.

Other refraction shooting methods. *Fan* or *arc shooting*, seen in Figure 4–4, has been quite successful in locating relatively shallow salt domes and plugs in areas where seismic velocity is normally rather low. Ideally, all shot-to-detector distances in fan shooting are kept equal. Hence; refractions that travel through the higher-velocity salt reach detectors ahead of those that do not.

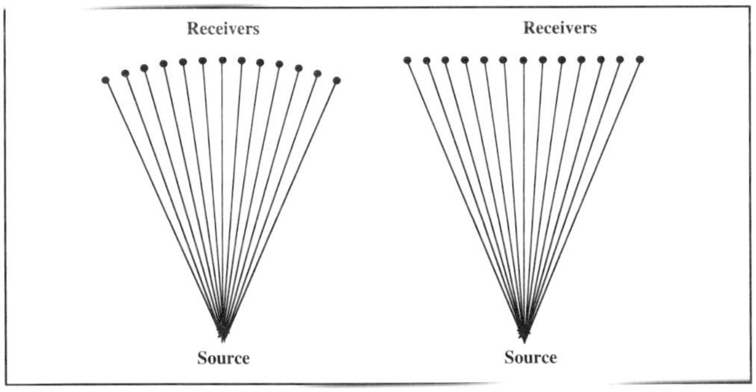

Fig. 4–4 *Arc and Broadside Shooting*

By shooting in mutually perpendicular directions across the salt mass, data are obtained from which the position and a rough outline of the salt mass can be determined. A general idea of the *normal structure*—away from salt intrusion—and locations of salt masses are needed to make the method feasible. Extensive fan shooting in the 1930s led to the discovery of nearly all relatively shallow salt domes in the Gulf Coast area of the United States.

Broadside shooting is well adapted to the mapping of subsurface relief of a particular refractor if the *overburden*—material between the surface and refractor—is reasonably well known. After adjustment for differences in shot-to-detector distances, refraction-time variations from trace to trace are interpreted as variations in refractor depth. By interspersing inline shots, a calibration of sorts is provided for mapping the subsurface along a straight line if spreads overlap sufficiently. Compared with most other methods, broadside shooting results in relatively low cost-per-mile subsurface coverage.

Advantages and disadvantages of the refraction method. Advantages of the refraction method of exploration are that it obtains velocity information directly, can be faster and often cheaper than reflection methods in terms of cost-per-mile of coverage, and, in a few cases, yields results where reflection methods fail. Refraction methods are especially useful for large-scale reconnaissance of new areas and for detailing intrusives such as salt domes and their flanks.

Disadvantages include interpretational ambiguities, cumbersome computations, possibly complex field operations, and the need for a higher degree of surveying accuracy and skill. Several surveying, shooting, and drilling crews may be required to keep pace with the recording crew so a large capital outlay may be needed for equipment and explosives.

See Appendix A for the derivation of calculations used in analysis and interpretation of refraction data.

Workshop

1. Use the *T-X* plot shown in Figure 4–5 to determine the dip angle *j*, the critical angle *qc*, and the true velocity of the second layer V_1.

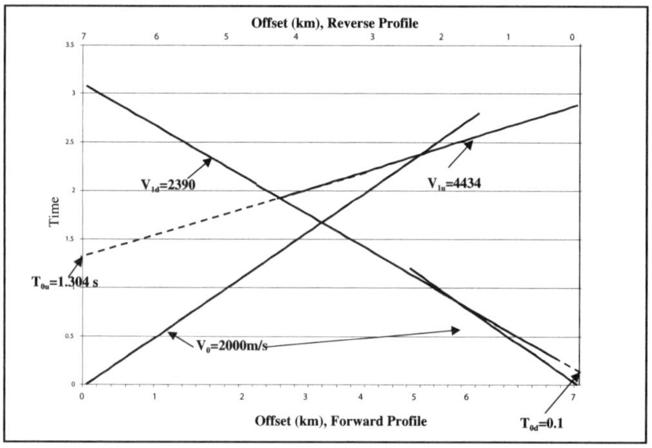

Fig. 4–5

5 Seismic Reflection Data Acquisition

Introduction

The functions required in any seismic reflection data acquisition program are the same, as seen in the following notes.

Permitting—Permission must be obtained from all owners or authorities prior to entering and working in the area. *Positioning*—Precise locations of source and receiver positions must be known. *Signal Generation* Seismic waves with appropriate amplitudes and frequency spectra must be generated. *Recording*—Seismic waves must be detected and converted to electrical signals. Electrical signals must be transmitted to the recording system with minimum attenuation and distortion. The way and the order in which these functions are performed depend to a great extent on the environment.

Land acquisition tends to be sequential. Positioning, usually called *surveying* on land, commences as soon as permitting has progressed sufficiently. If explosives are used for signal generation (energy source), then drilling and loading of holes precedes recording activities. If vibrators are used as the energy source, then signal generation and recording are performed simultaneously.

In marine acquisition, the source, receivers, and recording system are all constantly moving. The navigation system must determine the position of the vessel, sources, and receivers while recording data.

In transition zones between land and marine and in shallow water marine surveys, elements of both land and marine techniques are employed.

Land seismic surveys are conducted in many parts of the world with tremendous variations in topography, weather, surface conditions, population densities, and many other factors. There is usually little functional difference in how these surveys are conducted, but the kinds of equipment and instrumentation varies a great deal.

Another factor that causes variations in how seismic surveys are conducted is whether the survey is 2-D or 3-D. Up until the 1970s, all seismic surveys were 2-D. This means that lines of survey were on the order of one mile apart, and sources were usually in line with the receivers. A tacit assumption in 2-D surveys is that all subsurface reflection points lie in a vertical plane through the line of survey.

In 3-D seismic surveys, the sources and receivers are not all in line with each other. Receivers that record from a single source are laid out over an area instead of along a relatively straight line. Also, receiver lines are much more closely spaced in 3-D than in 2-D. This provides much denser sampling of the subsurface.

Permitting

The problems presented in permitting an area for seismic data acquisition differ considerably between land and marine and between those conducted in the United States and in other countries. In marine surveys, the property owners are nations or political subdivisions of them (states or provinces). For land surveys conducted in the United States, one must consider not only owners of the surface but also the subsurface. It is not uncommon in this country for a piece of land to be occupied by a person who leases it from the owner of the surface but does not own the mineral rights. In most, if not all, other countries the government owns all mineral rights, but the surface land may be owned privately.

Initial planning for a seismic acquisition program is done from maps. Before surveying or recording can begin, permission to enter the area must be obtained. Perhaps the most important thing for the permit agent to do is to plan ahead and start early. Neither a client nor a contractor can afford to have seismic surveying activities delayed as a result of not having secured required permits.

Identify all owners, whether private or government agencies. Failure to do this can result in lawsuits, skips in coverage, or even confrontations between field personnel and irate property owners. Once the owners are identified, contact the person or governmental agency responsible for granting permission to resolve any questions or clarify any uncertainty with respect to permit requirements.

International permitting presents special problems. Each country has its own unique set of permitting challenges. Import/export agents, who specialize in expediting the importation of seismic survey equipment and personnel, can aid in identifying permit issues for a particular country. Typically, the oil company requesting seismic services will be familiar with permitting requirements for the countries in which it operates.

Items to consider in international permitting are

- governmental permits
- environmental regulation and permits
- governmental import/export fees
- notification of military or coastal patrol authorities
- territorial water disputes
- requirements for use of local labor

The oil company contracting for seismic services is frequently in the best position to negotiate for and secure governmental permission.

In land operations, the permit agent is usually the first person to see the area of survey. The permit agent should inform the project manager and other interested parties about potential problems with access or obstructions that may require lines to be deviated or moved.

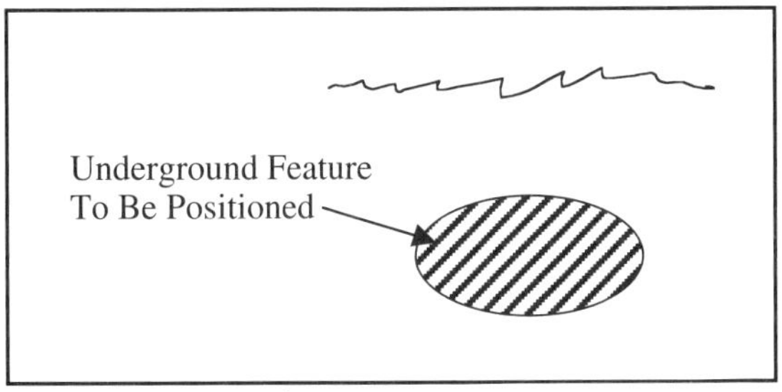

Fig. 5–1 *Target Reference*

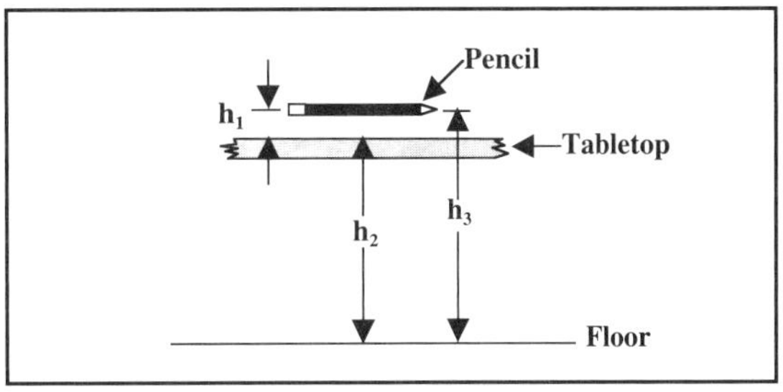

Fig. 5–2 *Simple Example of Vertical Reference Datum*

Positioning

Good seismic positioning provides accurate horizontal and vertical positions for the entire underground feature within the time that the client has allotted. One question that must be asked is to what will the positions be referred? (Fig. 5–1)

Datum is a very important concept in positioning. A datum is a well-defined reference surface to which measurements are referred. For example, assume you want to know the elevation of a pencil. To answer that question, the elevation must be defined. We can determine where the pencil lies in relation to the table it is on or in relation to its height from the floor. Figure 5–2 shows the situation. We can specify the height of the pencil above the tabletop (h_1) and the level of the floor below the tabletop (h_2). In this case, the tabletop is the datum. Alternatively, we could make the floor our datum and reference the heights of both the tabletop and the pencil to the floor. The height of the pencil depends on the datum—the height-above-the-tabletop datum is less than the height-above-the-floor datum. This is a simplified example of a *vertical reference datum*.

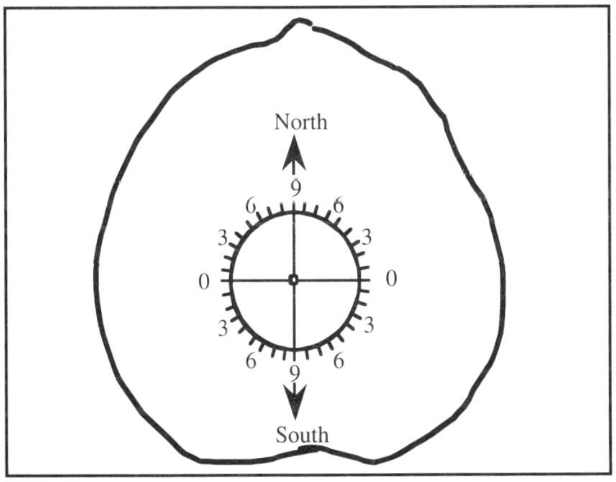

Fig. 5–3 *Geoid*

Unfortunately, the Earth is not flat! Therefore, surfaces that are close to spherical tend to be used to model the Earth's size and shape. One such surface is known as the *geoid*. The geoid is a vertical reference surface, and it is usually very close to mean sea level (MSL). See Figure 5–3. It must be noted that there are a number of different vertical reference datums, each with its particular strengths and weaknesses for a given region. Examples include GEOID '96 and OSU91.

Since the geoid is usable for heights only, we must introduce the concept of a *horizontal reference datum*. Consider this example—your house is the origin of a two-dimensional coordinate system. As you move away from your house, the relative positions of you and your car will change horizontally. Figure 5–4 shows a simple representation of a *two-dimensional horizontal datum* with its origin at your house. This datum does not take into account the fact that the Earth's surface is curved. A satisfactory horizontal datum for use in seismic exploration must recognize the Earth's curvature. *Geodetic latitude* (Fig. 5–5) is an angular measure of position. It is measured from the equator and reaches +90° at the North Pole and -90° at the South Pole.

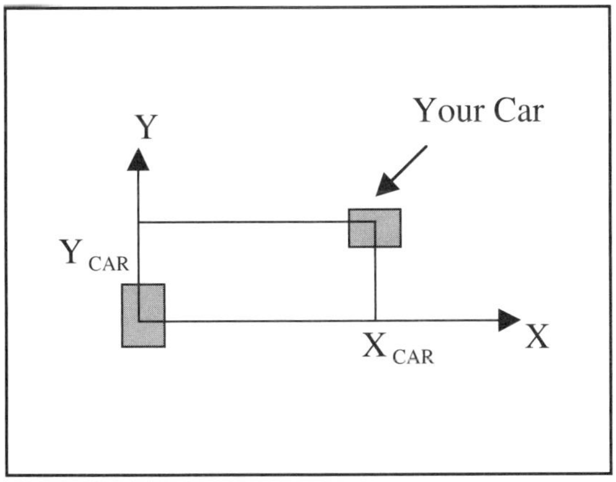

Fig. 5–4 *Horizontal Reference Datum Example*

Geodetic longitude (Fig. 5–6) is also an angular measure of position. It is measured in the equatorial plane. It is positive east of the Greenwich, England, Meridian of Longitude and negative west of the Greenwich, England, Meridian of Longitude. Note that the latitude (ϕ_G) and longitude (λ) values depend on the parameters of the reference ellipsoid.

Given the latitude, longitude, and ellipsoidal height, it is possible to calculate the *X*, *Y*, and *Z* Cartesian coordinates of a point.

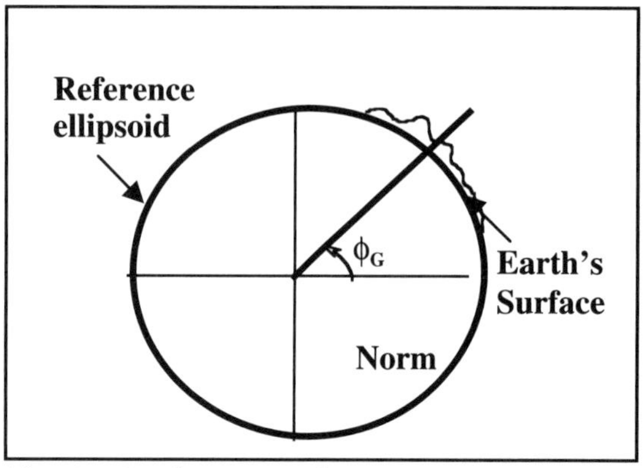

Fig. 5–5 *Geodetic Latitude*

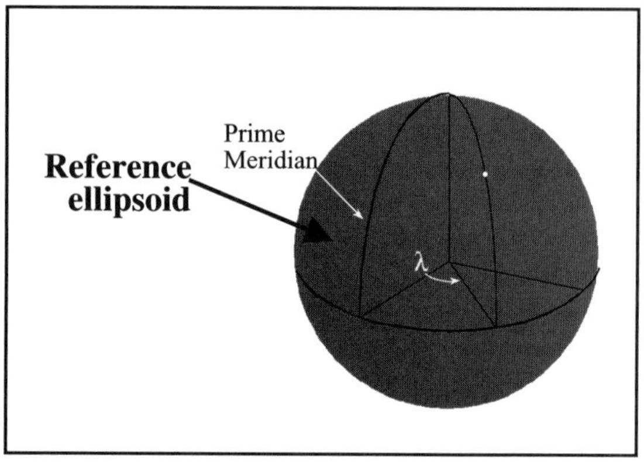

Fig. 5–6 *Longitude Measurement*

As implied previously, there is more than one kind of ellipsoid (Fig. 5–7). Geocentric ellipsoids are centered at the Earth's center of mass, while regional ellipsoids (such as Clarke 1866 or International) are not (Fig. 5–8). These differences (among others) lead to the need for datum transformations.

In Figure 5–9, Datum 2 is the *local* or *survey datum,* such as might be used in a marine seismic survey. It is desired to change from Datum 2 to Datum 1. The most complete datum transformation requires seven parameters. These are

- DX—datum shift in the X direction
- DY—datum shift in the Y direction
- D—datum shift in the Z direction
- e_x—rotation about the X axis
- e_y—rotation about the Y axis
- e_z—rotation about the Z axis
- DS—scale difference in parts per million

The critical thing is that these transformation parameters are entered into the survey system with the proper magnitude and sign.

Many clients will want the local datum positions of the subsurface feature to be expressed in a map projection. A map projection takes the latitude and longitude of a point and converts them into linear units of *northing* (N) and *easting* (E). Consider the simple example in Figure 5–10.

Begin at the origin and go due north for 1000.0 meters on the ellipsoid. This is about 33.3" of latitude. This will be at a northing of 999.6 meters. This northing would, of course, be much different on a different type of map projection.

Among the more common map projection types are the Lambert Conical and Universal Transverse Mercator. Each has its own particular parameters that the surveyor must carefully check to avoid costly positioning errors.

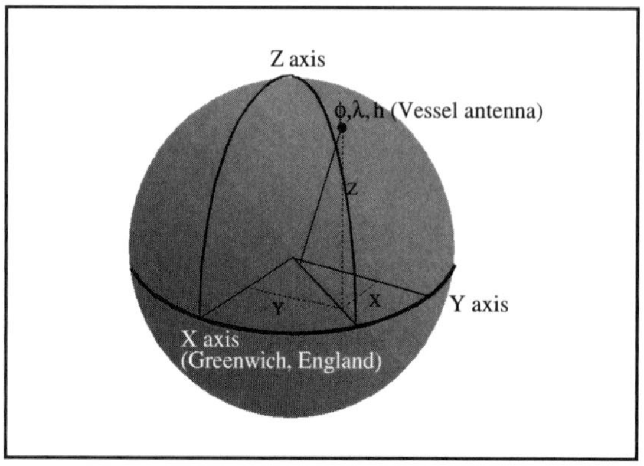

Fig. 5–7 *Horizontal Reference Datum*

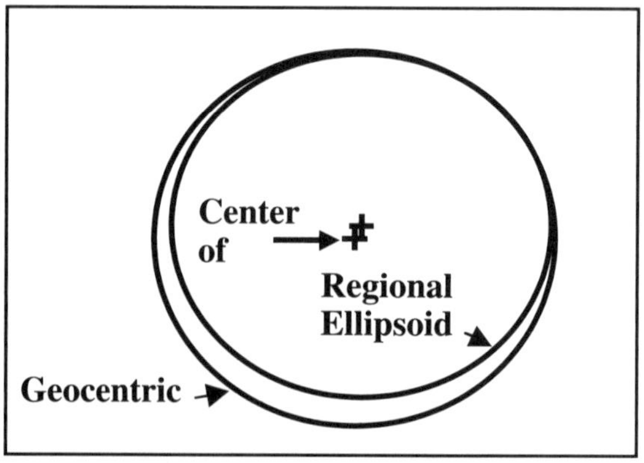

Fig. 5–8 *Geocentric and Non-geocentric Ellipsoids*

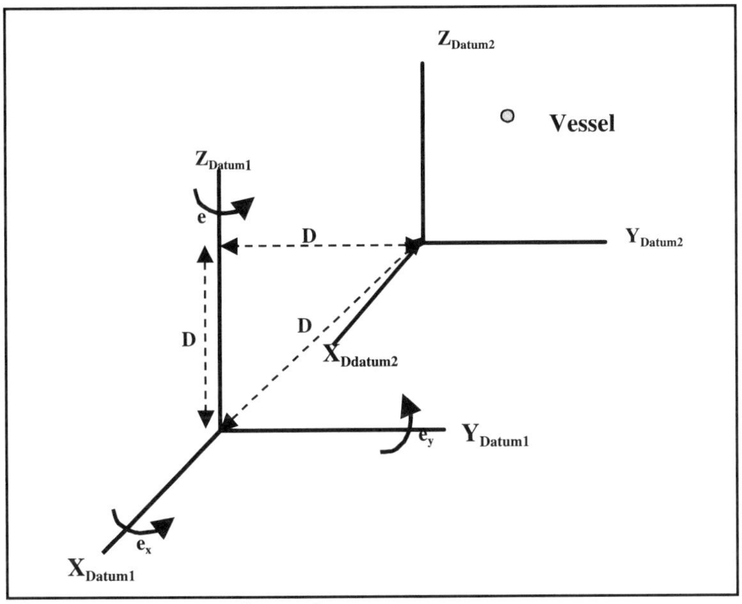

Fig. 5–9 *Datum Transformation*

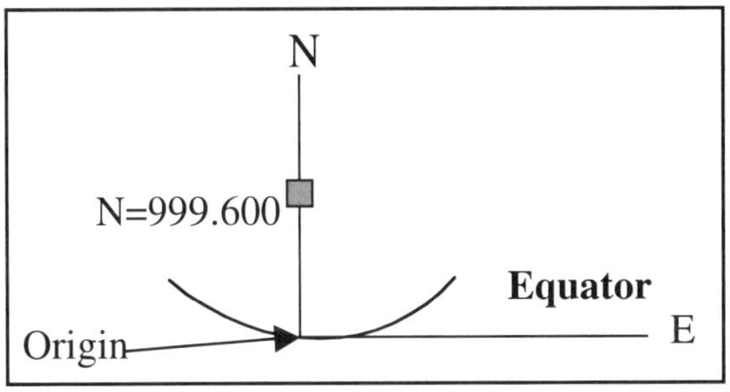

Fig. 5–10 *Map Projection*

Marine navigation. From a geophysical standpoint, the vessel is only located as an intermediate step in determining the location of the energy source array and seismic detectors. From a safety standpoint, accurate positioning and steering of the vessel is a necessity. The vessel is where all of the people are—this includes the clients and contractors. Except in rare cases, there are numerous hazards (surface facilities, buoys, reefs, shoals, and international boundaries) that must be avoided in order to execute a safe, incident-free marine 3-D survey. The vessel position is also used as a geodetic anchor point for vessel relative positioning systems such as laser or acoustic tracking systems.

Vessel navigation requires absolute positioning in order to relate the position of the vessel to other features (coastlines, platforms, shoals, or reefs) on the surface of the earth. All forms of absolute positioning commonly used in 3-D seismic survey rely on the observation of radio waves to determine the vessel position relative to precisely known reference positions known as *base stations*. There are two main types of radio positioning systems—surface-based and satellite-based systems. Surface-based systems rely on fixed-base stations located on the surface of the earth in the vicinity of the prospect site. Satellite-based systems use orbiting satellites as the base stations. The base stations may be *static (geosynchronous)* or *dynamic*. Table 5–1 provides a summary of surface-based positioning systems used in marine seismic acquisition programs.

Table 5–1 *Surface-based Navigation Systems*

System	Frequency	Geometry	Usable Range	Accuracy
Mini Ranger	5.4–5.9 GHz	Circular	25–50 km	5–20 m
Trisponder	8.8–9.5 GHz	Circular	25–50 km	3–12 m
Syledis	420–450 MHz	Circular	60–110 km	5–15 m
Maxiran	420–450 MHz	Circular	60–150 km	10–50 m
UF Transponder	430–440 MHz	Circular	60–80 km	5–15 m
ARGO	1.6–2.0 MHz	Circular	150–700 km	8–25 m
Microphase	1.6–2.0 MHz	Circular	150–700 km	10–30 m
SPOT	1.6–2.0 MHz	Circular	800–1000 km	10–500 m
Hyperfix	1.6–3.4 MHz	Circular	150–700 km	10–30 m
Geoloc	2.0 MHz	Circular	800–1000 km	10–50 m
Loran C	100 kHz	Hyperbolic	300–1500	50–500 m
Pulse/8	100 kHz	Hyperbolic	300–800 km	30–500 m

The basic concept of radio positioning systems is that radio signal transit times or phase measurements may by converted to equivalent distances by scaling by the propagation velocity of electromagnetic energy. Phase measurements have a cyclic ambiguity of an integer number of wavelengths.

Two primary positioning geometries for surface-based positioning are circular and hyperbolic. Circular systems measure the transit time or phase of a radio signal from a known location, called the *base station*, to the remote receiver resulting in a *circular line of position* (LOP), as seen in Figure 5–11. Hyperbolic systems measure the transit time difference or phase difference between radio signals sent from two known locations—master and slave base stations—to the remote receiver resulting in a hyperbolic line of position.

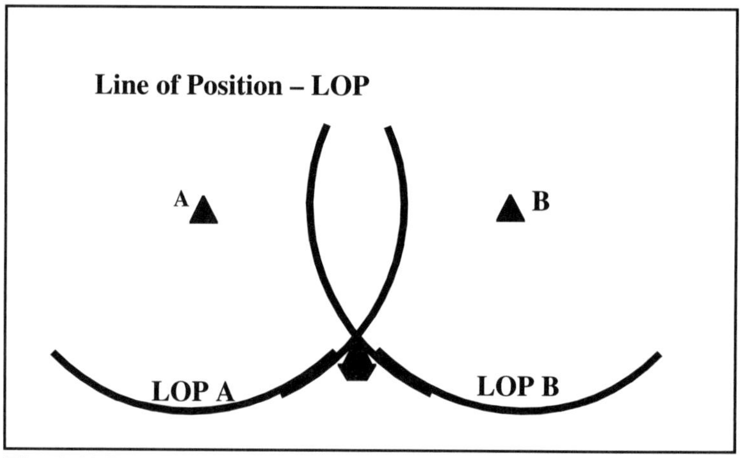

Fig. 5–11 *Range-Range LOPs*

Radio frequencies used in surface-based radio positioning systems are classified as:

- SHF (Super High Frequencies), 5 – 10 GHz
- UHF (Ultra High Frequency), 400 – 450 MHz
- MF (Medium Frequency), 1 – 4 MHz
- LF (Low Frequency), 100 kHz

In general, the higher the frequency, the greater the accuracy potential, and the lower the frequency, the greater the range potential.

Several satellite-based systems have been developed. The Transit or Navy Navigation Satellite System (NNSS) was developed by the U.S. Department of Defense. It is based on measurement of Doppler shift of transmitted frequency and uses hyperbolic geometry. It has been more or less superseded by the Global Positioning System (GPS). The Navstar GPS was also developed by Department of Defense. It is a ranging system that measures transit time of signal and uses spherical geometry. Starfix is a commercial system that uses geosynchronous communication satellites.

The wide acceptance of GPS positioning technologies, and in particular the wide application of differential GPS positioning techniques for high accuracy positioning, has transformed navigation and positioning in the seismic industry. Surface-based radio navigation systems have largely been superseded by differential GPS or similar satellite-based technologies on the basis of cost, accuracy, and simplified logistical considerations.

The GPS is a satellite-based positioning system that can be used night and day from virtually all corners of this planet. The benefits of GPS in seismic work follow:

- all-weather operation
- extremely good accuracy over long distances
- 24-hour per day availability virtually everywhere
- very good reliability
- much faster than conventional surveying techniques in many situations

There are currently 27 satellites in orbit around the world. These are mainly BLOCKII and BLOCK IIR satellites. GPS is made up of three segments:

- the space segment (the satellite hardware and software)
- the control segment (controls the quality of the space segment)
- the user segment (all of the users of GPS)

There are two types of measurements in GPS:

- **Pseudo-range:** measure an unambiguous range to the satellite.
- **Carrier phase:** an ambiguous measurement that has the benefit of higher resolution than the pseudo-range.

The satellite positions must be known to be able to position an object on the Earth's surface. The satellite positions are calculated from an ephemeris, which the user uses to calculate the satellite positions. Table 5–2 provides an overview of GPS systems.

Table 5–2 *Basic GPS Overview*

	Number of Reference Stations	Accuracy Positions	Minimum Satellites for 3-D	Accuracy Depends on	Accurate Enough for Seismic Work?
Stand-alone Position Non-Differential Point Position	None	100 m	4	Number of satellites	No
Differential Position or Relative Position	At least 1	Generally 1–5 m	4 (5 for some systems)	Number of satellites & distance from base station, etc.	Under proper conditions

In this context, a *reference station* is a GPS base station that is located on a very accurate survey point—not to be confused with the reference stations that the GPS control segment uses. There are several types of differential positions, each with its advantages and disadvantages in terms of accuracy, reliability, and cost.

GPS receivers are used to track the GPS satellites. These receivers do not send signals to the satellites; they receive radio waves from them. The GPS antenna receives the signals from the satellites. The receiver decodes these signals, calculates a position, and provides output (usually RS-232) to other systems (Fig. 5–12).

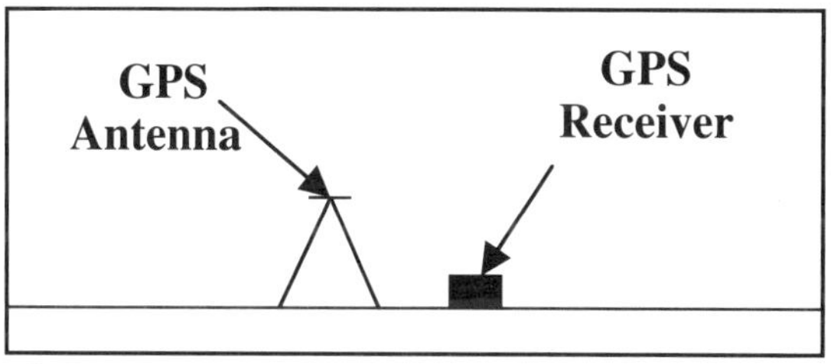

Fig. 5–12 *GPS Receivers and Antenna*

Every satellite sends out two carriers—*L1* at 1575.42 MHz and *L2* at 1227.6 MHz. The advantage is low noise. The disadvantage is an unknown number of cycles, which means range cannot be directly measured.

Both *L1* and *L2* carriers have a series of ones and zeroes modulated onto them. These are the pseudorandom codes. *L1* has the civilian access (C/A) and precise (P) code. The P code is also encrypted to form the Y code. Civilians can directly access the C/A code, but with only 100 meter accuracy. This effect is known as selective availability (S/A). S/A is thought to be caused by clock dithering in the satellites. This is the GPS standard positioning service. Only the military can directly access the P-code with special receivers. However there are some recent breakthroughs that simulate using the P-code with non-military receivers. *L2* has only the P-code.

Survey accuracy can be considered to have two components—absolute and relative. Absolute is important for interpretation, drilling etc. Relative is important for processing algorithms, frequency preservation, imaging, and more. Errors in absolute positioning accuracy are most commonly caused by incorrect datum transformations. The relative positioning accuracy necessary to preserve the required signal frequencies depends on application. For a *bin size* selected to preserve two samples per wavelength on steepest dips, a positioning error of 1/8 group interval or 1/8 of the bin dimension is required. (Bins are discussed further later in this chapter.)

Common Mid-Points (CMP) are an accounting tool used by geophysicists. It is important to know the positions of the sources and receivers. It is not important to have the locations in exact multiples of the group intervals to ensure that midpoints of traces are at bin centers.

Absolute positioning accuracy is required for determining drilling locations, reservoir monitoring/depletion studies, and survey extensions. Relative positioning accuracy for vessels, sources, and streamers is required for binning (fold and offsets), imaging, normal move out correction (NMO), dip move out (DMO), pre-stack migration, and location corrections.

Maximum error tolerance for land 3-D is 1/8 of the surface group interval (on surface). If this were extended to marine operations, it would imply 3 m accuracy for a 25 m group interval or line spacing, 1.5 m for 12.5 m group interval or line spacing, and 0.75 m for 6.25 m group interval or line spacing. None of these are fully realizable with today's technology.

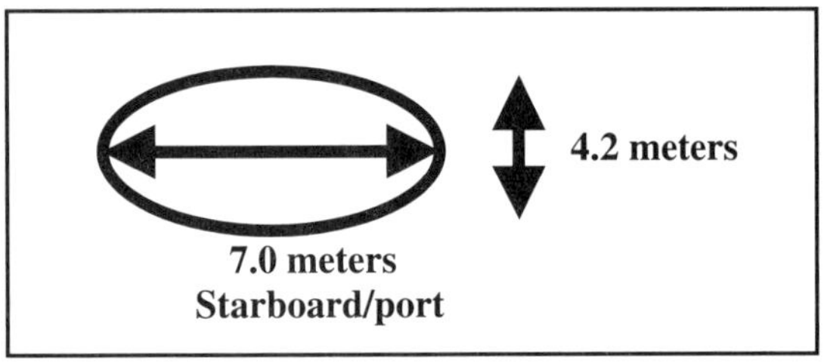

Fig. 5–13 *Antenna Motion*

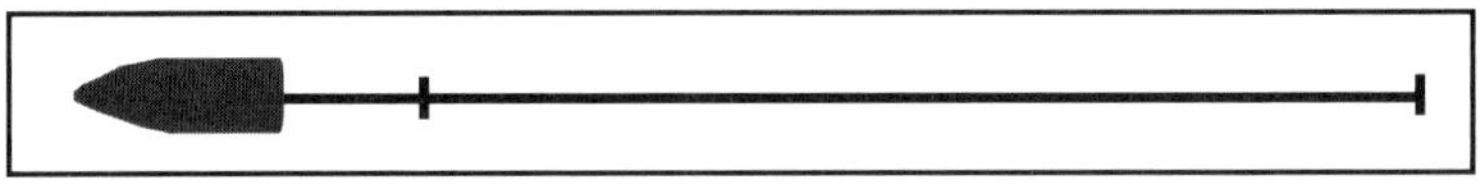

Fig. 5–14 *Source and Receiver Assumptions for 2-D Recording*

For any radio positioning system, the antenna position is what is actually determined. Typically, this antenna position is not coincident with the points of interest—vessel stern, center of source array, or center of near offset receiver group. Offset application errors, both for vessel positions and navigation buoy positions, are responsible for a surprisingly large number of major positioning busts. These are caused by offsets being ignored or not applied, offsets applied with wrong sign convention, offsets assumed to be the same as previous or incorrectly measured.

Antenna motion can introduce large positioning errors unless the antenna location is corrected for vessel pitch and roll using inclinometer data. For example (Fig. 5–13), for an antenna height of 20 m, if a vessel rolls 10 degrees to either side and pitches 6 degrees fore and aft, then the implied antenna movement is: *starboard/port* (right/left)= ±20 m x sin10° = ±3.5 m, *fore/aft* (front/back) movement = ±20 m x sin6° = ±2.1 m.

Relative positions of all elements of a marine acquisition system must be determined. Since some segments of the total system are below the water surface, in-water positioning devices are required. Vessel and platform attitude sensors include the gyrocompass, the *inclinometer* to measure pitch and roll, and the GPS vector systems. Remote target relative positioning systems include a short baseline (SBL) acoustics system, an ultra-short baseline (USBL) acoustics system, and laser ranging systems. Navigation head and tail buoys—radio navigation or GPS—provide remote target absolute positioning systems. Magnetic cable compasses—rotary card or flux gate—are streamer heading sensors. There are inter-cable, source, and vessel acoustics as well as ancillary devices like the velocimeter, cable tension, and depth indicators.

In 2-D recording, the streamer is assumed to be straight and that source and streamer follow boat track. Geometry is often assigned based on shot point and receiver group intervals without use of navigation data (Fig. 5–14).

These assumptions are rarely, if ever, satisfied. Sea currents, tides, and wind usually cause the streamer(s) to deviate from a straight line behind the vessel or *feather*. Figure 5–15 illustrates a simple example of feathering and defines the feathering angle. Feathering angle, α, is used to measure cable deflection due to crosscurrents. Cross-line cable deflection = cable length x sin (α) For a 6000 m streamer:

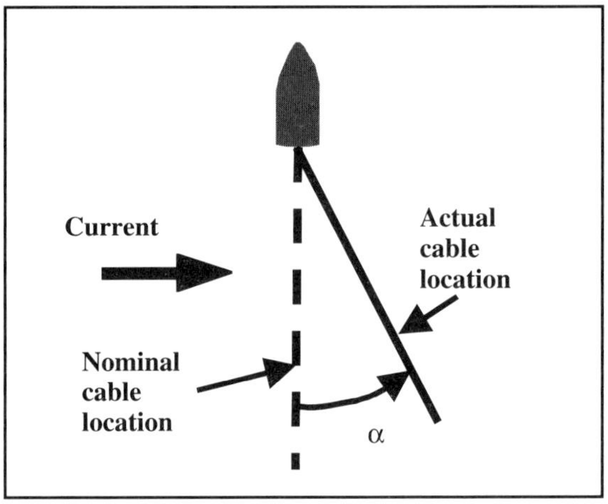

Fig. 5–15 *Cable Feathering*

1° = 105 m

2° = 209 m

3° = 314 m

4° = 418 m = 1/4 mile

8° = 835 m = 1/2 mile

16° = 1654 m = 1 mile

Variations in boat steering and currents result in the cable shape departing from simple assumptions of linearity. An example of streamer curvature caused by feathering is shown in Figure 5–16.

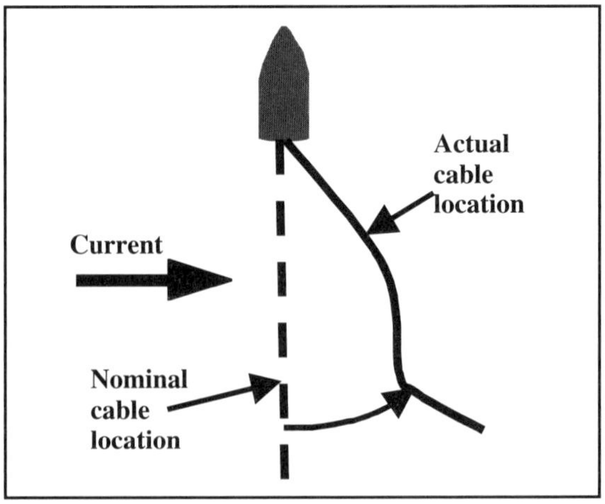

Fig. 5–16 *Streamer Curvature*

From about 1975 to 1977, the simple assumptions of Figure 5–14 were used in marine 3-D surveys. That is, the streamer shape was straight or a very simple curve, the source and streamer front end followed the boat track, and there was a constant streamer feather angle. Offsets from source to receiver were determined by measurement of deployed lengths of rope, hoses, and other towed equipment.

From 1977 to 1985, the streamer shape was defined by a curve fit to the cable compass readings. The tow fixture angle was used to define the departure angle of the streamer from the vessel. It was also assumed that the source followed the boat track or the gyrocompass.

By 1985, multiple streamers were standard. Up until 1991, streamer shapes continued to be defined by a curve fit to the cable compass readings. The near-vessel targets (source arrays and near detector group) were located using laser tracking or acoustic measurements. Radio navigation tail buoys at the end of the streamers augmented the cable compass measurements.

After 1991, the number of streamers being pulled continued to increase. Positioning measurements are made from a variety of sensors, including cable compasses, acoustic ranges, laser measurements, and navigation buoy positions. All are combined to form an integrated positioning network. Positioning accuracy indicators are now output along with coordinate estimates. Figure 5–17 illustrates such a network and identifies its components.

The vessel attitude sensors are the gyrocompass, the inclinometer, and the GPS vector system. A *gyrocompass* is an inertial device that measures ship's *heading*, the direction the vessel is pointing, as opposed to vessel *track* or *course made good*, which is the direction the vessel is traveling, with respect to true north. Unfortunately, gyrocompasses are subject to a number of errors that are speed, course, and latitude dependent. Overall accuracy for most systems is on the order of 0.5 to 0.7 degrees. An *inclinometer* is also an inertial device. It measures the ship's pitch and roll—along-axis and cross-axis vertical rotation—with respect to vertical.

Multiple GPS antennae are mounted on the vessel. GPS observations of these are used to solve for vectors between antennae. The vectors are, in turn, resolved into pitch, roll, and *yaw*—heading. Because carrier phase and carrier phase difference observations are used, this relative vector is an extremely accurate representation of vessel attitude. This is called the GPS vector system.

Figure 5–18 defines some vessel heading terms. The vessel track or course made good *t* is the direction that the ship is traveling. The vessel heading *h* is the direction that the ship is pointing. In general, vessel heading and vessel track are not equivalent because of wind and currents. Vessel heading is usually measured with a gyrocompass. The crab angle *c* is the angular difference between the direction the vessel is heading and the direction the vessel is traveling.

The remote target positioning system uses relative positioning sensors. The short baseline acoustics—Sonardyne SIPS, Syntron Multitrak, or DigiCOURSE Digirange—use acoustic travel time measurements to compute position fixes for remote targets. Ultra-short baseline acoustics—Simrad HPR or Ferranti ORE Trackpoint—use a phased array approach to measure distance and bearing to remote targets. Laser systems—MARS or LaserTrack—use laser range and angle measurements to locate remote targets. All relative positioning sensors measure the location of remote targets with respect to known points on the vessel and are strongly dependent on gyrocompass measurements to compute absolute positions. Consequently, the accuracy of these systems degrades as the distance from the vessel to the target increases.

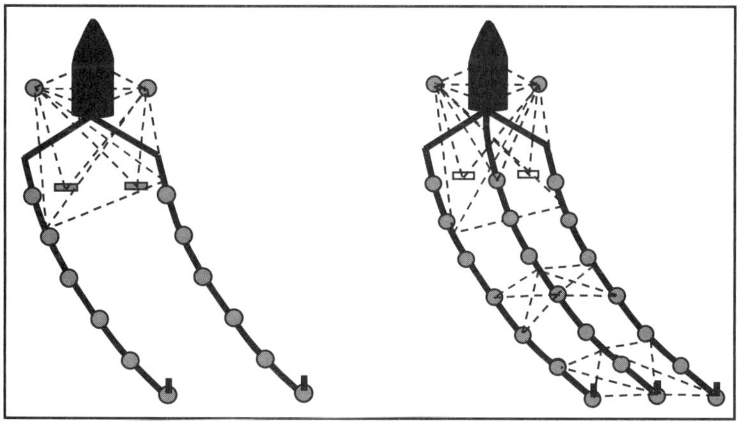

Fig. 5–17 *Source and Receiver Positioning, 1985–1991 and Current*

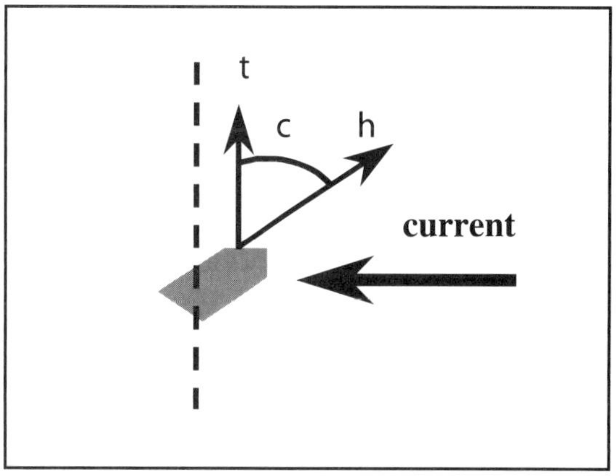

Fig. 5–18 *Vessel Heading Terms*

Acoustic systems have the advantage of being located beneath the water surface where the sources and receivers are. Acoustic systems also have high resolution. However, the signals they receive can follow multiple paths between sources and receivers, and the path followed may not be known. There is also some uncertainty in the velocity of water that can lead to error in distance computations.

All calculations are relative to the gyrocompass and, thus, include any errors in the gyrocompass reading. Pitch and roll of the vessel influence results obtained, and offsets to position of interest may be large. Geometry of the sensors may not be optimum.

The principal advantage of laser systems is their accuracy, with range at ±1 m and angle accuracy of ± 0.1. On the other hand, its surface location is a disadvantage. *Target acquisition* (locating the target) and *target discrimination* (differentiating among multiple targets) may be a problem, particularly in bad weather with fog or rain. The laser system locations are also relative to the gyrocompass.

Absolute positioning sensors used in remote target positioning include navigation tail and head buoys plus a GPS vector system. Navigation tail and head buoys may use either surface-based or satellite-based (GPS) radio positioning. Navigation tail buoys close the magnetic traverse at the end of the streamers and probably have the largest overall impact in improving positioning accuracy at the tail end of the streamers. Navigation head buoys are used to augment the front-end network by removing dependency on gyrocompass measurements associated with relative positioning systems. GPS-based vector systems that use GPS observations to compute vectors from a known location on the vessel to remote targets are NOT dependent on gyrocompass measurements to calculate absolute target coordinates. Unfortunately, navigation buoys measure surface, not subsurface, positions.

Streamer heading sensors (magnetic cable compasses) are of two types—rotary card and fluxgate magnetometer. Rotary card cable compasses have permanent magnets attached to a card that is gimbal-mounted and allowed to align with Earth's magnetic field. A *fluxgate magnetometer* is a toroid with two vertical coils placed orthogonally to measure the component of Earth's magnetic field in each horizontal axis. Cable compasses measure the *magnetic azimuth* of the streamer—the angle with respect to magnetic north. The difference between true or geographic north and magnetic north depends on declination. There are also deviations caused by local effects such as nearby magnetic bodies.

Declination is the difference between magnetic north and true north. Mathematical models of Earth's magnetic field and nautical charts are typically used to estimate true north. The Earth's magnetic field changes as a function of both position and time, so it is important to use a contemporary magnetic field model. Direction of magnetic north is determined by the horizontal component of Earth's magnetic field. The horizontal component of the magnetic field is a maximum at the equator and zero at the magnetic poles. The vertical component of Earth's magnetic field is zero at the equator and a maximum at the magnetic poles.

Fig. 5–19 *Bird/Compass on Test Rig*

Most modern cable compasses are housed as a component in an integrated compass/depth controller unit. Note that depth controllers are often referred to as *birds* (Fig. 5–19), so the integrated units are often termed *compass/birds*. These are typically spaced in 300 m or 400 m intervals along each streamer.

Compass data can be adversely affected by the sea state, which can translate rapid accelerations to the compass unit, thereby causing it to depart from a purely horizontal orientation. Magnetic compasses sense the total magnetic field, not just the Earth's field. Consequently, large metal objects (such us platforms or vessels) can adversely impact compass performance.

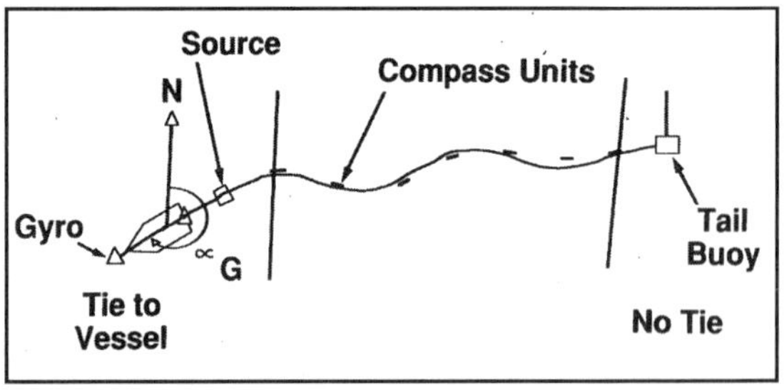

Fig. 5–20 *Traditional Applications–Cable Shaping*

A fifth-order polynomial fit is used for cable shape determination. Typically, up to 20 compasses may be used for each cable (Fig. 5–20). Compass calibration values can be used or zeroed. Real-time displays of cable shape and data are available. Multiple streamers use laser tracking data and acoustic data for front-end positions. Active navigation buoys are used to determine magnetic variation anomalies.

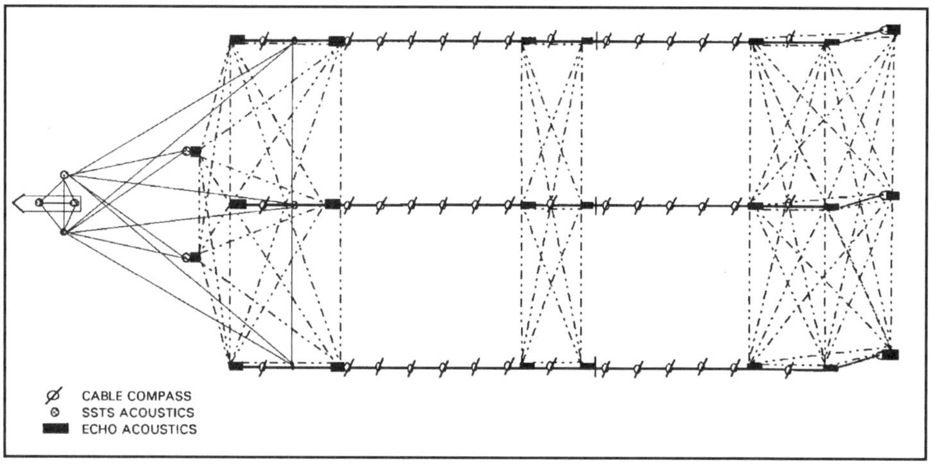

Fig. 5–21 *Acoustic Configuration Diagram*

The same or similar system used for remote target positioning is used for near vessel tracking of source arrays and near offset receiver groups as relative positioning targets. Acoustic devices are typically arranged in a braced quadrilateral arrangement with diagonal ranges included to measure streamer skew (Fig. 5–21). Acoustic devices tend to be distributed in sub-networks at the head, middle, and tail of the streamers. The cable head network is almost always coupled to the source arrays or a navigation head buoy or both, while the cable tail network is typically coupled to navigation tail buoys. Typical acoustic ranges are on the order of 500 m or less.

The speed of sound in water must be accurately estimated in order to convert measured acoustic travel times into corresponding distance measurements. The speed of sound in seawater depends on temperature, salinity, and pressure. It also varies as a function of depth, geographic location, and time. Devices used to measure water velocity include the *velocimeter* and the *expendable bathythermograph* (XBT). The velocimeter continuously measures speed of sound in water over a very short baseline. The XBT measures temperature as a function of depth in order to compute a sound velocity versus depth profile. Baseline measurements compute velocity from acoustic travel time measurements over known distances—along vessel hull transducers or streamer cable.

Depth indicators (DI) are typically deployed along the streamer as well as at the source array. All compass/bird units contain a depth indicator, which is used as a feedback mechanism to ensure that the streamer is at its target depth. In addition, some streamers have internal depth indicators. Depth indicators are also placed on each gun string in the seismic array. Typically, there is more than one depth indicator per gun string. Depth indicators function by sensing the ambient pressure and relating this pressure to depth.

Cable tension measurements can be used to estimate the amount of streamer stretch or elongation. Cable tension is also very useful in measuring the stresses the streamer cable is subject to during data acquisition.

Energy source location can be done several ways. *Stepback* methods apply nominal offsets from a known position on the vessel to the center of source array along an assumed azimuth—gyrocompass, vessel track, or nearest cable compass. This is extremely crude because the absence of any measurements to substantiate the selected offset and azimuth. It essentially amounts to guessing rather than measuring. *Vessel-relative* positioning methods measure distance and azimuth from vessel to energy source using laser or acoustic measurements. It still requires a smaller stepback from the measured position on the array to the center of the source. *Absolute* positioning methods employ a radio navigation receiver on or near the source array. This relaxes gyrocompass dependency, thereby improving position accuracy. In the *acoustically coupled* method, source position acoustic measurements are used to couple source array to relative or (preferably) absolute surface positions. This is the most accurate means of source positioning, short of a full network solution.

Various methods have been used for cable shaping. Piecewise linear is a simple but extremely crude method. It is not suitable for receiver locations. Curve fits include circular arc, polynomial fit, cubic spline, and streamer offset and arc length. Polynomial (fifth-order) fits are commonly used today.

Integrated positioning networks (Fig. 5–22) provide a comprehensive, integrated solution that fully uses all positioning measurements (absolute positions, relative positions, relative range and bearing, compass bearings, and cable arc length constraints) to derive optimum coordinate estimates for each network node.

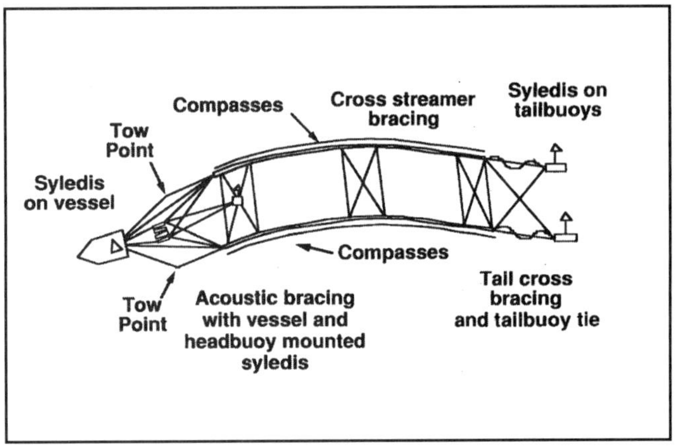

Fig. 5–22 *An Integrated Solution*

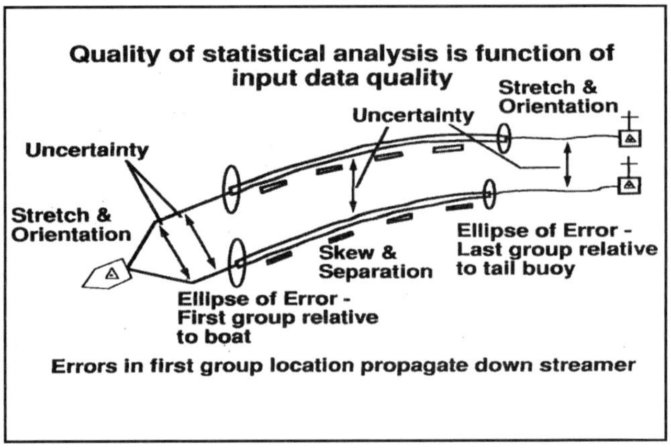

Fig. 5–23 *Statistical Analysis*

Redundant measurements can be exploited to improve coordinate reliability and to reduce statistical uncertainty. In addition to coordinates, integrated network solutions compute the coordinate uncertainties, which are vital for assessing positioning accuracy (Fig. 5–23). Most network algorithms are implemented as a *Kalman filter*. The Kalman filter can be thought of as a sequential implementation of a traditional least squares network adjustment algorithm that has been modified to predict future states of the network based on past behavior and a dynamic model.

Land surveying. Before any lines are laid out, a system must be developed for numbering source points, receiver points, and midpoints. The numbering system should work for all phases of operation—survey, data acquisition, data processing, and interpretation.

Some systems do not handle non-integers. For example, numbers assigned to source lines between receiver lines and/or receiver lines between source lines must usually be integers. Some systems expect regular Cartesian coordinates. Unique line numbers and point numbers work well for most systems except for Cartesian coordinates. Unique numbers make it easier to manage recovery shots. No two points should have the same number.

The surveyor is responsible not only for locating all source and receiver positions but also for laying out annotated markers on the ground to their locations. The surveyor must ensure that correct transformations are made between survey datums. This is the most frequent source of error.

The surveyor must provide x and y coordinates and elevations for all source and receiver positions. He or she must be aware of terrain limitations of crew equipment and permit restrictions—offsets from wells, springs and other similar elements. The surveyor must also under guidance from project manager, client, or geophysicist locate recovery shots to acquire the required subsurface coverage when source or receiver positions cannot be occupied.

The surveyor must also provide detailed maps for each line, swath, loop, or patch showing source and receiver locations, skips or offset source locations, safety hazards, permit problem areas, fences and drive-arounds, access routes to and from lines, and any wells (oil or water).

Surveyors use many of the same instruments as are used in marine navigation, particularly GPS. Electronic distance measuring devices are also frequently used. Conventional surveying instruments, such as transit and chain, still have a place in land surveying.

Surveying is done in *loops*—the surveyor returns to a particular starting point or previously located position. The difference between the first and second location of this point is called a *mistie*. The client company usually specifies the maximum error or mistie. There will be a maximum allowable error in both horizontal position and elevation. If the error is within the allowed amount, then it may be distributed around the loop followed by the surveyor.

Positioning summary

The final product for land and marine seismic surveys is an accurate three-dimensional position for the subsurface feature. To achieve this, the surveyor must check the following geodetic quantities carefully:

a. datums to be used

b. datum transformation parameters

c. ellipsoid parameters

d. map projection parameters

e. geoid model or incline plane

Survey quality control is an important aspect of positioning. For minimum GPS performance verification:

- take a position and record it at known location at start of the day
- take a position and record it at a different known location at the end of the day
- check positions against known locations

Conventional survey equipment should be checked for calibration on a regular basis, typically weekly, whenever possible damage has occurred, after instrument repair, whenever anomalous locations are detected, and whenever excessive misties and/or loop closures are observed. Good documentation of all instrument calibrations and checks should be maintained.

Despite modern positioning systems, errors can still occur. To prevent or minimize these, take special care with datum transformations. Use automated electronic data transfer whenever possible to reduce risk of errors. Surveyed positions plotted on maps should be checked against theoretical positions on program maps. Source and receiver position numbers for key locations (starts and ends of lines or wells) should also be checked against theoretical positions on program maps. Any differences should be checked and verified.

Final survey products include the following items.

- Source and receiver coordinates and elevations in a standard electronic format such as UKOOA or SEG P1. These are sent to the client and to the field crew seismologist for inclusion in files used in the data initialization stage of data processing.
- Post-plot maps showing source and receiver coordinates including key topographical features like coastlines, rivers, and lakes; survey benchmarks; and any wells, roads, or similar items that were surveyed.
- A final report that should include a description of personnel involved in the survey, method(s) and equipment used, datum transformations, and closure errors or misties.

Signal Generation

Seismic energy sources must have these characteristics:

- adequate strength for penetration to geologic targets
- broadband signature rich in both low and high frequencies
- balanced, flat spectrum—tuned source
- stable, repeatable signature
- mechanically and electrically reliable

For 3-D operations, energy sources should also have:

- compact spatial extent to minimize array effects
- spatial symmetry to minimize asymmetric array response in the in-line and cross-line dimensions

Air guns, water guns, and several others are or have been used as marine energy sources. Air guns are, by far, the most used for marine operations. In the transition zone, dynamite is used in shot holes or marine-type sources, depending on water depth and environment. Land energy sources include dynamite in shot holes, and Vibroseis. Source selection is usually based on availability, the ability to generate seismic energy required to image primary objective, cost effectiveness, and accessibility.

Airguns. Requirements for marine energy sources are

- produce a powerful pulse
- have capability of rapid firing (approximately 10 sec intervals)
- are simple to operate, consistent, and trouble-free
- remain at constant depth and have minimum drag when towed
- cause no injury to marine life
- minimize *bubble effect* (repeated expansions and contractions of volume of gas)

Airgun arrays satisfy all of these. Explosives satisfy all but one critical requirement—they do cause damage to marine life. Consequently, explosives have not been used in deepwater marine surveys for many years. Only in very shallow water, where holes can be easily dug, are explosives used in marine work.

Basic airgun operation is as follows (Fig. 5–24). Airguns of various sizes are combined to form arrays. The compressor on the back deck supplies compressed air. The navigation system sends a signal to the Airgun Controller when the source point is reached. The Airgun Controller fires each gun in the array and sends a signal to recording instruments to start recording.

Figure 5–25 illustrates major components of airguns and airgun operation. The solenoid valve is electrically connected to the airgun controller, and hoses to the compressor connect the air intake. Compressors operate at a pressure of 2000 psi. Pressures up to 4000 psi have been used in experiments, but it was found that the danger was too great to use in a production mode.

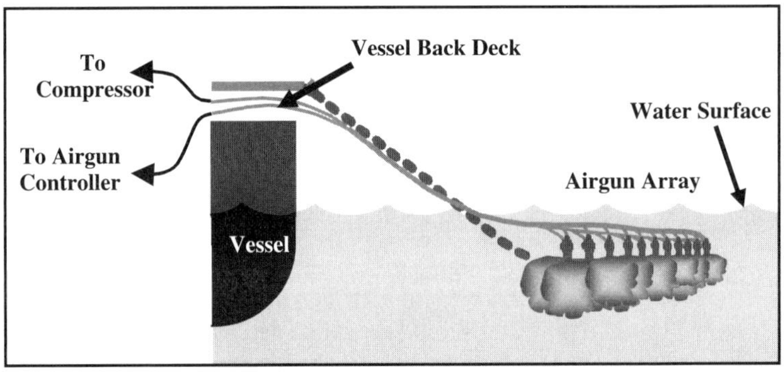

Fig. 5–24 *The Airgun Technique*

In the *charged* condition the solenoid is positioned to close the *firing line* and open the *fill passage*. Air is introduced through the air intake valve and the fill passage and enters the chamber via the *chamber fill orifice*. The airgun is *fired* when the pressure in the chamber reaches 2000 psi. On a signal from the airgun controller, the solenoid valve position is changed to open the firing line and airflows into the *firing chamber*. The pressure in the firing chamber causes the sleeve to move up and air begins *exhausting* into the water via the *exhaust ports*. The movement of the sleeve compresses the air in the *spring chamber* even further. As air exhausts from the chamber into the water, the pressure in the spring chamber becomes able to push the sleeve down, closing the exhaust ports and the air gun is now *exhausted*. The cycle can now begin again.

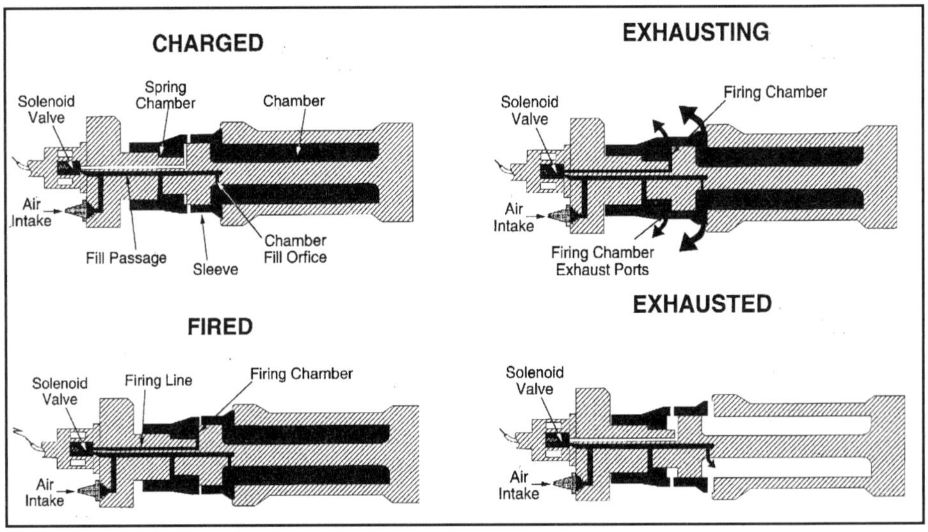

Fig. 5–25 *Airgun Operation*

A single airgun is a poor energy source because of its bubble effect. This is shown in Figure 5–26. When the compressed air is introduced into the water, it creates an air bubble that rapidly expands, and the pressure in the bubble becomes less than the surrounding water pressure. As a result, the bubble contracts. However, the contraction goes too far and air pressure is once

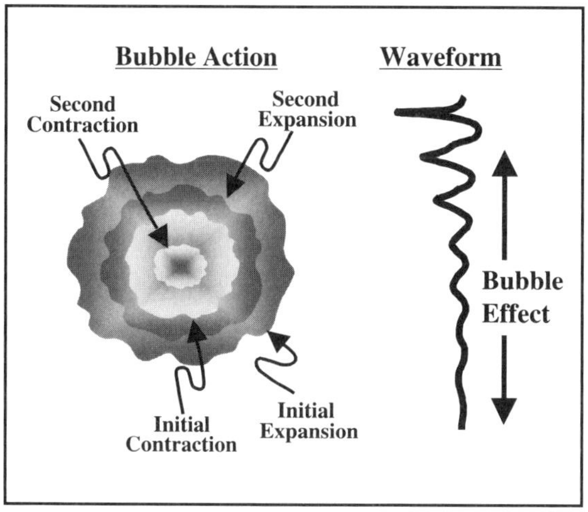

Fig. 5–26 *The Bubble Effect*

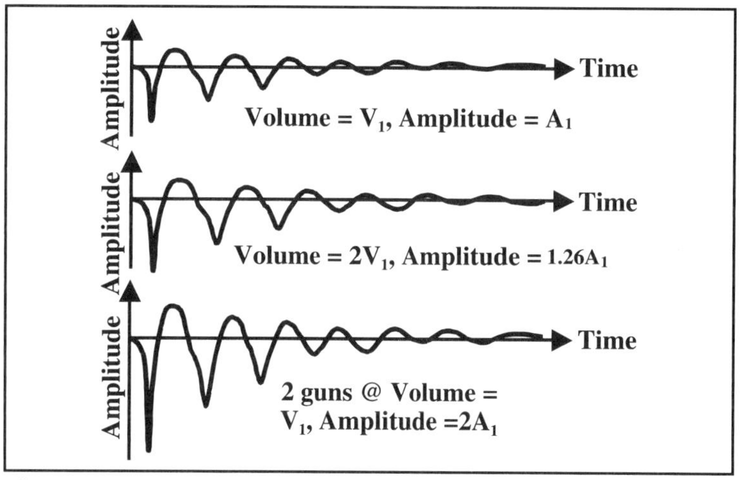

Fig. 5–27 *Airgun Pressure–volume Relationship*

again higher than water pressure and a second, although smaller, expansion occurs. Repeated contractions and expansions occur until all the energy is dissipated. Figure 5–26 illustrates this and the waveform produced by the bubble effect.

Pressure produced by a single airgun is proportional to the cube root of *gun volume*—the total space in the airgun occupied by air. Signal amplitudes are proportional to pressure, so amplitude is also proportional to $V^{1/3}$. So, as shown in Figure 5–27, increasing the volume of a single airgun gives only 26% larger amplitude. However, using two airguns of the same volume placed closely together produces twice the amplitude.

Airguns are used in arrays for two reasons—to increase signal amplitudes and to minimize the bubble effect. The latter is illustrated in Figure 5–28. In this very simple array, three different size guns (different volumes) are used—one large gun, three medium size guns, and three small guns. The guns of the same size are grouped closely together (clustered). Spacing between the one large gun and the two clusters is such that the bubbles interfere destructively except at the initial expansion. Note that the airguns in the array do not fire simultaneously. The smaller guns are delayed because their bubbles achieve maximum expansion earlier than the larger guns. Note also the lower frequency content of the larger guns. Large numbers of airguns of various sizes are grouped together to form tuned arrays. With proper spacing of single guns and gun clusters (array design), virtually any desired signal waveform can be achieved.

The two main objectives of airgun array design are to obtain adequate energy source strength and sufficiently broad frequency bandwidth. The best way to determine a source strength requirement is to conduct a field experiment using different strength sources to record a 2-D line and then process and analyze the results. For the obvious reasons (time and cost), this is almost never done. A review of previously acquired 2-D or 3-D data can aid in determining adequate source strength requirements. Amplitude decay analysis and time variant spectral analysis of previously gathered 2-D or 3-D data can help determine the depth (recording time) of penetration of useful seismic energy. It is possible to *overshoot* an area by using an energy source that is too strong.

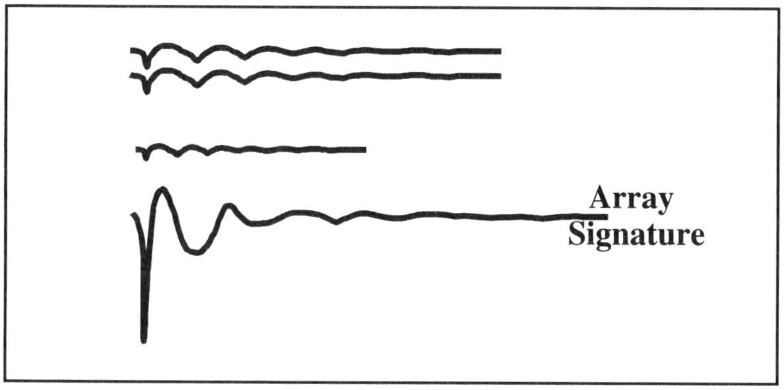

Fig. 5–28 *Waveform Synthesis by Airgun Arrays*

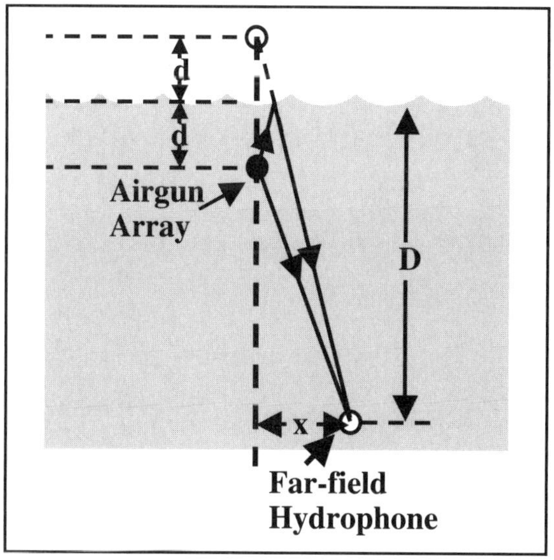

Fig. 5–29 *Measurement of the Far-field Signature*

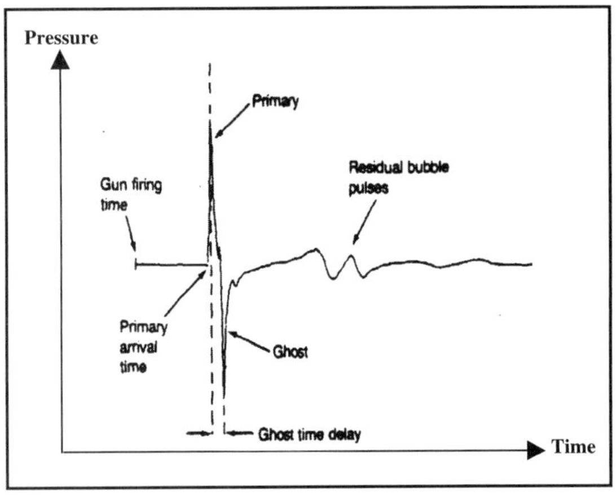

Fig. 5–30 *Time Domain Attributes*

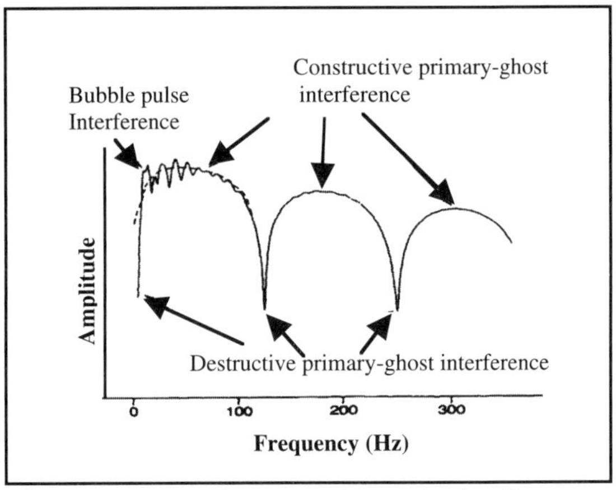

Fig. 5–31 *Airgun Frequency Domain Attributes*

Airgun arrays are towed below the water surface, usually at depths of around 10 m. As a consequence, upward-traveling energy from the airgun array is reflected at the surface and interferes with the downward-traveling energy to produce ghosts. In Figure 5–29 a hydrophone is shown at a depth *D* (usually about 100 m below the airgun array) below the water surface and displaced slightly behind the array. The airgun array is at depth d below the surface. When the array is fired, energy travels out in all directions. The ghost, produced by reflection at the water surface, appears to come from a source placed d above the water surface.

The angles of incidence of concern at the water surface are small enough that we can use the vertical incidence reflection coefficient from Equation 3.11 of Chapter 3.

$$R_{12} = \frac{\rho_2 V_2 - \rho_1 V_1}{\rho_2 V_2 + \rho_1 V} = \frac{\frac{\rho_2 V_2}{\rho_1 V_1} - 1}{\frac{\rho_2 V_2}{\rho_1 V_1} + 1}$$

In this case, medium 1 is water and medium 2 is air. The velocity of sound in water is about three times that of sound in air and the density of water is hundreds of times larger than that of air. So, with very little error, the ratio $\frac{\rho_2 V_2}{\rho_1 V_1} \cong 0$ and $R_{12} \cong -1$. The ghost is thus seen to be an inverted and delayed version of the primary signal, as shown in Figure 5–30.

Figure 5–31 shows the frequency domain attributes of an airgun array signature. In the frequency domain, ghosts appear as notches or zero points in the amplitude spectrum. The frequency at which the notches occur depends on depth of the airgun array and water velocity.

$$f = \frac{V_W}{2d} \tag{5.1}$$

where

f is the notch frequency

V_W is the water velocity

d is the source depth

Airgun array specifications, also called specs, may be given in the time domain or frequency domain. Time domain specs are strongly dependent on signature filtering and bandwidth. *Peak-to-peak strength (PTP)* measures, in bar-meters, the signature strength, and *primary-to-bubble ratio (PBR)*, a dimensionless quantity, measures signature shape. See Figure 5–32.

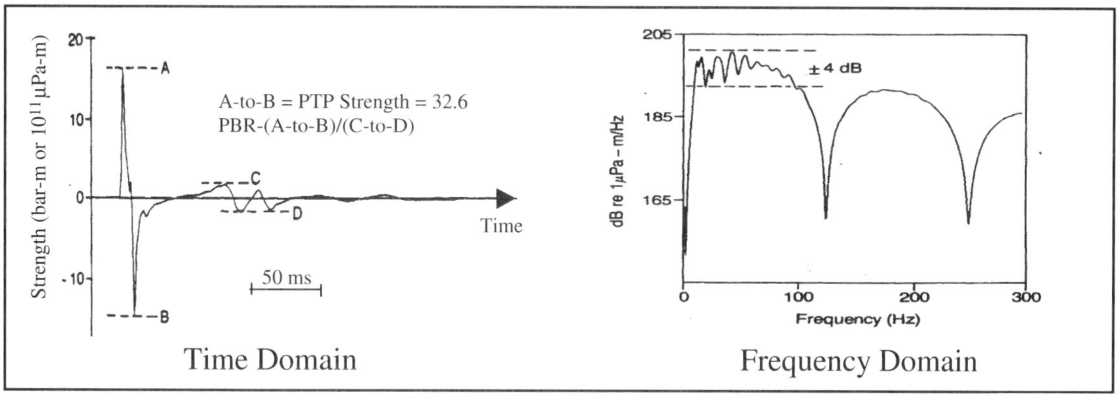

Fig. 5–32 *Airgun Array Signatures and Parameters*

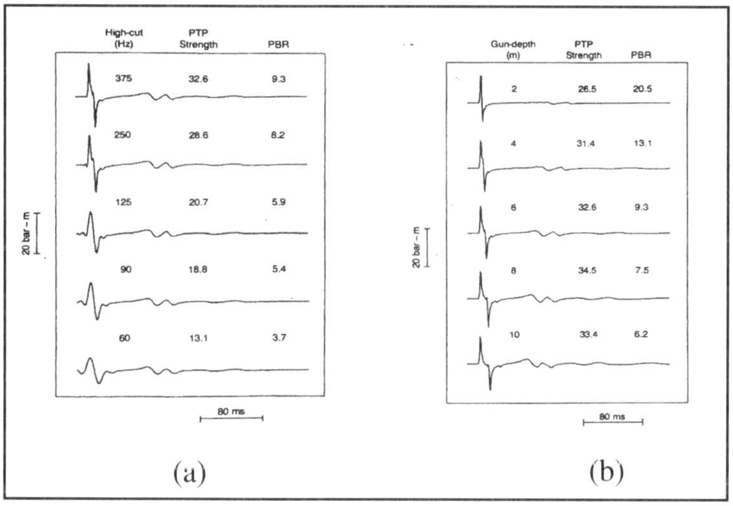

Fig. 5–33 *Signature Variation with Frequency and Depth*

Frequency domain specs are always preferred over time domain specs, since calibrated amplitude spectra provide absolute frequency-by-frequency strength information. *Spectral amplitude* (dB referenced to 1 micropascal-m/Hz) measures signature strength. *Spectral flatness* (dB); measures signature shape. Low and high cut filter effects are clearly evident and calibrated amplitude spectra also give a direct insight into ghost notch effects.

There are pitfalls in using time domain specs. The peak-to-peak strength and primary-to-bubble ratio of an airgun array signature are strongly dependent on the frequency bandwidth. Unless the bandwidth of a signature is known, it is impossible to validly compare time domain specifications for different source arrays. Because the depth of a source array has a profound effect on the frequency bandwidth of the array signature, time domain specifications are influenced by array depth (Fig. 5–33b). Low-cut instrumentation filters and high-cut anti-alias filters affect the time domain specs of an airgun array. The best way to compare the performance of different energy source arrays is by using calibrated amplitude spectra. Table 5–3 summarizes the effects of various array parameters on airgun specs.

The total volume of an airgun array is not, by itself, a good indicator of an energy source's strength. Source strength is roughly proportional to the *cube root* of the array volume. An airgun array's strength is roughly proportional to the number of elements (single guns or clustered guns) in the array.

Table 5–3 *Airgun Specs and Array Parameters*

Parameter	**Strength**	**Peak-to-Bubble Ratio**
Number of Elements	Proportional to the number of elements	No direct relationship
Element Volume	Proportional to the cube root of volume	Increases as volume increases
Element Depth	Increases initially, then decreases with increase in depth	Decreases as depth increases
Firing Pressure	Increases as firing pressure increases	Increases as firing increases
Port Area	Proportional to square root of port area	Increases as port area increases
Port Closure Pressure	No effect if closure pressure is < 0.5 of firing pressure	Decreases as port closure pressure increases
Synchronization	Decreases as spread in firing times increase	Decreases as spread in firing times increase
Horizontal Element Spacing	Decreases as elements move closer together	Increases when bubbles coalesce

Array parameters that depend on the type of airguns used in an array (firing pressure, port area, and port closure pressure) affect the fundamental characteristics of array signatures only weakly. An airgun array can be tuned to have a high primary-to-bubble ratio by towing the array at a shallow depth, but a shallow array has a much weaker low-frequency response than an array at a conventional depth.

Air guns in use today are the type called *sleeve guns*. Figure 5–34 shows a variety of sleeve guns.

Since airguns are used in arrays, some method must be used to assemble the array, maintain the separation between elements, and tow it as a unit at a fixed depth. Figure 5–35 shows a side view of a typical sub-array. The top part floats on the water surface and is attached to the boat by a tow cable or chain. The bottom frame is suspended by cables of a fixed length, according to the desired firing depth. Hoses and electrical connections run between the assembly and the back deck. Fittings allow airguns to be suspended by chains at any desired spacing.

Figure 5–36 shows a plan (top) view of dual airgun array. Each sub-array (heavy black lines indicated by numbers one through eight) has guns with a total volume of 750 cubic inches. The sub-arrays are 21.05 ft long and are six feet apart in each array. Location of other components (acoustic pods, GPS, MARS) used in source location are also shown in Figure 5–36.

Paravanes or other diverters are used to maintain the separation between the two arrays. It is important not only to maintain the desired separation between elements of the sub-arrays but also to maintain the required separation between sub-arrays. If this is not done, then the array response will not be as designed.

The depth at which the source array is towed determines the frequency of the ghost notches and affects the ghost response. Figures 5–37 shows the ghost responses for a 6 m and a 10 m source depth. Figure 5–38 gives the ratio, in dB, of the response at 10 m depth to that at the 6 m depth. Note that the comparison ends at the notch frequency for the 10 m depth.

Fig. 5–34 *A Single Sleeve Gun, a Disassembled Sleeve Gun, a Sleeve Gun with a Mounting Harness, and a Family of Sleeve Guns*

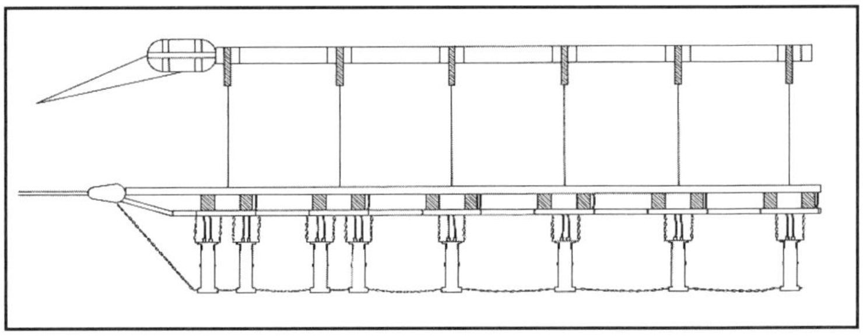

Fig. 5–35 *Side View of a Typical Sub-array Configuration*

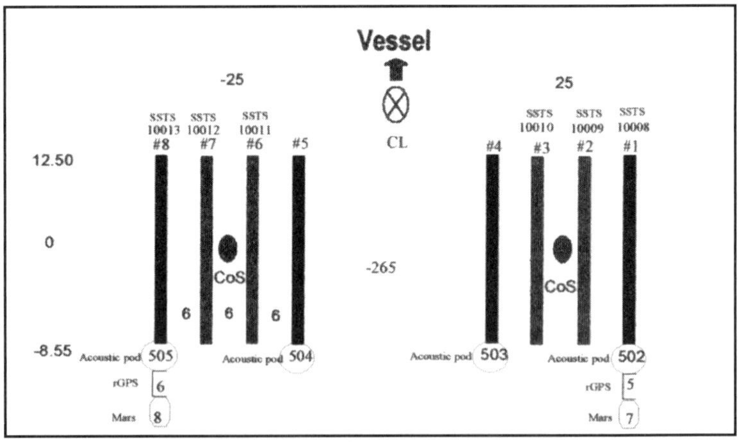

Fig. 5–36 *Plan View of a Typical Source Array*

Compact, symmetrical arrays are preferred to minimize filtering of seismic energy reflected from steeply dipping interfaces due to non-vertical arrival. Any filtering effects due to the spatial extent of the array will be uniform in both the in-line and cross-line directions. Wide or long source arrays, which are spatially extended in either the cross-line or in-line direction, are detrimental to high fidelity recording of steeply dipping energy in the direction of maximum spatial extent. Figure 5–39 illustrates energy transmission from compact and wide arrays. Compact is desirable for 3-D surveys and wide for 2-D surveys.

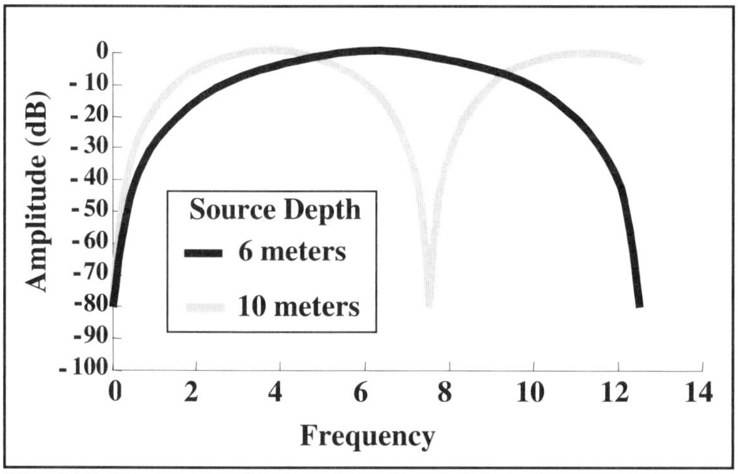

Fig. 5–37 *Streamer and Source Depth Ghost Notches*

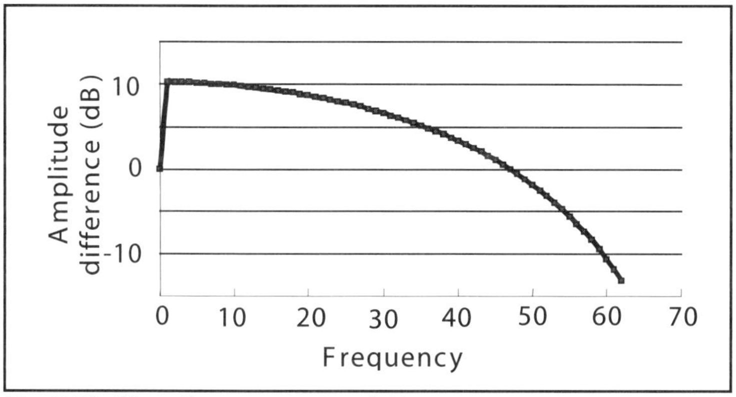

Fig. 5–38 *Ghost Response at 5 and 10 m*

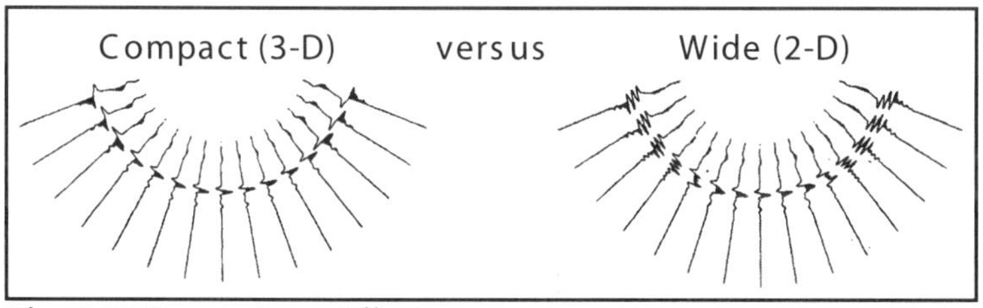

Fig. 5–39 *Source Array Effects*

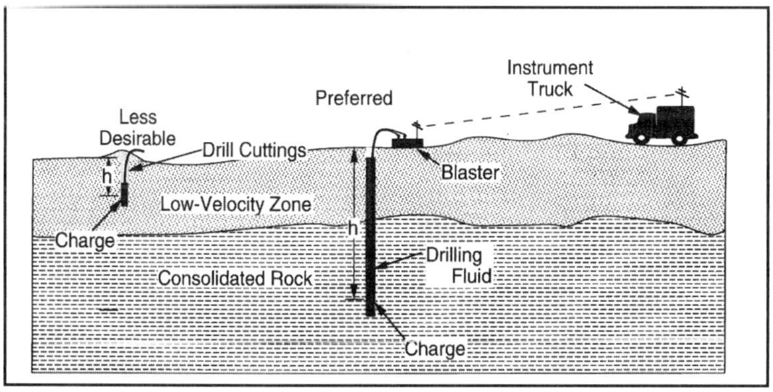

Fig. 5–40 *The Explosive Technique*

Explosives. Normally, on land the choice is between vibrators and explosives. The choice depends upon suitability to terrain, surface conditions, ease of drilling, availability, of personnel and equipment, environmental and governmental regulations, cost comparisons, and other factors. Explosives are not usually used in offshore data acquisition because of environmental regulations or in densely populated or industrial areas because of possible damage and permitting problems.

Explosives are an *impulsive* energy source because they transfer seismic energy into the ground in the form of a short duration impulse. Explosives are usually loaded and shot in drilled holes. However, some use has been made of explosive cord that is buried very shallowly in, usually, linear patterns. Further discussion follows that on explosives in drilled holes.

When loading explosives in drilled holes, it is desirable to drill deep enough to ensure that the explosive charge is below the base of the weathered layer. This gives the best signal level and bandwidth while minimizing shot-generated noise. Other factors to consider in the determination of hole depth follow.

- Is there a formation in which energy transfer is optimized?
- Can the surface ghost be used to reinforce the primary wavelet in the earth-air interface where the reflection coefficient equals -1?
- Should single deep holes be used or should multiple shallow holes be drilled in an array or pattern to cancel coherent noise?

The preferred technique (Fig. 5–40) is to load a charge or charges in a hole or holes drilled through the low-velocity zone (weathering). The charge may be dynamite, nitro-carbo-nitrate or ammonium nitrate fertilizer mixed with diesel fuel. The size of charge depends on depth *h* and shot medium. Principal advantages are that time through the low-velocity zone can be measured directly (via an up-hole geophone), surface wave generation is minimal, and having only one pass through the low-velocity zone reduces attenuation. In some cases, the time and cost of drilling dictate drilling shallow holes. When this is done, an array of several holes is drilled to enhance the signal and attenuate surface waves at the source.

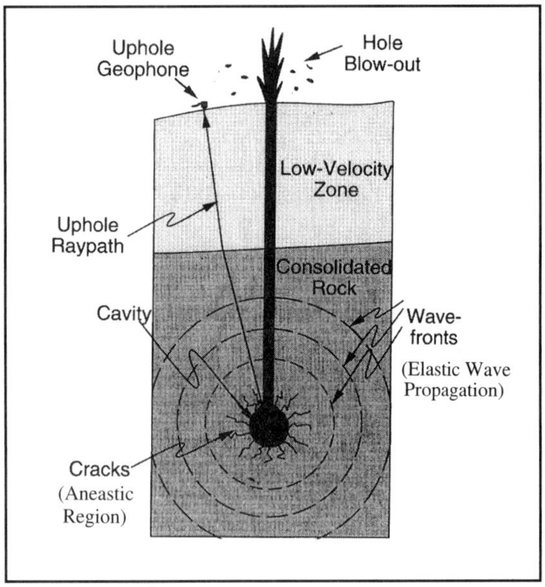

Fig. 5–41 *Explosive Source Operation*

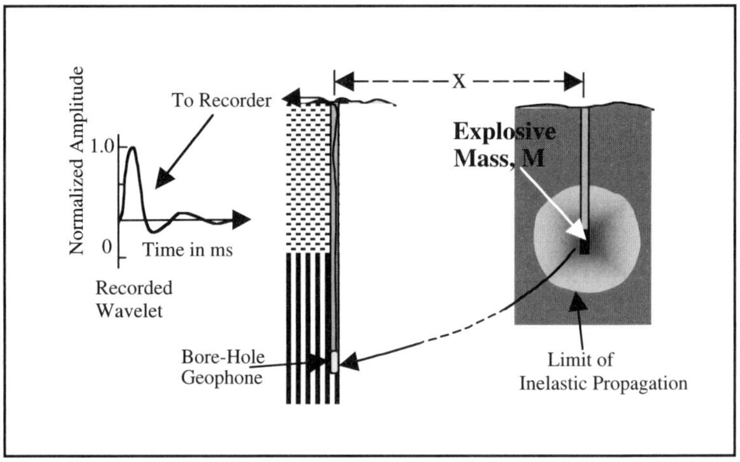

Fig. 5–42 *Measuring the Far-field Signature of an Explosive Source*

Figure 5–41 illustrates explosive source operation. A high-voltage pulse is applied to a blasting cap by a *blaster*. Heat from the cap initiates rapid combustion of the charge (explosive mass). Combustion produces a large volume of hot gas that expands at a very fast rate. Pressure against the hole walls is transmitted as seismic (elastic) waves in approximately spherical wave fronts radiating out in all directions. Some of this energy reaches the up-hole geophone via a minimum time path. Part of the energy produced by explosion may be expended in blowing out material (drilling mud, rocks) from the hole. This is called *hole blowout.* Some of the energy produces permanent deformation in the form of a cavity and cracks in the medium around the shot (an elastic zone).

Figure 5–42 shows a method of measuring the far-field signature of an explosive source, assuming a homogeneous medium. Two holes are drilled 100 m or so apart. A charge is loaded at the bottom of the shallower hole and a geophone (detector) connected to a recorder is placed at the bottom of the deeper hole. The charge is fired, and the signal received directly from the charge is recorded. The left side of Figure 5–42 indicates the shape of the recorded wavelet. The amplitude is *normalized*—amplitudes are divided by the maximum amplitude.

The amplitude, duration, bandwidth, and spectral amplitude of the pulse produced by an explosive source depend on charge size. Figure 5–43 shows wavelets and their amplitude spectra for two charges, one of mass *M* and one of mass *M/8*. The amplitude, duration, and bandwidth are proportional to the cube root of the charge mass. The amplitude of the spectrum is proportional to charge mass raised to the $^2/_3$ power.

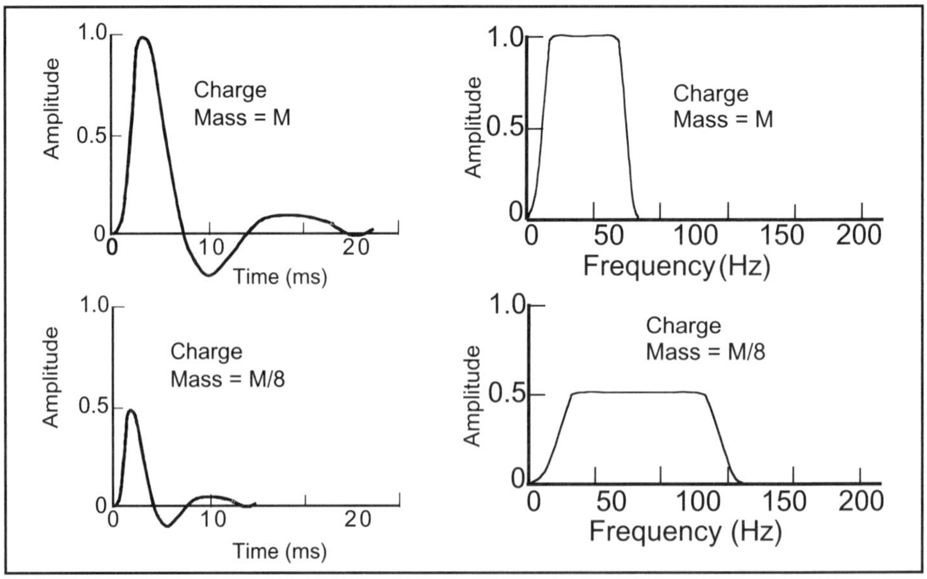

Fig. 5–43 *Effect of Charge Size*

The medium surrounding the charge also affects signal. Drilling deeper may place charges in a more favorable shot medium, producing higher amplitudes and wider bandwidth. However,

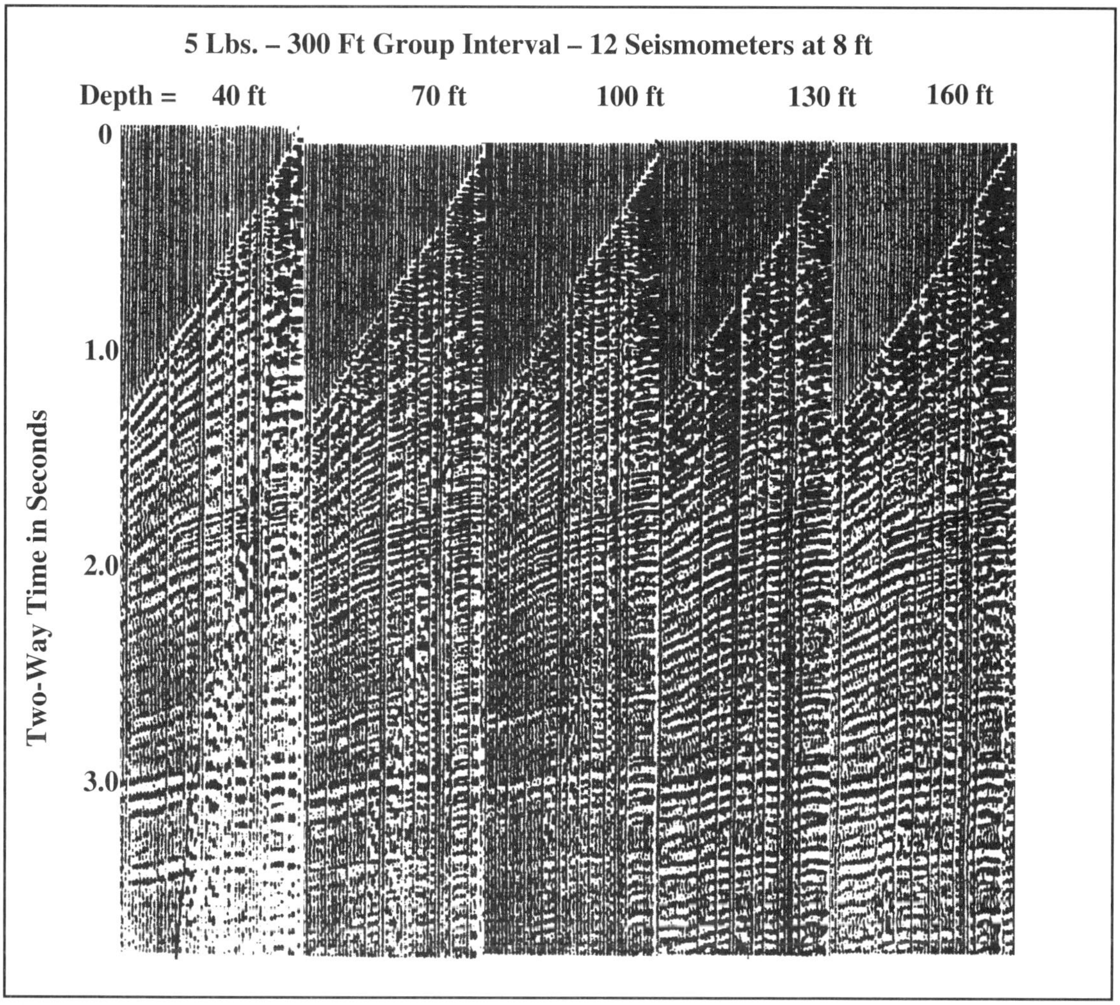

Fig. 5–44 *Charge Depth Test*

increasing either charge size or shot depth would not only increase cost but could adversely affect data quality. Increasing the number of holes would seem the best choice, but this would at least double the cost of drilling and the budget may not allow it. Field-testing is used to indicate an optimum charge size and shot depth.

Figure 5–44 shows the results of such a test. Usually, a hole is dug to a greater depth than that anticipated being used, and a constant charge size is shot at various depths, starting at the bottom of the hole. Once the optimum depth is determined, experiments with charge size are used to determine optimum charge.

Geoflex and *Primacord* are two commercial names for explosive cord. When used as a seismic energy source, cord is inserted into the ground with a tractor-mounted plow. As the plow turns over the soil to a depth of about 18 inches, the explosive cord is fed out of the rear of the tractor into the furrow. The cord is buried as the soil collapses over it. Usually, two or more lines of explosive cord are used. The lines are around 200 ft long and spaced about 15 ft apart. See Figure 5–45.

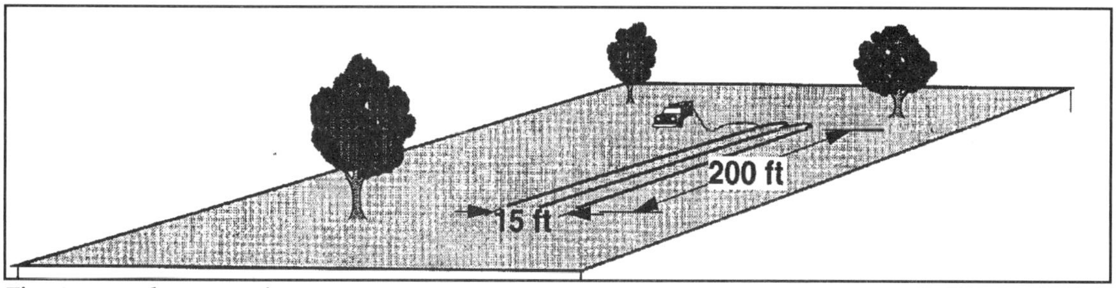

Fig. 5–45 *Explosive Cord as a Seismic Energy Source*

The cord detonates at 21,000 ft/sec, and being laid out in lines has some directivity in the energy propagation and attenuates horizontal noise. The advantages are that:

- charges are small but very efficient.
- it is cheap and fast because no drilling is required

It must be noted, however, that a soft surface is required for the method to be practical.

Vibrators. The vibrator (trade name Vibroseis) was developed by Conoco. A vibrator is a vehicle-mounted energy source that produces a vibratory or swept-frequency signal of relatively long duration (2 to 32 sec). Principal components of a vibrator system are illustrated in Figure 5–46, courtesy of WesternGeco. Major elements of a vibrator system include

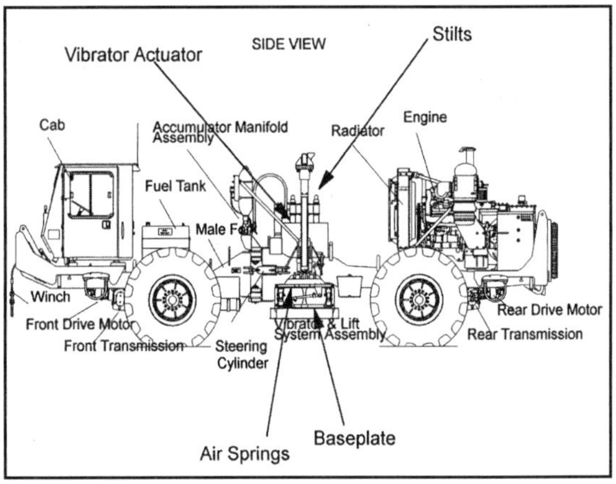

Fig. 5–46 *Truck-mounted Vibrator Components*

- diesel engine
  - converts fuel into mechanical energy
  - propels vehicle and powers hydraulic system (single engine)
- hydraulic pump
  - converts mechanical energy to hydraulic energy
  - compensates for pressure by providing flow necessary to maintain desired pressure
- hydraulic supply system
  - provides pulsation reduction
  - stores hydraulic energy on a duty cycle basis
  - filters and distributes hydraulic fluid
- vibrator control electronics
  - generates electrical signals to control vibratory system
- servovalve
  - modulates hydraulic flow to yield desired acoustic output under the electrical control of vibrator control electronics
- reaction mass
  - acts as an inertial element in actuator assembly for force to push or react against and is typically 4000 to 10,000 lb. in weight (Hold-down force holds the baseplate in contact with ground. It is sum of actuator assembly and vehicle weight applied through isolation spring.)
- base plate
  - applies generated vibratory force to ground
- isolation springs
  - decouples vibratory motion of base plate from vehicle—usually an air bag
- vibrator actuator
  - houses piston rod and servovalve
  - acts as inertial element
  - ported for oil passages
- stilts
  - provides mounting structure for vibrator actuator
  - delivers vibratory force to base plate
  - straddles vehicle drive line in some conventional truck mounts
- radius rods
  - provides lateral stability for isolation system
- accelerometer
  - measures phase and amplitude of base plate and/or reaction mass motion

Figure 5–47 is a photograph, courtesy of WesternGeco, of a vibrator used in West Texas.

Fig. 5–47 *Side View of a Truck-mounted Vibrator*

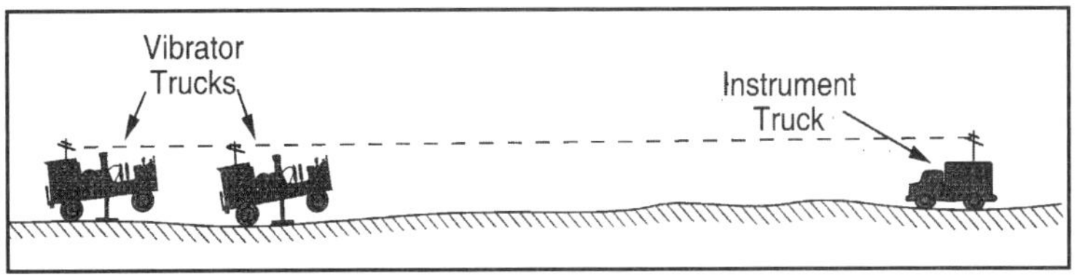

Fig. 5–48 *Vibrator Operation*

Vibrator operation is as follows (Fig. 5–48):

1. Two or more vibrator trucks are positioned at source points within a source array (VP).
2. The base plates are lowered to the ground and the trucks are jacked-up to place additional weight on the base plates, providing reactive mass.
3. An encoded swept-frequency signal (pilot sweep) is transmitted from the vibrator control unit in the instrument truck to similar units in each vibrator truck.
4. All vibrators input the prescribed signal into the ground and instruments begin recording simultaneously.
5. Recording continues for the length of the sweep plus the listen time.
6. All records obtained at a single source point are vertically stacked (summed) into one record.
7. Records are cross-correlated with the filtered pilot sweep to produce a record as long as the listen time.

Consider the following example. Reflections are recorded at times of 0.7s, 1.2 s, 1.95 s, 2.2 s, 2.86 s, 3.17 s, and 4.006 s. Sweep length is 10 s. The first reflection will be recorded from 0.7 s to 10.7 s, the second from 1.2 s to 11.2 s, the third from 1.95 s to 11.95 s, etc. As can be seen, there is considerable overlap among the reflections. In fact, the overlap is so great that raw vibrator records are unintelligible. This is why cross-correlation with the pilot sweep is required. An uncorrelated vibrator (Vibroseis) record using a 10 s sweep, and the corresponding correlated vibrator record is shown in Fig. 6–12.

Cross-correlation measures the similarity between two time series. In this case, the two time series are the pilot sweep that has had the instrument filters applied to it *(filtered pilot sweep)* and the raw vibrator record, which had the same filters applied while it was being recorded. In the frequency domain, cross-correlation is the product of the two amplitude spectra and the difference in phase spectra. Since the raw vibrator record has had the instrument filters applied to it, the filtered pilot must be used to cancel out the phase response of the recording instrument filters.

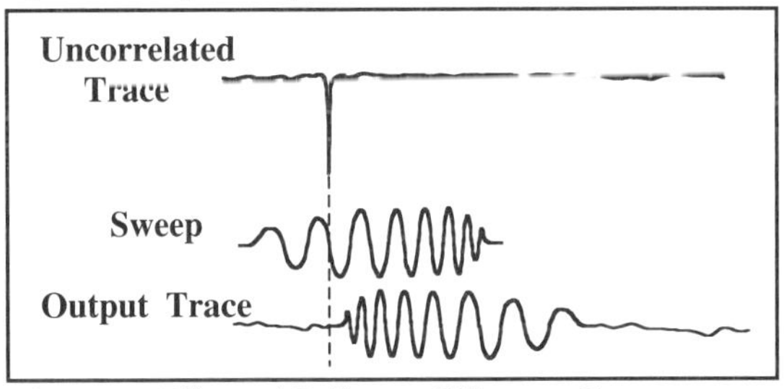

Fig. 5–49 *Effect of Spike on Vibrator Cross-correlations*

Noise spikes (static, parity) must be suppressed (*De-spike* applied) prior to correlation. Since correlation can be expressed as convolution of one time series with a time-reversed second time series, cross-correlation with a spike produces an inverted sweep that is combined with the output record. Figure 5–49 shows the effect of a single high-amplitude spike in an uncorrelated record. The amplitude is 1000 times larger than trace samples.

Figure 5–50 illustrates vibrator correlation, assuming an earth response with four primary reflections. The raw record is the convolution of the vibrator sweep and the earth response. Only the start of the first reflection can be determined.

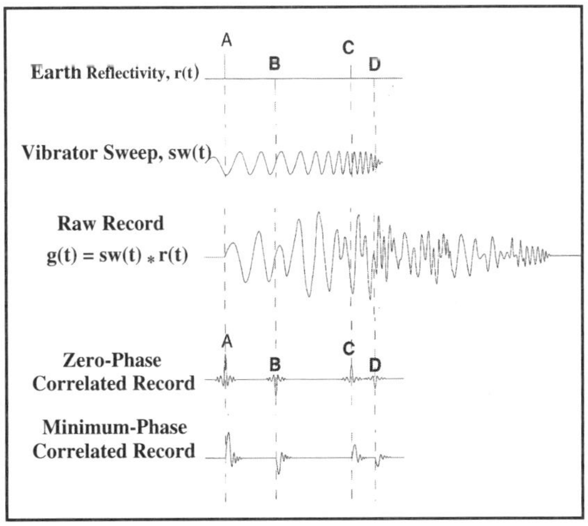

Fig. 5–50 *Vibroseis Correlation*

Since each reflection signal is the vibrator sweep scaled by the reflection coefficient, conventional cross-correlation produces a scaled version of the pilot sweep's autocorrelation, which is symmetrical and

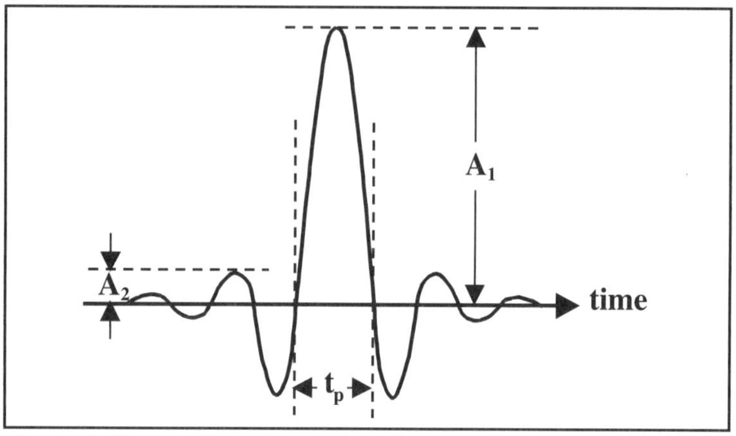

Fig. 5–51 *Klauder Wavelet and Resolution*

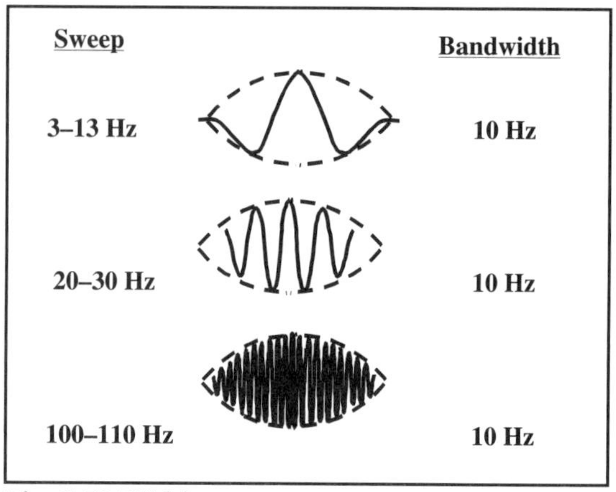

Fig. 5–52 *Width*

zero-phase. This is called *zero-phase correlation*. In many cases, *Wiener-type deconvolution* is applied in processing. This deconvolution assumes a minimum phase input. To provide for this, the option of minimum phase correlation is provided. Actually, *minimum phase correlation* uses an inverse filter technique to add the minimum-phase spectrum to the zero-phase correlation output. The difference between zero-phase and minimum-phase correlation is shown at the bottom of Figure 5–50.

The *Klauder wavelet* (Fig. 5–51) is used to represent the reflection wavelet produced by zero-phase correlation. The characteristics of a Klauder wavelet are described by definition, resolution, and width. *Resolution* may be thought of as the width of the peak (or trough) at the crossover point. In Figure 5–51, t_p = resolution. This assumes that all frequencies in a sweep are of the same amplitude and the sweep is linear. *Definition* may be thought of as the ratio of the amplitude of the largest peak (trough) to the amplitude of either adjacent peak (trough). This corresponds to the ratio A_1/A_2, in Figure 5–51.

The width is a measure of the wavelet's duration as measured in time units. It is inversely proportional to cycles of bandwidth and may be computed from 2/(bandwidth in Hz). Note that all three wavelets in Figure 5–52 have 10 Hz bandwidths, and, thus, the same width despite the differences in frequency content. As resolution and width get smaller and definition gets larger, the resemblance of the Klauder wavelet to a spike increases.

Good resolution depends more on bandwidth in octaves than in Hz. This is shown in Figure 5–53, where wavelets with zero through four octave bandwidths but different frequency ranges are compared. Infinite bandwidth gives a true spike or impulse function.

Parameters for vibrators are

- sweep type
- sweep length
- start and stop frequencies
- taper length
- number of vibrators
- number of sweeps/vibrator

Listen time depends on desired output record length.

A wide variety of sweep types are available. The original sweep was linear and frequency was a linear function of time. Other sweep types include T^n

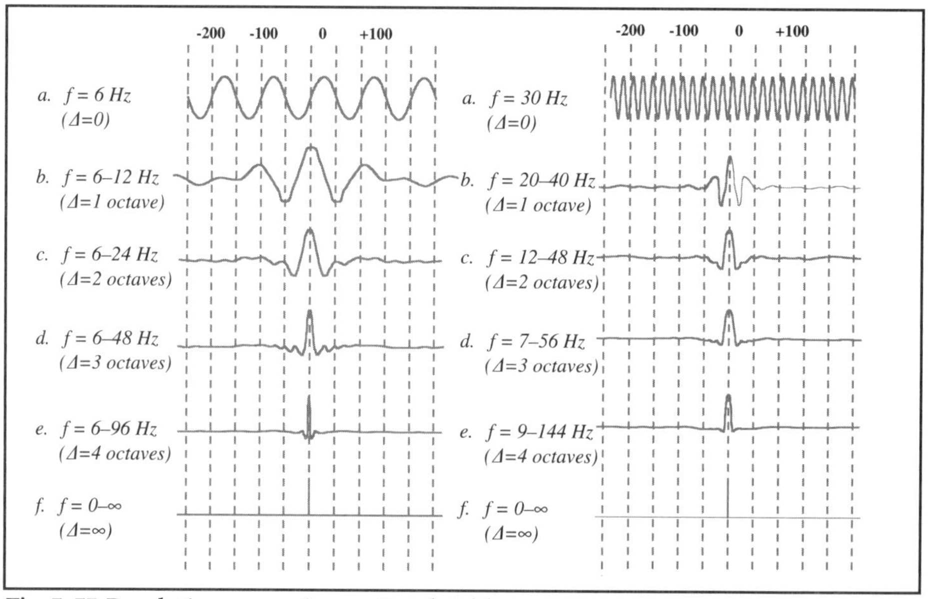

Fig. 5–53 *Resolution versus Sweep Bandwidth*

(usually $0.3<n<3$), exponential, and logarithmic. Figure 5–54 illustrates the change in sweep frequency with time for linear, $T^{0.3}$, T^3, logarithmic, and exponential sweeps. In each case, the start frequency is 10 Hz and the stop frequency is 80 Hz. Sweep length is 5 s in each case.

Of the five sweeps shown in Figure 5–54, the $T^{0.3}$ sweeps longest at the higher frequencies. This results in more high-frequency energy that is desirable to compensate for inelastic attenuation. The T^3 sweeps longest at the lower frequencies. This is desirable for penetration, since the lower frequencies are less affected by inelastic attenuation. The linear sweep is intermediate between the other two. The logarithmic sweeps slightly more in the higher frequencies than the linear, and the exponential sweeps slightly less in the lower frequencies than the T^3.

Figure 5–55 shows the pilot sweeps corresponding to Figure 5–54. It is clear that the $T^{0.3}$ sweep has more high-frequency content than the others. Likewise, it is clear that the T^3 sweep has more low-frequency content than the others. The logarithmic sweep has an appearance closer to the linear sweep than to the $T^{0.3}$ sweep, and the exponential sweep has an appearance closer to the $T^{0.3}$ sweep than to the linear sweep.

The number of sweeps and sweep length are selected based on acceptable signal-to-noise ratio (S/N). Both the number of sweeps and sweep length affect the production rate. Fewer long sweeps may result in better data and an increase in production if reduction in the number of sweeps/pattern results. A larger move-up distance between sweeps (a larger separation between source points in the source array or more groups between source arrays for lower fold) makes it more difficult to maintain source array configuration and increases time between sweeps.

The time between sweeps is equal to sweep time + listen time + move-up time (time required to move a vibrator from one source point to the next). The current tendency is to move vibrators during listen time, but moving during listen time causes coherent noise on receiver spread.

Signal-to-Noise ratio (S/N) for vibrators is proportional to:

- square root of number of sweeps, vN_S
- square root of number of receivers, vN_R
- square root of sweep length in seconds, vL
- number of vibrators, N_V

Hence, $S/N \propto (N_V) \, x \, (N_S \, x \, L)^{1/2}$

The number of sweeps to compensate for lost vibrators N_{S2} is given by:

$$N_{S2} = (N_{V1}/N_{V2})^2 \, x \, N_{S1} \tag{5.2}$$

where

N_{V1} = original number of vibrators

N_{V2} = number of vibrators remaining after loss

N_{S1} = original number of sweeps

The change in sweep length to compensate for lost vibrators L_2 is given by:

$$L_2 = (N_{V1}/N_{V2})^2 \, x \, L_1 \tag{5.3}$$

where

L_1 = original sweep length

Both the number of sweeps and the sweep length can be changed to compensate for lost vibrators, as indicated here.

$$N_{S2} \, x \, L_2 = (N_{V1}/ N_{V2})^2 \, x \, N_{S1} \, x \, L_{1V} \tag{5.4}$$

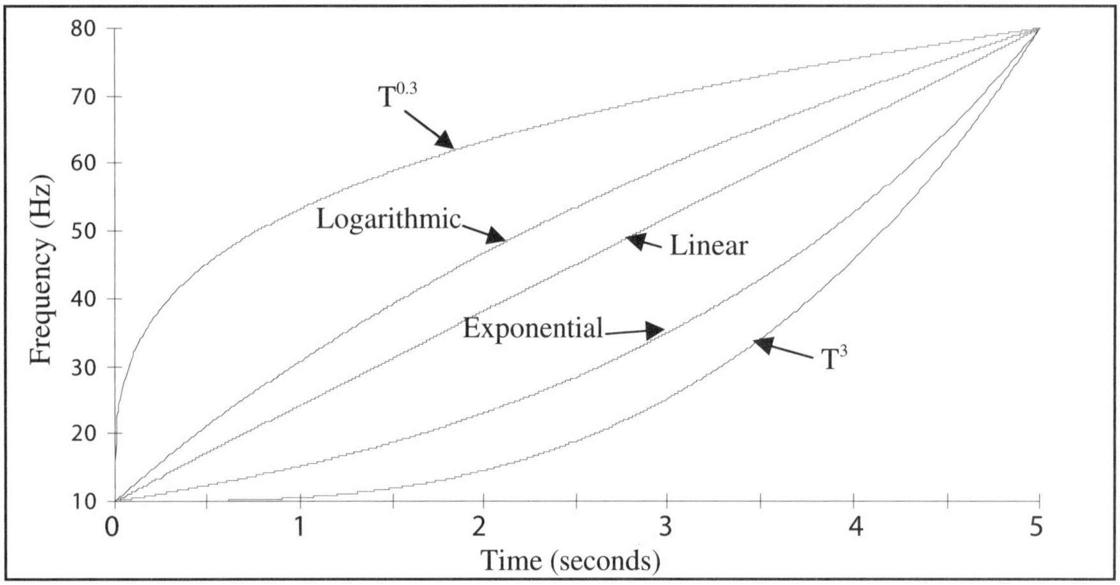

Fig. 5–54 *Frequency versus time for Various Sweep Types*

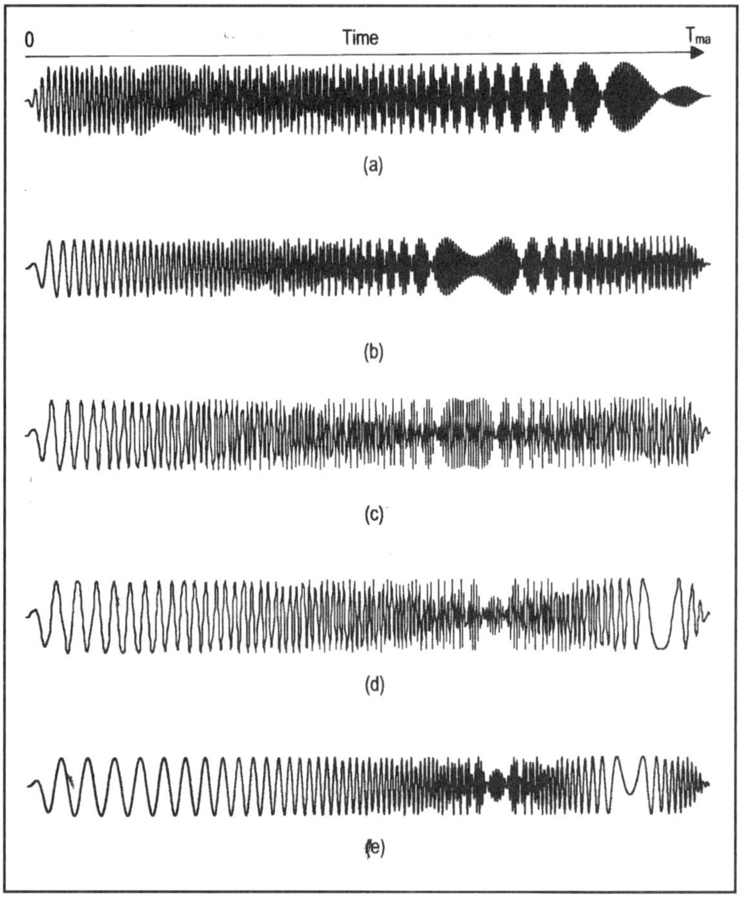

Fig. 5–55 *Pilot Sweeps for $T^{0.3}$, Logarithmic, Linear, Exponential, and T^3*

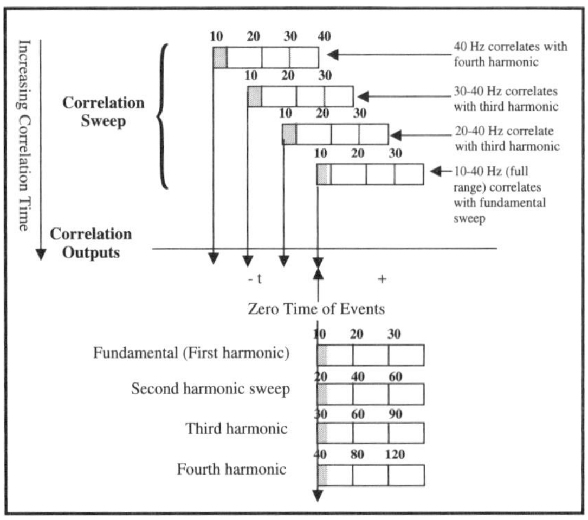

Fig. 5–56 *Vibroseis Correlation Ghosts*

It follows from the preceding that doubling the amplitude of a vibrator signal requires increasing the number of vibrators by two, the number of sweeps by four, the sweep length by four, or the number of sweeps *and* the sweep length by two.

The bandwidth of a vibrator is established by initial and final frequency values for the sweep. The bandwidth, in turn, depends on the required resolution to solve the particular geophysical problem presented by the survey. The start frequency may depend on the type of surface material encountered. Hard surface material results in production of higher amplitude harmonics.

Harmonics are always produced when vibrators sweep. Only the amplitudes vary. One effect of vibrator harmonics is the generation of *Vibroseis correlation ghosts*—a type of noise on correlated vibrator records. Figure 5–56 demonstrates the generation of harmonics for an *upsweep*—start frequency lower than stop frequency.

The top part of the figure illustrates how the pilot sweep is moved past the sweep recorded at the start of a reflection event. The bottom part of the figure illustrates the position of the fundamental (pilot) sweep plus the second, third and fourth harmonics. When the pilot sweep first encounters the signal, the only correlation is the 40-Hz component of the pilot sweep with the 40-Hz component of the fourth harmonic.

The vertical arrow drawn from the left of this position indicates the time of this correlation is earlier than the event actually occurs. As the pilot moves through the recorded event to the time of the 30-Hz component, the 30–40-Hz range of the pilot correlates with the same frequencies in the third harmonic. The vertical arrow from this position also indicates the correlation occurs earlier than the event actually occurs. Next is shown the shift position at the time of the 30-Hz component of the pilot. Now the 20–40-Hz range of the pilot correlates with the same frequencies in the second harmonic. This also occurs earlier than the event actually occurs.

The last position is where the start of the sweep is aligned with the start of the reflection event. Here the full range, 20–40 Hz, correlates only with the fundamental sweep and is coincidental with the event time.

The duration of any ghost is, of course, the difference between highest and lowest frequency arrival times. For a down-sweep, these times are given by:

$$t_1 = \frac{T(f_1 n - f_1)}{\Delta} \text{ and } t_2 = \frac{T(f_2 n - f_2)}{\Delta} \tag{5.5}$$

For an up-sweep, these times are given by:

$$t_1 = \frac{T(f_1 - f_1 n)}{\Delta} \text{ and } t_2 = \frac{T(f_2 - f_2 n)}{\Delta} \tag{5.6}$$

where

t_1 = arrival time of lowest ghost frequency

t_2 = arrival time of highest ghost frequency

T = sweep length

Δ = sweep bandwidth $(f_1 - f_2)$

n = order of harmonic (second = 2, third =3, etc.)

It follows from Equation 5.5 that for a down-sweep, times t_1 and t_2 are positive, while from Equation 5.6, it can be seen that times are negative for an up-sweep. This is the reason up-sweeps are generally preferred over down-sweeps.

Figure 5–57 shows a set of correlated vibrator records with sweep length, T, ranging from 7 seconds to 2 seconds. The sweep is from 35 Hz to 10 Hz, making the bandwidth, Δ, 25 Hz. The first break refraction is labeled PP_1, and the correlation ghosts are labeled GG_1, on each record.

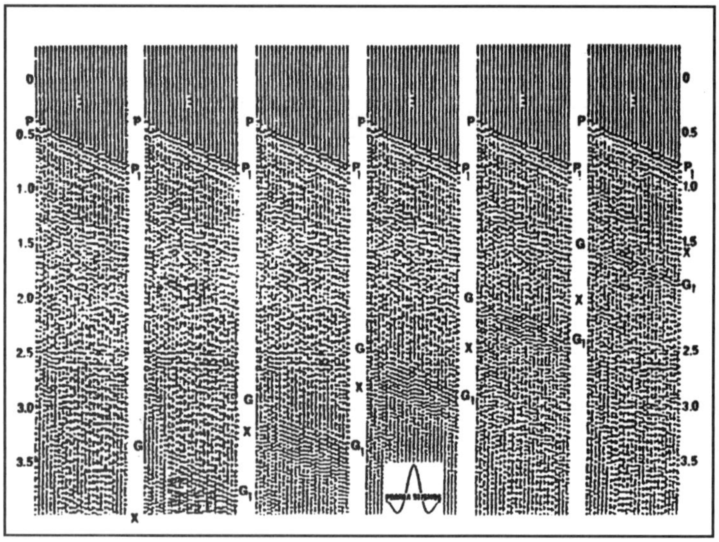

Fig. 5–57 *Vibroseis Correlation Ghost Examples*

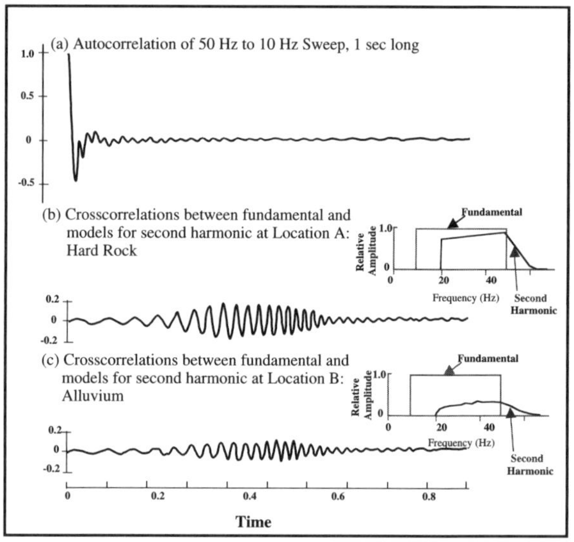

Fig. 5–58 *Effect of Surface Material*

Table 5–4 lists the start and stop times of each harmonic relative to PP_1. The worst problems occur for short sweeps and lower order harmonics.

Table 5–4 *Correlation Ghost Start and Stop Times*

	Second Harmonic		Third Harmonic		Fourth Harmonic	
T(s)	$t_1(s)$	$t_2(s)$	$t_1(s)$	$t_2(s)$	$t_1(s)$	$t_2(s)$
7	9.8	2.8	19.6	5.6	29.4	8.4
6	8.4	2.4	16.8	4.8	25.2	7.2
5	7	2	14	4	21	6
4	5.6	1.6	11.2	3.2	16.8	4.8
3	4.2	1.2	8.4	2.4	12.6	3.6
2	2.8	0.8	5.6	1.6	8.4	2.4

Figure 5–58 illustrates the effect of surface material on the generation of harmonics. Figure 5–58a is the autocorrelation of a 50-Hz to 10-Hz sweep that is 1 second long. Figures 5–58b and 5–58c are crosscorrelations between the fundamental and models—shown at right of crosscorrelations—for second harmonics at locations *A* and *B*. Location *A* has a hard rock surface, and location *B* has an alluvium surface. Note the lower level of harmonic generation at the softer surface.

Another parameter affecting vibrator correlation waveforms is the end taper. This is usually a cosine ramp function of the form:

$$s_s(t) = 1/2\left[1 - \cos\left(\pi \, {}^{t}/_{t_t}\right)\right] \text{ for } 0 = t = t_1$$

and (5.7)

$$s_s(t) = 1/2\left[1 - \cos\left(\pi \, [\, \text{T-1}] / t_t\right)\right], \; for\ T - t_1 = t = T$$

where

t_t = the taper length in seconds

T = is the sweep length in seconds

The effect of the taper on the sweep can be seen in Figure 5–59. Figure 5–60 shows the effect of the taper by comparing pilot autocorrelations. With no taper, the side lobes have much higher amplitudes.

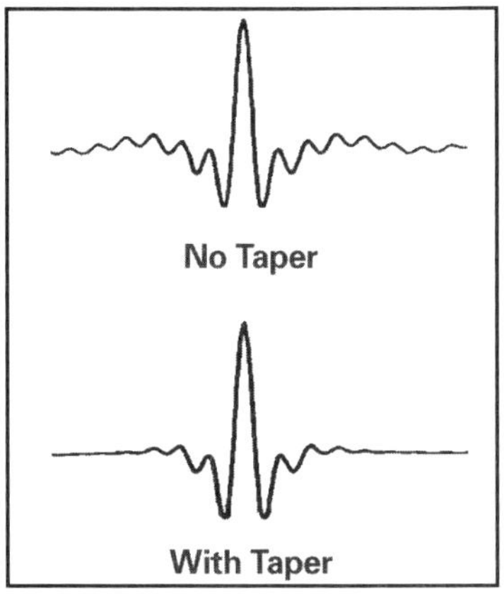

Fig. 5–59 *Effect of Taper on Pilot Autocorrelation*

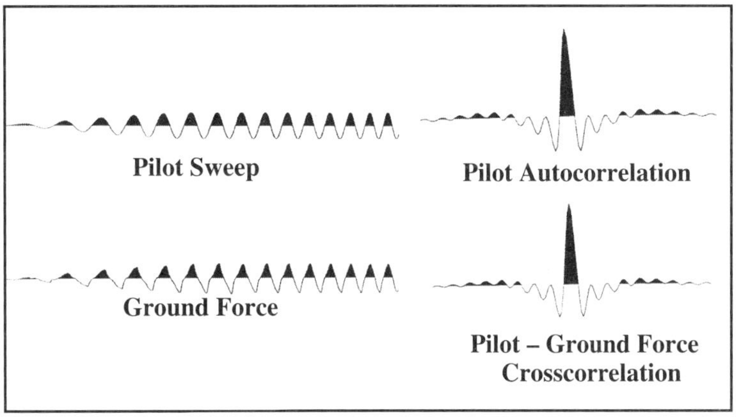

Fig. 5–60 *Attenuation of Sweep Distortion by Crosscorrelation*

Even though the sweep put into the ground may be considerably distorted, the crosscorrelation between the filtered pilot and the sweep will not necessarily show this distortion. Remember that crosscorrelation behaves as a filter, and its output can contain no frequency components not in the pilot. The distortion seen on sweeps is usually at higher frequencies than in the pilot sweep and, thus, do not appear on the correlated sweep. The early part of a 5–65 Hz, 8 s pilot sweep is shown at the top left of Figure 5–60. At the top right is the autocorrelation of this sweep. The measured sweep put into the ground (ground force) is shown at the lower left, and to its right is the cross-correlation between the two sweeps. Note the strong similarity between the autocorrelation and the cross-correlation in Figure 5–60.

In summary, vibrators produce a controlled frequency-content signal, the only source that does so. Maximum frequency range is from about 5 Hz to 511 Hz. Maximum sweep length is 32 seconds. A wide variety of sweep types are available: linear, T^n, logarithmic, exponential, and others. Sweeps may be up (increasing frequency) or down (decreasing frequency). A cosine taper is applied at each end of a sweep to limit side-lobe amplitudes. A vibrator signal is low *power* but not low *energy*. Cross-correlation of output record with pilot sweep collapses long duration signals into impulsive-appearing signals and increases signal power. Two or more (usually more) vibrator units are used together. This increases cost but speeds operation. Multiple sweeps per shot point increase signal levels and signal-to-noise ratios. Vibrators can be used anywhere heavy vehicles can operate, including towns and cities, but cannot be used in marshy or mountainous areas or in jungles. Very hard surfaces may limit vibrators' effectiveness, as this type of surface tends to distort signals.

Recording

The seismic waves generated by the seismic energy sources propagate as mechanical energy. Recording of seismic waves presents the following problems.

1. Ground motion on land and changes in water pressure in the marine environment, resulting from seismic waves generated by the source, must detected.
2. The detected seismic energy must be converted to a form that can be stored, displayed, and manipulated in such a way as to allow optimum information extraction from the data.
3. Implemented such that fidelity and precision are maximized, distortion is minimized, and retrieval of the stored information for processing and analysis can be readily accomplished.

Seismic detectors are special forms of *electro-mechanical transducers*—devices that convert mechanical energy into electrical energy). This is necessary because seismic energy is a form of mechanical energy. Mechanical methods of detecting and storing the amplitude and arrival time of seismic events are difficult and imprecise, and computer processing requires an electrical input.

More than one type of detector is required since in land exploration, both P- and S-waves are detected as motion of the earth's surface while in marine exploration, P-waves are detected as pressure changes in the water. No single detector type can detect both ground motion and pressure change. In the marine environment detectors called *hydrophones* are used. On land, *geophones* are used.

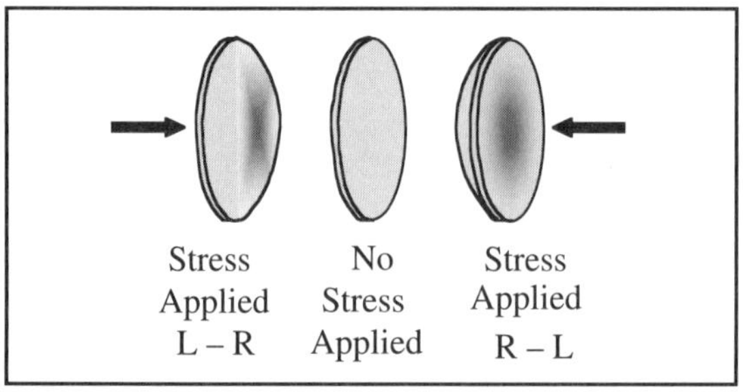

5–61 *Piezoelectric Crystals*

Hydrophones. Most hydrophones make use of the *piezoelectric effect*, which is that the application of stress to the surface, in some materials, causes a flow of electrons in the direction of the stress. Change in pressure is the stress for marine operations. Figure 5–61 illustrates the reaction of piezoelectric crystals to applied stress. Since current is the flow of charge, a current is produced proportional to the applied stress. Current across a load produces a voltage. The voltages so produced are then input to a seismic recording system.

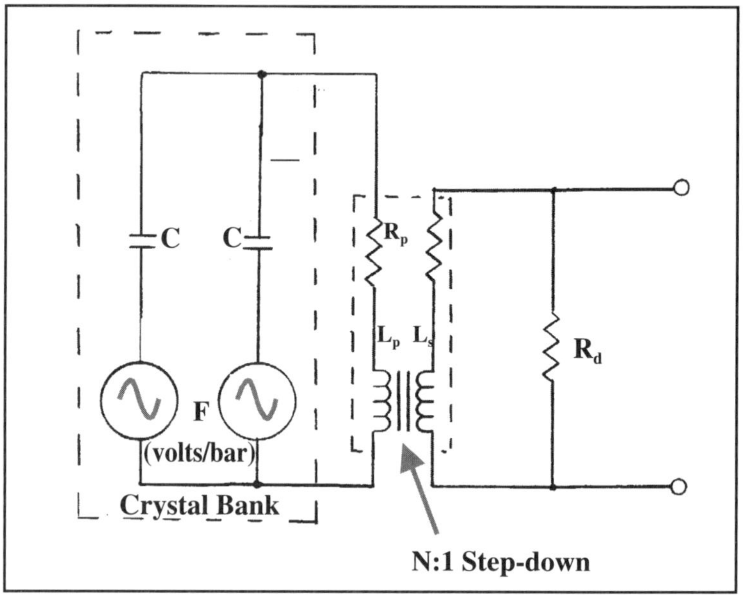

Fig. 5–62 *Hydrophone Equivalent Circuit*

Most hydrophones used today are transformer-coupled, piezoelectric hydrophones. An equivalent circuit (simplified) for a generic hydrophone group is shown in Figure 5–62. Here, the hydrophones are represented as a voltage source in series with a capacitor. There are usually 14 or 16 crystals in a bank. R_p and L_p are the resistance and inductance, respectively, of the transformer primary. R_s and L_s are the resistance and inductance, respectively, of the transformer secondary. R_D represents the damping resistance.

A modern transformer-coupled hydrophone used for geophysical prospecting has an amplitude frequency response represented by the following (simplified) equation, which is based on the equivalent circuit of Figure 5–62.

$$H_{Bd}(\mathrm{F}) = H_{Bo}\left[\frac{\dfrac{\mathrm{R_d}}{\mathrm{R_d}+\mathrm{R_z}}}{\sqrt{\left[\left(\dfrac{F_n}{F}\right)^2+1\right]^2+\left[2b_t\left(\dfrac{F_n}{F}\right)\right]^2}}\right] \tag{5.8}$$

where:

$H_{Bd}(F)$ is the sensitivity of the damped hydrophone at the test frequency F in Volts/Bar

H_{Bo} is the open circuit sensitivity of the hydrophone, measured unloaded at two or three octaves above resonance in Volts/Bar

R_d is the total effective damping resistance, which includes the influence of the acquisition system in ohms

R_z is the output impedance of the phone in ohms

F_n is the natural frequency of the phone, based on the input inductance of the transformer and the capacitance of the sensor bank in Hz

F is the test frequency in Hz

b_t is the total damping—the sum of the open circuit damping and the damping due to the damping resistor—expressed as a decimal fraction

Note that *resonance* and *damping* will be discussed later in this chapter.

Equation 5.8 is valid as long as the RMS amplitude of the differential pressure signal source is constant, and the frequencies of interest are within normal geophysical limits. Additional terms must be added to the equation to account for the high frequency characteristics of the transformer and the sensing crystals, as well as any frequency-dependent characteristics of the mechanical housing. The phase versus frequency response for a constant pressure drive is defined by the following equation.

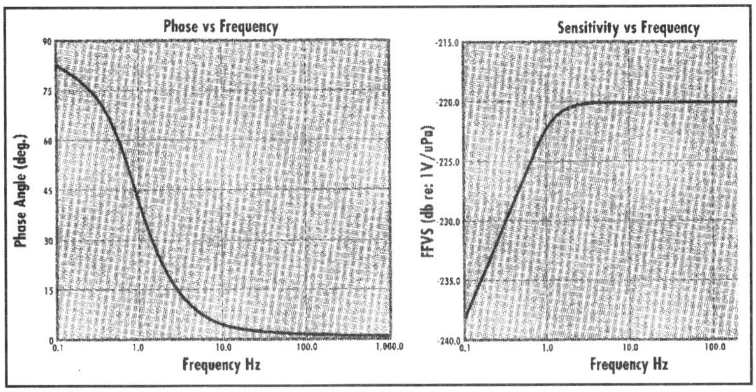

Fig. 5–63 *Input/Output Model 2522 Near-field Hydrophone*

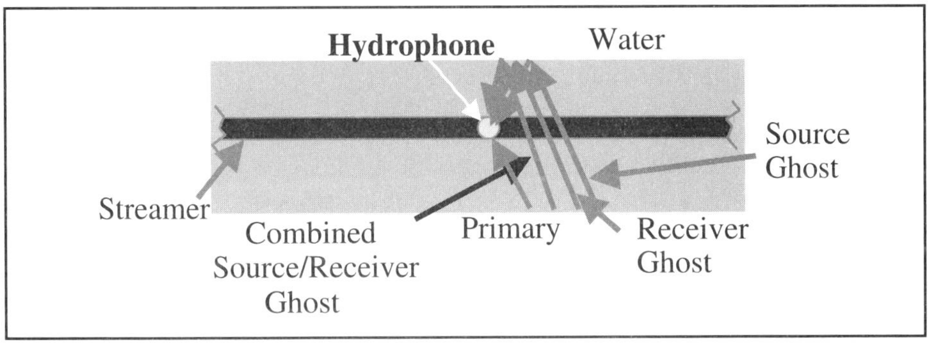

Fig. 5–64 *Hydrophone Ghosts*

$$\phi = \left[\frac{2b_t \dfrac{F_n}{F}}{\left[\left(\dfrac{F_n}{F}\right)^2 + 1\right]} \right] \tag{5.9}$$

Figure 5–63 shows the phase and amplitude response of an Input/Output Model 2522 near-field hydrophone. The response is flat from about 1 Hz.

Hydrophones in marine streamers are always used in closely spaced groups or arrays. This is because an individual hydrophone output is small. Current systems employ 14 to 16 hydrophones/group and group lengths are usually 6.25 m, 12.5 m, or 25 m.

Marine streamers are towed beneath the water surface. Shallower depths give higher frequencies but, usually, more noise. Greater depths give lower frequency but less noise. Also, as shown in Figure 5–64, the fact that the hydrophones and the source are towed beneath the surface allows energy to reach the hydrophones via undesired ray paths. Every reflection event is recorded four times.

- **Primary**. Energy travels from the source to the reflecting interface and then upward to the hydrophone group.
- **Source ghost**. Initially upward-traveling energy is reflected down at the water surface to the reflecting interface and then up to the hydrophone group.
- **Receiver ghost.** Energy travels from the source to the reflecting interface and then up to the water surface where it is reflected down to the hydrophone group.
- **Combined source-receiver ghost**. Initially up-traveling energy is reflected down at the water surface to the reflecting interface and then up to the water surface where it is reflected down to the hydrophone group.

Ghosts introduce notches in hydrophone group responses as shown in Figure 5–65.

Geophones. Seismic energy sources generate waves that travel mostly down, initially, and are returned to the surface by reflection and refraction. On land, these waves cause the earth's surface to move and this motion can be described in terms of ground particle displacement, velocity (speed), and acceleration. Of the three parameters of motion, the most readily measured is velocity, using moving coil seismometers.

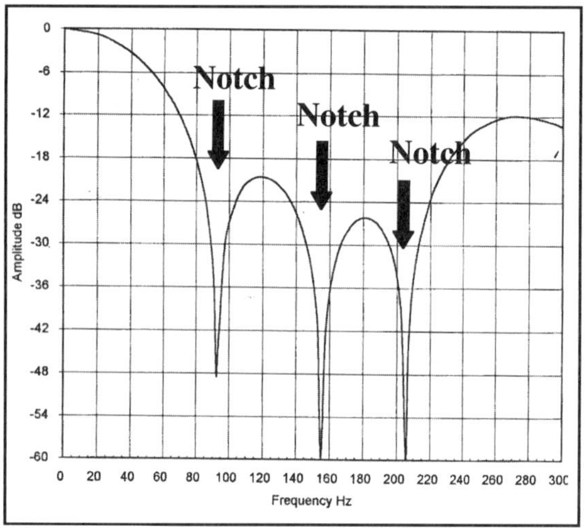

Fig. 5-65 *Amplitude Response for 12.5 m Group, 14 Phone Array*

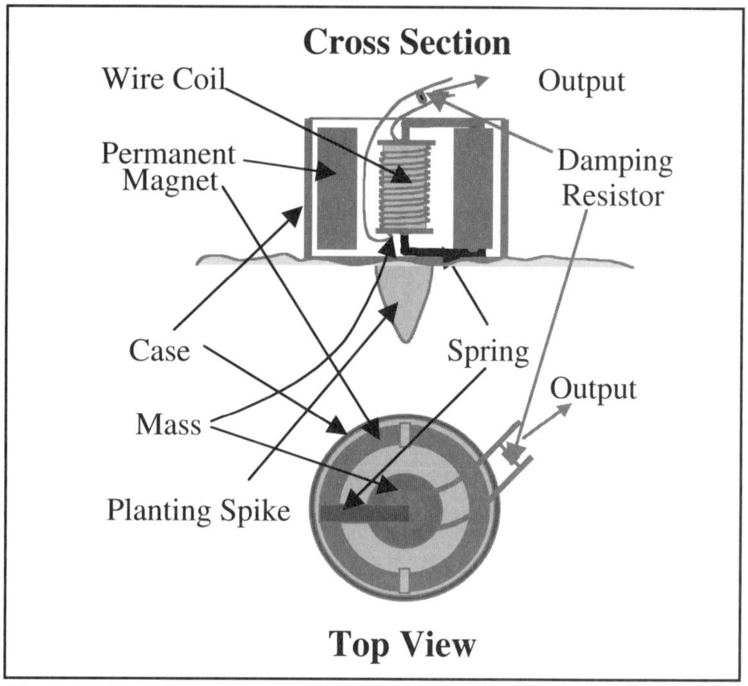

Fig. 5–66 *Geophone Components*

Moving coil seismometers make use of the property that an electrical voltage is generated in a conductor moving through a magnetic field (Lenz' Law). Mathematically, this is written:

$$e = \frac{d\phi}{dt} \tag{5.10}$$

where

e is the induced voltage

Φ is magnetic flux (lines of force)

Figure 5–66 is a simplified representation of an exploration seismometer, more commonly called a geophone. Principal components are

- case
- planting spike
- permanent magnet
- mass
- wire coil
- spring

The planting spike couples the case to the ground, so that we can assume that the case faithfully follows the ground motion. The magnet is attached to the case and moves with it. The mass, with a wire coil wrapped around it, is suspended by a spring so that it remains fixed with regard to the earth as a whole. Seismic waves arriving at the geophone cause the case and magnet to move up and down.

The case and mass lie within the magnetic field of the permanent magnet, which is described by lines of force or flux, $\tilde{\Phi}$ (Fig. 5–67). The relative motion between mass and case causes the coil to be cut by lines of force. This is described by $d\Phi//dt$.

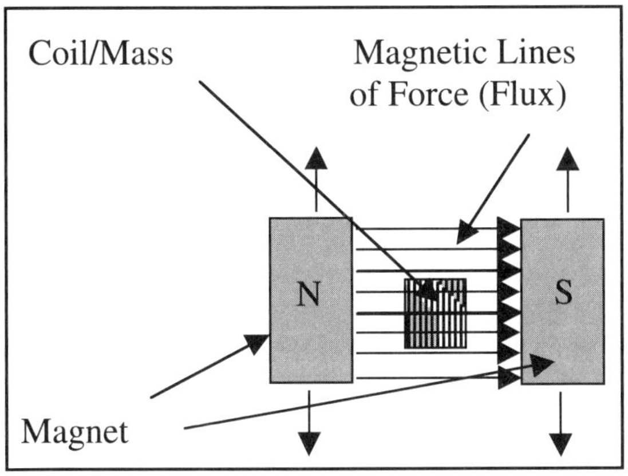

Fig. 5–67 *Geophone Magnetic Field*

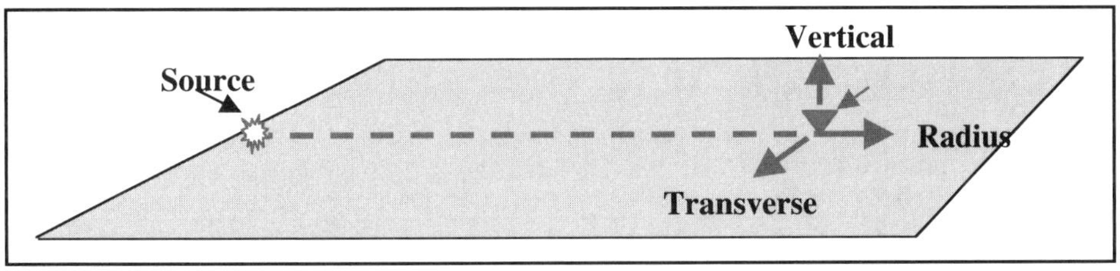

Fig. 5–68 *Three-component Geophones*

Magnetic flux can be defined by the flux density B and the area A through which the flux flows—$\Phi = BA$. Assuming a constant flux density, the change in flux with time is caused by a change in the total area across which lines of force flow through the coil.

$$e = -\frac{d\phi}{dt} = -B\frac{dA}{dt} \tag{5.11}$$

The coil is assumed to be constrained to motion perpendicular to the magnetic field, and the change in area can be described by $dA = l\ ds$, where l = circumference of the coil and ds = instantaneous displacement of the case and magnet. Substituting for dA in the previous equation yields:

$$e = -Bl\frac{ds}{dt} \tag{5.12}$$

Since the case moves with the ground, $\frac{ds}{dt} = v$, where v = ground particle velocity. Thus, we have, finally:

$$e = -Blv \tag{5.13}$$

The previous series of equations shows that $e \propto v$ or that the voltage output by a geophone is directly proportional to the ground particle velocity.

Geophones can be made to respond to vertical, radial (in the direction of the receiver line) and transverse (orthogonal to the direction of the receiver line) motion of the ground by rotating the mounting of all active elements of the geophone 90° from the vertical position. Three-component geophone systems (Fig. 5–68) can used to record both P- and S-waves. This procedure allows more information to be extracted from seismic data but adds to the expense in both data acquisition and processing.

Damping is used in geophones to retard the tendency to oscillate at the natural frequency that depends on mass and spring constant. Damping used to be accomplished with some fluid, such as oil, being placed within the case. Damping, today, is accomplished by placing shunt resistors across the output of a phone that provides a back *electromotive force* (emf)—an opposing emf field.

The parameter h is used to describe damping, with $h = 1$ corresponding to *critical damping*, $h < 1$ meaning the geophone is *under damped*, and $h > 1$ meaning the geophone is *over damped*. For the case of no damping, $h = 0$, the geophone oscillates continuously but, in practice, gravity and air friction would eventually bring the mass to rest (Figure 5–69).

For seismic exploration, geophysicists require as short a signal as possible. Hence, some damping is required. Critical damping means that after the case is displaced, it returns to the rest position with no oscillations and no over-swing. Normal damping for geophones is 0.7 (70% of critical). This gives a slight (only 5%) over-swing past the zero position. It is a compromise between considerations of over-swing, output amplitude, and phase shift.

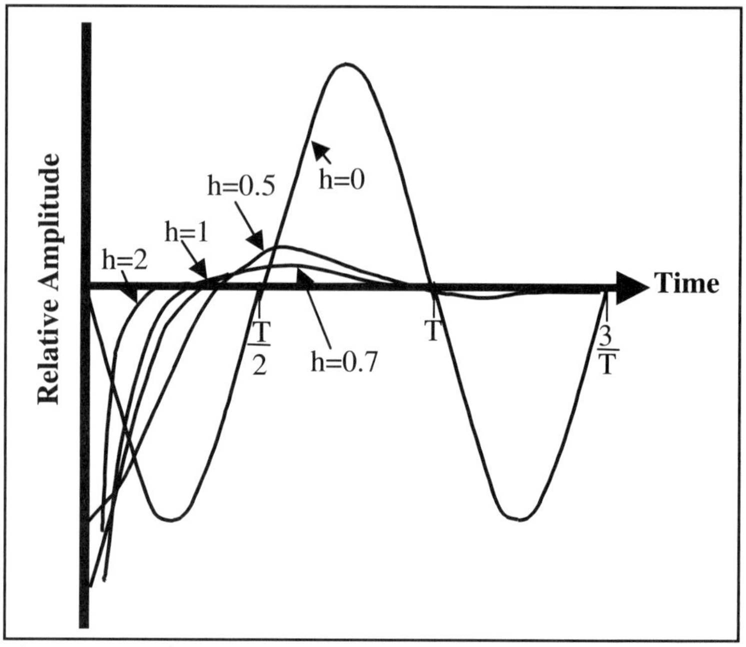

Fig. 5–69 *Geophone Damping*

For no damping, $h = 0$, the amplitude peaks at the resonance or natural frequency. Low frequencies are preferred to extend width—a low resonant frequency is desirable. This can be accomplished through design with a heavier mass or a stiffer spring. High geophone sensitivity is desirable. (Geophone *sensitivity* or *transduction* is measured in volts/inch/second.) Heavy damping provides a shorter impulse but at the cost of reduced sensitivity. Light damping tends to promote oscillations and enhance low frequencies such as those at which ground roll occurs. Typical resonant frequencies for reflection work are 14 Hz and 10 Hz. These choices are sometimes made in lieu of low-cut system filters.

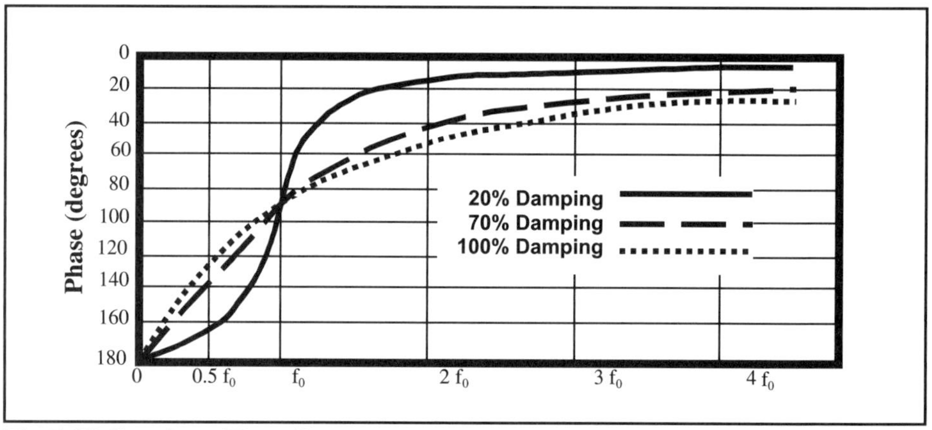

Fig. 5–70 *Geophone Phase Response*

The geophone behaves as a filter since it attenuates frequencies lower than its natural frequency. With 70% damping, the geophone acts as a low-cut filter with cutoff equal to the natural frequency and a slope of 12 dB/octave. The geophone also introduces phase shifts between the input and output signal. *Input* is the physical up and down movement of the geophone. A phase shift of 90°always occurs at the natural frequency regardless of the degree of damping. Phase shift below the natural frequency is greater than 90°. Above the natural frequency the phase shift is less than 90°. See Figure 5–70.

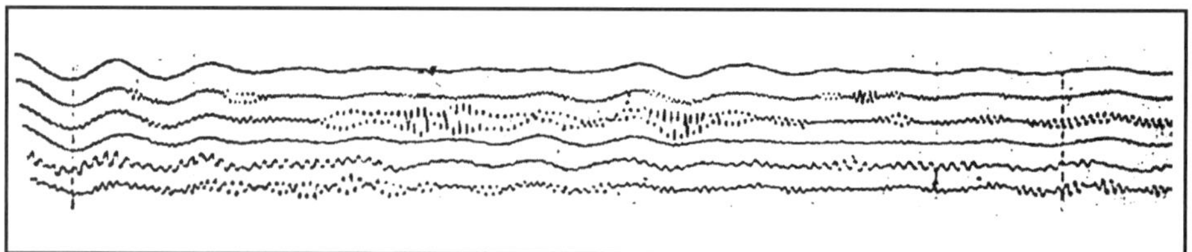

Fig. 5–71 *Spurious Resonance or Parasitics*

A geophone is a mechanical system designed to allow a single degree of freedom in the vertical direction. Also, a geophone has other *vibrational resonances* that couple loosely into its primary motional mode, including a horizontal oscillatory mode in which the mass behaves as though it were attached to the case via a horizontal spring. This produces a *spurious resonance*, or *parasitic*. Figure 5–71 is a recording of parasitic resonances on a single-geophone-per-trace spread of Geospace's GSC 11D, 8-Hz geophones at distances of 180 ft to 415 ft from a 1-lb. Nitramon charge. While the noise has rather high amplitude, it can be easily recognized because it is much higher frequency than the nearby ambient noise.

Geophone parameters that can be varied are natural or resonant frequency, open circuit damping (some damping is always present), and coil resistance. Table 5–5 provides some guidelines in selecting these parameter values.

Table 5–5 *Selecting Geophone Parameter Values*

Parameter	Goal	Approach
Natural Frequency	Reduce low frequency signal or ground roll	Select a higher natural frequency detector to suppress low frequencies (28 to 30 Hz or higher)
	Enhance the high frequency data for higher resolution	
	Enhance low frequency signals	Select a low natural frequency (8 to 14 Hz)
Open Circuit Damping	Have total damping equal to 0.6	Select a seismometer with an open circuit damping approaching but not greater than 0.6
	Maintain high sensitivity under load	
Coil Resistance	Increase sensitivity	Choose large coil resistance
	Decrease noise pickup	Choose low coil resistance
	Reduce effect of load impedance	

The normal assumption is that geophone case motion is a replica of ground velocity. However, this assumption is invalid in loose soil. Since the soil is elastic, geophone amplitude and phase are distorted. These effects can reduce within seismic frequencies by increasing the mass, increasing the radius of the geophone, using longer spikes to provide firmer contact with compacted soil, or using buried phones for better plants in compact soil. These solutions are expensive and require careful consideration. Small phones are generally more economical and all that is required is a good vertical plant in compact soil. If compact soil is not available, the cost of burying has to be considered.

Geophones are directional, sensing only the vertical component of motion. If a geophone is planted (placed in the ground) at an angle θ, then amplitude is proportional to cosθ. See Figure 5–72. If the geophone is at an angle of ± 20° from the vertical, then there is a loss of 5% in amplitude. Far offsets add an angle of emergence to a tilted or poorly planted phone. For example, an emergence angle of 30° adds to the geophone tilt of 20°. The amplitude is now proportional to cos (20° + 30°) giving a 35% loss in amplitude. Geophones are generally rated as operational within ± 20 degrees. Spurious harmonics may also contaminate the signal within the seismic bandwidth outside this angle.

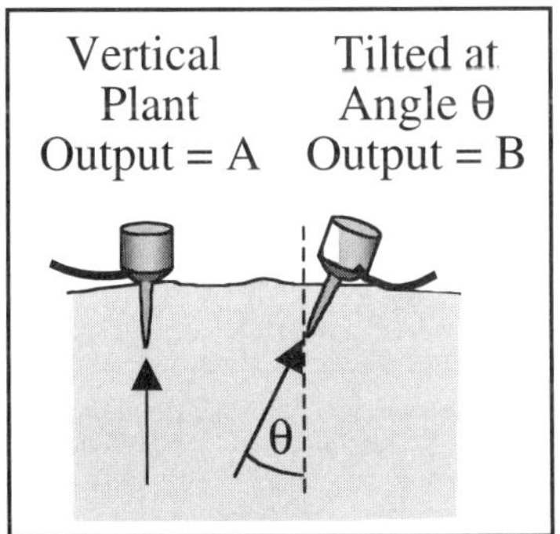

Fig. 5–72 *Geophone Tilt*

It is not only important that geophones be planted vertically but also firmly in contact with the ground. Figure 5–73 shows four planting conditions that rate from best to poor with regard to earth coupling. The best plant is either buried (Fig. 5–73a) or a geophone with its 5½-inch planting spike solidly placed in firm ground with all loose soil first scraped away (Fig. 5–73b). A geophone with its 3-inch planting spike solidly placed in firm ground with all loose soil first scraped away (Fig. 5–73c) is a good plant. A geophone with its 3-inch planting spike solidly placed in firm ground with only vegetation, such as pine needles, brushed away is a fair plant. A geophone planted on top of vegetation, such as pine needles or very loose soil, is a poor plant. Figure 5–74 shows the effect of these planting conditions by displaying the amplitude and phase responses predicted from impulse response measurements.

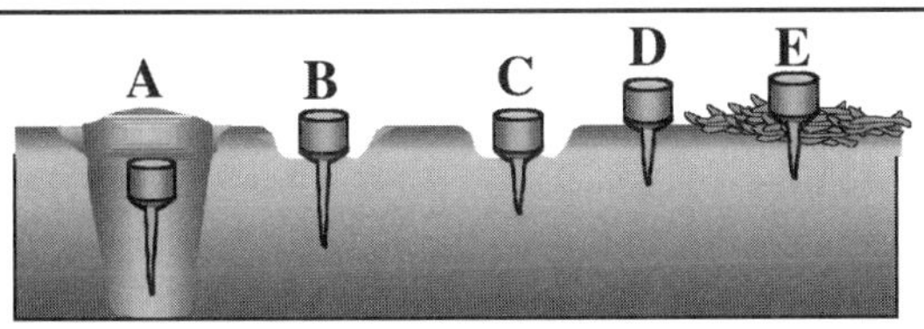

A or B: Best plant – Geophone with 5½-inch spike, buried or scraped surface

C: Good plant – Geophone with 3 inch spike, scraped surface

D: Fair plant – Geophone with 3 inch spike, brushed surface

E: Poor Plant – Geophone with 3 inch spike and *no* preparation of surface

Fig. 5–73 *Geophone Planting Conditions*

Noise

Noise on a seismic record is, basically, anything you don't want to record. There are two general types of noise—ambient and source-generated. *Ambient noise* is naturally occurring (wind, rain, lightning, thunder, surf, earthquakes) or cultural (vehicular and foot traffic, grazing animals, pumps). *Source-generated* noise includes ground roll, air blast, guided waves, and others. Since noise is undesirable, although unavoidable, measurement of noise characteristics (frequency, wavelength propagation velocity) aids in design of techniques to minimize noise recording. Such measurements are done in noise tests.

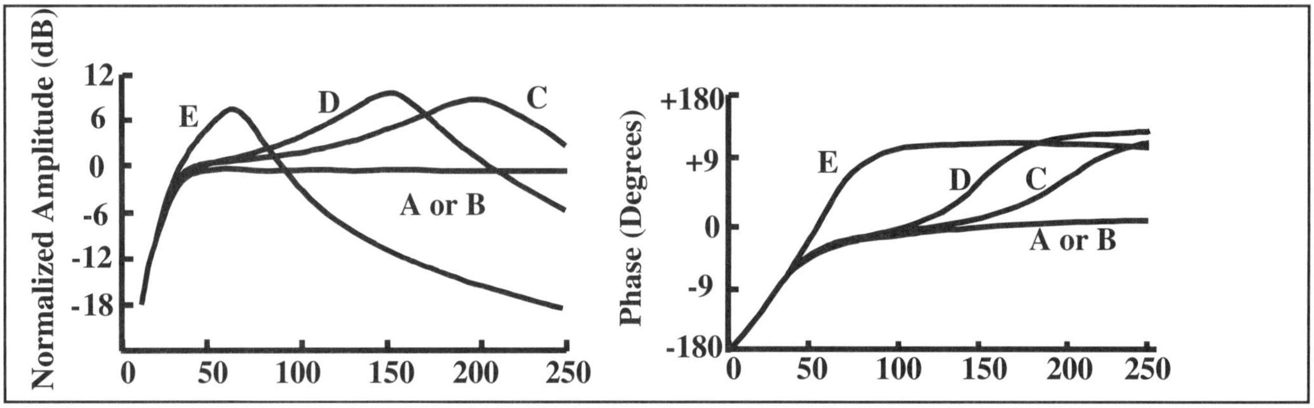

Fig. 5–74 *Effect of Geophone Planting Conditions on Amplitude and Phase Responses*

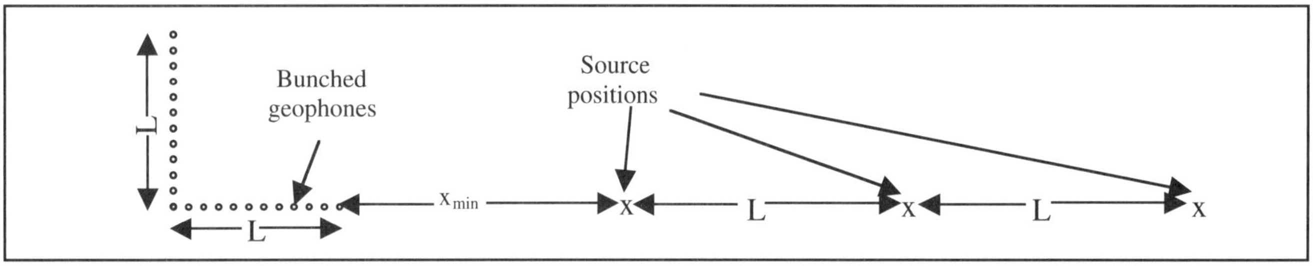

Fig. 5–75 *Noise Spread and Shooting Procedures*

A useful method of conducting noise tests is to lay out about 12, or more, groups of several bunched geophones, with total length L, and a similar set of 11 geophones perpendicular to these. The length L should be equal to the planned group interval in the seismic survey. Figure 5–75 shows the suggested layout and shooting procedure. Shoot (or sweep) into these geophones starting at the minimum offset x_{min} and continuing at intervals L as shown in Figure 5–75. The minimum offset should also be equal to that planned for the survey. One record is made at each source position. Offsets for the first record are x_{min} to $x_{min}+ L$. Offsets for the second record are $x_{min}+L$ to $x_{min}+ 2L$, etc. Records are combined with traces offset-ordered.

Figure 5–76 is an example of a noise test record. It is important in conducting noise tests that the same source be used as in the seismic survey. While the same information is present in both the records shot with explosives and the records obtained from vibrator sweeps, there are some differences as well.

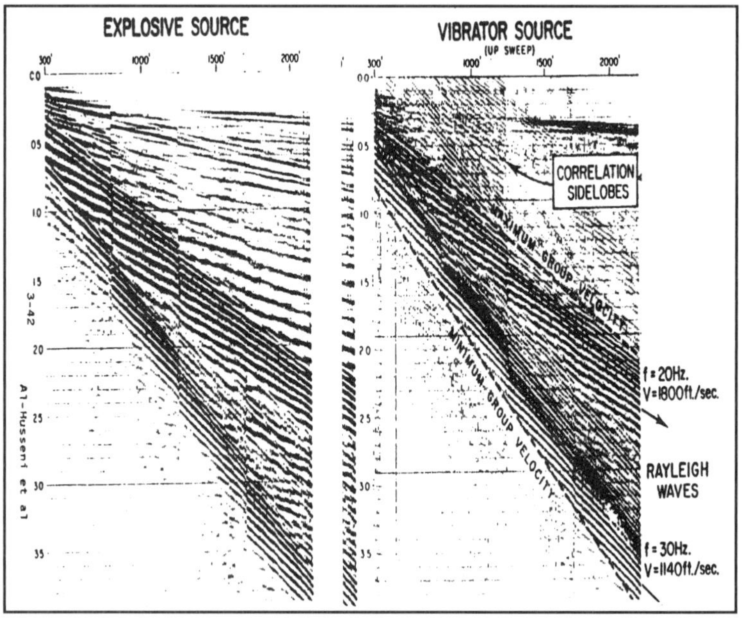

Fig. 5–76 *Noise Test Example*

The various noise types are recognized by velocity and frequency. Note that ground roll is not a single event but is, instead, a dispersed noise train. Ground roll normally contains all frequencies released by source, and it penetrates into earth as a function of wavelength. Low frequency equals long wavelength, which means deeper penetration into the surface layer. High frequency equals short wavelength, which means shallow penetration. Since velocity in earth normally increases with depth, the low frequency, long wavelength ground roll travels faster than shallower-penetrating high frequencies. This causes dispersion of ground roll. Velocity is a function of wavelength or frequency.

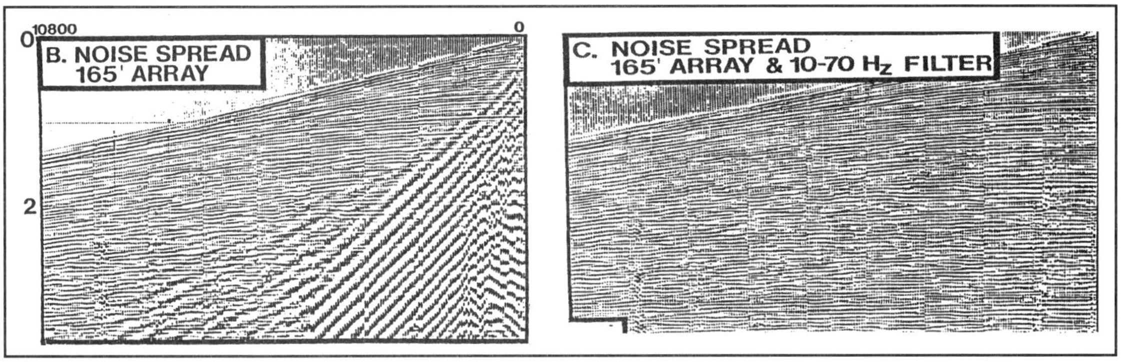

Fig. 5–77 *Attenuation of Ground Roll with Band-pass Filter*

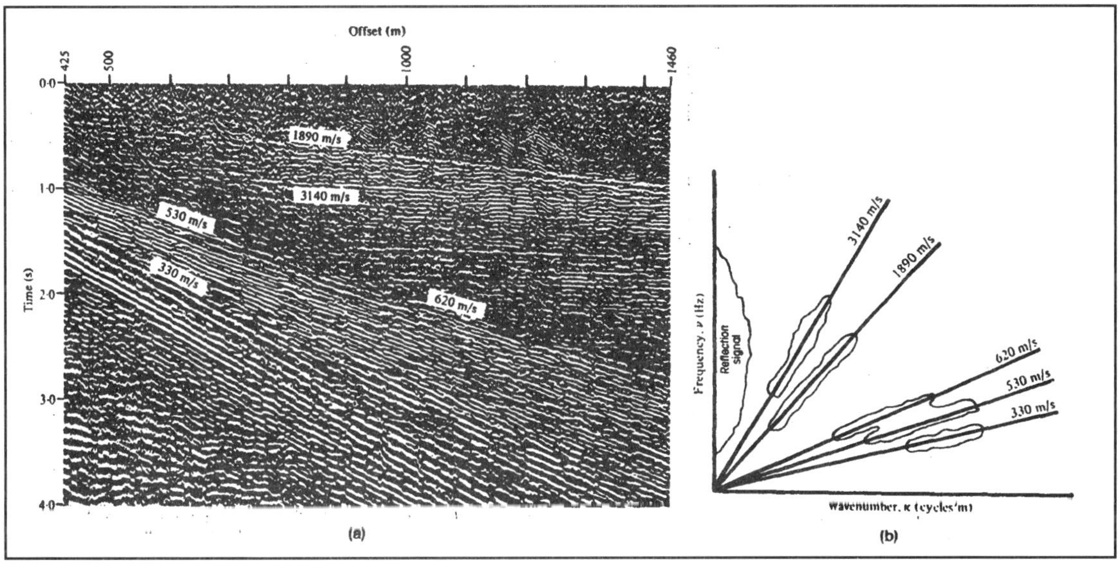

Fig. 5–78 *Noise Analysis or Walk Away*

A variety of methods are available for attenuating noise. In many cases, ground roll has significant amplitudes only at frequencies lower than the signal. In such cases, a band-pass filter will provide adequate noise attenuation. Figure 5–77 illustrates this by showing a noise test record with high amplitude, low frequency ground roll on the left and the same record after application of a 10 Hz low-cut filter. The filter essentially eliminates the ground roll.

Figure 5–78 shows a noise test record on the left and its *F-K* domain representation on the right. Noise is well separated from signal. A velocity or *F-K* filter should give good attenuation of the linear noise.

From measurements of noise wavelength λ and velocity V, calculate frequency f from $f = 1/T$ and wavelength λ from above relationship. Calculate wave number k for signal and noise from the relationship of $k = 1/\lambda$. Use these values to design the optimum receiver array, considering availability of geophones, cost, time, and difficulty of lay out.

In seismic exploration, arrays comprise two or more elements in a geometric pattern. The elements may be sources or receivers. In land exploration, arrays are used primarily to attenuate source-generated, horizontally propagating noise. In marine exploration, arrays are used primarily to increase signal levels. Arrays can be linear (usually in line with or parallel to the receiver lines) or areal (with both in-line and cross-line components).

There are different kinds of linear arrays—equal weights and equal spacing, variable weights and equal spacing, and equal weights and variable spacing. Areal arrays include rectangular, parallelogram, chevron, and star. See Figure 5–79.

Consider a theoretical receiver that responds continuously and equally along its length L. Figure 5–80 shows the response of this array in the space x and wave number k domains. However, seismic data are not recorded continuously but at discrete points. Therefore, seismic data are *space-sampled* data. Arrays are made of discrete elements, and thus they are discrete spatial filters. Figure 5–81 shows a 6-element, linear, equally spaced, equally weighted array. Spacing between elements is d, ft, or m.

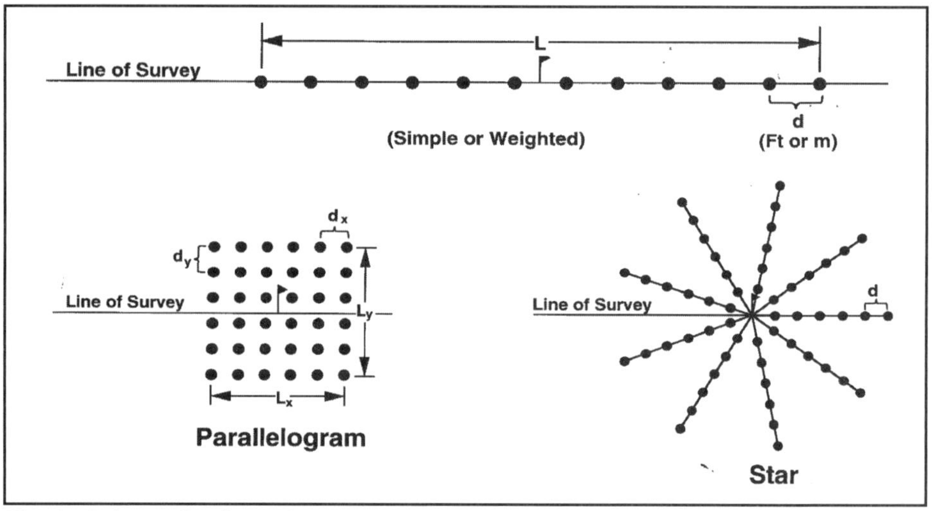

Fig. 5–79 *Array Configurations*

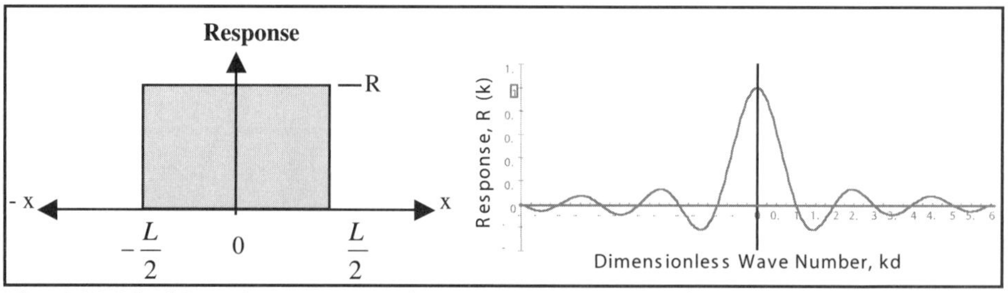

Fig. 5–80 *Ideal Receiver Response for Space Domain x and Wave Number Domain K*

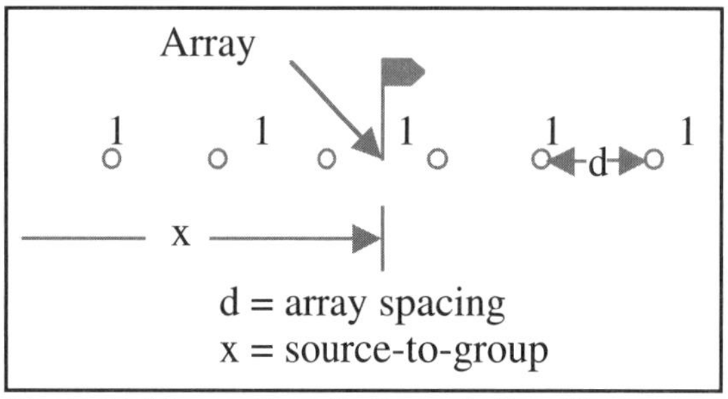

Fig. 5–81 *Linear, Equally Spaced, Equally Weighted Array*

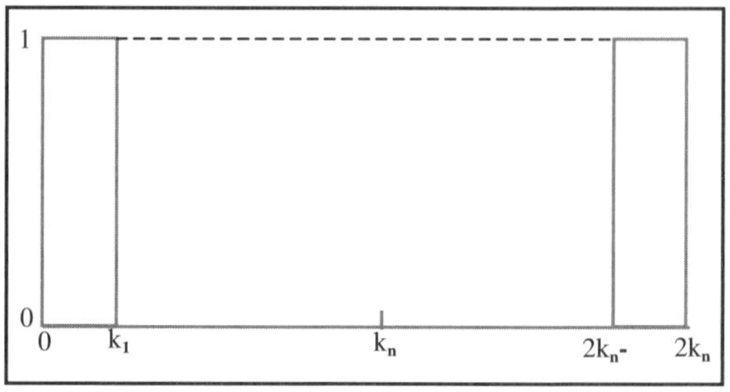

Fig. 5–82 *Conventional Representation of Desired Response of Linear, Equally Weighted, Equally Spaced Array*

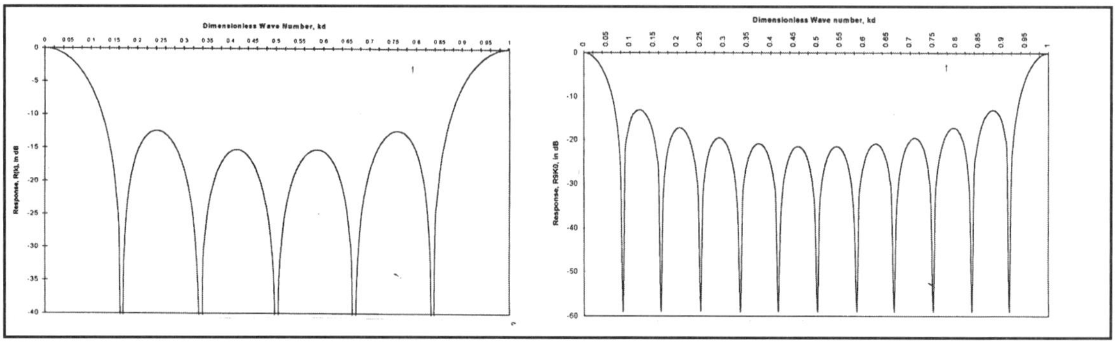

Fig. 5–83 *Response of Linear, Equally Weighted, Equally Spaced Arrays, 6 Element and 12 Element*

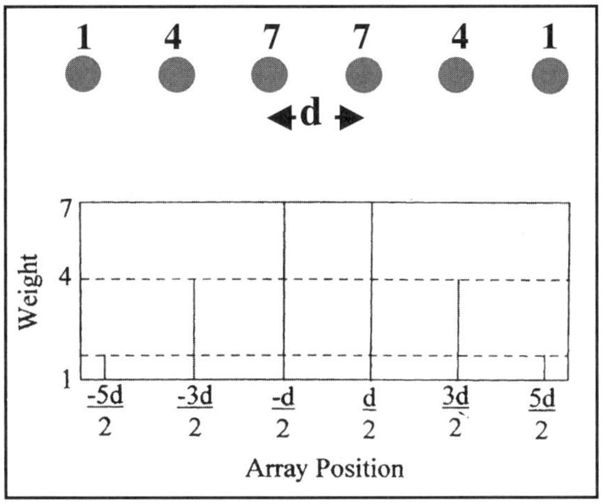

Fig. 5–84 *Alternative Methods of Representing the Spatial Response of Linear, Unequally Weighted, Equally Spaced Arrays*

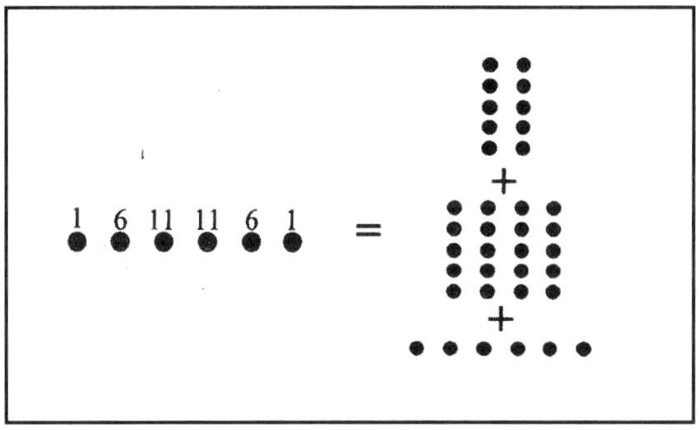

Fig. 5–85 *Decomposing Linear, Unequally Weighted, Equally Spaced Arrays*

If k_1 is the wave number at which the signal-to-noise ratio = 1, then the desired array response is as shown in Figure 5–82. Note that array responses are usually shown for the range $0 < k < 2k_n$ or $0 < kd < 1$. The response of an equally weighted, linear array having N equally spaced elements is given by:

$$R(k) = \frac{\sin(N\pi kd)}{\sin(\pi kd)} \tag{5.14}$$

Actual responses, thus, differ from the desired response as shown in the example of Figure 5–83.

Linear, variably weighted, equally spaced arrays can be used for both receiver and source arrays. Receiver arrays can be weighted by clustering geophones with the number of geophones per cluster determining the weight. Source arrays can be weighted by loading different charges or by varying number of sweeps. Array weights can be shown as numbers corresponding to weights, above array positions, or as lines whose heights represent weights and separations represent array spacing (Fig. 5–84).

Responses of weighted arrays are found by decomposing them into equally weighted linear arrays. The array shown in Figure 5–85 is decomposed into one 6-element, five 4-element, and five 2-element arrays. The k-domain responses of each component is calculated from Equation 5.14 and summed. The result is shown in Figure 5–86. Note that *wi* is the weight of each array element.

While arrays can be an effective tool for attenuating horizontally propagating noise, they can also be harmful to recording signal. One such situation is when there are large elevation changes over the area where the array is to be placed. It is generally accepted that the maximum elevation difference across an array should not exceed one-fourth of the smallest wavelength expected λ_{min}. Since $V = f\lambda$, the signal having the smallest wavelength will have the highest frequency f_{max} and the restriction on elevation difference can be written as:

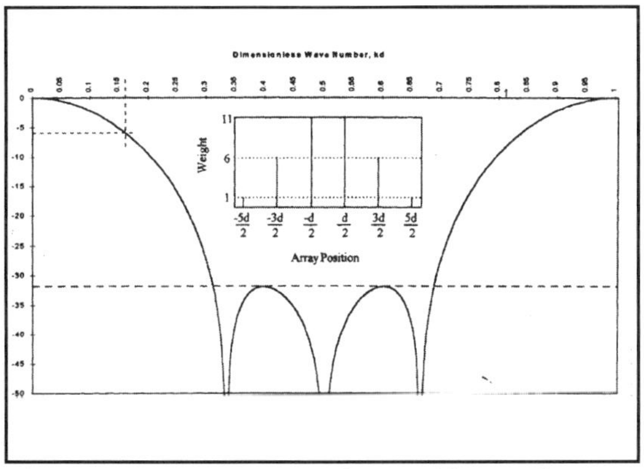

Fig. 5–86 *Linear Array with Variable Weights*

$$x \le \frac{V}{4f_{max}} \tag{5.15}$$

On the other hand, given an elevation difference x, the highest frequency that can be recorded satisfactorily is given by:

$$f_{max} \le \frac{V}{4x}$$

Figure 5–87 shows alternative methods of laying out geophones in the presence of excessive elevation differences.

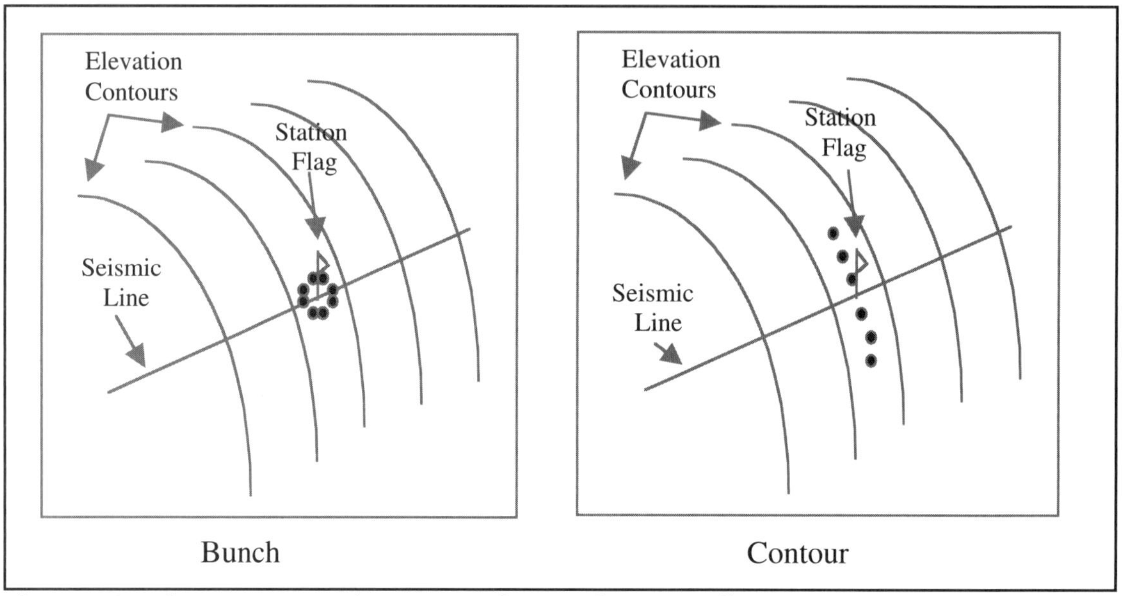

Fig. 5–87 *Geophone Layouts in the Presence of Large Elevation Differences*

Recording systems

Seismic recording systems measure and record electrical inputs from detector groups with maximum fidelity and precision, minimum distortion, and the capability to be readily retrieved for subsequent processing and analysis. The environment under which this must be done is challenging.

First-break amplitudes are usually the largest inputs and may be 140 to 160 dB higher than amplitudes of the deepest reflections. Source-generated noise amplitudes are often considerably larger than signal received at the same time. Amplitudes of multiples and reverberations may also be larger than reflections that are recorded at about the same time. Amplitudes of the earliest reflections can be 100 dB or more—larger than those of the latest primary reflections. Ambient noise is usually about 70 to 80 dB below first-break amplitudes but larger than later reflection amplitudes. Amplitudes of signals and various noises received at the same time can cover a range of up to 100dB. All recording systems produce internal noise, and the level of this noise, expressed as an equivalent input noise, establishes a floor for signal recording. Figure 5–88 shows the situation graphically.

System performance requirements include a large dynamic range, wide bandwidth, large amplitude range, low harmonic distortion, and minimal system noise. System operational requirements include:

- large channel capacity at all sampling rates
- multi-line/multi-streamer capacity
- large number of channels/line or channels/streamer
- large roll capacity
- adequate auxiliary channels
- wide range of group and line intervals
- correlation and stack capability (land)
- internal testing capability
- remote status monitoring
- long battery life (land)
- good status and QC displays

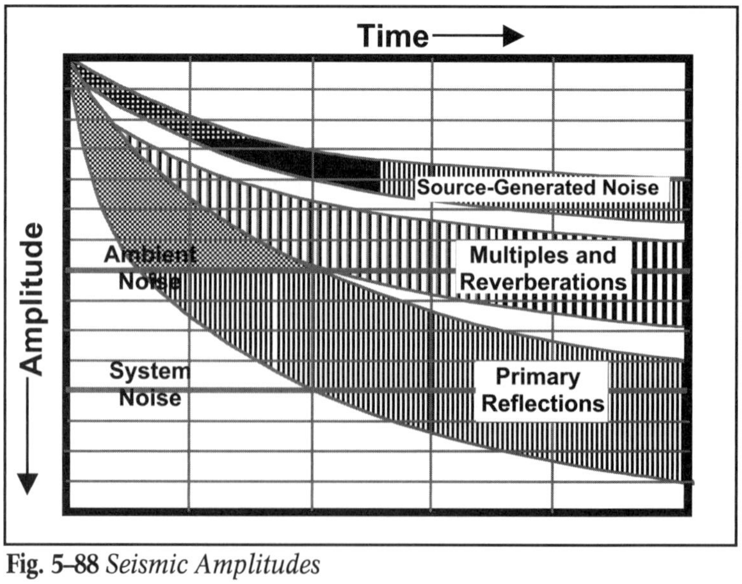

Fig. 5–88 *Seismic Amplitudes*

Modern seismic recording systems consist of two distinct parts—ground (land) or in-water (marine and OBC) systems and truck or onboard systems.

Ground and in-water system modules are located near detector groups. Major components are

- preamplifier (preamp)
- low-cut and notch filters
- sigma-delta modulator
- finite impulse response (FIR) module

Modules also include components for testing and status reporting. Figure 5–89 is a block diagram for a typical 24-bit module. The number of channels (*N* = number of geophone group inputs) varies considerably, ranging from one to eight. Each channel is identical up to the serial data bus.

Preamplifiers increase amplitude of seismic signal received from the cable by applying a constant gain. Usually, preamplifier gain can be selected as 0 dB (x1), 12 dB (x4), 24 dB (x16), 36 dB (x64), or 48 dB (x256). In some 3-D surveys, preamp gain may be set differently for different offset ranges to compensate for amplitude variation with distance. It is very important that preamplifiers operate with the least possible addition of noise and distortion.

The filters here are analog filters. Low-cut filters should be used with caution. Do not kill low-frequency noise with these filters if it damages data. Emphasis today is on preservation of signal rather than attenuation of noise. Notch filters can be used to attenuate highline noise but see the following discussion first. Note the absence of anti-alias filters. In 24-bit systems, they are not needed at this point because data are sampled at rates far above the seismic frequency range.

To minimize phase and amplitude distortion, use wide data band pass, avoid steep filter slopes, use notch filters only as last resort, and do not change recording filter settings once survey has begun, although playback filters can be changed as required. Amplitude distortion can only be partially corrected by seismic processing, but some signal amplitude information lost due to the S/N ratio of recorded seismic data. Phase distortion can theoretically be removed in seismic processing, but this requires the impulse response of the recording system, including filters used in data acquisition. Practical limitations are imposed by the non-linear phase response at filter limits and steep cut-off filter slopes found in notch filters.

The sigma-delta modulator is a single-bit, analog-to-digital converter. That is, one bit—0 or 1—is output each sample period, which is on the order of 4 nsec (1 nsec = 10^{-9} seconds). The output stream of bits goes to the FIR where it goes through two stages of averaging and application of digital anti-alias filters appropriate to the selected output sample period (0.5, 1, 2, or 4 ms). The averaging process results in the final value being output as a 24-bit binary number.

Marine in-water recording systems. Marine recording systems have two parts—the in-water system and the onboard system. The in-water system is consists of a set of streamers that are towed behind the vessel. Figure 5–90 shows a typical streamer configuration. Cable reels on the stern of the vessel are used to hold the streamers when not deployed. Electrical connection between the streamer electronics and the onboard system are made at these reels.

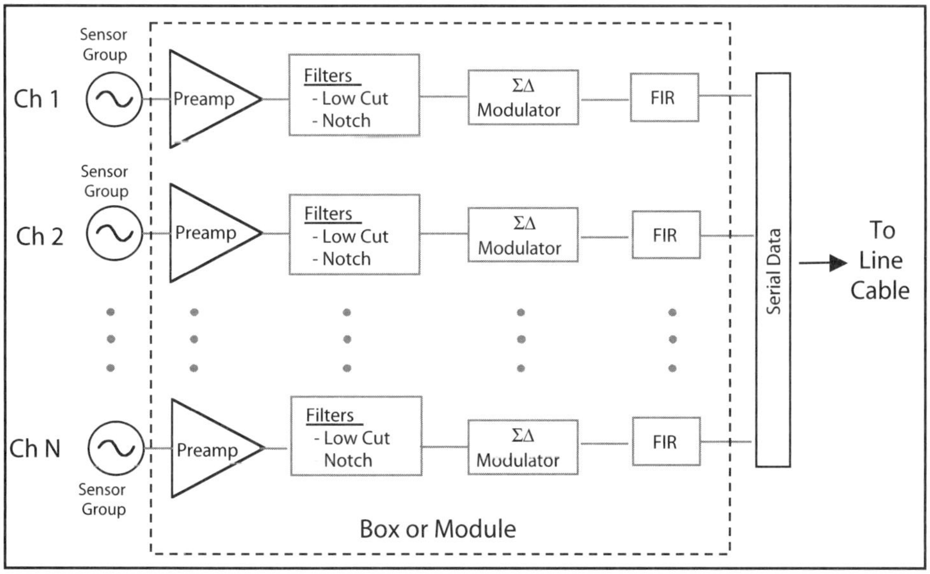

Fig. 5–89 *24-Bit Electronics*

The physical connection between the vessel and the streamer proper is via the *towing bridle* and the *lead-in section.* Streamers are made in sections of various types. *Stretch sections* must be deployed at the front and tail ends of streamers being towed from lateral separation devices (diverters) in order to attenuate vibration attributable to the motion of the separation device and from tail buoy jerks. The depressor *paravane* acts as a vertical diverter to place the streamer at the required depth. Depth sensors and depth controllers are also put at various points

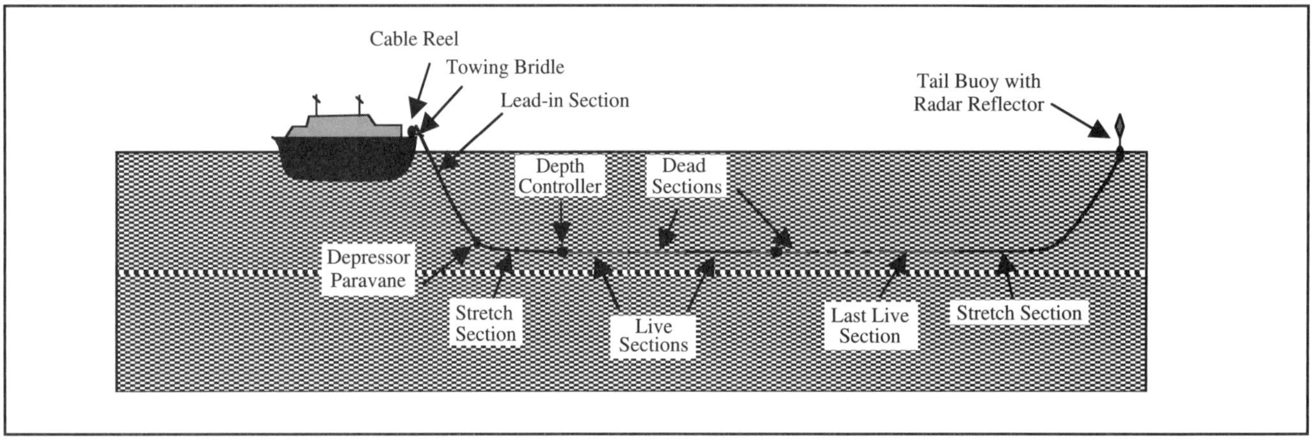

Fig. 5–90 *Streamer Configuration*

in the streamer to maintain the streamer at a constant depth throughout. The balance of the streamer is composed of *live* sections and *dead* sections.

Live sections contain hydrophone groups and electronic components that apply filters to the data, convert the analog outputs of the hydrophone groups into digital signals, perform instrument tests, and monitor various aspects of the electronics modules' operation. Dead sections act as spacers. All sections from the last live section to the lead-in section have either twisted pair copper wires or optical fiber cables to carry data from the hydrophone groups to the onboard system and receive commands from the onboard system. A tail buoy with a radar reflector is placed at the end of each streamer. Devices used by the positioning system are also in or on the streamers.

Live sections vary in length (Table 5–6). The group interval should be sufficiently small to ensure that data are adequately sampled in the common shot domain—that the data are not spatially aliased. Most systems allow selection of group intervals that are subintervals of section length. The group length, or length of the hydrophone array, should be as small as practical in order to minimize attenuation of high-frequency components of steeply dipping in-line reflection events. Ideally, the source array and receiver array lengths should be comparable.

Table 5–6 *In-water Systems Comparison*

Parameter	I/O System Two MSX	Syntron SYNTRAK 480–24 MSTS	Hydroscience CenturyMux
Section Length	100 m	75 m	100 m
Streamer OD	6.35 cm	5.6 or 6.2 cm	5.1 cm
Groups/Section	8	12, 6, 3	32, 16, 8, 4
Group Spacing	12.5 m	6.25, 12.5 or 25 m	3.125, 6.25, 12.5 or 25 m
Channels/Module	16	12	Distributed electronics
Max Length	12,000 m	7200 m expandable to 10,000 m	7900 m
Max Ch/Streamer	704 at 1 ms	480 at 1 ms	640 at 1 ms
Hydrophones/Group	14	16	16

Streamer depth should be consistent with project bandwidth requirements in terms of the frequencies at which ghost notches occur. For shallow depths (< 8 m), noise specifications need to be relaxed due to the increased sea noise near the surface.

Table 5–6 lists significant parameter values for I/O System Two MSX, Syntron SYNTRAK 480-24 MSTS, and Hydroscience CenturyMux in-water systems.

There have been many improvements in marine systems in recent years. Among these are an increase in number of channels per streamer, ability to tow and record from more streamers, reduction in streamer noise, lower instrument noise, less distortion, and improved dynamic range. Much of the improvement stems from conversion to 24-bit systems from 16-bit systems. Figure 5–91 compares Syntron 16-bit with 24-bit steamers and electronics.

Land ground systems. Land ground systems have similar components to marine in-water systems. Major components are

- electronics modules or boxes
- sectionalized line cables
- line interface modules
- jumper cables

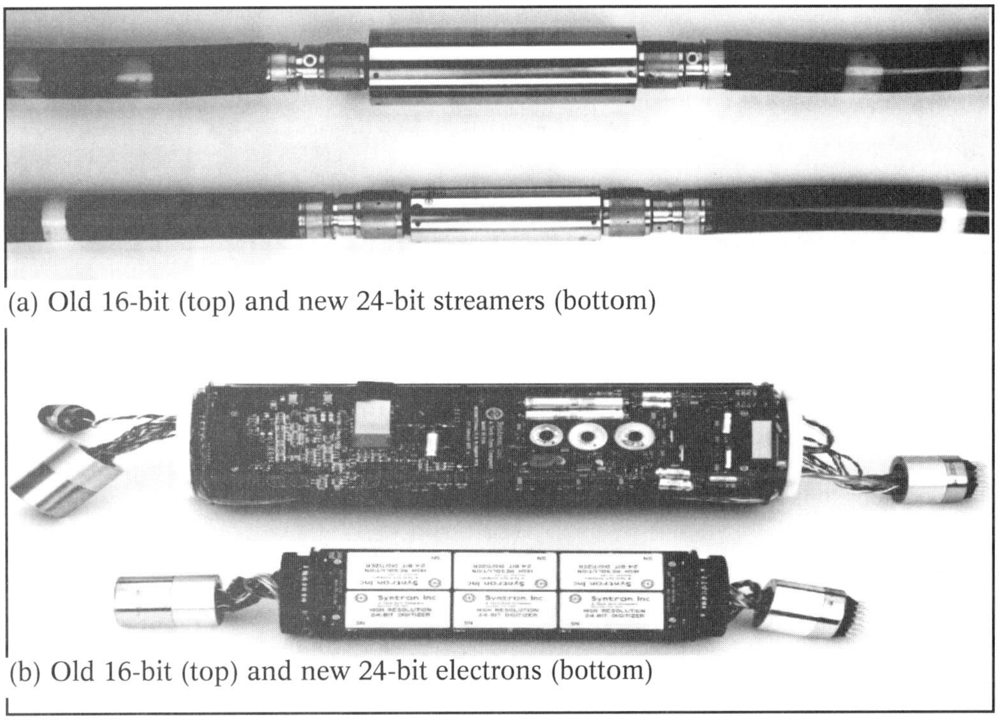

(a) Old 16-bit (top) and new 24-bit streamers (bottom)

(b) Old 16-bit (top) and new 24-bit electrons (bottom)

Fig. 5–91 *Comparison between 16-Bit and 24-Bit Systems*

Every manufacturer has its own special names for these components, but all systems have the same basic configuration. Figure 5–92 shows the typical ground system configuration.

The boxes receive analog inputs from geophone groups. There are usually six or eight geophone groups, but Sercel has only one geophone group input per box. The boxes then convert these to digital signals and transmit these as data channels through the line cables, line interface module, and jumper cable to the truck-mounted system. Boxes also receive digital data from other boxes downstream from them and retransmit them at required voltage levels. They receive and process commands for configuration, testing, and status (conditions within the box); and transmit results of tests and status inquiries back to the truck-mounted system. For systems that use fiber optic cables to transmit data, boxes must also have optical-electric converters since data are transmitted as light but processed in the box as electrical signals.

Land sectionalized cables can be connected to boxes at each end. There will be as many connectors, called *takeouts*, for geophone groups as there are channels in the boxes. In the case of six- or eight-channel systems, half of the geophone inputs go to the box at one end of the cable and half to the other.

Line interface modules receive all geophone group inputs from a single receiver line, order them, and transmit them to the truck-mounted system. They also may receive and retransmit data from other line interface modules.

Jumper cables provide connections between line interface modules and from a line interface module to the truck-mounted system,

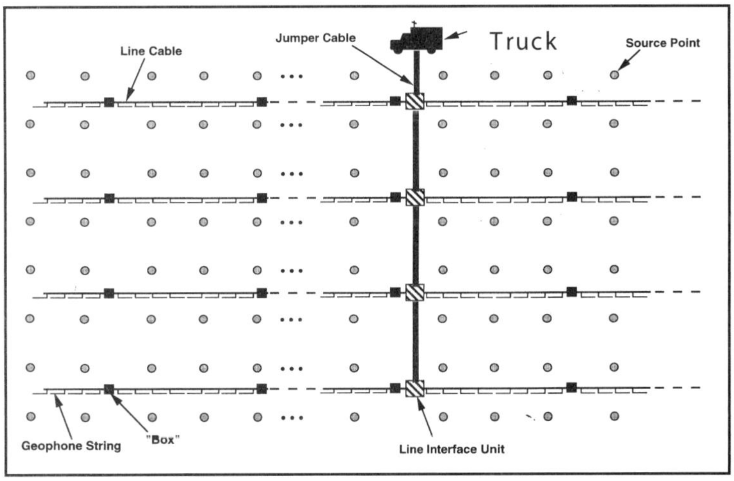

Fig. 5–92 *Land Ground System Configuration*

Onboard or truck-mounted systems. Major components include

- system control unit
- line/streamer interface modules
- correlator/stacker module
- operator console modules
- tape transports
- printers and/or cameras

Figure 5–93 shows these components as they appear in the recording on WesternGeco's marine vessel, the Western Monarch.

Fig. 5–93 *Recording Room on the Western Monarch*

In marine operations, the system control unit interfaces with the navigation system and the airgun controller to assure that recording begins at the proper location and that this is coordinated with airgun array firing. The system control unit also controls the flow of data from the streamer interface modules and to the tape transport units as well as the functions of peripheral devices.

Streamer interface modules function much the same way as the line interface modules in land ground systems but are located onboard the vessel rather than in the streamers.

The correlator/stacker module is optional land equipment used with vibrators. It is specially designed to perform cross-correlation between the filtered pilot sweep and the raw vibrator records and to do vertical stacking (including the diversity stack option).

The system operators use the operator console modules to configure and monitor the recording system. Tape transports, along with tape transport control units, format the data received from the system control unit and record it on magnetic tape. Printers and/or cameras are used to produce monitor records—hard copy or visual versions of data recorded on tape—status reports and other necessary records.

Some definitions of terms are required before continuing the discussion of recording.

Dynamic range is an amplitude ratio. It has the general definition of the ratio of the largest signal that can be recorded to rms instrument noise or the largest recoverable signal divided by the smallest recoverable signal.

Total dynamic range is the maximum system gain divided by the system noise level.

Instantaneous dynamic range is the ratio of the smallest signal that will cause a change in the presence of a larger signal. Instantaneous dynamic range is a function of the number of bits in either the A/D converter or the mantissa of the seismic word. (A 14-bit converter gives an instantaneous dynamic range of $20\log 2^{14}$ or 84 db). It controls amplitude capabilities of system's digital recording of seismic data.

All of these ranges are usually expressed as dB above system noise level.

Harmonic distortion is the ratio of the sum of the recorded amplitudes of second and higher-order harmonics to the amplitude of the input fundamental, which is usually stated in dB.

Cross-feed is the ratio of the amplitudes of electromagnetically induced signal in one channel to the amplitude of that input on another channel.

These parameters are used to evaluate recording system performance and capability.

Recorded data includes noise and signal superimposed on each other. Seismic processing may enhance data if the seismic signal is recorded within the precision of the A/D converter (instantaneous dynamic range - IDR). Seismic signals may have amplitude range of 120 dB or more when considering the amplitudes of direct arrivals and ground roll versus those of deep, faint reflections.

All recording systems have definable phase and amplitude (impulse) response for each component. The recording system fidelity is less than perfect because of amplitude, phase, and bandwidth distortion combined with the corruption of seismic wavelet with system noise. The recoverable dynamic range of seismic data is restricted by noise introduced by the recording system. Seismic recording systems have limited dynamic range. Signal and noise are superposed over the desired bandwidth, and a distorted wavelet is obtained because each component of the recording system tends to operate on or filter the seismic energy.

A magnetic tape is a strip of plastic coated with iron oxide particles. In magnetic tape recording, data are retained by magnetization of the iron oxide particles. In analog magnetic tape recording, data are represented by varying magnetization intensity along each *track*, with one track per *channel.* In digital magnetic tape recording, the direction of magnetization at small, discrete areas of tape indicates 0 or 1 values. There is no connection between channel number and track number.

Digital magnetic tape recording and related operations make use of number systems other than decimal. Data are actually recorded on tape using *binary* numbers. Some use has been made of *octal* numbers, and *hexadecimal* numbers are frequently used in tape dumps as well as in other applications of digital data. The top part of Table 5–7 shows the *radix*, or base, of these number systems and the digits used by each. Decimal numbers are included for reference. Note that the hexadecimal system requires six more digits than the decimal system has, so capitals *A* through *F* are used. The bottom part of Table 5–7 shows how to count from zero to decimal 15 in all four number systems. Note that three binary digits or bits are required to count from zero to decimal seven. This means that a single octal digit can represent a set of three bits. Four bits are required to count from zero to decimal 15. In hexadecimal, only one digit is required. Thus, a single hexadecimal digit is often used to represent sets of four bits.

Table 5–7 *Number Systems*

Name		**Radix**		**Digits Used**			
Binary		2		0.1			
Octal		8		0, 1.2, 3, 4, 5, 6, 7			
Decimal		10		0, 1.2, 3, 4, 5, 6, 7, 8, 9			
Hexadecimal		16		0, 1.2, 3, 4, 5, 6, 7, 8, 9, A, B, C, D, E, F			
Hex.	**Dec.**	**Oct.**	**Binary**	**Hex.**	**Dec.**	**Oct.**	**Binary**
0	0	0	000	8	8	10	1000
1	1	1	001	9	9	11	1001
2	2	2	010	A	10	12	1010
3	3	3	011	B	11	13	1011
4	4	4	100	C	12	14	1100
5	5	5	101	D	13	15	1101
6	6	6	110	E	14	16	1110
7	7	7	111	F	15	17	1111

All the four of these number systems use positional representation. That is, the position of a digit in the number indicates its magnitude with regard to power of the radix. The following illustrates this.

Hexadecimal - $92D_{16} = 9 \times 16^2 + 2 \times 16^1 + D \times 16^0 = 2304 + 32 + 13 = 2349_{10}$

Decimal - $2349_{10} = 2 \times 10^3 + 3 \times 10^2 + 4 \times 10^1 + 9 \times 10^0 = 2000 + 300 + 40 + 9$

Octal - $4455_8 = 4 \times 8^3 + 4 \times 8^2 + 5 \times 8^1 + 5 \times 8^0 = 2048 + 256 + 40 + 5 = 2349_{10}$

Binary - $100100101101_2 = 1 \times 2^{11} + 0 \times 2^{10} + 0 \times 2^9 + 1 \times 2^8 + 0 \times 2^7 + 0 \times 2^6$

$+ 1 \times 2^5 + 0 \times 2^4 + 1 \times 2^3 + 1 \times 2^2 + 0 \times 2^1 + 1 \times 2^0$

$= 2048 + 256 + 32 + 8 + 4 + 1 = 2349_{10}$

This example shows how to write the same number in four different number systems.

The advantage of the binary system is that its two digits—0 and 1 can be represented by systems with only two states—off-on, up-down. In magnetic tape recording, the iron oxide particles can be magnetized in either of two directions, so one can represent 0 and the other 1.

Computers can be programmed to do many things but these many things are done with only a few operations. All logic operations are simply combinations of three basic logic operations—or, and, and not. (Not is also called *complementation*). These operations are summarized in the following example.

0 OR 0 = 0	0 AND 0 = 0	NOT 0 = 1
0 OR 1 = 1	0 AND 1 = 0	NOT 1 = 0
1 OR 0 = 1	1 AND 0 = 0	
1 OR 1 = 1	1 AND 1 = 1	

Note that there are no + and – signs in computers, so the first (from left to right) bit of binary numbers is the *sign bit,* with a 0 indicating a positive number and a 1 indicating a negative number. The second bit is the *most significant bit (MSB)* and the last bit is the *least significant bit (LSB)*.

Two methods of complementation are used in binary arithmetic—*one's complement and two's complement*. These are illustrated using the binary equivalent of decimal 872. The one's complement is formed by changing all 0s to 1s and vice versa. The two's complement is formed by adding a 1 to the least significant.

$+872_{10}$ = 000001101101000

-872_{10} = 111110010010111 = one's complement

$\underline{\qquad\qquad\qquad\qquad +1}$

= 111110010011000 = two's complement.

The only arithmetic operation a computer can perform is addition! Multiplication is done by repeated addition. Division is done by repeated subtraction. To subtract, add the *complement* of the number to be subtracted.

The following shows how a computer subtracts decimal 872 from decimal 872, using the one's complement.

872_{10} = 000001101101000

$\underline{-872_{10}}$ =+$\underline{111110010010111}$

= 0 = 111111111111111

The result is all 1s! This is sometimes called negative zero. Intuitively, this does not seem to be a correct answer. However, it works.

The following shows how a computer subtracts decimal 872 from decimal 872, using the two's complement.

872_{10} = 000001101101000

$\underline{-872_{10}}$ = +$\underline{111110010011000}$

= 0 = $\underline{1}$|000000000000000

In the above operation, a 16th bit is created by the carry of one in adding the two numbers. Since in this example, only 15 bits can be retained in the register, the 16th bit overflows and only the 15 zeros are present. This is a result with which we are, intuitively, more comfortable. Either complement yields valid results but it must always be stipulated which is being used.

Write heads in the tape transport unit do the writing of data on magnetic tape. *Write heads* are, basically, electric coils maintained in fixed position a very small distance above the magnetic tape. The magnetic tape passes under the write heads and current in the write heads causes the iron oxide particles to be magnetized in the up or down direction depending on direction of current through the head. As the magnetic tape moves, tracks of data are recorded along the length of the tape. For many years, the standard for magnetic tape recording was nine tracks. Now, magnetic tape cartridges commonly have 36 tracks.

A variety of methods have been used to encode data on magnetic tape. One of the earliest methods was called *non-return-to-zero* (NRZI). This is illustrated in Figure 5–94. Note that the data are being recorded along a single track (along the length of the tape, not across the tape). When a one is recorded, the direction of current in the write head changes and so does the direction of magnetization on tape. When a zero is recorded, the direction of current in the write head does not change and neither does the direction of magnetization on tape. When the tape is being read, changes of direction in tape magnetization cause changes in direction of current in the *read heads.*

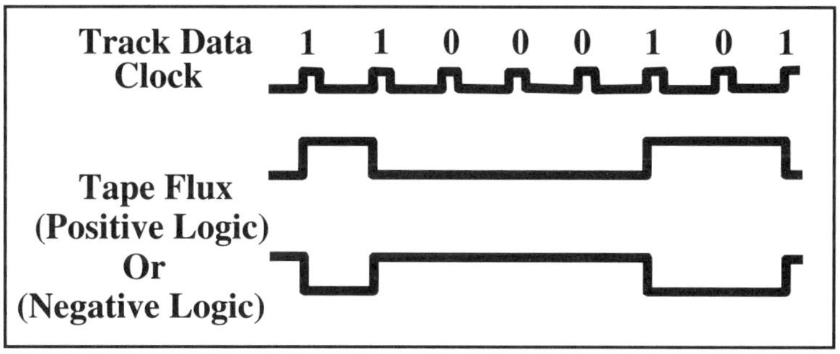

Fig. 5–94 *Non-return-to-zero Encoding*

A more recent and more often used method of encoding data on magnetic tape is called *group coded recording* (GCR). In GCR, data are initially encoded in NRZI with resynchronization bursts at 1600-bit intervals. Data are recorded as follows:

- track data are split into groups of seven bits—DDDDDDD
- A horizontal parity *E* is added to form an 8-bit group—DDDDDDDE
- the 8-bit group is split into two 4-bit subgroups—DDDD DDDE
- the 4-bit subgroup is translated into a standard 5-bit code

Translation table examples:

1000 → 11010

1001 → 01001

1111 → 01111

0001 → 11001

Table translation is a table look-up procedure, not application of an algorithm or formula.

- Two 5–bit subgroups are combined to form a 10-bit storage group.
- Then, 158 storage groups are written in NRZI followed by an all ones resynchronization group.

There are some tape format terms that need to be defined. Figure 5–95 provides a visual aid in understanding these terms. Near the beginning of a magnetic tape there is a physical, magnetic marker, called the *beginning of tape* (B.O.T.). The tape transport must sense the B.O.T. before recording can begin. At the end of the tape is a similar physical, magnetic marker, called the *end of tape* (E.O.T.). The tape transport looks ahead for the E.O.T. and, when it is detected, determines whether there is enough room on the tape to continue recording. If not, the tape is stopped, and the operator is informed that a tape change is required. A *magnetic tape record* is quite different from a seismic record. A magnetic tape record is a volume of data separated by an internal record gap (I.R.G.). A *file* is the largest division on tape, and is separated from other files by an end-of-file (E.O.F.) code that is all ones.

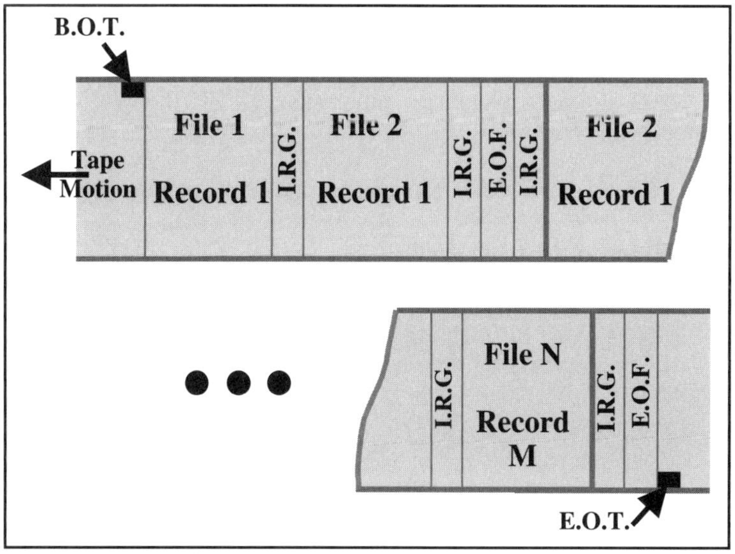

Fig. 5–95 *Tape Schematic*

A *byte* has eight data bits plus a parity bit. The byte is the basic index of tape format. A *parity* bit (P) is used for quality control in magnetic tape recording. One bit in a group is used to assure that the number of logical ones written on the tape is odd (*odd parity*) or even (*even parity*). Odd parity, illustrated here, is most used in seismic exploration.

Data Value: 0100110 (Three 1s)

Parity Bit Value: 0

S MSP LSP P

Value Written to Tape: 0 1 0 0 1 1 0 0 0 (Three 1s)

Data Value: 0101 1001 (Four 1s)

Parity Bit Value: 1

S MSP LSP P

Value Written to Tape: 0 1 0 1 1 10 1 1 (Five 1s)

Here, *S* identifies sign bit, *MSB* is most significant bit, *LSB* is least significant bit and *P* is parity bit.

A *word* is, essentially, a computer term. It is important when interrogating tape by dumping. Dumps are indexed by words. Word size varies from computer to computer.

Multiplexed data are written on tape in sample order. That is, all samples obtained at the same time are written together in data channel order. *De-multiplexed data* are written in channel order. That is, all samples for the same channel are written together, ordered by sample times. See Figure 5–96.

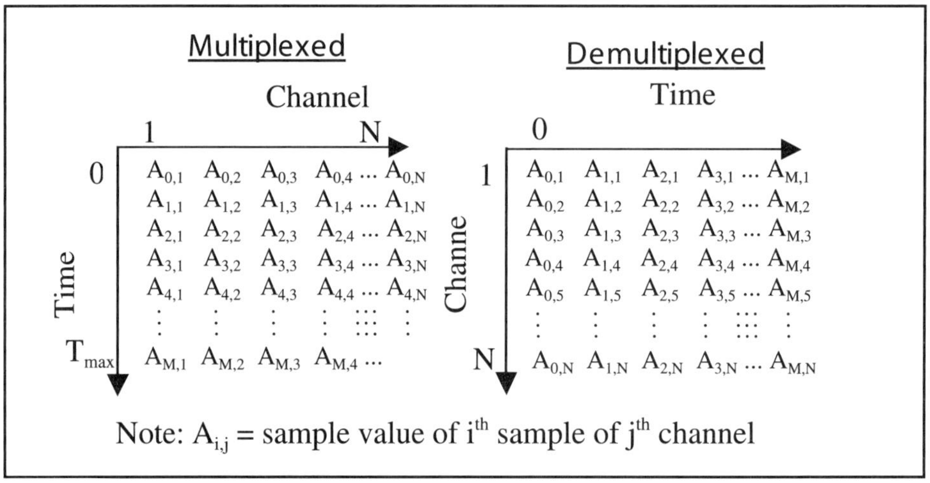

Fig. 5–96 *Multiplexed and De-multiplexed Data*

At the beginning of digital recording in the early- to mid-1960s, a great variety of tape formats were in use by contractors and clients. To eliminate the chaos, the Society of Exploration Geophysicists (SEG) adopted standard formats. These were:

1967 — *SEG A* (field data, multiplexed, NRZI), *SEG B* (field data, multiplexed, NRZI), and *SEG X* (data exchange, de-multiplexed, NRZI)

1972 — *SEG C* (field data, multiplexed, NRZI or PE) introduced to accommodate IFP recorders

1975 — *SEG Y* (de multiplexed, NRZI or Phase Encoded) introduced as new data exchange format to accommodate computer field equipment and newer processing hardware

1980 — *SEG D* (multi-purpose, multiplexed or de-multiplexed, details in header) introduced to accommodate further advances in data acquisition and processing. SEG D was revised in 1994 to accommodate other developments including 24-bit recording

Details on tape formats are found in Appendix B.

2-D Acquisition Parameters and Operations

Figure 5–97 shows typical geometry for a 2-D seismic survey. Survey parameters include

- lines
- number, orientation, and spacing
- offsets
- maximum (x_{max}) and minimum (x_{min})
- receiver parameters
- number of receiver groups, group interval, type, and arrays
- source parameters
- type, configuration, number of units/holes, etc.
- fold

The number of lines depends on spacing and areal extent of the target zone. Spacing shown in Figure 5–97 is fairly common. In English units, spacing usually ranges from 0.5 miles to 1 mile for lines in the direction of regional dip and from 1 to 2 miles perpendicular to regional dip. As implied in the preceding sentences, orientation of lines is usually in the direction of regional dip and perpendicular to it. In some cases, fault orientation may determine line orientation.

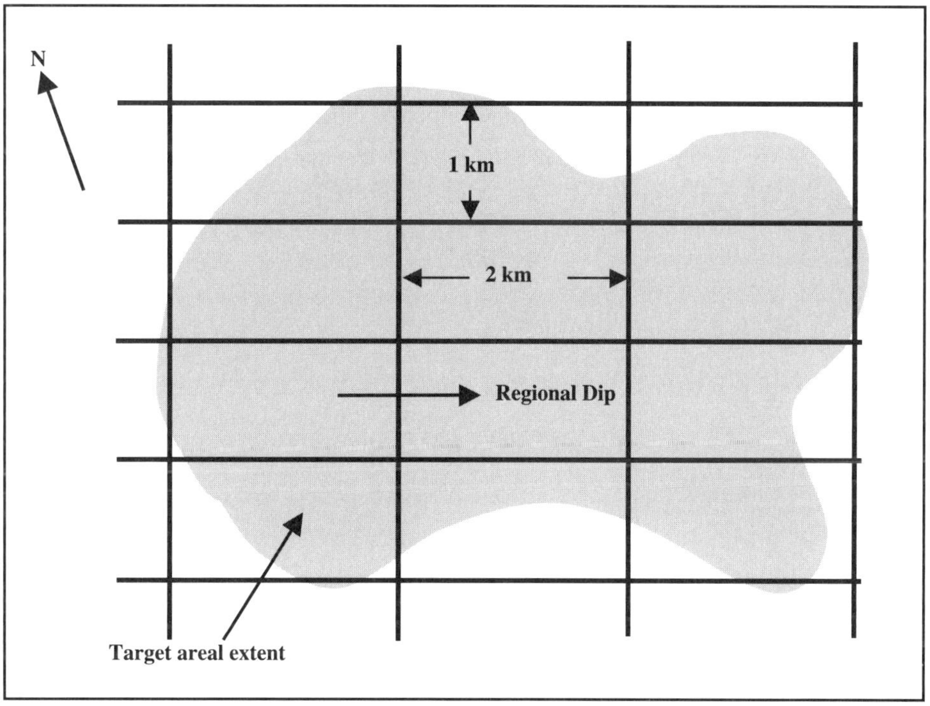

Fig. 5–97 *Typical 2-D Geometry*

The minimum offset should be less than depth to shallowest layer of interest (for geological, data processing or data interpretation reasons) but far enough from the source to avoid interference of signal with source-generated noise and to prevent *clipping* of near-channel data. Clipping occurs when amplitudes exceed the maximum the recording system can handle.

The maximum offset should be compatible with the depth to deepest horizon of interest. Normally, this means less than or equal to the depth to the main target.

The maximum offset should be long enough to give adequate differential move out for velocity determination and multiple attenuation but short enough that wavelet stretching due to NMO corrections not be excessive at target times—25 to 30% for normal processing. See Figure 5–98.

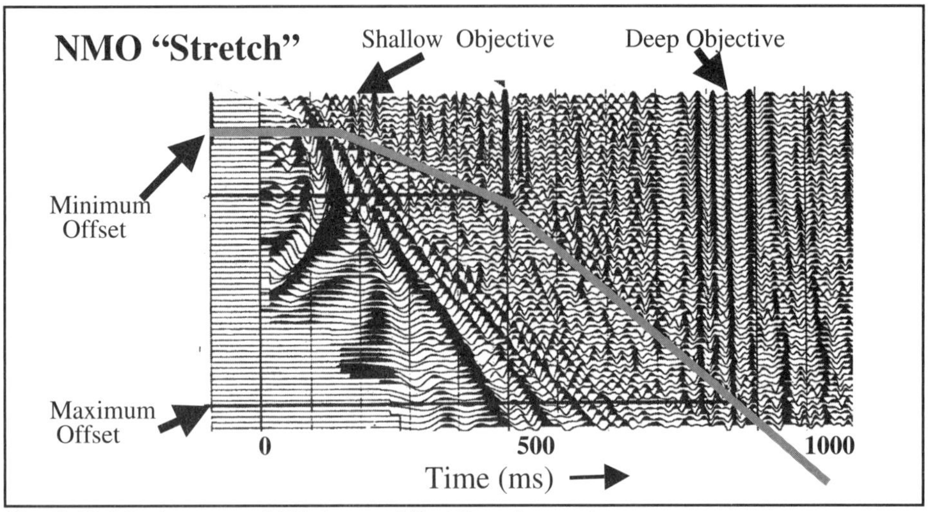

Fig. 5–98 *Maximum and Minimum Offset Requirements*

The maximum offset should be compatible with fold requirements at shallow and deep horizons of interest.

$$F_1 = F\left(\frac{V_1T_1}{2x_{max}}\right) \qquad (5.16)$$

where

F = fold at deepest target

F_1= fold at shallow target at depth =

x_{max} = maximum offset = depth to deepest target

V_1T_1 = rms velocity and zero-offset reflection time to the shallow target

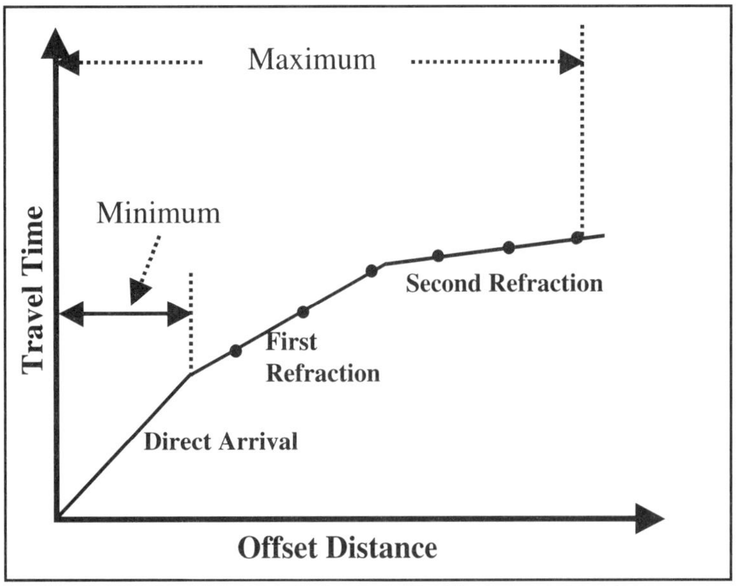

Fig. 5–99 *Maximum and Minimum Offset Requirements for Refracted Arrivals*

The maximum offset should also be large enough to allow at least three observations of the deepest refraction arrival. The minimum offset should be small enough to allow at least three observations of the shallowest refraction arrival (Fig. 5–99).

The receiver group interval should be adequate for the number of groups required to get the desired fold but large enough for desired receiver array. The receiver group interval must be large enough to prevent spatial aliasing.

If the maximum dip angle expected for the target is ϕ, and the highest frequency to meet resolution requirements is f_{max}, then the group interval dx is given by:

$$dx = \frac{Vdt}{4f_{max}\ \sin\phi} \tag{5.17}$$

where

V is the rms velocity at the target

dt is the maximum time dip per group interval

Since x_{max}, x_{min}, and dx are known, the number of receiver groups NG in the spread can be calculated as:

$$NG = \frac{(x_{max} - x_{min})}{dx} \tag{5.18}$$

If this number does not provide the required fold or adequate number of traces for velocity analysis, then a split spread may be required to increase the number of traces.

An off-end spread (Fig. 5–100) is preferred if it is compatible with the maximum offset and fold requirements when considering the group interval. In off-end spreads, all receivers in the spread are on one side of the source. If progression along the line in Figure 5–100 is to the right, then the source is said to be *pulling the spread*. If progression along the line is to the left, then the source is said to be *pushing the spread.*

Figure 5–101 shows a symmetric split spread—it is symmetric about the source point. The *gap* is the number of group intervals between the two near groups. The gap should always be odd to avoid duplicating offsets in CMP gathers. No receivers in the gap are connected to the recording instruments.

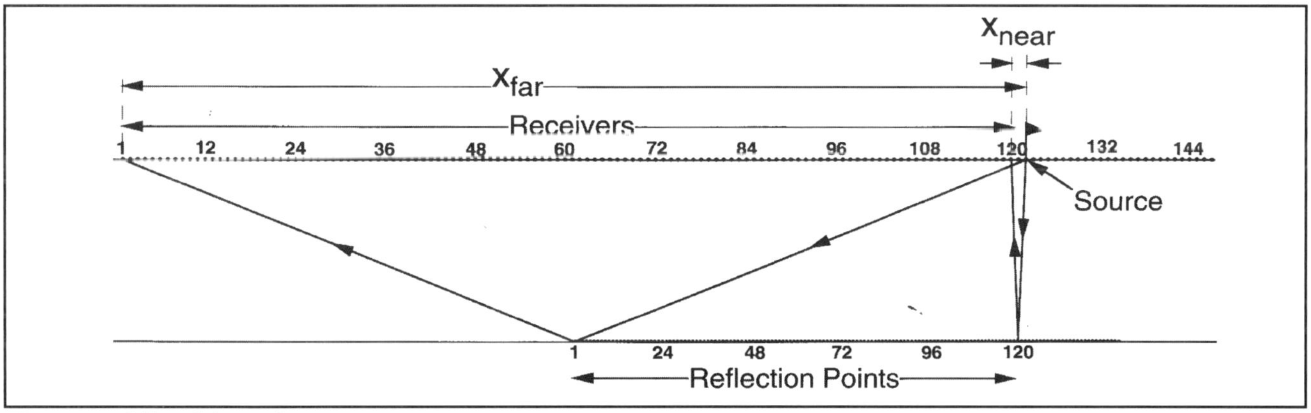

Fig. 5–100 *Off-end Spread*

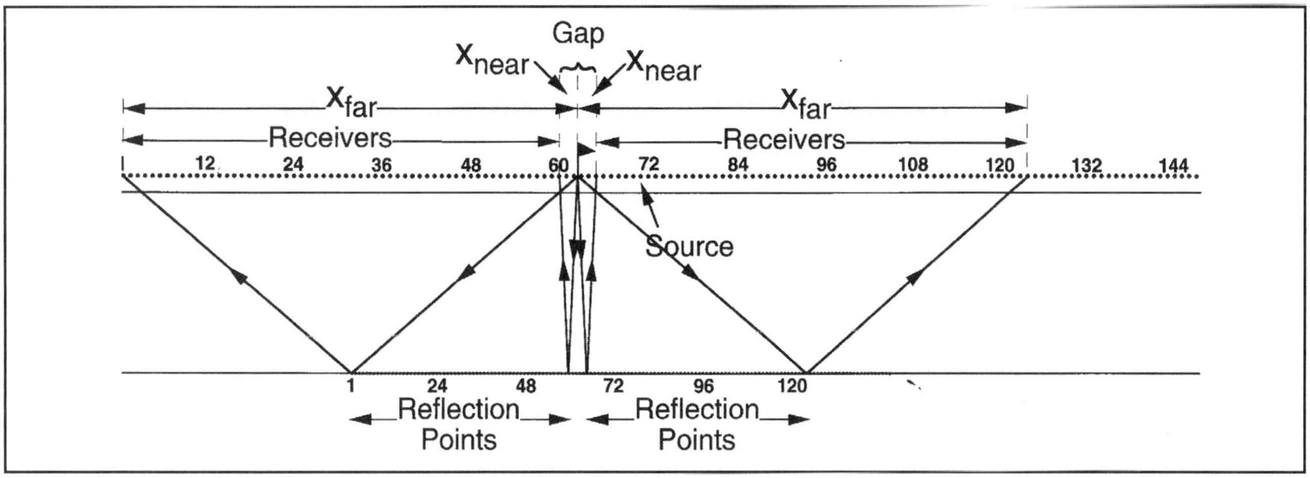

Fig. 5–101 *Symmetric Split Spread*

The asymmetric split spread is often used when there is significant dip. By placing fewer groups on the down-dip side, the angle of emergence is kept relatively small, to minimize signal attenuation by the source and receiver arrays.

A rule of thumb for array length L is

$$L = \frac{0.44}{k_{\text{pass}}} = \frac{0.44V_x}{f_{\max}} \tag{5.19}$$

where

k_{pass} is the wave number at which signal is attenuated 3 dB by the array

V_x is the apparent horizontal velocity at offset equal depth ($x = Z$)

$f_{\max}$ is the required high frequency at the depth of interest Z

The apparent horizontal velocity V_x can be calculated for flat dip from

$$V_x = \sqrt{5}V_{NMO} \tag{5.20}$$

where

V_{NMO} is the velocity required for NMO at the zone of interest

For dip at angle ϕ, V_x can be calculated from:

$$V_x = \frac{\sqrt{5 \pm 4\sin\phi}}{1 \pm 2\sin\phi} V_{\text{NMO}} \tag{5.21}$$

In Equation 5.21, use + for down-dip and – for up-dip.

The maximum wave number for the required signal k_{pass} is calculated from

$$K_{\text{pass}} = \frac{f_{\max}}{V_x} \tag{5.22}$$

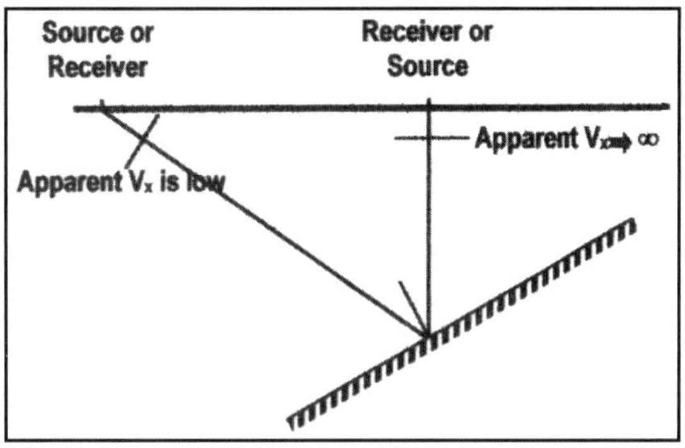

Fig. 5–102 *Up-dip or Down-dip?*

According to tradition, up-dip shooting yields higher frequency signals. This tradition is based on array theory for receivers and point sources (single-hole dynamite). Array theory says the longer array should be up-dip. Figure 5–102 shows that the down-dip part of the path will give the most signal attenuation because of time difference across the array. Since most surface sources (vibrators, land airguns) are deployed in arrays, the array with the longest effective length should be up-dip. Thus, traditional rules are broken by shooting down-dip from off-end spreads. Note that when shooting split spreads, one would need to process up-dip and down-dip traces separately, to avoid breaking the rule.

In marine operations, the source array is generally much larger than the hydrophone array. Thus, the boat shooting in the up-dip direction should result in less attenuation of high frequencies. This assumes good knowledge of geologic structure, which may not be the case but, in any event, always shooting in one direction would be very inefficient use of boat time.

The fold required depends on the signal-to-noise ratio (S/N), the velocity analysis requirements, and the multiple attenuation needed. The S/N improvement from N-fold CMP stack is $\sqrt{N}$, where N is the fold for ambient noise. Fold N is given by:

$$N = \frac{NG}{2GR} \tag{5.23}$$

where

NG is number of groups in spread

GR is group roll or number of group intervals the spread is moved between shots

It is desired to have *full-fold coverage* in the target zone so a minimum distance (number of shots) is required to obtain minimum fold. This minimum distance must be added to the line lengths to assure that full-fold coverage is obtained over the entire target zone.

Some deviations from normal procedures may be required in land surveys. This includes

- start-of-line variations
- source array variations (obstacle avoidance)
- recovery shots (corrections for shot points that cannot be occupied)

Figure 5–103 illustrates the normal or conventional method of ramping on to full-fold coverage. For a 24-group, 3-group gap, split spread and 12-fold shooting, the source and spread move up one group interval from shot-to-shot. Thus, each spread overlaps the previous spread by 22 groups. After 13 shots, 12-fold is obtained at a distance of 18.25 group intervals from the start of the line. A similar ramp-off occurs at the end of the line.

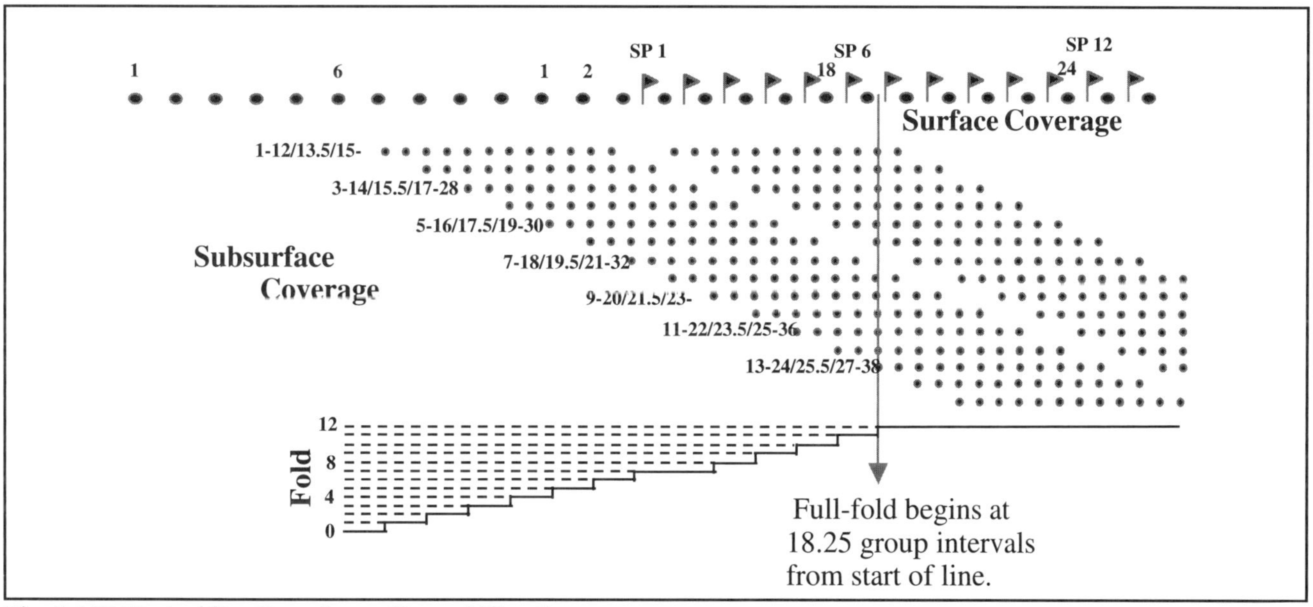

Fig. 5–103 *Start-of-line Procedures, Normal Shooting*

Figure 5–104 shows an alternative, faster method of achieving full-fold called *shooting through the spread.* Instead of beginning normally, the first shot point is placed between receiver positions 1 and 2 and shot into 12 groups at positions 3 through 14. The shot point is moved one group interval and shot into one additional receiver group each time. Full-fold begins at 6.25 group intervals from start-of-line, 12-group intervals before normal procedures. Similar procedures are followed at the end of the line, resulting in the total line length being reduced by 24-group intervals. Larger savings are made when longer spreads with many more receiver groups are used. Data processors must be fully aware of the different procedures so that correct traces are gathered for CMP stack.

Many times obstacles of various types that prevent using a normal source array are encountered. Figure 5–105 illustrates this and the methods that might be used to avoid the obstacle. In this example the obstacle is a river. *A* is the normal pattern, *N* holes in line, equal spacing, and equal charges. *B* is an acceptable method, the pattern is offset parallel to the line. *C* is an acceptable method, it is a normal pattern at an angle to the line. *D* is an acceptable method. The pattern is in line with the receiver line but with reduced spacing between holes. This could affect attenuation of shot-generated noise. *E* is an acceptable method. The pattern is in line with the receiver line and has normal spacing, but the reduced number of holes can affect signal

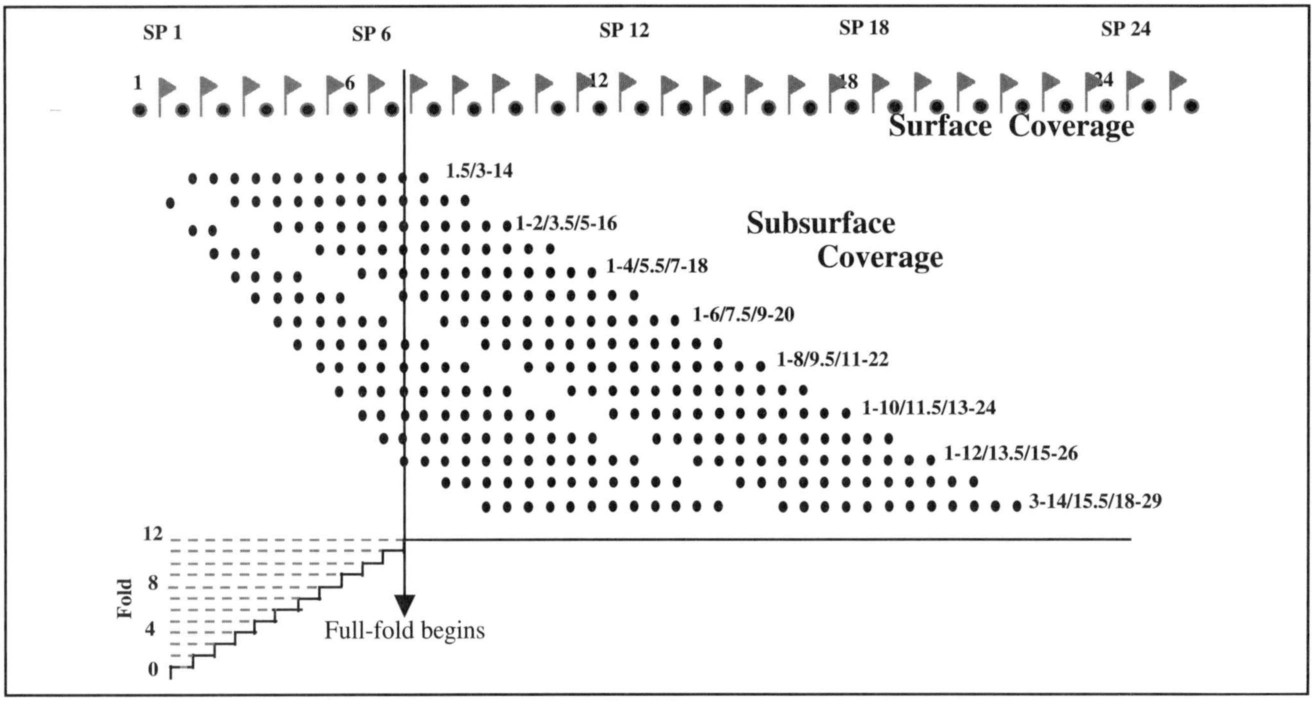

Fig. 5–104 *Start-of-line Procedures, Fast Ramp On to Full Fold*

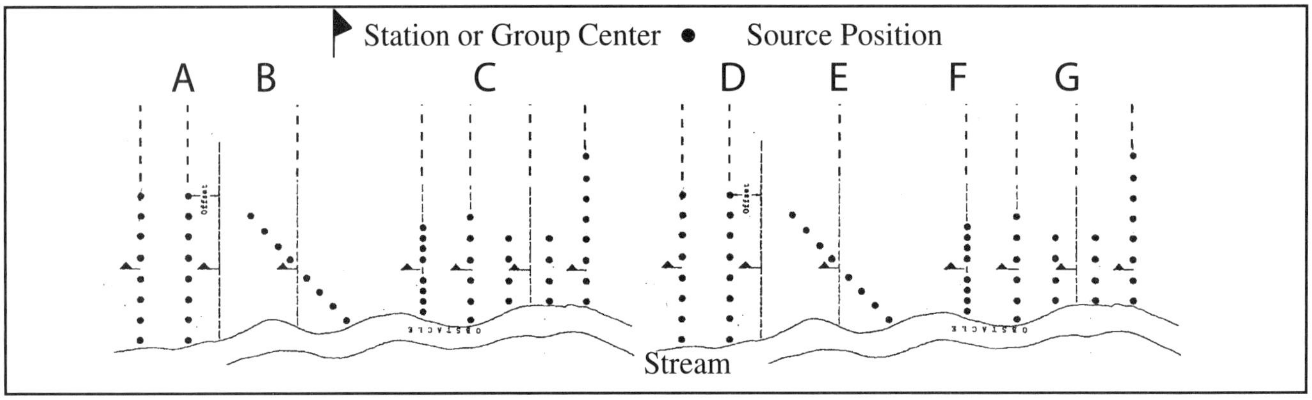

Fig. 5–105 *Obstacle Avoidance at the Source Array*

strength. *F* is an acceptable method; two parallel lines with normal spacing. This could, however, also affect attenuation of source-generated noise because of the reduced number of elements in the in-line direction. *G* is an unacceptable method, the pattern center is offset from station in the in-line direction. This means that all offsets are now changed because offsets are measured from the center of the source array.

Sometimes, in land acquisition, a source point cannot be occupied. It may fall in a body of water, or there may be a structure in the way, or any of a a variety of other reasons. In such cases, it is still desirable to obtain the subsurface coverage a shot at that position would have obtained. This is done by shooting recovery shots. Figure 5–106a shows a shot point that cannot be occupied and the corresponding subsurface coverage. The shot point is at position 155 and the spread occupies positions 130 through 153. Figure 5–106b shows one recovery shot. The shot point is moved from surface position 155 to 156 and the spread is moved in the opposite direction on group interval (positions 129 through 152). Figure 5–106c shows another recovery shot. The shot point is moved from surface position 155 to 157 and the spread is moved in the opposite direction one group interval (positions 128 through 151). Note that all offsets increase in both recovery shots.

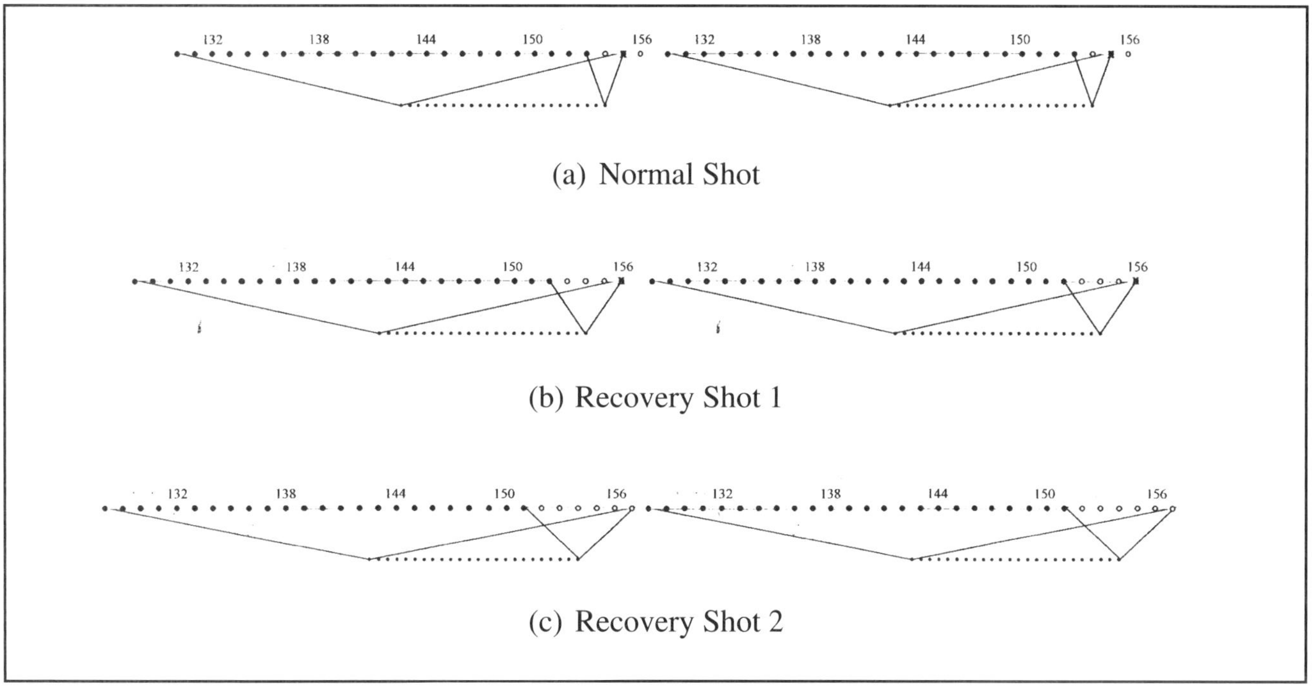

Fig. 5–106 *Recovery Shots*

3-D Acquisition Parameters and Operations

Figure 5–107 shows typical 3-D geometry. Note how closely spaced the receiver lines are. Also note that the source lines are closely spaced and orthogonal to the receiver lines. Source lines are not always orthogonal to receiver lines and even when they appear to be, they may not be shot that way. In 3-D, multiple receiver lines are shot from a single source so that subsurface reflection points are distributed over a surface instead of along a line. Along each of these lines, receiver groups are laid out in spreads similar to 2-D. Sources and spreads are moved along these lines until all sources are shot. Then some of the lines are moved laterally (rolled) such that there is overlap in surface and/or subsurface coverage. This continues until all sources have been shot and the entire survey area covered.

In 3-D *maximum fold* can be obtained with a lower *in-line* fold than in 2-D because, in general, fold is obtained in both the in-line and *cross-line* directions and total fold is the product of the two. For 3-D, the *in-line fold* may be calculated using the following formula.

$$\text{In-line fold} = \left(\frac{NC_i \times RI_i}{2SI_i} \right) \tag{5.24}$$

where

NC_i = number of receiver recording channels used in the in-line (receiver line) direction

RI_i = spacing between receivers in the in-line (receiver line) direction

SI_i = spacing between source lines in the in-line (receiver line) direction

Cross-line fold may be calculated using the following formula.

$$\text{Cross-line fold} = \left(\frac{NRL_x \times NS_x \times SI_x}{2R_x} \right) \tag{5.25}$$

where

NRL_X = number of receiver lines in the cross-line direction (perpendicular to the receiver line)

NS_x = number of sources in cross-line direction

SI_x = spacing between sources in the cross-line direction

R_x = distance rolled in the cross-line direction

Effective fold plots are used to illustrate poor offset distributions. An offset window is defined and any trace that is within the offset window of another trace is ignored in the fold computation. In order to understand the quality of the offset distribution, compare the effective fold with the full fold. If effective fold is ≥ 80% of full fold, offset distribution is probably good. Careful selection of the offset window is required. If the window is too small, the effective fold will be very similar to full fold. If the window is too large, the effective fold will be very small. Note that effective fold ignores the azimuths of the traces. It is useful for offset evaluations for the purposes of multiple attenuation, velocity analysis, and others. However, effective fold is not useful for understanding the imaging issues. Identical offsets with different azimuths contribute different information.

A regular offset distribution is preferable for many reasons. It gives the ability to determine a stable velocity field, provides structural stability after application of NMO, enhances coherent noise cancellation in CMP stack, and improves attenuation of ground roll and multiples. Regular offset distribution also provides consistency of reflection character for stratigraphic work, makes NMO stretch more uniform, and gives consistent response of residual static computations to different wavelengths of near surface velocity anomalies. Regular offset distribution is especially important in DMO application. DMO is discussed fully in the next chapter.

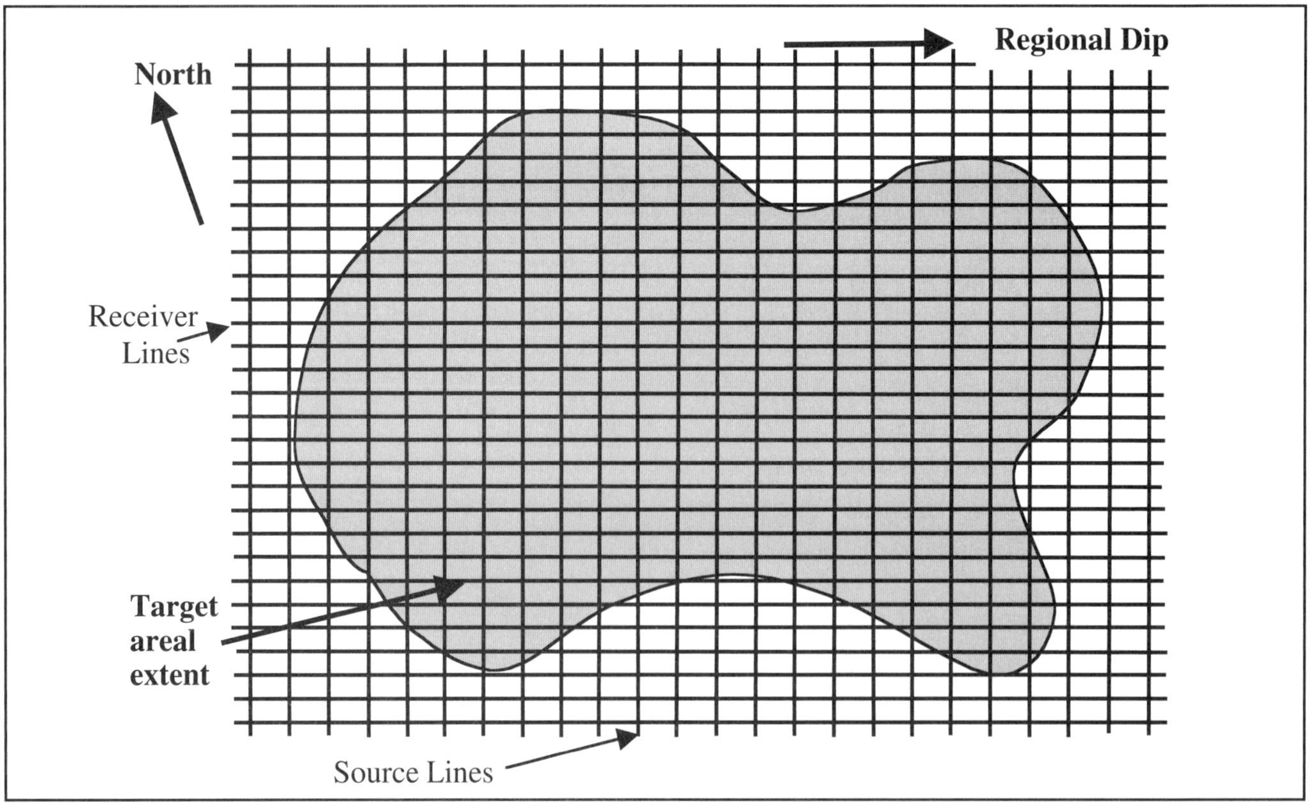

Fig. 5–107 *Typical 3-D Geometry*

Distribution of fold and offsets is determined by the requirements for both the resolution in the zone of interest and the data processing sequence. Small short offsets are needed to image shallow objectives. Short offsets are also needed for shallow velocity control, in refraction statics, and for near-surface depth modeling. Large far offsets are required to accurately measure the velocity field for use as stacking velocities, for migration, and for depth conversion. Consistent arrivals from far offsets are also needed for refraction modeling.

Azimuthal distribution is the most controversial of 3-D survey attributes. Differences of opinion regarding the benefit of wide versus narrow azimuths continue. The presence of anisotropy and fractures and the velocity variations related to them affect the desirability of particular azimuthal distributions. Surveys usually do not have uniform azimuthal sampling because of techniques adopted for cost savings, equipment limitations, access problems, etc.

A restricted range of azimuths combined with recording along in-line dip lines could significantly reduce processing difficulties, especially for DMO, velocity analysis, migration, and refraction analysis. Reflection azimuths also affect the amplitude and phase response of geophone arrays. In-line orientation is usually along the line of steepest dip in the target zone. The in-line subsurface sampling interval is designed to sample the steepest dips without spatial aliasing. Cross-line sampling may be coarser than in-line. This is usually selected to reduce acquisition costs. Interpolation to finer CMP sampling may be required to avoid dip aliasing in the cross-line direction.

Migration is a process that needs a large number of statistics. Hence, some minimum number of traces is required in order to get adequate constructive and destructive interference of wave fronts. Migration moves reflection energy to its correct location. This energy may have been recorded from reflectors that are outside of the area to be imaged and energy reflected within the area to be imaged may have to be recorded outside the area. Thus, correct imaging requires recording of data over an area on the surface that is larger than the subsurface target area. The additional area is known as the *migration aperture* or *halo*.

The migration process focuses blurred images. Scattered energy is gathered and focused in the correct locations. This requires a large spatial aperture. In determining the dimensions of the migration aperture, the following need to be considered:

- dip effects
- the Fresnel zone or minimum aperture
- diffraction energy

The migration aperture depends upon reflection time, velocity, and dip at the target. The area of the migration aperture should, theoretically, be recorded with full CMP fold.

Migration aperture defined by dip is the most commonly used method. When calculating the migration aperture from dip, one needs to consider both the constant velocity (straight ray-path) equations and the depth variable velocity (curved ray-path) equations. Straight ray-path equations have been used, traditionally, and work reasonably well for small to medium dips and shallower record times. Straight ray-path equations always give a larger aperture and, hence, provide a safety margin. The Fresnel aperture should, theoretically, be added to dip aperture calculations but this is not often done.

Figure 5–108 shows a dipping reflector and the zero-offset reflection ray-path from source *S* in a medium that has a constant velocity v. Since the reflector is dipping at an angle θ, the line *SP* (along which propagation occurs) also makes an angle θ with the vertical. The right triangle *SPP'* can be formed, and it follows from simple trigonometry that:

$$x_{\min} = \frac{vt}{2}\sin\theta \tag{5.26}$$

where t is the two-way, zero-offset time to the reflector from source *S*

Alternatively, the geologic depth *Z* and geologic dip θ can be used, as shown in Figure 5–109 and Equation 5.26.

$$x_{\min} = Z\tan\theta \tag{5.27}$$

Since $Z = \frac{vt}{2}\cos\theta$, Equations 5.26 and 5.27 are equivalent.

When velocity increases with depth, the ray path is curved. Figure 5–110 illustrates this for a situation where the velocity increases linearly with depth z—$V = V_0 + kz$. In this case, the migration aperture is given by:

$$x_{min} = \frac{V_0 T_0 \sin\theta}{2}\left[\sinh\left(\frac{kT_0}{2}\right)\right] \tag{5.28}$$

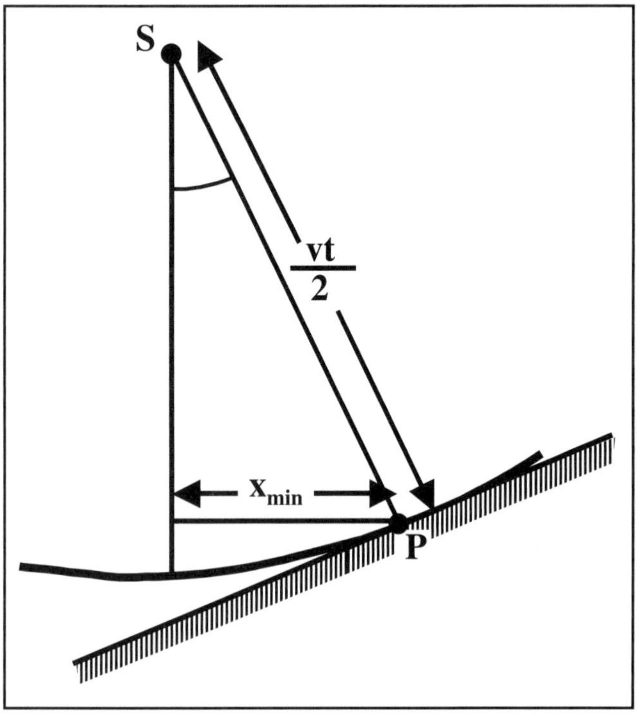

Fig. 5–108 *Migration Aperture from Dip for Constant Velocity (Straight Ray Paths)*

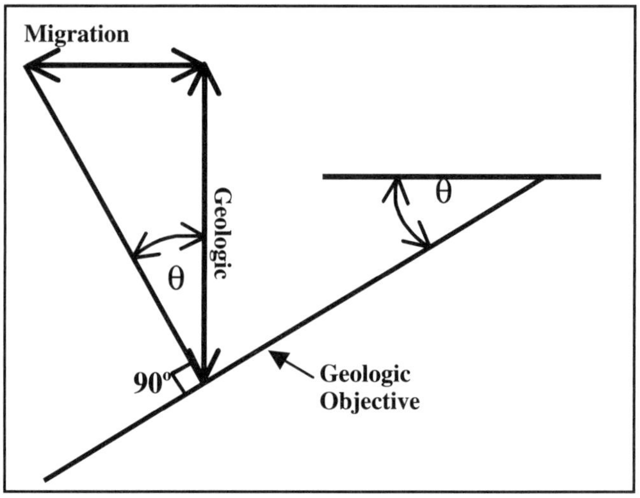

Fig. 5–109 *Migration Aperture from Dip for Constant Velocity (Straight Ray Paths) Using Geologic Depth*

where

T_0 is the two-way, zero-offset reflection time to the geologic objective

sinh is the hyperbolic sine

k is a constant of units sec^{-1}

The magnitude of the aperture calculated as above is always less than that determined by the straight ray approximation.

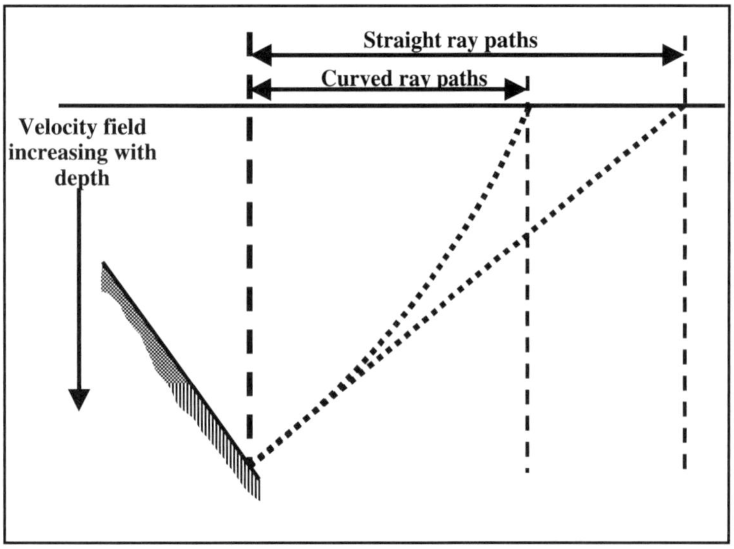

Fig. 5–110 *Migration Aperture from Curved and Straight Ray Paths*

Where there is no dip, the dip calculation implies no additional aperture is required. However, migration requires some minimum aperture to obtain constructive and destructive interference of wave fronts to attenuate diffractions. This minimum aperture is based upon Fresnel Zones. The half-aperture (one side) is defined in Figure 5–111. Here a spherical wave front from point *S* has reached the point discontinuity *P* at a time *t*. Since the velocity is *v*, the distance *SP* is $vt/2$. Point *P'* is at the intersection of a vertical line from *S* and a horizontal line from *P*. The distance from *P'* to the wavefront is one quarter of the wavelength, $\lambda/4$. The wave is assumed to be at the lowest signal frequency of interest f_{min}. Since $\lambda = V/f$, $\lambda/4 = v/4f_{min}$.

Noting that SPP' is a right triangle whose sides are x_{min}, $vt/2$, and $vt/2 - v/4f_{min}$, an expression for xmin can be derived. Applying the Pythagorean theorem:

$$x^2_{min} = \left(\frac{vt}{2}\right)^2 - \left(\frac{vt}{2} - \frac{v}{4f_{min}}\right)^2 = \left(\frac{v}{2}\right)^2 \left(\frac{t}{f} - \frac{1}{4f^2_{min}}\right)$$

$$\therefore x_{min} \approx \left(\frac{v}{2}\right) \sqrt{\frac{t}{f}} \quad (5.29)$$

The computed Fresnel aperture is added to the required image area. Small changes in the value selected for f_{min} can make significant differences in the Fresnel aperture.

When accurate delineation of faults is required, diffraction energy may be used to define sampling requirements. Diffraction energy should also be considered to define migration aperture. The same equations used to define sampling can be used for migration aperture.

The Fresnel aperture should handle diffraction energy up to about 15° of apparent dip. However, as shown in Figure 5–112, getting 95% of diffraction energy requires recording out to apparent dips of 30°. The larger the amount of diffraction energy recorded, the more successfully will migration collapse it.

The migration aperture exists in time as well as space so there is a need to consider recording length as well as spatial aperture. Diffractions can extend to very large record times, particularly in low-velocity media. One should look for evidence of diffraction energy on un-migrated time sections.

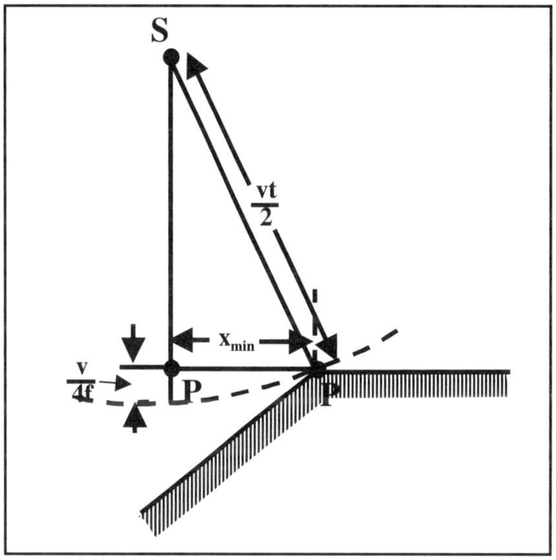

Fig. 5–111 *Migration Aperture from Fresnel Zone*

From Figure 5–112, it follows that if the deepest diffraction to be collapsed is from a point diffractor at depth $Z = VT_0/2$, then

$$x_{min} = \frac{VT_0}{2} \tan 30° \quad (5.30)$$

The minimum aperture can exceed the maximum aperture. The maximum aperture equation may result in a calculated aperture lower than that from the minimum aperture equation because of low dip values and/or low frequency values. It is usually better to use the larger of the two calculated apertures especially for *Kirchhoff time migration.*

The migration aperture may (and probably should) be different along each edge of a survey. The migration aperture selected should be the one that extends the farthest distance of that determined from dip calculations, Fresnel zones, and diffraction energy. Dip apertures should be measured from locations where the dips were evaluated and several measurements should be made to assure steepest local dip has been measured. The aperture from gentle dip at the edge of a survey may extend farther than that calculated for steep dip at center of the survey. See Figure 5–113.

The final full-fold survey area should go out to the farthest limits in all directions. The CMP fold taper should be added on all sides. See Figure 5–114.

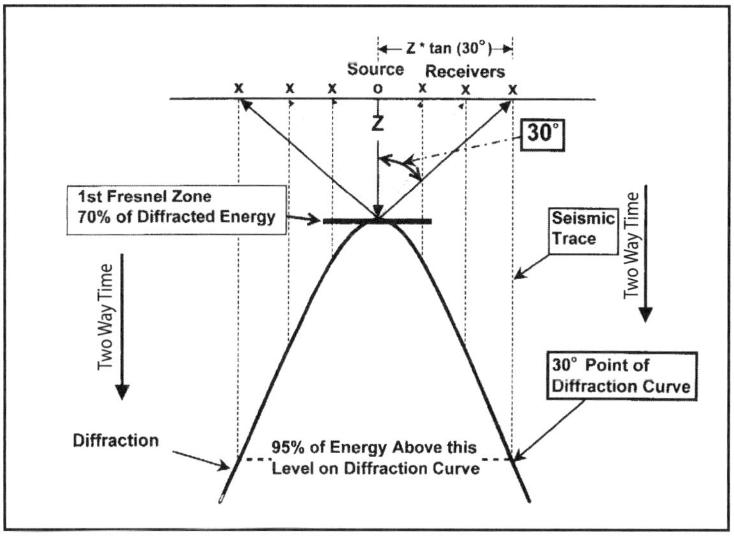

Fig. 5–112 *Distribution of Diffraction Energy*

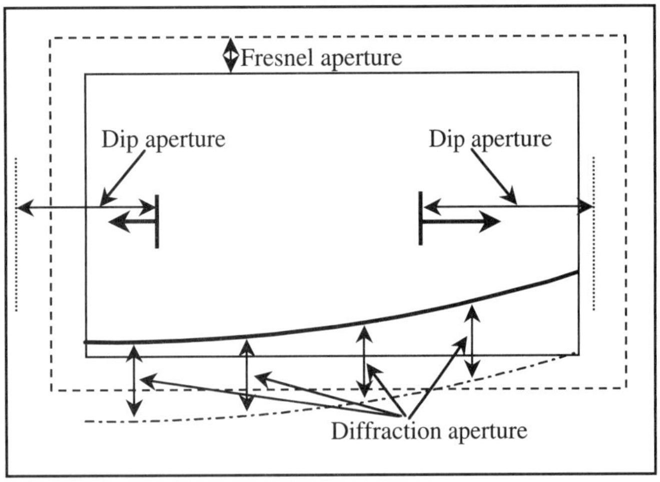

Fig. 5–113 *Determining Total Migration Aperture*

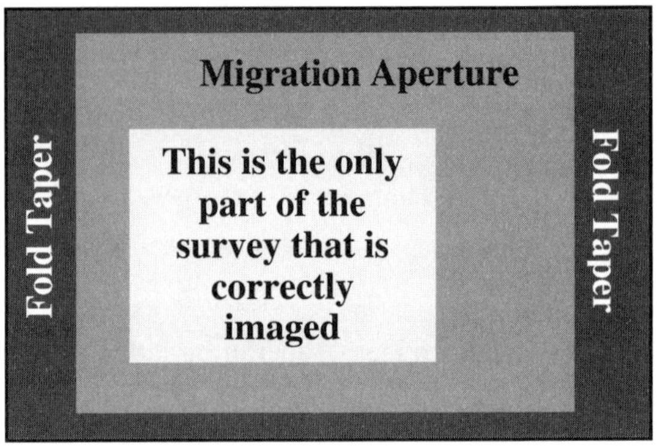

Fig. 5–114 *Total Survey Area Required to Correctly Image Target*

Marine 3-D data acquisition. Common midpoint or common depth point processing techniques require gathering of data from some defined area of subsurface. Some survey geometries, such as land surveys in unobstructed areas, have genuine natural common midpoints. Most surveys, however, have irregular mid-point distributions. This includes marine surveys and land surveys with surface obstructions. Such surveys use a regular array of bins (Fig. 5–115). All seismic traces with midpoints that fall within the bin boundaries are gathered for CMP stacking.

Bins may be square, rectangular, circular, or elliptical but normally are square or rectangular (Fig. 5–116). Different in-line and cross-line spatial sampling is handled by use of rectangular bins.

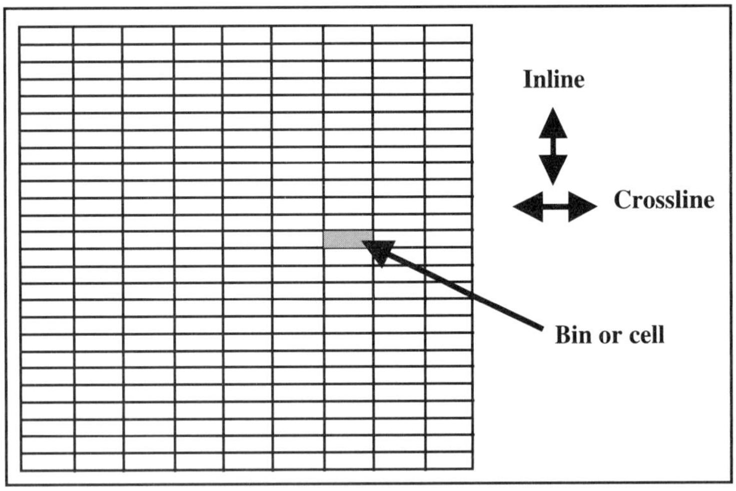

Fig. 5–115 *Bins or Cells*

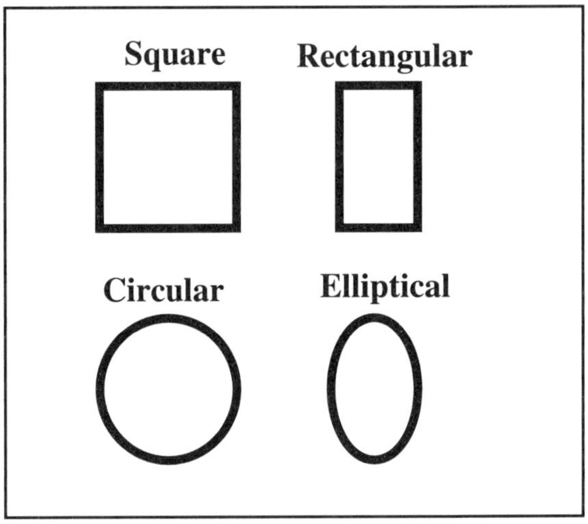

Fig. 5–116 *Bin Shapes*

Marine 3-D surveys recognize that the streamers are curved (Fig. 5–117). Cable compass readings and other in-water devices are used to define cable shape. The sources are located precisely, and midpoints between source and hydrophone groups in the streamers are determined in real time.

Binning serves two functions—monitoring subsurface coverage during the acquisition of a 3-D survey and collating data for seismic processing (in particular, CMP stacking). Properties of binned information are referred to as attributes, which include

- mid-point location
- source to receiver offset
- source to receiver azimuth
- average streamer depth
- RMS noise
- horizontal midpoint (HMP) uncertainty
- feather angle

For 3-D data acquisition, subsurface coverage is monitored in real time by the onboard quality control (QC) geophysicist, who steers the subsurface midpoint coverage within the 3-D grid. The vessel course is adjusted such that the subsurface midpoint coverage falls along the pre-plotted traverse (Fig. 5–118). Not all offsets will fall within the planned traverse because of cable feathering, so the onboard QC geophysicist must decide which offset range to steer into the desired bins.

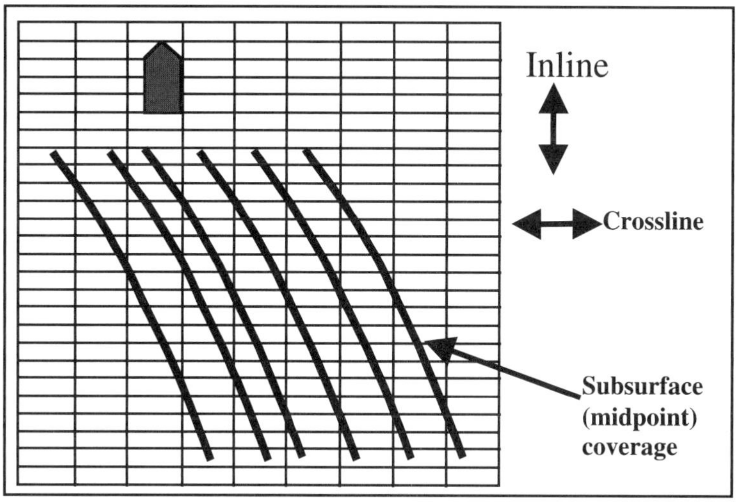

Fig. 5–117 *Subsurface Coverage in Marine Surveys*

A binning system should always be used for real-time monitoring of the subsurface coverage during data acquisition and must have the following features:

- ability to monitor offset distribution of data within bins
- ability to monitor fold of coverage, excluding duplicated offset traces
- capability to display steering graphically and to perform re-binning

Desirable features of a binning system include the ability to:

- monitor azimuth distribution of data
- exclude duplicated offset traces on the basis of azimuth, distance from bin center, etc.
- monitor fold of coverage as a function of azimuth
- exclude individual traces, elements (sources or streamers), or shots based on quality indicators

Static binning is a simple way to do CMP binning. The bin positions are defined prior to the start of acquisition. The bins are uniformly spaced and have constant in-line and cross-line dimensions. In static binning, there is neither overlap of adjacent bins nor gaps between bins.

Overlapping or wide binning (Fig. 5–119) is similar to static binning except that bins are allowed to overlap in the cross-line direction. Overlap can be as much as 200%. Advantages of wide binning are that prospect infill is significantly reduced and it allows optimum CMP distribution. Traces that fall in the overlap zone are used twice in the two adjacent bins.

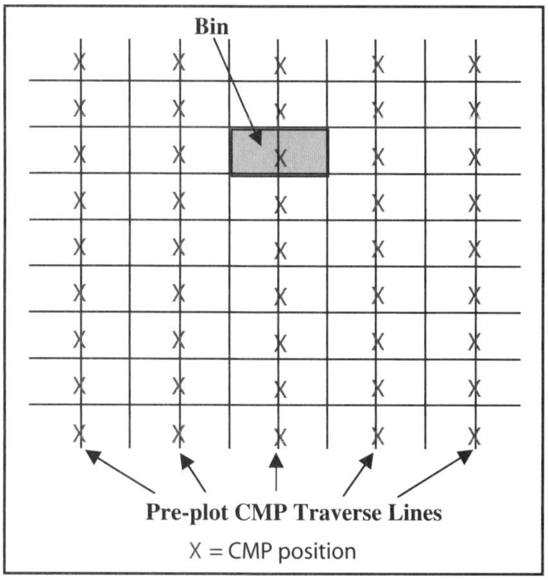

Fig. 5–118 *Static Binning*

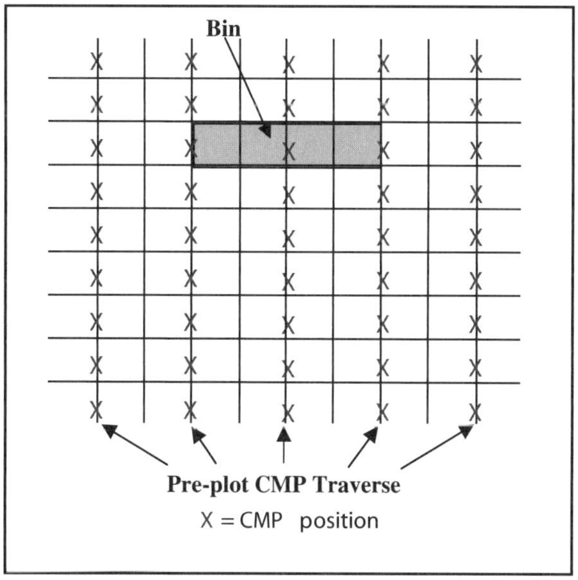

Fig. 5–119 *Overlapping or Wide Binning*

In tapered or flex binning, a static cross-line bin dimension is defined for the near offset trace and another, larger bin dimension for the far offset trace. Cross-line bin dimension increases linearly with offset from the near offset to the far offset. The advantage of the method is that it provides better far offset trace distribution than static binning in areas that have large feathering angles. It is assumed that far offset data will be muted at earlier recording times, and that the deeper data that is not muted will be lower frequency.

Multiple subsurface lines are now commonly recorded for each boat pass. This provides the finer spatial sampling required for 3-D, reduces cost, and decreases acquisition time. Technological changes that made the increase in subsurface lines feasible include

- more recording channels
- bigger boats
- better airgun arrays
- larger compressor capacity
- improved navigation and in-water positioning

Figure 5–120 illustrates the effect of multiple streamers and multiple sources. A single streamer and single source produce only one set of subsurface reflection points for each pass of the boat along a seismic line. A boat using this type of operation typically covers around 4800 km of surface distance per month (boat track km/mo) and also provides about 4800 km of subsurface coverage per month (subsurface km/mo).

Adding another streamer reduces the boat track km/mo to 4050, but since there are two subsurface lines per pass, subsurface production increases to 8100 km/mo. Two streamers and two sources further slow the boat track km/mo to 3750, but with four subsurface lines per pass, subsurface production increases to 15,000 km/mo. Adding a third streamer reduces the boat track km/mo to 3000 but there are now six subsurface lines per pass and subsurface production increases to 18,000 km/mo. When using multiple streamers and multiple sources, the spacing between streamers and sources can be adjusted to give the desired spacing between subsurface lines.

Note, however, that when dual sources are used, they are not fired simultaneously but alternately. As a consequence, fold for each subsurface line is only half as much as when a single source is used. This is shown in Figure 5–120 by representing subsurface coverage from the left source as open circles and those from the right source by filled circles. This is usually an acceptable tradeoff, as a lower fold is required for 3-D than 2-D in most cases.

As many as 12 streamers, with two sources, have been used to acquire data in recent years. This can provide 24 tracks of CMPs per boat pass.

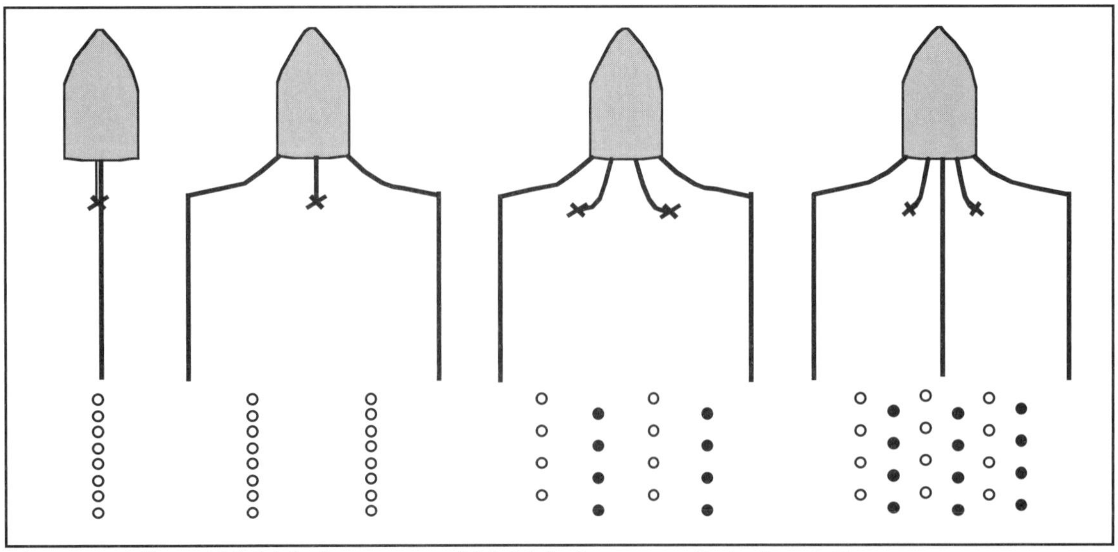

Fig. 5–120 *Effect of Multiple Streamers and Multiple Sources on Data Acquisition*

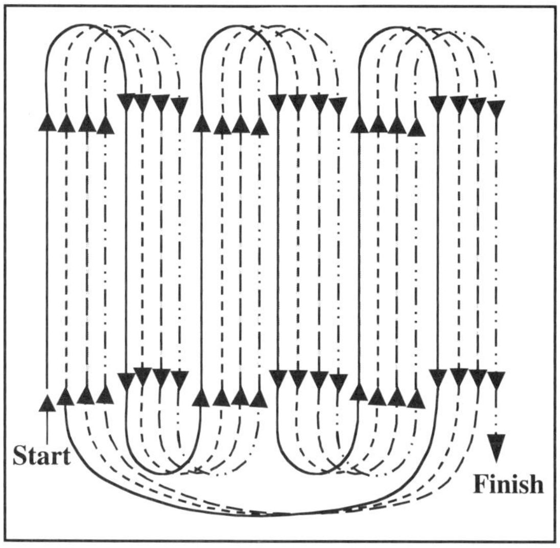

Fig. 5–121 *Conventional Straight Line or Racetrack Shooting*

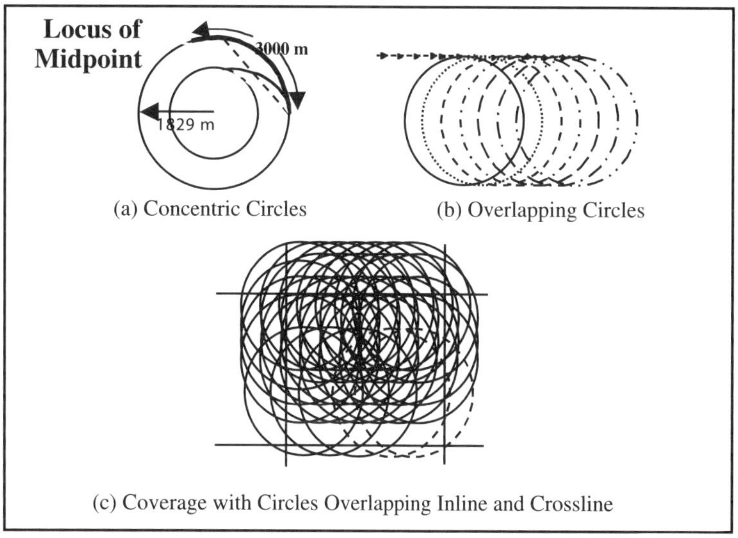

Fig. 5–122 *Circle Shooting*

The paths followed by the boat(s) are also very important for both data quality and acquisition efficiency. Lines are usually orthogonal to the steepest dips encountered. Theoretically, all data should be acquired in the up-dip direction, but this would be horribly inefficient for marine data acquisition. The boat would have to travel from the end of one line back to the start of the next line with no acquisition in transit. The most common method, called conventional straight line or *racetrack* shooting, is shown in Figure 5–121.

The boat begins at the first line on one side of the prospect area. When the line is finished, the boat makes a turn to another line. The turning radius is at least half the length of the streamers, so the next line would be on the order of one streamer length from the first line. Note that the second line is shot in the opposite direction from the first. A similar turn is made at the end of this line, and shooting proceeds in the same direction as the first line. The pattern continues in this way so that the boat is following a continuous looping path until there are no more lines within the turning distance. At this point, the boat turns and travels back to the start of the line next to the starting point and follows another looping path similar to the first. It is only at the end of one loop and the start of the next that the boat travels further than the minimum turning distance without acquiring data. Note that there is an alternation between up-dip and down-dip shooting on successive lines shot but it is very efficient.

Circle shooting provides a wide range of azimuths and minimizes line change time. In Figure 5–122a, the outside circle has a radius of 1829 m. That means that the circumference is about 11,492 m or not quite four times the length of the 3000 m long streamer. The chord connecting the source and the far hydrophone group is on the order of 60°. Midpoints from one shoot are distributed within this chord. The next shot produces a similarly distributed set of midpoints farther along the circle. Completing the circular path then produces midpoints fairly distributed over a circular ring having an outside radius of 1829 m and an inside radius of about 1574 m.

Moving to the interior circle (half the radius of the outer) is done easily over a short distance but the circumference now is less than two streamer lengths and the tighter circle increases the turbulence caused by the constantly changing direction of the streamer, which means more noise is recorded. This is one of the main problems with circle shooting—increased levels of streamer noise.

Figure 5–122b is an alternative approach to circle shooting. Line changes are made following the tangent to the set of overlapping circles. Since midpoints are grouped toward the outside of each circle sufficient overlap is required to fill in the gaps. Figure 5–122c carries the overlapping circle approach further by adding additional lines of circles that overlap the first in the opposite or cross-line direction. This can produce a high density of midpoints and large azimuthal variation at many different offsets.

Figure 5–123 shows two methods used for salt domes. Figure 5–123a shows concentric circle shooting. This gives, more or less, continuous strike shooting since salt domes are somewhat cylindrical in shape. A benefit of this method is that line changes are minimized.

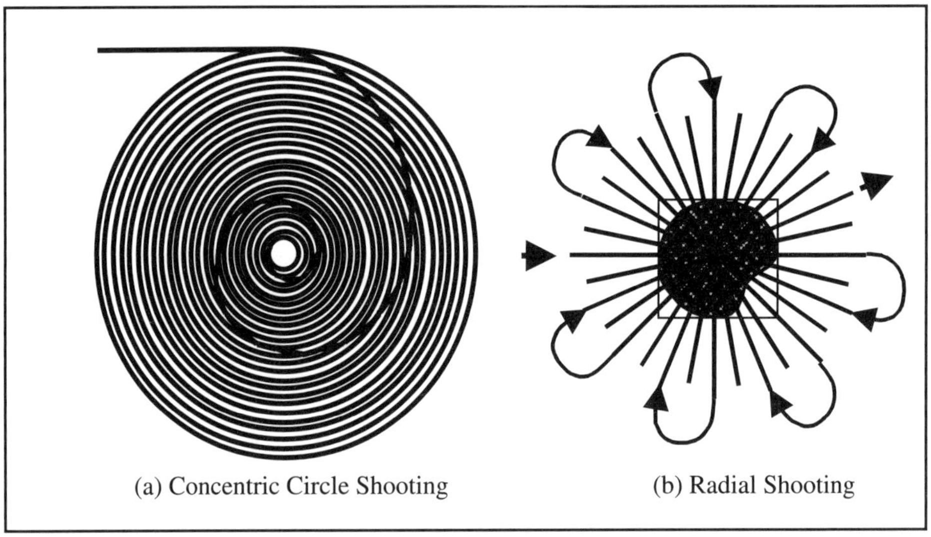

Fig. 5–123 *Approaches to Salt Dome Shooting*

However, as circle radius decreases, streamer noise increases and steering becomes more difficult. The method of Figure 5–123b, called radial shooting, essentially gives all times in the dip direction. This approach does not minimize line change times but does reduce total recording time because of wider spacing between lines away from the dome.

The use of multiple streamers and multiple sources means that the distribution of azimuth varies systematically as a function of source to receiver offset and cross-line swath position. Azimuthal variation can lead to amplitude artifacts during pre-stack imaging (DMO, MZO). Undershooting with a two-boat acquisition method also results in a similar azimuthal variation. Both situations are shown in Figure 5–124.

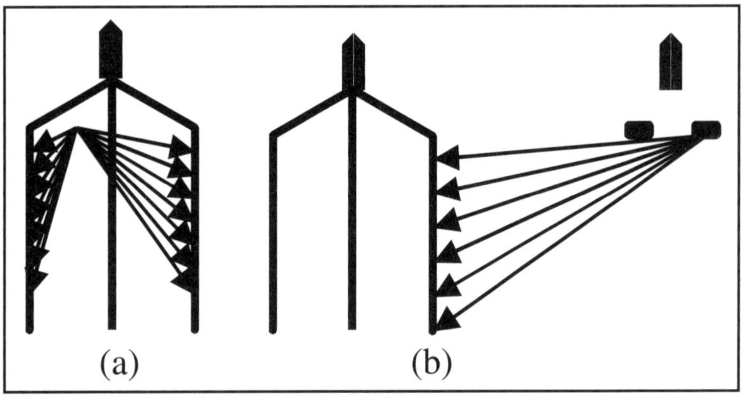

Fig. 5–124 *Azimuthal Variation in Conventional One-boat Operation and Two-boat Undershoot Operations*

Many surveys are conducted in areas where there are drilling rigs, production platforms, shipping lanes, and other obstacles that restrict surface access. When using one boat in such areas, the boat is forced to make a turn well before reaching an obstacle and continue the line on the other side of the obstacle, resulting in recording many truncated lines. This is very expensive. A two-boat operation is a common approach to avoiding truncated lines by under-shooting obstructions. In this method, the shooting boat is on one side of the obstacle and the recording boat is on the other. Midpoints under the obstacle are obtained. The two approaches are contrasted in Figure 5–125.

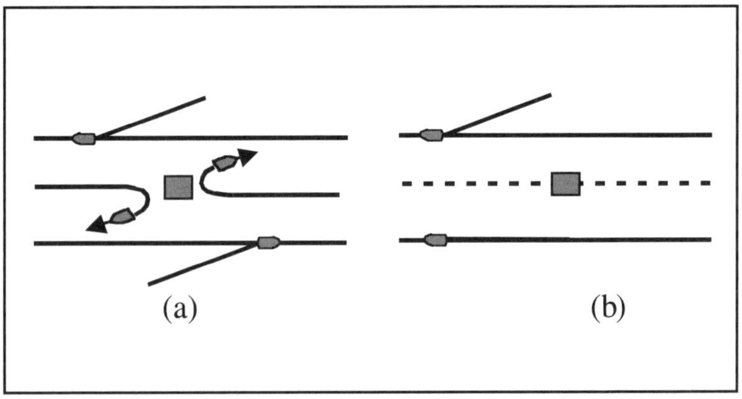

Fig. 5–125 *Obstacle Avoidance, One-boat Operation and Two-boat Undershoot*

Figure 5–126 shows how two boats can be used to obtain data between obstacles. In Figure 5–126a, the shooting boat leads the recording boat by some specified distance. The objective here is to obtain far offset data. In Figure 5–126b, the recording boat leads the shooting boat. In this latter situation the streamer is often shortened to provide better access through the obstructed area.

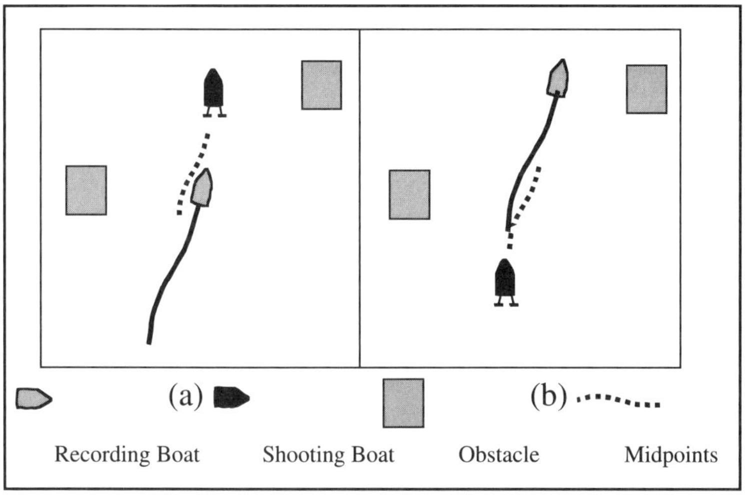

Fig. 5–126 *Obstacle Avoidance with Two Boats*

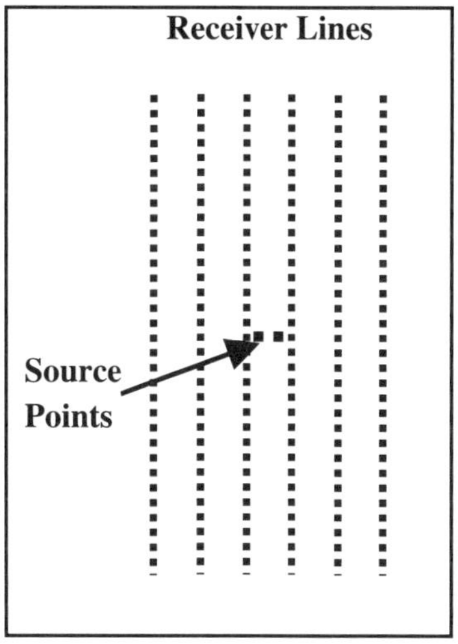

Fig. 5–127 *Acquisition Template*

Land 3-D data acquisition. Before discussing land 3-D acquisition in detail, some basic definitions of commonly used nomenclature are in order.

A *patch* consists of all live stations (positions where geophone groups are input to the recording system) for a particular source point in the 3-D survey. It usually consists of a rectangular area formed by several parallel receiver lines.

A *template* is the combination of a particular receiver patch and the associated source points that are shot into and recorded by the receiver patch. The source points can be inside, outside, or both inside and outside the patch.

The receivers shown on the six receiver lines in Figure 5–127 constitute a patch. The addition of the sources shot into that patch makes a template.

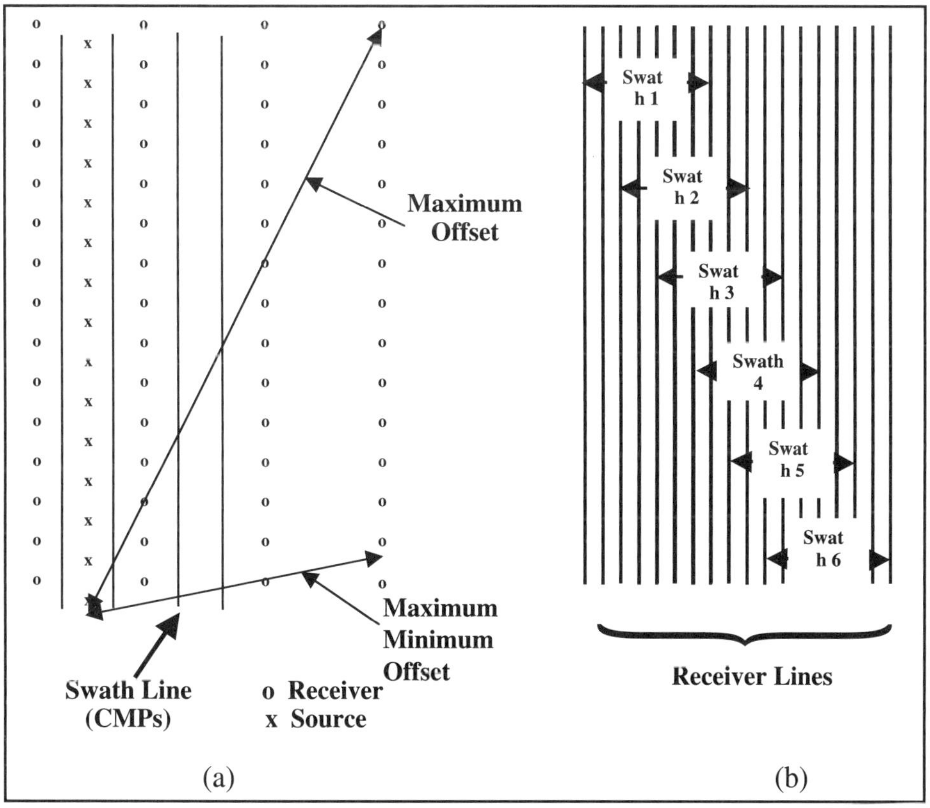

Fig. 5–128 *Swath Layout Method, Swath Shooting Method*

The term *swath* is used in two ways. One definition (Fig. 5–128a) refers to the swath layout method in which source lines are parallel to receiver lines. The other definition (Fig. 5–128b) refers to the progress of the template when shooting a survey—the number of receiver lines being rolled in-line at one time.

Fold is equal to the number of midpoints that fall within a bin.

The *maximum minimum offset* is the shortest distance between source and receiver on a line farthest from the source in a 3-D survey. See Figure 5–128a. A relatively small maximum minimum offset is required to record shallow reflectors. This may be the most neglected 3-D survey parameter.

The *maximum offset* is the largest consistently recorded source to receiver offset. It depends on the shooting scheme and patch size (Fig. 5–128a). A large maximum offset is required to record deeper reflections with sufficient normal move out for processing.

The *migration aperture* is the width of the fringe area that must be added to a 3-D survey to assure that reflections from dipping reflectors within the target area are recorded.

Fold taper is the additional surface width that must be added to a project area to assure that the target area and migration aperture are recorded full fold.

Shot index (SI) is the number of source lines between receiver lines. If SI = 1, then the subsurface line interval is one-half the surface line interval. If SI = 2, then the subsurface line interval is one-fourth the surface line interval. If SI = 4, then the subsurface line interval is one-eighth the surface line interval. See Figure 5–129.

The choice of source is usually between dynamite (or some other explosive) and vibrators. When explosives are used, shot holes are normally predrilled and loaded. Misfires must be expected and allowance made for make up shots. In most cases, no special line discipline (limiting crew activity) during shooting is required. Drill trucks require relatively easy access to shot points but this depends on specific equipment. The order of shot point shooting should provide the most efficient movement of the recording spread. Having two, three, or even four shooters can enhance efficiency if many shots are fired into a particular patch. Usually the speed at which the geophone crew lays out and moves geophone groups determines speed of acquisition. Larger crews are desirable if labor cost is less than time cost. Fold is a function of the number of shots and the number of recording channels. Planning an optimum 3-D survey requires balancing equipment, labor, and shot hole costs.

Vibrator and/or other surface sources provide optimum production (lower cost) when geophone crew layout speed is matched to recording time of vibrators. If there are too many recording channels, the geophone layout can be slowed down and the amount of time spent troubleshooting bad groups increased. A combination of too few channels and long sweeps may cause the geophone layout crew to be idle too much.

The direction of vibrator shooting can be restricted by geophone cable locations. Because of their size and weight, vibrators may have limited access. The kind of topography, ground conditions, environmental concerns, and other factors can also limit use of vibrators. It is possible to mix vibrator and dynamite sources by using data processing techniques to match the source signatures of one to the other or to modify both to some other desired signature.

Both geophone groups and source points may not always be accessible because of such things as hazardous waste sites, refineries, wildlife refuges, wilderness areas, harvesting or planting of crops, lakes, and uncooperative land or mineral owners. In such cases, acquisition of subsurface information below these points requires undershooting. Loops, squares, or very wide swath may be used. This could mean that shallow data are absent and that structural and wavelet stability are degraded.

In some areas, geophone arrays and cables may be allowed while sources are not permitted. Areas restricted because of springs, water wells, oil and gas wells, crops, cattle, houses, buildings, and terrain are usually not such a problem. Production facilities and refineries, however, usually result in considerable increase in recorded noise. The scope of the problem caused by lakes and rivers depends on their sizes. Subsurface coverage can sometimes be improved by adding geophone groups and careful selection of recovery source points. Shallow coverage and offset distribution may still be poor.

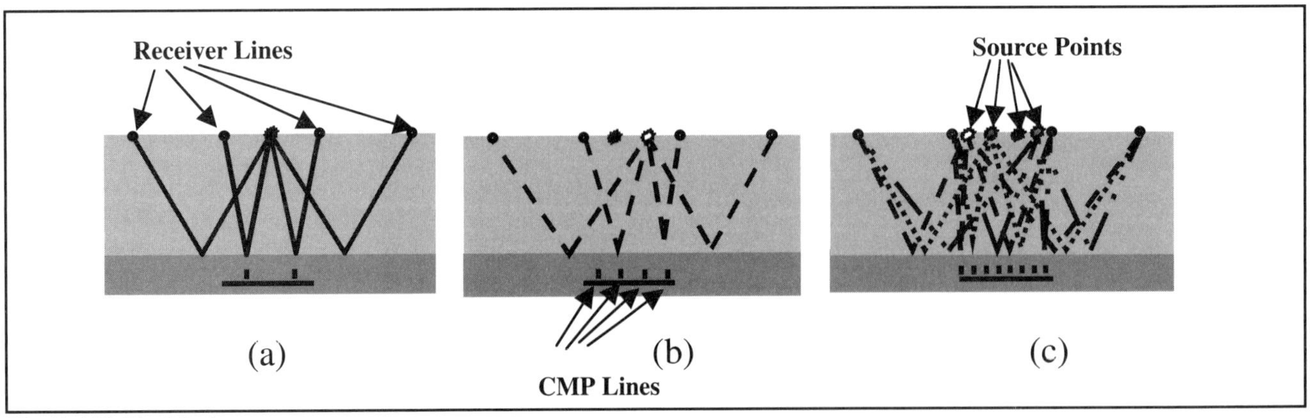

Fig. 5–129 *Shot Index*

There will be a variety of azimuths from each shot and to each receiver in 3-D acquisition (Fig. 5–130). Coherent source-generated noise and back-scattered noise can be recorded with very large amplitudes. When using swaths with a restricted range of azimuths, linear arrays can be used to attenuate coherent, source-generated noise. However, areal arrays (stars, wagon wheels, etc.) are needed to attenuate scattered noise. In 3-D surveys with a wide range of azimuths, both scattered and coherent source-generated noise can be present and areal arrays should be considered. Note that errors in laying out the areal arrays may reduce array efficiency, particularly for the more elaborate arrays. Another option for surveys with a wide range of azimuths is using orthogonal source and receiver arrays to attenuate coherent source-generated noise.

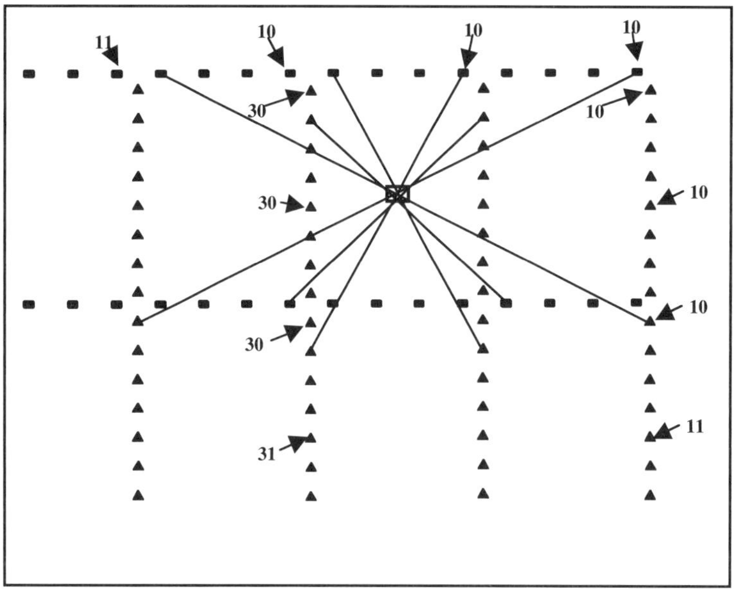

Fig. 5–130 *Offset and Azimuth Variation in a Bin*

In 3-D recording, source and receiver arrays are rarely parallel, and noise does not always travel directly from source to receiver. Areal arrays can be used to attenuate directional, coherent noise but areal array polar responses should be investigated before selecting a particular array. These show theoretical attenuation for all azimuths by means of color plots of amplitude or by contour plots. Figure 5–131 is an example of a polar response for a star array. For surveys with a wide range of azimuths, arrays should have reasonably symmetric responses.

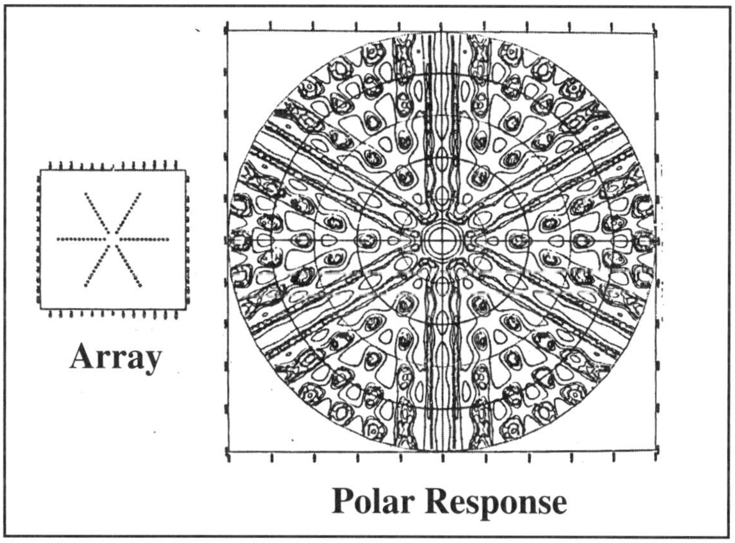

Fig. 5–131 *A Star Array and Its Polar Response*

In designing or selecting arrays to attenuate coherent source-generated noise that travels directly from the source to the receiver array it should be remembered that total attenuation is the product of source and receiver attenuation (or sum expressed in dB). Only the receiver array, however, can be used to attenuate scattered noise.

There are a great variety of ways to shoot a land 3-D survey. The swath shooting shown in Figure 5–128 is very efficient in open country, using vibrators. A variation called *full swath roll* is extremely efficient for large land 3-D surveys. Cross-line roll consists of all receiver lines. No receiver positions are reoccupied, but each source point is occupied twice. Full swath roll is cost-effective in areas where source points are relatively inexpensive and when the crew has a limited number of recording channels. Cross-line fold is obtained even though all receiver lines are rolled. Source positions are reoccupied in such a way that source-to-receiver ray paths are not duplicated. A rarity for the full swath roll method is that an odd number of receiver lines can be used.

Figure 5–132 illustrates the *straight line* method. Receiver lines and source lines are orthogonal. As indicated by the two rectangles, after one template is completely shot, the template is moved diagonally one receiver line and one source line.

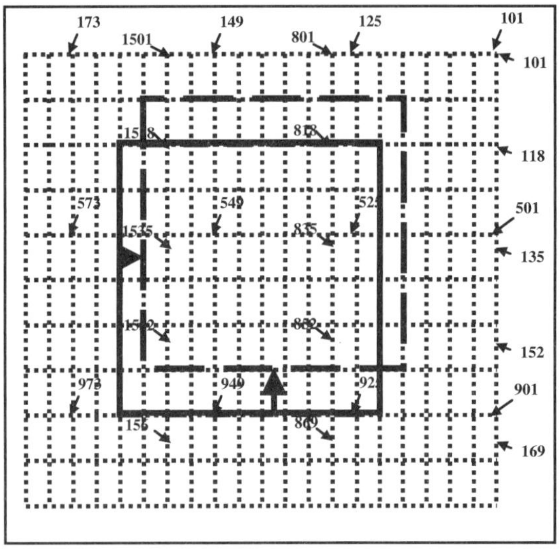

Fig. 5–132 *Straight Line Method*

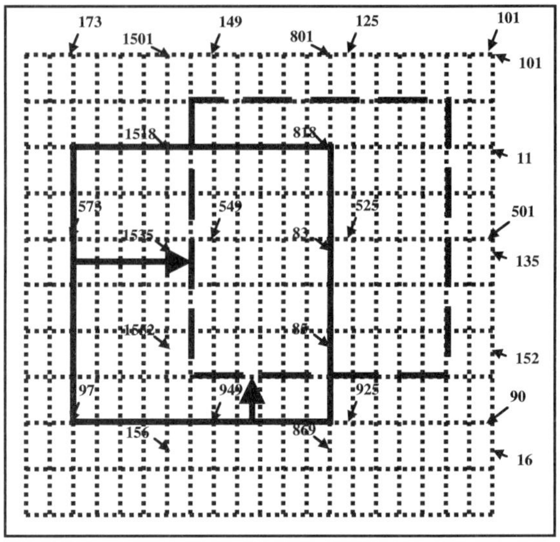

Fig. 5–133 *Another Type of Swath Shooting*

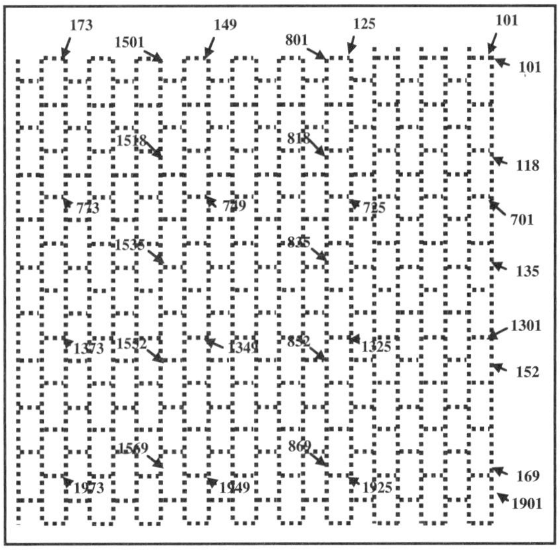

Fig. 5–134 *The Brick Pattern*

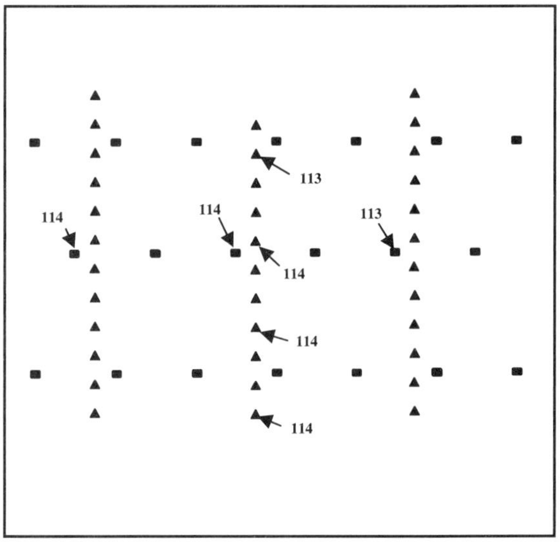

Fig. 5–135 *Odds and Evens*

Figure 5–133 shows a variation of the straight-line method that is also called swath shooting. All shots between receiver lines that are several lines apart are fired. The receiver patch is rolled over several lines and the process is repeated. The operational advantages are attractive, but the resulting offset/azimuth mix is quite poor.

The *brick* pattern (Fig. 5–134) has one tremendous advantage. For a typical template, the largest minimum offset will be equal to the receiver line interval. In contrast, the straight-line method (Fig. 5–132) will have a largest minimum offset equal to the diagonal of one template. This means that the line interval can be increased yet still retain the required minimum offset. Spacing the receivers farther apart results in cheaper acquisition costs.

The brick pattern can be shot very efficiently in a similar manner to swath shooting. Vibrators move from source position to source position parallel to a receiver line. On reaching the end of the swath, the vibrators move laterally to the next source position and shoot the next set of sources parallel to the receiver lines in reverse order.

Odds and evens (Fig. 5–135) is really a variation on the brick pattern. In this case, there is a new brick every shot. Operationally, twice as many lines must be traversed as in the straight-line method but only every second shot is fired on each line.

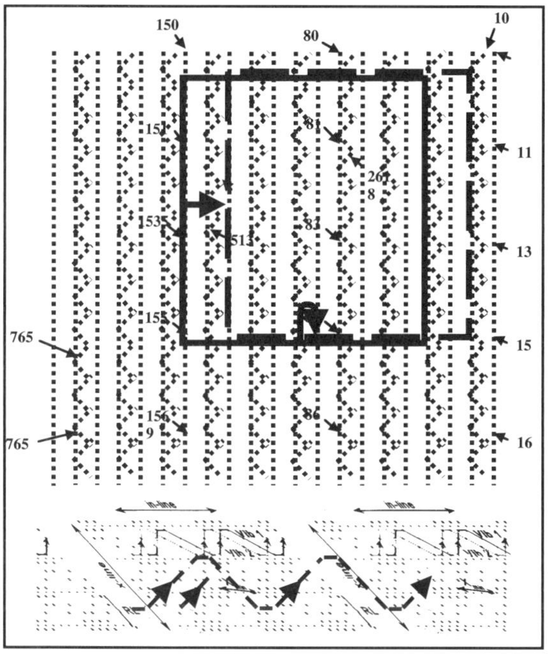

Fig. 5–136 *The Zig-zag and Double Zig-zag Patterns*

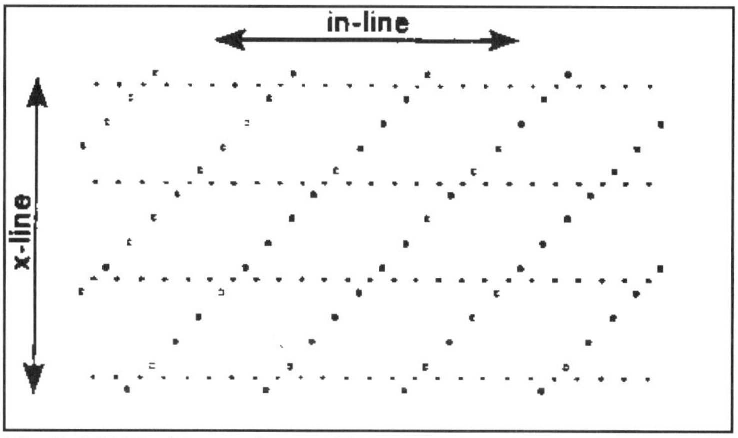

Fig. 5–137 *The Non-orthogonal Pattern*

The *zig-zag* pattern, shown at the top of Figure 5–136, is very popular in desert areas or anywhere there is good access between receiver lines. In operations, vibrator trucks drive up and down between receiver lines. This method has the same largest minimum offset values as in the brick method. The bottom of Figure 5–136 shows a variation on the zig-zag pattern called the double zig-zag. Two sets of vibrators are used. One set sweeps while the other moves.

The non-orthogonal pattern shown in Figure 5–137 has source lines that make an angle other than 90 degrees with the receiver lines.

The *bin fractionation* method (Cordsen 1993) is shown in Figure 5–138. This pattern can be laid out in many different ways as long as the receiver and shot-line spacing are non-integral with respect to the group interval. For example, a group interval of 50 m, a receiver line interval of 220 m, and a source line interval of 440 m satisfies the requirement. This arrangement causes midpoints to fall in a regular pattern inside each CMP bin. Instead of being clustered near the center of each bin, this method has midpoints evenly distributed inside each bin.

The *button patch* method, shown in Figure 5–139, was developed by ARCO and is now routinely used in all of its 3-D surveys. Each button contains a tight pattern of receivers, typically six by six or eight by eight, although there is no reason to keep them square.

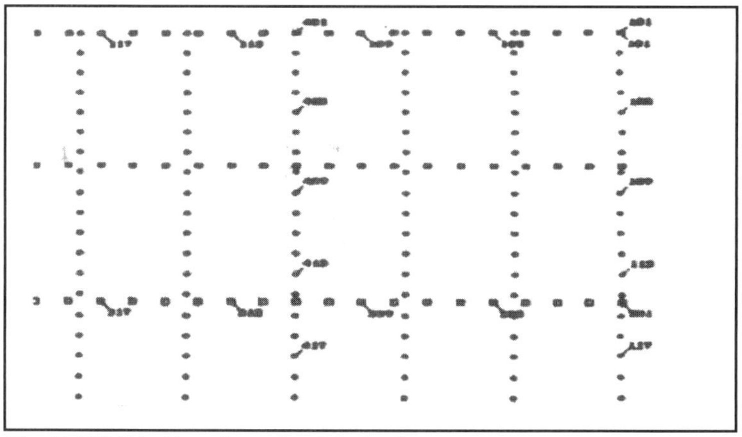

Fig. 5–138 *Bin Fractionation Method*

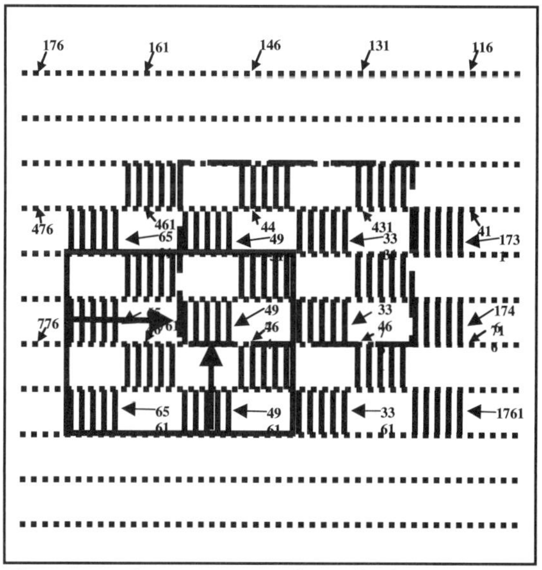

Fig. 5–139 *The Button Patch Method*

Several buttons are combined in a checkerboard pattern to form the receiver patch. Often shots all over the survey are fired before moving the receiver patch exactly half its dimension.

Table 5–8 summarizes the advantages and disadvantages of the acquisition methods described above.

Table 5–8 *Pros And Cons Of Various Layout Strategies*

Layout	Pros	Cons
Straight	Simple geometry	Large maximum, minimum offset
Swath	Simple geometry, cost efficient, good offset distribution, minimum equipment movement	Poor azimuth distribution, statics coupling
Brick	Smaller Xmin may allow a wider RLI, reasonable offsets and azimuths	Access can be a problem
Odds/Evens	Special case of brick without such severe access problems, better offsets and azimuths	Twice the source lines as conventional straight shoot, only half the shots are taken on each line
Flexibin or Bin	High resolution with low fold or low resolution with high fold, super bins for normal use have good offset and azimuth mix, good statics coupling	Same as straight
Button Patch	Efficient use of large channel systems, good offset and azimuth distribution require detailed planning	Can require large number of source points over a wide area for each patch, needs large channel capacity, static coupling hard to accomplish
Zig-Zag	Same as brick, efficient for equipment moves	Must have very open access
Non-Orthogonal	Simple geometry	Same as straight
Hexagonal Patch	Layout hexagonal receiver patches and use hexagonal CMP bins, more efficient use of equipment since more channels will be at useful offsets	Difficult to lay out
Circular	Consistent offset	Operationally difficult
Source: M. Galbraith, *3-D Survey Design by Computer,* 1994.		

Irregular, chaotic, or *goat trail* designs are needed where access is limited by topography, lakes, rivers, etc. Other applicable situations involve environmental limitations, surface facilities, towns, refineries, production facilities, and existing oil fields. Aerial photos, digital topographic maps, and satellite images are useful tools in the design of such surveys.

If a source cannot be located at the desired location, it is best to shift it in-line (parallel to the geophone lines). When it is shifted, it should be *moved an integer number of groups* to produce midpoints at the bin centers. The fold for bins at the ends of the spread will be reduced or increased. The shift will not affect fold for the other bins.

When an in-line move has been made, any other required shifts along the same source line should be in the same direction if they have CMP bins in common. In Figure 5–140, the first source position move was a shift to the right so the second source location was also shifted to the right.

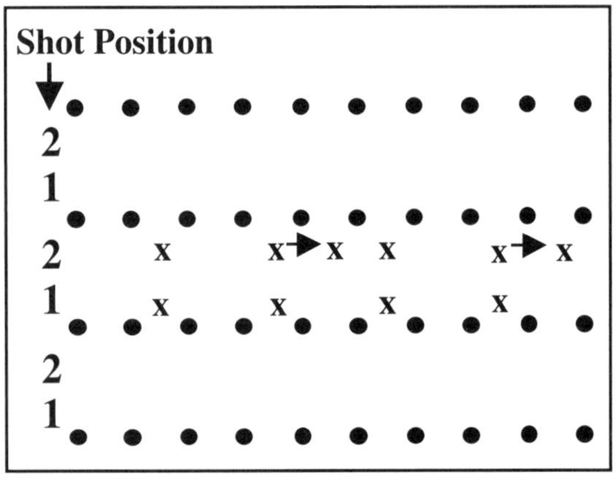

Fig. 5–140 *Recovery Shots with In-line Shift*

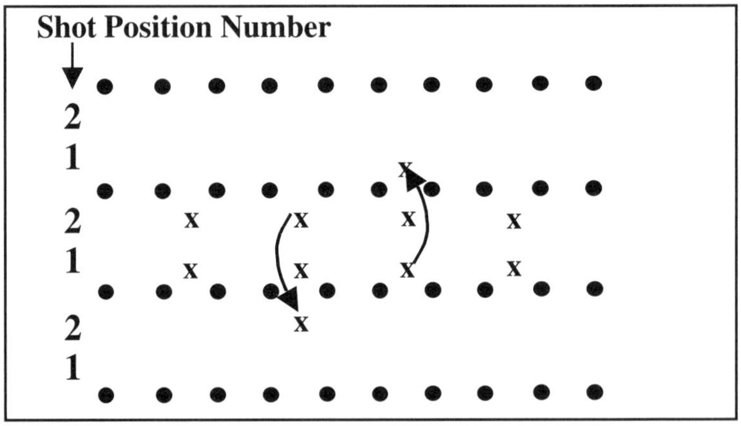

Fig. 5–141 *Recovery Shots with Cross-line Shifts*

If an in-line shift cannot be made, then a cross-line move may be the only alternative (Fig. 5–141). In this case, the source location must be consistent. That is, if a source is moved in a cross-line direction, then it must occupy a similar position between two other geophone lines.

In Figure 5–141, the source that was shifted in-line in Figure 5–140, is now shifted from a source position two to another source position two. Also shown in Figure 5–140 is a source along line one that has been moved from its desired location to another line one between the top two geophone lines. It could have been moved to the one source line between the bottom two geophone lines and still satisfied the requirements for a recovery shot.

Workshop

1. A marine vessel is pulling a 6 km long streamer at a speed of 6 knots along a boat track of N75°E (Fig. 5-142). There is a current of 1.5 knots in the direction S30°W. What will be the angle between the vessel heading and the streamer? How many m of cross-line deflection will result at the end of the streamer? (Assume the current acts uniformly on the streamer.)

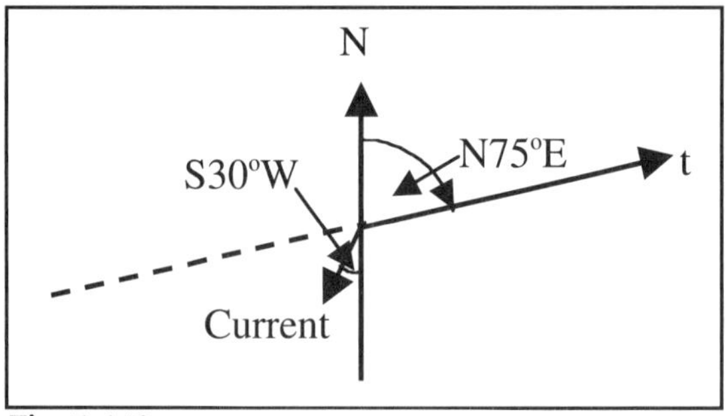

Fig. 5–142

2. How would an air gun array signature be affected if the volume of every gun in the array were doubled?

3. Shown below in Figure 5-143 are sketches representing the far-field signatures (left) and amplitude spectra (right) for a one-pound charge (top) and an eight-pound charge of dynamite. Based on these, sketch the signature and amplitude spectrum for two one-pound charges fired simultaneously in two holes.

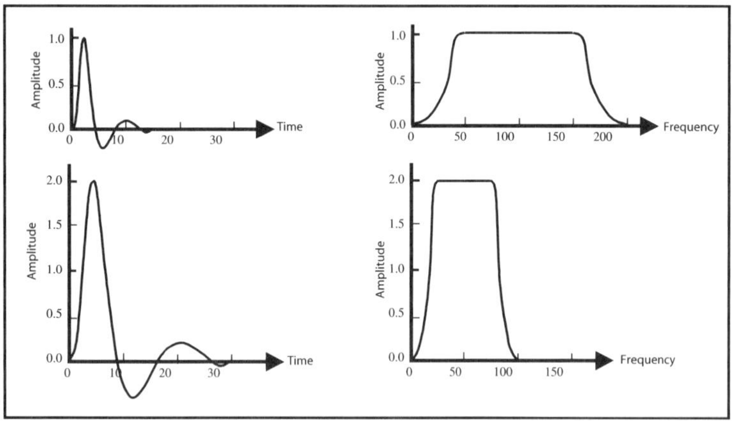

Fig. 5–143

4. If six vibrators, each making four sweeps of four seconds, provide an adequate signal-to-noise ratio, what changes could preserve this S/N if one vibrator were lost?

5. If the airgun arrays are at a depth of 6 m and the streamer is being towed at a depth of 10 m, at what frequencies would the ghost notches be seen? (Assume water velocity is 1500 m/s.)

6. Identify the natural frequency of the geophones from the amplitude responses shown in Figure 5-144:

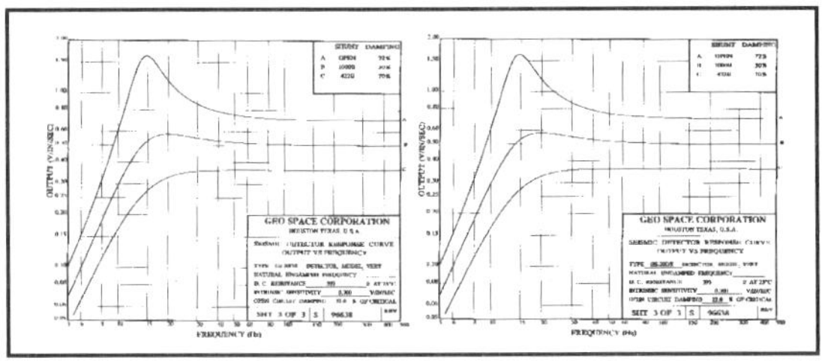

Fig. 5–144 a

A. ____ Hz

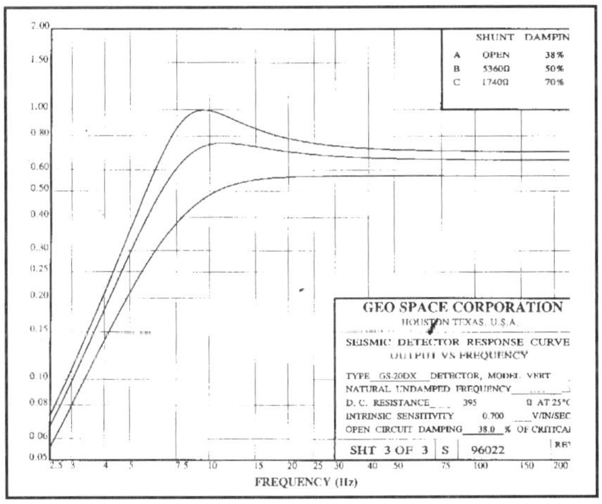

Fig. 5–144 b

B. ____ Hz

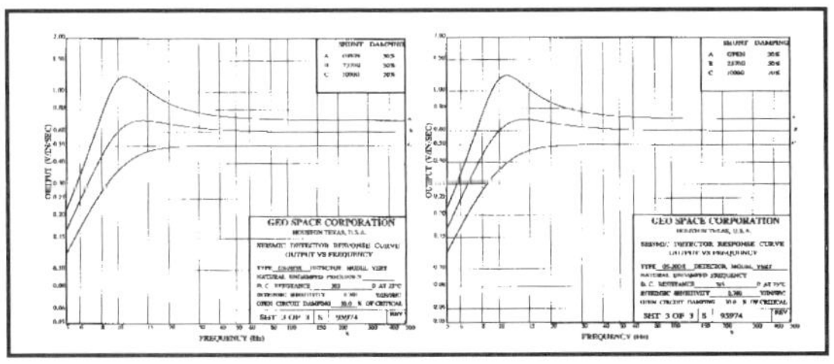

Fig. 5–144 c

C. ____ Hz

7. A 4-element source array and an 18-element receiver array are to be used. The group interval is 110 ft. If the source-generated noise shown in Figure 5–76 is expected, what spacing between elements in the two arrays will maximize noise attenuation?

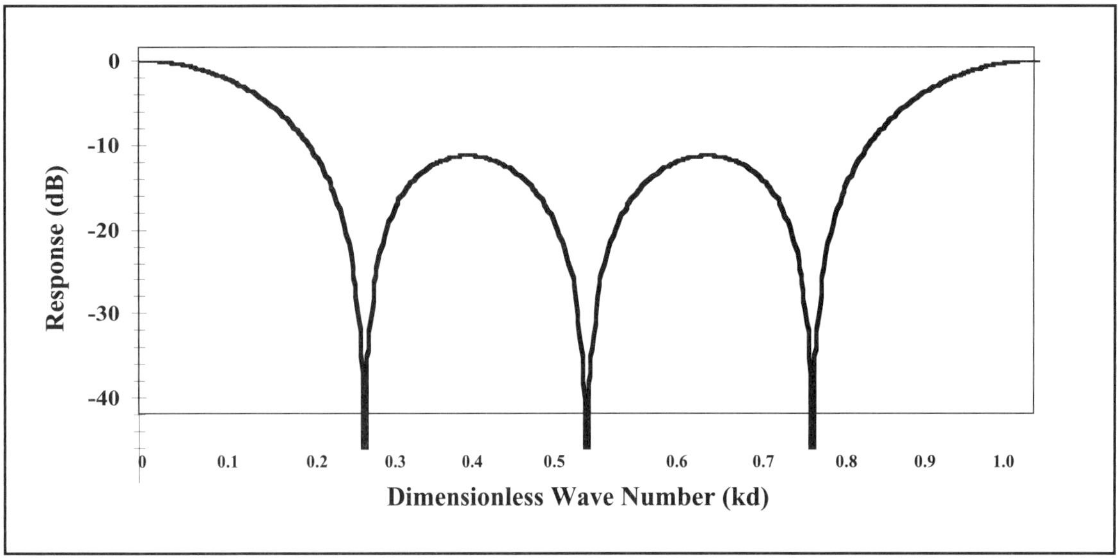

Fig. 5–145 a

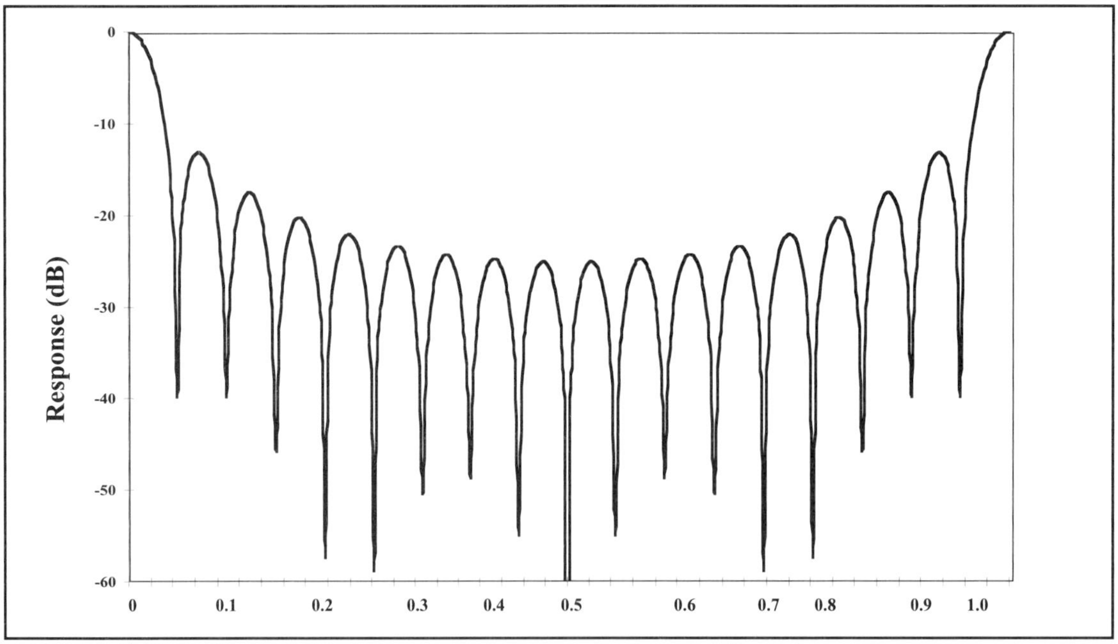

Fig. 5–145 b

8. Show how decimal 583 is subtracted from decimal 1947 in computers.

9. A tape dump yields the following information:

    first four bytes of general header: 0371 8048

    bytes 101–104 of first data block: 43F8 8000

    What amplitude is written in bytes 101–104 of first data block?

10. A 2-D line is to be shot over a target about 10 km long and 5.5 km deep. Maximum dip over the target is expected to be about 15°. Strike is N25°E. A secondary, shallow target is at about 1.8 km deep with a maximum dip of 10°. A fold of at least 30 is needed on the deep target and 16 on the shallow target. Assume an rms velocity of 3770 m/s for the deep target and 2800 m/s for the shallow target. Determine optimum parameter values for the following:

    spread type _______ .

    maximum offset _______ .

    minimum offset _______ .

    group interval _______ .

    shot interval _______ .

    line length _______ .

    direction of shooting _______ .

11. The primary target for a 3-D survey is at 2.2 seconds, and a shallow target is at about 900 ms. A velocity function for the area is provided by Figure 5-146. The surface velocity is 600 m/s and the interval velocity just above the target is 3600 m/s. Maximum frequency required is 50 Hz. Nominal fold for the survey is 48. Shown in Figure 5–146 is a plot of the direct arrival, two refractions and the shallow reference reflection. Also shown is a sketch indicating target size, orientation, and dips (including maximum dip) at various points.

Use this information to determine the following parameters:

a. maximum offset

b. maximum minimum offset

c. in-line spatial sampling

d. cross-line spatial sampling

e. migration aperture width on all four sides

f. number of groups/line in the patch

g. group interval

h. group roll

i. number of receiver lines in the patch

j. line interval

k. line roll

l. number of sources between lines

m. number of source lines in template

n. in-line fold taper distance

o. cross-line fold taper distance

p. total area of survey

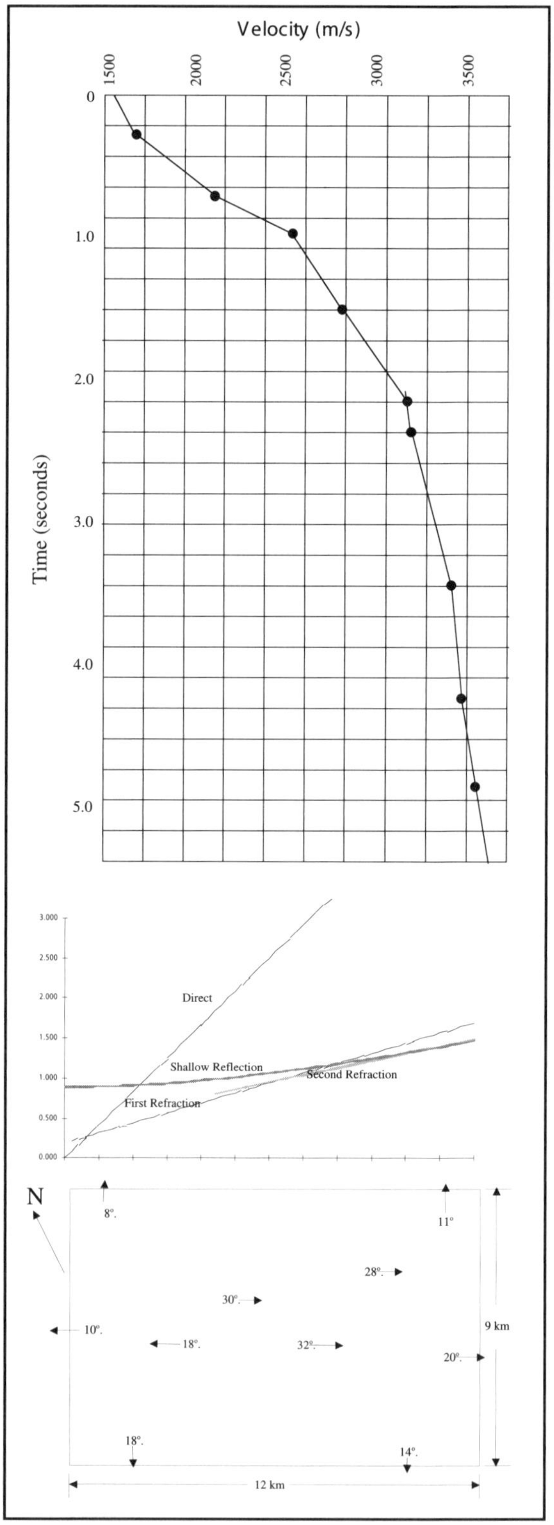

Fig. 5–146

6 Seismic Reflection Data Processing

Introduction

There are some variations in processing related to acquisition environment (land, marine, or ocean bottom cable) and source type. Land and ocean bottom cable (OBC) processing require *static corrections* but marine does not. Marine processing requires integration of navigation processing with initial seismic processing. OBC requires calculation of constants needed to sum hydrophone and geophone data. Land vibrator data must be *vertically stacked* and cross-correlated with the sweep.

Preliminary processing for vibrator data

Vertical stack is applied to vibrator data to attenuate coherent noise when sweeps are made at separate points in a source array and to attenuate random noise by $\sqrt{N}$, where N is the number of records stacked together. Vertical stack reduces data processing cost through reduction of data volume. Thus, vertical stack outputs fewer records with better signal-to-noise ratios. Modern recording systems integrate vertical stacking with acquisition.

Figure 6–1 illustrates the vertical stack process. In this example, sweeps are made at four locations in the source array resulting in four records, which are shown here schematically. The four records are summed (sample values at like times on like-numbered channels added together) and divided by four. Since each record was acquired from different effective source points, source-to-receiver offsets differ, and, thus, arrival times are different on each record. The time differences are much more significant for the ground roll since its propagation velocity is much slower than that of primary reflections. Time differences translate into phase differences producing attenuation of the ground roll. Time differences between primary reflections are insignificant and no appreciable attenuation occurs.

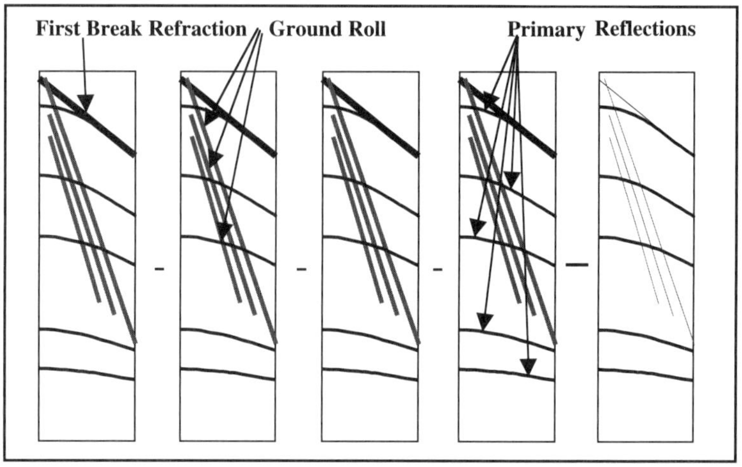

Fig. 6–1 *Vertical Stack*

In *simple* or *straight stack* (described previously) there is no scaling of traces before stack. In *diversity stack*, scalers that depend on trace power are applied before the traces are stacked. The same samples are stacked in each type of stack. Differences among the stacks are the way the amplitudes are scaled prior to stack.

Straight stack (Fig. 6–2) is used when signal and noise levels are about the same on all input records, and the noise level is equal to or less than signal level (S/N ratio ≥ 1). The figure uses synthetic traces to demonstrate straight stack, which is a two-step process.

1. Sum the samples at the same time on traces having the same number on records at the same shotpoints.

2. Divide the sum trace by the number of contributors to keep amplitudes within the dynamic range of the instruments and to maintain trace modulation.

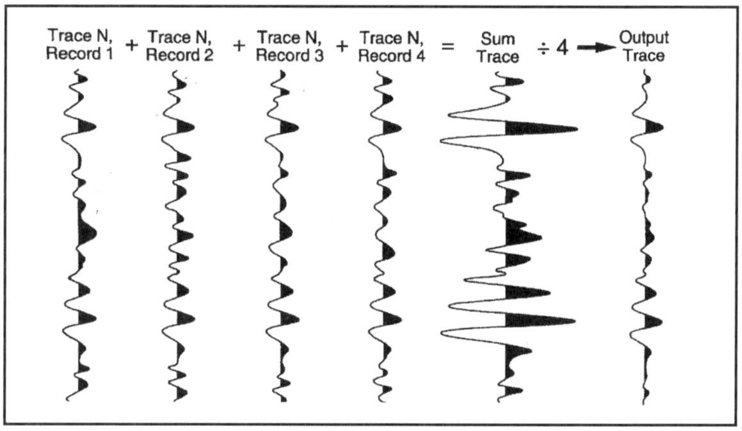

Fig. 6–2 *Straight Stack*

Summarizing straight stack

- no scaling is applied before stack
- the output trace is the average of the input traces
- signal is coherent and stacks in phase are reinforced
- noise is random and stacks out of phase are attenuated
- signal-to-random noise improvement is $\sqrt{N}$ where N is the number of contributing traces

Diversity stack (Fig. 6–3) is used when signal amplitudes are weak but are consistent from record to record and noise level is greater than signal level—S/N ratio < 1. The condition that diversity stack best handles is when noise is not coherent but, instead, appears as noise bursts of limited time duration (city traffic, animals, line truck).

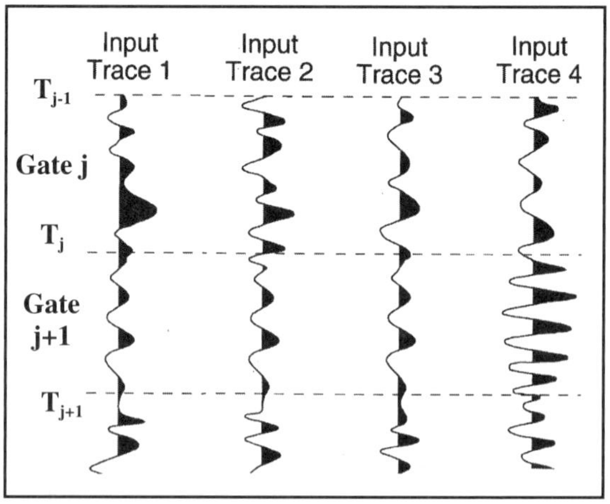

Fig. 6–3 *Diversity Stack Time Gates*

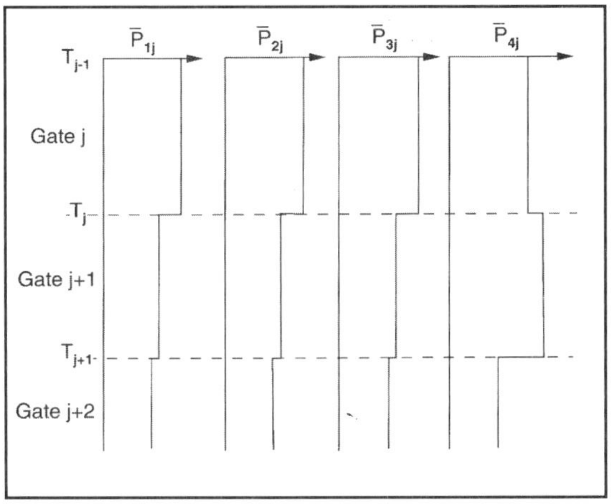

Fig. 6–4 *Average Power of Each Trace in Each Gate*

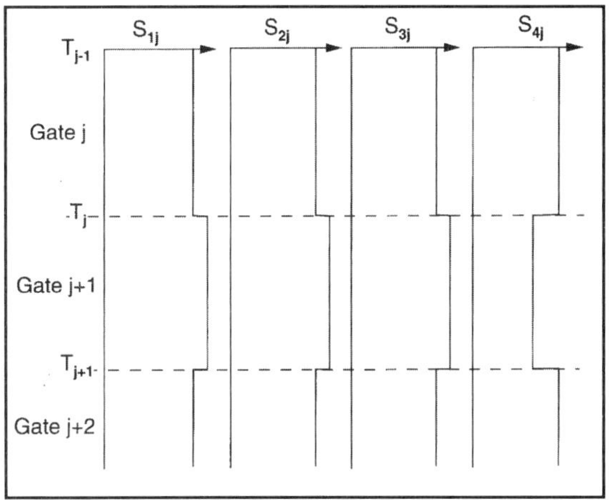

Fig. 6–5 *Scalers for Each Gate*

Application of diversity stack requires the following steps.

1. Divide traces into time gates. Gates should be no longer than noise bursts.
2. Compute average power in each gate of each trace.

Average power in j^{th} gate of ith trace $\overline{P_{ij}}$ is given by $\overline{P_{ij}} = \frac{1}{K}\sum_{j=1}^{K} A_{ijk}^2$

where

A_{ijk} is the amplitude of the kth sample in the jth gate of the ith input trace

K is the number of samples in the gate (Fig. 6–4)

3. Compute scalers for each gate of each trace. To maintain correct amplitude balance, make scalers relative to average gate power for all traces in stack.

Average gate power, $\overline{P_j}$ is given by:

$$\overline{P_j} = \frac{1}{N}\sum_{i=1}^{N} \overline{P_{ij}}$$

where

$\overline{P_{ij}}$ is the average power of the in the jth gate of the ith input trace

N is the number of traces to be stacked

The scaler (Fig. 6–5) for the jth gate of the ith input trace S is given by:

$$S_{ij} = \frac{\overline{P_j}}{P_{ij}}$$

4. Compute scaling function $Sj(t)$ for each trace by linear interpolation between gate center times (Fig. 6–6).
5. Apply scaling functions to traces (Fig. 6–7). $Oj(t) = Aj(t) \; x \; Sj(t)$.
6. Stack scaled traces (Fig. 6–8).

$$O(t) = \frac{1}{N}\sum_{j=1}^{K} O_j(t)$$

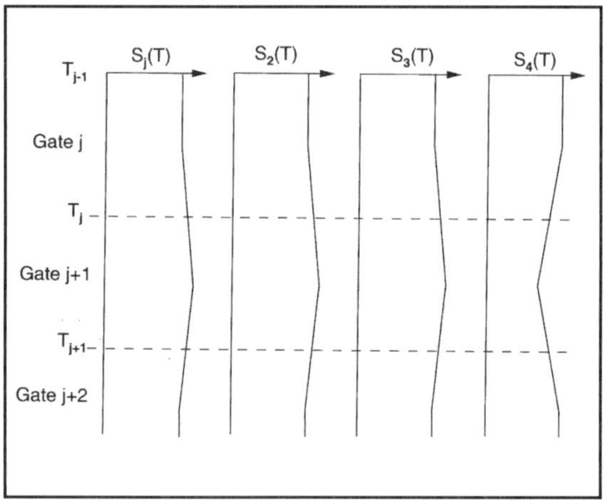

Fig. 6–6 *Diversity Stack Scaling Functions*

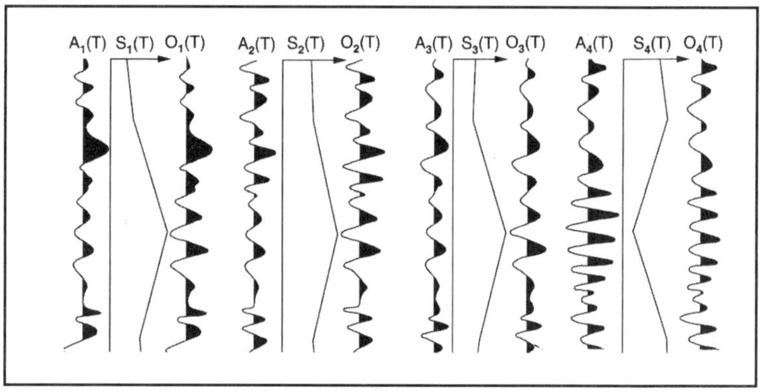

Fig. 6–7 *Application of Diversity Scalers*

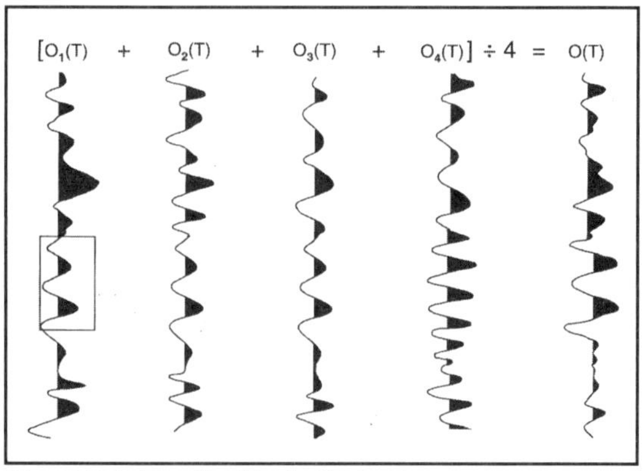

Fig. 6–8 *Diversity Stack Output*

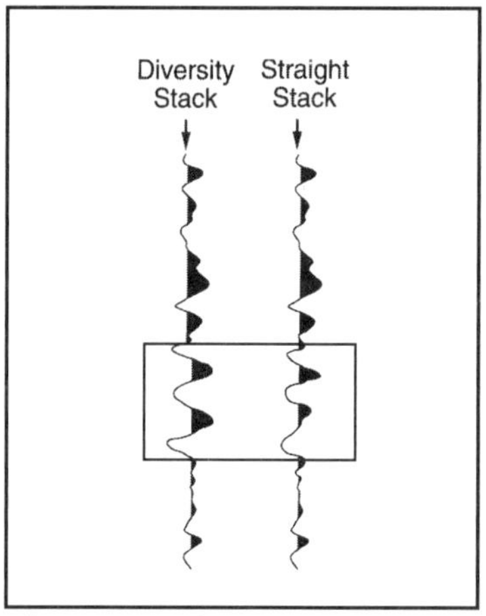

Fig. 6–9 *Stack Comparison*

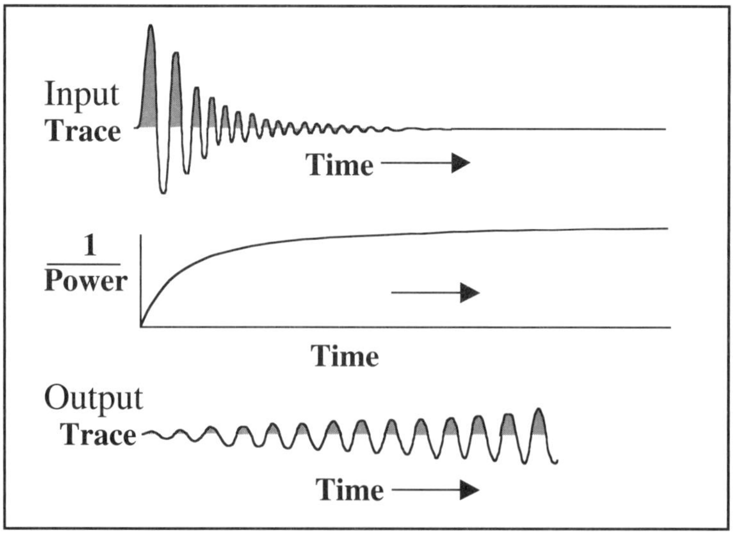

Fig. 6–10 *Need for Scaler Normalization*

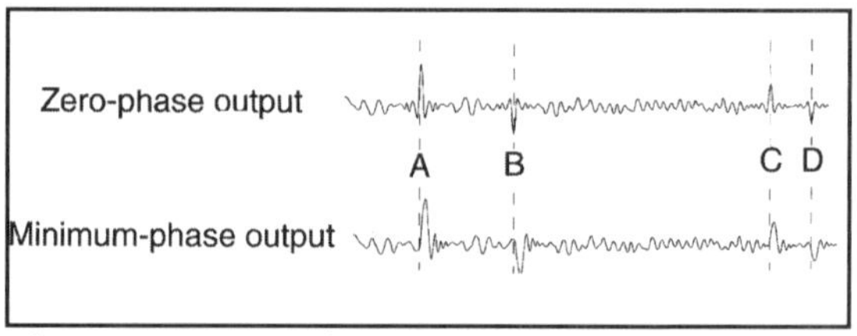

Fig. 6–11 *Zero- and Minimum-phase Correlation*

As shown by the synthetic trace example (Fig. 6–9), diversity stack, when used in proper conditions, is superior to straight stack. The boxed region of Figure 6–10 demonstrates the superiority of diversity stack in revealing signal in the presence of incoherent noise.

Since recorded amplitudes decay rapidly with time, scalers inversely proportional to power would increase rapidly with time, and scaling would make amplitudes increase with time. Thus, diversity stack scalers need to be normalized (Fig. 6–10).

A weak, noisy record (from an event such as a misfire) would be scaled very high and cause distortion in stack output. An amplitude threshold should be set and records with amplitudes below this threshold omitted from the stack to prevent this.

When vibrators are used as the energy source, the pilot sweep is recorded on a separate (auxiliary) trace of each record or on a special record as is the pilot sweep that has been passed through the analog filters in the recording. The sweep to use in correlation must be specified by the user, and the filtered sweep should always be specified so that the phase of the filters will be removed from the output wavelet. Otherwise, the output will not be zero phase.

Vibroseis correlation involves the following steps:

- align the pilot sweep and trace at time zero
- cross multiply pilot and trace sample values and sum the product
- store sum of products as output value at time zero
- shift the pilot sweep one sample period in the positive direction and repeat previous steps to obtain output value for next sample period
- continue the procedure until desired output record length (listen time) is obtained

This was shown by Figure 5-50, using a synthetic example. The input trace (top) is unintelligible. After the filtered pilot sweep is shifted past primary reflection *A*, a short duration, zero-phase wavelet can be seen. Subsequent shifts disclose the presence of primary reflections *B, C,* and *D*.

When the pilot sweep passes over a reflection event the output is, essentially, an autocorrelation scaled by the reflection coefficient, making the output wavelets zero-phase. Deconvolution, which is applied in later processing, usually requires a minimum-phase input for optimum results. Minimum-phase correlation output is an option in modern Vibroseis correlation programs. Zero-phase wavelets are converted to minimum phase by inverse filtering, and sweep frequencies are input to permit correct inverse filter design.

Figure 6–11 compares zero-phase and minimum-phase Vibroseis correlation. Note that the time for a reflection event is at the peak (or trough for a negative reflection coefficient) for zero-phase wavelets but is where the wavelet first breaks down, or up, for minimum phase wavelets.

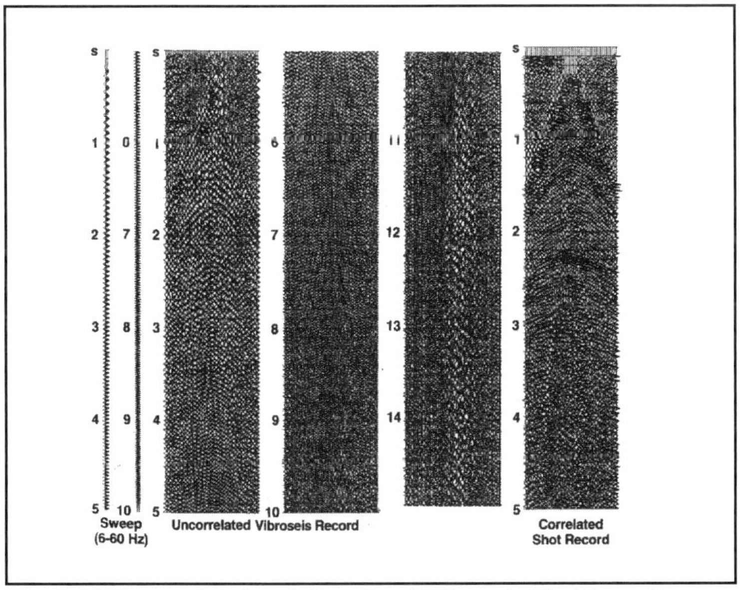

Fig. 6–12 *Uncorrelated and Correlated Vibroseis Shot Records*

In summary, Vibroseis correlation collapses a long-duration sweep into a short-duration wavelet similar to impulsive source wavelets. Either zero- or minimum-phase wavelets can be output as specified by the user. Record length is reduced by an amount equal to the duration of sweep.

Figure 6–12 shows a 10 second sweep, an uncorrelated Vibroseis record made with that sweep, and the correlated version of the same record. The uncorrelated record is 15 seconds long and the correlated record is five seconds long.

Seismic data processing sequences. The objective of seismic data processing is to produce an image of the subsurface target that is as accurate as possible. Inputs are field records on magnetic tape, survey/navigation data on magnetic tape, reports, and maps

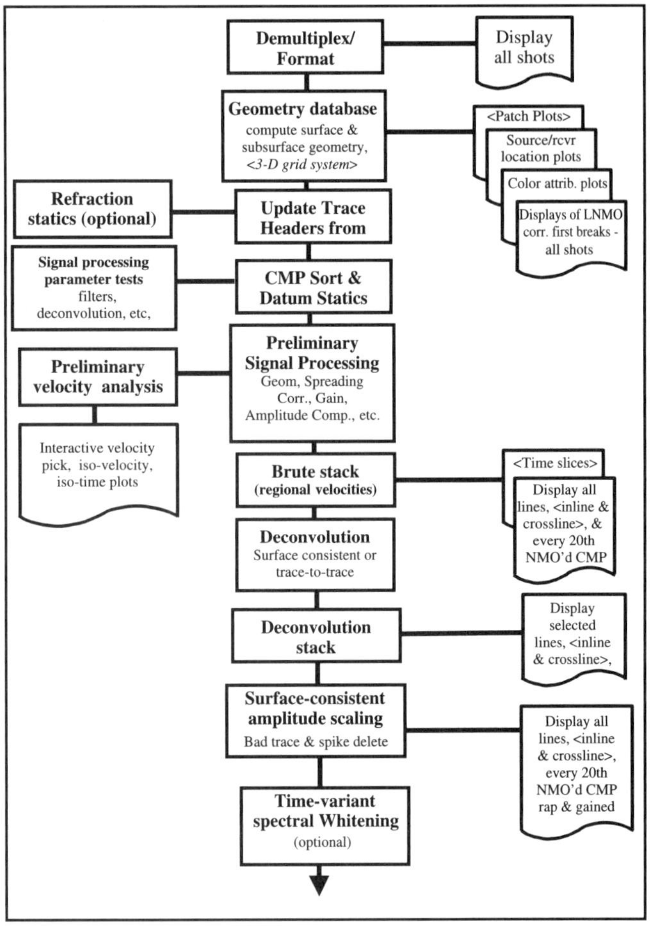

Fig. 6–13 *Typical Processing Sequence*

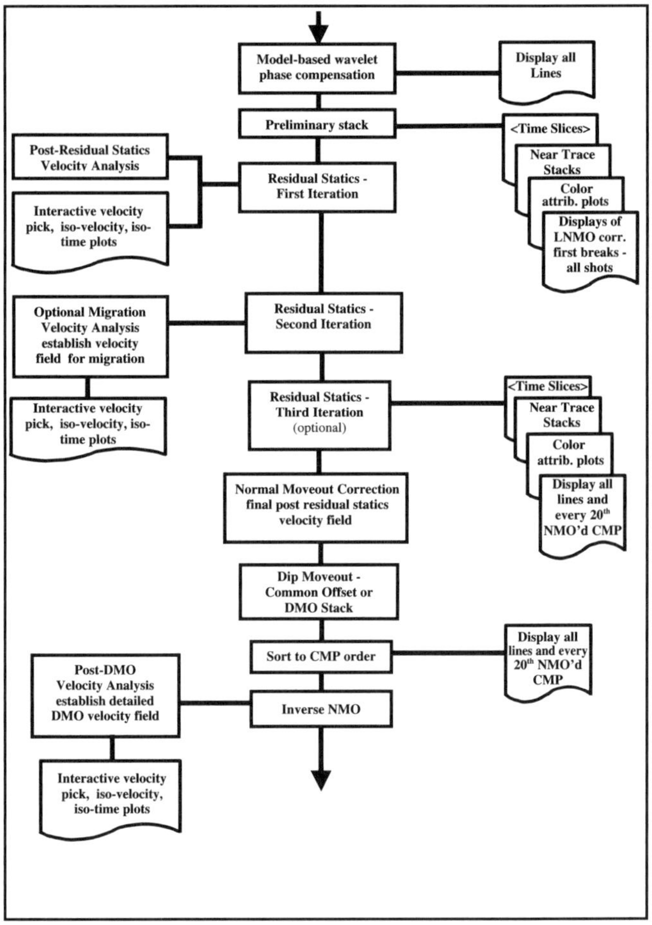

Fig. 6–14 *Typical Processing Sequence (continued)*

on a variety of media. Outputs are *seismic (cross) sections* with the same value displays and *iso-time* (amplitude variations over the data volume at constant times) displays. Iso-velocity displays (countours of equal velocities) are also produced and used to evaluate velocity picks.

The amplitude processing done to this point has had no effect on signal bandwidth. Usually, deconvolution of some sort is applied for this purpose. Note that Wiener type deconvolution is statistical in nature and should not be applied if AVO analysis is to be performed. Surface consistent deconvolution should be applied when AVO is planned.

The same basic processes are applied to both 2-D and 3-D data, but the nature of the 3-D data volume is such that some processes are applied only to 3-D data and other processes are applied quite differently in 3-D. Figures 6–13 through 6–15 comprise a flow chart for a typical processing sequence. Processes, displays, and other elements that apply only to 3-D are shown as <3-D only>.

Processing begins with data initialization. This is shown in the first three boxes of Figure 6–13. The first process shown, *demultiplex*, should only be necessary when reprocessing very old data. Modern seismic recording systems supply field data in *demultiplexed* form. These data will, however, be in one of the standard SEG formats and will have to be converted to the format specified by the processing system being used. It is best to display all shots at this stage to ensure data correspond to documentation and to identify problems in the data.

The *geometry database* is extremely important. All source and receiver positions must be assigned unique numbers and given coordinates in the *x-y* coordinate system adopted. Extensive and intensive quality control must be applied to assure that the geometry described by field documentation is correct and, if not, corrections made. Geometry errors can produce velocity

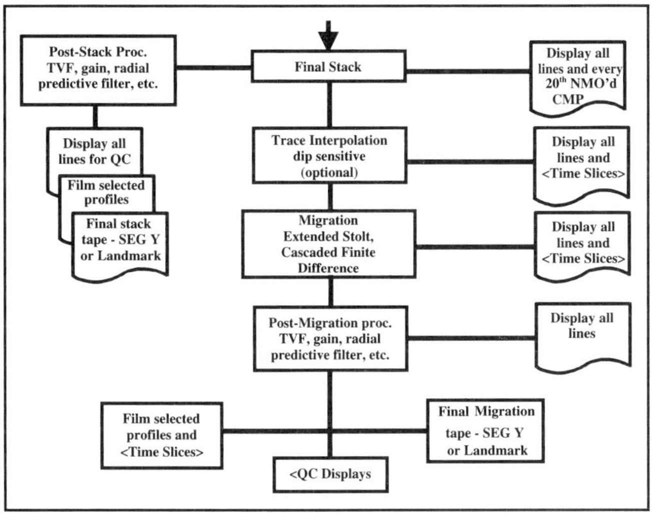

Fig. 6–15 *Typical Processing Sequence (continued)*

and statics errors that cannot be removed except by geometry correction. The displays shown at the right of the geometry database box in Figure 6–13 are used for this purpose. The results of the geometry quality control (QC) are used to update the geometry database.

At this point, a sort into *common midpoint* (CMP) format is made for use in parameter testing and selection. In land processing, datum statics are also calculated and applied. If refraction statics are used, then calculations are performed separately for later use. The updated geometry database should provide data needed for both datum and refraction statics.

Since each trace in the final output is to be a representation of the earth impulse response at each midpoint position, amplitudes must be corrected for losses along the reflection ray paths.

Geometrical spreading correction and some sort of *time-variant amplitude correction (gain)* are applied for this purpose. The output of this preliminary signal processing provides input to a *brute stack*, in which a *regional velocity function* is used for *NMO correction*. Brute stack sections for all lines are displayed. In 3-D processing, this means not only all the lines parallel to the receiver lines (in-lines) but also all lines created by selecting CMPs perpendicular to the receiver lines (cross-lines). *Horizontal sections,* or *time slices*, at times in the target zone are displayed as well.

CMP gathers with preliminary signal processing are used to perform an initial *velocity analysis*. Results of the analysis are interpreted via an interactive velocity picking process. The effect of velocity choices on CMP gathers and on temporal and spatial variation of the *velocity field* can be monitored in real time. *Iso-velocity* (contours or color-coding to show times at which the velocities have the same value) displays and *iso-time* (amplitude variations over the data volume at constant times) displays are also produced and used to evaluate velocity picks.

The amplitude processing done to this point has had no effect on signal bandwidth. Usually, *deconvolution* of some sort is applied for this purpose. After deconvolution is applied, data should be *stacked* using the same regional velocity function. If the velocities obtained in the initial velocity analysis are used, changes caused by the differences in velocities cannot be separated from those caused by deconvolution.

Amplitude variations dependent on surface location of sources and receivers may still be present. Corrections for such amplitude variations should be determined and applied—*surface-consistent amplitude correction.* Amplitude thresholds (minimum and maximum) should be established and traces with rms amplitudes below the threshold (weak or dead traces) and those above the threshold (noisy traces) should be zeroed and not used in subsequent processing. Traces should be checked for *spikes*—abrupt, large positive or negative changes in amplitude—and any found should be removed. Since many processes are applied as convolutions, spikes will cause the impulse response of the process to be superimposed on the trace. Data should again be stacked and all lines displayed. In addition selected CMP gathers (every 20th gather, for example) should be displayed.

If data are not sufficiently *white*, a process called *time-variant spectral whitening* can be applied. This has the effect of making the amplitude spectrum flat over a desired bandwidth and does not attempt to increase amplitudes outside this bandwidth. If used, data should be stacked and all lines displayed for evaluation of process effect.

Model-based phase compensation (Fig. 6–14) can be applied to remove residual phase components and provide the desired zero-phase wavelets. Time-variant spectral whitening can be used as an alternative to statistical deconvolution when coupled with model-based phase compensation. Again, display all lines after a process is applied.

At this point, the velocities obtained in the initial velocity analysis are used for NMO corrections and a preliminary stack performed. Quality control displays at this stage consist not only of the stack sections for all lines but also time slices in the target zone, near trace stacks or gathers, color attribute (amplitude) displays, and selected CMP gathers (same set as previously selected). Comparison with stack sections from the previous process shows the effect of the new velocities.

In land processing, at least two iterations of *residual statics* are run, with a third iteration optional. The number of iterations depends on how quickly residual statics converge to a stable value. Each residual statics iteration is followed by a velocity analysis. The combinations of residual statics and velocity analyses better separate time variations in the near surface from those along reflection ray paths and those caused by geological structure variations.

In marine processing, no residual statics analysis is performed, and usually at least one less velocity analysis is done. However, in both land and marine processing, DMO is applied to remove the effect of dip from velocities and to move data to their true zero-offset position. This requires application of NMO corrections and, in some cases, sorting into the common offset domain. After the DMO correction is applied, the NMO correction is backed out—inverse NMO corrections are applied.

After this, a final stacking velocity field is obtained via velocity analysis and interactive velocity picking. This velocity field is used to apply NMO corrections to CMP traces and then do the final CMP stack. The usual QC displays are also produced. Post stack processes such as time–variant filtering (TVF), attenuation of residual multiple reflections, and attenuation of ambient noise are applied to produce a CMP stack with the best possible signal-to-noise ratio.

A variety of migration algorithms are available. The choice depends on cost, maximum dip, and frequency requirements. In 3-D, a choice must be made between one-pass and two-pass migration. One-pass migration is true 3-D migration while two-pass migration applies 2-D migration in-line followed by another 2-D migration cross-line. In Figure 6–15, two-pass migration using *extended Stolt migration* for the in-line migration and cascaded (multi-stage) *finite difference migration* for cross-line migration is indicated.

The final stack is then input to migration. It is usually best to input a stack that does not have TVF applied to it. Interpolation to a smaller spatial sampling interval may be advisable prior to migration to limit the harmful effect of spatial aliasing on migration. This is particularly the case in 3-D when the cross-line sampling interval is quite large compared to in-line. It may also be necessary to apply some sort of *multiple attenuation* processing other than deconvolution. Such processing can be done pre-stack, post-stack, or both. Multiple attenuation before the final velocity analysis can improve the quality of the velocity field.

Data Initialization

Field tape processing has two basic functions—reformatting magnetic tape data and editing traces.

Reformatting magnetic tape may, but usually does not, involve demultiplexing. Most data, today, are recorded in demultiplex format. In earlier times, the conversion from multiplex to demultiplex (shown in Fig. 5–96) was the first process applied. However, data must be put in the format required by the processing software.

Trace editing includes *despiking* (elimination of high amplitude anomalies), *polarity reversal* (change to correct polarity), *trace zeroing* (set trace amplitudes to zero if average amplitude is outside amplitude thresholds), and *source/receiver coupling compensation* (correct trace amplitudes for poor coupling at specific surface locations).

After field tape processing is complete, geometry must be checked. This is usually done by applying *linear moveout* (LMO) *corrections* and displaying first breaks. Linear moveout corrections take the form:

$$T_c = T - \frac{x}{V_R} \tag{6.1}$$

where

T_c = first break time corrected for linear moveout

T = raw first break time

x = offset for trace being corrected

V_R = refraction velocity of first refractor

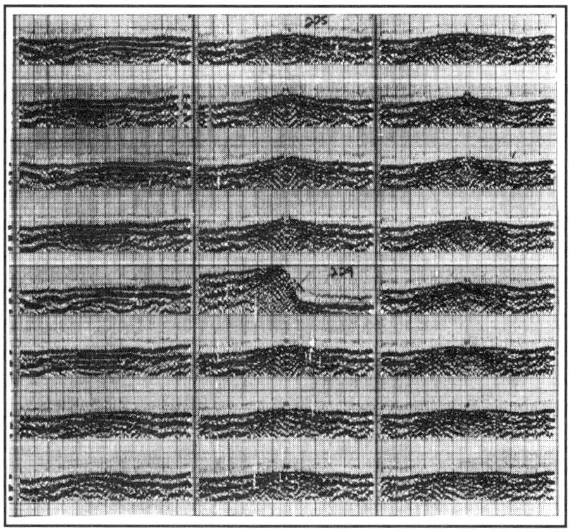

Fig. 6–16 *LMO Traces Display*

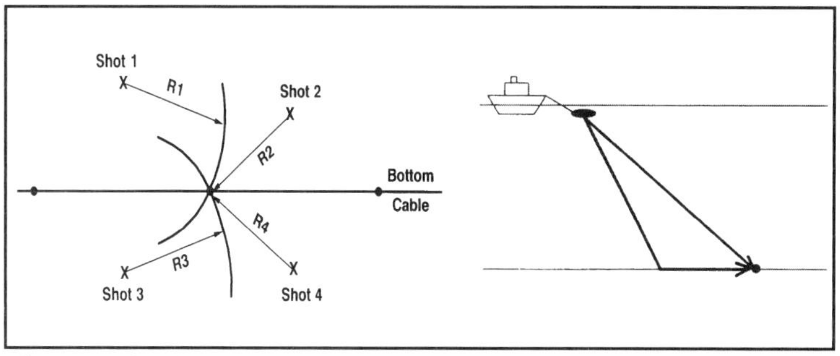

Fig. 6–17 *OBC Receiver Location*

If geometry is correct, traces corresponding to arrivals from the first refractor will have first breaks aligned. Short offset traces will be overcorrected if direct arrivals are first breaks. Far offset traces will be under corrected. This is shown in Figure 6–16. Most records appear to be normal, indicating geometry provided from the field is correct for these shots. However, shot number 329 (fifth record from top in the center) shows anomalous first breaks. Traces on the left are under corrected and traces on the right are overcorrected. This indicates that the shot point location is incorrect and should be moved to the right.

In OBC operations, receiver locations are not known precisely. Source positions are located using differential GPS positioning (GPS antenna on source array) and first arrivals from the seismic data are picked. For direct arrivals, the times picked multiplied by water velocity give the acoustic ranges or source-to-receiver distances. For first break refractions, calculate ranges by subtracting delay times from first break times and multiplying the difference by the refractor (sub-water layer) velocity. Since records are shot from multiple sources into the same receivers, receiver locations can be computed using a multi-range acoustic positioning method analogous to conventional radio positioning with the known source positions acting as base stations (Fig. 6–17).

Data are input as *shot* or *common source* records. After data initialization is completed, data are output in common midpoint or CMP format. Other applications require *common offset* or *common receiver* format. Reflection point or stacking diagrams, such as in Figure 6–18, facilitate collecting or gathering traces for various purposes. Depth points for each field record are shifted down, vertically, but are in correct horizontal positions.

Stacking diagrams form a trace matrix. Lines drawn through the matrix indicate which traces to collect for specific gathers, as shown in Figure 6–19. Vertical lines connect traces that have common midpoints. Horizontal lines connect traces that have common sources. A line drawn through the shortest offset traces connect traces in a *near-trace gather* (NTG). Lines parallel to the NTG line connect traces that have common offsets. Another set of slanting lines, with a steeper slope, connect traces that have common receivers.

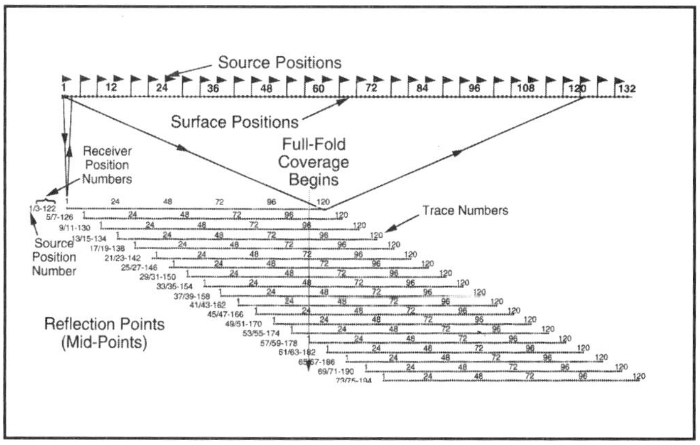

Fig. 6–18 *A Stacking Diagram*

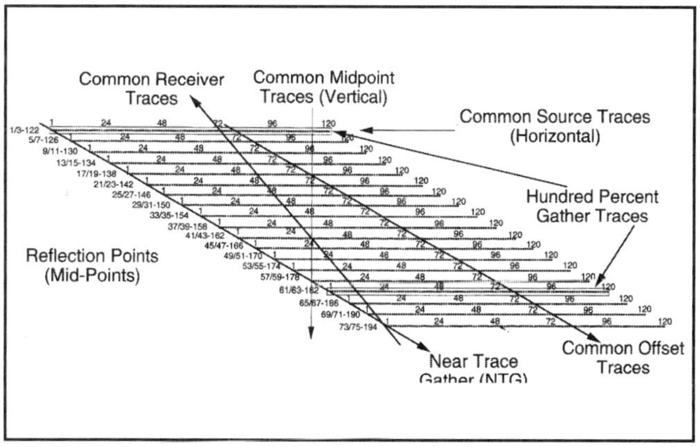

Fig. 6–19 *Trace Gathers*

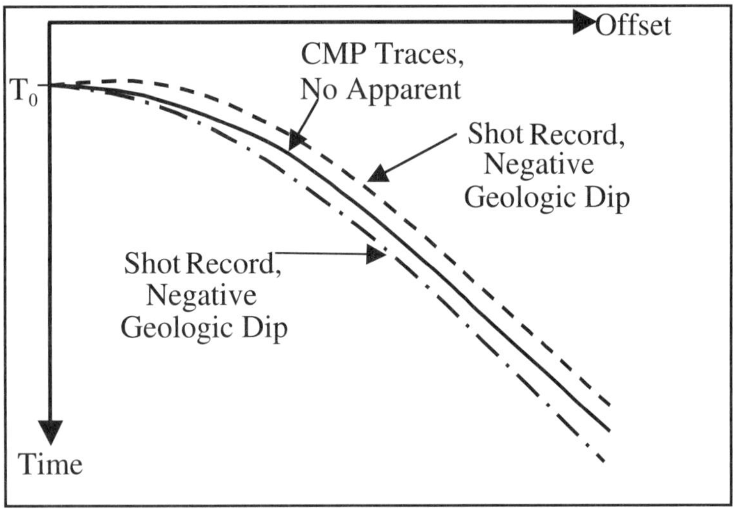

Fig. 6–20 *Shot and CMP Traces with Geologic Dip*

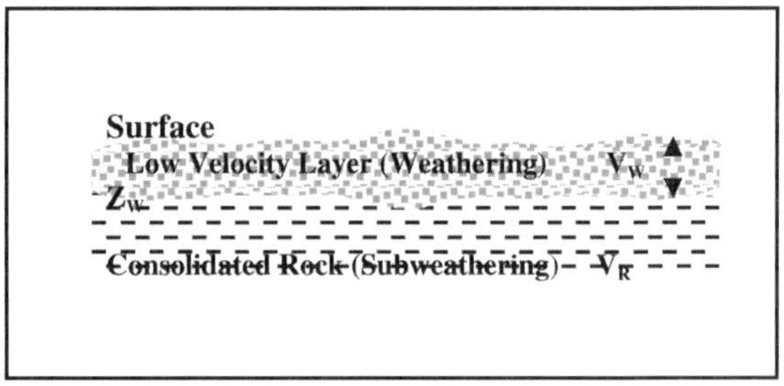

Fig. 6–21 *Near-surface Model*

Common source (shot) traces differ in moveout, offset, receivers, and may differ in dip. This is the format for input to amplitude recovery and velocity filtering. Common receiver traces differ in moveout, offset, sources, and may differ in dip. This format may be used in surface consistent processing. Common midpoint traces differ in moveout, offset (split spread shooting may result in duplicate offsets), receivers, and sources. This format is the input to velocity analysis and CMP Stack (after NMO/DMO and static corrections, if the latter are done). Independence from effect of dip on NMO curves is an important attribute for the just-mentioned processes.

Figure 6–20 shows moveout on CMP traces and common source (shot) traces. The moveout on each trace is the sum of NMO and geologic dip. If dip is negative, then reflection times initially decrease and minimum reflection time may not be at zero offset. If dip is positive, then moveout is greater than NMO alone at all offsets. On CMP traces however, the effect of dip on reflection times is negligible for dips up to about 10°.

Common offset traces differ in receivers, sources, and reflection points. Trace-to-trace time differences are caused by differences in dip and/or velocity. This is the format for input to DMO. It is also used to determine residual moveout in residual statics analysis and other special analyses.

An NTG is a set of common offset traces with minimum offset. It is used for QC and parameter determination. It is a good indicator of horizon times and dips. A 100% gather (HPC) is a collection of records that provide single-fold coverage of the subsurface. The HPC is obtained by selecting every N^{th} shot record, where N = maximum fold. It may be used in conjunction with the NTG to analyze spatial variation of velocity.

There is a layer of unconsolidated rock just below the surface in virtually all land areas. See Figure 6–21. This layer, or layers, has quite low and variable seismic velocities. It is sometimes called the *low-velocity layer* or *weathering*. The variable velocity and/or thickness produces trace-to-trace time differences which give false pictures of subsurface structure. The effect of variations in surface elevations and the near-surface thickness and velocity can readily be seen on a reflection from a flat reflector.

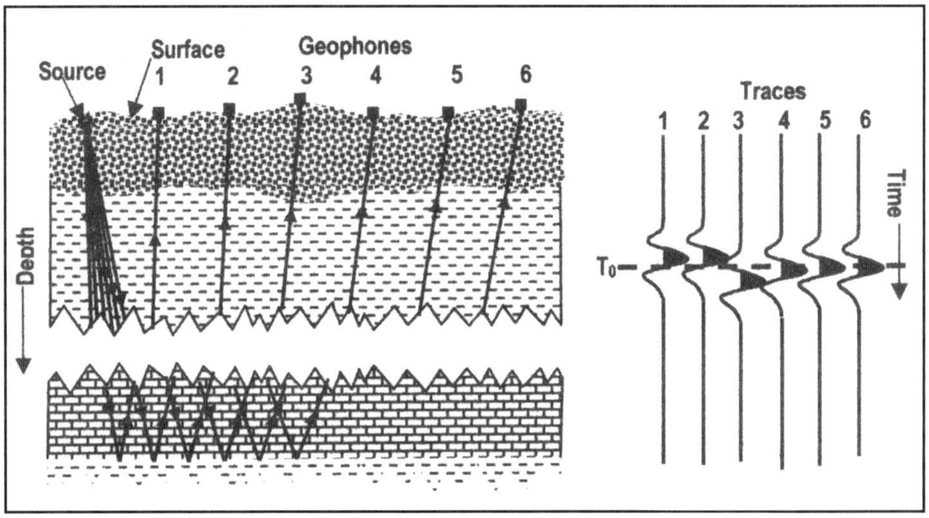

Fig. 6–22 *Effect of Variable Near Surface*

Ray paths from a source to a flat reflector and upward to six geophones through a variable near surface are shown on the left of Figure 6–22. On the right are the reflection events recorded at geophones one through six on traces one through six. If it were not for the variation in the near surface, all events would be recorded at time T_0. A variable near surface causes reflection time variations that, if not properly accounted for, can give false impressions of geologic structure and/or seismic velocity.

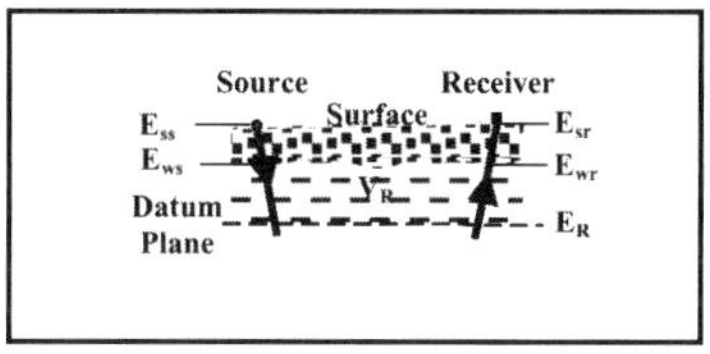

Fig. 6–23 *Static Corrections, Case I: Surface Source*

Static corrections are applied to remove the effect of elevation and near-surface variations on reflection times. In effect, datum statics strip off the near surface and materials down to the datum, which is usually flat. This moves sources and receivers from the actual surface to a new reference plane underlain by material having a higher velocity V_R than weathering.

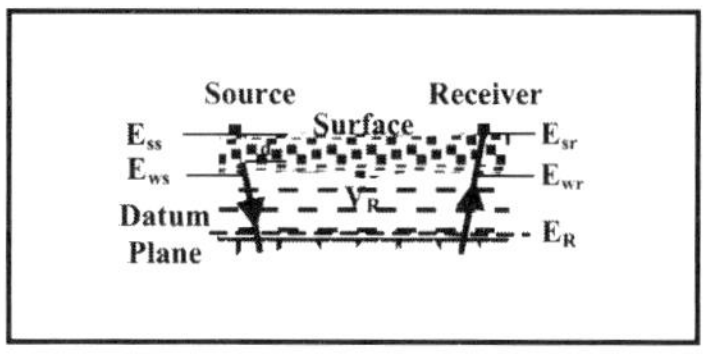

Fig. 6–24 *Static Corrections, Case IIa: Source in Weathering*

Datum static corrections are calculated separately for sources *SSC* and receivers *RSC*. The total static *DSC* for a particular trace is the sum of the source static and the receiver static.

$$DSC = SSC + RSC \tag{6.2}$$

There are two basic situations for datum statics computations. Case I, for a surface source, and Case II, for sources shot beneath the surface. Figure 6–23 shows Case I. The source static *SSC* is calculated from:

$$SSC = -\left[\frac{E_{ss} - E_{ws}}{V_W} + \frac{E_{ws} - E_R}{V_R}\right] \tag{6.3}$$

The receiver static *RSC* is calculated from:

$$RSC = -\left[\frac{E_{sr} - E_{wr}}{V_W} + \frac{E_{wr} - E_R}{V_R}\right] \tag{6.4}$$

where

E_{ss} = elevation of the surface at the source position

E_{sr} = elevation of the surface at the receiver position

E_{ws} = elevation of the base of the weathering at the source position

E_{wr} = elevation of the base of the weathering at the receiver position

E_R = elevation of the datum plane

V_w = velocity of the weathering

V_R = replacement velocity

Case II has two subclasses. The source, for Case IIa (Fig. 6–24), is within the weathering or near surface layer. In this case the source static correction is computed from Equation 6.5 or Equation 6.6.

$$SSC = -\left[\frac{E_{ss} - d - E_{ws}}{V_W} + \frac{E_{ws} - E_R}{V_R}\right] \quad (6.5)$$

$$SSC = -\left[\frac{E_{ss} - E_{ws}}{V_W} - T_{uh} + \frac{E_{ws} - E_R}{V_R}\right] \quad (6.6)$$

where

E_{ss}, E_{sr}, E_{ws}, E_{wr}, E_R, V_w, and V_R are as previously defined

d = depth of source

T_{uh} = up-hole time

Up-hole time is the time required for energy to travel from the source to a geophone at the surface near the top of the hole. The receiver static is calculated from Equation 6.4.

In Case IIb (Fig. 6–25), the hole is drilled through the weathering. The source static correction is calculated from Equation 6.7a or Equation 6.7b.

$$SSC = -\left[\frac{E_{ss} - d - E_R}{V_R}\right] \quad (6.7a)$$

$$SSC = -\left[\frac{E_{ss} - E_R}{V_R} - T_{uh}\right] \quad (6.7b)$$

Sometimes a less detailed approach, called elevation statics, is used. In this method, the base of the weathering is ignored and an average velocity to the datum Vavd is used. Computation of the source elevation static (SESC), the receiver elevation static (RESC), and the total elevation static (ESC) is shown below for Cases I and II.

Case I

$$SESC = -\frac{E_{ss} - E_R}{V_{avd}}$$

$$RESC = -\frac{E_{sr} - E_R}{V_{avd}}$$

Case II

$$SESC = -\frac{E_{ss} - E_R}{V_{avd}} + T_{uh} = \frac{E_{ss} - d - E_R}{V_{avd}} \quad (6.8)$$

$$RESC = -\frac{E_{sr} - E_R}{V_{avdd}} \quad (6.9)$$

Datum statics, however they are calculated, are saved as separate source and receiver statics and total datum statics are added to trace headers.

Refraction static corrections are computed in a four-step process:

1. pick refraction travel times
2. decompose refraction times
3. construct refraction elevation model
4. compute static corrections

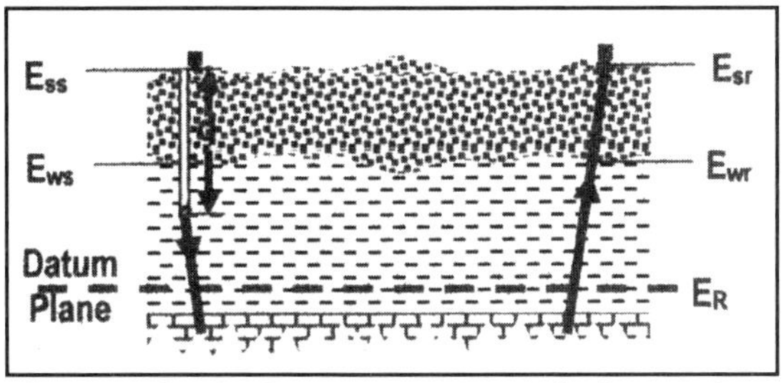

Fig. 6–25 *Static Corrections, Case IIb: Source in Subweathering*

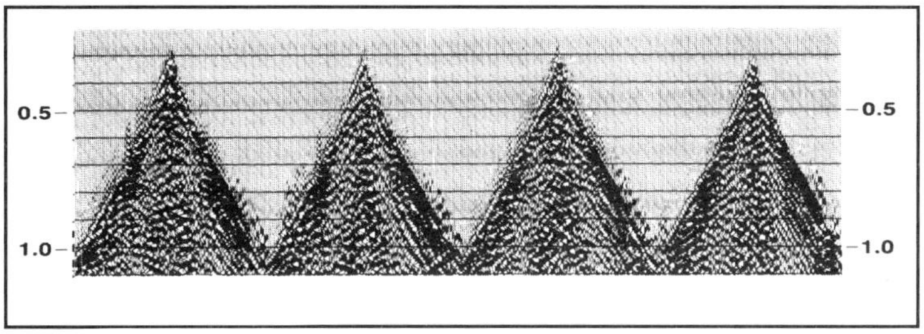

Fig. 6–26 *First Breaks of Selected CMP Gathers*

Figure 6–26 shows first breaks of a few CMP gathers. First break times are picked as the time at which the traces first break downward. The short offset times, in general, will correspond to direct arrivals. Later picks will be refraction arrivals from the base of the weathering or a deeper refractor. Figure 6–27 shows ray paths and a time-distance or *T-X* plot of first break times versus offset. Velocities V_0, V_1, and V_2 are calculated from the inverse slopes ($\Delta d/\Delta t$) of the lines fitted to the data in the *T-X* plot.

Refraction arrivals can be decomposed into three parts:

$$T_{ij} = t_i + t_{ij} + t_j \tag{6.10}$$

where

T_{ij} = refraction time from the source *i* to receiver *j*

t_i = time from the source down to the refractor

t_{ij} = time along the refactor beneath the source to beneath the receiver

t_j = time from refractor up to the receiver

In Figure 6 28, $ti = SP/V_0$, $t_{ij} = PQ/V_1$, and $t_j = SP/V_0$.

In multifold shooting, many estimates of t_i and t_j are obtained. For example, in Figure 6–29 traces are recorded at receiver group R_1 from sources S_1, S_2, S_3, and S_4. Similarly, traces are recorded at receiver groups R_A, R_B, R_C, and R_D from source S_4. Folds much greater than four are common, and a particular source position is often also a receiver position. The redundancy of information about surface positions allows application of statistical methods to extract delay times of t_i and t_j.

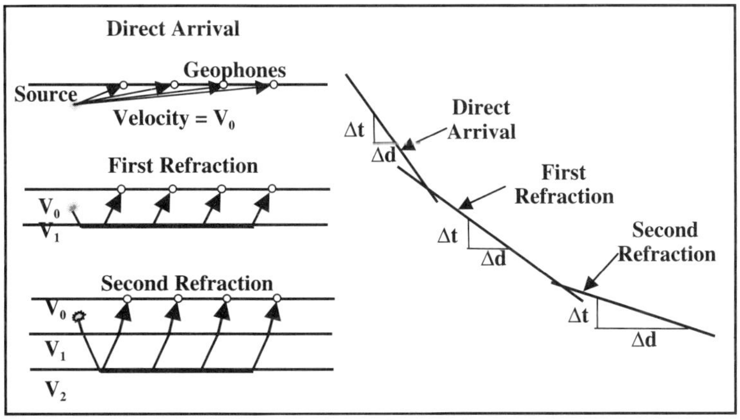

Fig. 6–27 *First Break Ray Paths and T-X Plot*

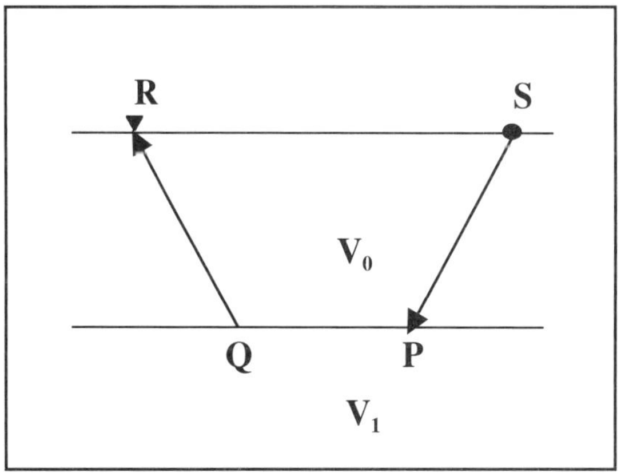

Fig. 6–28 *Refraction Ray Path Segments*

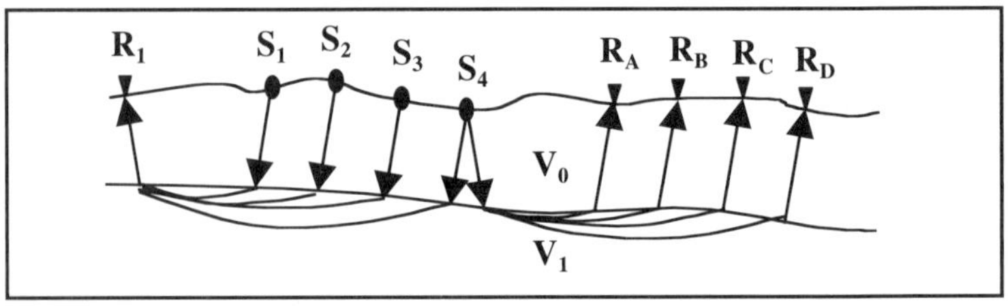

Fig. 6–29 *Redundancy of Data in Refraction Statics Determination*

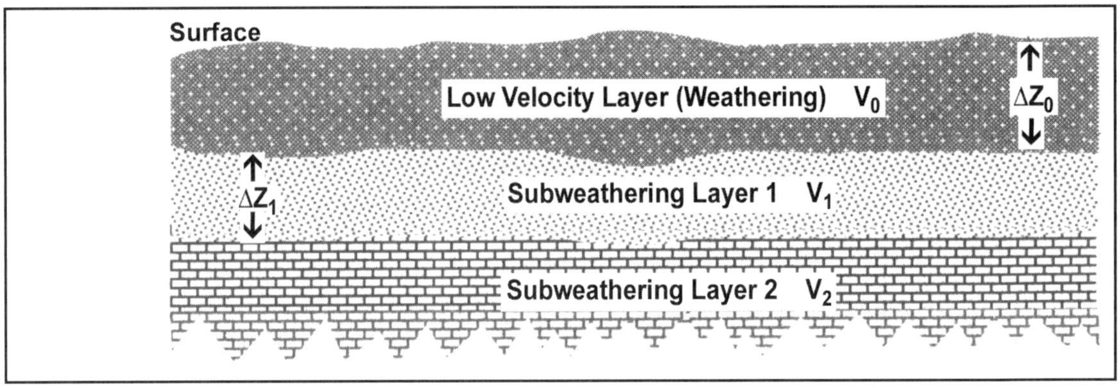

Fig. 6–30 *Refraction Elevation Model*

If the refraction velocity V_R is known, then the time along the refractor can be calculated from:

$$t_{ij} = \frac{x_{ij}}{V_R}$$

where

x_{ij} is the distance between source i and receiver j.

Substituting this into Equation 6.10 and solving for t_i yields:

$$t_i = T_{ij} - \frac{x_{ij}}{V_R} - t_j \qquad (6.11)$$

The Gauss-Seidel method is an iterative method of obtaining surface consistent solutions from redundant data. In Equation 6.11, initial values of t_j factors (e.g. = 0) are selected and a least-square error estimate of t_j is obtained. The error is used to update the t_j factors and a second iteration, third, and more iterations are performed. The iterations are done very quickly and there is usually a rapid convergence to a particular value of t_j. The same approach is used to determine all t_i and t_j values.

Using the methods described in chapter 4, layer thicknesses ΔZ_0 and ΔZ_1 below the various surface positions (stations) can be calculated from the delay times t_i and known velocities V_0, V_1, and V_2. This allows construction of a refraction elevation model, such as that shown in Figure 6–30. Datum statics can then be calculated from relevant Equations 6.4 through 6.9.

Preliminary Signal Processing

There are two main objectives in this stage of processing—eliminating amplitude variations caused by factors other than reflection coefficient variations and attenuating coherent noise.

Amplitude processing. If a field record recorded by a 24-bit system or one recorded by an FTP system with gain removed is displayed, it has the appearance of the record of Figure 6–31. Amplitudes rapidly decay primarily due to geometrical spreading and inelastic attenuation.

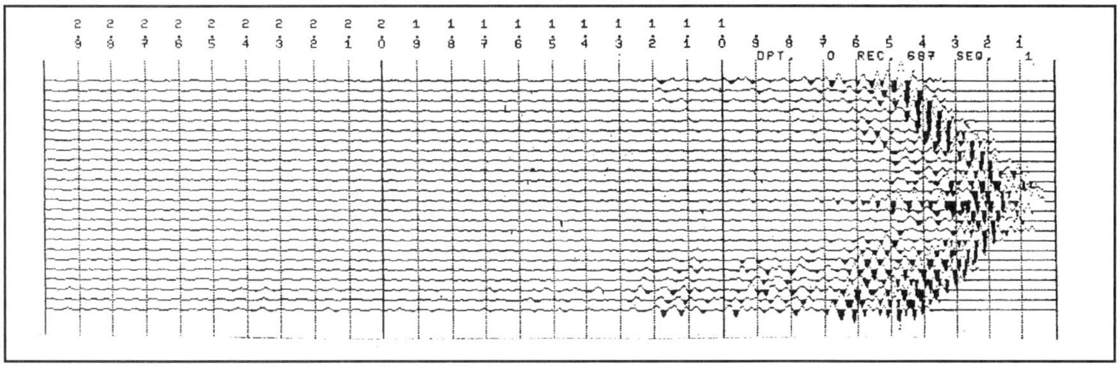

Fig. 6–31 *Field Record with No Gain Applied*

Geometric spreading. In a constant velocity medium, energy spreads out from the source at the same rate in all directions. Wavefronts in such a medium are spherical. At a time t_1, the surface area of the wavefront in a medium of velocity V is $\pi(Vt_1)^2$. At a later time t_2, the surface area of the wavefront is $\pi(Vt_2)^2$. Amplitude is proportional to the square root of energy per unit area. Assume no energy is lost in propagation and the total energy is E. The amplitude at time t_1 is A_1, and the amplitude at time t_2 is A_2. The ratio of amplitudes is

$$\frac{A_1}{A_2} = \sqrt{\frac{\frac{E}{\pi(Vt_1)^2}}{\frac{E}{\pi(Vt_2)^2}}} = \frac{t_2}{t_1} \tag{6.12}$$

It follows from Equation 6.12 that $A_1t_1 = A_2t_2$ = constant. In the case of a constant velocity medium, amplitude losses caused by geometric spreading, also called *spherical divergence*, can be corrected by multiplication by time t.

In real cases, however, velocities are not constant. In fact, velocities usually increase with depth so actual wavefronts are not spherical, and their areas increase at a faster rate than in spherical divergence. This is illustrated in Figure 6–32.

To compensate for the faster increase in wavefront surface area when velocity increases with depth (and time), the geometric spreading correction is defined as $g(t) = (V/V_0)^2t$ where V is a smoothed regional velocity function and V_0 is the initial velocity or velocity at time zero. Figure 6–33 shows the effect of the Geometric Spreading Correction.

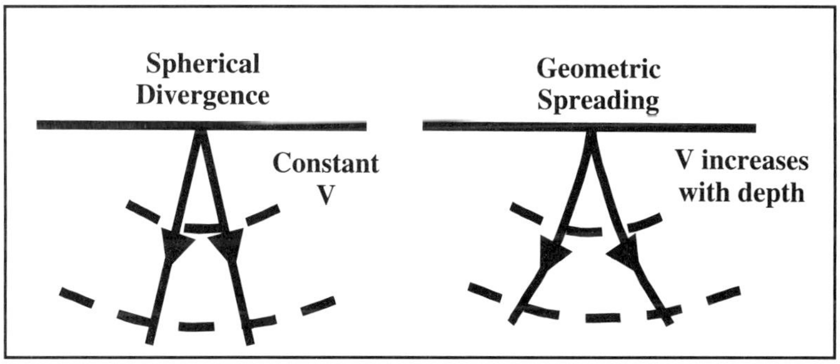

Fig. 6–32 *Spherical Divergence and Spherical Spreading*

Gain functions. The geometric spreading correction does not compensate for all amplitude losses. Usually the most severe losses are caused by inelastic attenuation—absorption and scattering—along the path. These losses depend on path length (and record time) plus the Q-factors of subsurface rocks through which the seismic waves propagate. A number of approaches (often called gain or scaling) are used to correct amplitudes for inelastic attenuation. These can be broadly classed as either *empirical* or *statistical*.

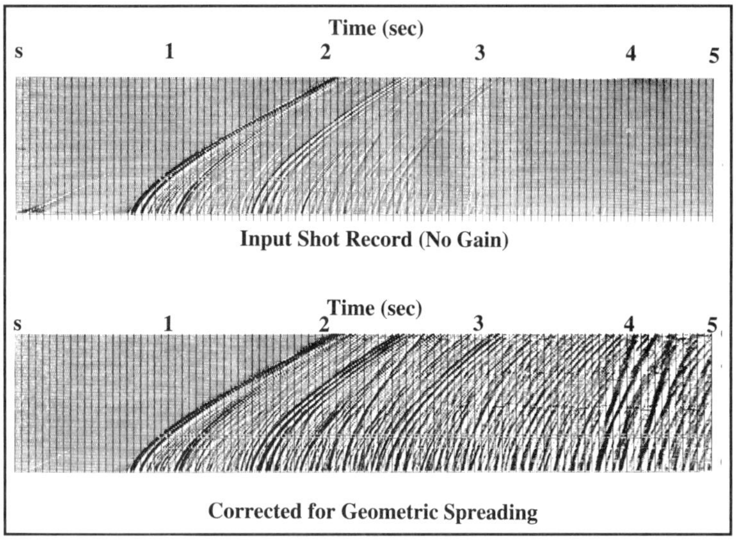

Fig. 6–33 *Geometric Spreading Correction Example*

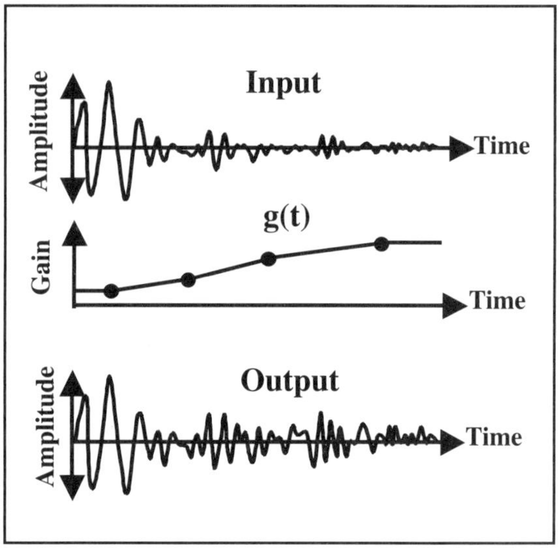

Fig. 6–34 *Programmed Gain Control*

Programmed gain control (PGC) is one of the empirical methods. It is a rather straightforward approach. One simply analyzes the data and selects two or more (usually more) pairs of times and gains. Gain is simply a number by which the input amplitudes are multiplied. Gain between the selected times is determined by linear interpolation. This is illustrated in Figure 6–34.

Designating $A_i\,(t)$ the input trace, $g\,(t)$ the PGC gain function, and $A_0\,(t)$ the output trace, input and output are related by:

$$A_0(t) = A_i(t)g(t) \tag{6.13}$$

Usually, a variety of functions are applied to representative CMP gathers. These outputs are then displayed and visually inspected to determine the best result. This evaluation is very qualitative since knowing what the result *should be* is required.

Another empirical approach is exponential scaling. This seems a good approach, intuitively, since it is known that inelastic attenuation is a type of exponential decay. The relation between input $A_i(t)$ and output $A_o(t)$ in this case is

$$A_0(t) = A_i(t)10^{at} \tag{6.14}$$

where a is a constant, usually expressed in dB/sec.

Similar to the procedure for PGC, various values of the constant a are used to apply exponential scaling to representative CMP gathers. These outputs are then displayed and visually inspected to determine the best result. Again, this evaluation is very qualitative since knowing what the result *should* be is required.

A frequently used method of amplitude processing is called automatic gain control (*AGC*). This is a statistical approach that is based on methods of volume control used in electronic audio systems. In such systems, a part of the output from an audio amplifier is fed back to the input. The effect is to maintain audio outputs within a narrow range, avoiding very loud or very faint volumes. Some electronic systems have fast AGC and others have slow AGC. In fast AGC the time between output and feedback action to limit output variation is very short. As a result, there is little variation in output sound volume. In slow AGC, the time between output and feedback action to limit output variation is longer and more variation in output sound volume is allowed.

One approach to AGC, called instantaneous AGC (IAGC), is as follows:

1. select a time gate or number of samples N
2. calculate the *average absolute amplitude* $\bar{A}$ within the time gate

$$\bar{A} = \frac{1}{N}\sum |A_n| \tag{6.16}$$

where

$|A_n|$ = absolute value of amplitude samples in the time gate, $n = 1, 2, \ldots, N$

3. determine the scaler from $S = \frac{A_{desired}}{\bar{A}}$, where $A_{desired}$ is the desired average absolute amplitude for the entire trace—A value of 1800 mv for $A_{desired}$ is frequently used
4. select the position within the gate at which the scaler will be applied
5. multiply the amplitude in the selected position by the scaler—this is the first output
6. move the gate down the trace one sample and repeat steps one through five
7. continue until all samples are used

The critical parameter is *gate length*. A very short gate is similar to fast AGC and results in elimination of almost all amplitude variation. Note that amplitude variations caused by subsurface geology must be preserved. A very long gate is similar to slow AGC and allows too much amplitude variation. Another negative effect of a long gate is a high-frequency filtering effect. The usual approach is to try different gate lengths and evaluate by visual inspection of CMP displays. Better results are sometimes obtained by allowing gate length to linearly increase. Signal-to-noise ratios decrease with record time so less gain variation is desirable at later times. Also, bandwidth decreases on the high side as record time increases, making high-frequency filtering less important.

Another AGC is *RMS AGC*. This is probably used more often than IAGC and may be the most-used form of amplitude processing. The method involves the following steps:

1. select a time gate or number of samples N
2. calculate the *root mean square (rms) amplitude* $\hat{A}$ within the time gate

$$\hat{A} = \sqrt{\frac{1}{N}\sum_{n=1}^{N} A_n^2} \tag{6.16}$$

where

A_n = amplitude samples in the time gate, $n = 1, 2, \ldots, N$

3. determine the scaler from $S = \frac{A_{desired}}{\hat{A}}$, where $A_{desired}$ is the desired rms amplitude for the entire trace—a value of 2000 mv for $A_{desired}$ is frequently used
4. assign this scaler to the center of the gate
5. move to the next gate down the trace and repeat steps one through four
6. continue until all samples are used
7. develop the RMS AGC function $g(t)$ by extrapolating the value of the first scaler back to start time, linearly interpolating between scalers up to the center of the last gate, and extrapolating the value of the last scaler forward to end of trace
8. multiply the input trace by the rms scaler function $g(t)$ as in Equation 6.13 and illustrated in Figure 6–34

As with instantaneous AGC, gate length is the critical parameter in RMS AGC. However, gates do not overlap in RMS AGC so the AGC action is not so fast. Gate lengths increasing with time are very common in RMS AGC.

OBC scaling. Ocean bottom cables are used in relatively shallow water. A major problem in this type of seismic acquisition is the presence of multiple reflections in the water layer. In OBC operations, these are called *OBC ghosts*. Figure 6–35 shows ray paths for a primary, plus the first and second ghosts. A third ghost is produced by additional reflections within the water layer.

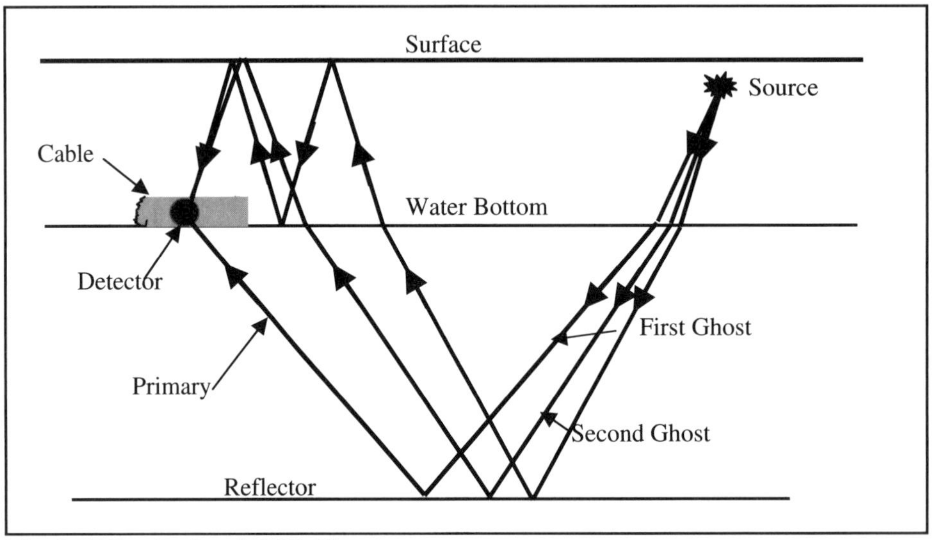

Fig. 6–35 *Ray Paths of OBC Ghosts*

The first ghost has opposite polarity from the primary because the reflection coefficient at the surface equals -1. The ghost arrivals have alternating polarity and decreased amplitude because the reflection coefficient of the water bottom is less than 1. As a result, a long source wavelet is produced, as shown in Figure 6–36.

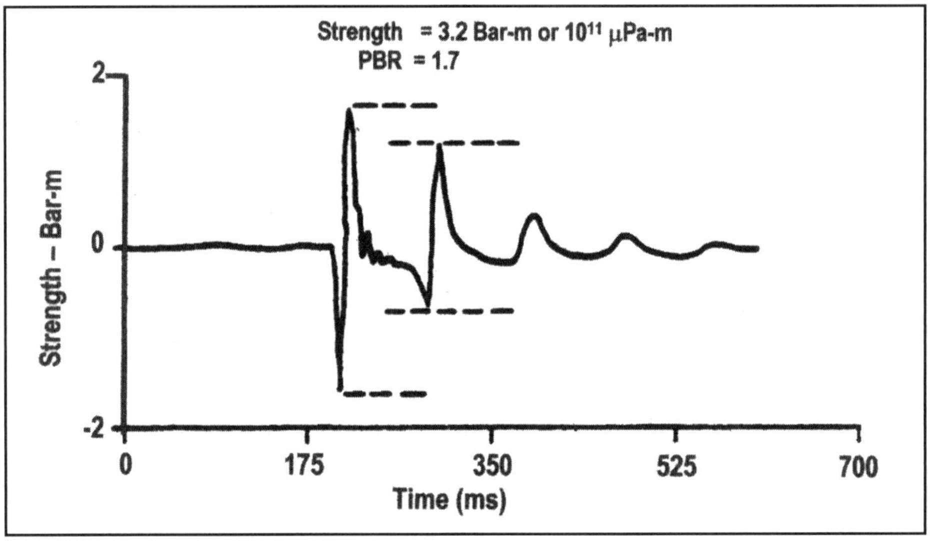

Fig. 6–36 *OBC Primary Plus Ghosts*

A wavelet such as that in Figure 6–36 is unacceptable. The ghosts must be attenuated so that wavelets are reduced to acceptable durations. The OBC method records data with both hydrophones and geophones and takes advantage of the differences in their responses to ghosts.

Figure 6–37 illustrates the generation of hydrophone ghosts. The numbers in the figure correspond to the following:

- the amplitude incident on the water bottom after reflection (primary reflection amplitude) = 1
- reflection coefficient of the water bottom = R
- reflection coefficient of the surface = -1
- the amplitude transmitted into the water from below = $1 + R$
- the amplitude reflected downward at the water surface = $-(1 + R)$
- the amplitude reflected upward at the water bottom = $-R(1 + R)$
- the amplitude reflected downward at the water surface = $R(1 + R)$

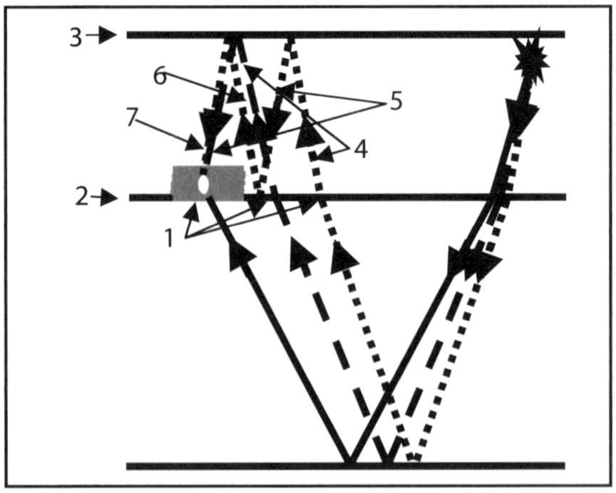

Fig. 6–37 *Hydrophone Ghosting*

Continued reflection at water bottom and surface result in the amplitude of successive ghosts being $-R$ times that of the previous ghost.

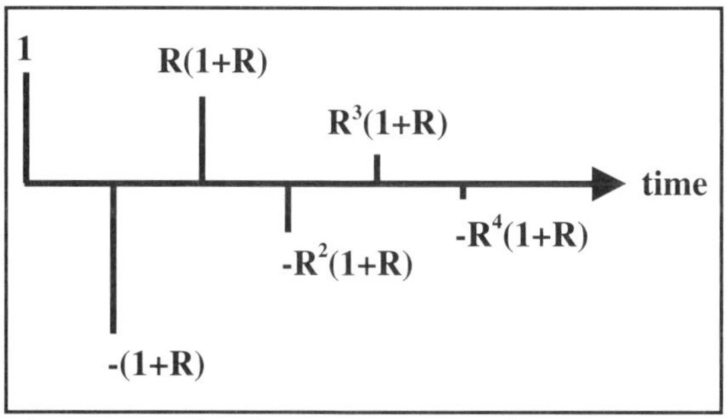

Fig. 6–38 *Hydrophone Ghost Impulse Response*

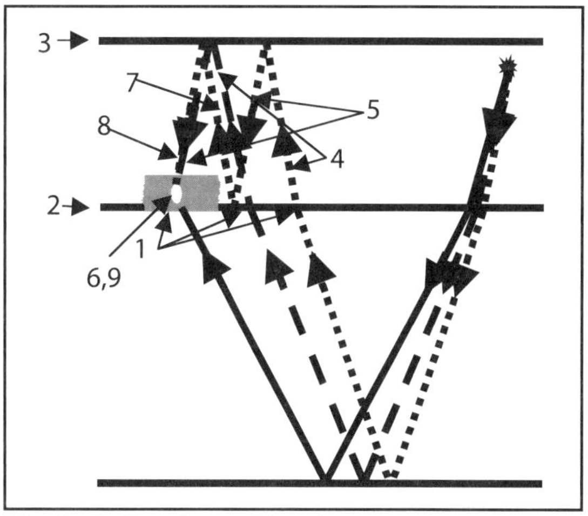

Fig. 6–39 *Geophone Ghosting*

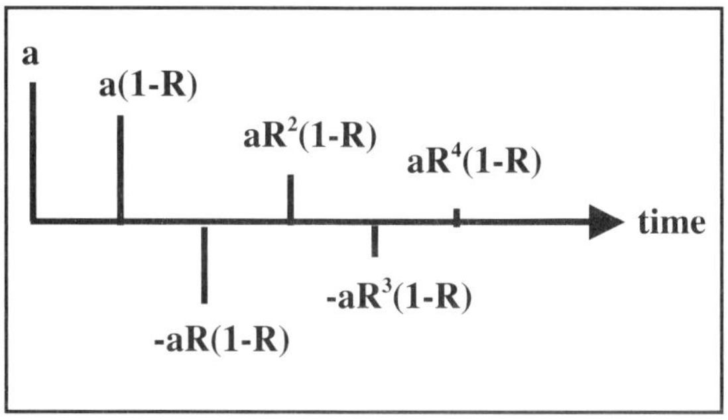

Fig. 6–40 *Geophone Ghost Impulse Response*

The effective wavelet for the hydrophone ghost is the convolution of the source wavelet with the hydrophone ghost impulse response, which is shown in Figure 6–38.

Outputs of geophones differ from those of hydrophones for two reasons:

- the polarity of hydrophone signals do not depend on the direction in which energy is incident on them but that of a geophone does
- the geophone sensitivity differs from that of the hydrophone

This means that the geophone ghost differs from the hydrophone ghost, as shown in Figures 6–39 and 6–40. The numbers in Figure 6–39 refer to the following explanations.

1. the amplitude incident on the water bottom after reflection (primary reflection amplitude) = 1
2. reflection coefficient of the water bottom = *R*
3. reflection coefficient of the surface = -1
4. the amplitude transmitted into the water from below = 1 - *R*
5. the amplitude reflected downward at the water surface = -(1 - *R*)
6. the amplitude recorded by the geophone is a (1-*R*) where *a* is the ratio between geophone and hydrophone sensitivity. The downward travel causes the change in signal polarity
7. the amplitude reflected up at the water bottom = -*R* (1 + *R*)
8. the amplitude reflected down at the water surface = *R* (1 + *R*)
9. the amplitude recorded by the geophone is -*a* (1-*R*)

Continued reflection at water bottom and surface result in the amplitude of successive ghosts being $-R$ times that of the previous ghost.

The effective wavelet for the hydrophone ghost is the convolution of the source wavelet with the hydrophone ghost impulse response that is shown in Figure 6–40.

The problem is to determine the scaling factor that will result in ghost cancellation when the scaled outputs are combined. The ratio of hydrophone ghost amplitude to geophone ghost amplitude is

$$\frac{(1+R)}{a(1-R)}$$

Since neither a nor R are known, some method of determining the proper scaling is needed. A frequently used method is

1. do a common receiver sort
2. select a window where both sensors have signal
3. balance the amplitudes in the windows of each sensor
4. guess at scale factor and apply to the geophone window
   a. sum traces in the window
   b. autocorrelate the summed traces
   c. compute *varimax* over region where the side lobe due to ghost energy appears

The varimax is defined as follows:

Let xj, $j = 1, 2, 3, ..., N$ represent the digital samples of a trace, then

$$\text{Varimax} = \frac{x_1^4 + x_2^4 + x_3^4 + \ldots + x_N^4}{(x_1^2 + x_2^2 + x_3^2 + \ldots + x_N^2)^2} \qquad (6.17)$$

5. increment the scale factor and go back to step four until the factor range is covered
6. select the scale factor that has the best varimax value
7. correct that factor for amplitude balance in step three
8. apply the corrected scale factor to the entire geophone gather and sum
9. repeat steps two through eight for each receiver gather
10. re-sort data to common shot gathers

Figure 6–41 illustrates the variation in the ratio $\frac{(1+R)}{(1-R)}$ across 360 receiver locations.

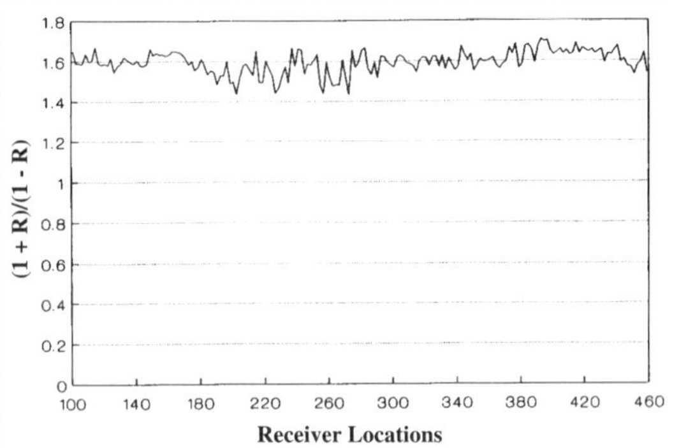

Fig. 6–41 *A Plot of Scale Factor (1+R)/(1–R) for 360 Receiver Locations*

Dual sensor recording significantly reduces water column reverberations and the associated spectral ghost notches and increases the spectral bandwidth of the data.

Other methods of determining the appropriate scalars are based on different responses to S/N ratio and different responses to variation in water bottom reflection coefficient. Some people do not acknowledge reverberations at the source and use calibration shooting based on receiver ghosts only. Calibration shots are expensive to record.

Varimax scalars are very robust and work well for areas with small reflection coefficients at the water bottom. They are relatively insensitive to variations in signal-to-noise ratio.

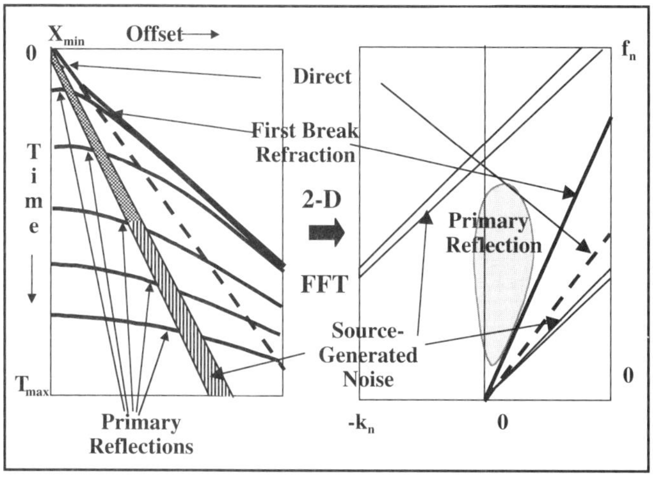

Fig. 6–42 *Transformation from T-X to F-K Domain*

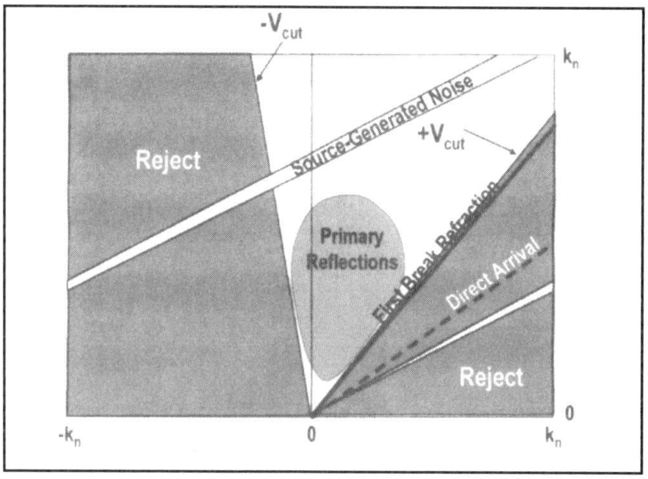

Fig. 6–43 *Positive and Negative Cut Lines*

Noise Suppression

The position of noise suppression in the processing sequence is somewhat controversial. The controversy is over the order in which spiking deconvolution and noise suppression should be applied. The authors' opinion is that noise suppression should precede deconvolution in order that the source wavelet, which spiking deconvolution is to remove, can be more clearly seen and its inverse better designed. It is generally true that in deconvolution the noise amplitudes are often increased by the deconvolution filter, but the objectionable noise components can usually be removed by band pass filtering.

Velocity filtering. The objective of velocity filtering is to eliminate or, at least, significantly attenuate source-generated noise and other undesired linear events. The procedure for applying velocity filtering is as follows:

1. Transform field records from *T-X* to *F-K* domain, separating signal from noise on the basis of dip or apparent velocity differences (Fig. 6–42).
2. Define positive and negative velocity cut lines. Data between cut lines will be passed, and everything else will be rejected (Fig. 6–43).
3. Multiply data in the pass zone by 1 and in the reject zone by 0. Linearly increase multiplier from 0 to 1 in narrow ramp zones just outside the two cut lines (Fig. 6–44). Velocities are close to $+V_{cut}$ (Fig. 6–45).
4. Transform *F-K* domain back to *T-X* domain. The first-break refraction is still evident but attenuated. All other linear events are eliminated. Far offset part of early reflections may be partially attenuated because their velocities are close to $+V_{cut}$ (Fig. 6–45).

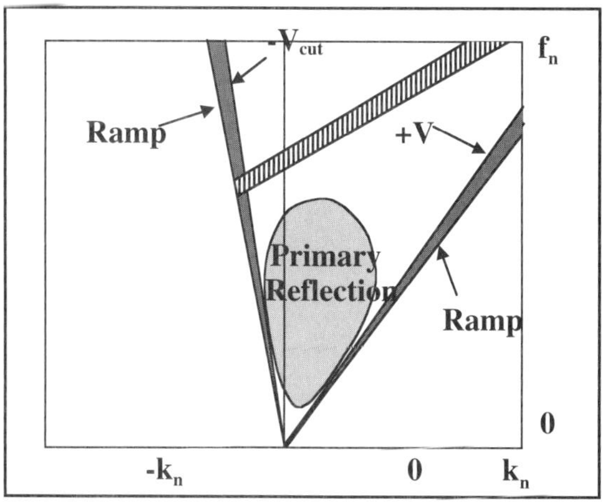

Fig. 6–44 *Filtering in the F-K Plane*

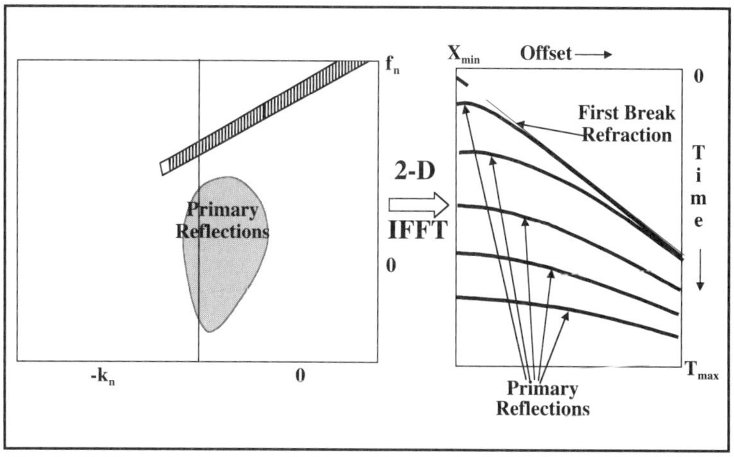

Fig. 6–45 *Transformation of Filtered Record from F-K to T-X Domain*

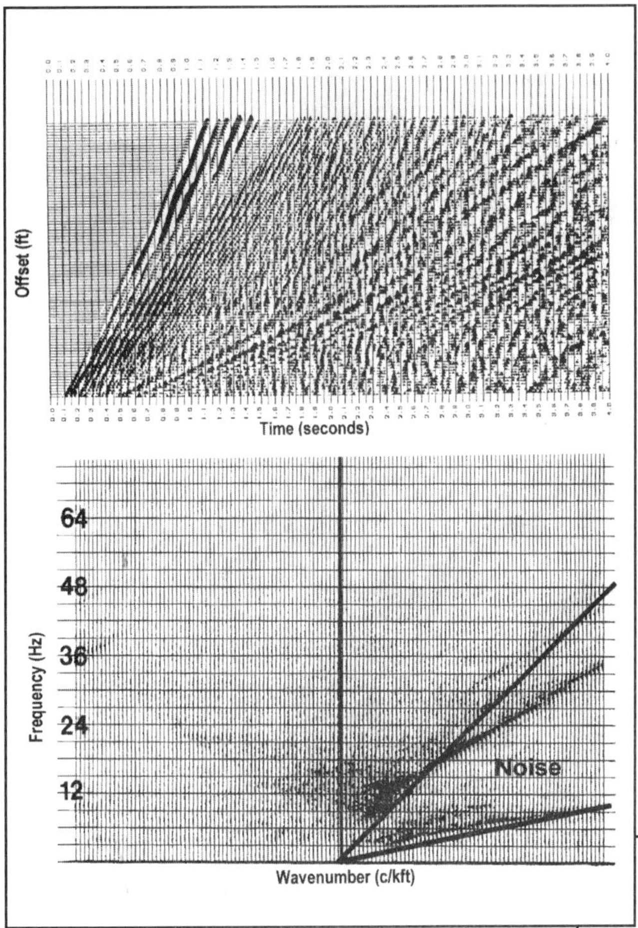

Fig. 6–46 *Noise Record in the T-X Domain and the F-K Domain*

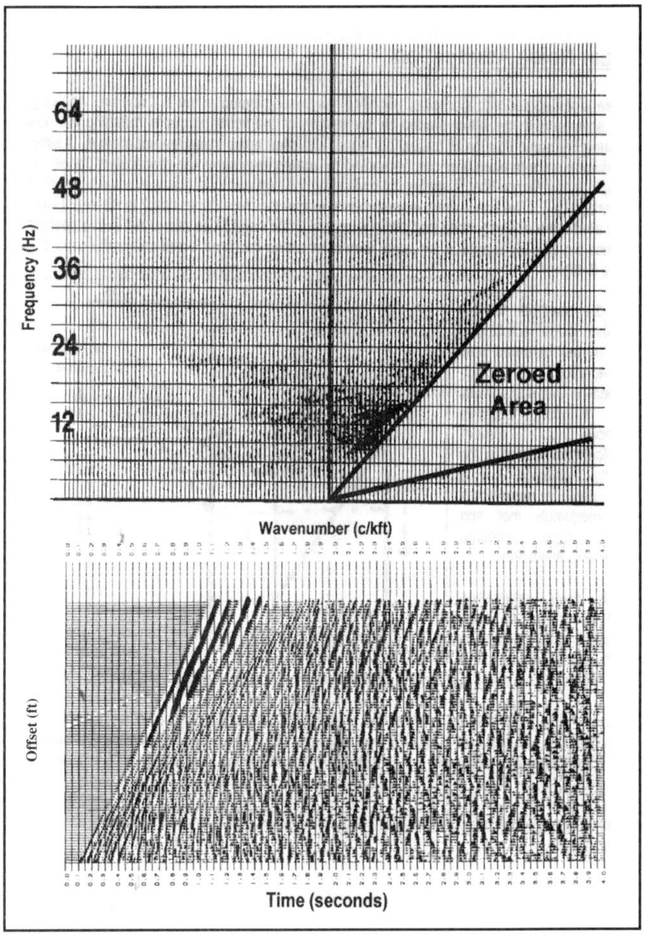

Fig. 6–47 *Filtered Field Record in the F-K Domain and the T-X Domain*

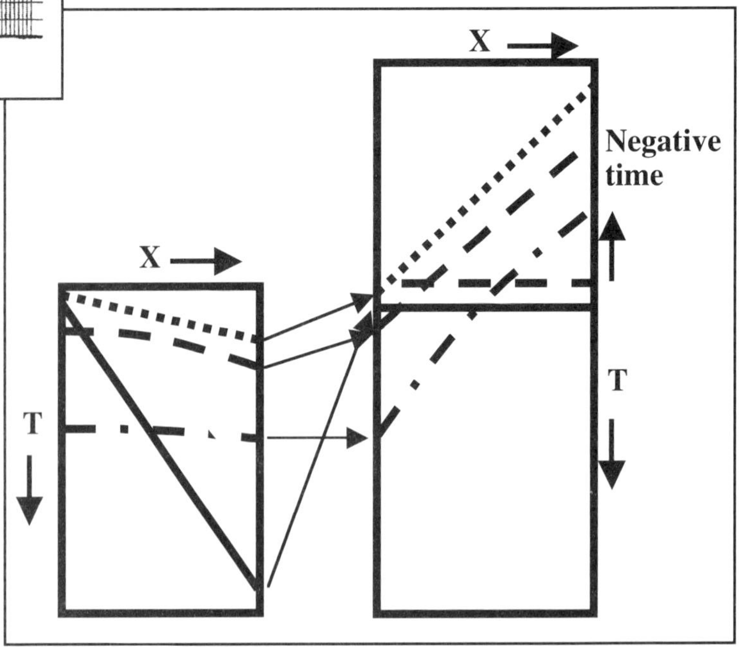

Fig. 6–48 *Linear Moveout Applied to Linear Noise*

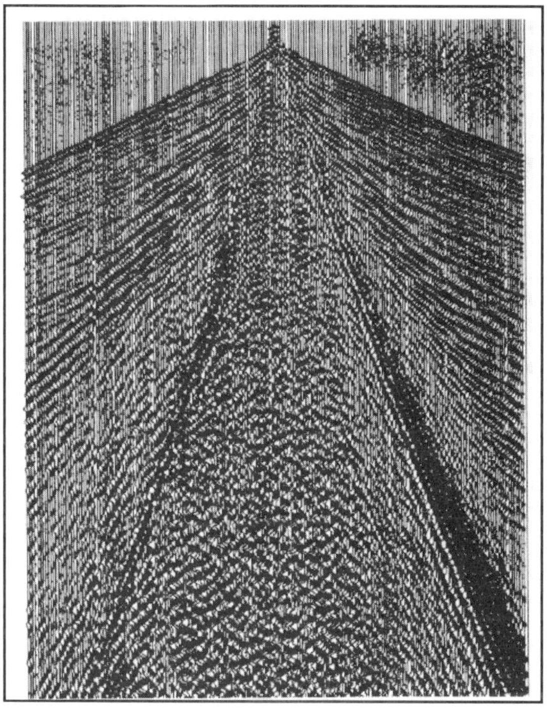

Fig. 6–49 *Field Record Dominated by Linear Noise*

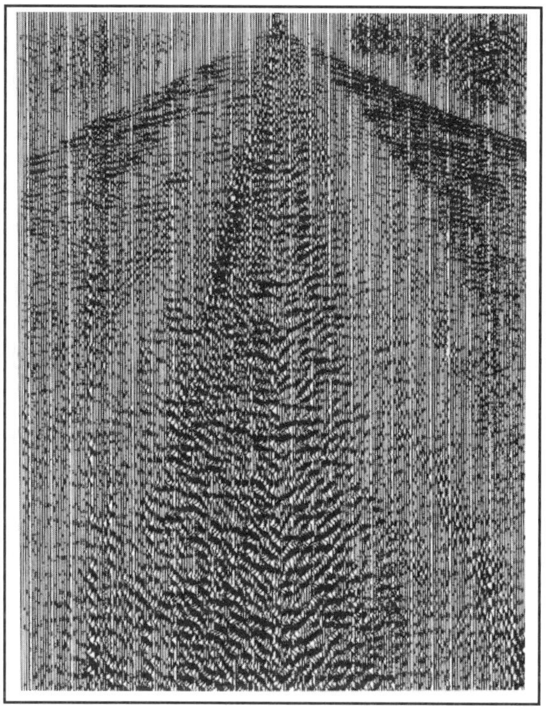

Fig. 6–50 *Record of Fig. 6–49 after Six Passes of Linear Noise Attenuation*

Figure 6–46 shows a noise record in both the T-X and *F-K* domains. In the *F-K* domain, lines are drawn that include the minimum and maximum noise velocities. These are used as the velocity cut lines for velocity filtering. The area inside these two lines is zeroed. Figure 6–47 shows the *F-K* domain after this is done and the inverse transform back into the T-X domain.

Linear noise attenuation. If noise is propagated with a constant velocity, it has linear moveout in the T-X domain. Applying a static shift equal to the offset divided by the velocity of the noise results in flattening the noise—occurs at the same time on all traces—but signal and other noise trains are not aligned. A simple time domain trace-to-trace filter can be applied by scaling then summing consecutive traces. For example, a three-trace filter is applied by multiplying the first and third traces in a set by -0.5 and the second by 1. The traces are then summed to output the trace at position two in the set. This is also called trace mixing. Note that negative record times need to be preserved in this operation because the time shifts applied are larger than moveout on first breaks and early reflections (Fig. 6–48).

For the method to be effective, noise amplitudes must be well balanced from trace to trace. Another problem is that signal must have sufficient negative moveout to prevent significant attenuation in the filtering process. This assumption is sometimes not adequately met, and signal does undergo significant attenuation.

Linear noise attenuation must be applied to every noise train that significantly interferes with signal. This may require a great many applications when noise trains at many different velocities are present.

Figure 6–49 is presented as an example of a record with many distinct noise trains. Noise propagating at 500 m/s, 666 m/s, 1000 m/s, 1500 m/s, 2100 m/s, and 2800 m/s can be identified. The combination of many noise trains with high amplitudes causes signal to be almost totally obscured.

Figure 6–50 shows the same record after the linear noise process has been applied for all six noise velocities.

Velocity Analysis

Seismic velocity means, in general, the rate of seismic wave propagation—distance/time. It is usually measured in meters per second or feet per second. The term velocity rarely occurs alone in seismic exploration because there are so many *kinds* of seismic velocities, such as instantaneous velocity, interval velocity, average velocity, rms velocity, NMO velocity, stacking velocity, migration velocity, and apparent velocity.

Instantaneous velocity is the velocity at which a seismic wave propagates at a point within a medium. It varies with wave type or propagation mode. P-wave velocity V_P is always the fastest. S-wave velocity V_S is the second fastest. The ratio V_P/V_S is greater than or equal to $\sqrt{2}$. Rayleigh wave velocity is usually about 92% of S-wave velocity and Love wave velocity is slightly slower than Rayleigh wave velocity. Instantaneous velocity also depends on rock type, age, depth of burial, porosity, pore shape, pore fluid, confining pressure, and other elements.

The closest thing to instantaneous velocity measurement is the sonic log or *continuous velocity log* (CVL). This is done in a borehole using a special tool.

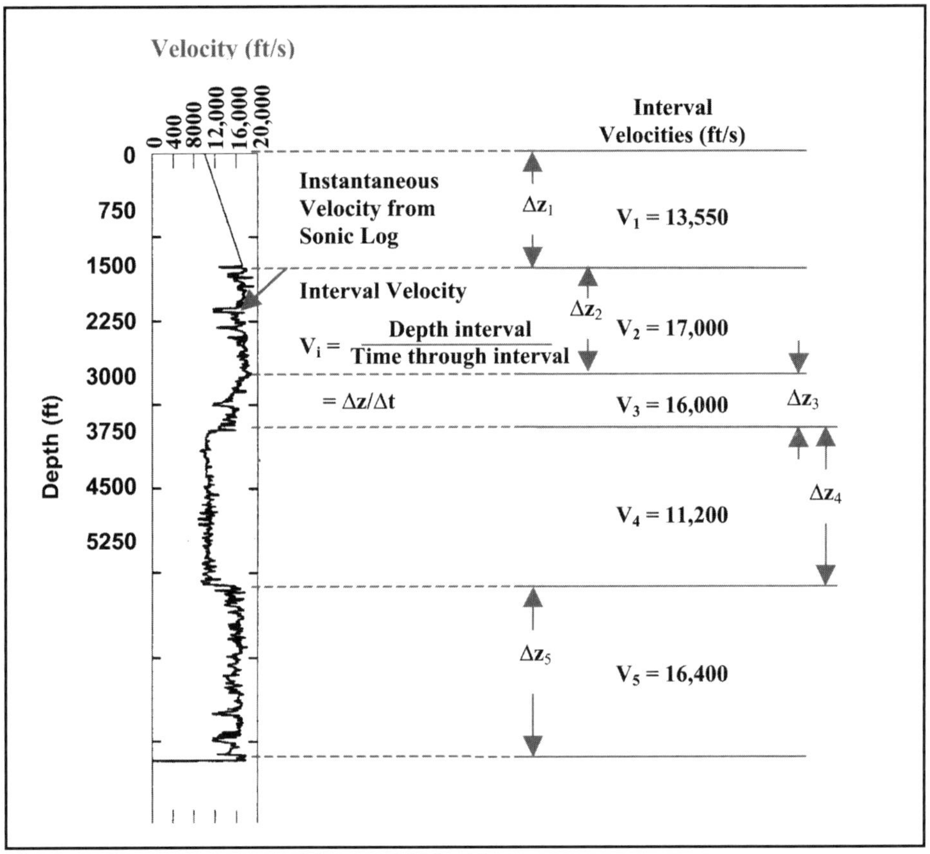

Fig. 6–51 *Instantaneous and Interval Velocities*

Interval velocity V_i is the average propagation velocity through a depth or time interval, and it equals the thickness of the depth interval divided by vertical time through the interval. Figure 6–51 shows a well log on the left and determination of interval velocity from the log on the right. It can be seen on the log that there are definite velocity trends. That is, over certain depth intervals, velocity tends to be about the same, and between these depth intervals velocity changes abruptly. Denote these depth intervals by Δz_i, $i = 1, 2, \ldots, N$. The direct measurement by the CVL is μ s/ft. Summing these times over the indicated depth intervals gives the times Δt_i, i = 1, 2, …, N. Interval velocities are obtained from:

$$V_i = \frac{\Delta z_i}{\Delta t_i} \tag{6.18}$$

Average velocity $\overline{V}$ is total depth to a reflector Z_r divided by time to the reflector or twice the depth to the reflector divided by two-way, zero-offset reflection time T_i.

$$\overline{V} = \frac{2Z_i}{T_i} \tag{6.19}$$

Average velocity can be calculated from interval velocities using:

$$\overline{V}_n = \frac{\sum_{i=1}^{n} V_i \Delta t_i}{\sum_{i=1}^{n} \Delta t_i} \tag{6.20}$$

Root mean square velocity $\tilde{V}$ is the square root of the average squared velocity and is calculated from:

$$\tilde{V}_n = \sqrt{\frac{\sum_{i=1}^{n} V_i^2 \Delta t_i}{\sum_{i=1}^{n} \Delta t_i}} \tag{6.21}$$

NMO velocity is the velocity used to correct for NMO—to make primary reflections on CMP gather records occur at the same time on all traces. For plane horizontal or gently dipping layers, NMO velocity is equal to the RMS velocity.

Stacking velocity is the velocity that gives the optimum CMP stack output when used for NMO corrections.

Migration velocity is the velocity that optimizes the output of a migration algorithm—repositioning of the reflected energy to the correct locations.

Apparent velocity is the velocity at which seismic waves appear to propagate along the surface of the earth.

Velocities of most concern, usually, are stacking velocity, migration velocity, and average velocity—used to convert from time to depth. Sonic logs, when available, give direct measures of interval velocities from which the desired velocities can sometimes be calculated but this can only be done where wells exist and must be interpolated between wells or extrapolated beyond wells. Sonic logs measure vertical velocity that, because of anisotropy, may differ from horizontal velocity and also uses higher frequency signals than surface seismic surveys that may give different velocities.

Normal moveout. In most cases, it is necessary to determine the required velocities from the data that are to be processed. In one way or another, this is done from measurements of *moveout*—the change in reflection time across a set of CMP traces. In general moveout is the algebraic sum of NMO and geologic dip.

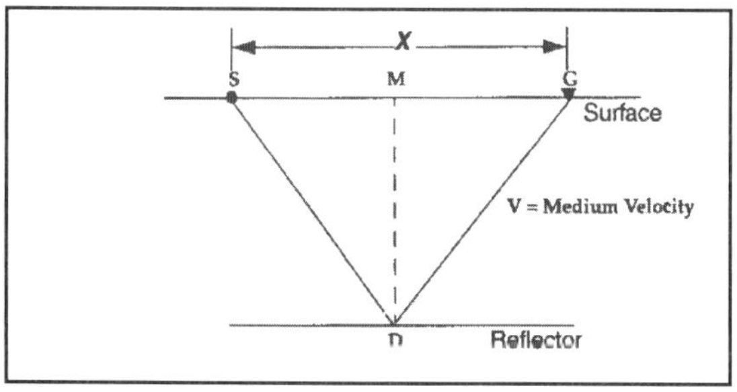

6–52 *Single Horizontal Layer*

NMO is the extra time required for energy to travel from a seismic energy source to a seismic receiver group at some distance from the source via reflection from a subsurface reflector. This is illustrated in Figure 6–52.

In Figure 6–52 the source is at *S* and the receiver is at *G*. Since the reflector is horizontal and the velocity *V* is constant, reflection occurs at point *D*, vertically below *M*, the midpoint between *S* and *G*. The time from *S* to *D* to *G* is:

$$t_x = \frac{(\overline{SD} + \overline{DG})}{V}$$

The time from *M* to *D* and back to *M* (zero-offset time) is:

$$t_0 = \frac{2\overline{MD}}{V}$$

From the above, $\overline{SD} = \overline{DG} = \frac{Vt_x}{2}$ and $\overline{MD} = \frac{Vt_0}{2}$ Applying the Pythagorean theorem yields:

$$\left[\frac{V_{NMO}t_x}{2}\right]^2 = \left(\frac{x}{2}\right) + \left[\frac{V_{NMO}t_0}{2}\right]^2 \tag{6.22}$$

Solving Equation 6.22 for t_x:

$$t_x = \sqrt{\left[\frac{x}{V_{NMO}}\right]^2 + [t_0]^2} \tag{6.23}$$

NMO Δt_{NMO} is the difference between the zero offset time t_0 and the reflection time t_x.

$$\Delta t_{NMO} = t_x - t_0 = t_0 \{ \sqrt{\left[\frac{x}{V_{NMO}t_0}\right]^2 + 1 - 1} \} \tag{6.24}$$

It can be seen from Equation (6.24) that Δt_{NMO} increases as offset x increases and decreases as V_{NMO} and/or $t(0)$ increase. This is also shown in Table 6–1. Equation 6.22 is actually the equation of a hyperbola, and Figure 6–53 shows that reflection events have hyperbolic shapes.

Table 6–1 *Variation of DtNMO with Time, Velocity, and Offset*

		Δt_{NMO} **for offset x =**	
Time (s)	**Velocity (m/s)**	**1000m**	**2000 m**
0.25	2000	0.309	0.780
0.5	3500	0.140	0.443
1.0	3000	0.054	0.201
2.0	3500	0.020	0.080
4.0	4000	0.008	0.031

NMO correction is usually done trace by trace. Each sample is mapped from a particular trace offset to zero offset. The time for each sample is set equal to T_0, and T_x is calculated for a given offset for that trace and velocity for that time. The sample value at the calculated T_x is moved up to T_0 by subtracting Δt_{NMO}. See Figure 6–54.

Figure 6–55 gives a more detailed picture of the NMO correction process for a particular trace for an offset x. On the left is shown that the NMO velocity is different for each sample. It actually increases with time. In the center is a table representing sample amplitudes versus sample time on the input, shot-organized record before NMO. On the right is shown the sample values that have been shifted up in time. In the figure, Δt_{NMO} = integral multiples of sample period Δt.

Note that Figure 6–55 assumes that the NMO corrections are integral multiples of the sample period. When this is not the case (the usual situation), there is no valid sample amplitude to shift. The solution is to interpolate between known values using an interpolation filter. This is shown in Figure 6–56, where Δt_{NMO} = nonintegral multiples of sample period Δt.

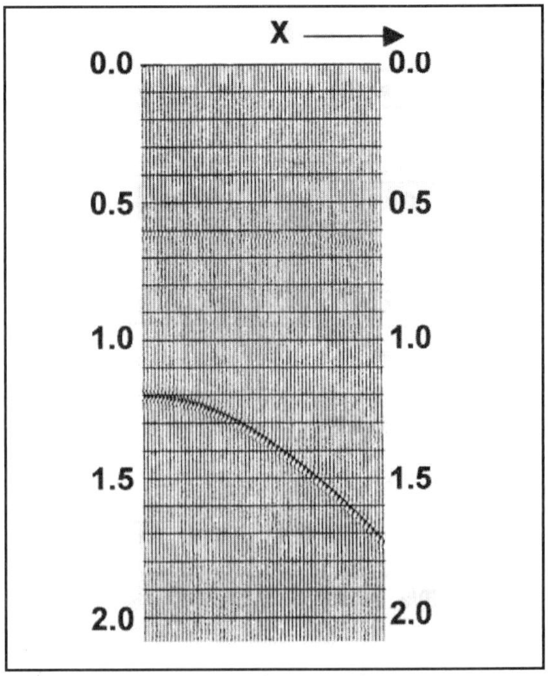

Fig. 6–53 *Reflection with Normal Moveout*

Reflection events will be *flattened—arrive* at same time on all traces of a CMP record—only if the correct NMO velocity is used. The effect of using incorrect velocities is shown in Figure 6–57. On the left

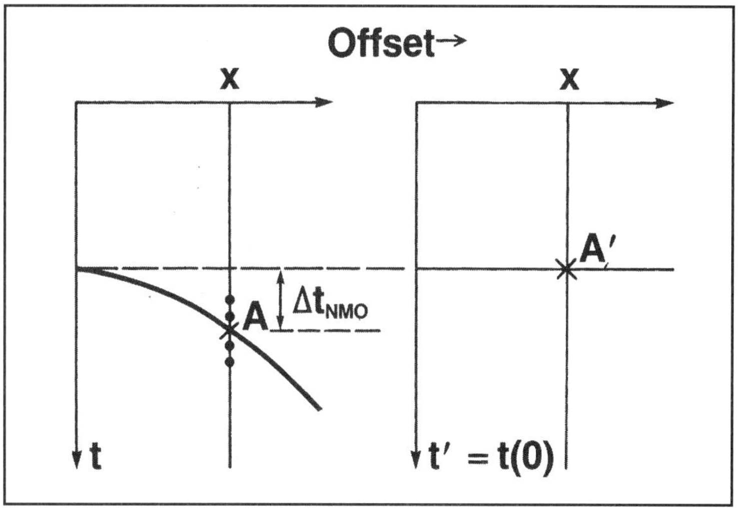

Fig. 6–54 *NMO Correction*

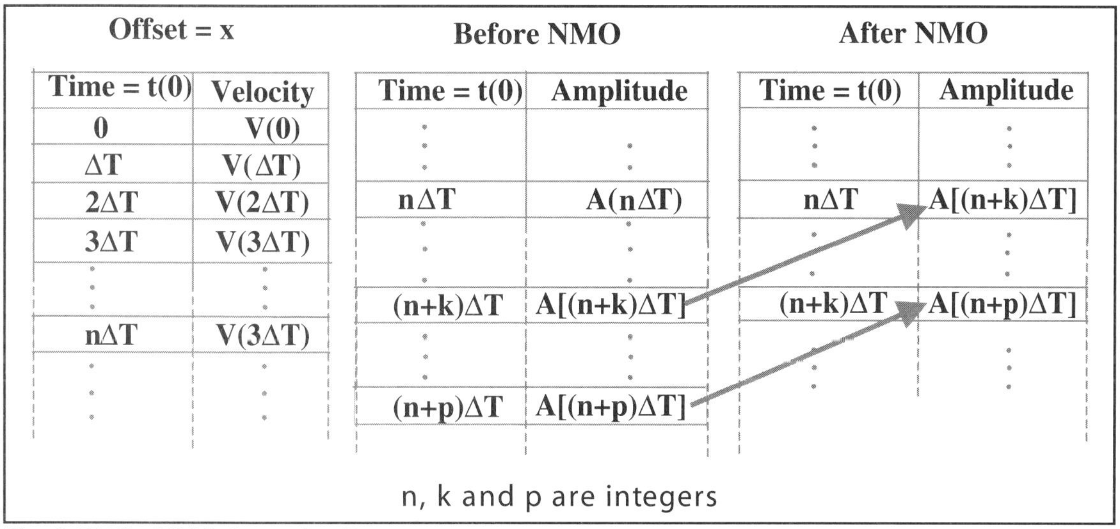

Fig. 6–55 *The NMO Correction Process*

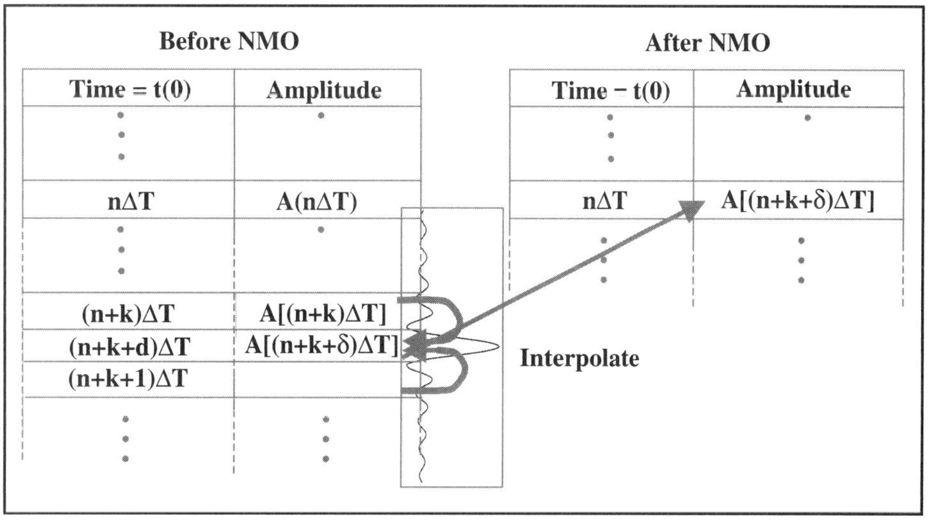

Fig. 6–56 *The NMO Correction Process*

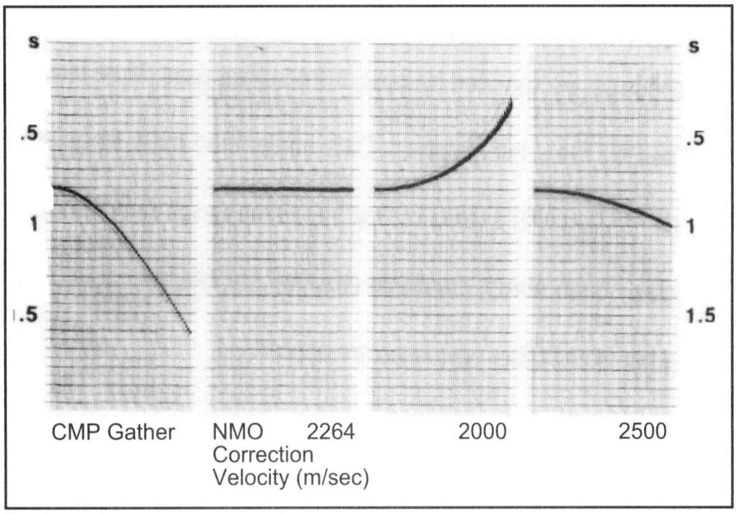

Fig. 6–57 *Effect of Velocity on NMO Correction*

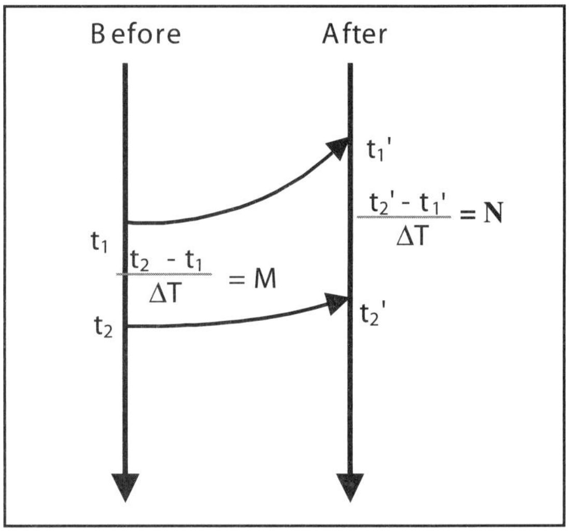

Fig. 6–58 *NMO Stretch*

of the figure is the input gather record indicated by a single hyperbolic event. Using the correct velocity of V_{NMO} = 2264 m/s results in a flat event. Using a velocity of 2000 m/s—slower than V_{NMO}—results in the event being *overcorrected*. The correction exceeds NMO and the event curves upward. Using a velocity of 2500 m/s—faster than V_{NMO}—the event is *under-corrected*. Not all of the NMO has been removed.

NMO correction is done sample by sample. Each sample time is used as t_0 in the NMO equation and the velocity at that time used for V_{NMO}. As previously noted, V_{NMO} increases as t_0 increases. As a result, NMO corrections decrease as t_0 and V_{NMO} increase. Thus, the mapping process places output samples farther apart than they were on the input, causing an apparent increase in wavelet duration or decrease in frequency. This effect, called *NMO stretch*, is greatest at early times and long offsets.

Figure 6–58 demonstrates this. Before the NMO corrections are applied, there are M samples at a sample interval ΔT, between times of t_1 and t_2. The sample at time t_1 maps into t_1' and time t_2 maps into t_2'. Between a time of t_1' and t_2', there are N samples at a sample interval ΔT. The amount of stretch is indicated by the increase in the number of samples N compared to the original M samples. Data are muted (zeroed) at times and offsets where the stretch exceeds the maximum allowable. The allowed stretch is some-times measured by the ratio $N{:}M$. In which case, a maximum ratio of 2:1 is the usual default value.

Mute schedules are specified by input of time-offset pairs. Mute times between time-offset pairs are determined by linear interpolation. Data earlier than mute times are zeroed. Note, also that the mutes are applied with a ramp so that trace amplitudes do not change abruptly from zero to some large value. Such abrupt amplitude changes are interpreted as "spikes" and can have unwanted effects when some other processes are applied.

Mute times should be picked by inspection rather than using a default. Figures 6–59 and 6–60 show why this is so. Figure 6–59 shows three CMPs as they appear on input, after NMO correction (note the heavy distortion at early times and far offsets caused by NMO stretch), and after default NMO mutes are applied. Figure 6–60a shows the center NMO-corrected record of Figure 6–59. Figure 6–60b is the stack of the two near traces of the original gather record. The dotted line is the hand-picked mute schedule. Figure 6–60c is the full-fold stack obtained from the original gather with a poor mute choice based on picks from the original gather. The mutes in Figure 6–60b are preferred over those in Figure 6–60c because more data are preserved. Muting is necessary to remove data overly distorted by NMO correction, but, it must be remembered, muting also reduces fold at

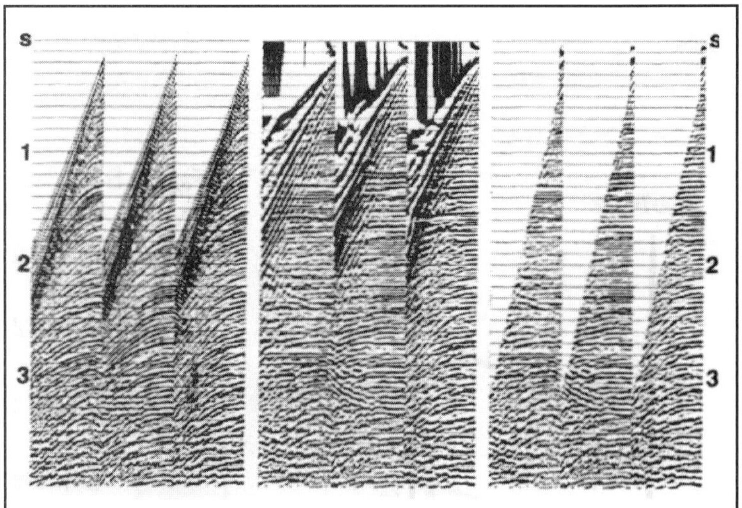

Fig. 6–59 *Input CMP Records, the Same Records after NMO Correction, and after NMO Mute*

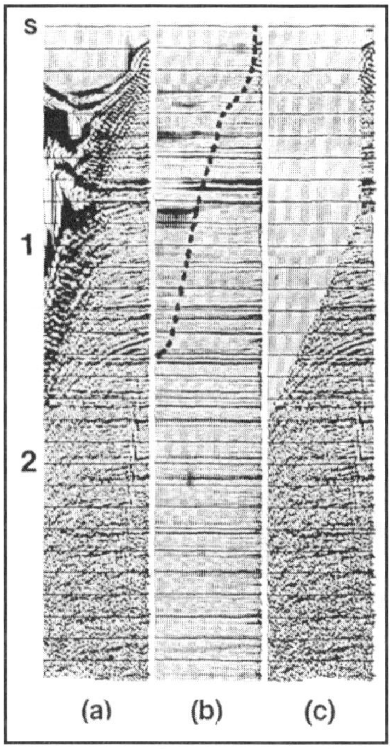

Fig. 6–60 *Picked versus Default Mutes*

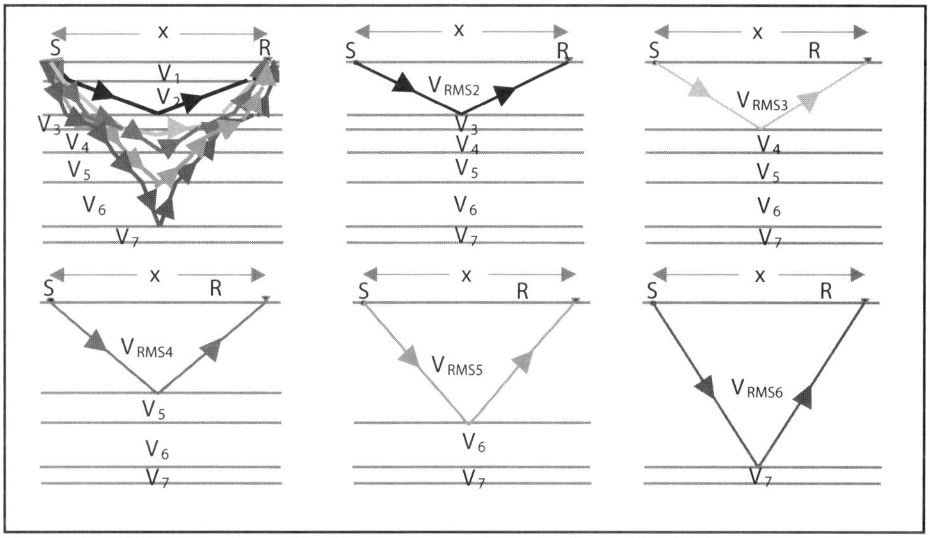

Fig. 6–61 *RMS Velocity as the Replacement Velocity in the Multi-layer Situation*

early record times. The choice of mutes is, thus, a compromise between optimizing fold and minimizing the effect of distortion caused by NMO correction.

The NMO equations, Equations 6.21 and 6.22, assume a constant velocity and a horizontal reflector. This is not the situation encountered in seismic exploration. Instead, there are multiple layers with different velocities. The solution is to "scrape off" the layers one at a time and replace these layers with a single layer and a constant velocity. If the reflectors are flat or gently dipping, then the RMS velocity can be used.

Figure 6–61 illustrates this approach to NMO correction. The top left drawing shows ray paths for reflections from the bottoms of layers two through six. For a reflection from layer one, velocity V_1 is used for NMO correction. For a reflection from the base of layer two, layers one and two are replaced by a single layer with the RMS velocity V_{RMS2}. The remaining drawings in Figure 6–61 show how the pattern continues for subsequent reflections. For NMO calculation purposes, all layers above the reflecting interfaces are replaced by a single layer having the appropriate RMS velocity.

The use of the RMS velocity is subject to some restrictions:

- reflecting interfaces must be horizontal or gently dipping
- trace offset must be less than or equal to reflector depth

Figure 6–62 shows the effect of offset on the applicability of RMS velocity to NMO calculation by comparing NMO obtained from the earth model and by use of RMS velocity in equation (6.21). Note that the errors between model times and NMO times calculated by using V_{RMS} are no more than 1 ms when the offset x is less than reflector depth Z. Even at an offset of 11,000 ft and depth of 5500 ft, $x = 2Z$, the error is only 0.014 s, which is less than 1%.

Earth Model

V_1 = 6000 ft/s Z_1 = 1500 ft

V_2 = 10,000 ft/s Z_2 = 4000 ft

V_3 = 9000 ft/s Z_3 = 3375 ft

	Z = 5500 ft		Z=8875 ft	
Offset (ft)	**T_{model}**	**T_{VRMS}**	**T_{model}**	**T_{VRMS}**
0	1.300	1.300	2.050	2.050
1000	1.305	1.305	2.053	2.053
3000	1.345	1.345	2.078	2.078
5000	1.421	1.422	2.127	2.127
7000	1.527	1.530	2.198	2.199
9000	1.655	1.663	2.290	2.291
11,000	1.801	1.815	2.398	2.401

Fig. 6–62 *NMO for Plane, Horizontal, and Iso-Velocity Layers*

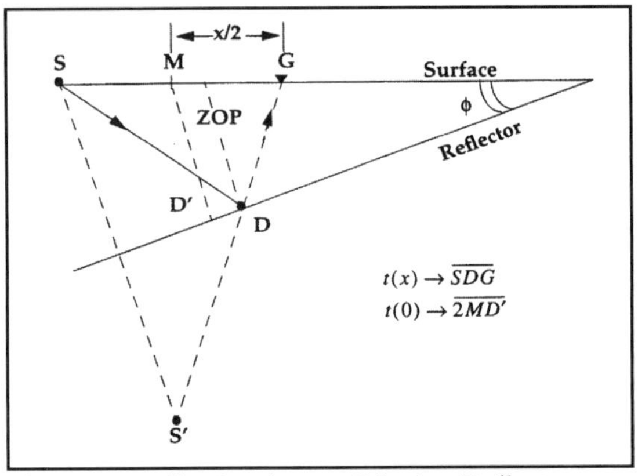

Fig. 6–63 *Reflection from a Dipping Reflector*

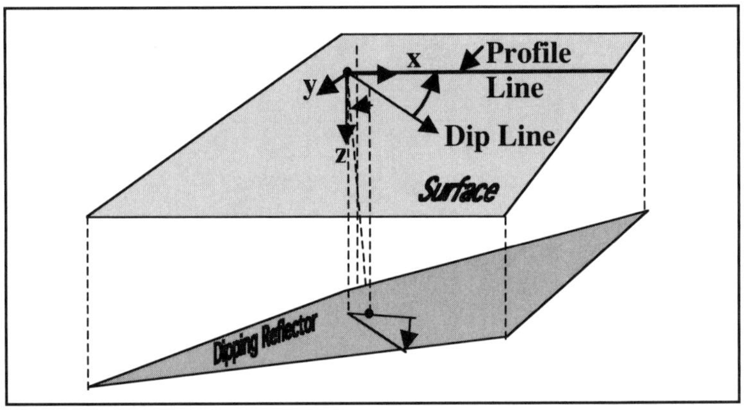

Fig. 6–64 *A Dipping Reflector in 3–D*

Of greater interest to petroleum exploration are dipping layers. Figure 6–63 shows a single layer with the reflector dipping at angle ϕ. The reflection ray path from source *S* to receiver *G* reflects at point *D* on the dipping reflector. The midpoint between source and receiver is at *M* but a line from *M* perpendicular to the reflector intersects the reflector at *D*' not *D*. Thus for the dipping case, the midpoint is not the zero offset position (ZOP) for the reflection.

In Figure 6–63, *S*' is the image point of the source *S*. From triangle *SGS*':

$$V^2t_x^2 = V^2t_0^2 + x^2 \cos^2 \phi$$

Dividing by V^2 gives:

$$t_x^2 = t_0^2 + \frac{x^2 \cos^2 \phi}{V^2}$$

$$= t_0^2 + \frac{x^2}{\left(V/\cos\phi\right)} = t_0^2 + \frac{x^2}{V^2_{DIP}} \qquad (6.25)$$

where

$$V_{DIP} = \frac{V}{\cos\phi}$$

For CMP gathers with a dipping reflector, the velocity needed to correct for NMO is always greater than the RMS velocity or that used for flat reflectors. Also, as dip increases, the velocity required to flatten data increases and the amount of NMO decreases. Dip decreases travel time from source to receiver because of the shorter path from source to receiver.

In 3-D operations, the azimuth between source and receiver must also be considered. Figure 6–64 shows a reflector dipping at angle ϕ, but the receiver line is at an angle θ—the azimuth—from the dip direction. In this situation:

$$V_{NMO} = \frac{V}{\sqrt{1 - \sin^2\phi + \cos^2\theta}} \qquad (6.26a)$$

or

$$V_{NMO} = \frac{V}{\cos\phi'} \qquad (6.26b)$$

where

$$\cos \phi' = \sqrt{1 - \sin^2\phi + \cos^2\theta}$$

Table 6–2 *NMO Velocities*

Model	NMO Velocity
Single horizontal layer	Layer Velocity
Horizontally stratified	RMS velocity if small spread layers assumption is valid
Single dipping layer	Layer velocity/cos(dip angle)
Multi-layered with arbitrary dips	RMS velocity if small spread and small dip assumptions are valid

If shooting is along the on dip azimuth, then $\theta = 0°$, and apparent dip equals true dip. If shooting is in the strike direction azimuth, then $\theta = 90°$, and apparent dip equals 0°. That is, the reflector appears to be flat with no dip! At intermediate azimuths, the apparent dip is less than true dip.

Stacking velocities do not tie at line intersections for dipping reflectors. This is because the lines are shot at different azimuths, causing differences in apparent dip angle and, hence, different velocities and different NMO calculations.

Table 6–2 summarizes the choices of NMO velocity for various subsurface situations. With regard to Table 6–2, the small-spread assumption is that offset is less than reflector depth. The small-dip assumption is that dip angles are less than 10°.

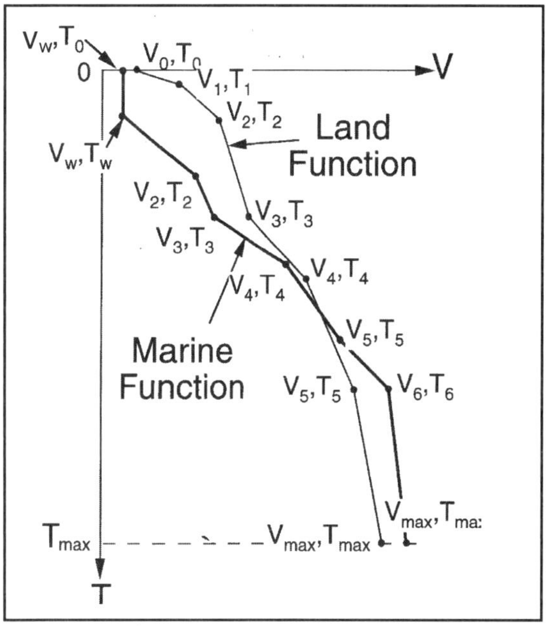

Fig. 6–65 *Velocity Functions*

Velocity analysis techniques. The purpose of velocity analysis is to determine stacking velocity, called *velocity functions* and seen in Figure 6–65, that, when used to implement NMO corrections, yield the optimum stack. In general, approaches to this problem include the following:

- time-velocity space restricted to velocity values that can be reasonably expected
- NMO correction evaluated statistically, using small time gates and many estimates of stacking velocity
- both time and velocity adequately sampled
- results of velocity calculations displayed in forms that allow ccurate velocity interpretation

The initial result of velocity analysis is a set of velocity functions determined at specific points within the survey. Velocity functions are sets of time and velocity pairs that define velocities for every sample in the CMP traces. Interpolation between and extrapolation from these functions yields a *velocity field* that defines a velocity for every sample in the data volume.

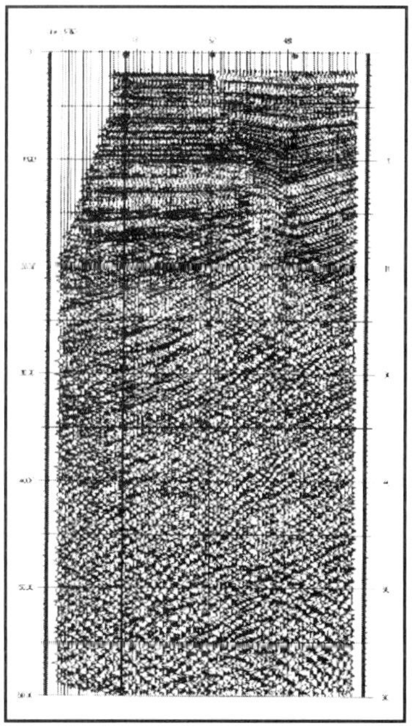

Fig. 6–66 *Velocity Analysis Location Used for Example*

Velocity analysis requires the following:

- selection of velocity analysis locations
    - this is often done by regular sampling (such as every 20th CMP) of every seismic line
    - locations should be chosen to comprehend structural variation in areas of greater structural variation—Figure 6–66 shows a portion of a seismic section and the CMP selected for the velocity analysis example that follows
- determination of the velocity sweep, which is defined by a low velocity trend and a high velocity trend that depends on the area as seen in Figure 6–67
    - a constant velocity stack (CVS) display is sometimes used to generate the velocity trend. (*Constant velocity stack* is a process in which a set of constant velocities, equally incremented between the lowest and highest velocities expected, is used to apply NMO corrections and then stack the CMPs.) In Figure 6–70, the open diamonds indicate picks made for the trend function
    - a velocity fan consisting of this trend function plus three to five slower and three to five faster functions may also be generated

- within the velocity sweep, a total of 40 to 60 velocities are used to calculate and apply NMO corrections. Seven to 11 CMPs (center CMP plus 3 to 5 on either side) are used in each analysis
    - selection of time gate width
    - time gates should contain at least four to five samples
    - they usually overlap
    - choice of statistical parameter used to determine the velocities that provide the best stack
    - *semblance* is frequently used for this purpose. Semblance is a determined by dividing the square of summed stack amplitudes by total gate power. Semblances that are plotted in velocity displays are often smoothed in time and velocity

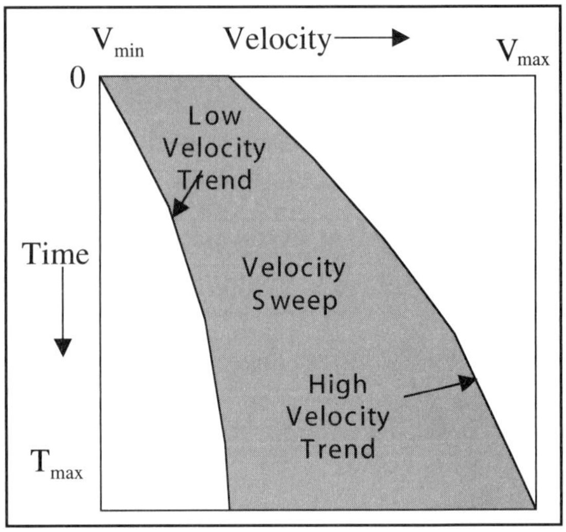

Fig. 6–67 *The Velocity Sweep*

Figure 6–68 is a semblance plot in gray scale—the darker the gray, the higher the semblance value. The central velocity trend is shown as a heavy broken line, and the four lower and higher velocities are on each side. Figure 6–69 shows the stack of CMPs used in this analysis for each of the nine velocity functions of Figure 6–68.

Figure 6–70 is a CVS display used to generate the velocity trend for the IVP session. In CVS, NMO corrections are made using a single constant velocity at all times for each trace in the CMP, then stacked together. Stack response is the criterion for velocity picks. The open diamonds indicate picks made for the trend function.

Usually, velocity analysis and interpretation are done interactively. Primary interpretation is based on the semblance display but other displays such as the center CMP before and after NMO correction can be viewed to confirm picks of velocity and time pairs. Figures 6–71 through 6–76 illustrate interactive velocity processing.

Figure 6–71 is the semblance plot generated from the input velocity fan. The white circles are the picks (time and velocity pairs) selected from the CVS display. The event at 1729 ms (indicated by the white arrow) is used to demonstrate the effect of interactive velocity picks.

Shown at the left in Figure 6–72 is CMP 61 corrected for NMO using the velocity function shown in white on the semblance plot. The right display is a stack panel with CMP 61 at the center. It can be seen that the event at about 1729 ms—see black double arrow—has a doubled appearance.

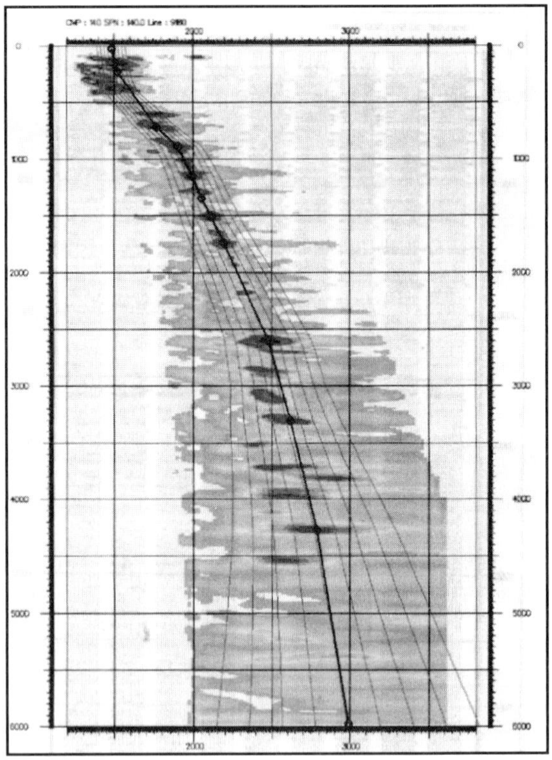

Fig. 6–68 *Semblance Plot*

Figure 6–73 is the semblance plot with a slower velocity picked for the event at 1729 ms, as indicated by the white arrow. Figure 6–74 is the same as Figure 6–72 except that CMP 61 corrected for NMO using the modified velocity function (too slow at 1729 ms) shown in white on the semblance plot of Figure 6–73. The stack panel with CMP 61 at the center shows the event at about 1729 ms—see black double arrow—has lost its doubled appearance.

Figure 6–75 is the semblance plot with a faster velocity picked for the event at 1729 ms, as indicated by the white arrow. Figure 6–76 shows CMP 61 corrected for NMO using the second modified velocity function (too fast at 1729 ms) shown in white on the semblance plot of Figure 6–75. The stack panel with CMP 61 at the center shows the event at about 1729 ms—see the black double arrow—has, again, lost its doubled appearance.

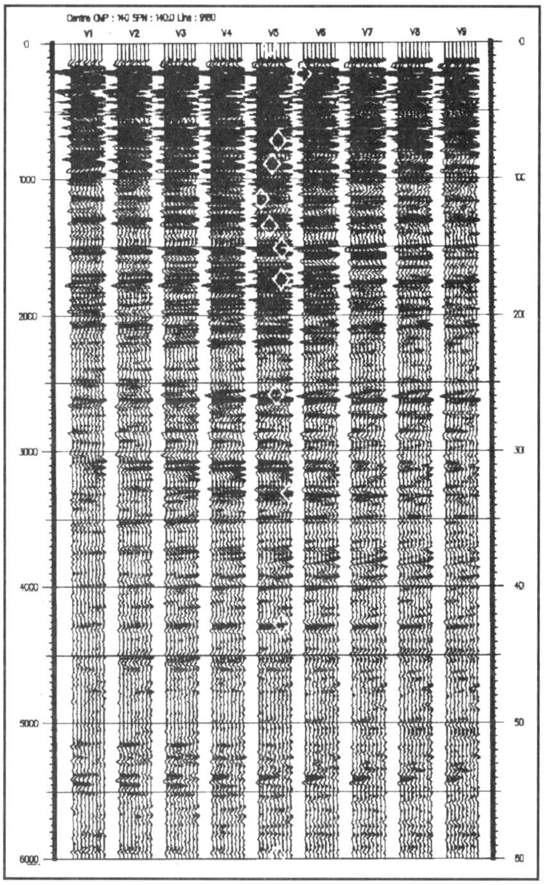

Fig. 6–69 *Stack of CMPs Using the Nine Velocity Functions of Figure 6–68*

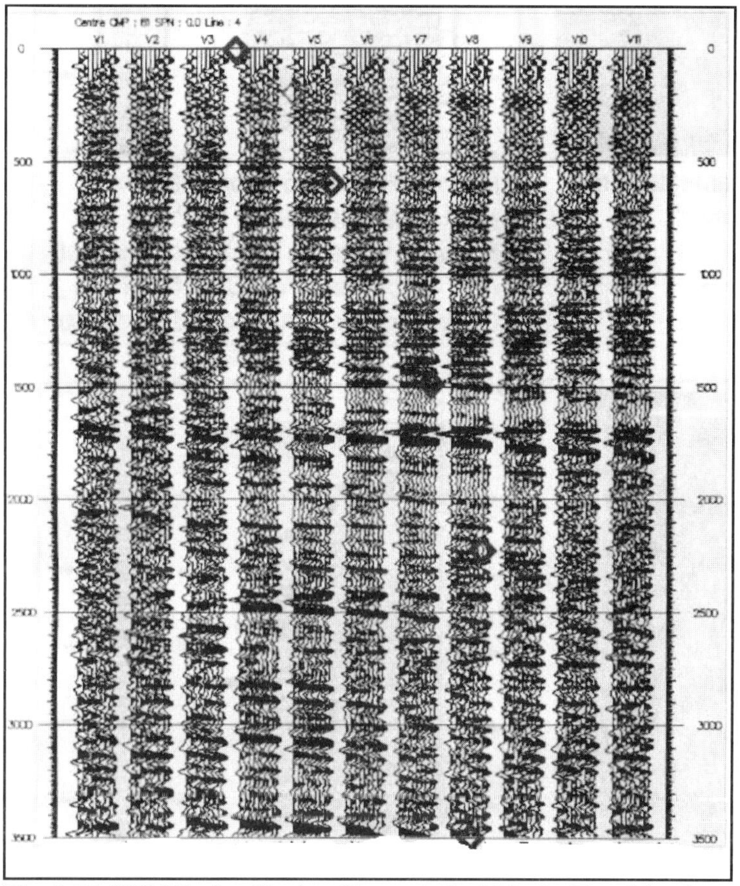

Fig. 6–70 *CVS Display Used to Generate the Velocity Trend for the IVP Session*

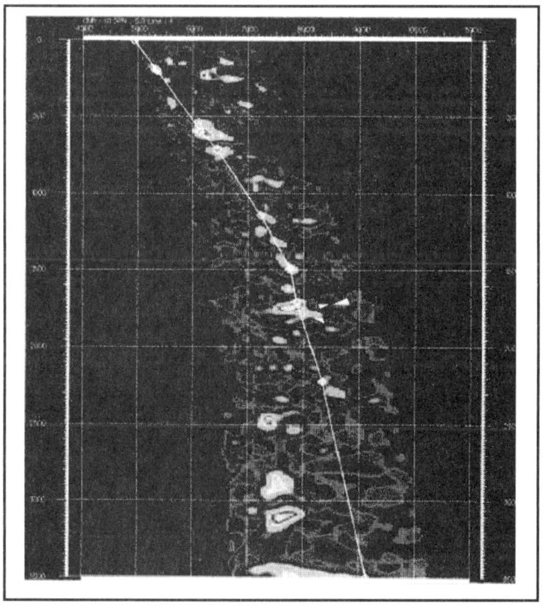

Fig. 6–71 *Semblance Plot Generated from the Input Velocity Fan*

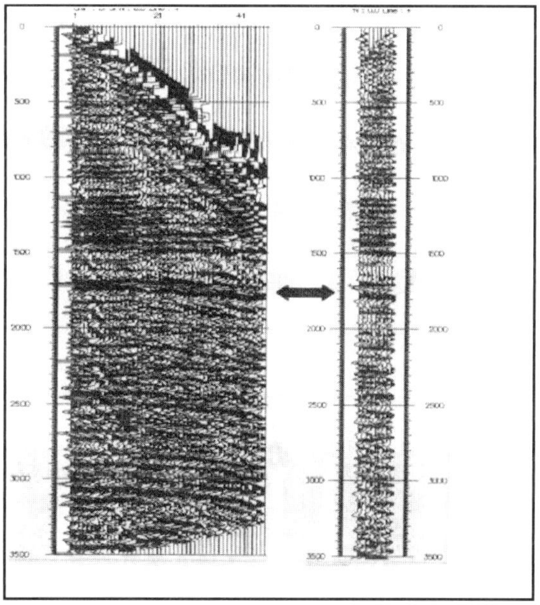

Fig. 6–72 *CMP 61 Corrected for NMO Using Velocity Function*

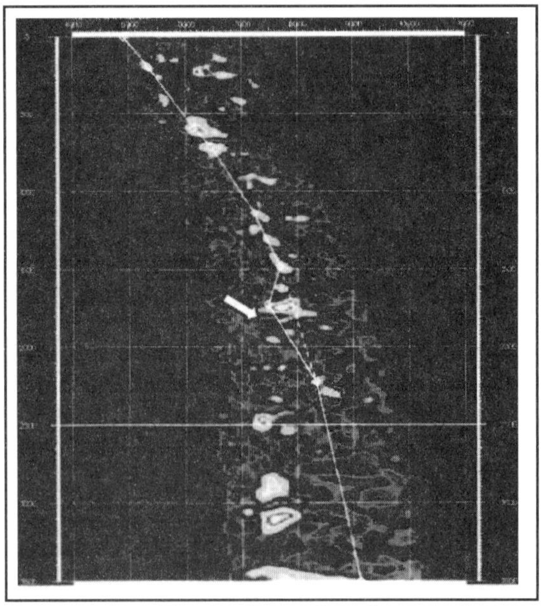

Fig. 6–73 *Semblance Plot with a Slower Velocity Picked for Event at 1729 ms*

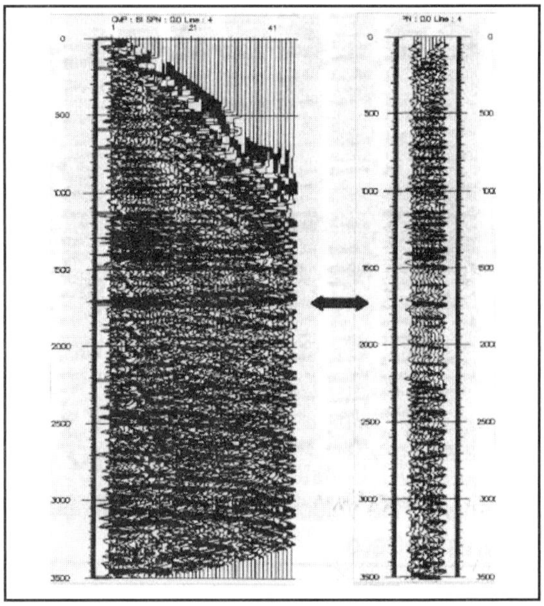

Fig. 6–74 *CMP 61 Corrected for NMO Using the Modified Velocity Function*

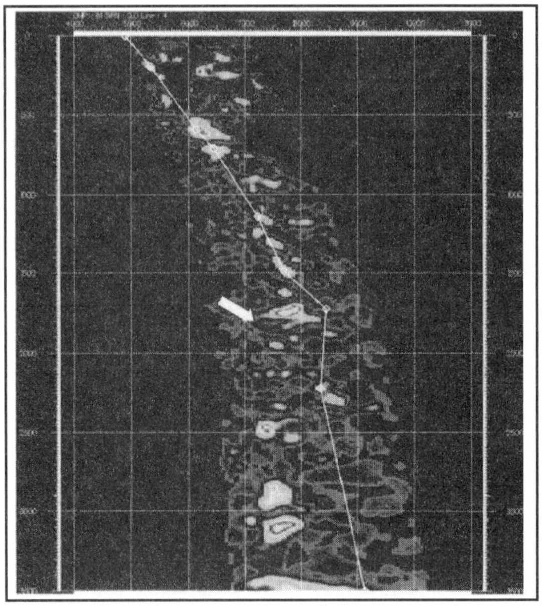

Fig. 6–75 *Semblance Plot with a Faster Velocity Picked for Event at 1729 ms*

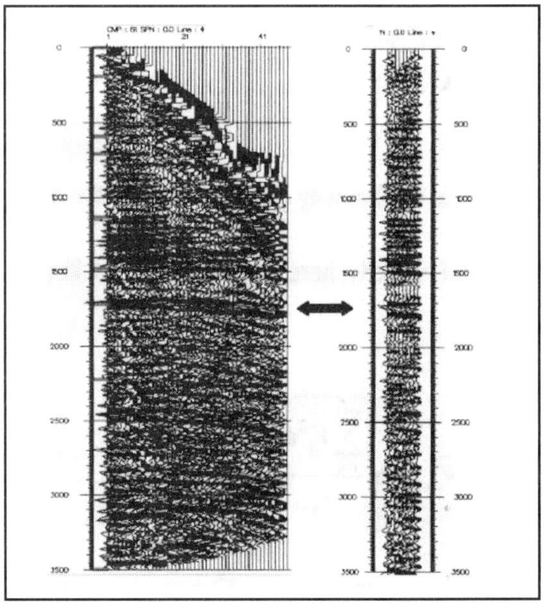

Fig. 6–76 *CMP 61 Corrected for NMO Using the Modified Velocity Function*

When the too slow and too fast picks of the event at 1729 ms were made in Figures 6–73 and 6–75, a *velocity inversion* of the stacking velocity resulted—the stacking velocity decreased from one pick to the next. This is never a valid interpretation of a velocity analysis.

Velocity inversions for interval velocities are expected because of changes in lithology of sedimentary strata. The *Dix Equation* can be used to show that stacking velocities must always increase with time. The Dix equation can be derived from Equation 6.21. To do that, substitute $T_n = \sum_{i=1}^{n} \Delta t_i$ into that equation, to give:

$$\tilde{V}_n = \sqrt{\frac{\sum_{i=1}^{n} V_i^2 \Delta t_i}{T_n}} \qquad (6.27)$$

where

T_n is the two-way reflection time for the nth event picked in the velocity interpretation

$\tilde{V}_n$ is the RMS velocity, here equated to the stacking velocity.

The $n+^{th}$ stacking velocity pick is thus

$$\tilde{V}_{n+1} = \sqrt{\frac{\sum_{i=1}^{n+1} V_i^2 \Delta t_i}{T_{n+1}}} \qquad (6.28)$$

From Equation 6.27, $\tilde{V}_n^2 T_n = \sum_{i=1}^{n} V_i^2 \Delta t_i$. Substituting this into Equation 6.28 yields:

$$\tilde{V}_{n+1} = \sqrt{\frac{V_n^2 T_n + V_{n+1}^2 \Delta t_{n+1i}}{T_{n+1}}}$$

Solving this equation for V_{n+1}, the implied interval velocity between the reflection times T_n and T_{n+1}, gives the Dix equation:

$$V_{n+1} = \sqrt{\frac{\tilde{V}_{n+1}^2 T_{n+1} - \tilde{V}_n^2 T_n}{T_{n+1}}} \qquad (6.29)$$

If $\tilde{V}_n > \tilde{V}_{n+1}$, which is the case for a stacking velocity inversion, the quantity inside the square root sign is likely to be either negative or too small to give a realistic value for the interval velocity $\tilde{V}_{n+1}$.

An example will further illustrate the point. Let $\tilde{V}_n$ = 2800m/s, $\tilde{V}_{n+1}$ = 2500m/s, T_n = 2 s, and T_{n+1} = 2. 25 s. This means that ΔT_{n+1} = 0.25 s. Substituting these values into Equation 6.29 gives:

$$V_{n+1} = \sqrt{\frac{2500^2 \times 2.25 - 2800^2 \times 2}{0.25}} = \sqrt{\frac{14962500 - 15680000}{0.25}} = \sqrt{-2870000}$$

So, our example gives an imaginary number for interval velocity, and the pick at 2.25 s is too slow or the pick at 2 s is too fast!

Computational procedures. In performing velocity analysis, it should be kept in mind that the velocity field is the only output of velocity analysis. Processes and computational procedures can be employed that enhance the accuracy and resolution of the velocities but are not passed on to later stages of processing. There are two basic aspects of this—reduction of data quantity and enhancement of signal.

Data reduction. The obvious thing in the first category is to resample to a larger sample period—from 2 ms to 4 ms or even from 4 ms to 8 ms. This results in about a 50% reduction in data to be processed. The anti-alias filter must be applied, but usually the trend of the low frequency part of signal is the most significant.

Partial stacking—adding together a selected number of adjacent offsets in CMP gather—is another approach to data reduction. By adding traces corresponding to four offsets, 64-fold data is reduced to 16 fold.

Calculate coherences for each gather individually and average coherence for multiple CMP gathers or calculate coherence for the sum of CMP gathers. Remember to sum CMPs in the presence of dip with great caution—the time dip across the CMPs stacked must be less than one-fourth of the dominant period for CMPs in the sum. When summing adjacent CMP gathers, the sum should have every possible shot-to-receiver offset!

Use a preliminary stack mute to remove data outside the NMO stretch mute, but be conservative, and do not mute too much. Use different velocities with different mutes.

Use only as high velocity values as absolutely necessary. High velocities are difficult to resolve because residual moveout is very small and very high velocities are rarely encountered.

Signal enhancement. Apply differential NMO before stacking. Too much summation causes deterioration of velocity resolution. Time dip between adjacent CMPs limits the number of adjacent CMPs that can be summed.

Apply band pass filtering to restrict data to those frequencies having the best signal-to-noise ratios As noted previously, only the low frequency velocity trend is required for most processing.

AGC decreases high amplitude noise, such as ground roll and multiples, relative to data, improving ability to pick correct velocities.

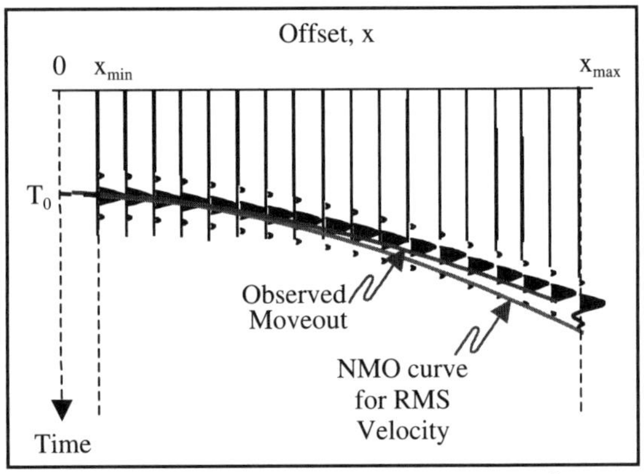

Fig. 6–77 *Observed Moveout versus NMO for a Flat Layered Earth*

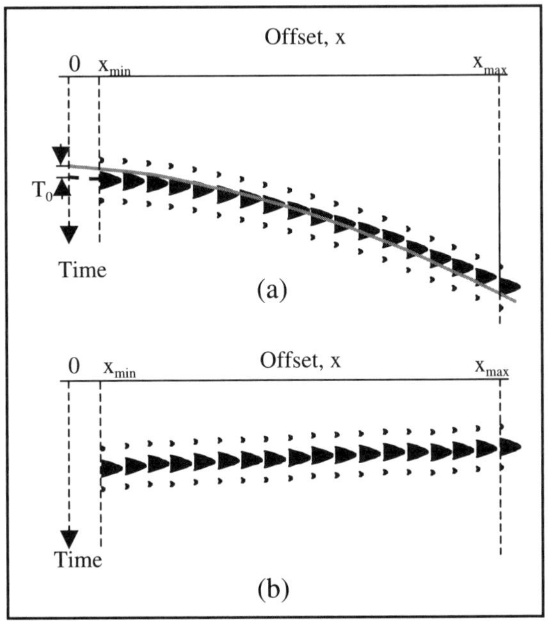

Fig. 6–78 *Best Fit Hyperbola to Observed Moveout, NMO-corrected Traces Using Velocity Obtained from Best Fit Hyperbola*

Velocity problems. A number of things can be and are encountered that affect velocity analysis negatively. Among these are

- dip
- offsets
- near-surface anomalies (statics)
- multiple reflections

Dip-related problems. Stacking velocities do not tie at line intersections for dipping reflectors. This is because the lines are shot at different azimuths, causing differences in apparent dip angle and, hence, different velocities and different NMO calculations. The solution for this problem is the application of DMO. The DMO process will be discussed later in this chapter.

Offset-related problems. Offsets that do not satisfy the short spread requirement—$X \leq Z$—can produce problems. This is demonstrated in Figure 6–77. Here the reflection times are those observed for a flat layered earth model with the maximum offset of $x_{max} > Z$, where Z is the depth of the reflector. As offset increases, the calculated NMO diverges further from the actual moveout.

Velocity analysis is, basically, the fitting of NMO curves to observed moveout. Figure 6–78a shows a best fit to the observed moveout of Figure 6–77 and Figure 6–78b shows the NMO-corrected event using this best fit hyperbola. The event is not flattened, resulting in a less than optimum stack, and the zero-offset time T_0 is too early.

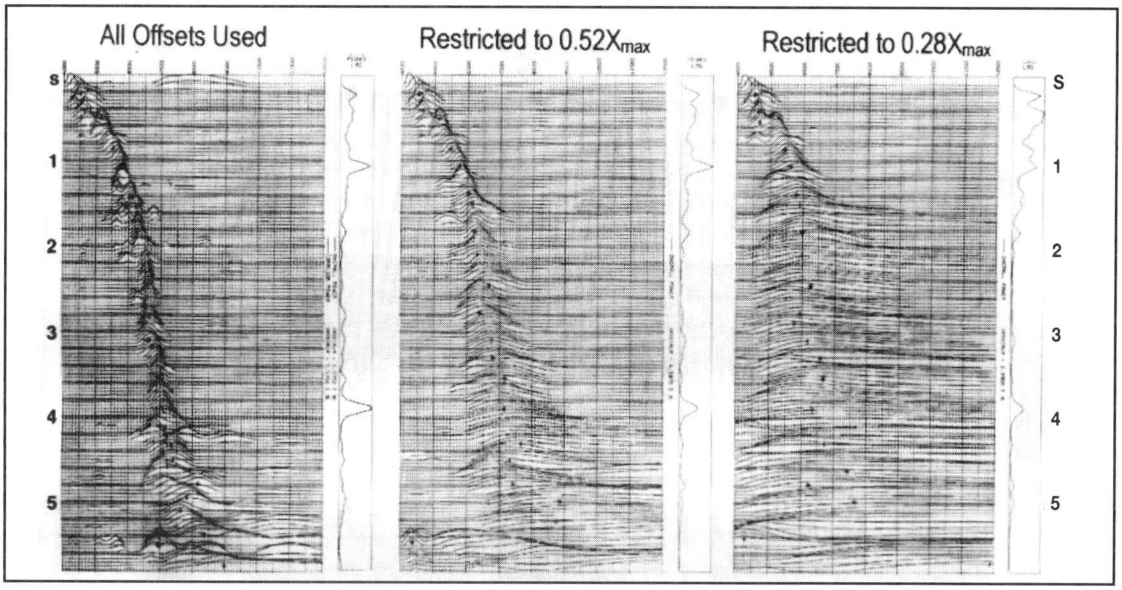

Fig. 6–79 *Effect of Restricting Offsets on Velocity Analysis*

Not using a full range of both long and short offset traces can also have a very negative effect on velocity analysis. This is demonstrated in Figure 6–79. In this figure, coherence is plotted as a wiggly trace with amplitude indicating coherence values. It is desirable to have the velocity trend defined by large coherence maxima with small breadth. The display on the left of Figure 6–79 satisfies this requirement. The center display shows what happens when traces are restricted to those corresponding to 52% of the maximum offset. Clearly, velocity resolution has decreased. The display on the right shows the effect of further offset restriction. Here it is only out to 28% of the maximum offset. The velocity trend is very difficult to see.

Effect of near-surface anomalies on velocities. Near-surface anomalies produce erroneous velocities if proper account of them is not taken. Figure 6–80 presents an earth model with a near-surface anomaly.

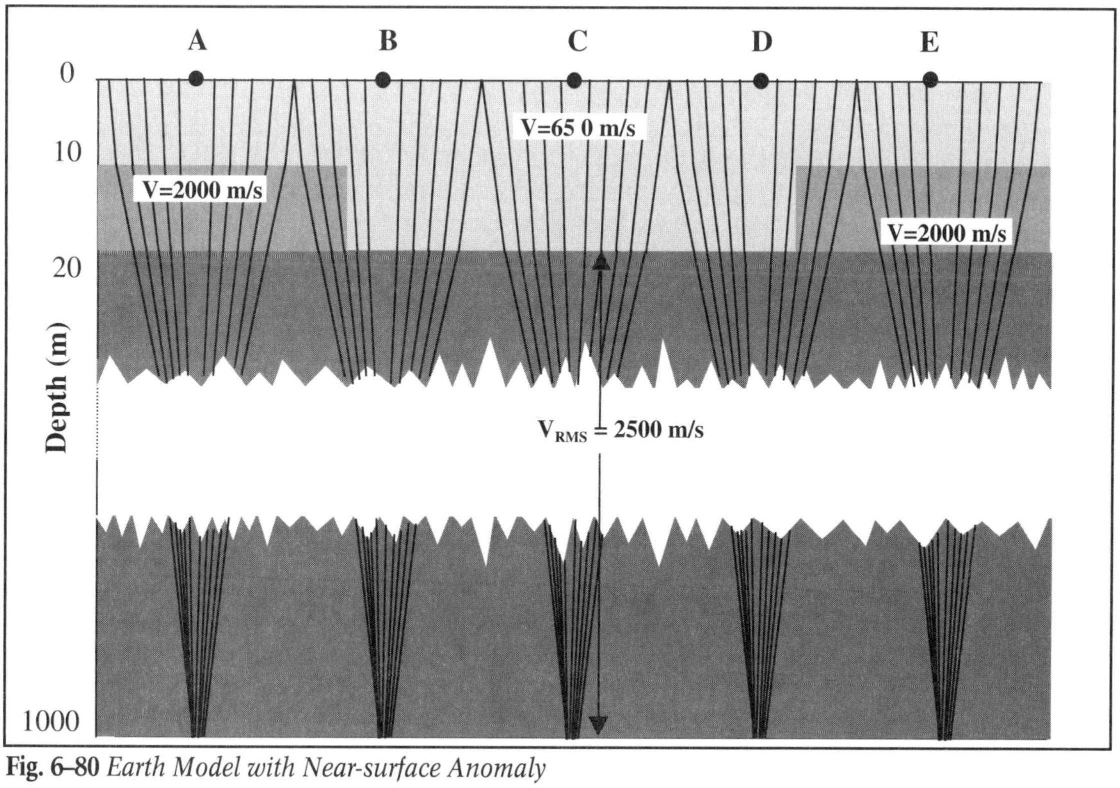

Fig. 6–80 *Earth Model with Near-surface Anomaly*

In this model, the unconsolidated rock extends to a depth of 10 m, abruptly increases to 20 m depth, then abruptly returns to the 10 m depth. Velocity of the unconsolidated rock is 650 m/s. The 10 m depth sections of the near-surface are underlain by 10 m layers of consolidated rock having a velocity of 2000 m/s. From a depth of 20 m, there are layers of consolidated rock with an RMS velocity of 2500 m/s.

Five CMPs, designated *A, B, C, D*, and *E*, and five ray paths corresponding to offsets of 200 m, 400 m, 600 m, 800 m, and 1000 m for each CMP are shown. Because of the near-surface anomaly some ray paths go through the 2000 m/s layer and others do not. Figure 6–81 shows the effect of this by displaying simulated recordings of the reflection from 1000 m deep reflector at CMPs *A* through *E*.

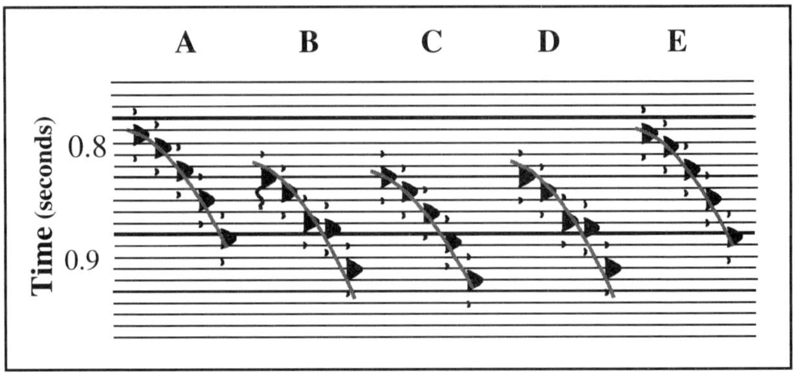

Fig. 6–81 *Synthetic Traces Based on Model of Figure 6–80*

The events at CMPs *A, C*, and *E* show hyperbolic moveout, although the hyperbola at *C* differs from those at *A* and *E*. Events at *B* and *D* do not exhibit hyperbolic moveout, but best fit hyperbolas are shown for both events.

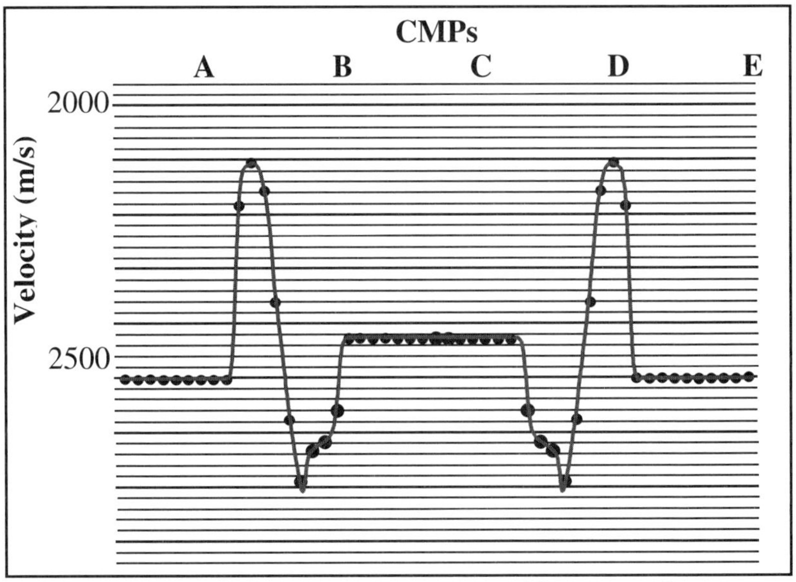

Fig. 6–82 *Variations in Stacking Velocity Caused by Near-surface Anomaly*

The model of Figure 6–80 was used to calculate arrival times for CMPs at 100 m intervals. From these arrival times, velocities were determined. Figure 6–82 displays the results of these calculations. Note that the variation in apparent stacking velocity across the model does not exactly follow the near surface anomaly. At the edges of the anomaly, velocities rapidly decrease, increase, then again decrease to a stable value where the all ray paths of the CMPs are in the thick portion of the unconsolidated rock.

Figure 6–83 is a simulated CMP stack for the model of Figure 6–80. While the reflector is actually flat, the near-surface anomaly has produced apparent structure.

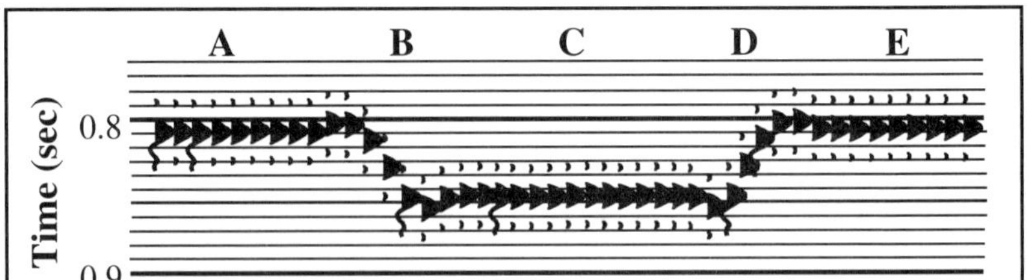

Fig. 6-83 *Stack of Synthetic Traces Based on Model of Figure 6–80 and Velocities of Figure 6–82*

Problems caused by multiple reflections. When multiple reflections are present in the data, semblance maxima are produced for both primary and multiple reflections. In some cases, the two sets of semblance maxima can be easily distinguished with the multiples showing a lower velocity trend than the primaries. In other cases, multiple amplitudes are so strong that maxima for primary reflections are obscured.

An example of the latter situation is shown in Figure 6–84. The center CMPs at locations 224 and 684—the coherence displays for velocity analyses at these locations—are shown before (left) and after (right) multiple attenuation is accomplished. Note that before multiple attenuation, the primary velocity trend at times later than about 3 seconds cannot be discerned. After multiple attenuation, the primary velocity trend becomes obvious.

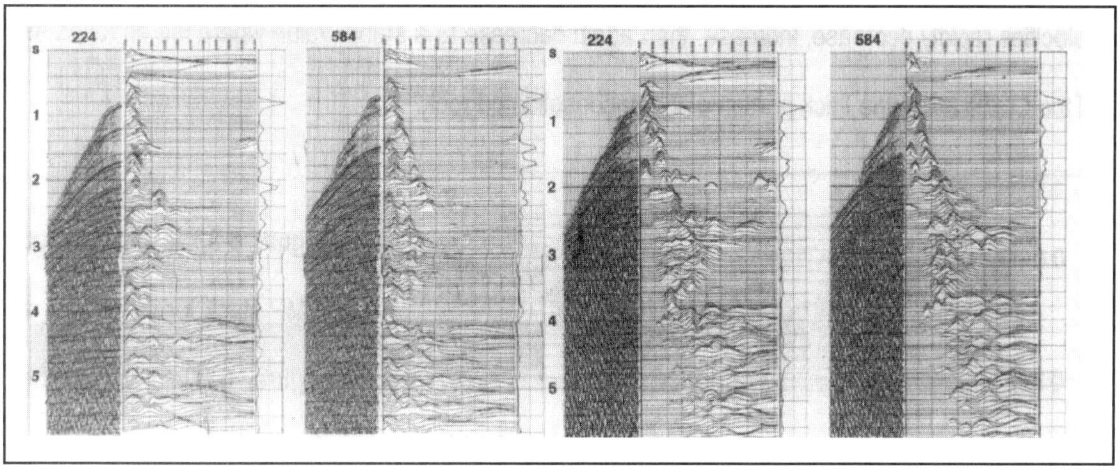

Fig. 6–84 *Velocity Analysis before and after Multiple Attenuation*

As shown in Figure 6–84, it is often desirable to attenuate multiples before the final velocity field is defined. Methods used to attenuate multiples will be presented later in this chapter.

Velocity analysis summary. The accuracy and resolution of velocity estimation from seismic data depends upon the following factors:

- spread lengths
- stacking fold
- muting
- time gate
- velocity sampling
- choice of coherency measure
- true departures from hyperbolic moveout
- bandwidth of data

Most of these are under the control of the processor, who should carefully select parameter values for velocity analysis.

Brute Stack

Brute stack is simply CMP stack with the preliminary velocity field being used to perform NMO corrections. The steps in CMP stack are:

- apply NMO corrections to all CMP gathers
- mute traces
- sum all traces in each CMP gather (add amplitudes on each trace at each sample time)
- divide summed amplitudes by number of live traces—traces not zeroed by mutes—to normalize output amplitudes

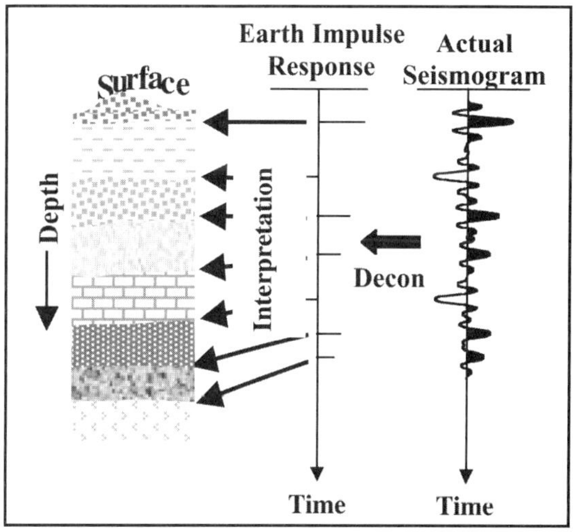

Fig. 6–85 *Deconvolution Objectives*

Deconvolution

The objectives of deconvolution are to:

- shorten reflection wavelets
- attenuate ghosts, instrument effects, reverberations and multiple reflections

In so doing, vertical resolution is increased and earth impulse response or reflectivity is approximately recovered. This is demonstrated in Figure 6–85.

If deconvolution were entirely successful, reflection wavelets would be collapsed into spikes. Their amplitudes would be proportional to reflection coefficients at reflecting interfaces. After NMO corrections, times of occurrence would be two-way times to these interfaces.

If deconvolution is broadly defined as those processes that operate on seismic wavelets to produce a more desirable wavelet shape, then there are many deconvolution types: These types include:

- time-domain deconvolution
- signature deconvolution
- time-variant deconvolution (TVD), with subtypes of
- whitening or spiking
- predictive or gapped
- surface consistent deconvolution
- frequency domain deconvolution
- *F-K* domain deconvolution

A variety of other, special-purpose deconvolutions are also available. TVD may be used in post-processing as well as in pre-processing.

Signature deconvolution. The simplest trace model is

$$x(t) = w(t) * e(t) + n(t) \tag{6.30}$$

In general, the only known element is the recorded trace $x(t)$. The source wavelet $w(t)$, the reflectivity $e(t)$, and noise $n(t)$ are unknown. In some circumstances, however, the source wavelet or far-field signature is known. For example, it is common procedure in marine seismic acquisition to record the airgun signature at a depth of about 100 m beneath the airgun array (Fig. 5–29). In the case of vibrators, the source autocorrelation of the sweep can be used as the source wavelet on cross-correlated vibrator records.

If, in addition to having a known source wavelet, it is assumed that noise is zero or negligible, then the trace model can be modified to:

$$x(t) = s(t) * w'(t) * e(t) \tag{6.31}$$

In the airgun model:

$s(t)$ = source signature measured near source

$w'(t)$ = earth's effective wavelet (the effect that causes the waveform to be time-variant)

$e(t)$ = reflectivity

Since the source wavelet $s(t)$ is recorded, its inverse $s^{-1}(t)$ can be designed. The inverse is defined by:

$$s(t) * s^{-1}(t) = \delta(t) \tag{6.32}$$

where

$\delta(t)$ is the Dirac delta function or unit impulse (commonly called a spike)

Note that $s^{-1}(t)$ does not equal $\frac{1}{s(t)}$. Convolving the wavelet inverse with the trace:

$$x(t)*s^{-1}(t) = s(t)*s^{-1}(t)*\ w(t)*\ e(t) = \delta(t)\ *\ w(t)*\ e(t) = w(t)*\ e(t) \tag{6.33}$$

Reflectivity is obtained but with modified, time-variant, waveforms (not spikes).

In Equation 6.31 for the vibrator model, $x(t)$ is the uncorrelated vibrator trace, $s(t)$ is the known vibrator sweep, $w'(t)$ is the earth's effective wavelet, and $e(t)$ is reflectivity. It is common practice to cross-correlate the sweep with each trace, and since cross-correlation is equivalent to convolving with the time-reversed version of the sweep, the correlated record $x'(t)$ is:

$$x'(t) = x(t)*\ s(-t)= s(t)*\ w(t)* = s(t)*\ w(t)*\ e(t)\ *s(-t)$$

$$= s(t)*s(-t)*\ w(t)*\ e(t)$$

But $s(t)*s(-t)$ is the sweep autocorrelation, which is described by the Klauder wavelet $k(t)$. Thus, the correlated vibrator record is described by:

$$x'(t) = k(t)*w(t)*e(t)$$

Signature deconvolution for Vibroseis records is

$$x'(t) = x(t)*k^{-1}(t) = \delta(t)*\ w(t)*\ e(t) = w(t)*\ e(t) \tag{6.34}$$

The result is identical to that for the airgun. Reflectivity is obtained but with modified, time-variant, waveforms (not spikes).

Signature deconvolution is only partially successful. The source wavelet is removed but the earth's effect on the wavelet has not been determined nor removed. Usually, some other kind of deconvolution must be applied after signature deconvolution.

Time-variant deconvolution. Equation 6.35 is a more realistic mathematical model of a seismic trace:

$$y(t) = w(t)*e(t)*\ i(t)*\ g(t)*\ m(t) + n(t) \tag{6.35}$$

where

$y(t)$ = seismic trace

$w(t)$ = source wavelet

$e(t)$ = earth impulse response

$i(t)$ = instrument filter response

$g(t)$ = ghosting operator

$m(t)$ = multiple reflection operator

$n(t)$ = additive noise

A whitening deconvolution filter is a filter $h(t)$ such that:

$$y(t)\ *\ h(t) = e(t) \tag{6.36}$$

Comparing Equations 6.35 and 6.36, it can be seen that with $n(t) = 0$, the whitening deconvolution filter must be:

$$h(t) = w^{-1}(t)*\ i^{-1}(t)*\ g^{-1}(t)*\ m^{-1}(t) \tag{6.37}$$

That is, the deconvolution filter, if it perfectly accomplishes its objective, must contain the exact inverses of the source wavelet, the instrument filter response, the ghosting operator, and the multiple reflection operator. This is so, because, as noted in chapter 3, when a time series is convolved with its inverse, the result is an impulse function and a time series convolved with an impulse is the same time series.

Deconvolution filter design is based on the Weiner-Hopf optimum filter equation:

$$\phi_{ii}(\tau) * h(\tau) = \phi_{id}(\tau), \tau > 0 \qquad (6.38)$$

where

$\phi_{ii}(\tau)$ = autocorrelation of the input trace or gate

$h(\tau)$ = the deconvolution filter

$\phi_{id}(\tau)$ = cross-correlation of the input with the desired output

Thus, only the input itself and the desired output are needed to design the filter. For practical filter design, it is also necessary to specify the length of the filter to be designed.

For sampled data, $\tau = \lambda\Delta T$, where $0 \le \lambda \le N$, $\phi_{id}(\tau) = \{\phi_\lambda^{id}\} = \{\phi_0^{id}, \phi_1^{id}, \phi_2^{id}, \ldots, \phi_N^{id},\}$ and ΔT = sample period, Equation 6.30 becomes:

$$\sum_{j=0}^{N} \phi_j^{ii} h_{j-\lambda} = , \phi_\lambda^{id}, \; 0 \le \lambda \le N \qquad (6.39)$$

Remembering that the autocorrelation is symmetrical— $\phi_n = \phi_{-n}$—Equation 6.39 can be written as the set of linear, simultaneous equations:

$$\begin{aligned}
&\phi_0^{ii} h_0 + \phi_1^{ii} h_1 + \phi_2^{ii} h_2 + \phi_3^{ii} h_3 + \ldots + \phi_N^{ii} h_N = \phi_0^{id} \\
&\phi_1^{ii} h_0 + \phi_0^{ii} h_1 + \phi_1^{ii} h_2 + \phi_2^{ii} h_3 + \ldots + \phi_{N-1}^{ii} h_N = \phi_1^{id} \\
&\phi_2^{ii} h_N + \phi_1^{ii} h_{N-1} + \phi_2^{ii} h_{N-2} + \phi_3^{ii} h_{N-3} + \ldots + \phi_{N-2}^{ii} h_0 = \phi_2^{id} \\
&\quad\vdots \\
&\phi_N^{ii} h_0 + \phi_{N-1}^{ii} h_1 + \phi_{N-2}^{ii} h_2 + \phi_{N-3}^{ii} h_3 + \ldots + \phi_0^{ii} h_N = \phi_0^{id}
\end{aligned} \qquad (6.40)$$

For whitening deconvolution, the desired output is a spike at time 0. Thus, the cross-correlation between input and desired output is a one at zero-lag and zero at all other lags since only positive lags can be considered. Thus Equation 6.40 can be written in matrix form as:

$$\begin{bmatrix}
\phi_0^{ii} & \phi_1^{ii} & \phi_2^{ii} & \cdots & \phi_N^{ii} \\
\phi_1^{ii} & \phi_0^{ii} & \phi_1^{ii} & \cdots & \phi_{N-1}^{ii} \\
\phi_2^{ii} & \phi_1^{ii} & \phi_0^{ii} & \cdots & \phi_{N-2}^{ii} \\
\phi_3^{ii} & \phi_2^{ii} & \phi_1^{ii} & \cdots & \phi_{N-3}^{ii} \\
\vdots & \vdots & \vdots & & \vdots \\
\phi_N^{ii} & \phi_{N-1}^{ii} & \phi_{N-2}^{ii} & \cdots & \phi_0^{ii}
\end{bmatrix}
\text{X}
\begin{bmatrix} h_0 \\ h_0 \\ h_0 \\ h_3 \\ \vdots \\ h_N \end{bmatrix}
=
\begin{bmatrix} 1 \\ 0 \\ 0 \\ 0 \\ \vdots \\ 0 \end{bmatrix} \qquad (6.41)$$

It can also be written as the matrix algebra equation:

$$\Phi H = \Psi \tag{6.42}$$

where

Φ is the N+1 x N+1 matrix of autocorrelation values

H is the N+1 point column matrix of filter points

Ψ is the N+1 point column matrix 1, 0, 0, ..., 0. The Wiener-Levinson iterative method allows relatively fast solutions of Equation 6.42.

The autocorrelation function contains information needed for deconvolution filter design, such as source wavelet shape, periods of multiples and reverberations, and amplitude relationships between multiples and primaries, as shown by the autocorrelogram of Figure 6–86. Note that the filter should be somewhat longer than the period of the multiple to be attenuated.

The whitening deconvolution filter includes approximate inverses of the:

- source wavelet
- recording instrument filter responses
- ghosting function
- reverberation operator
- multiple operators

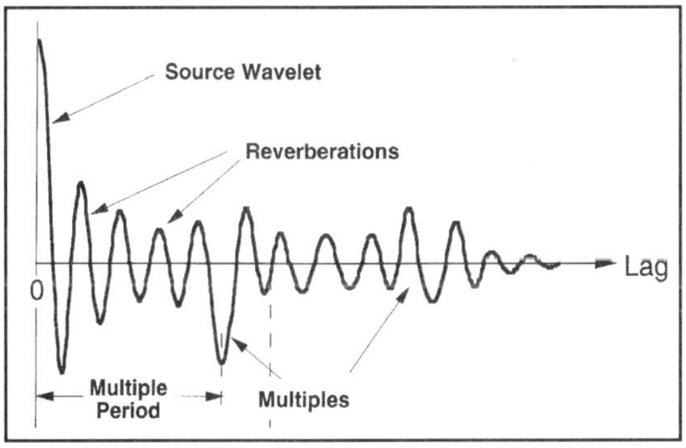

Fig. 6–86 *Information in a Correlogram*

To be exact inverses, the filter would have to be infinitely long.

The effect of whitening deconvolution is easily seen in the frequency domain. Figure 6–87 shows this by displaying the input amplitude spectrum of a trace with multiples present, the amplitude spectrum of the filter, and the amplitude spectrum of the deconvolved trace. In this ideal case, the filter is the exact inverse of the input, and the output spectrum is perfectly flat from 0 to the Nyquist frequency.

Deconvolution filter design often fails because, in approximating the inverses, a divide by zero is attempted. To avoid or correct for this, white noise is added to the input. Figure 6–88 illustrates this. Note that when white noise is added to the amplitude spectrum, the filter does not completely flatten the output amplitude spectrum. This is because the inverse of the whitened amplitude is applied, not the inverse of the actual input.

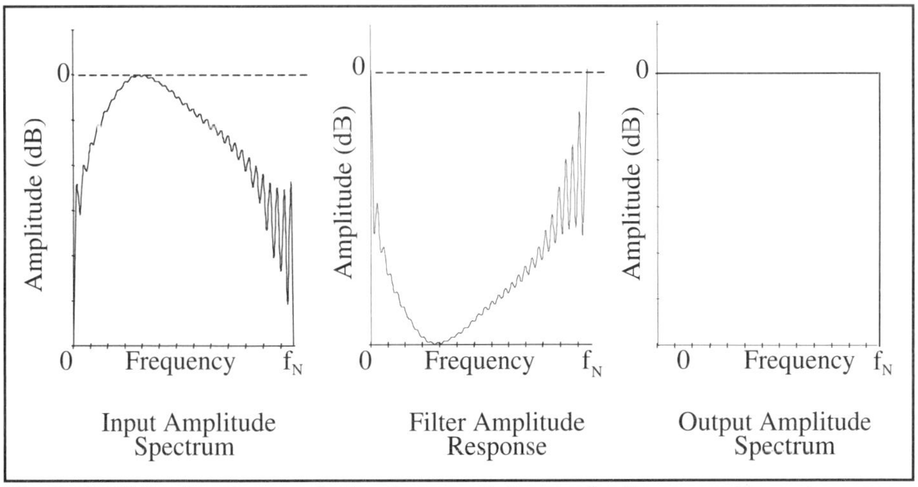

Fig. 6–87 *Whitening Deconvolution Representation in the Frequency Domain*

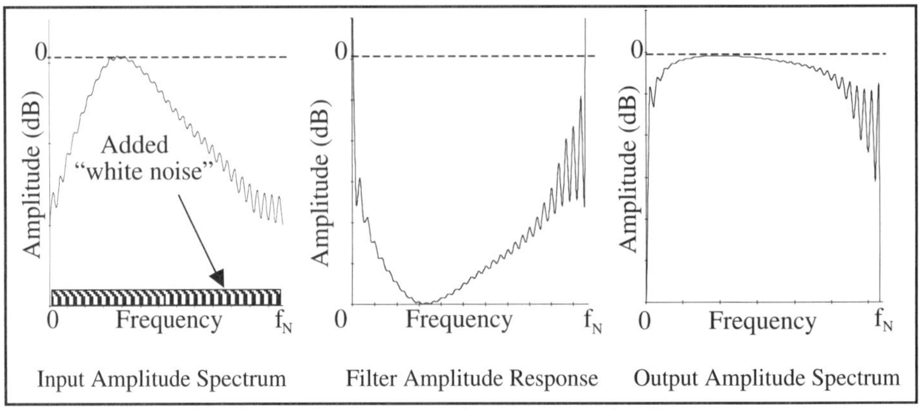

Fig. 6–88 *Effect of Adding White Noise to the Input Amplitude Spectrum*

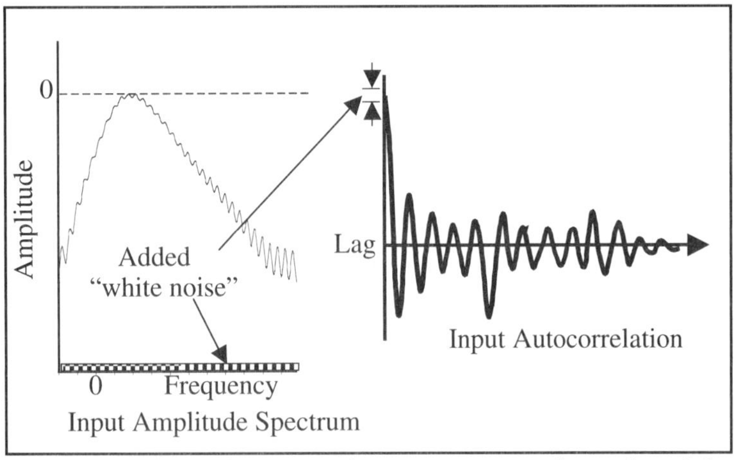

Fig. 6–89 *Adding White Noise by Increasing Zero-lag Value of the Input Autocorrelation*

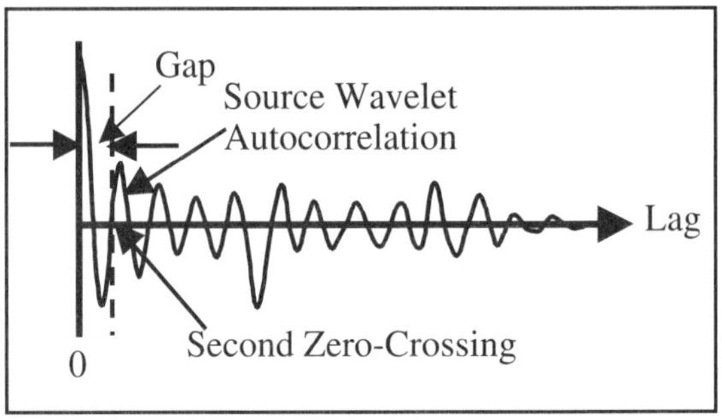

Fig. 6–90 *Input Autocorrelation and the Information It Contains*

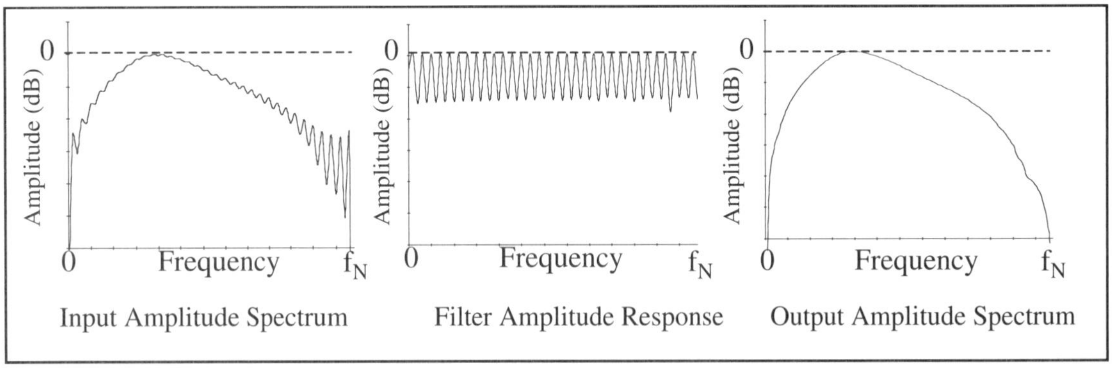

Fig. 6–91 *Representation of Gapped Deconvolution Representation in the Frequency Domain*

Addition of white noise is accomplished by increasing the zero-lag value of the autocorrelation by a small percentage. This is equivalent to adding a constant amount to the amplitude spectrum for all frequencies. This is illustrated in Figure 6–89. Adding white noise, or pre-whitening, should never be done unless necessary to get a filter designed. Even then, the minimum amount of white noise should be added.

For predictive or gapped deconvolution, the desired output consists of the input values at some time in the future, say a samples ahead. The crosscorrelation between desired output and input is, thus, the input autocorrelation beginning at a lag of $\tilde{\alpha}$. The basic filter equation takes the form:

$$\begin{bmatrix} \phi_0^{ii} & \phi_1^{ii} & \phi_2^{ii} & \cdot & \cdot & \cdot & \phi_N^{ii} \\ \phi_1^{ii} & \phi_0^{ii} & \phi_1^{ii} & \cdot & \cdot & \cdot & \phi_{N-1}^{ii} \\ \phi_2^{ii} & \phi_1^{ii} & \phi_0^{ii} & \cdot & \cdot & \cdot & \phi_{N-2}^{ii} \\ \phi_3^{ii} & \phi_2^{ii} & \phi_1^{ii} & \cdot & \cdot & \cdot & \phi_{N-3}^{ii} \\ \cdot & \cdot & \cdot & \cdot & \cdot & \cdot & \cdot \\ \cdot & \cdot & \cdot & \cdot & \cdot & \cdot & \cdot \\ \cdot & \cdot & \cdot & \cdot & \cdot & \cdot & \cdot \\ \phi_N^{ii} & \phi_{N-1}^{ii} & \phi_{N-2}^{ii} & \cdot & \cdot & \cdot & \phi_0^{ii} \end{bmatrix} \text{X} \begin{bmatrix} h_0 \\ h_0 \\ h_0 \\ h_3 \\ \cdot \\ \cdot \\ \cdot \\ h_N \end{bmatrix} = \begin{bmatrix} \phi_{\alpha}^{id} \\ \phi_{\alpha+1}^{id} \\ \phi_{\alpha+2}^{id} \\ \phi_{\alpha+1}^{id} \\ \cdot \\ \cdot \\ \cdot \\ \phi_{\alpha+N}^{id} \end{bmatrix} \qquad (6.43)$$

The filter of Equation 6.43 predicts the amplitude of the trace samples ahead. The prediction filter can be used to form a prediction error filter as shown in the following:

$$\{p_n\} = \{1, \overbrace{0, 0, \ldots, 0, 0}^{\alpha \text{ zeros}}, -h_0, -h_1, -h_2, -h_3, -h_4, \ldots, -h_N\}$$

where

$\{p_n\}$ = the prediction error filter

h_k = filter points from solution of Equation 6.43

$k = 0, 1, 2, \ldots, N$

The output of the prediction error filter is the difference between the actual and predicted value. The filter points calculated from the set of equations just shown, predict the values of the input a sample periods ahead.

The length of the gap (α) in a gapped deconvolution filter determines the amount of wavelet shortening, or whitening, the filter does. A popular choice for gap length is the time of the second zero crossing of the autocorrelation function. See Figure 6–90. It is, however, often desirable to try other values for the gap in order to select an optimum value for this parameter.

In the frequency domain (Fig. 6–91), it can be seen that gapped deconvolution does not completely whiten the amplitude spectrum. The filter amplitude response does not contain the inverse of the wavelet, hence it only removes the ripples in the input amplitude spectrum caused by multiples, reverberations, etc.

Figure 6–92 compares whitening and gapped deconvolution. On the left, whitening deconvolution collapses wavelets and attenuates multiples. On the right, gapped deconvolution applied to the same trace attenuates multiples but does not remove the source wavelet.

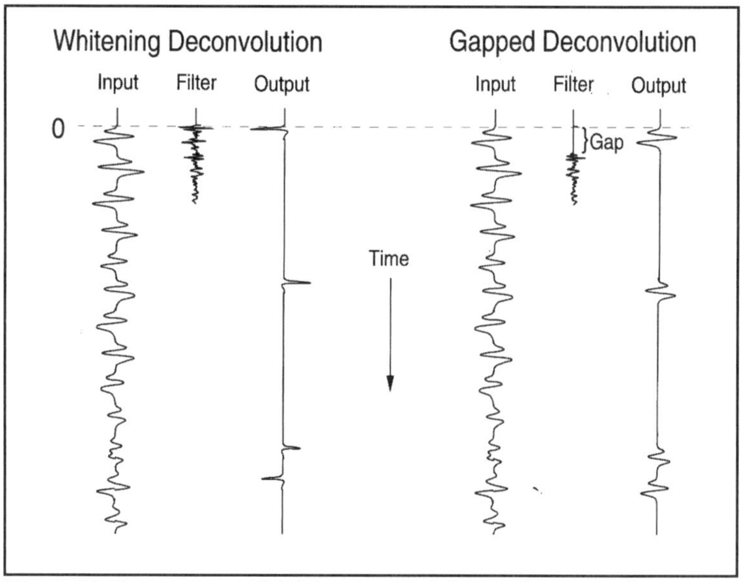

Fig. 6–92 *Comparison of Whitening and Gapped Deconvolution*

From another point of view, the distinction between gapped and whitening deconvolution does not seem so great. This is because whitening deconvolution can be thought of as a prediction error filter with a prediction distance of one sample. Thus, the gap shrinks to zero. All deconvolution filters are, thus, prediction error filters with prediction distances ranging from one upward.

Filter design and application are time-variant because loss of higher-frequency components makes reflection wavelets become progressively lower frequency, as shown in Figure 6–93. Because of this, no single filter will adequately collapse all wavelets. Since wavelets are time-variant, deconvolution filters must also be time-variant. Thus, traces are divided into time gates and different filters are designed and applied in each gate. First breaks are excluded from the first analysis gate. As far as reflection sequences and multiples are concerned, first breaks are anomalous events.

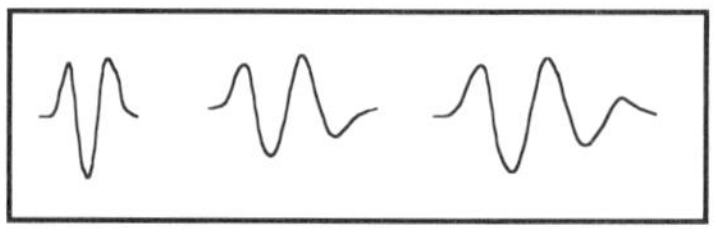

Fig. 6–93 *Wavelet Shapes at Early, Middle, and Late Record Times*

In essence, each filter is applied to the trace, and scaling functions for each filter output are applied such that the filters are applied to the gates for which they were designed but with overlap. In the overlap regions, the earlier filter is ramped on and the later filter is ramped off with scalers. This is done to avoid abrupt amplitude changes at gate boundaries and to give smooth transitions between filters. Selection of design gates and time-variant filter application are illustrated in Figure 6–94 for the three-filter case.

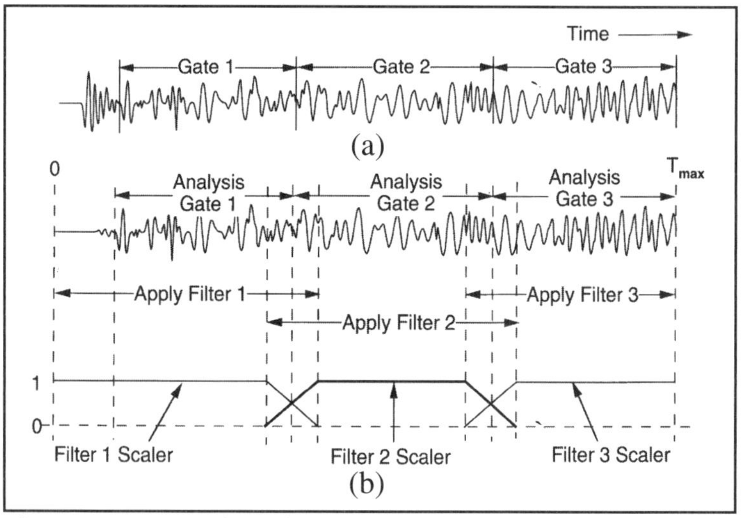

Fig. 6–94 *Design and Application Gates for Time-variant Deconvolution*

Design, or analysis, gates should be long enough for multiple sequences to be well-defined. In general, the gates should be at least 10 times as long as the filters. Since filters should be a bit longer than the multiple periods, this means each gate should be seven to 10 times the multiple period. Gates may overlap.

First breaks may also be excluded from the first application gate. If the filter is applied to first breaks, then some of their high amplitudes will be smeared into parts of the traces just past the first breaks. Except for short offset traces, this is not serious as these early parts of traces will be eliminated when NMO corrections are made.

A tacit assumption in Wiener deconvolution is that reflectivity $e(t)$ is random. If this assumption is correct, the autocorrelation of reflectivity is an impulse function or spike. Cross multiplication at zero lag is the sum of the squares of all amplitudes, but at all other lags the sums cancel out to zero. Remembering also that noise is assumed to be zero, represent the trace by:

$$x(t)= w(t)^* e(t)$$

Then, noting that autocorrelation of the trace is equivalent to convolution with the time-reversed version of the trace:

$$\varphi^{xx}(\tau) = x(t) * x(\tau-t) = [w(t)]*e(t)]*[w(\tau-t)*e(\tau-t)]$$
$$= w(t) * w(\tau-t) = [e(t)]*e(\tau-t)] = \varphi^{ee}(\tau) \quad (6.44)$$

where

$\varphi^{xx}(\tau)$ is the trace autocorrelation

$w(t)$ is the wavelet

$e(t)$ is reflectivity

$\varphi^{ww}(\tau)$ is the autocorrelation of the wavelet

$\varphi^{ee}(\tau)$ is the autocorrelation of the reflectivity

If $e(t)$ is truly random, then $\varphi^{ee}(\tau) = A^2\varphi(\tau)$, where A^2 is a constant representing reflectivity amplitudes squared. Substituting this into Equation 6.44 gives:

$$\varphi^{ee}(\tau) = \varphi^{ww}(\tau)*[A^2\delta(\tau)] = A^2\varphi^{ww}(\tau) \quad (6.45)$$

Equation 6.45 states that if noise is zero and reflectivity is random, then the trace autocorrelation is a scaled version of the wavelet autocorrelation. If this is true, or it is sufficiently correct, then the Wiener filter Equation 6.38 is valid.

In practical terms, Wiener deconvolution yields good results when the S/N ratio is good and traces have many reflections. In the case of sparse reflections, results may not be so good. This is demonstrated in Figure 6–95, using synthetic data.

Trace *b* in Figure 6–95 is the input to a spiking deconvolution. The trace is noise-free and multiple-free. The numbers on the left indicate the length of the deconvolution filter applied. The desired output is trace a. Based on the autocorrelograms, deconvolution outputs get better as filter length increases. However, as filter length increases, more and more spurious signals are introduced. The filters are trying to change reflectivity function into random reflectivity.

The amplitude spectra also become progressively whiter (flatter) as filter length increases. In this case it is undesirable. The amplitude spectrum desired is that of trace a, not a white spectrum. It is obvious, then, that spiking decon is not always the best approach to extracting reflectivity from recorded traces.

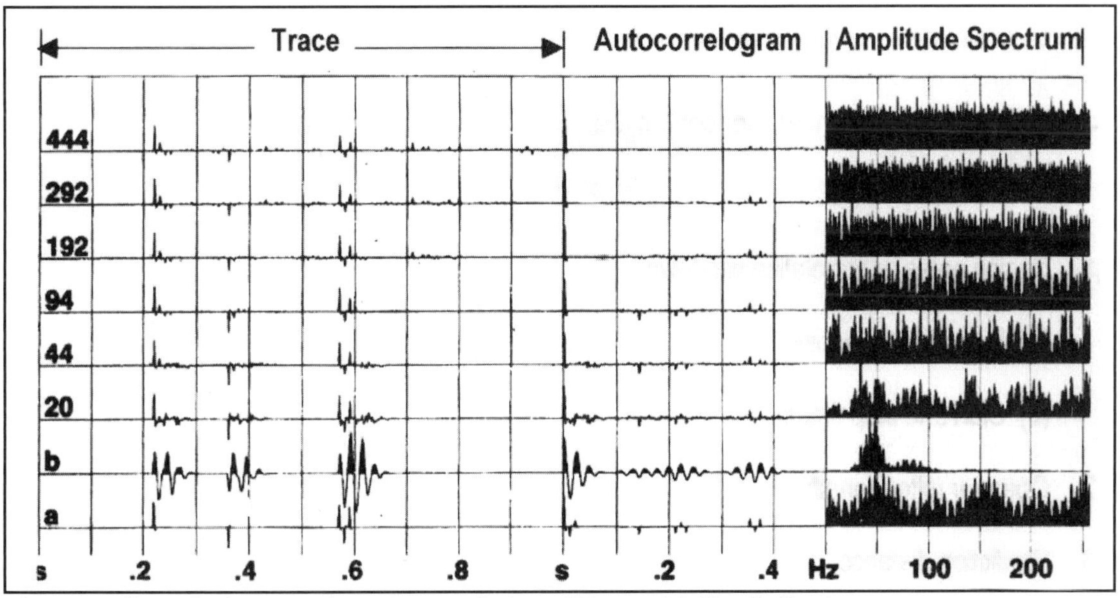

Fig. 6–95 *Effect of Filter Length on Deconvolution of Trace with Five Reflections*

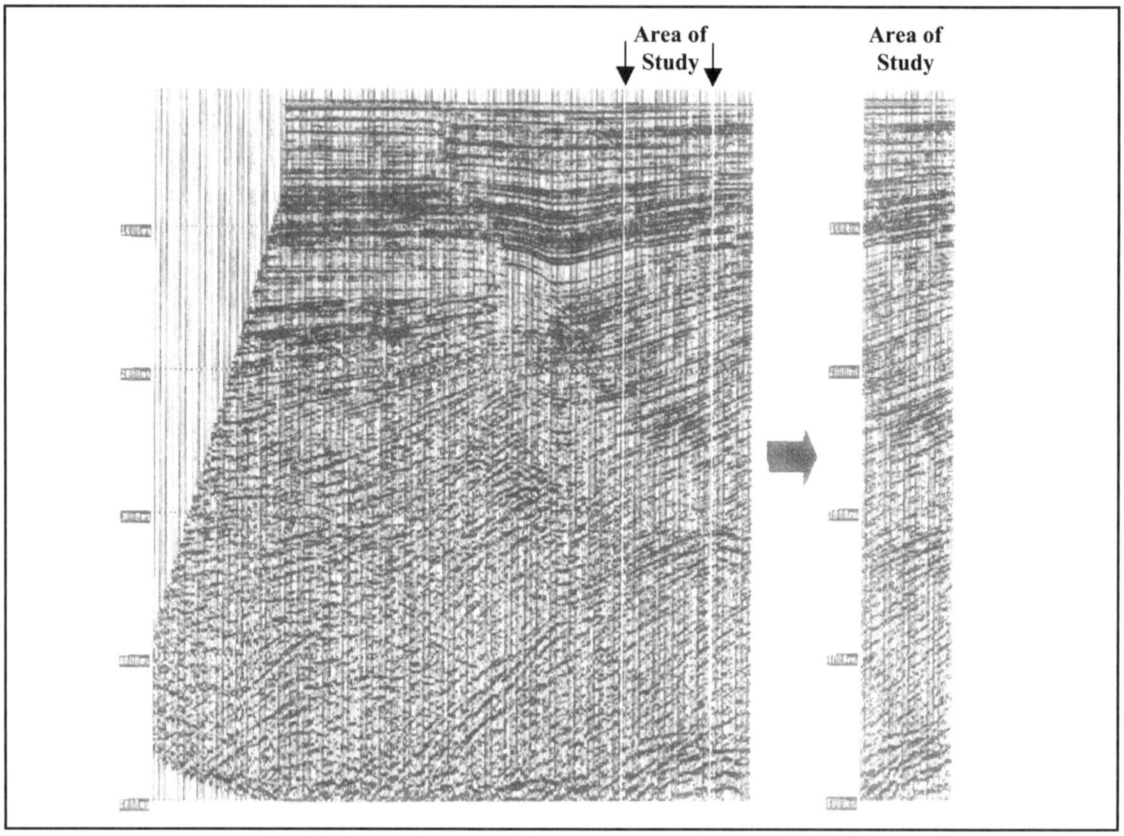

Fig. 6–96 *Data Used to Illustrate TVD Parameter Testing*

Time-variant deconvolution parameter selection. Deconvolution parameters under user control are:

1. design gates or correlation windows
    (a) number of windows
    (b) start and stop times
2. operator (filter) length
3. prediction distance
4. white noise level

Parameter testing is recommended before selecting final TVD parameters. The marine field data of Figure 6–96 will be used to illustrate deconvolution parameter testing. Each parameter is tested separately and illustrated by two panels—a set of CMP panels and a set of stack panels centered on the selected CMP. Only one parameter is varied at a time. *Never* change more than one parameter at a time!

Figure 6–97 shows three autocorrelation windows selected for a CMP near the center of the study area. On the left, the window extends from just below the first breaks to nearly the end of the record. Note that window start and stop times are extrapolated from the near offset trace to the far offset trace along NMO hyperbolas. Another single window selection is shown at the center. Start time is unchanged but stop time is just slightly below 3 s. This selection includes the parts of the record having the best S/N ratio. The selection on the right has two windows that overlap. If multiple windows are chosen an overlap of 400 ms or more is often required by the software.

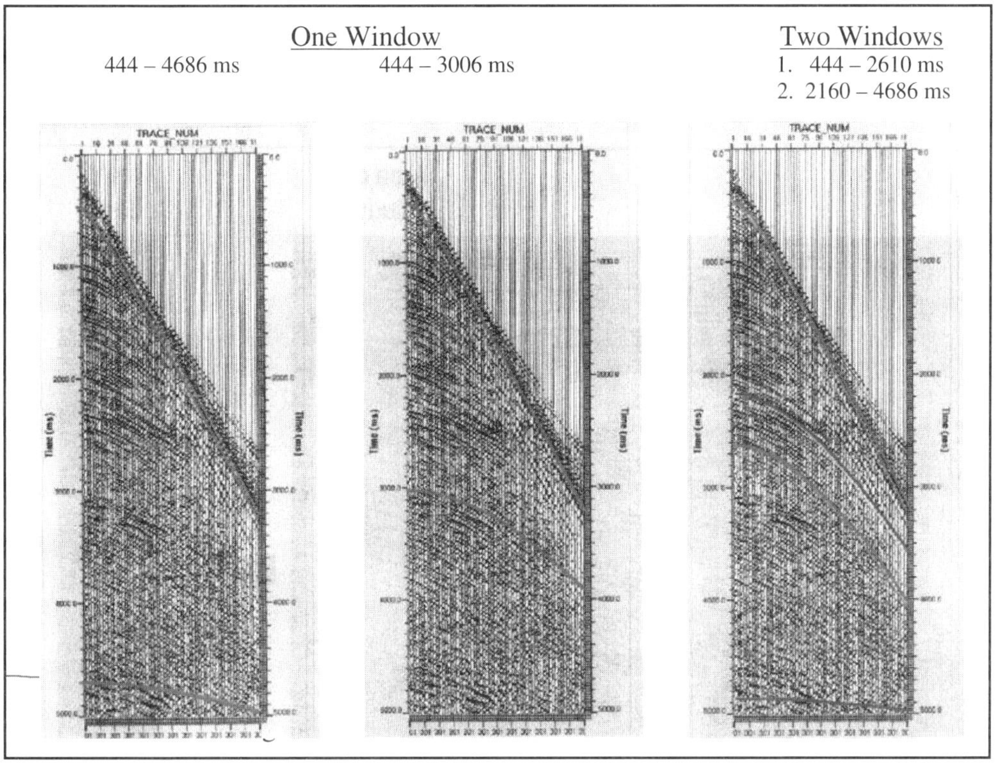

Fig. 6–97 *Autocorrelation Window Selection*

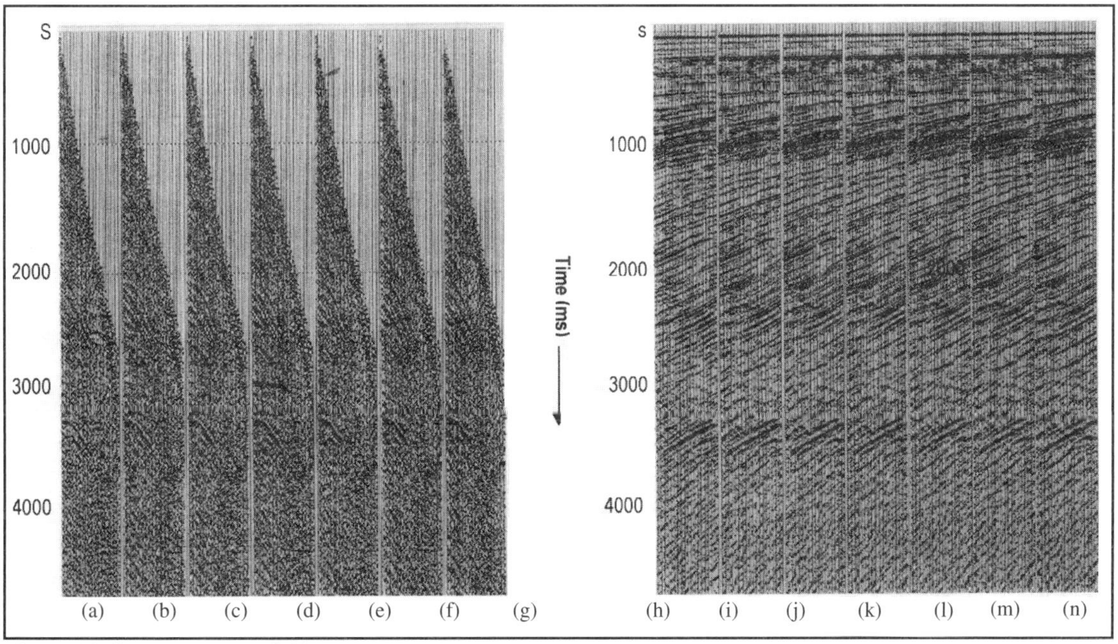

Fig. 6–98 *Operator Length Tests*

Figure 6–98 shows outputs, with CMP gathers on the left and stack panels on the right. In each case, the operator length was 120 ms, the prediction distance was 2 ms (spiking decon since sample period is 2 ms), and the white noise level 0.01%—minimum value for the software used. The decon filters were applied over the same times as the design gates (autocorrelation window) for the long single window and the two window cases. The decon filter was applied to the end of the record for the short window case. The longer window seems to give a somewhat better result at later times than does the shorter window. The two-window result does not appear to appreciably differ from the long single window result so the 444–4868 ms window was chosen.

Figure 6–98 shows the results of the operator length test. All outputs were obtained using a single correlation window from 444 ms to 4868 ms, 2 ms prediction distance, and 0.01 percent white noise. Results using operator lengths of 320 ms, 240 ms, 160 ms, and 120 ms do not significantly differ. The 100 ms operator does not provide as much wavelet collapse as the others. The 120 ms operator length was chosen since, in general, the shortest operator that gives acceptable results is used.

Figure 6–99 shows the results of the prediction distance test. All outputs were obtained using a single correlation window from 444 ms to 4868 ms, 120 ms operator length, and 0.01 percent white noise. The 2 ms prediction distance output retains the most character—a recognizable wavelet shape, usually requiring high frequency content—and wavelet collapse. Longer prediction distances give more high frequency filtering. The choice for prediction distance is, thus, 2 ms, or spiking decon.

Figure 6–100 shows the results of the white noise test. All outputs were obtained using a single correlation window from 444 ms to 4868 ms, 120 ms operator length, and 2 ms prediction distance. A filter was designed with no white noise added (white noise level = 0), so this could well be the choice. There is no observable difference between white noise levels of 0 and 0.01%, and 0.01% is the default value. The default (0.01%) is the choice. Note that 99.99% and 0.01% are interpreted as the same value.

Time-variant deconvolution summary. TVD is:

- performed in the time domain
- applied to every trace of every record
- expects minimum-phase inputs and results may be less than optimum if not minimum-phase
- has two options—whitening and gapped
- shortens reflection wavelets
- attenuates ghosts, reverberations, and multiples if filter is long enough

Note that TVD may be applied in both pre-processing and post-processing. Whitening TVD is usually done in pre-processing and gapped in post-processing.

Figure 6–101 is an example of a whitening deconvolution application. The output of whitening deconvolution often looks very noisy, as in the center record of Figure 6–101. This is because the filter whitens the whole trace, both signal and noise. Application of a band pass frequency filter usually improves the appearance of the output and allows better evaluation of deconvolution effect. Note that the filtered output is not passed on to the next stage of processing.

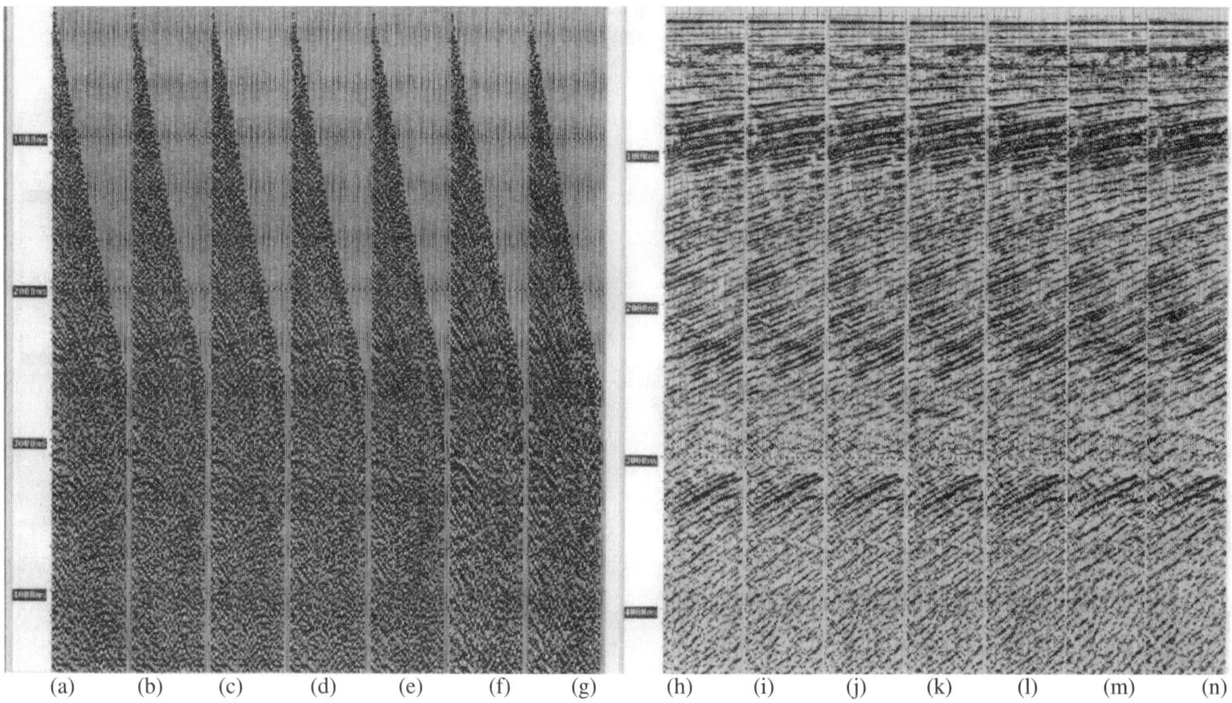

Fig. 6–99 *Prediction Length Tests*

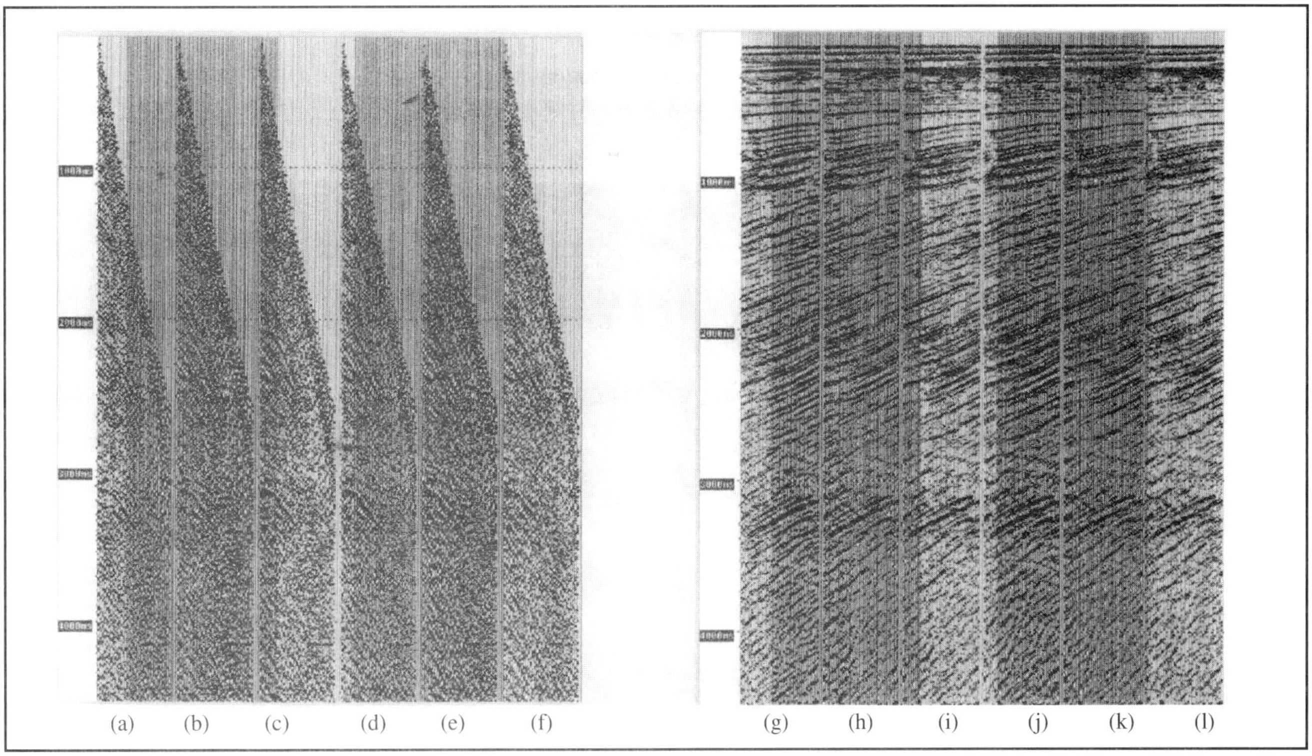

Fig. 6–100 *White Noise Tests*

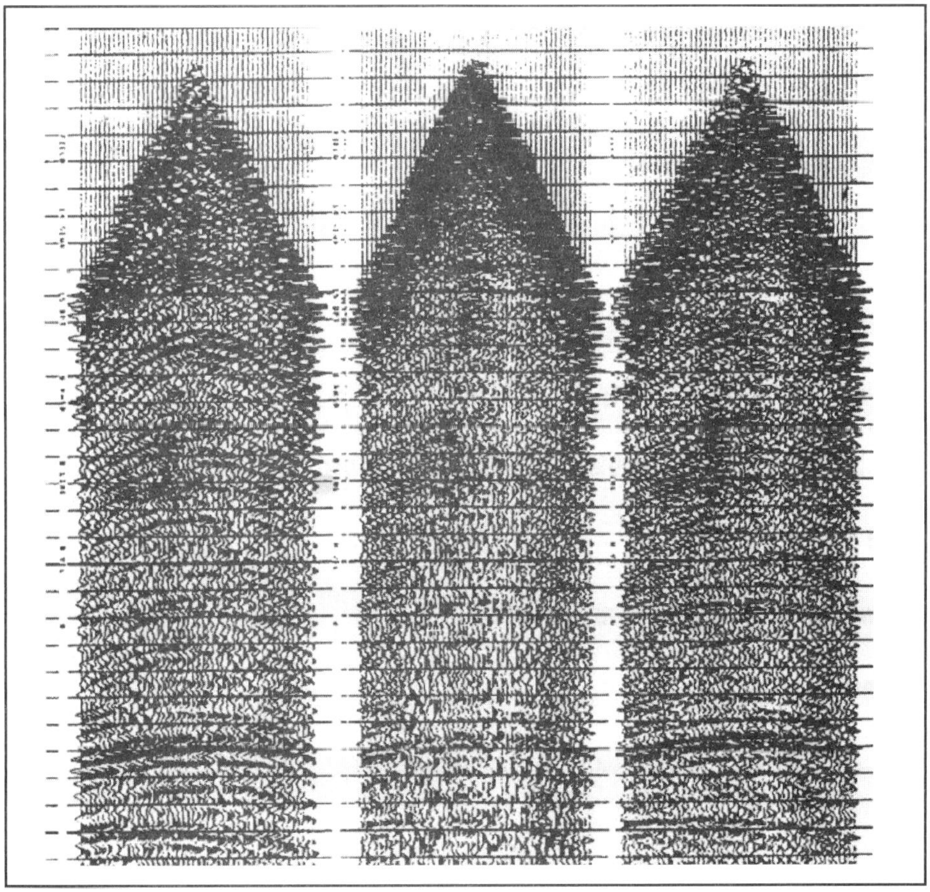

Fig. 6–101 *Example of TVD*

Surface-consistent deconvolution. Surface-consistent deconvolution is a method of decomposing the seismic trace *x(t)* into components that reflect the effects of source, receiver, offset, and the earth's reflectivity. Inverses of source, receiver, and offset effects can be designed and applied to recover the earth's reflectivity from the input trace.

The classic convolutional model is

$$x(t) = w(t)*y(t) + n(t) \tag{6.46}$$

The surface-consistent convolutional model is, however:

$$x(t) = s_i(t)*h_{(i-j)/2}(t)*y_{(i+j)/2}(t)*r_j(t) + n(t) \tag{6.47}$$

where

$s_i(t)$ = effect of source i

$h_{(i-j)/2}(t)$ = effect of offset from source i to receiver j

$y_{(i+j)/2}(t)$ = effect of midpoint between source i and receiver j

$r_j(t)$ = effect of receiver j

$w(t)$ = combined effects of $s_i(t)$, $h_{(i-j)/2}(t)$, and $r_j(t)$

$n(t)$ = noise

Surface-consistent deconvolution assumes that the source wavelet depends, primarily, on locations of source and receiver. It also assumes that traces can be decomposed into source, receiver, offset, and midpoint components and that any error between components obtained and trace is caused by noise. As with TVD it assumes that inputs are minimum phase.

Trace decomposition involves the following steps.

- Apply FFT to trace, yielding:

  $$X(f) = S_i(f)H_{\frac{i-j}{2}}(f)Y_{\frac{1+j}{2}}(f)R_j(f)$$

  Note that noise is ignored and is assumed to be minimum phase.

- Separate amplitude and phase spectra:

  $$A_x(f) = A_s(f)A_h(f)A_y(f)A_r(f)$$

  $$\phi_x(f) = \phi_s(f) + \phi_h(f) + \phi_y(f) + \phi_r(f)$$

Take logarithm of amplitude spectrum:

$$\ln[A_x(f)] = \ln[A_s(f)] + \ln[A_h(f)] + \ln[A_y(f)] + \ln[A_r(f)]$$

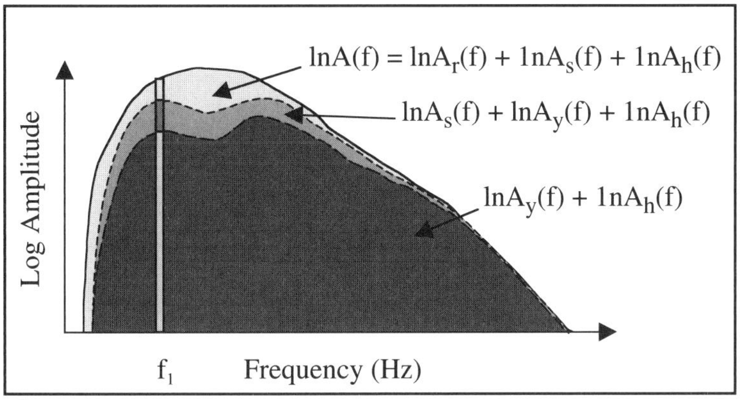

Fig. 6–102 *Trace Decomposition*

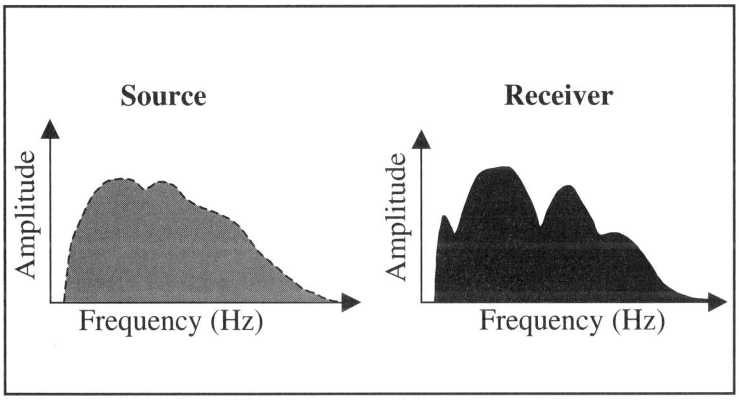

Fig. 6–103 *Source and Receiver Amplitude Spectra Extracted from Trace Decomposition*

Apply the Gauss-Seidel iterative method to extract $\ln[A_s(f)]$ and $\ln[A_r(f)]$ for all sources and receivers. Source and receiver values are found for individual frequencies, such as f_1 in Figure 6–102, within the pass band. The individual frequency values combine to define $\ln[A_s(f)]$ and $\ln[A_r(f)]$. The anti-logarithms of $\ln[A_s(f)]$ and $\ln[A_r(f)]$ give their amplitude spectra. Combining these amplitude spectra with their corresponding minimum phase spectra, inverse FFTs yield $s_i(t)$ and $r_j(t)$ for all i and j. See Figure 6–103.

Inverse filters $s_i^{-1}(t)$ and $r_j^{-1}(t)$ are determined such that $s_i^{-1}(t)s_i(t) = r_j^{-1}(t)r_j(t) = \delta(t)$. There will be a unique combination of source and receiver for any trace out of the data volume, so for a given trace, the appropriate $s_i^{-1}(t)$ and $r_j^{-1}(t)$ are selected and convolved with the input trace. The result should be relatively uniform and shortened wavelets on all traces.

Surface-consistent deconvolution summary. Filters can be designed in the presence of noise but they do not attenuate noise. No assumption is made about reflectivity, so the spectrum will not be forced to be white as in Wiener filters. Surface-consistent deconvolution can be used on traces that are input to AVO analysis, whereas Wiener filters should not be used if AVO analysis is to be done.

After deconvolution—either TVD or surface-consistent—has been applied, another CMP stack should be done. This is to evaluate the effect of deconvolution. Sections should be displayed for lines (in-line and cross-line) selected on the basis of being typical or critical to accomplishing the program objective(s).

Surface-consistent Scaling

By taking advantage of the high surface redundancy in most data acquisition geometries, many amplitude variations can be adequately corrected. Amplitude variations may be caused by factors related to surface locations. Such factors include source and/or receiver coupling variations and absorption and transmission effects in the near surface. Components of amplitude variations that have relatively *short spatial wavelengths* compared to the maximum source to receiver distance can usually be adequately handled. However, those components that have a relatively *long spatial wavelength* compared to the maximum source to receiver distance will not be adequately corrected because they are not sampled.

Before applying surface-consistent scaling, all traces should be investigated to detect anomalous amplitudes. Spikes—large abrupt positive or negative changes in amplitude—should be deleted. Amplitude thresholds—minimum and maximum RMS amplitude values—should be established. Traces with amplitudes that exceed the upper limit or that are below the lower limit should be zeroed. Traces with anomalous amplitudes will have a negative effect on surface-consistent scaling and other processes. It is better to eliminate them from the data volume.

The surface-consistent approach to amplitude correction makes the assumption that the signal from a particular horizon on a particular trace can be related to source, geology, offset, and receiver by:

$$x_{ijkh}(t) = s_i(t)*g_{kh}(t)*m_h(t)*r_j(t) \tag{6.48}$$

where

$x_{ijkh}(t)$ = signal recorded from source i, reflected at horizon k, below midpoint h, by receiver j

$s_i(t)$ = source effect at surface position i

$g_{kh}(t)$ = geology effect at midpoint h and horizon k

$m_h(t)$ = offset effect for an offset x_h, $h = i - j$

$r_j(t)$ = receiver effect at surface position I

From the Fourier Transform of the preceding equation, it is seen that:

$$A_{ijkh} = S_i G_{kh} M_h R_j \tag{6.49}$$

where

A_{ijkh} = amplitude of signal recorded from source i, reflected at horizon k below midpoint h by receiver j

S_i = effect of source at surface position i on signal amplitude

G_{kh} = effect of geology at midpoint h on signal amplitude

M_h = effect of an offset = x_n on signal amplitude

R_j = Effect of receiver at surface position j on signal amplitude

In the absence of anomalies, $S_i = M_h = R_j = 1$. Observed amplitude variations of a long-period nature should result only from G_{kh}.

Taking the logarithm of Equation 6.49 allows decomposition of the recorded amplitude into a sum instead of a product—G_{kh}.

$$\log(A_{ijkh}) = \log(S_i) + \log(G_{kh}) + \log(M_h) + \log(R_j) \tag{6.50a}$$

If it is assumed that the offset effect is different for each horizon, this equation is modified to:

$$\log(A_{ijkh}) = \log(S_i) + \log(G_{kh}) + \log(M_{kh}) + \log(R_j) \tag{6.50b}$$

Equation 6.50a or 6.50b produces a set of linear equations, one for each valid combination of j, k, and i. On the surface, it appears that there will be more equations than unknowns but, in fact, the number of *independent* equations is less than the number of unknowns. As a consequence, a statistical approach to solution of the set of linear equations is required.

Recognizing that some error is inevitable, we require a solution that minimizes the square error. An expression for the squared error is developed and, from it, a set of equations—one each for $\log(S_i)$, $\log(M_{nh})$, $\log(G_{kh})$, and $\log(R_j)$—describing conditions for error minimization. The set of equations are then solved by the Gauss-Seidel iterative method, which starts with an initial guess of:

$$\log(S_i) = \log(M_{kh}) = \log(R_j) = 0, \text{ for all } i, j, k, h, \text{ and } n$$

This initial assumption implies that $\log(A_{ijkh}) = \log(G_{kh})$, but when this is substituted into the equations, errors are found. These errors are used to generate a new set of values for $\log(S_i)$, $\log(M_{kh})$, $\log(G_{kh})$, and $\log(R_j)$. This process continues with the set of values obtained in one iteration used as the initial values for the next iteration. Iterations continue for a specified number or until the error is reduced to some predetermined value, such as 0.1%. One last iteration may be performed after this error level is reached.

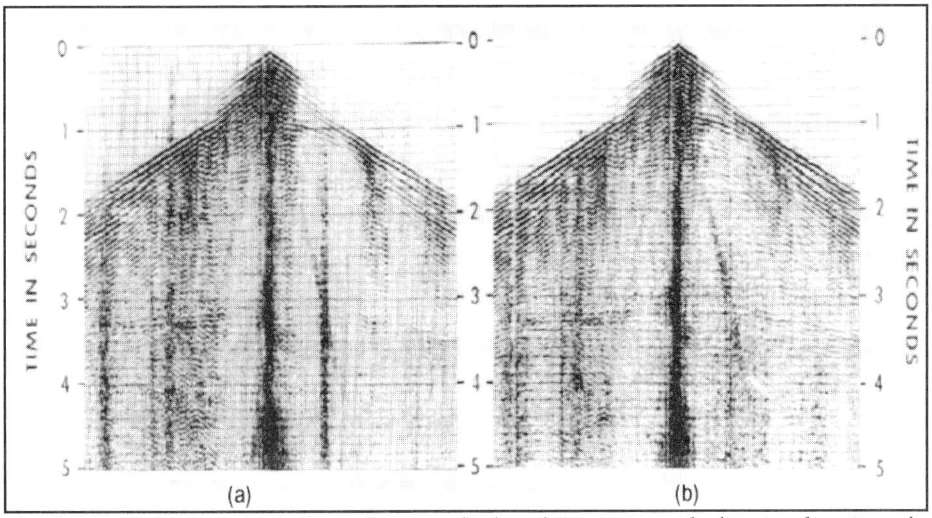

Fig. 6–104 *Only Geometric Spreading Applied and Same Record after Application of Surface Consistent Amplitude Correction*

Figure 6–104 illustrates surface consistent amplitude correction. Note the weak amplitudes on the right side of the record in *a* and improvement in *b*.

After surface-consistent scaling, all CMPs should be stacked and sections displayed for lines—in-line and cross-line—to evaluate the effect of surface-consistent scaling.

Time-variant Spectral Whitening (TVSW)

TVSW compensates for frequency-dependent attenuation. It can be applied pre- or post-stack, but post-stack is generally preferred. TVSW does a better job of flattening the amplitude spectrum than statistical deconvolution. However, it provides no attenuation of multiples and has no effect on phase. The combination of TVSW for spectrum whitening and model-based wavelet processing to obtain zero-phase wavelets can be much more effective in achieving the desired output wavelets than TVD.

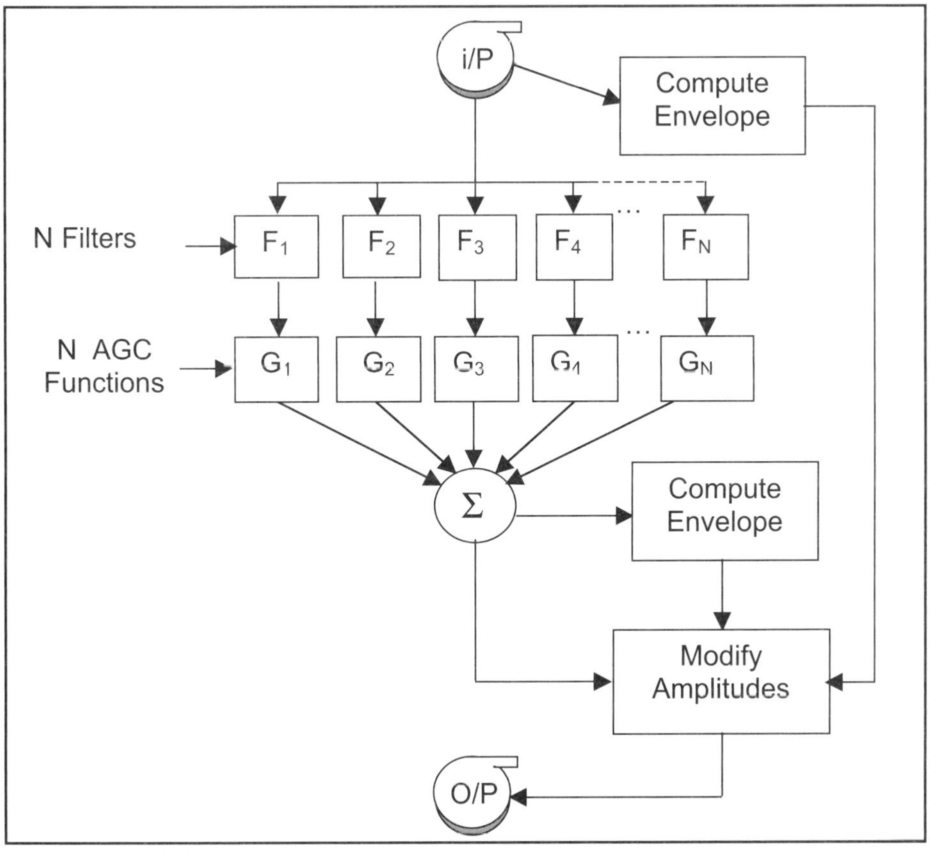

Fig. 6–105 *TVSW Flow Chart*

TVSW operates as a time- and frequency-variant gain function on each input trace. Input data are decomposed into spectral components by band pass filters (F_1, F_2, ... , F_N). Spectral components are gained by application of AGC functions (G_1, G_2, ..., G_N). Outputs of gain are summed to form a composite trace. Amplitudes of the composite trace are modified to make the output envelope match the input envelope. Figure 6–105 is a flow chart for TVSW.

The filters may be designed automatically or input manually. If designed automatically, define the bandwidth by the input of the four frequency response pairs: f_L, a_L; f_A, a_A; f_B, a_B; and f_H, a_H. Frequencies f_L and f_A should be chosen such that $\frac{f_L + f_A}{2}$ is equal to the lowest frequency for which $\frac{S}{N}$ =1. Similarly, select f_B and f_H such that $\frac{f_B + f_H}{2}$ is equal to the highest frequency for which $\frac{S}{N}$ =1.

At frequencies f_L and f_H, the responses equal .01—$a_L = a_H$ = .01. At frequencies f_A, and the responses equal 1—$a_A = a_B$ = 1.

These four frequencies, if selected as above, define signal pass band. Next, specify the number of filters to design. A good option is to require all filters to have equal bandwidth in octaves. Figure 6–106 illustrates automatic selection of five filters.

It may be desirable, instead, to define filters by supplying sets of four frequencies for n = 1, 2, ..., N, where N is the number of filters. That is, supply [f_{L1}, a_{L1}; f_{A1}, a_{A1}; f_{B1}, a_{B1}], [f_{L2}, a_{L2}; f_{A2}, a_{A2}; f_{B2}, a_{B2}], ..., [f_{LN}, a_{LN}; f_{AN}, a_{AN1}; f_{BN}, a_{BN}]. In this case, frequencies f_{L1} and f_{A1} should be chosen such tha $\frac{f_{L1}+f_{A1}}{2}$ is equal to the lowest frequency for which $\frac{S}{N}$ =1 and frequencies f_{BN} and f_{HN} selected such that $\frac{f_{BN}+f_{HN}}{2}$ is equal to the highest frequency for which $\frac{S}{N}$ =1. At frequencies f_{Ln} and f_{Hn}, the responses equal .01—$a_{Ln} = a_{Hn}$ = .01, for n = 1, 2, ..., N. At frequencies f_{An} and f_{Bn}, the responses equal 1—$a_{An} = a_{Bn}$ = 1, for n = 1, 2, ..., N. Figure 6–107 illustrates user-supplied filters for TVSW.

Filters must be designed so that $f_{B1} = f_{L2}$, $f_{H1} = f_{A2}$, $f_{B2} = f_{L3}$, $f_{H2} = f_{A3}$, $f_{B3} = f_{L4}$, $f_{H3} = f_{A4}$, etc. This is to assure that the total filter response=1, from f_{A1} to f_{BN}. Frequency f_B (or f_{BN}) should be less than 75% of the Nyquist frequency or equal to the highest sweep frequency for vibrators. No filter bandwidth should be less than 5 Hz.

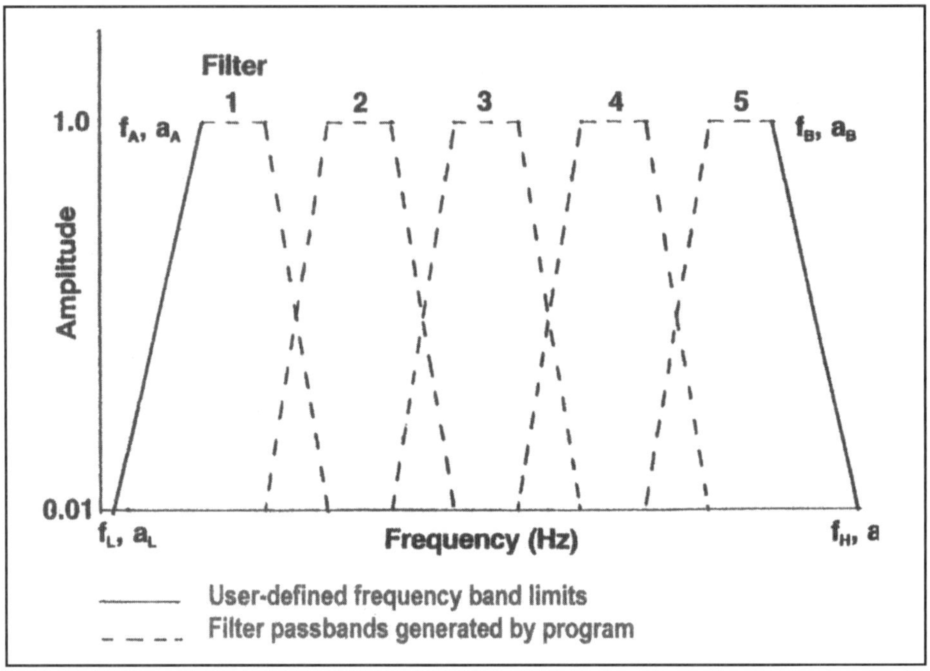

Fig. 6–106 *Automatic Filter Design in TVSW*

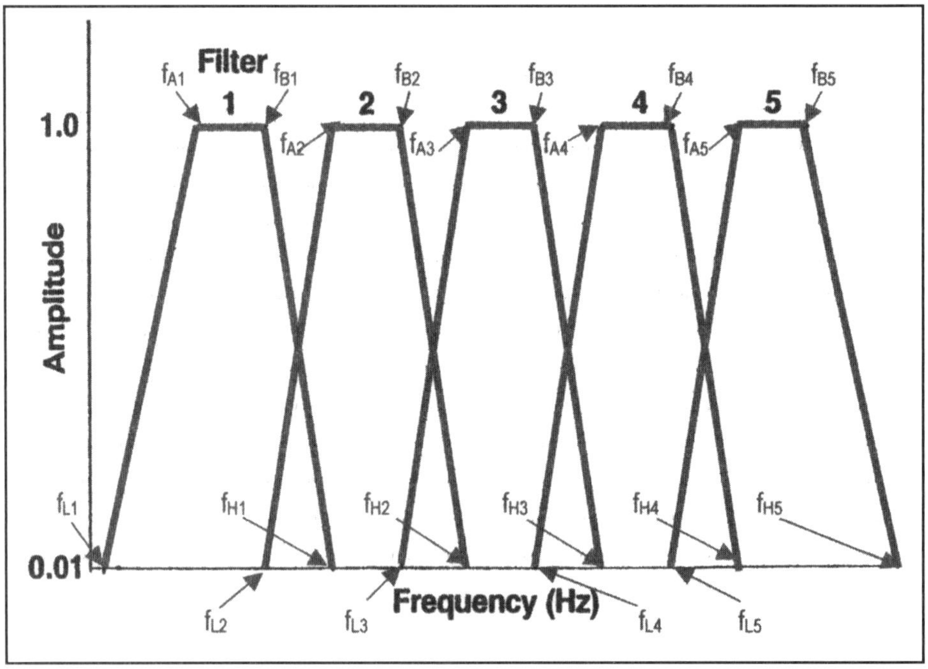

Fig. 6–107 *User-Defined Filters for TVSW*

Figure 6–108 shows filters used in a three-filter example. Figure 6–109a shows the outputs of the three filters and amplitude envelopes. Figure 6–109b shows the three outputs of AGC and the composite. Dashed lines are the amplitude envelopes.

Figures 6–110 and 6–111 show how the number of filters affects the output of TVSW. Figure 6–110 shows the same CMP stack without TVSW, with a one-filter TVSW, with a three-filter TVSW, and with a seven-filter TVSW. The seven-filter TVSW output appears to be the best of the four sections.

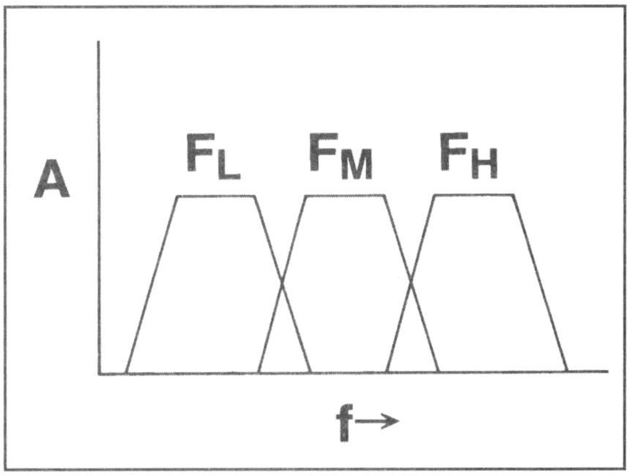

Fig. 6–108 *Three-Filter TVSW Example*

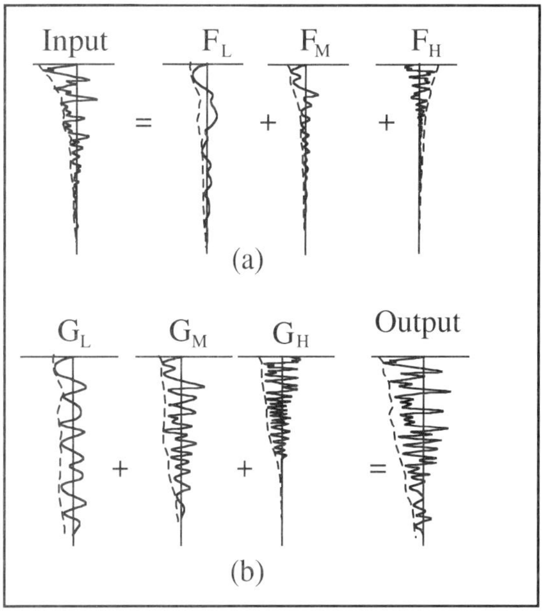

Fig. 6–109 *Filter Output, Gain Output*

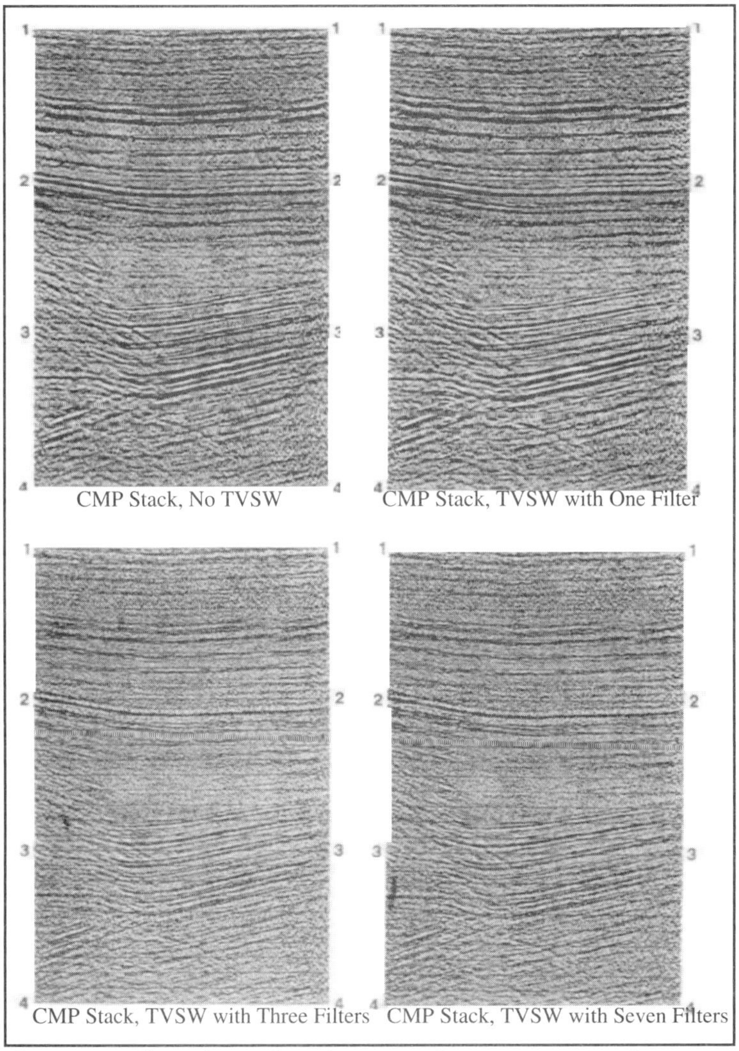

Fig. 6–110 *Effect of Number of TVSW Filters on CMP Stack*

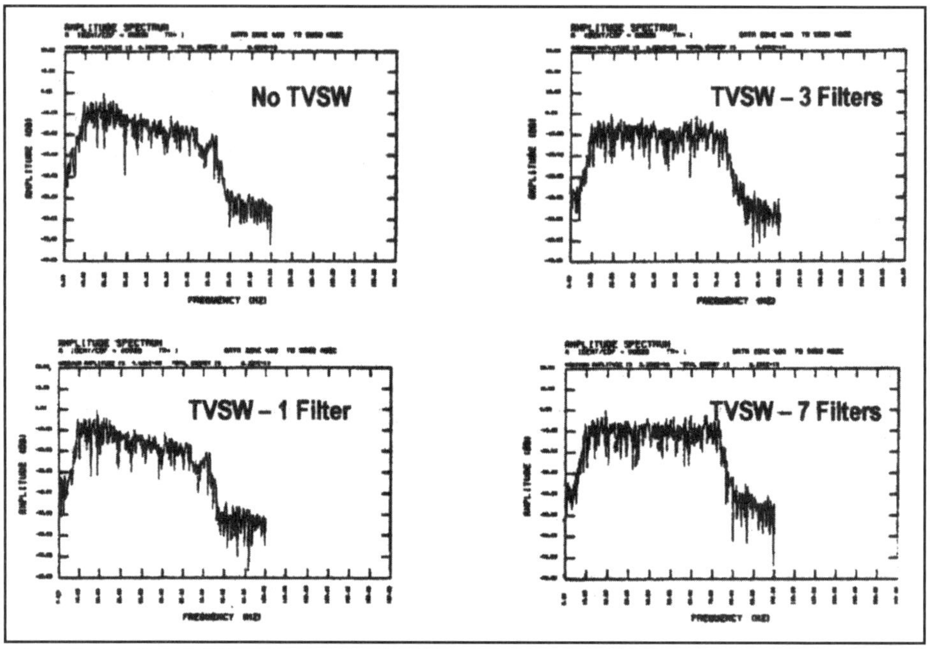

Fig. 6–111 *Effect of Number of TVSW Filters on Amplitude Spectra*

Figure 6–111 shows amplitude spectra corresponding to the four CMP stacks of Figure 6–110. The one-filter TVSW is actually only a band pass filter. Its amplitude spectrum is the same as the one without TVSW. This is because there is no decomposition into frequency components. The three-filter TVSW amplitude spectrum is considerably flatter over the signal pass band and the seven-filter TVSW is even flatter.

TVSW summary. TVSW may do a better job of whitening data than spiking deconvolution. One advantage is that it whitens data only within the signal pass band rather than the entire spectrum as TVD does. TVSW compensates for frequency dependent amplitude decay by:

- decomposing data into frequency bands
- applying AGC gain to component frequency bands
- recombining gained components consistent with input amplitudes

TVSW does not affect phase or attenuate multiples.

Model-based Wavelet Processing (MBWP)

If the input to spiking or whitening deconvolution is minimum phase, then the output wavelets are zero-phase. However, inputs may not be minimum phase or differ from the predicted phase. In such cases, deconvolution does not output the desired zero-phase wavelet. MBWP estimates and minimizes phase errors resulting from use of standard deconvolution algorithms. It does this by examining traces to see how they differ from ideal traces assumed by the deconvolution algorithm and then develops phase-correction filters to apply after conventional deconvolution to shape the wavelet to the desired waveform.

MBWP does this by modeling the various elements that produce the observed wavelet and the desired wavelet. For wavelets produced from a vibrator source, the observed wavelet is described by:

$$s(t) = i(t) * g(t) * q(t) * k(t) \tag{6.51}$$

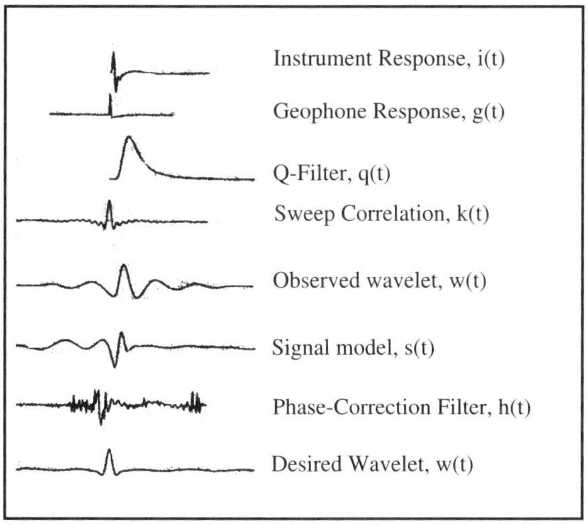

Fig. 6–112 *MBWP Models for Vibrator Records*

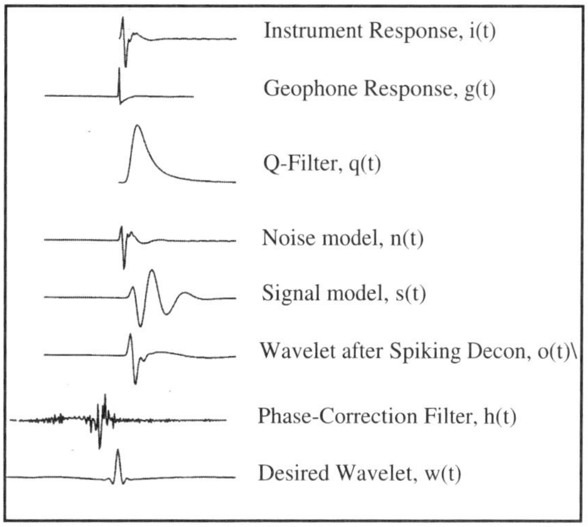

Fig. 6–113 *MBWP Models for Dynamite Records*

where

$i(t)$ = instrument impulse response

$g(t)$ = geophone impulse response

$q(t)$ = Q-filter

$k(t)$ = sweep correlation

$s(t)$ = signal output from spiking deconvolution

MBWP develops mathematical models for all the components of Equation 6.48 and uses these to develop the phase-correction filter $h(t)$ needed to obtain the desired wavelet, $w(t)$.

$$w(t) = h(t) * s(t)$$

The phase-correction filter $h(t)$ is an inverse filter that subtracts out the input phase and replaces it with the desired phase. That is

$$h(t) = s^{-1}(t) * w(t) = i^{-1}(t) * g^{-1}(t) * q^{-1}(t) * k^{-1}(t) * w(t) \quad (6.52)$$

Figure 6–112 illustrates the models developed for the components of Equation 6.52, the signal model, the phase-correction filter, and the desired wavelet.

When an impulsive energy source such as dynamite is used, a slightly different set of models is required. In this case:

$$o(t) = i(t) * g(t) * q(t) * n(t) * s(t) \quad (6.53)$$

Figure 6–113 illustrates components of Equation 6.53, the wavelet after spiking decon, the phase-correction filter, and the desired wavelet. In this case:

$$w(t) = h(t) * o(t)$$

and

$$h(t) = o^{-1}(t) * w(t) = i^{-1}(t) * g^{-1}(t) * q^{-1}(t) * n^{-1}(t) * s^{-1}(t) * w(t) \quad (6.54)$$

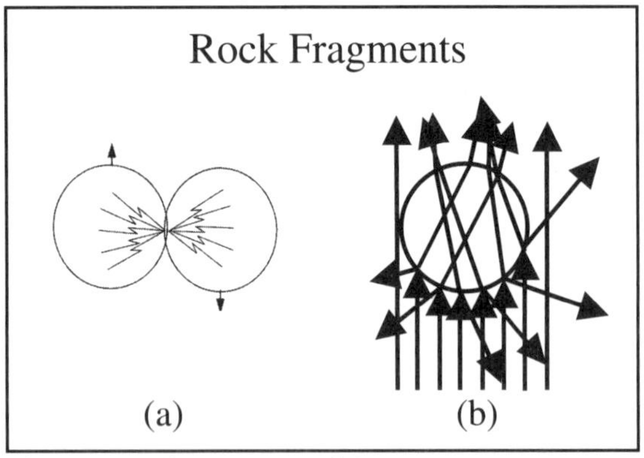

Fig. 6–114 *Absorption and Scattering*

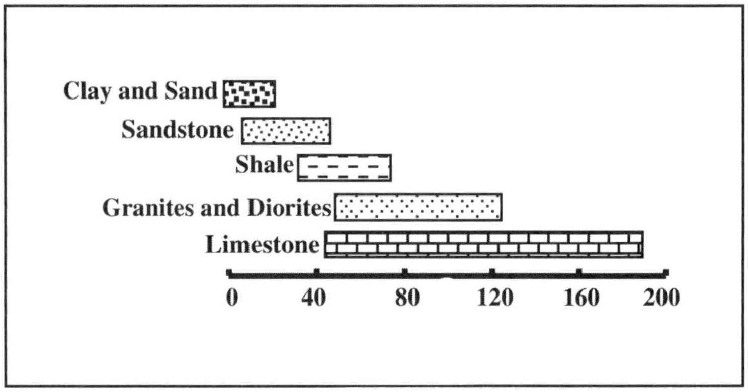

Fig. 6–115 *Representative Q Values*

The Q-filter describes the effect of propagation through inhomogeneous and anisotropic rocks. Propagation through such media produces frequency-dependent attenuation that change in the amplitude and phase spectra of recorded signal. The loss of energy during propagation through rocks is caused by absorption and scattering. See Figure 6–114.

The *Q-factor* (Q) of the rocks determines the amount of attenuation that will occur during propagation through those rocks.

To show what Q is, start with a general expression for inelastic attenuation:

$$A(x) = A_0 e^{-ax} \tag{6.55}$$

where

$A(x)$ = amplitude as a function of distance

A_0 = reference amplitude

x = seismic path length

e = base of natural logarithms = 2.718 ...

a = attenuation constant

The attenuation constant $a = \frac{\pi}{Q\lambda}$. Since $\lambda = \frac{V}{f}$, the attenuation constant also = $\frac{\pi f}{QV}$.

where

Q = quality factor.

λ = wavelength

f = frequency

V = velocity

For each wavelength of travel through a rock there is an amount of attenuation equal to $1/Q$. Since $\lambda = V/f$, higher frequencies correspond to shorter wavelengths. It follows that a higher frequency completes more wavelengths over the same distance than does a lower frequency.

Since attenuation is inversely proportional to Q, large values of Q imply low attenuation and low values of Q imply high attenuation. Q depends on rock type but there are other factors, so each rock type has a range of values. Some values of Q for common rock types are shown in Figure 6–115.

It is necessary to evaluate Q in order to model the Q-filter To see how to do this, substitute $a = \frac{\pi f}{QV}$ and $x = VT$ in Equation 6.55. After these substitutions, it becomes:

$$A(T) = A_0 e^{-(\frac{\pi f}{QV})(VT)} = A_0 e^{-\frac{\pi f T}{Q}} \tag{6.56}$$

Expressing amplitude in dB, Equation 6.56 becomes:

$$A_{dB} = 20\log\left[\frac{A(T)}{A_0}\right] = -\left(\frac{20\pi fT}{Q}\right)\log e = -28.2875\left(\frac{fT}{Q}\right) \tag{6.57}$$

Taking the derivative of Equation 6.57 with respect to f yields:

$$\frac{dA_{dB}}{df} = -28.2875\left(\frac{T}{Q}\right) \tag{6.58}$$

Solving Equation 6.58 for Q:

$$Q = -28.2875\left(\frac{T}{\frac{dA_{dB}}{df}}\right) \tag{6.59}$$

Q is evaluated from Equation 6.59 using the following steps.

- Calculate the amplitude spectrum of the trace over a narrow window centered at time T. As shown in Figure 6–116, amplitudes are small at the lowest frequencies, increase to a maximum, then decrease in a more or less linear trend, and then flatten out. The linear portion part corresponds to the signal pass band. The flat part represents noise outside the pass band. Signal is weaker than noise.
- Fit a straight line to the spectrum where the amplitude (in dB) has approximate linear decay.
- Determine the slope of the line. The slope = $\frac{dA_{dB}}{df}$.
- The center time of the window over which the spectrum was calculated is T, in Equation 6.59.

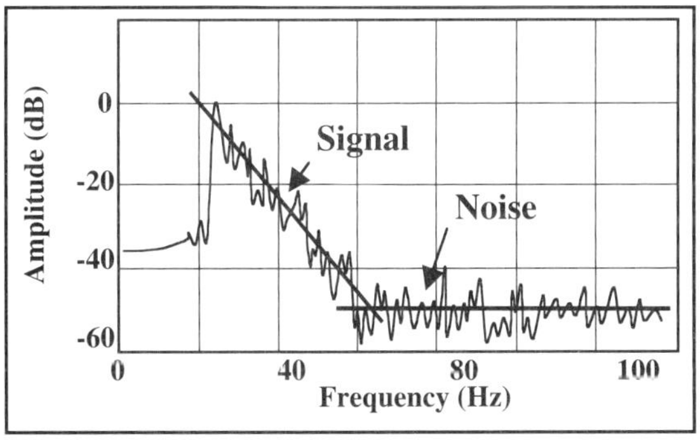

Fig. 6–116 *Evaluation of Q*

Substitute for T and $\frac{dA_{dB}}{df}$ in Equation 6.59 to determine Q.

Figure 6–117 provides an example of model-based wavelet processing. Data were recorded in a transition zone area. One part of the line was acquired using Vibroseis. The other, near and in shallow water was acquired with dynamite because Vibroseis could not operate there. The right side of Figure 6–117 is the preliminary CMP stack section of the line prior to MBWP. The left side of Figure 6–117 is the CMP stack section of the line after MBWP.

Note the apparent mistie at the transition position (indicated by the triangle at the top of the section) before MBWP. This is because of phase differences between wavelets produced by the different sources. Note that the phase of both dynamite and Vibroseis data were changed, not one corrected to the other.

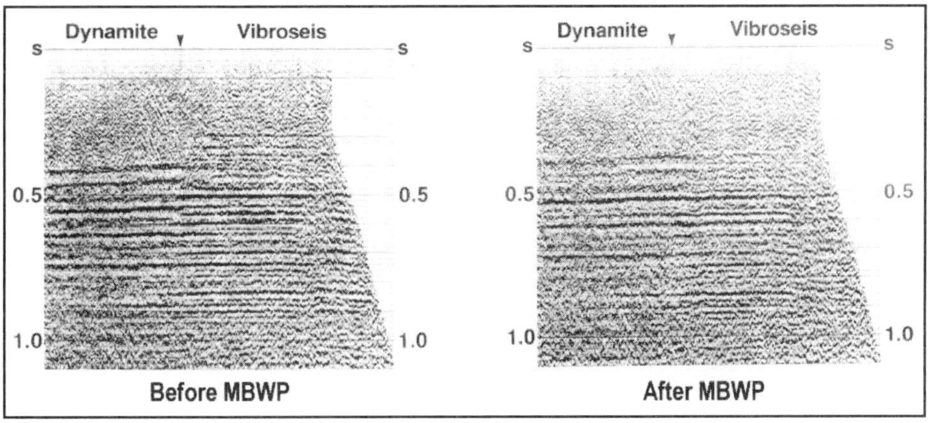

Fig. 6–117 *Transition Zone Example of MBWP Application*

Figure 6–118 shows that, with additional residual statics and velocity analyses, continuity of data across the transition zone is excellent and no indication of different source types can be seen.

The effectiveness of MBWP depends on signal-to-noise ratio and proper selection of Q values. This is demonstrated in Figure 6–119. The left side of Figure 6–119 shows wavelets after deconvolution but before MBWP. The right shows wavelets after deconvolution and MBWP. The desired wavelet is zero phase. While MBWP improves wavelet shape in all cases, the best result is with S/N = 24 dB and $Q = 30$.

MBWP summary. The effectiveness of model-based wavelet processing in correcting phase errors introduced by departures from assumptions inherent in statistical deconvolution depends upon the accuracy and realism of the models used for systems and processes. Variability of signal and noise from trace to trace also strongly influences MBWP effectiveness. Note that MBWP does not, and is not intended to, correct for frequency-dependent attenuation.

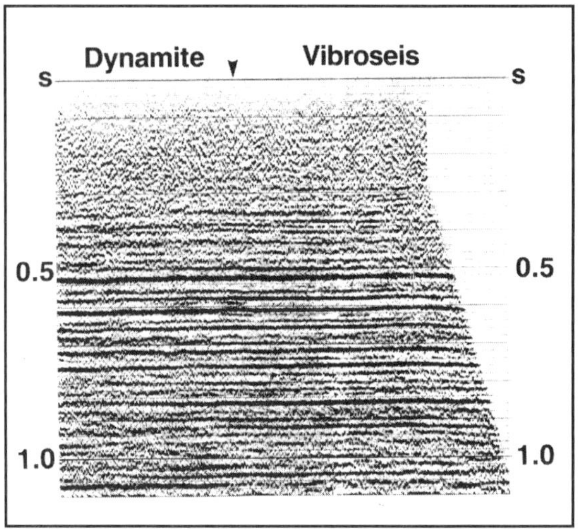

Fig. 6–118 *Data of Figure 6–117 after MBWP*

Inverse-Q Filtering

Inverse-Q filtering is another alternative to TVD, or statistical deconvolution. It does, essentially, what the combination of MBWP and TVSW do. The following introduces and describes Inverse-Q Filtering.

The geologic sections through which seismic waves propagate are absorbing media. The result of this absorption in subsurface rocks is that seismic waves are attenuated in a time- and frequency-dependent manner and that velocity dispersion occurs (velocity varies with frequency), causing phase distortion and time delays.

Figure 6–120 demonstrates the effect of distortion by showing a synthetic sonic log that is a set of equally spaced, equal amplitude spikes. The corresponding synthetic seismic trace shows non-zero phase wavelets of decreasing amplitude and bandwidth plus increasing delays in reflection time relative to the sonic.

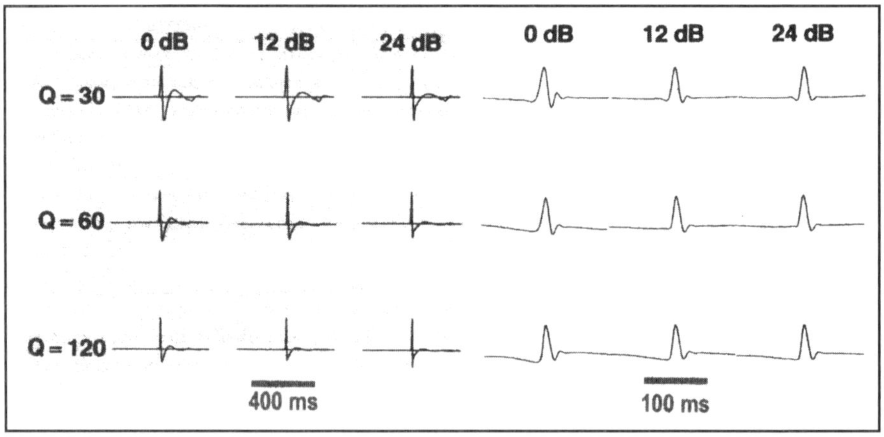

Fig. 6–119 *Wavelets after Decon but before MBWP and after Both Decon and MBWP*

The objectives of spiking deconvolution are to correct for earth filtering by flattening the amplitude spectrum and to apply a minimum-phase correction to the wavelet, producing a zero-phase output wavelet. Spiking deconvolution is a statistical process that assumes random reflectivity, hence the autocorrelation defines the wavelet shape and the output amplitude spectrum should be flat. It also assumes *stationarity* of waveforms (no change in wavelet shape) over the design window plus zero or negligible noise.

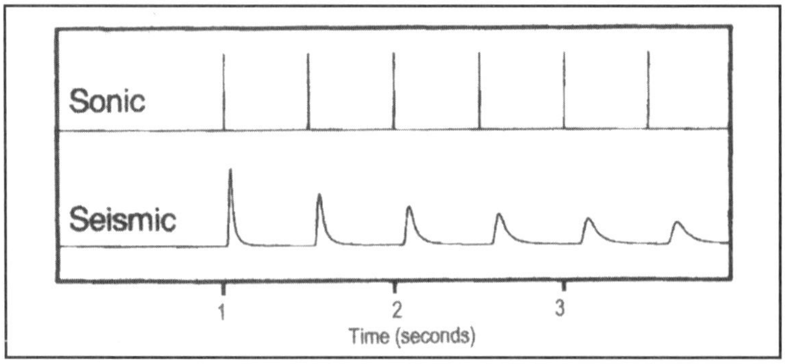

Fig. 6–120 *Effects of Absorption*

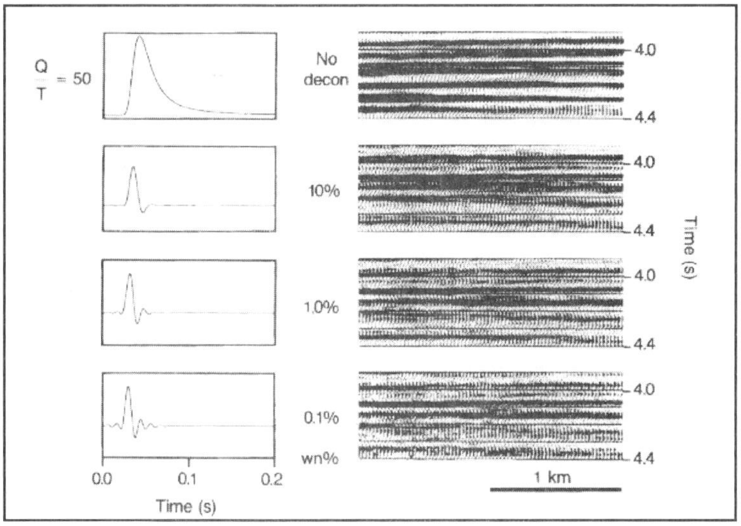

Fig. 6–121 *Effect of Noise on Deconvolution*

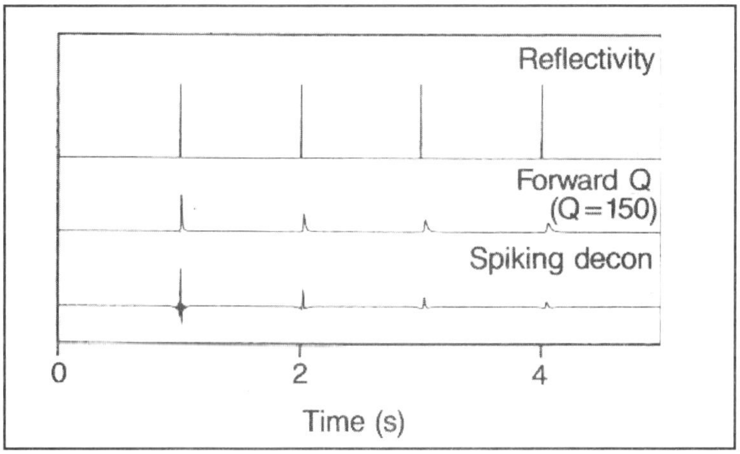

Fig. 6–122 *The Forward Q-Filter and Deconvolution*

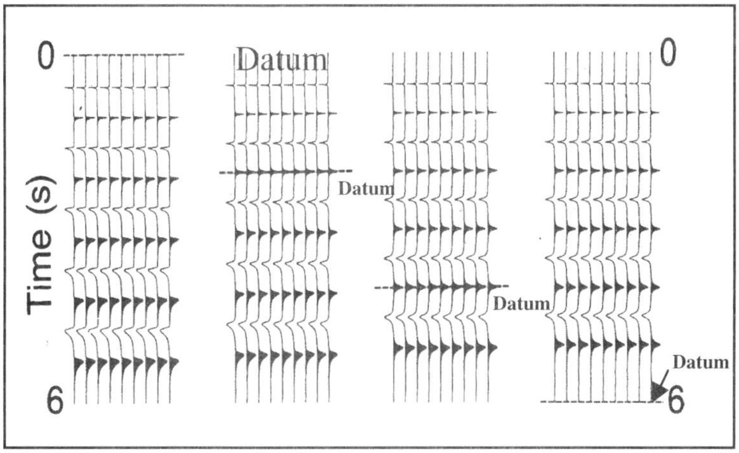

Fig. 6-123 *Phase Compensation*

Spiking deconvolution works on the spectrum of the signal-plus-noise in the data. This may cause the signal spectrum to be deficient in those frequencies that are obscured by the noise. The noise may also lead to inaccurate estimates of the minimum-phase spectrum of the wavelet. This limits the deconvolution filter's ability to convert the wavelet to zero phase. Another possibility is inconsistency in the amplitude and phase treatment of the wave, especially if the noise varies from trace to trace.

Figure 6–121 illustrates how noise modifies deconvolution effectiveness. Changes on the wavelet shape are shown on the left and part of a CMP stack on the right. The value of $Q/T = 50$ is equivalent to a very absorptive medium at early times and becoming less so with increasing time. This is the general trend in the earth. At the top of Figure 6–121, the wavelet and section are shown without deconvolution. The lower parts of the figure show the wavelet and section after deconvolution is applied to traces with white noise levels of 10, 1, and 0.1% of signal levels added. Note how a change in the noise level can produce a marked change in the character of the wavelet after deconvolution.

Figure 6–122 shows a synthetic reflectivity function that is a set of equal amplitude spikes at equal time intervals, the wavelets resulting from the forward Q-filter, assuming a value of 150 for Q, and the wavelets after spiking decon is applied. The forward Q-filter, which describes the time- and frequency-dependent attenuation in the earth, is time-variant. The wavelets seen in the design window of spiking decon filter vary over the length of the window. Conesquently, wavelets at early times are overcorrected and wavelets at later times are under-corrected by spiking decon.

Inverse Q-filtering is a deterministic approach to solving the problem of phase distortion and high-frequency amplitude loss. Earth filter models (Q-values) are assumed and an algorithm based on Stolt (*F-K*) migration is employed for phase correction to zero-phase. Amplitude correction is done separately.

Phase compensation is accomplished by *backward propagation*. This is similar to migration and uses an algorithm based on Stolt migration and is similar to downward continuation migration. Migration is discussed later in this chapter. Constant-phase corrections are applied in multiple overlapping windows.

Figure 6–123 illustrates how phase compensation is done in inverse-Q filtering. On the left, the datum is at time zero and no phase compensation has been done. When the datum is lowered to two seconds, Q-phase compensation is applied to data below the datum. Data above the two-seconds datum have been converted to zero-phase. The datum is lowered again, and the next two seconds of data are converted to zero phase. With the next lowering of the datum, the last two seconds of data are converted to zero phase.

The rock parameter Q, which generally varies with depth, determines the amount of attenuation. The Q-filter effect is described by $e^{-\pi ft}/Q$, where f = frequency and t = time.

The simple approach to inverse Q-filtering involves multiplying by $e^{\pi ft/Q}$, using specific Q values within time windows corresponding to the earth model selected. In cascaded application (Fig. 6–124), $e^{\pi ft/Q_1}$ is applied at all times, $e^{\pi ft(1/Q_2 - 1/Q_1)}$ is applied at times later than the end of the first window, $e^{pft(1/Q_3 - 1/Q_2)}$ from the top of the third window, etc.

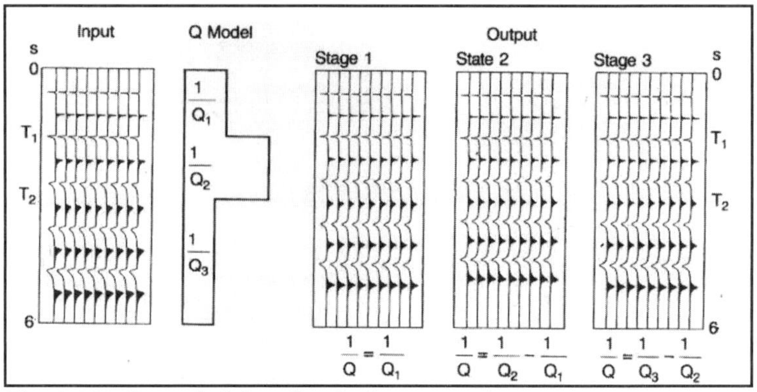

Fig. 6–124 *Cascaded Amplitude Compensation*

The problem with this simplistic approach is that there is the tendency to over-amplify noise. Figure 6–125 illustrates a more sophisticated approach. On the left of the figure are shown amplitude spectra and traces for three time windows before any amplitude compensation. In the center are the inverse filters that limit the maximum gain level and roll off smoothly from this maximum level above the signal frequency band. On the right of the figure are shown amplitude spectra and traces for three time windows after amplitude compensation. Considerable spectral flattening is accomplished in all windows. The later windows are not flattened as much at the higher frequencies because of the poorer signal-to-noise ratios.

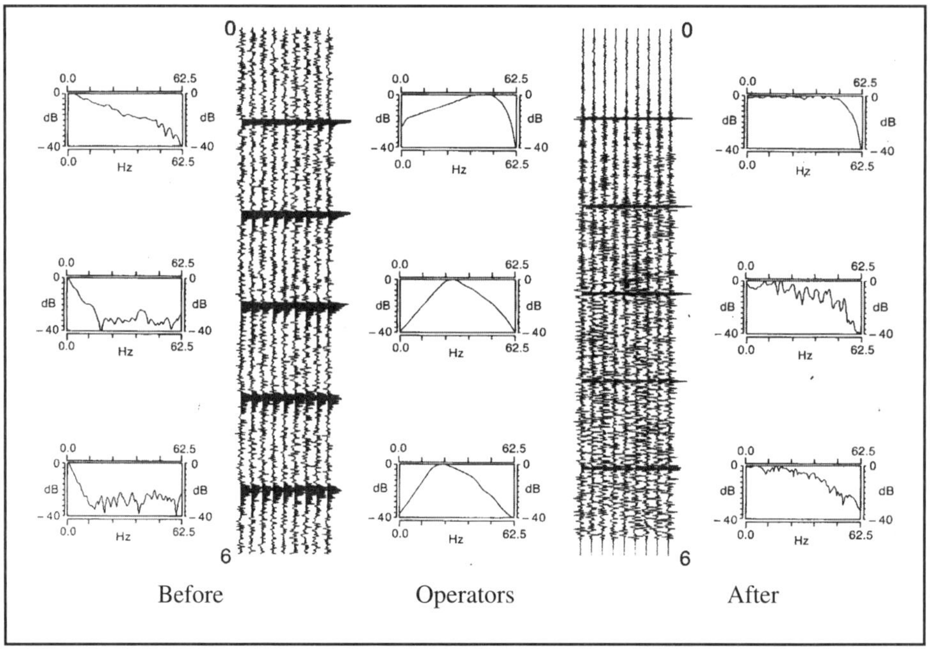

Fig. 6–125 *Amplitude Compensation*

Figure 6–126 is an example of inverse-Q filtering. Both sections show vibrator data. Conventional processing includes pre-stack, windowed spiking decon. At right is the same data set with inverse-Q filter and gapped decon applied pre-stack. Arrows and the circle indicate areas in which inverse-Q filtering has made significant improvement.

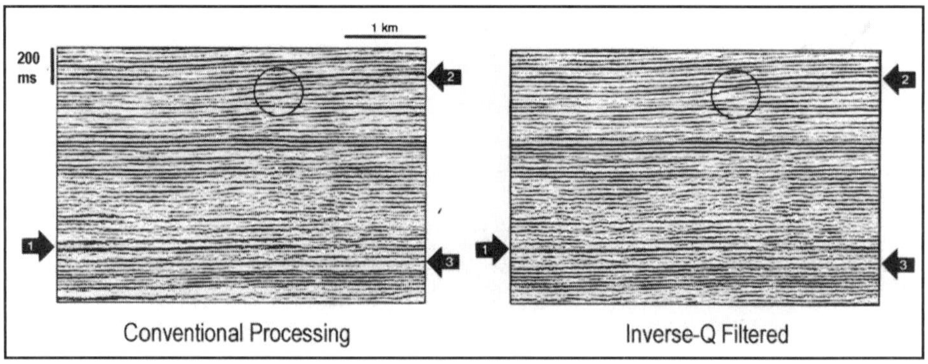

Fig. 6–126 *Inverse-Q Example*

In conclusion, Inverse-Q filtering can be used when reflectivity is non-random, it handles time-variance of forward Q filter much better than statistical decon, it corrects for phase distortion and timing delays, and produces better result than statistical decon in the presence of noise, in many cases.

Residual Statics

Often, after both NMO and static corrections are applied, reflections on CMP traces are still not exactly aligned in time. This is usually due to small errors in static corrections called *residual statics*. There may also be errors in NMO corrections, called *residual NMO*, due to incorrect velocities.

Figure 6–127 shows what is meant by residual statics. Even though NMO corrections have been applied, the reflection events do not line up across the CMP. When the traces are stacked two bad things happen—there is an error in T_0 and amplitude and frequency are lowered.

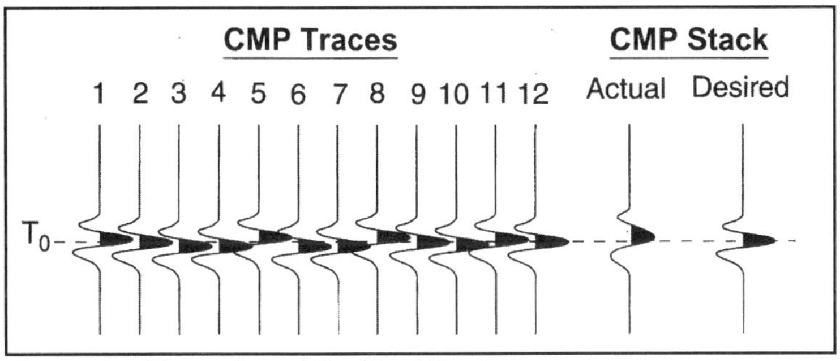

Fig. 6–127 *Illustration of Residual Statics*

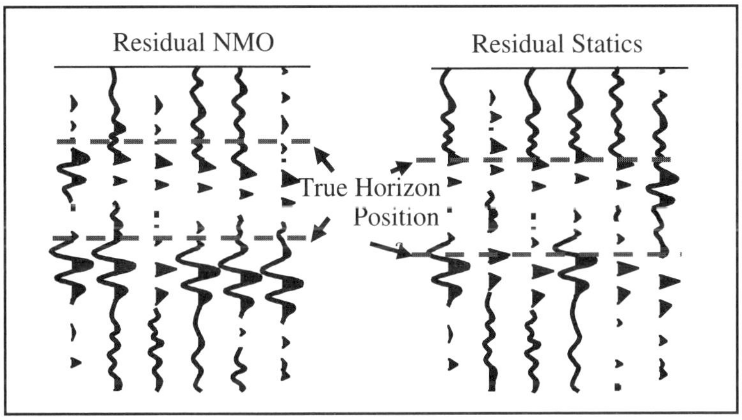

Fig. 6–128 *Residual NMO and Residual Statics*

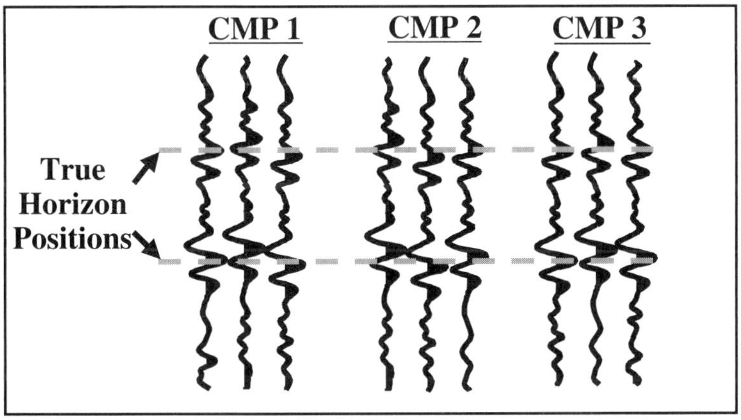

Fig. 6–129 *Three CMPs with Residual Statics*

As shown in Figure 6–128, residual NMO is characterized by trace-to-trace time differences that increase, or decrease, with offset. Residual statics, by contrast, are characterized by trace-to-trace time differences that are random in nature.

Figure 6–129 further demonstrates the variability of residual statics where three CMPs are represented. Not only is a trace-to-trace pattern lacking but also no pattern is seen from CMP to CMP.

Residual statics analysis can be done with refraction or reflection data. Figure 6–130 illustrates the general reflection residual statics method.

Conventional reflection-based statics. NMO corrections are applied to the CMP traces, which are then stacked. This stacked trace is the model or reference trace. Residual statics are picked by comparing times on individual traces and the model trace. The residual statics are applied to the input traces and these are stacked to obtain an improved stack trace.

Picking residual statics involves the following steps.

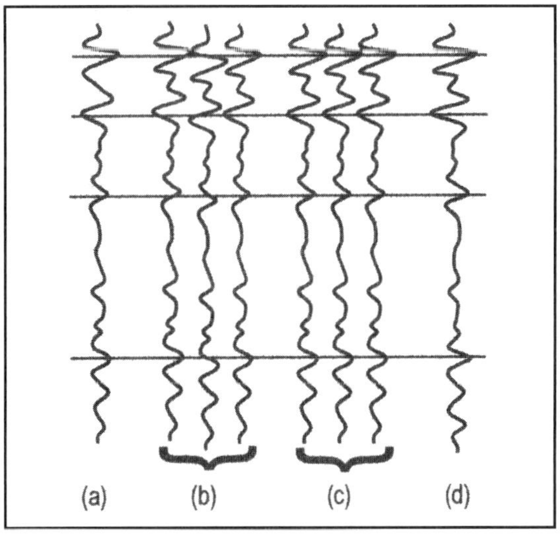

Fig. 6–130 *The Reflection Residual Statics Method*

1. Apply NMO corrections using velocities from the most recent iteration of velocity analysis.
2. Select correlation window(s).
3. Build the preliminary model trace.
4. Cross-correlate traces from the gather with the model trace.
5. Align the traces in accordance with the calculated time deviations. Build a new model trace from these traces.
6. Cross-correlate again with the new model trace. Calculate new static time deviations and quality factors. Generate the final model trace.
7. Cross-correlate the final model trace with traces in the next gather.
8. Make a new model trace for the second gather. This new model trace is built from all the time-adjusted traces in the second gather plus the final model trace from the previous gather (so there are now twice as many traces).
9. Cross-correlate again, then repeat this sequence for the rest of the line. The final product is a set of static time deviations and quality factors for each trace.

Correlation window selection. Insofar as possible, windows used for residual statics analysis should avoid steep dip and complex structures. Window(s) are usually defined by specifying spatial coordinates—CMP numbers—plus start and stop times of the window(s). This procedure allows window boundaries to follow structure to some extent. Some people prefer a single window in all cases. The authors believe two windows are usually better. There are two reasons for this.

- By definition, static corrections are time invariant. Thus, basing estimates of residual statics on two different time intervals makes it more likely that only residual statics are determined rather than being combined with residual NMO and structure components.
- Often deep and shallow structures are markedly different, making it impossible to eliminate structure from the residual statics estimate. At best, this situation will make the cross-correlation maxima more difficult to pick.

Figure 6–131 illustrates the selection of window boundaries. The vertical blue lines represent spatial *control points* or spatial coordinates used to define the two windows. The crooked red lines represent the first or shallow window. The crooked yellow lines represent the second or deep window. The times at which these lines cross the blue lines are the time coordinates of the widows. Window lengths should be about 8 to 10 times the *maximum correlation shift.* The maximum correlation shift will be explained later in this chapter.

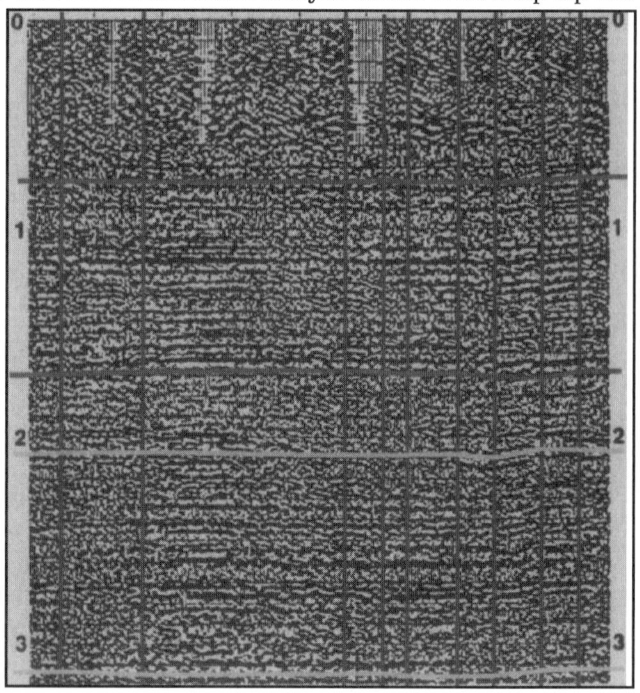

Fig. 6–131 *Defining Windows for Residual Statics Analysis*

Model trace building. Model traces must be built for both time windows if two windows are used. The first step is to scale all trace amplitudes to the same RMS amplitude. The preliminary model for the first CMP, $M_{p,1h}$, is the stack of all traces in the first CMP.

$$M_{p,1h} = \sum_{k=1}^{N} f_{1kh}(t) \tag{6.60}$$

where

$f_{1kh}(t)$ is the kth trace in the first gather of time gate h

N is the number of traces in the gather

Cross-correlate traces from the gather with the model trace. Cross-correlate $f_{1kh}(t)$ with $M_{p,1h}(t)$ and determine the maximum normalized cross-correlation quality factor $Q1lh$, and corresponding time shift Δt_{1lh}. From Figure 6–132 it can be seen that the quality factor is the maximum value of the normalized cross-correlation between the individual trace and the model, and Δt_{jlh} is the time shift from time zero to the time of the cross-correlation maximum.

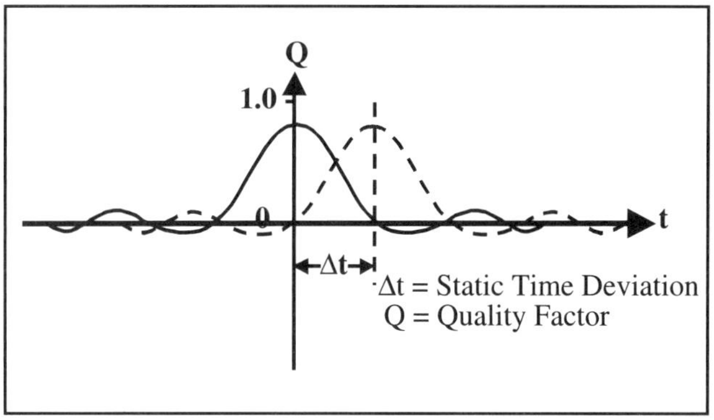

Fig. 6–132 *Definition of Quality Factors Q_{jlh} and Δt_{jlh}*

Build a new model trace. Construct the new model trace by applying the time deviations found from the cross-correlations to each trace, obtaining a weighted trace average and averaging it with the models determined for a designated number of previous gathers. This process is described by Equation 6.61.

$$M_{1h} = \frac{\sum_{k=1}^{N} W_{1kh} \times f_{1kh}(t - \Delta t_{1kh})}{\sum_{k=1}^{N} W_{1kh}} \tag{6.61}$$

where

W_{1kh} is the trace weight

$f_{1kh}(t - \Delta t_{1kh})$ is the kth trace in the first gather of time gate h to which time deviations determined by cross-correlations are applied

One recommended trace weight is $W_{jkh} = \frac{Q_{jkh}}{1 - Q^2_{jkh}}$. This results in large weights for high values of quality factor and low weights for poor quality factors. A high quality factor implies a good S/N and lack of conflicting events or rapid structure changes.

Cross-correlate again. Cross-correlate as before, but use the new model trace. New values of quality factors, Q_{jlh} and time shifts Δt_{jlh} are obtained and used to determine the final model using Equation 6.61.

Cross-correlate the final model trace with traces in the next gather. The quality factors, Q_{jlh} and time shifts Δt_{jlh} obtained are used in Equation 6.61 to determine the preliminary model for the second CMP.

Make a new model trace for the second gather. Construct the new model trace M_{2h} by applying the time deviations found from the cross-correlations to each trace, obtaining a weighted trace average, and averaging it with the model determined for first gather. This process is described by Equation 6.62.

$$M_{2h} = \frac{1}{2}\left[\frac{\sum_{k=1}^{N} W_{2kh} \times f_{2kh}(t - \Delta t_{2kh})}{\sum_{k=1}^{N} W_{2kh}} + M_{l,h}(t)\right] \tag{6.62}$$

$M_{l,h}(t)$ is the final model for the first gather.

Continue as discussed previously to obtain the final model for the second CMP. The same procedure is followed for the third CMP with final models for both the first and second CMPs averaged with the third CMP model.

Construct models for each CMP by averaging the individual trace models with the models determined for a designated number of previous gathers. The general form of process is described by Equation 6.63.

$$M_{jh} = \frac{1}{L+1}\left[\frac{\sum_{k=1}^{N} W_{jkh} \times f_{jkh}(t-\Delta t_{jkh})}{\sum_{k=1}^{N} W_{jkh}} + \sum_{k=1}^{L} M_{j-l,h}(t)\right] \tag{6.63}$$

$M_{j-l,h}(t)$ is the model obtained for the $(j\text{-}l)^{th}$ gather, and L is the number of gathers to be averaged with the current model to obtain the new model.

The number of prior CMPs to use in model building depends on signal-to-noise ratio and dip. If S/N is very good, then few CMPs are needed. If the dip is steep, then only a few CMPs should be used. So, with very good S/N and steep dip, no more than two to six prior CMPs should be used in model building. With average S/N and moderate dip, perhaps 10 to 100 CMPs can be used. With poor S/N, there are no guidelines since no conventional technique has worked.

Parameters to apply in reflection residual statics are correlation window, maximum correlation shift, and NMO correction. Maximum correlation shift is simply the largest time shift for which cross-correlations will be calculated. The value chosen should be larger than the largest expected residual static but care should be taken avoid cycle skipping.

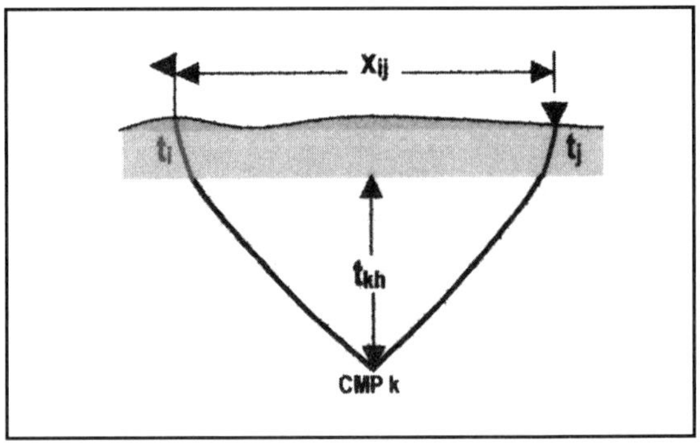

Fig. 6–133 *Surface Consistent Travel Time Model*

Surface consistent statics are based on a travel time model shown in Figure 6–133 and mathematically by Equation 6.64.

$$t_{ijkh} = t_i + t_{kh} + t_{NMOh} + t_j \tag{6.64}$$

where

t_{ijkh} = observed reflection time horizon h from source point I to receiver point j at CMP k

t_i = time from surface to base of weathering at source point I

t_{kh} = two-way time from base of weathering to reflector at normal incidence

t_{NMOh} = additional reflection time caused by separation between source and receiver by the distance x_{ij}

t_j = time from surface to base of weathering at receiver point j

If t'_{ijkh} is the true reflection time at horizon h from source point i to receiver point j at CMP k, then:

$$t_{ijkh} - t'_{ijkh} = \Delta t_{ijkh} = \Delta t_i + \Delta t_{kh} + \Delta t_{NMOh} + \Delta t_j \tag{6.65}$$

In Equation 6.65, Δt_{NMOh} is residual NMO or RNMO and is attributed to velocity errors. Normally, this is ignored (set to zero). Similarly, Δt_{kh} is assumed to be without error. Thus, residual static picks are resolved only into components associated with sources and receivers. If two correlation gates are used, then a single residual static is picked for each trace, so the h counter can be ignored. Equation 6.65 becomes:

$$\Delta t_{ijkh} = \Delta t_{ij} = \Delta t_i + \Delta t_j \tag{6.66}$$

Long and short wavelength statics. Reflection-based residual statics can only detect relatively *short wavelength statics* or time variations caused by near-surface anomalies with spatial dimensions less than (usually much less than) the maximum offset for the CMPs. Near-surface anomalies with spatial dimensions larger than the maximum offset for the CMPs are called *long*

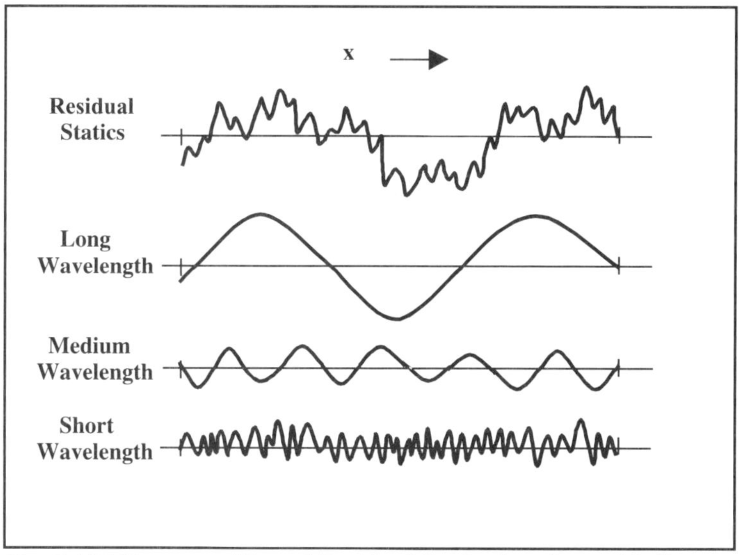

Fig. 6–134 *Decomposition of Residual Statics into Long, Medium, and Short Wavelength Statics*

wavelength statics. Note that short wavelength statics are also called *short period statics* and long wavelength statics are also called *long period statics.* In general, short and long wavelength statics, as well as intermediate or medium wavelength statics are present. See Figure 6–134.

To optimize the ability to detect and correct for all wavelengths of statics, both reflection-based and refraction-based residual statics should be determined and applied. Figure 6–135 demonstrates the value of combining reflection-based and refraction-based residual statics.

Residual statics summary. Residual statics analysis is required to identify and measure deviations in reflection times not associated with subsurface structure and stratigraphy. These deviations are generally caused by anomalous conditions in the shallowest parts of geological sections (mostly in the near surface) and velocity errors. Residual statics are used to correct for the first problem. Separating the picked values of residual statics into source and receiver components (surface consistent statics) allows residual statics to be added to datum statics calculated in the data initialization stage. This is often called statics update.

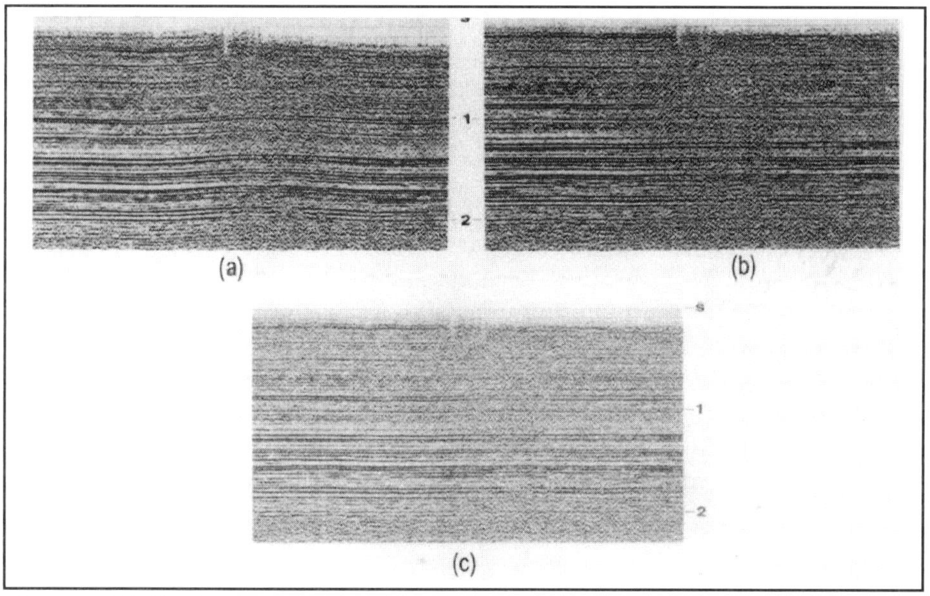

Fig. 6–135 *Reflection-based and Refraction-based Residual Statics (a) Brute Stack, (b) After Refraction-based Statics, and (c) After both Refraction-based Statics and Reflection-based Statics.*

Additional iterations of velocity analysis following application of residual statics, are used to reduce velocity errors. It is usually best to run at least two iterations of residual statics analysis, each followed by velocity analysis.

The presence of large amplitude multiple reflections can degrade the accuracy of reflection-based residual statics, particularly when the multiple generating horizon(s) have significant dip. Because such events cross through primary reflections and do not align after application of NMO corrections that use primary velocities, cross-correlation maxima tend to be broad, have low quality factors, and fall at incorrect times. If strong multiples are present, some sort of pre-stack multiple attenuation is suggested, at least before the second iteration of residual statics.

Pre-stack Multiple Attenuation

To attenuate multiples, advantages must be taken of those characteristics that distinguish multiple reflections from primary reflections. These characteristics are

- predictability
  - multiple reflections are predictable but primaries are not
- multiples occur at regular intervals (at least at near offsets)
- multiple amplitudes and polarities show a pattern of variation
- moveout/velocity/dip
  - multiple reflections always have more moveout than a primary occurring at about the same time
  - the NMO velocity required to flatten multiple reflections is lower than NMO velocities for primary reflections
  - a multiple associated with a dipping reflector has a greater apparent dip than the corresponding primary and the apparent dip increases, linearly, with multiple order (Fig. 6–136)
- amplitudes of primaries and multiples recorded at about the same time are often different

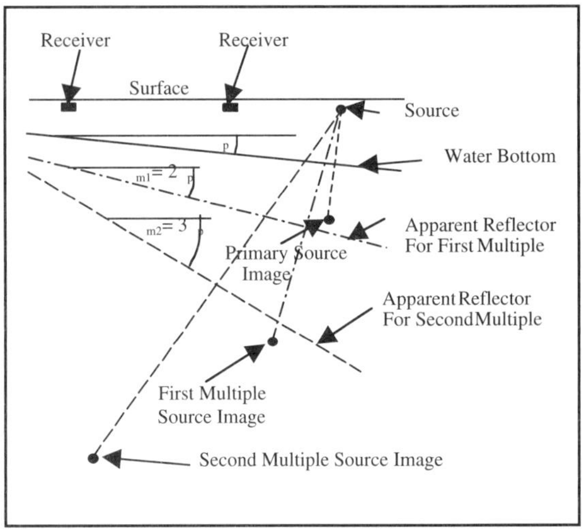

Fig. 6–136 *Increase in Dip of Multiples*

Here are methods of multiple attenuation grouped by the characteristic upon which they are based:

Predictability

- deconvolution
- time domain
- τ-p domain
- wave equation multiple attenuation

Moveout/velocity/dip

- CMP stack
- *F-K* Filtering
- velocity filtering in the τ-p domain

Amplitude

- median stack

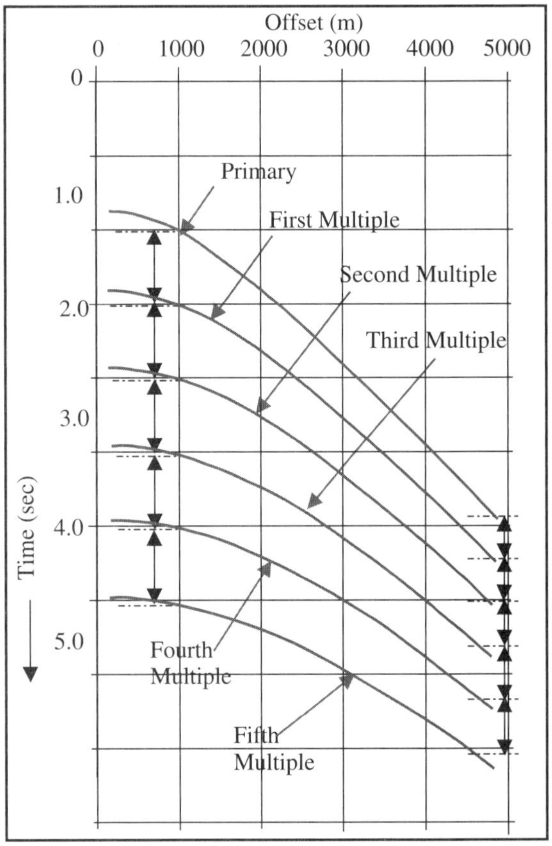

Fig. 6–137 *Change in Multiple Period with Offset*

All of the methods based on predictability are appropriate for pre-stack application. CMP stack is, of course, not a pre-stack process, but *F-K* filtering and velocity filtering in the τ-p domain can be applied pre-stack. Median stack is an alternative to conventional CMP stack.

Methods based on predictability. *Time-variant deconvolution* is a multiple attenuation method that uses the predictability or periodic nature of multiples to attenuate them. Since TVD was covered earlier in this chapter, this multiple attenuation method will be only briefly reviewed.

While either spiking (whitening) or gapped (predictive) deconvolution can be used to attenuate multiples, gapped deconvolution is frequently used when multiple attenuation is the primary objective. In this case, the gap is usually chosen to be longer than wavelet duration. In either case, the total filter length (gap plus filter) *must be longer than the period of the multiple to be attenuated.*

Deconvolution is most effective on short offset traces. Because of NMO, multiple periodicity decreases with offset resulting in a loss of effectiveness in attenuating multiples. The change in time between multiples, or loss of periodicity, with offset is shown in Figure 6–137.

Deconvolution in the τ-p domain. A synthetic example is presented to demonstrate the method. Figure 6–138 is generated from a model of the water layer.

Events shown are:

A – first break refraction from sub-water layer

B – water bottom reflection

C – direct arrival

D – noise generated by the modeling process

M_1, M_2, M_3 – multiples of sub-water refraction

m_1, m_2, m_3 – water bottom multiples (reverberations)

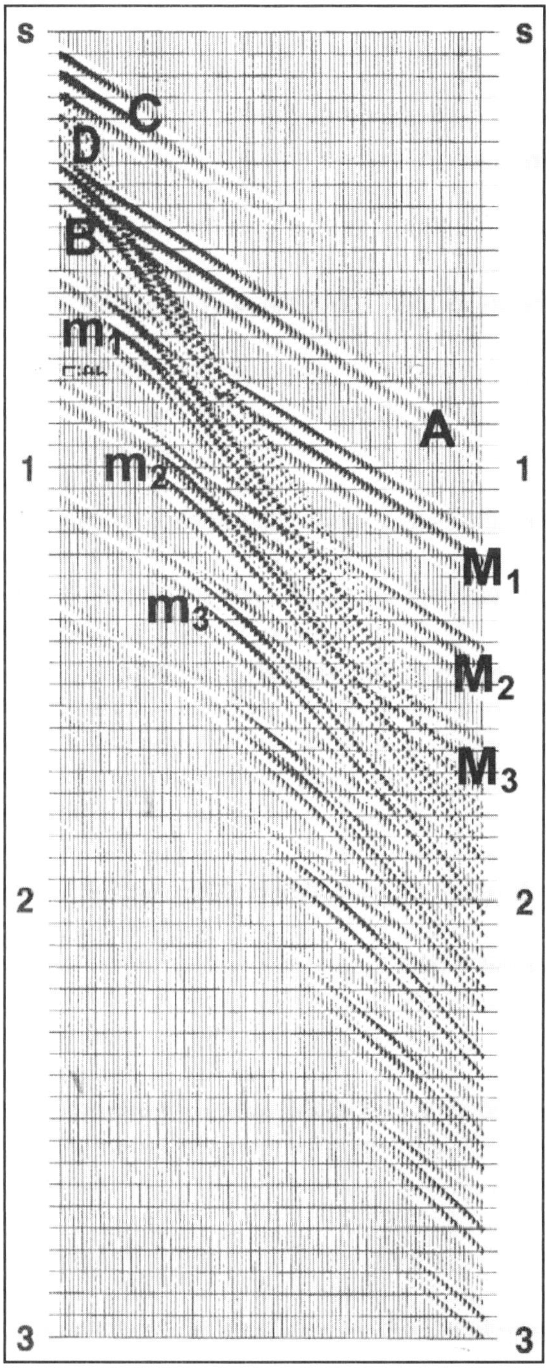

Fig. 6–138 *Synthetic Example*

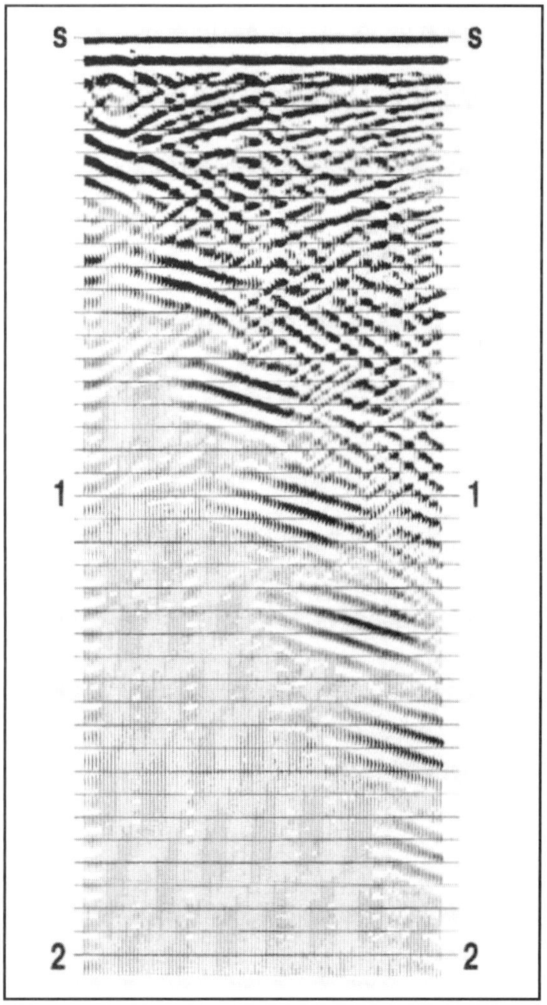

Fig. 6–139 *Autocorrelation of Data in Figure 6–138*

Figure 6–139 is the autocorrelation of the data in Figure 6–138. The change in periodicity of the multiples with offset is clearly evident.

Figure 6–140 is the radon transform of the data in Figure 6–138. Note that events *A*, M_1 M_2, and M_3 are very restricted in both τ and *p*. In theory, all should map to points since they have a single velocity and angle of emergence. For events *B*, m_1, m_2, and m_3 we see that $\Delta\tau$ along any p-trace is nearly constant.

Figure 6–141 is the autocorrelation of the data in Figure 6–140. The regularity at which the multiples occur is even more evident here.

Figure 6–142 is the result of applying a gapped deconvolution filter to the data of Figure 6–138. All multiples are severely attenuated.

Figure 6–143 is the autocorrelation of the data in Figure 6–142. Figure 6–144 is the inverse radon transform of the data in Figure 6–142. Multiple attenuation is even more obvious here than in Figure 6–143.

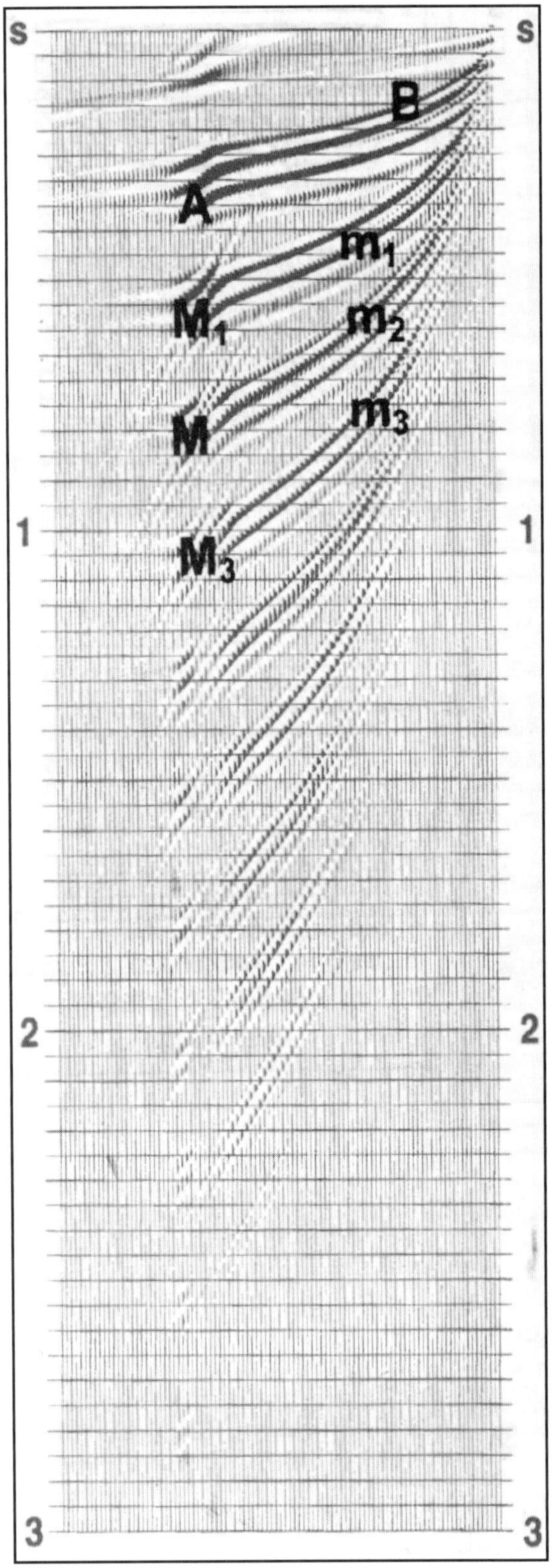

Fig. 6–140 *Radon Transform of the Data in Figure 6–138*

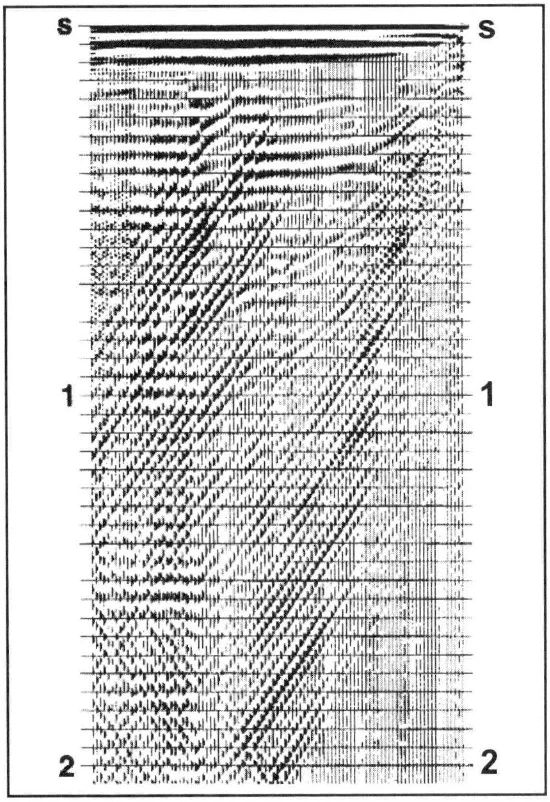

Fig. 6–141 *Autocorrelation of Data in Figure 6–140*

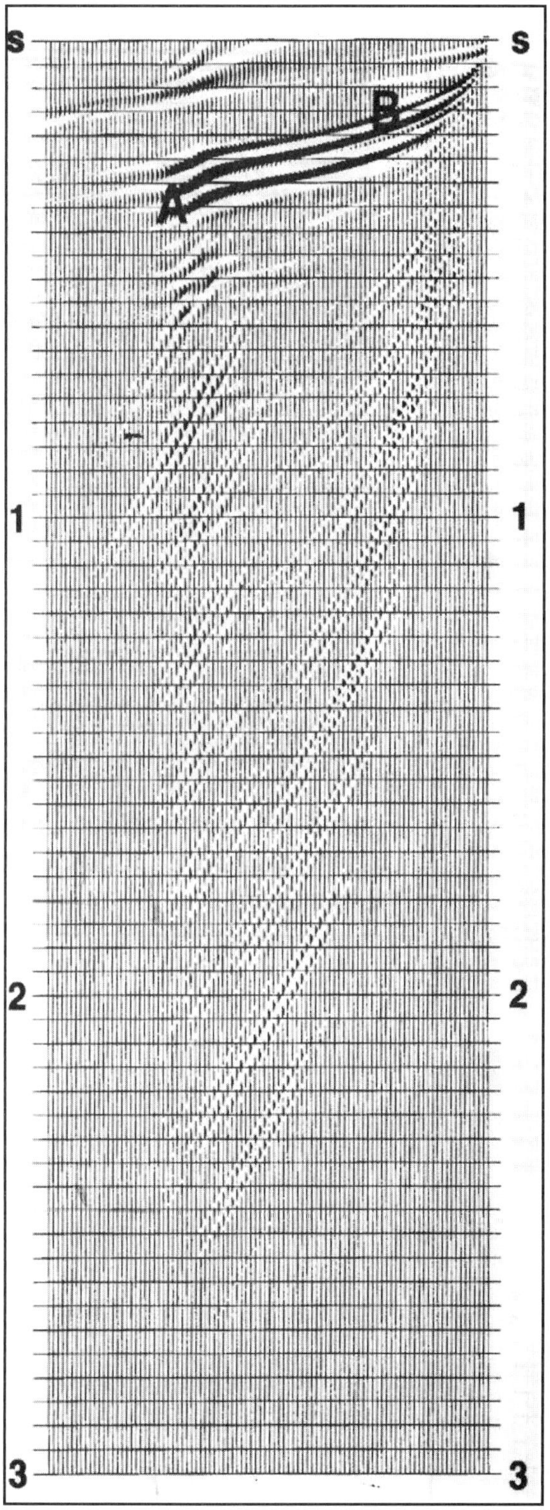

Fig. 6–142 *Data of Figure 6–140 after Gapped Deconvolution*

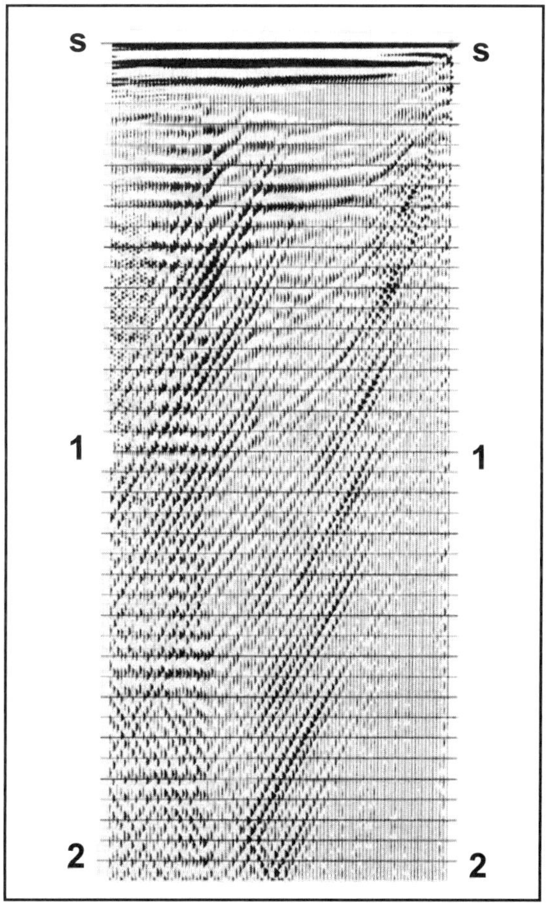

Fig. 6–143 *Autocorrelation of Data in Figure 6–142*

Wave equation multiple attenuation (WEMA). Transformation of data into the τ-p domain to observe the characteristics of the multiple reflections is most effective when the multiple-generating interface is horizontal or nearly so. For dipping or irregular interfaces multiples will not be periodic even in the τ-p domain. In such cases, another approach is required.

When seismic data contains strong, long-period water-bottom multiples—water-bottom multiple amplitudes are larger than pegleg multiple amplitudes. WEMA can be effective in attenuating water bottom multiples (reverberations). WEMA can also be applied to peg-leg multiples, but may not be as effective on them as on water-bottom multiples. WEMA uses knowledge of the multiple generation process to predict the occurrence of multiples in time and space as well as predicting multiple amplitudes.

It is not necessary for the predictive process to span more than one cycle of the multiple because the recorded data provides information about the reverberating wave field each time the wave arrives at a receiver. This is shown in Figure 6–145.

Here, the energy from the source reflected at the water bottom and is recorded at receiver 7. This energy is reflected at the water surface and the down-going energy is again reflected at the water bottom and recorded at receiver 11. There is a second reflection at the water surface, another reflection from the water bottom, and recording at receiver 13. Thus, information obtained from the recording at receiver 7 allows a prediction to be made of the recording at receiver 11. Similarly, the recording at receiver 11 allows a prediction of the recording at receiver 13.

Figure 6–146 shows the sequence of events that are modeled to predict a water bottom multiple from a previous occurrence.

1. surface reflection (at offset x and depth 0)
2. propagation to bottom (at offset x' and depth z_B)
3. bottom reflection
4. propagation to receiver (at offset x'' and depth 0)

Event 1, in the previous list, is the determination of surface reflectivity. Event 2 is the determination of the time to the water bottom, the position of the reflection point, and the effect of geometrical spreading. Event 3 is the determination of the water-bottom reflectivity. Event 4 is the determination of the time from the water bottom to the receiver plus the effect of geometrical spreading.

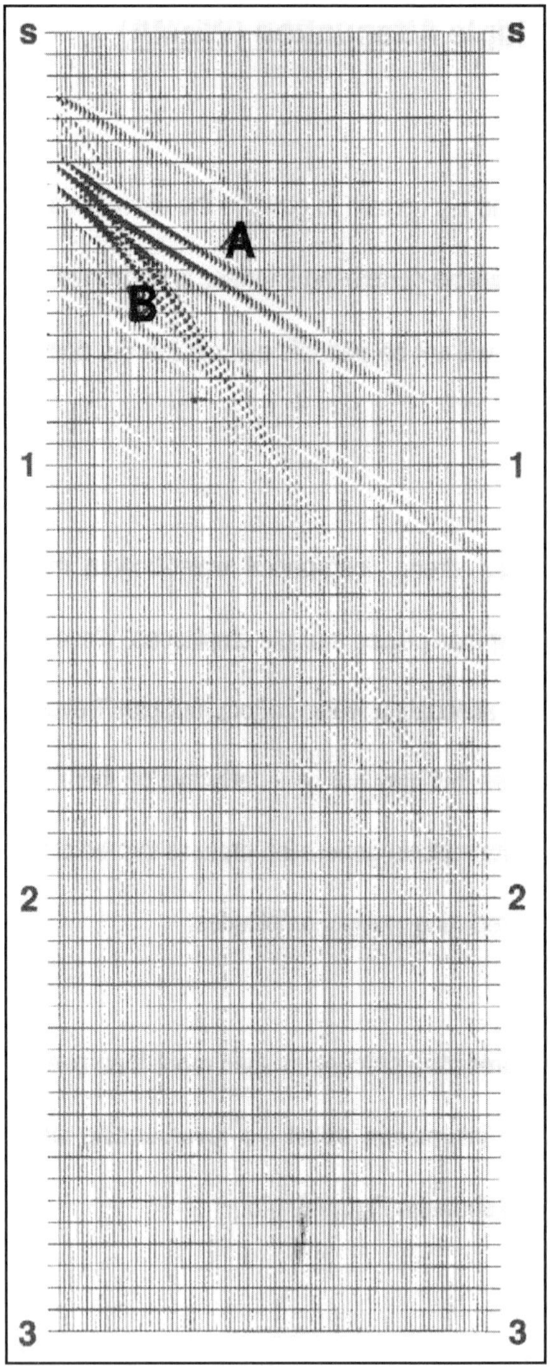

Fig. 6–144 *Inverse Radon Transform of the Data in Figure 6–142*

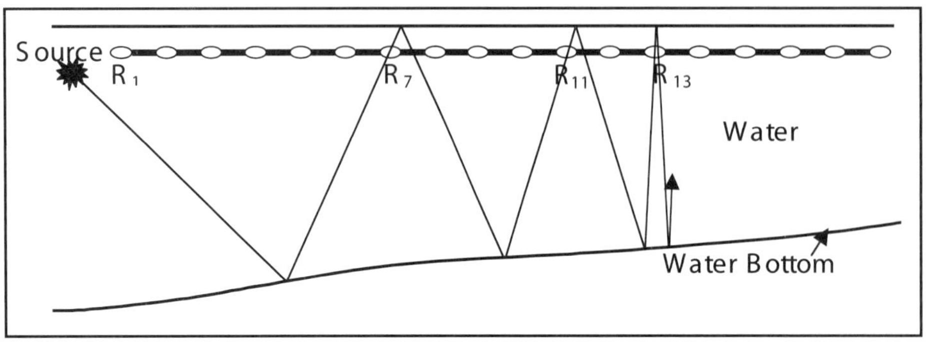

Fig. 6–145 *Reverberation Sampling by Receivers*

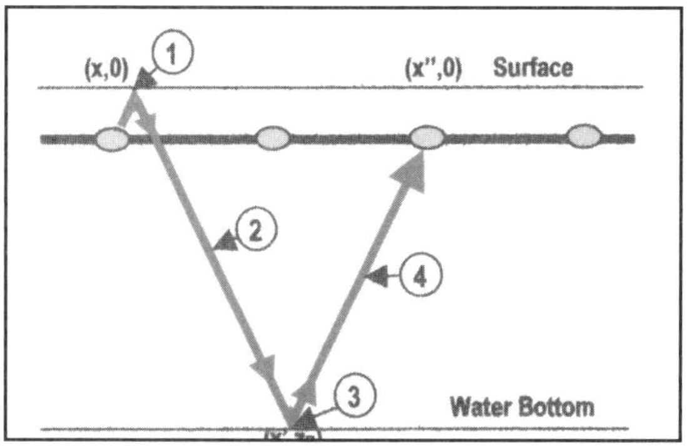

Fig. 6–146 *Modeling of Water Bottom Multiple Sequence from Previous Occurrence*

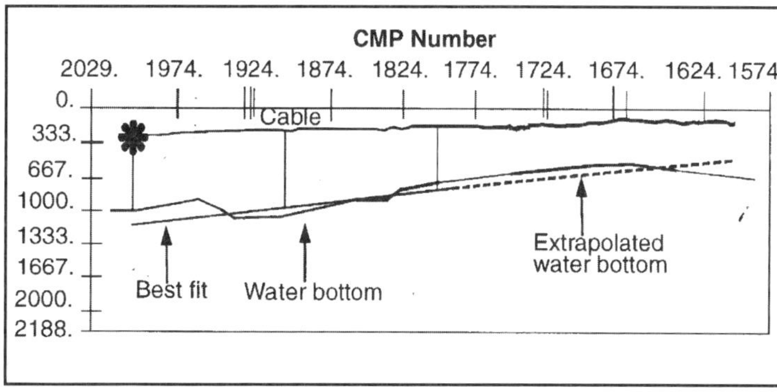

Fig. 6–147 *Approximating the Water Bottom by a Straight Line*

Events 3 and 4 require the definition of the water bottom in time and space. To simplify this, it is assumed that the water bottom can be approximated by a straight line for any one shot or receiver record.

Figure 6–147 shows how a best fit straight line is used to approximate an irregular water bottom. Note that this line is dipping in Figure 6–147. The line used can have any dip.

To simplify the first step, it is assumed that the surface reflectivity is a constant over a single shot or receiver record. It is also assumed that the water-bottom reflectivity can be estimated and applied after propagation to the surface. This implies that the reflectivity changes slowly within a shot or receiver record. The last two assumptions also allow combination of the surface and water bottom reflectivity into a single operator.

A multi-valued operator that is convolved with the extrapolated data best represents reflectivity. Further, there is no way to directly measure water bottom reflectivity so it must be found statistically.

If WEMA is successful in its objective of subtracting multiple energy from the input records, its optimum output will be that which has minimum record energy. Since seismic energy is measured by amplitudes squared:

$$[s_o(t, x) - r(x)*s_e(t, x)]^2 = \text{minimum} \tag{6.67}$$

where

$s_o(t,x)$ = original data

$s_e(t,x)$ = extrapolated data

$r(x)$ = combined reflectivity operator

$*$ = mathematical symbol for convolution

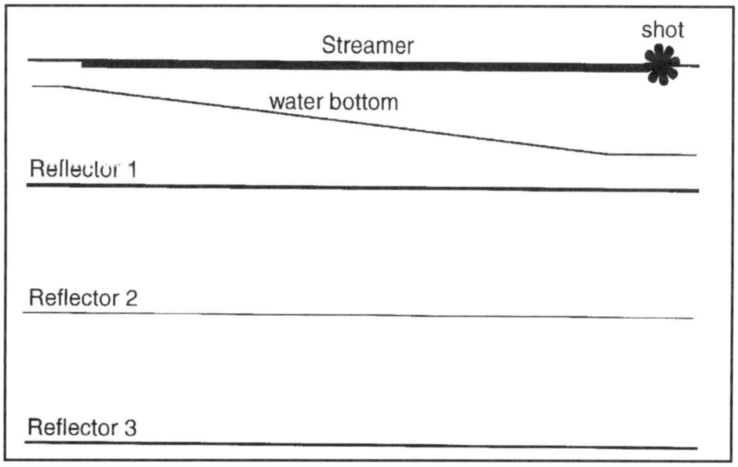

Fig. 6–148 *Model Used to Generate Synthetic Data*

Synthetic data example. To demonstrate WEMA, synthetic data were generated using the model shown in Figure 6–148. The water bottom slopes toward the shot position at an angle of 6°. Below the water bottom, there are three flat reflectors that generate primary reflections at 1100 ms, 2100 ms, and 3450 ms. Figure 6–149 shows data generated from the model in Figure 6–148. This synthetic shot record contains 14 cycles of the water-bottom multiples—the events that occur with a little less than 500 ms interval at zero offset and exhibit increasing reverse moveout as the cycle number increases. In addition to the water-bottom multiples, the data contain three sub-bottom primary reflections and the associated peg-leg multiples resulting from reverberation in the water layer at the receiver end following reflection at the primary reflector. Peg-leg multiples resulting from reverberations of the source wave before transmission below the water bottom are not included in the data.

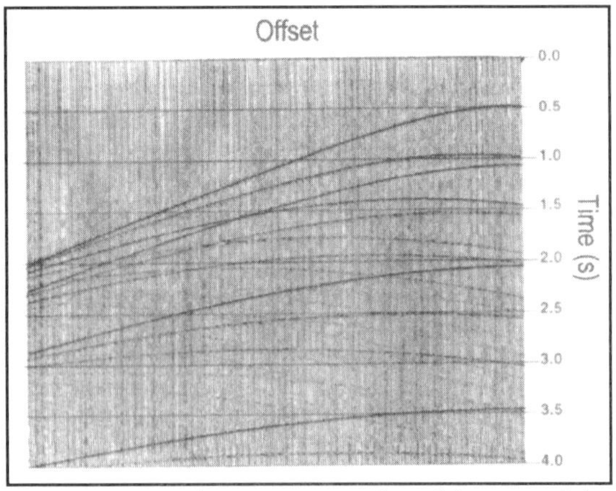

Fig. 6–149 *Synthetic Data Produced from Model of Figure 6–148*

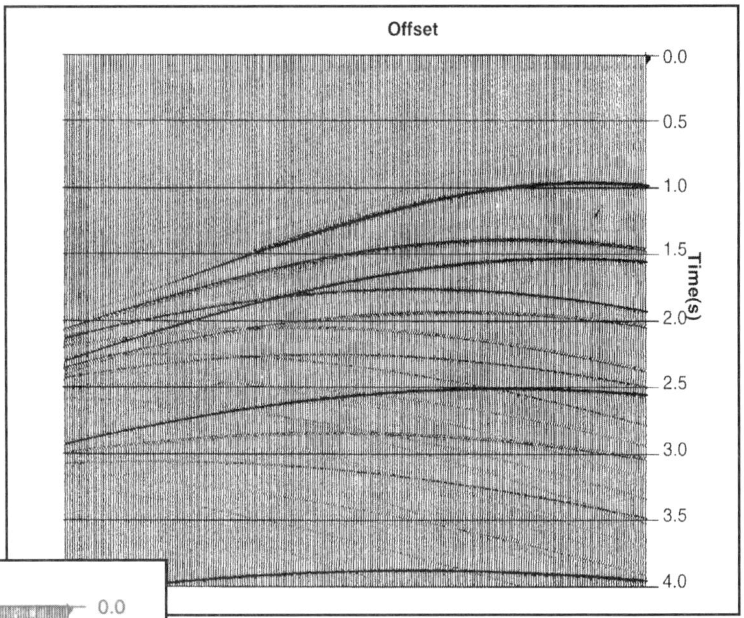

Fig. 6–150 *The Data of Figure 6–149 after Forward Extrapolation of One Round Trip Through the Water Layer*

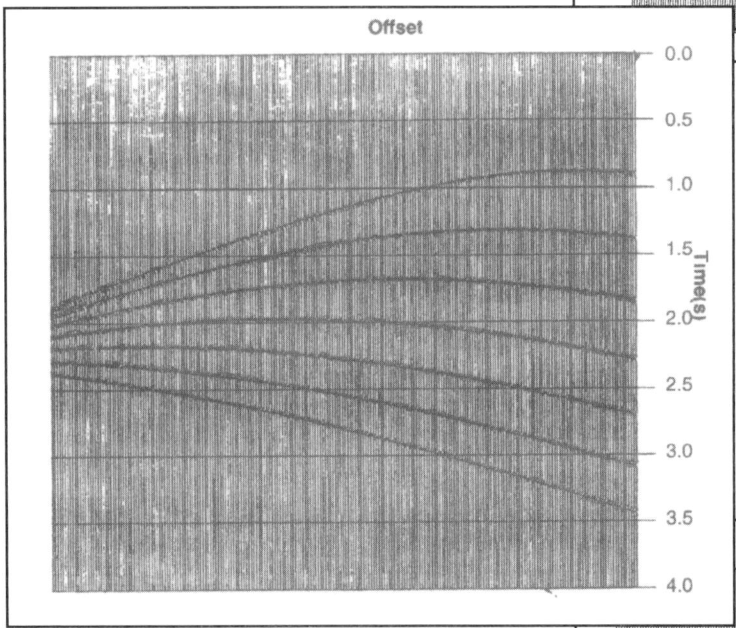

Fig. 6–151 *Reflectivity Operators Designed for Seven Different Gates of Data in Figure 6–149*

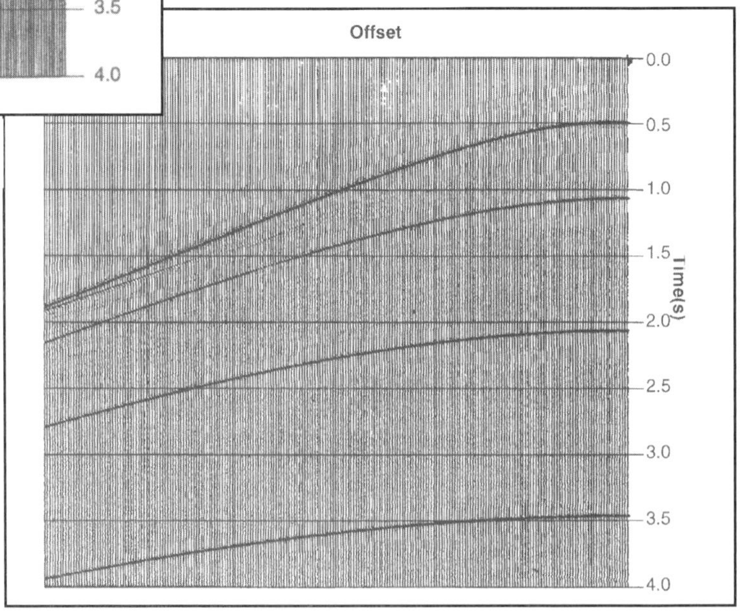

Fig. 6–152 *Result of Subtracting Predicted Multiples from Data of Figure 6–149*

It should be noted that the data in Figures 6–149, 6–150, 6–151 and 6–152 have been scaled proportional to time as a correction for spherical spreading. No geometric spreading correction is applied before WEMA.

Figure 6–149 shows the data after the wave extrapolation step. Note that the water-bottom primary reflection now occurs at the same time as the first water-bottom multiple occurred in Figure 6–149. Likewise, each multiple in Figure 6–150 occurs at the same time as the succeeding multiple in Figure 6–149. The wave extrapolation operator has numerically simulated one complete cycle through the water layer—surface reflection, down to water bottom, reflection at the bottom, and up to the receiver.

Figure 6–151 shows combined surface and water-bottom reflectivity operators for seven time windows. There is some change in reflectivity as multiple order—number of bounces in the water—increases because angles of incidence change. Reflectivity depends on acoustic impedance contrast across a boundary and angle of incidence.

The reflectivity operators are 50 ms long with a 20 ms lead time—a parameter that helps compensate for error in picking water bottom reflection times. The relatively long reflectivity operators are used to compensate for the behavior of a complex water bottom. The bottom may be laminated—many thin strips of material—or have three or more layers of equal magnitude. Such water bottoms often exhibit frequency-dependent reflectivity. The operators in Figure 6–151 are negative because the surface reflectivity is negative and the water bottom reflectivity is positive. This is the normal situation.

The reflectivity operators of Figure 6–151 are convolved with the extrapolated data of Figure 6–152 to produce the multiple predictions. The predicted multiples are then subtracted from the input data.

Figure 6–152 shows that WEMA was successful in attenuating all multiples for this synthetic case as only primary reflections can be seen after subtraction of the predicted multiples. One reason for this, besides its being synthetic data, is that peg-leg multiples were not completely defined. In real cases, peg-leg multiples occur at both the shot and receiver ends of the travel path. Thus, to attenuate peg-leg multiples, it is usually necessary to apply WEMA twice. For a relatively complex water bottom or where there is significant dip, peg-leg applying the second pass of WEMA to receiver records may better attenuate multiples.

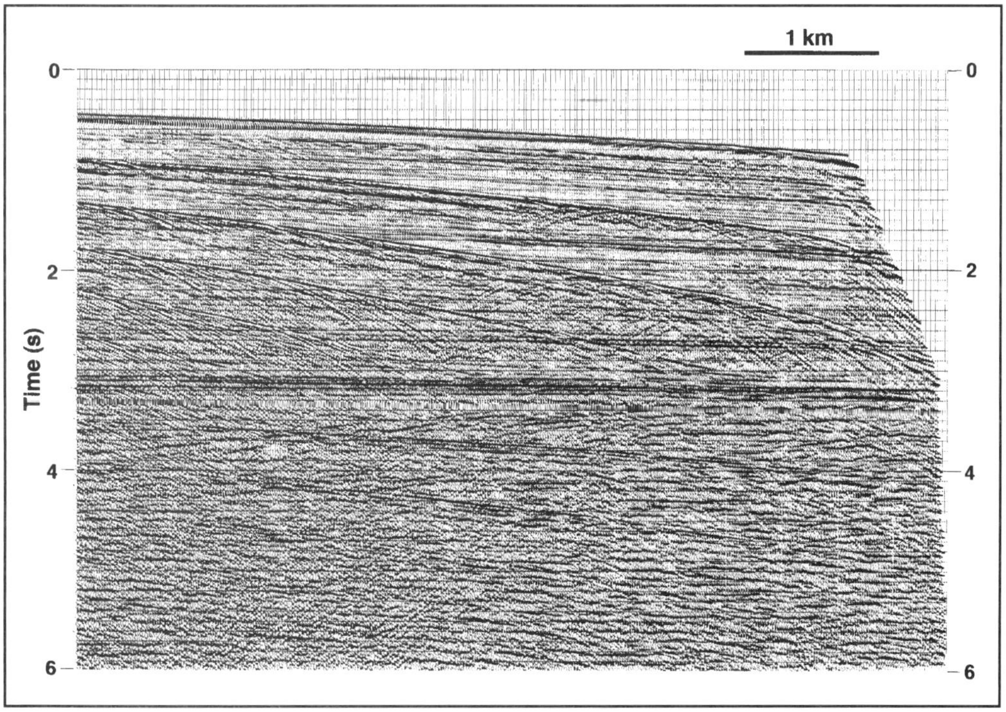

Fig. 6–153 *A CMP Stack Showing Strong Multiples*

Field data example. Figure 6–153 shows a CMP stack section to which no pre-stack multiple attenuation has been applied. The water bottom dips at a nearly constant angle from left to right. Several water bottom multiples can be seen with progressively increasing dip. The steep dip on the multiples causes multiple velocities to approach primary velocities. As a result, there is inadequate moveout differential for CMP stack to appreciably attenuate the multiples.

Figure 6–154 shows the data of Figure 6–153 after application of WEMA.

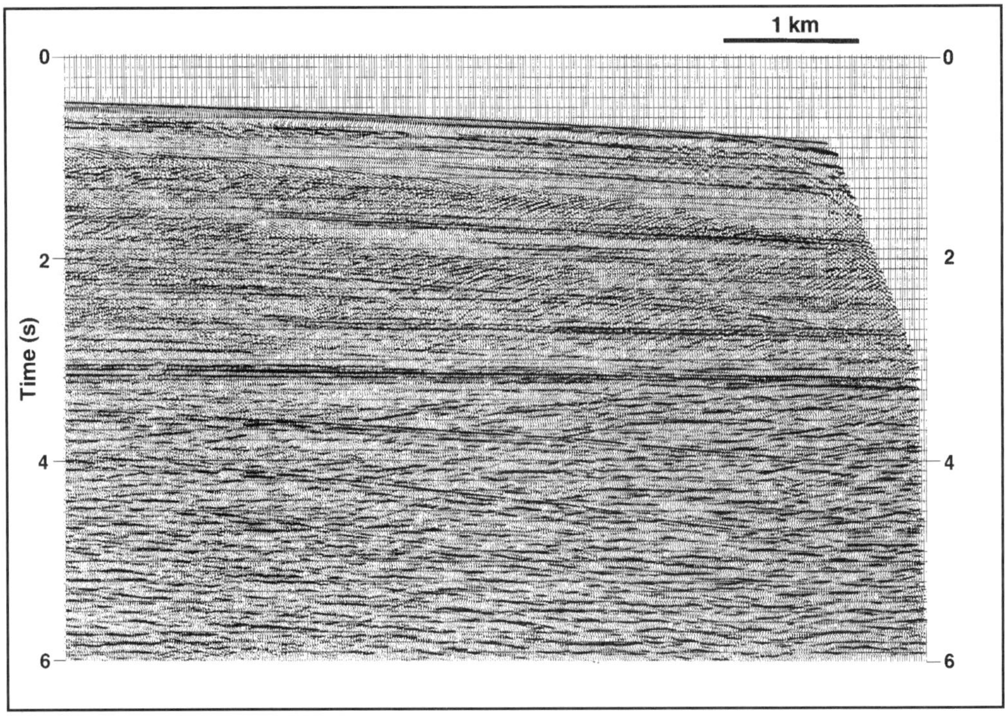

Fig. 6–154 *CMP Stack Section of Figure 6–153 after Application of WEMA*

Despite some residual noise from the early multiples, the processed stack reveals primary events that were previously obscured by the multiples. Notice, in particular, the primaries between 1.0 and 1.4 s and between 2.2 and 2.4 s near the left side of Figure 6–154. At the far right of the unprocessed section, the first and second water-bottom multiples cut across primary events. The wave equation processing has removed these multiples and their associated diffractions, uncovering the underlying primaries.

Below the strong primary at about 3.2 s, the water-bottom multiples are weak even in the conventional section because of the attenuating action of the stacking process. Nevertheless, the conventional and multiple-suppressed sections differ noticeably in that the multiple-suppressed data appear free of the background noise present in the conventional stack.

As with other wave-extrapolation processes, the data must be adequately sampled in space to prevent artifacts arising from spatial aliasing. The data here were recorded with a receiver group interval of 50 m. This sampling interval is larger than typically used in acquisition today and is not sufficient to avoid spatial aliasing for waves traveling at angles greater than 17°. To avoid problems of insufficient sampling with these data, the angular aperture was restricted in the wave extrapolation to eliminate the waves propagating at wide angles. This restricting of aperture requires a tradeoff between the temporal frequency retained in the data and the wave angles passed by the aperture. Notice that the unprocessed stack contains diffraction tails following each multiple.

Because of the restricted aperture, the right dipping half of each diffraction has been removed by the wave-equation processing while the left dipping half has been left in the stack. One-half of each diffraction fell within the extrapolation aperture, which is asymmetrical because of the off-end recording geometry, while the other half was outside the aperture and was, therefore, not predicted or subtracted.

Though a group interval that is small enough to avoid spatial aliasing at all angles recommended for wave-equation processing, restricting the extrapolation aperture allows processing of more coarsely sampled data. It does so, however, at the expense of a tradeoff among steep-dip signal retention, multiple suppression, and economy.

Processing considerations. Wave extrapolation is a process that models the actual propagation of the seismic wave through the water layer in a manner determined by the geometry of the streamer and the water bottom. Amplitudes are handled in a way that automatically compensates for spherical spreading. While estimation of the water-bottom reflection operator is adaptive, it works best when the effects that it must compensate for are simple. Trace balancing, statistical deconvolution, spectral whitening, or geometric spreading corrections are effects that do not match the multiple model definition, and they complicate the model-building job. Do not perform these operations before using WEMA. If there is a good measure of the source signature, apply signature deconvolution before applying this process.

Data supplied to WEMA should be checked to see that the early arrivals are not clipped. Pre-processing should include the following items:

- survey geometry placed in trace headers
- time correction for recording equipment
- band pass filter, especially low-pass filtering
- signature deconvolution to remove instrument response
- bad trace processing
- resampling in time (optional)

It is extremely important that noisy and bad traces are dealt with before applying WEMA because the wave extrapolation step will spread the bad data from one trace across many traces of the shot gather. When the data are CMP-sorted later, one bad shot gather will contaminate many CMPs.

Summary and conclusions. A list of processing considerations follows.

- No gain or balancing, geometrical spreading compensation, or statistical deconvolution should be applied before wave equation prediction.
- The effective angle increases with higher order multiples for dipping water bottom so there is a need to window the data. Window lengths should be 2, 3, or 4 times the water-bottom cycle.
- Constant velocity Stolt migration of the near group may be required to accurately model the water bottom.
- Closely spaced layers just below the water may cause computational errors that result in a failure to discriminate between multiples.
- Lack of near offset data means that there will not be data from which to forward predict multiples to the nearest recorded offsets.
- Spatial sampling must be small enough to record the bottom irregularities and dip. It is especially important in the case of receiver sorts for peg-leg multiple attenuation that the shot interval be small.

Peg-leg multiples may be described as having shot-generated reverberations attached to a primary and receiver-related reverberations after the primary. See Figure 6–155. One pass of WEMA only attenuates the shot-generated reverberations. To attenuate the associated receiver-related reverberations requires sorting to common receiver gathers and repeating the procedure. This doubles the computer effort.

The procedure is normally applied to shot gathers, but, as noted previously, attenuation of peg-leg multiples may require a second pass on receiver gathers.

WEMA is effective even when the water bottom is complex—dip, curvature, and locally varying reflectivity—but accurate water depths must be specified. It exploits the predictability of the multiples. It is possibly the only suitable multiple attenuation technique if pre-stack data analysis (AVO) is considered.

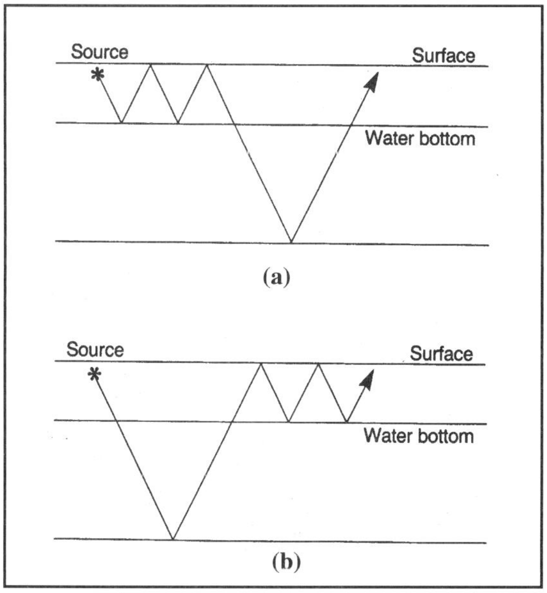

Fig. 6–155 *Source-generated and Receiver-generated Reverberations for Peg-leg Multiples*

Methods based on velocity, moveout, or dip. Methods of multiple attenuation that depend on velocity differences may take advantage of the effect of velocity on moveout—CMP stack, *F-K* filtering, and velocity filtering in the t-p domain. In any of these, it is the difference between primary and multiple velocities that underlies the approach.

CMP stack attenuated multiples because multiples tend to have residual moveout after NMO corrections if good primary velocities are obtained. This is true if multiple and primary velocities differ considerably, but in some cases there is inadequate differential moveout for satisfactory multiple attenuation. CMP stack is obviously not a pre-stack process and will be discussed later in this chapter.

***F-K* filtering.** Reverberations and multiple reflections may be recorded at nearly the same time as primary reflections, but they have lower normal moveout velocities than primary reflections. Thus, NMO corrections that flatten reverberations or multiples overcorrect primaries. When NMO-corrected CMP records are transformed into the *F-K* domain, all primaries are placed in the negative half of the *F-K* plane and the NMO-corrected reverberations or multiples along the $K = 0$ axis. A velocity filter that passes everything except a narrow reject band centered at $K = 0$ eliminates the reverberations or multiples.

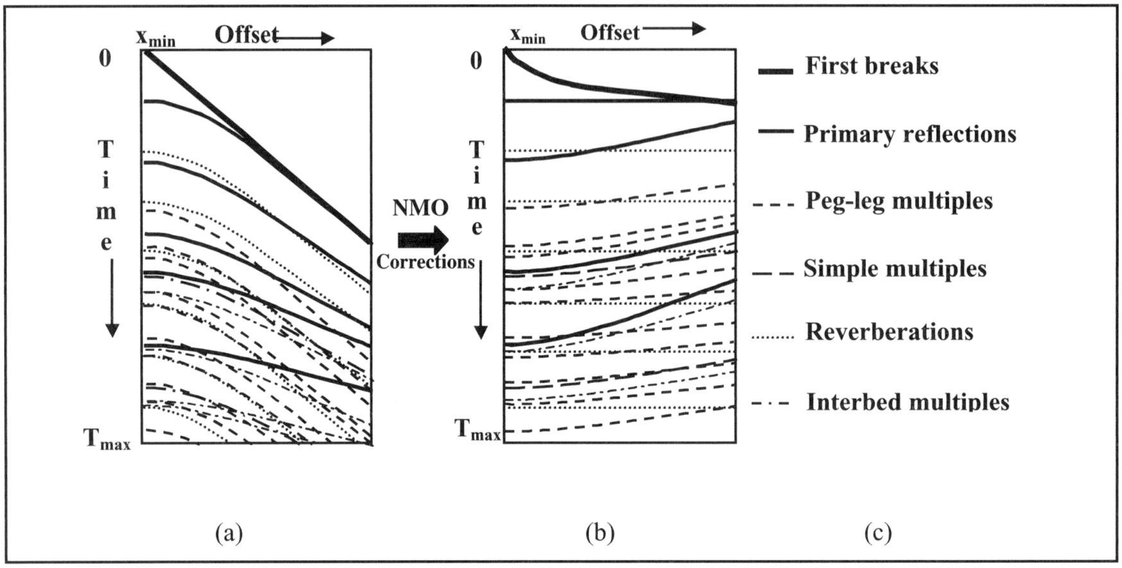

Fig. 6–156 *Input Record and Corrected Record for NMO Using Reverberation Velocity*

The same basic approach is applied to reverberations, simple multiples and peg-leg multiples, but different velocity functions are required for each of these. Reverberations are flattened by using a constant velocity equal to water velocity (about 1500 m/s). Simple multiples and peg-leg multiples require picking appropriate events in a velocity analysis similar to that used to determine primary velocities. Simple multiple velocities are faster than water velocity but slower than peg-leg velocities, which are somewhat slower than primary velocities.

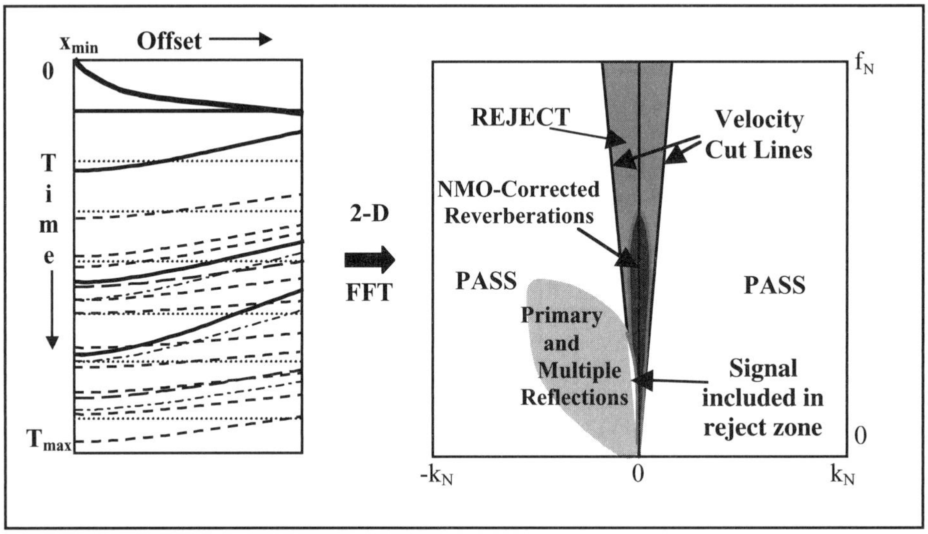

Fig. 6–157 *NMO-corrected Record of Figure 6–156 and Record Transformed into F-K Domain*

Figure 6–156a shows a schematic record with primaries and a variety of multiples. In Figure 6–156b, the record has been corrected for NMO using the water velocity. As a result, the reverberations are flattened.

Figure 6–157 illustrates the next step, transformation from the *T-X* domain into the *F-K* domain via a two-dimensional FFT. Note that, in the *F-K* domain, reverberations are along the $k = 0$ axis, and all other events are in the negative half-plane, except for a slight overlap at low frequencies.

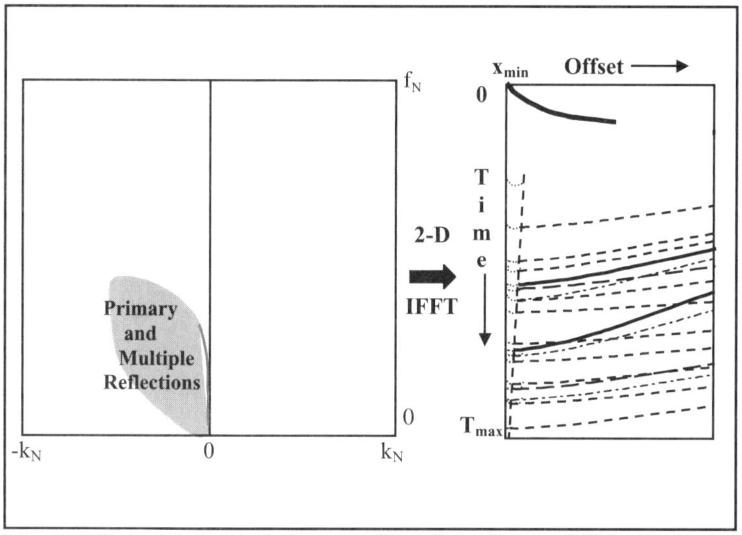

Fig. 6–158 *F-K Record of Figure 6–157 after Velocity Filter Applied and Record Transformed into T-X Domain*

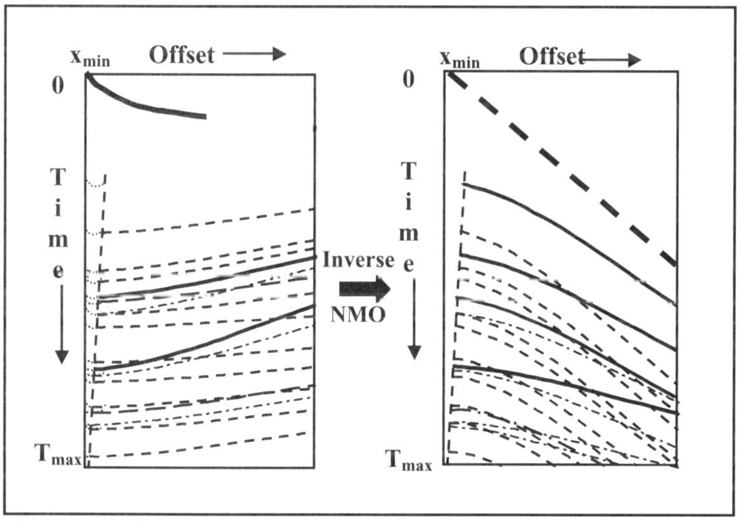

Fig. 6–159 *The T-X Domain Record of Figure 6–158 and Record After Inverse NMO Corrections*

Figure 6–158 continues the illustration of *F-K* domain multiple attenuation by showing the *F-K* domain record after the velocity filter is applied to attenuate the reverberations. The right side of Figure 6–158 shows the transformation back into the *T-X* domain via an inverse FFT (IFFT). The reverberations are absent but there has also been some attenuation of events at short offsets because the apparent velocities were very close to the negative velocity cut line. The effect is most evident at early record times. The result is distortion of these events.

Figure 6–159 completes the illustration of *F-K* domain multiple attenuation. The *T-X* domain record of Figure 6–158 is repeated with the short offset events zeroed. The left side of Figure 6–153 shows the record after inverse NMO is applied—NMO correction times that were subtracted from record times have been added back. The distorted events must be zeroed. This can be accomplished by applying a wedge-shaped scaler or ramp function. The zeroed segments are shown on the left of Figure 6–159.

Simple, peg-leg and inter-bed multiple attenuation is accomplished by similar means. However, to attenuate these multiples, it is necessary to run a velocity analysis, basing interpretation on whichever type of multiple is to be attenuated. Use velocity function(s) so obtained to apply NMO corrections. Primaries and faster multiples, if any are present, will be overcorrected. From this point the method is as described for reverberation attenuation.

If a velocity function intermediate between the primary and multiple velocity functions is used for NMO corrections, the primaries are overcorrected—although not as much as when the multiple velocity function is used—and plot in the negative half of the *F-K* plane. Multiples are under-corrected, although not as much as when the primary velocity function is used. In this situation, an *F-K* filter that rejects most of the positive half of the *F-K* plane is used to pass the primaries and reject the multiples. As with the other approaches, this is followed by inverse 2-D transformation and inverse NMO correction.

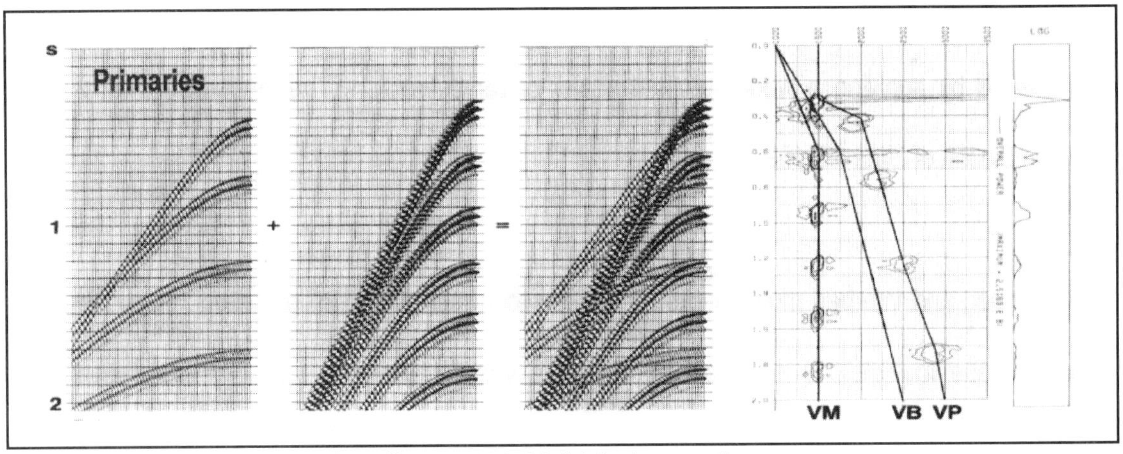

Fig. 6–160 *Synthetic Data Used to Illustrate F-K Multiple Attenuation*

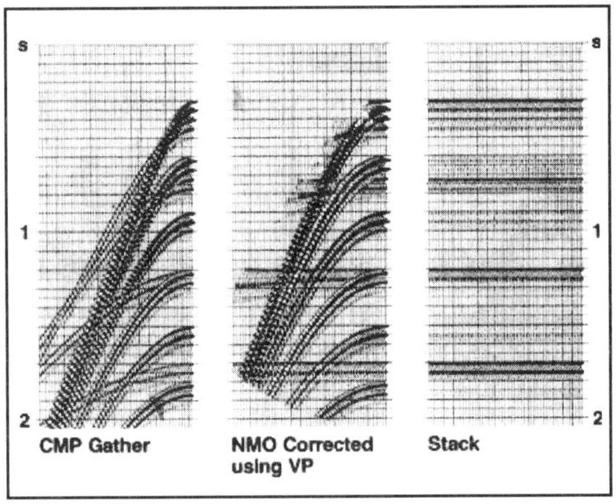

Fig. 6–161 *Multiple Attenuation from CMP Stack*

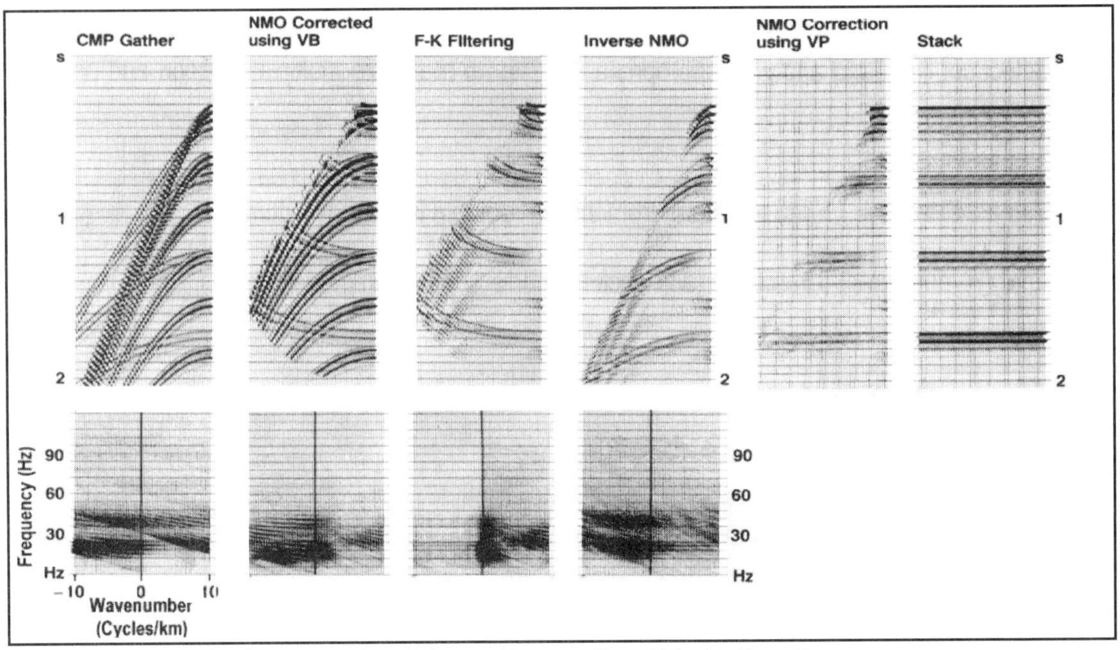

Fig. 6–162 *F-K Multiple Attenuation Using an Intermediate Velocity Function*

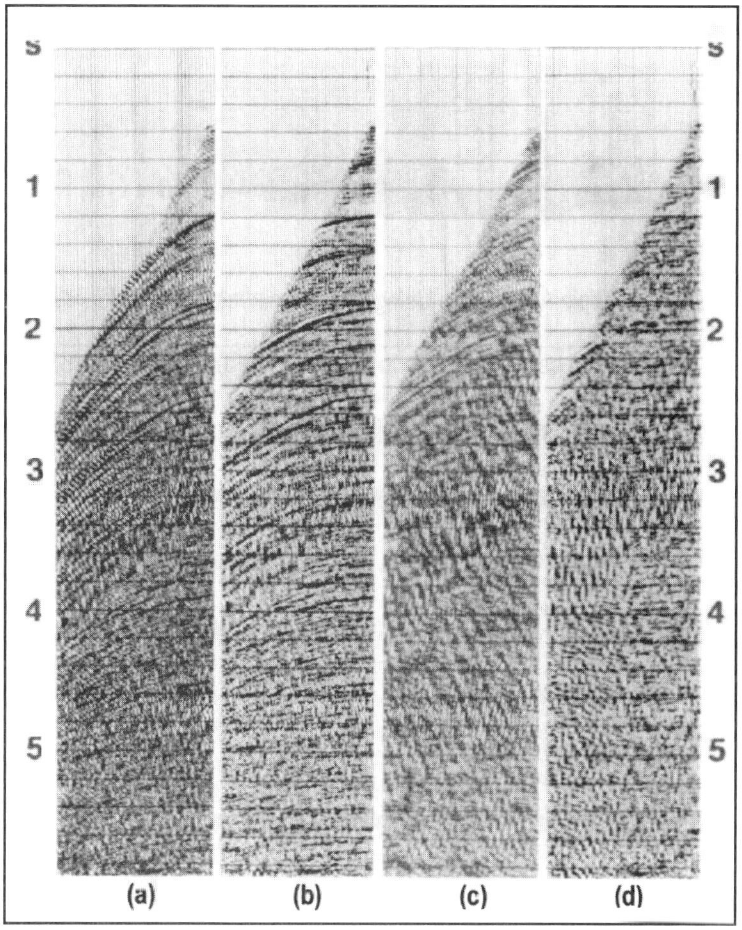

Fig. 6–163 *F-K Multiple Attenuation Example*

Figure 6–160 shows the synthetic data used to illustrate the *F-K* multiple attenuation. These data consist of three sub-water primary reflections, the water-bottom reflection and five water-bottom multiples. On the far right of Figure 6–160 is a velocity analysis display with interpretation of interpretation of a multiple velocity function V_M and a primary velocity function V_P. A velocity function V_B intermediate between the multiple and primary velocities is also shown.

Figure 6–161 uses the synthetic data of Figure 6–160 to illustrate the conventional approach of using CMP stack to attenuate multiples. Residual multiples can be seen from the stack of near-offset data. There is also a smearing of energy across the stack.

Figure 6–162 illustrates the use of the intermediate velocity function, V_B, for NMO correction. From left to right, see an input CMP gather, the gather after NMO correction with the intermediate velocity function, the gather after *F-K* filtering, the gather after inverse NMO correction, the gather after NMO correction using the primary velocity function, and, finally, the stack. This stack is obviously superior to that of Figure 6–161.

Figure 6–163 demonstrates the effect of *F-K* multiple attenuation with a field data example. Figures 6–163a and 6–163b shows the same CMP gather record before NMO corrections. *F-K* multiple attenuation has been applied in Figure 6–163b but not in Figure 6–163a.

Note the many events present in Figure 6–163b that cannot be seen in Figure 6–163a. Figures 6–163c and 6–163d show the same CMP gather as in Figures 6–163a and 6–163b, but NMO corrections have been applied but before *F-K* multiple attenuation. Figure 6–157c shows the CMP gather before *F-K* multiple attenuation, and Figure 6–163d shows the CMP gather after *F-K* multiple attenuation. That many multiples are present in Figure 6–163c is evident from the events with residual NMO. In Figure 6–163d these events are not seen.

Summarizing, *F-K* velocity attenuation involves the following steps.

1. Perform velocity analyses to determine primary and multiple reflection velocities.
2. Interpolate an intermediate velocity between these two velocity functions.
3. Apply NMO corrections to CMP traces using one of the velocity functions, with the intermediate velocity usually preferred.
4. Use a 2-D forward FFT to transform from the *T-X* domain into the *F-K* domain.
5. Select cut lines and apply an appropriate *F-K* filter to maximize multiple rejection and primary preservation.
6. Use a 2-D inverse FFT to transform from the *F-K* domain into the *T-X* domain.
7. Apply inverse NMO corrections (add back the corrections subtracted in the second step).

Radon transform (τ-p) filtering. This approach to multiple attenuation involves the following steps.

1. Apply NMO corrections to CMPs.
2. Use a radon transform to change from the *T-X* domain into the τ-p domain.
3. Apply a low and high velocity mute to retain only the multiples.
4. Inverse radon transform from the τ-p domain back to the *T-X* domain.
5. Subtract the record obtained from step 4 from that of step 1.

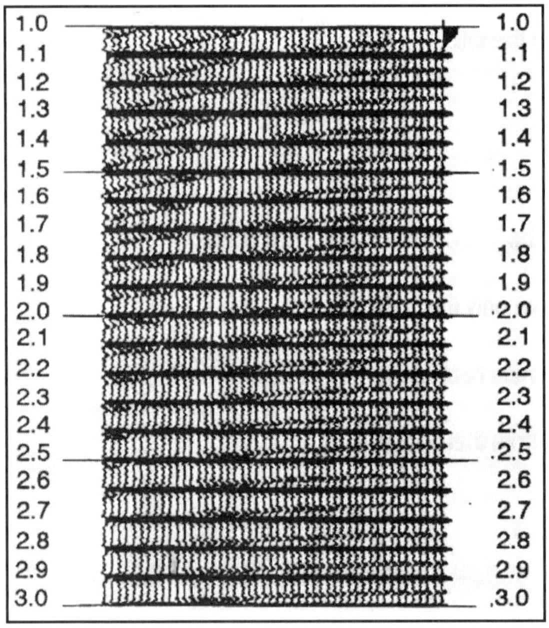

Fig. 6–164 *Synthetic Data Used to Illustrate Radon Transform Filtering*

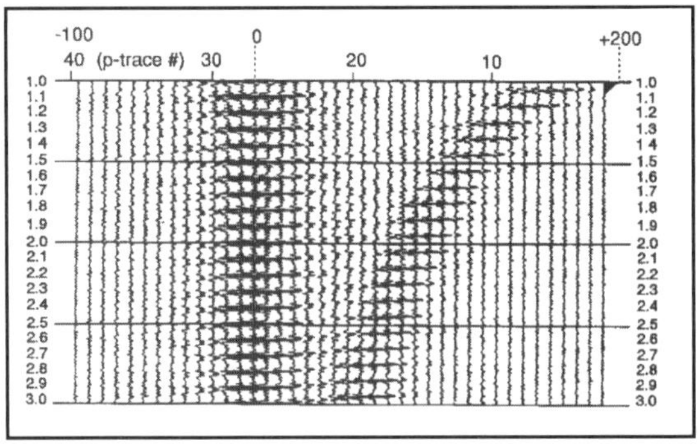

Fig. 6–165 *Data of Figure 6–164 Transformed into τ-p Domain*

After step 4, records should contain only multiple reflections with Primary NMO corrections removed. Subtraction of these data from the NMO-corrected gathers should leave only flattened primaries.

Figure 6–164 shows a synthetic gather record with two types of events—those with an 1800 m/s stacking velocity (multiples) and those with a 2000 m/s stacking velocity. NMO corrections have been applied using 2000 m/s. The 1800 m/s events have residual moveout and retain a somewhat hyperbolic shape.

Figure 6–165 is the data of Figure 6–164 after transformation into the t-p domain via the radon transform. Note that the 2000 m/s events line up along 0 dip (vertical event) while the 1800 m/s events are curved events in τ-p.

Figure 6–166 shows the velocity functions used to define the mutes and the data of Figure 6–159 after the mutes are applied. All of the 2000 m/s energy has been removed, leaving only the curved 1800 m/s events. Figure 6–167 is the inverse radon transforms of the data in Figure 6–166.

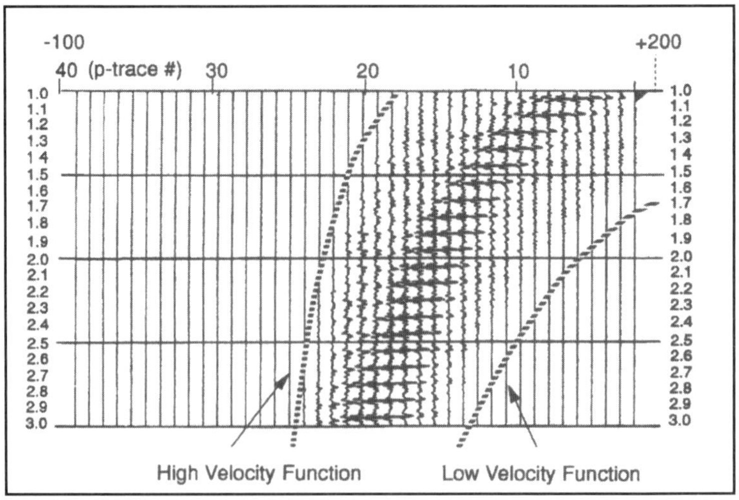

Fig. 6–166 *Data of Figure 6–165 after Velocity Mutes Are Applied*

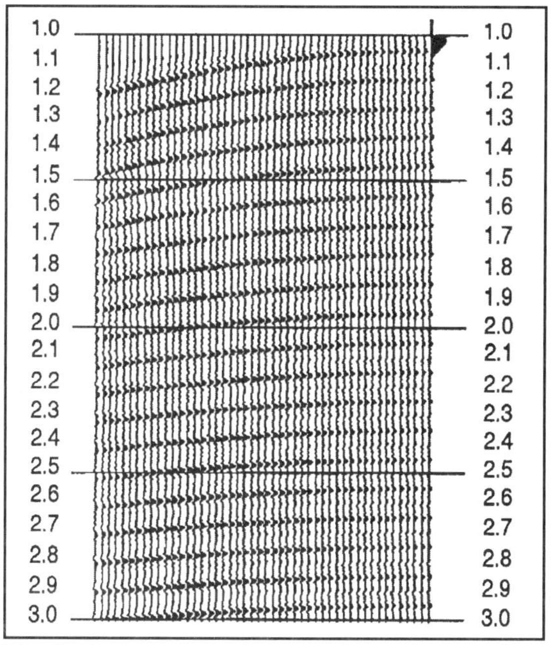

Fig. 6–167 *Inverse Radon transform of Data of Figure 6–166*

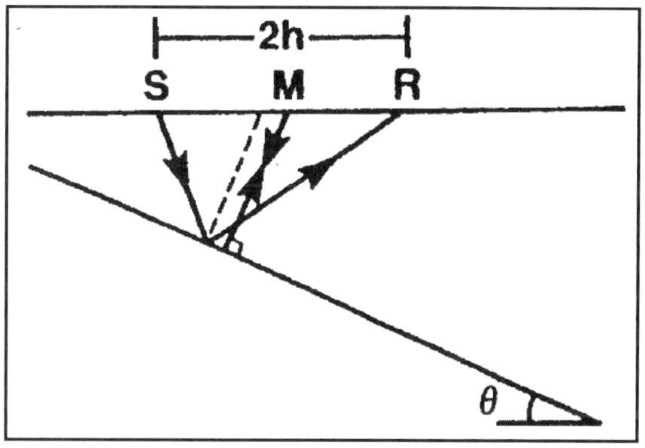

Fig. 6–168 *Reflection from a Dipping Horizon*

Comparing this with Figure 6–164, it can be seen that the under-corrected 1800 m/s events on the two figures appear to be the same, but Figure 6–167 lacks the flattened 2000 m/s events. Thus, subtracting the data of Figure 6–167 from that of Figure 6–164 will leave only the 2000 m/s events.

Pre-stack multiple attenuation summary. A variety of pre-stack techniques are available for attenuation of multiple reflections. Use of these techniques can improve results of velocity analysis and residual statics analysis. Their use can also produce a cleaner CMP stack.

The choice of method to use depends on the attribute that best distinguishes multiples from primaries.

Dip moveout

DMO, also called pre-stack partial migration:

- removes dip dependence of stacking velocity (stacking velocities are higher when dip is present)
- corrects for reflection point smear in CMP gathers caused by dip on the reflector
- allows focusing of crossing reflections (different dips mean different stacking velocities)

Moveout for a dipping reflector. The time t at which the reflection from source S to receiver R is related to half offset h, velocity V, and dip angle θ, (Fig. 6–168) can be expressed as:

$$t^2 = t_0^2 + \frac{4h^2 \cos^2 \theta}{V^2} \tag{6.68}$$

Letting $V_s = V/\cos\theta$ = the stacking velocity, Equation 6.68 can be written as:

$$t^2 = t_0^2 + \frac{4h^2}{V_s^2} \tag{6.69}$$

Since $\cos \theta < 1$, $V_s > V$.

Substituting $\cos^2 \theta = 1 - \sin^2 \theta$ into Equation 6.69 gives:

$$t^2 = \underbrace{t_0^2 + \frac{4h^2}{V^2}}_{\text{NMO}} - \underbrace{\frac{4h^2 \sin^2 \theta}{V^2}}_{\text{DMO}} \tag{6.70}$$

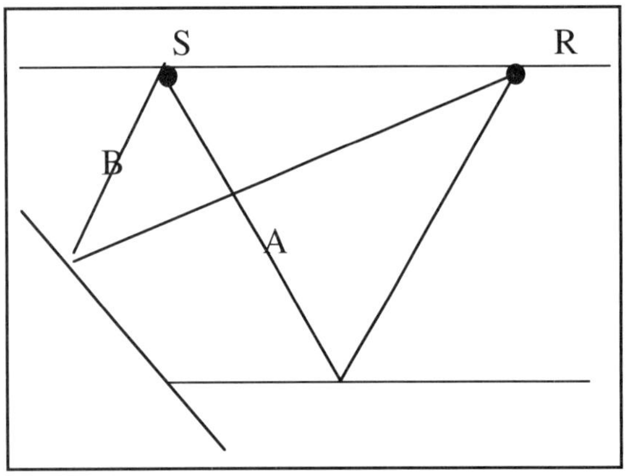

Fig. 6–169 *Conflicting Dips*

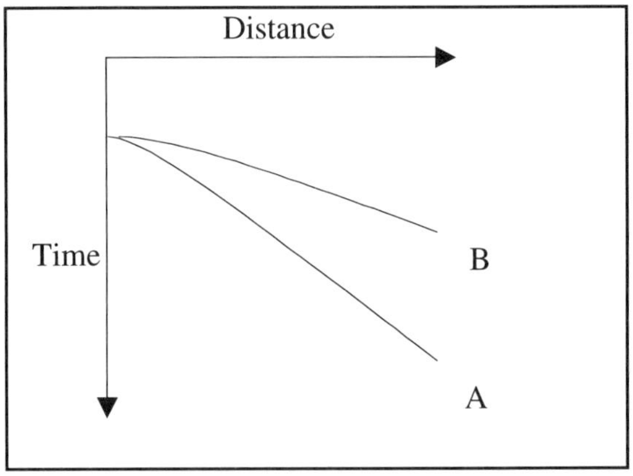

Fig. 6–170 *Moveout from Conflicting Dips*

It can be seen that the effect of dip is to *decrease* moveout. That is, the DMO term $4h^2\sin^2\theta/V^2$ is *always* negative. The DMO time correction increases as offset $2h$ and dip θ increase and as velocity V decreases. The DMO time correction is, thus, stronger for lower velocity structure (generally for shallow data) and for large offsets.

Figure 6–169 shows a situation in which two reflections are recorded at R from a source at S. The first A, is reflected from the flat horizon and the second B is reflected from a dipping horizon (possibly a fault plane). Reflection A will have greater moveout because the NMO velocity is less than that for reflection B—V versus $V/\cos\theta$ This is shown in Figure 6–170.

Depth point smear is another undesirable result of recording CMP data from dipping reflectors. Figure 6–171 illustrates depth-point smear in the presence of dip. The assumption made in CMP stack is that the point of reflection, for all traces corresponds to the zero-offset reflection point—where a line from the CMP perpendicular to the reflector intersects it. However, Snell's law requires that the angle of reflection equal the angle of incidence. As a result, the reflection point moves farther *up-dip* as offset between source and receiver increases so that, instead of a single reflection point for all CMP traces, the reflection points are spread out over a distance up-dip from the assumed common reflection point. In fact, it is possible that reflection points from one CMP set overlap those of another set.

In the case of deep-seated synclines, it is possible that energy reaches the receiver from more than one ray path at different times—this is in accord with Fermat's Principle. Figure 6–172a shows true zero-offset ray paths for coincident sources and receivers at surface points A through G. The curvature of the reflecting horizon is such that there are multiple perpendicular ray paths from surface positions B through F. On the stack section traces, it appears as two reflection events with an apparent anticline below them. This is called a *buried focus* or *bow tie effect* and is illustrated in Figure 6–172b.

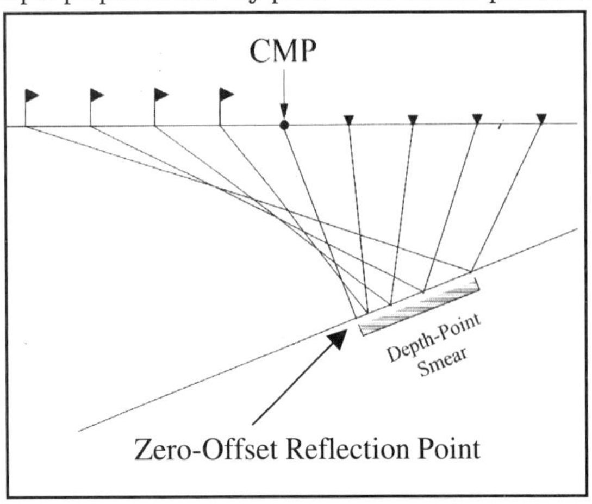

Fig. 6–171 *Depth-point Smear in the Presence of Dip.*

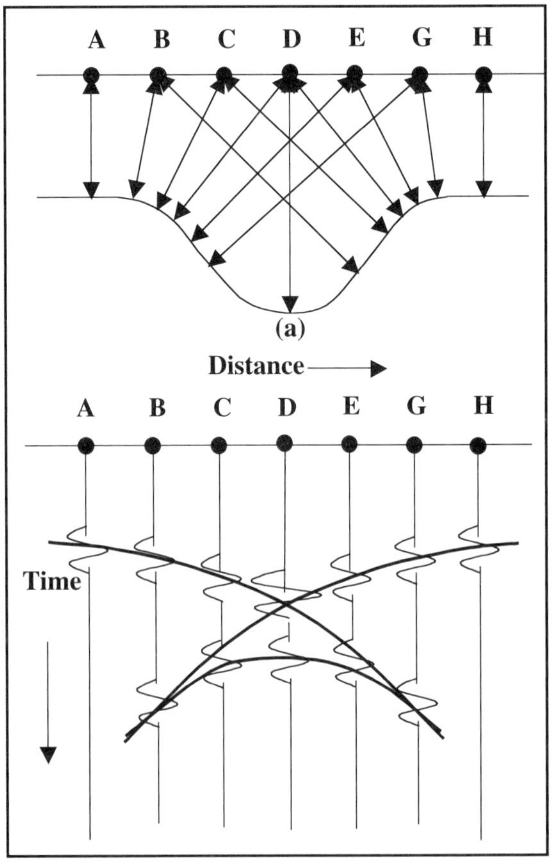

Fig. 6–172 *Zero-offset Ray Paths for a Synclinal Reflector and Appearance on CMP Stack*

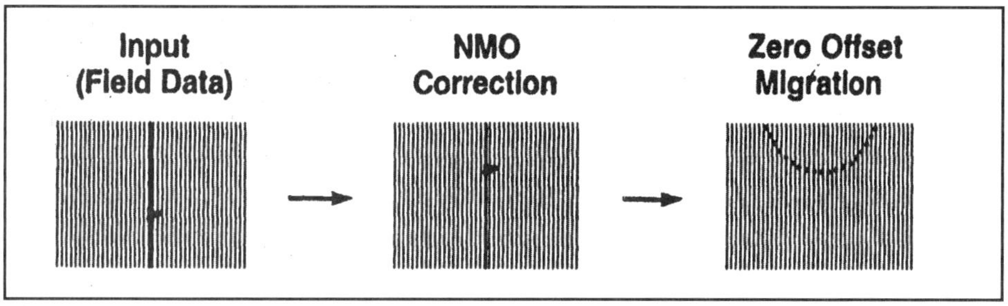

Fig. 6–173 *Conventional Processing*

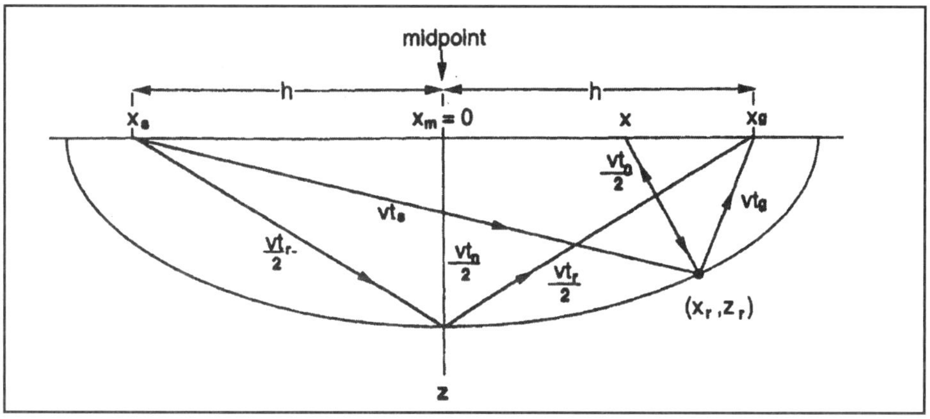

Fig. 6–174 *Locus of Non-zero Reflection Points*

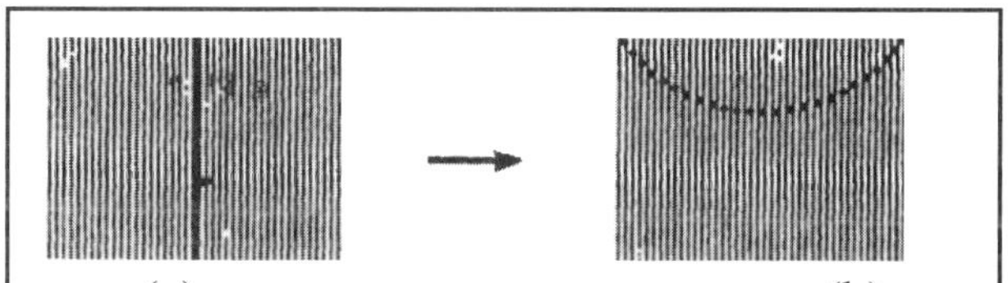

(a) (b)

Fig. 6–175 *Pre-stack Migration of a Spike Showing Input and Proper Migration*

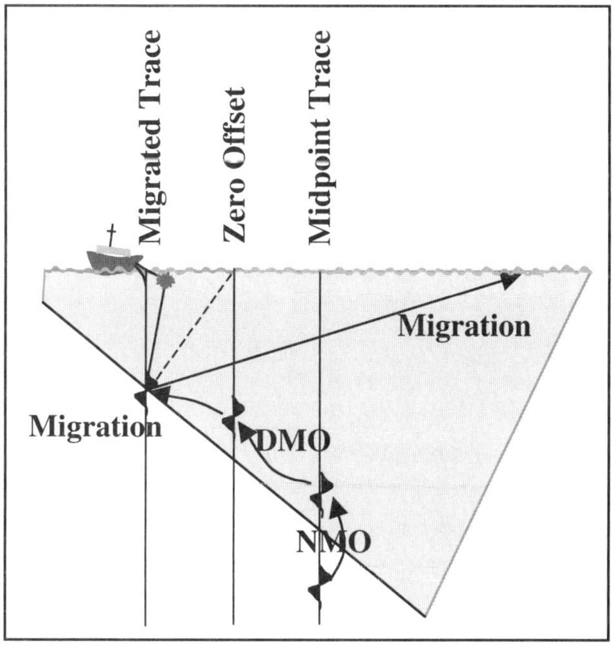

Fig. 6-176 *Three-step Imaging*

Data processing with DMO. Conventional data processing applies NMO correction (which really assumes zero dip) and then stacking in the offset plane. This is followed by zero-offset migration in the midpoint plane. Conventional processing is illustrated in Figure 6–173 for a constant-offset section and a constant velocity medium.

Field data are represented by a single event on the center trace. NMO correction moves the event to an earlier time on the same trace. Zero-offset migration results in convolution of the migration operator with the input event. The result is a semicircular pattern of events since zero-offset migration assumes the event lies somewhere on a semicircular *isochron.*

A problem with this approach is that data were not recorded at zero offset. Whereas the locus of all reflection points for zero-offset data is a semicircle, for non-zero offset, the locus of all reflection points is an elliptical isochron. See Figure 6–174.

Figure 6–175 illustrates proper migration—migration that comprehends that data were collected with non-zero offset and thus, the locus of all reflection points is an ellipse. Application of the migration impulse response to the spike in Figure 6–175a results in the elliptical event of Figure 6–175b. This is pre-stack migration.

Pre-stack migration algorithms are relatively slow and computationally more difficult than post-stack migration. An alternative approach is shown in Figure 6–176.

In this approach, the complete imaging process has three steps. Before stack the reflection point is assumed to be vertically below the midpoint *M* at time T_x. The NMO correction changes the reflection time to zero-offset time but leaves the apparent reflection below the midpoint. DMO moves the apparent reflection to the zero-offset trace. After stack, migration moves the reflection to the trace that represents the correct position and at the correct vertical time.

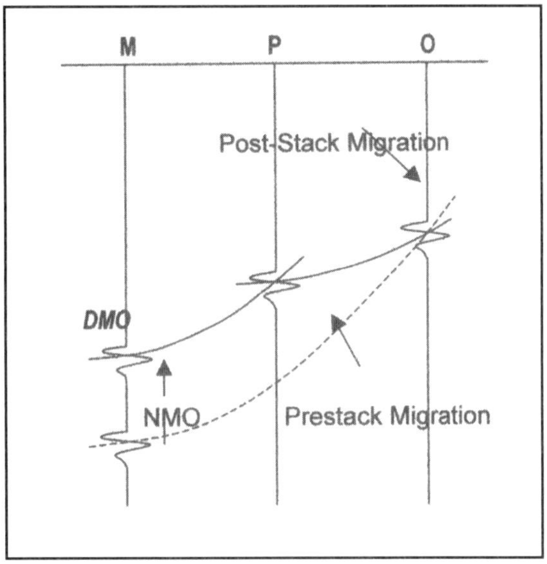

Fig. 6–177 *Post-stack and Pre-stack Migration*

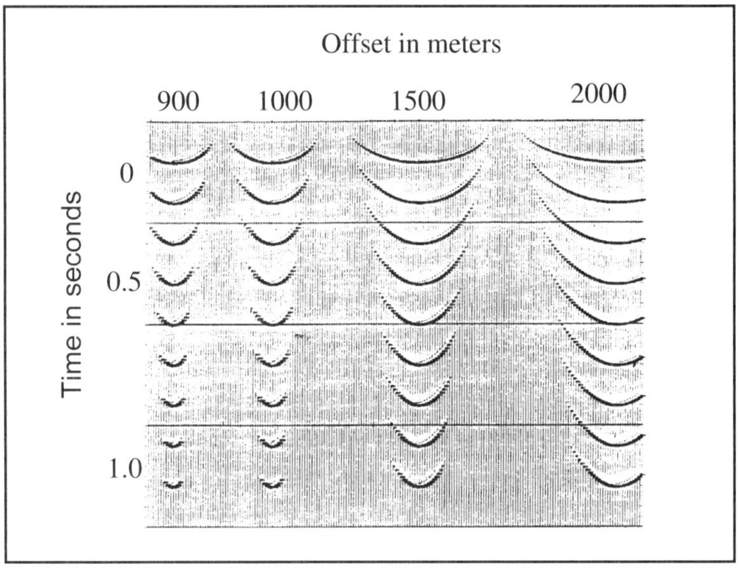

Fig. 6–178 *Kirchhoff Impulse Response*

Figure 6–177 compares the two imaging techniques, Pre-stack partial migration, also called DMO, followed by post-stack migration and full pre-stack migration.

Figure 6–178 relates the DMO impulse response to acquisition geometry. The DMO ellipse is given by:

$$\frac{t_0^2}{t_x^2} + \frac{x^2}{h^2} = 1 \tag{6.67}$$

Where t_0 is the NMO-corrected time at zero offset t_x = NMO-corrected time at offset x, and h is the half-offset.

The three-step approach is a desirable alternative approach because pre-stack migration does not produce an intermediate stack. Pre-stack migration also needs very accurate velocities for correct imaging and lacks a procedure to improve the velocity data. DMO offers opportunity for improved (migration) velocity analysis. It should also be noted that DMO is a small correction.

The DMO impulse response varies with time and offset. This is illustrated in Figure 6–178, which shows the DMO impulse responses created by the Kirchhoff algorithm for offsets of 900, 1000, 1500, and 2000 m. Note that the response broadens laterally with increasing offset and shrinks with increasing time.

DMO is usually performed in the common offset domain. The NMO correction moves data upward in time to the (assumed) CMP position, and DMO moves data to the zero offset position by moving to another (up-dip) trace at an earlier time. See Figure 6–179.

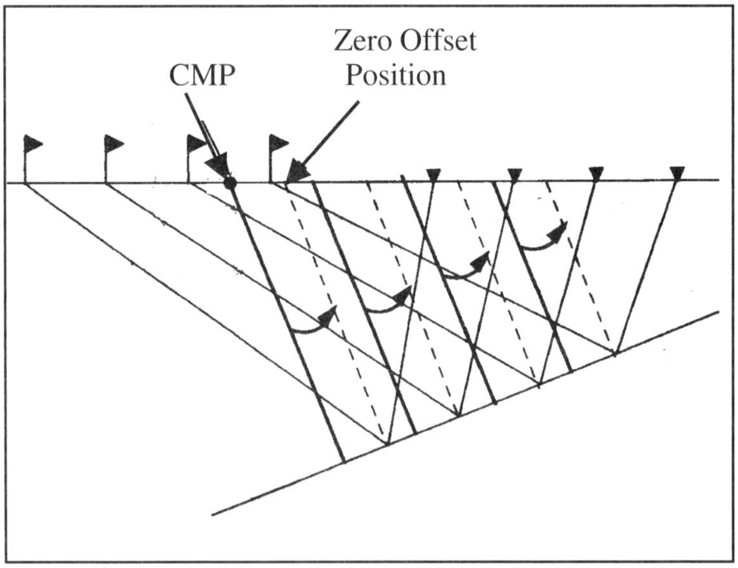

Fig. 6–179 *DMO in the Common Offset Domain*

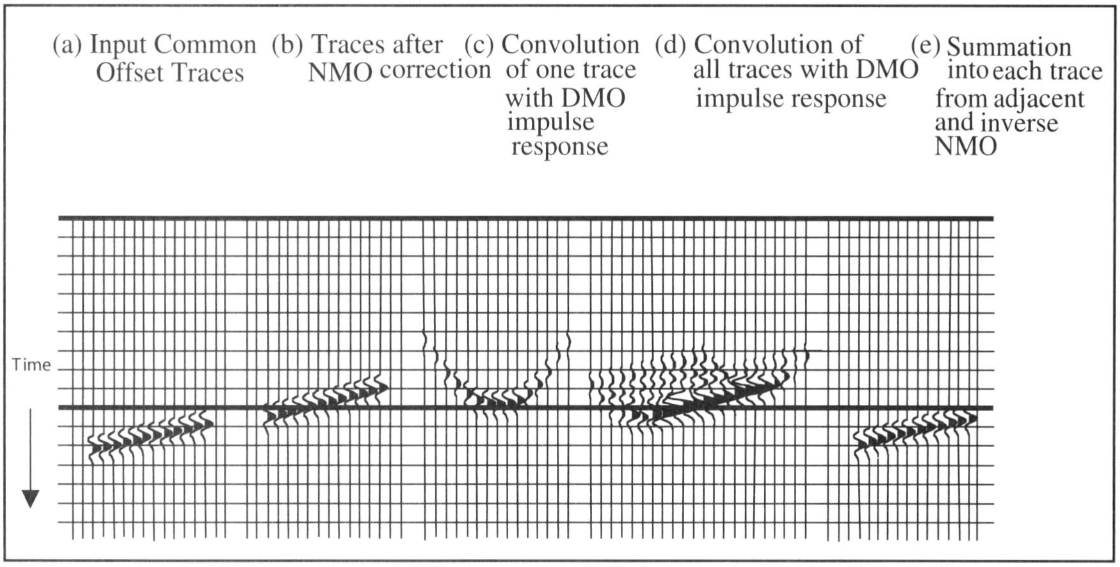

Fig. 6–180 *The DMO Process*

Figure 6–180 illustrates the Kirchhoff DMO approach. This is performed in the time domain, usually on traces in the common-offset domain, and involves the following steps:

- sort into common offset domain
- apply NMO corrections
- convolve each trace with the Kirchhoff DMO impulse response to distribute amplitude contributions along the DMO ellipse
- sum adjacent traces into the trace being processed, one trace at a time

DMO can also bc applicd in thc *F-K* domain.

If DMO is applied after pre-processing is completed, then the following processing sequence is recommended:

1. CMP sort
2. preliminary velocity analysis
3. NMO correction (using preliminary velocities)
4. common offset sort
5. DMO
6. CMP sort
7. inverse NMO correction (add back those times subtracted in step 3)
8. velocity analysis
9. NMO correction (using revised velocities)
10. stack
11. migration

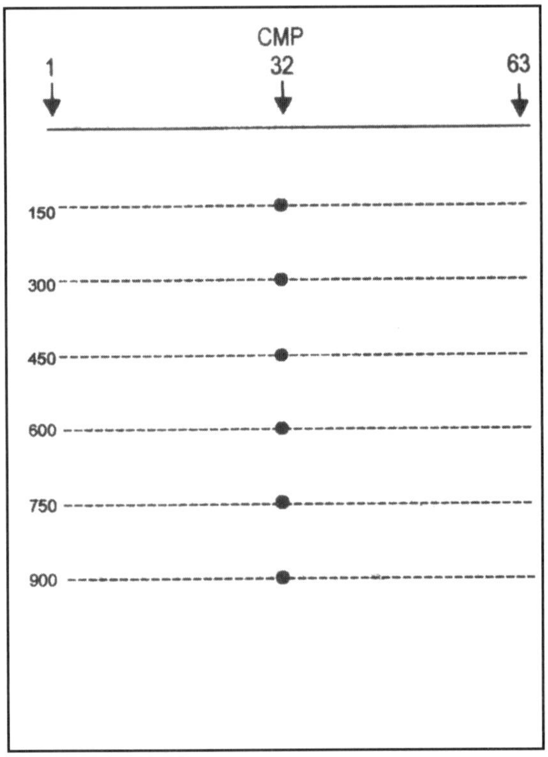

Fig. 6–181 *Depth Model of Six Point Scatterers Buried in a Constant Velocity*

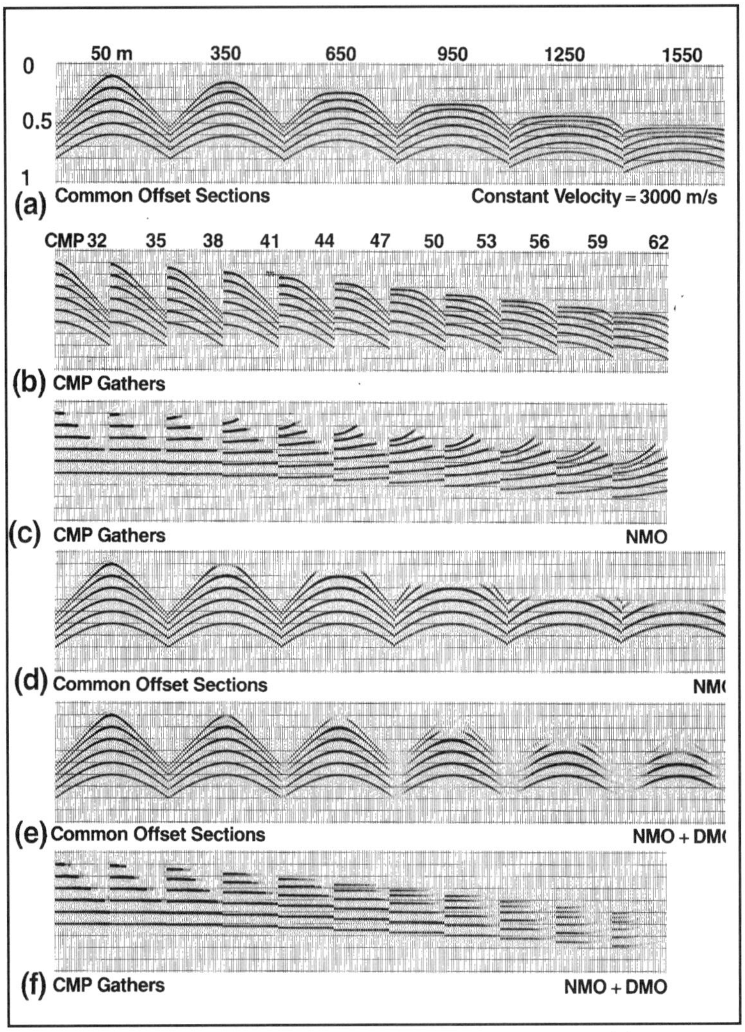

Fig. 6–182 *DMO Example Using Synthetic Data Derived from Model of Figure 6–181*

Synthetic example. The first illustration of DMO uses the depth model of Figure 6–181. The point scatterers are located vertically below CMP number 32, as indicated by the filled circles. These will result in the recording of diffractions.

Selected common offset sections from the model of Figure 6–181 are shown in Figure 6–182a. With increasing offset, the diffraction curves become more and more flat. CMP gathers are shown in Figure 6–182b. Note that the travel times at the center midpoint are perfectly hyperbolic while the travel times at CMP gathers away from the center CMP become more non-hyperbolic with distance from the center CMP. The remaining parts of Figure 6–182 show the results of DMO processing of these data.

Figure 6–182c shows the NMO-corrected and muted CMP gathers. Since the medium velocity (3000 m/s) was used for NMO correction, diffractions near the center midpoint (CMP 32) are flat after NMO correction and those recorded away from the center mid-point are increasingly over-corrected.

Figure 6–182d shows the NMO-corrected gathers of Figure 6–182c after sorting into common offset sections for DMO processing. Figure 6–182e shows the result of DMO correcting each common offset section and Figure 6–182f is the NMO- and DMO-corrected data re-sorted to CMP gathers. Figure 6–182e, showing the data after NMO and DMO corrections, is approximately equivalent to the zero-offset section.

Unlike conventional migration, the effect of DMO is greater for shallow depths. (Recall Equation (6.70),

$$t_x^2 = t_0^2 + \frac{4h^2}{V^2} - \frac{4h^2 \sin^2 \theta}{V^2}$$

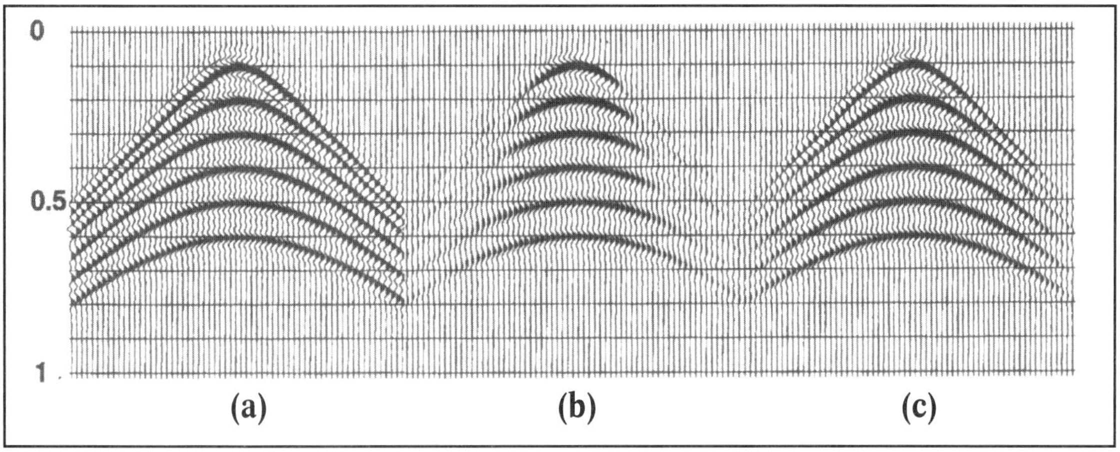

Fig. 6–183 *Zero-offset Section Associated with the Depth Model in Figure 6–181 is Seen in Here (a), (b) Shows a Stack of CMP Gathers from Figure 6–182c, and (c) Shows a Stack of DMO-Corrected CMP Gathers from Figure 6–182f. (Courtesy of WesternGeco.)*

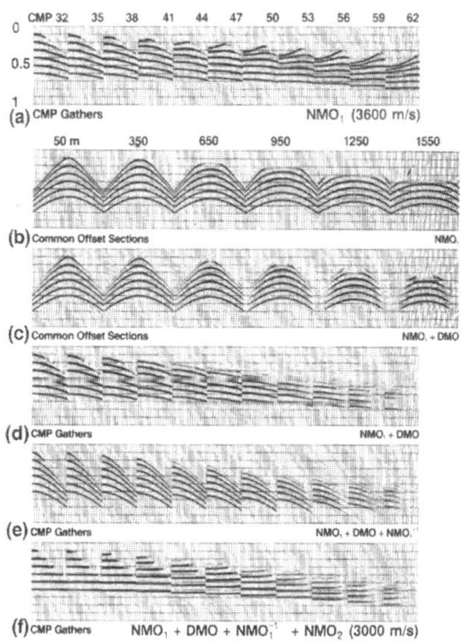

Fig. 6–184 *DMO Processing Stages for Synthetic Data of Figure 6–181 Using Incorrect Velocity for NMO Correction*

The effect of DMO is also greater at increasingly large offsets, and it does nothing to the zero-offset section. DMO is similar to conventional migration in that the steeper the event, the greater the partial migration that takes place. Flat events are not affected.

Figure 6–182f shows data re-sorted into CMP gathers after DMO correction. Comparison of this to the CMP gathers without DMO correction in Figure 6–182c reveals that the combination of DMO and NMO correction results in flattened events rather than overcorrected events.

Use of a low velocity for NMO correction results in considerable over-correction, especially at CMPs located farther from the point scatterers. In Figure 6–182f, DMO has eliminated the effect of dip on velocity, resulting in Figure 6–182c, which is the stack with DMO applied, resembles the desired zero-offset section of Figure 6–183a than does the stack without DMO in Figure 6–183b.

Figure 6–184 also shows DMO processing stages on the synthetic data obtained from the depth model of Figure 6–181. In this case, however, the medium velocity (3000 m/s) was not used. Instead, a velocity of 3600 m/s was used for NMO correction. This higher velocity corresponds to the kind of stacking velocity that might be obtained from steeply dipping data. Thus, in Figure 184a, data near CMP 32 and shallower data are under-corrected while the deeper and more steeply dipping data are flattened or overcorrected. The effect of using the higher velocity is quite apparent from the comparison of Figures 6–184a through 184d with Figures 6–183c through 183f.

Figure 6–184e shows the CMP gathers of Figure 6–184d with the original NMO correction backed out. In a real situation, these data would be used for another velocity analysis that would yield dip-independent velocities. In the case of these synthetic data, the equivalent is to use the medium velocity of 3000 m/s for NMO correction, yielding flattened data. Note that a stack of NMO and DMO corrected data of Figure 6–184d would not give a stack any superior to that obtained from the stack of Figure 6–182c.

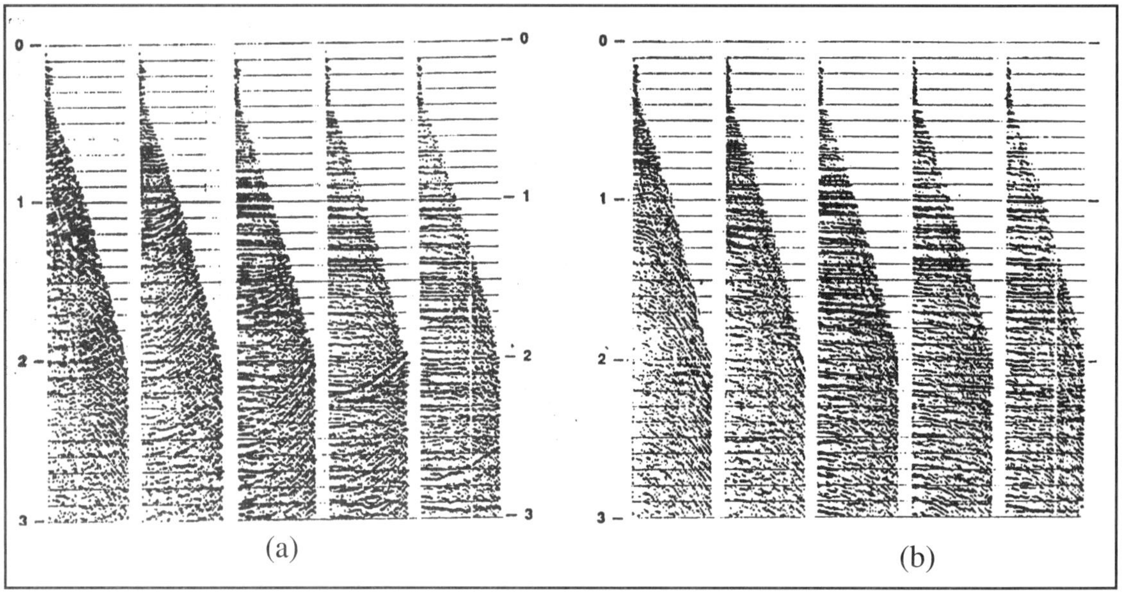

Fig. 6–185 *CMP Gathers (a) without DMO and (b) with DMO*

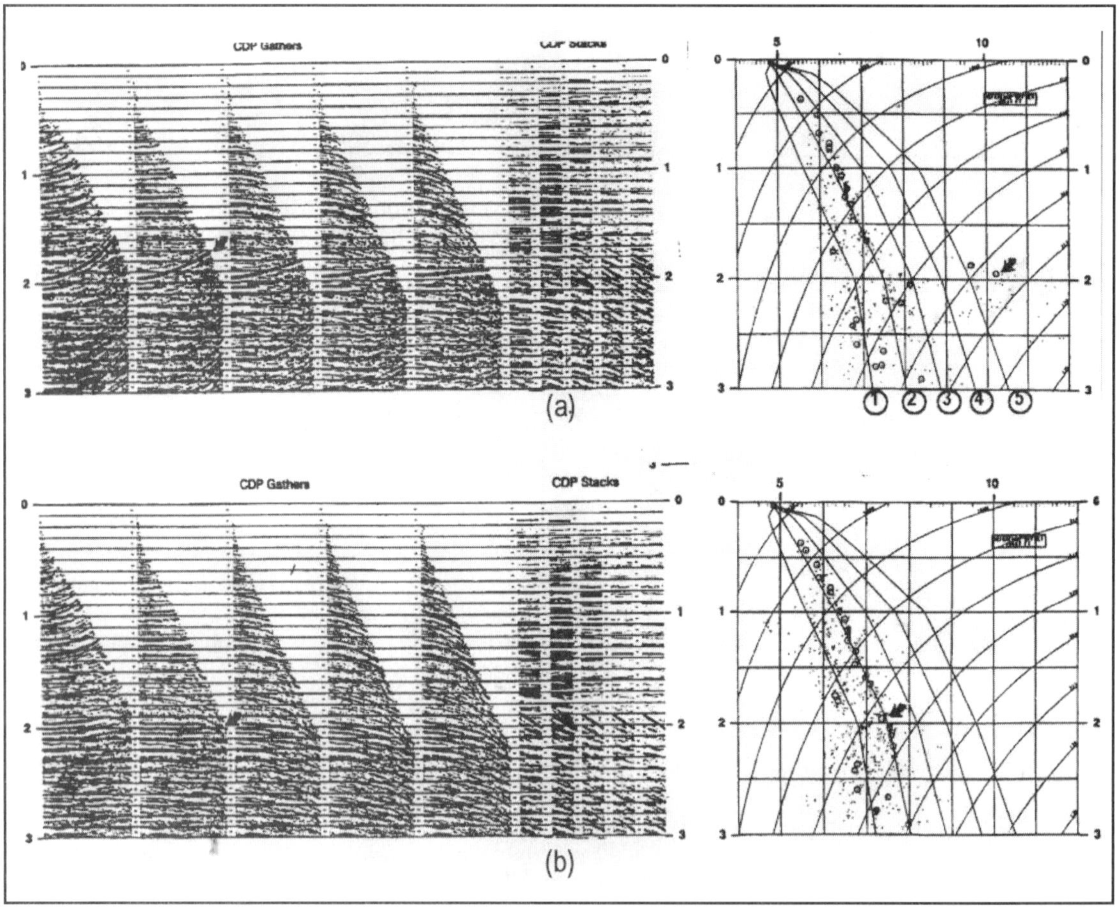

Fig. 6–186 *Velocity Analysis Displays (a) without DMO and (b) with DMO*

Field data examples. Figure 6–185 shows a set of CMP gathers that apparently require two different NMO velocities at the same time. In Figure 6–185a, DMO has not been applied, and many overcorrected events can be seen, especially at the larger offsets. The problem is *conflicting dips*. That is, we have a situation similar to that of Figure 6–169. Events recorded from reflectors with different dips will have NMO velocities $V/\cos\theta_1$ and $V/\cos\theta_2$, where θ_1 and θ_2 are the two different dip angles, and V is the velocity for a flat reflector. When DMO is applied, events with conflicting dips are separated and the need for different NMO velocities is eliminated, as shown in Figure 6–185b.

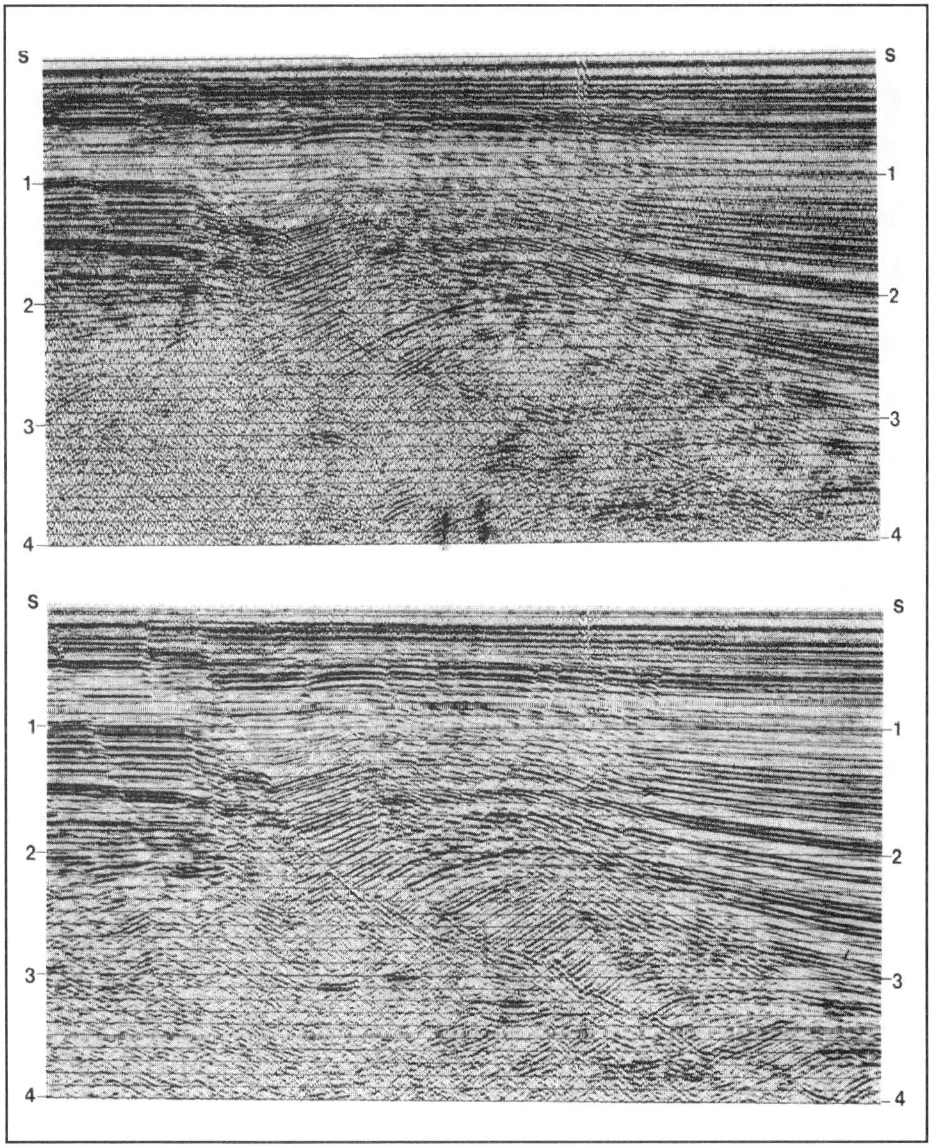

Fig. 6–187 *Data Acquired near a Major Fault, Conventional CMP Stack (top), and Post-stack Migration (bottom).*

Figure 6–186 shows the effect of DMO on velocity analysis. Figure 6–186a shows the traces and scattergram display of an analysis done without DMO. The arrows indicate an event on the center CMP at about two seconds that is under-corrected when using velocity function two. The stack traces for this velocity function show weak flat and dipping events at about this time. The CMP traces for velocity function five show a nearly flat event and an overcorrected event at this time. The stack traces for velocity function five show a rather strong dipping event at about two seconds. The arrow on the scatter-gram of Figure 6–186a indicates this event has an anomalously high velocity.

Figure 6–186b shows the same trace and scatter-gram displays after DMO has been applied to the traces. The arrows, again, point to the event at about two seconds. Note that the event is now flat when velocity function two is used for NMO correction. Also, the event on the scatter-gram has moved from the very high value to the velocity function two values. Note that the stack traces for velocity function two show a strong dipping event at about two seconds now. Thus, it is clear that DMO has eliminated the effect of dip on velocity by moving the dipping event to its zero offset position.

Figure 6–187a is a conventional CMP stack of data shot in the vicinity of a major fault, which dips down from left to right. There are also many auxiliary faults. Figure 6–187b is the migration of this stack. While evidence of the faults can be seen, the fault planes are not well-defined. This is because the fault planes dip quite steeply so reflections from them appear to have high stacking velocities. In picking the velocities for the conventional stack, these high velocities are ignored, and, as a consequence, reflections from the fault planes are not seen on the section.

Figure 6–188 illustrates the improvement in velocities obtained from application of DMO by showing displays of the analysis done at the position indicated by A in Figure 6–187a. Figure 6–188a and Figure 6–188b show the display of the analysis before and after DMO, respectively. DMO not only eliminates the peaks at anomalously high velocities but also clarifies the velocity trend, making interpretation easier and more accurate.

Figure 6–189a shows the stack of the same data as in Figure 6–187a but with DMO applied before stack. Figure 6–189b is the migration of the stack in Figure 6–189a. Applying DMO resolves the conflicting dip problem, which accomplishes at least two things:

- reflections from the steeply dipping fault planes are now evident
- better velocities are obtained since velocities are now single-valued (no more conflicting dips after DMO)

The migration in Figure 6–187b does improve the picture over the stack in Figure 6–187a, but, because reflections from fault planes were lacking, the fault planes are still not as well-defined as in the migration in Figure 6–189b. Here, the presence of fault plane reflections and the better-defined velocities give far better definition of the faults.

Figure 6–190b is a pre-stack migration of Figure 6–187a. The post-stack migration of Figure 6–189 is repeated in Figure 6–190a for ease in comparison. This comparison shows that, at least in this case, pre-stack migration and post-stack migration of data to which DMO has been applied results in imaging of nearly the same quality.

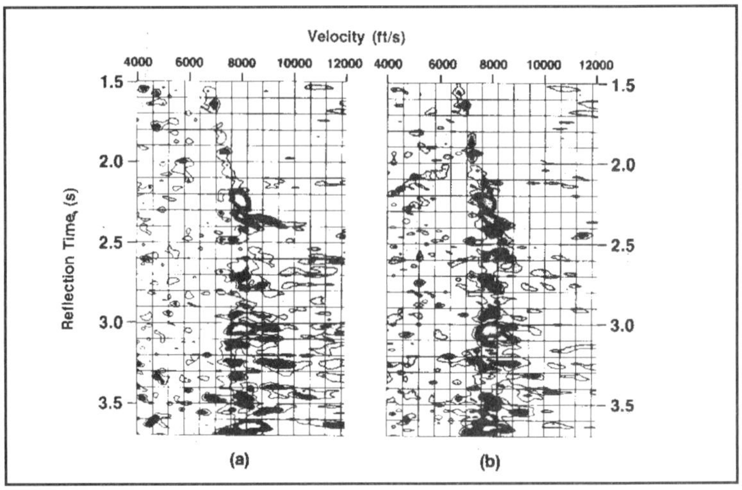

Fig. 6–188 *Velocity Analysis Displays Contoured in Correlation Coefficients*

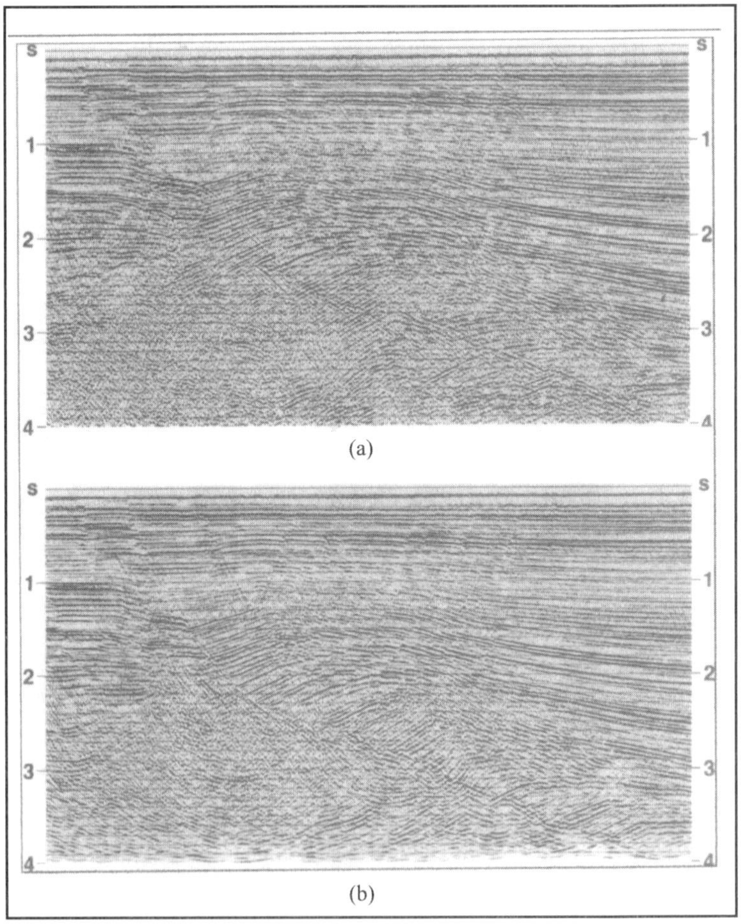

Fig. 6–189 *(a) Data of Fig. 6-187a after DMO and (b) Migration of Data in Fig. 6-189a.*

CMP Stack

Before the actual stack begins, apply NMO correction using the final velocity model or function and final statics including final round of residual statics. Next, sum all traces in CMP gather, sample time-by-sample time. Divide each sum by the number of live traces in sum, including fractional parts resulting from mute ramp application. Figure 6–191 illustrates how event alignment or misalignment affects CMP stack output.

Figure 6–191a shows the situation desired for primary reflections—total *constructive* interference. Figure 6–191b is the situation desired for noise attenuation—total *destructive* interference.

Figure 6–191c is the situation that gives partial attenuation. The presence of residual statics or residual NMO causes this to happen—the output is slightly lower amplitude, appears to be lower frequency and T_0 is at an intermediate time. The situation also often occurs at near offsets for multiples, especially peg leg or interbed multiples because residual NMO is small.

The situation in Figure 6–191d provides 6 dB of attenuation. At higher folds, greater attenuation results. For example, 30-fold provides almost 30 dB of attenuation.

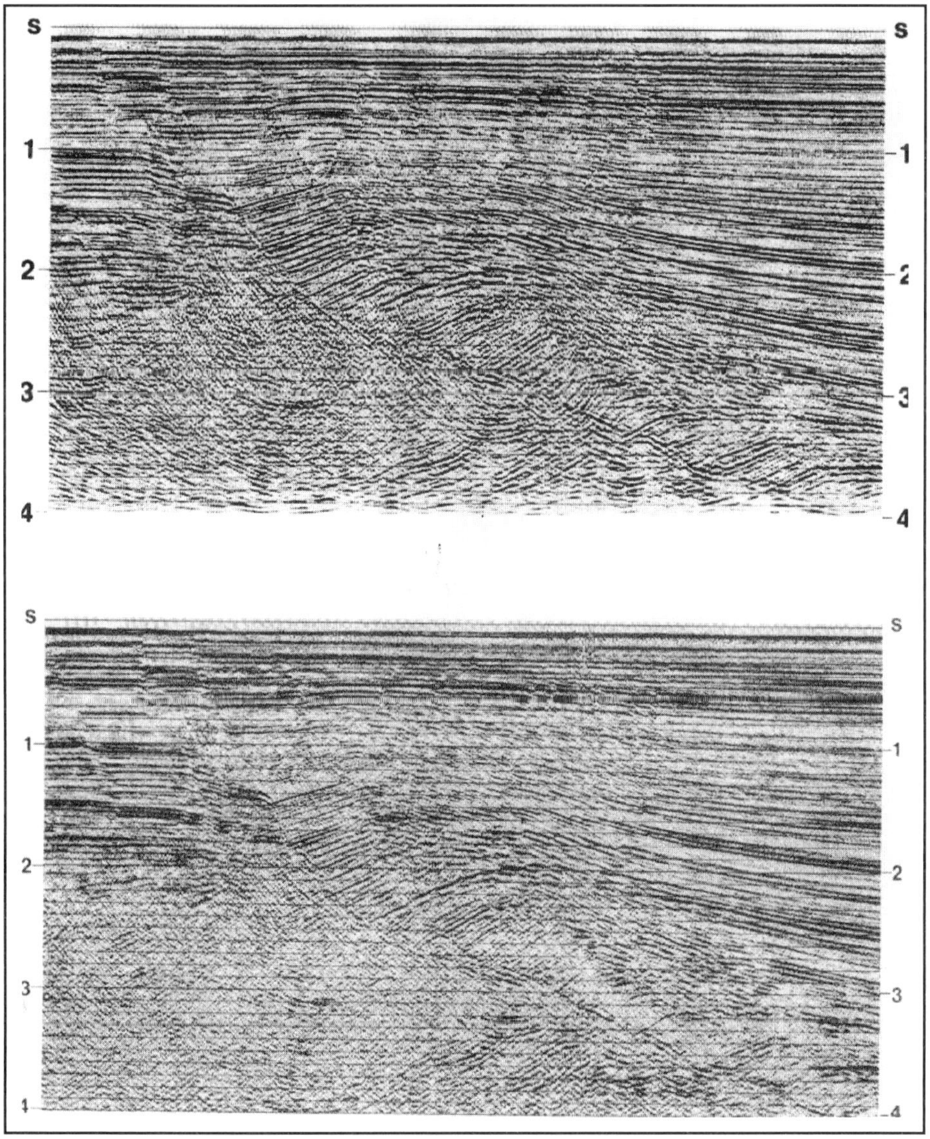

Fig. 6–190 *Data of Figure 6–187 after DMO, CMP Stack, Post-stack Migration (top), and DMO and Pre-stack Migration (bottom).*

Figure 6–153 shows a CMP stack that has residual multiples. After NMO corrections were applied to the CMP gathers, residual NMO was small, at least at near offsets. Only partial attenuation was obtained, as in Figure 6–191c. Figure 6–190a is an example of a good CMP stack. Signal-to-noise ratio is good and structure can be seen. However, some structural distortion is inherent in CMP stacks because all reflection points are assumed to be vertically below the CMP positions. Migration is required to remove these distortions and show reflection ponts in their true positions.

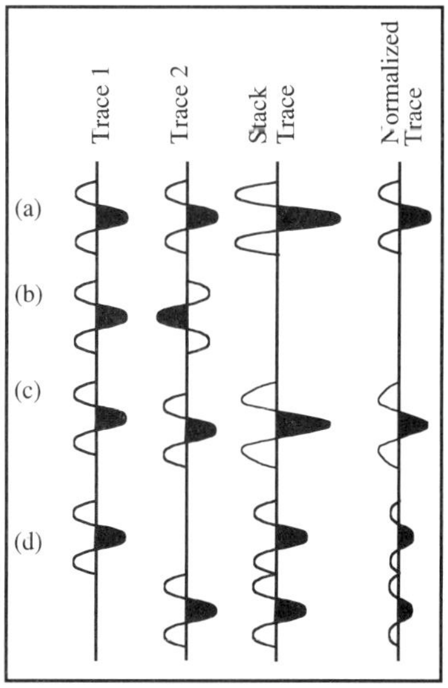

Fig. 6–191 *Effect of Event Alignment on CMP Stack*

Post-stack Multiple Attenuation

Deconvolution, particularly gapped TVD, is applicable to post-stack multiple attenuation and is frequently used for that purpose. Other methods, based on amplitude differences between primary and multiple reflections, can be used. These are not, strictly speaking, post-stack methods but are altyernative stack methods. These are:

- median stack
- optimum trace-weighted stack

Median stack will be described later in some detail, but optimum trace-weighted stack will be discussed only briefly as it is rarely used today.

Median stack. The basis of median stack is statistical. That is, it has been found that in any sample of any CMP set of trace, primary reflection amplitudes cluster around the median value of all the amplitudes for that sample. *Median* is a statistical parameter. The median value for a set of numbers is that value for which the number of values that are smaller equals the number of values that are larger. Note that the median value may differ from the average value, which is simply the sum of all trace amplitudes in the data set divided by the number (47 in the example) of traces summed.

Table 6–3 shows the recorded amplitudes for two consecutive sample periods for a particular CMP. Data were recorded at 48-fold but trace 75 was bad and, therefore, zeroed. Amplitudes have been sorted from smallest (most negative) to largest (most positive).

Table 6–3 *Sorted Trace Amplitudes—Median and Average Values*

Time =4816 ms				Time = 4820 ms			
Trace	Amplitude	Trace	Amplitude	Trace	Amplitude	Trace	Amplitude
81	-32768	7	1811*	73	-32768	51	1165*
85	-32382	15	1937	77	-31202	9	1175
83	-31719	13	1959	79	-24774	59	1356
79	-28984	59	2074	71	-15522	39	1559
87	-25051	33	2135	81	-14637	43	1647
77	-22897	41	2466	47	2857	15	1648
89	-13719	43	2545	85	-2114	23	1890
93	-6321	61	2693	13	-1273	31	1996
91	-955	63	2812	55	-1034	21	2068
9	-119	11	2828	1	-997	27	2197
1	269	45	3189	35	-422	33	2604
19	743	49	3203	37	-300	41	2714
21	753	17	3398	25	-189	19	3140
51	810	35	3437	45	-116	3	3398
31	821	23	3449	57	109	69	5662
39	1035	53	4223	5	125	61	5683
27	1097	3	4691	53	418	87	8161
47	1358	73	4724	29	468	63	12670
25	1415	55	5138	7	548	65	18656
5	1482	95	5682	11	618	67	19238
57	1497	65	7097	17	957	89	26557
29	1595	71	15541	49	1034	91	30070
37	1609	67	16341	83	1064	93	30416
		69	19	664		95	32767
		Average	-3839			Average	-2614

For each sample time, the median value is indicated by an asterisk to the right. The median value for the 4816 ms sample is 1811 millivolts (mv) compared to the average value of -3839 mv. For the 4820 ms sample the median is 1165 mv and the average is –2614 mv. The average value is the value that would be obtained in a conventional CMP stack. If the assumption that true signal amplitude is close to the median value is correct, then the conventional stack amplitudes do not represent the desired output.

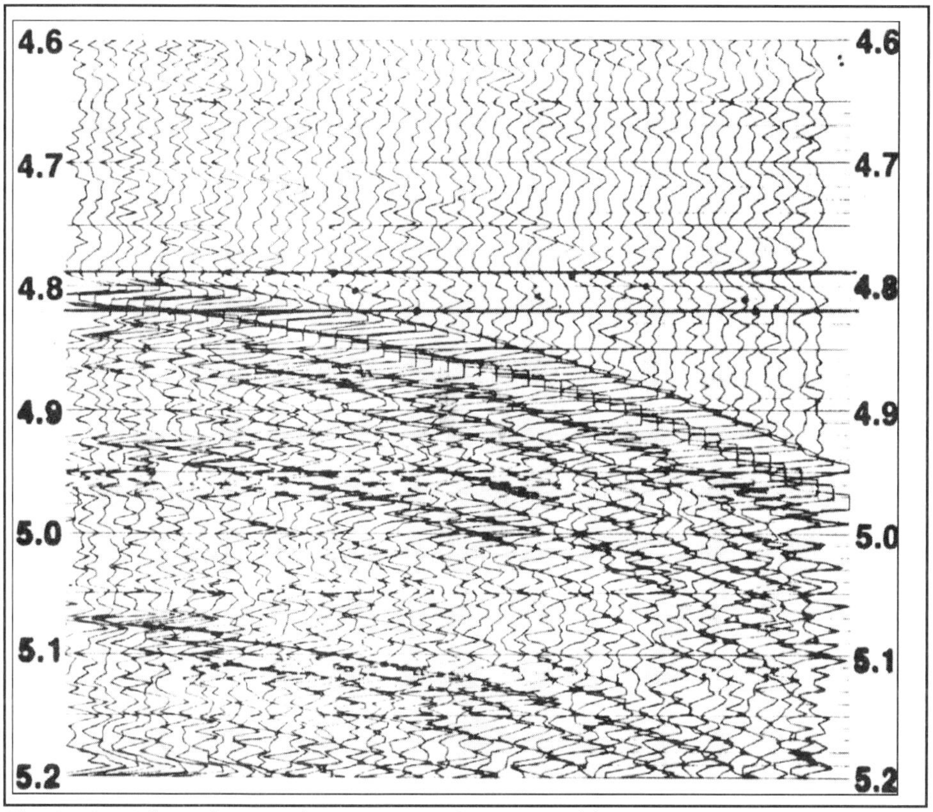

Fig. 6–192 *Enlarged View of Time Scans Shown in Table 6–3*

Figure 6–192 shows an enlarged view of the time scans shown in Table 6–3.

An alternative approach is to select only those amplitudes near the median value, sum them and divide by the number of amplitudes summed. There are two variations to this approach.

- *Two-Sum*—The three values closest to the median for each sample of each CMP are summed and divided by three. All other amplitudes are discarded. In Table 6–3, these values are printed in boldface and underlined.
- *Eight-Sum*—The eight values closest to the median for each sample of each CMP are summed and divided by eight. All other amplitudes are discarded. In Table 6–3, the five values printed in boldface are used in addition to the three that are also underlined.

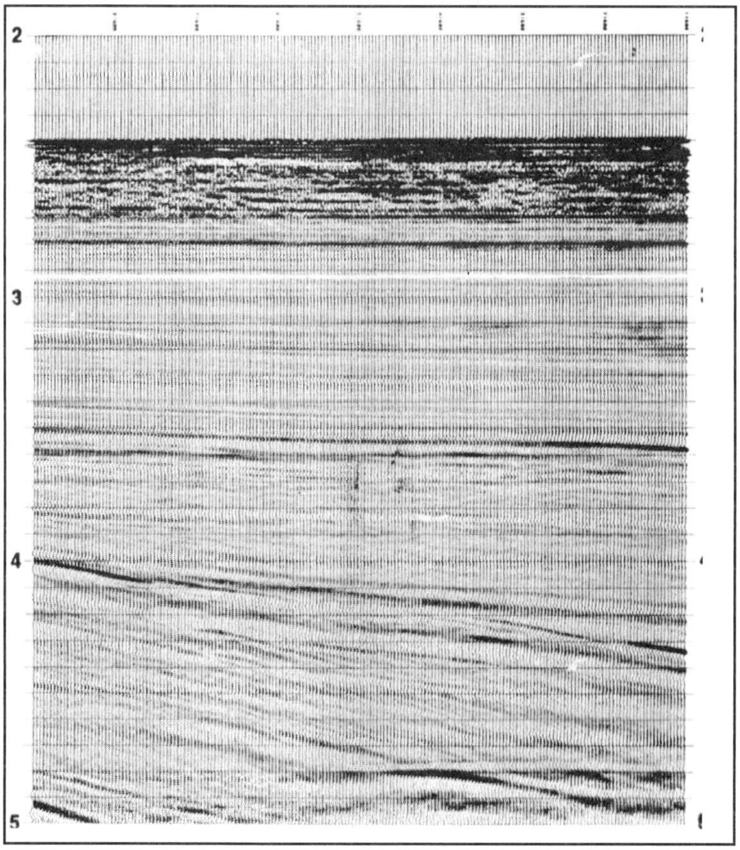

Fig. 6–193 *Two-sum Median Stack*

Figure 6–193 shows the output of a two-sum median stack. This was in relatively deep water as the water bottom reflection is at about 2.4 seconds. There is no evidence of the water bottom multiple at about 4.8 seconds. Notice the rather sharp reflection wavelets that provide good resolution.

Figure 6–194 shows the eight-sum output for the same data. This output is quite similar to the two-sum. Between 4 and 5 s signal is stronger and random noise appears to be slightly lower amplitude because of the increased averaging (sum of eight versus sum of three). Some reflection wavelets, however, do not appear to be quite as sharp.

Figure 6–195 shows another output for the same data. In this case, there was a partial stack (summing of four adjacent traces) before the two-sum median stack was done. This mixing process smeared noise, multiples, and primaries together. As a result, the basic assumption in the median stack was invalidated. A strong water-bottom multiple at just past 4.8 seconds can be seen on this section.

Figure 6–196 compares a conventional CMP stack with a median stack in *a* and in *b* for cases where there is strong water-bottom multiples from a steeply dipping water bottom. These are clearly evident on the conventional stack but virtually absent on the median stack.

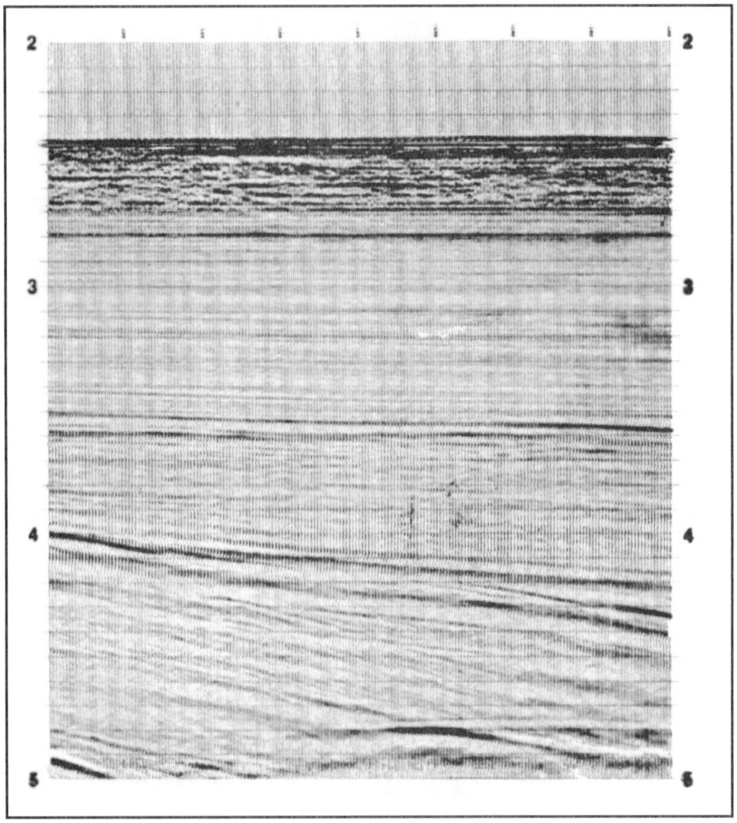

Fig. 6–194 *Eight-sum Median Stack*

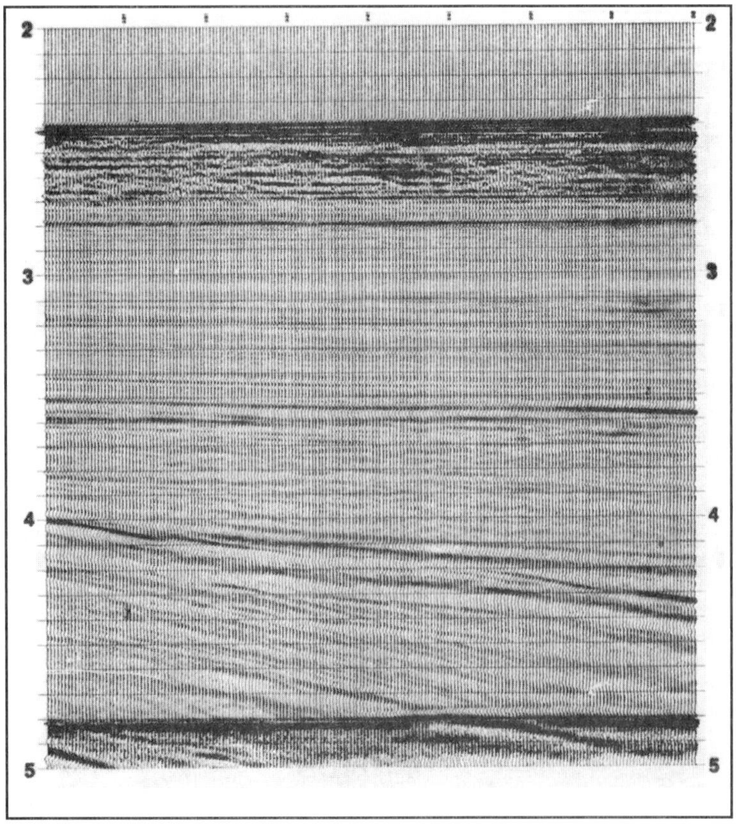

Fig. 6–195 *Partial Stack to 12–Fold Then Two-sum Median Stack*

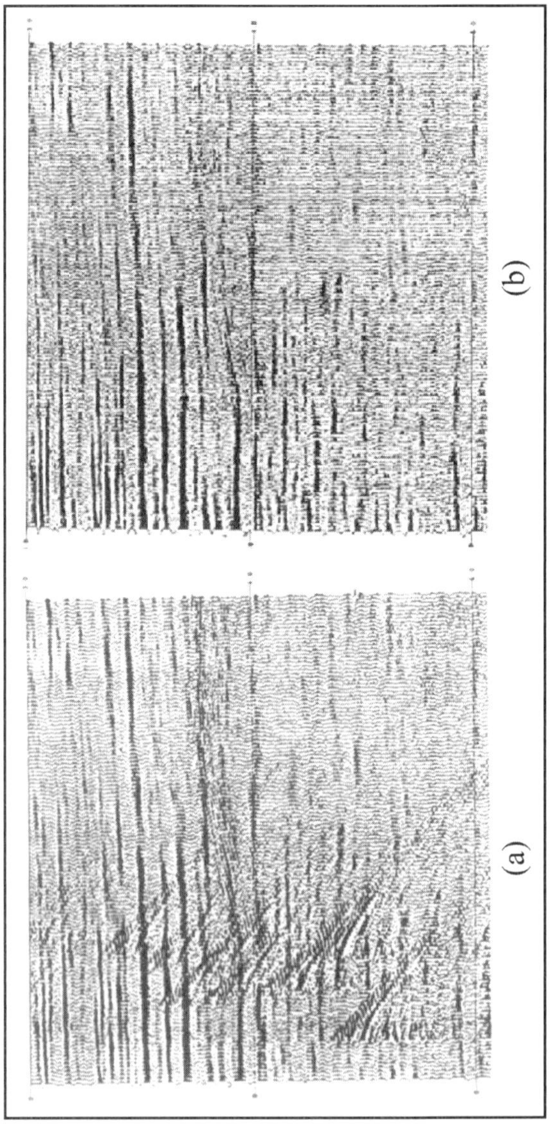

Fig. 6–196 *Comparison of Conventional CMP Stack and Median Stack*

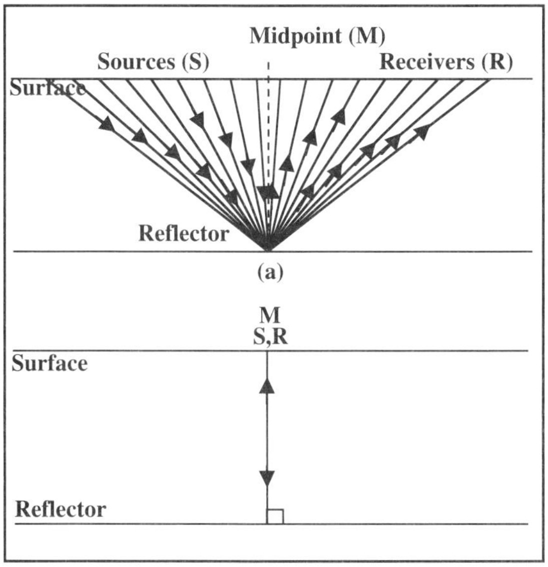

Fig. 6–197 *CMP Traces before NMO Correction and Stack – Flat Reflector and CMP Traces before NMO Correction and Stack*

Optimum trace-weighted stack. It has been found that on CMP gathers with both primaries and multiples present; multiple amplitudes tend to be larger than primary amplitudes at some offsets (generally at the shorter and longer offsets) while the reverse is true at other offsets. Optimum trace-weighted stack takes advantage of this by developing a set of weights that are multiplied times the recorded sample values before stacking.

The output of the Optimum trace-weighted stack can be expressed, mathematically as:

$$o_w(t) = \frac{\sum_{j=1}^{N} w_j x_j(t)}{\sum_{j=1}^{N} W_j} \qquad (6.71)$$

where

$o_w(t)$ = output of optimum trace-weighted stack at time t

w_j = weight for j^{th} trace, $0 < w_j < 1$

$x_j(t)$ = recorded sample at time t on trace j

N = number of traces

This may be compared to the mathematical expression for a conventional CMP stack:

$$o_s(t) = \frac{1}{N}\sum_{j=1}^{N} w_j x_j(t) \qquad (6.72)$$

where

$o_s(t)$ = output of CMP stack at time t and the other parameters are as described for Equation 6.68

Migration

The problem. Figure 6–197a shows CMP ray paths from a flat, horizontal reflector. As shown in Figure 6–197b, NMO corrections followed by CMP stack in effect move both source *S* and reflectors *R* to the midpoint position *M*. In this case, midpoint and zero-offset point are identical. Only when reflectors are flat and horizontal is the geologic situation shown in true perspective.

Figure 6–198a shows another set of CMP ray paths but from a dipping reflector. In Figure 6–198b, S_1, $R_1 - S_9$, R_9 are the combined source and receiver points without DMO. P_1 – P_9 are the reflection points without DMO. Note that the reflection points move progressively farther up-dip as offset increases. *Z* is the coincident source and receiver point, or zero-offset point, with DMO applied and *P*' is the coincident reflection point with DMO applied.

The assumption of true zero-offset condition in the case of a dipping reflector is true only if DMO is applied before stack. The combination of NMO and DMO corrections places all sources and receivers at the midpoint position.

At this point in the processing sequence, reflection times are known but not the location of the reflection points. After stack, it is assumed that all reflection ray paths are perpendicular to the reflectors.

This assumption results in the presence of geometric distortions unless all reflections arise from flat reflectors—a most uninteresting situation for petroleum exploration. Figure 6–199 shows the distortion on the CMP stack section for a dipping reflector segment.

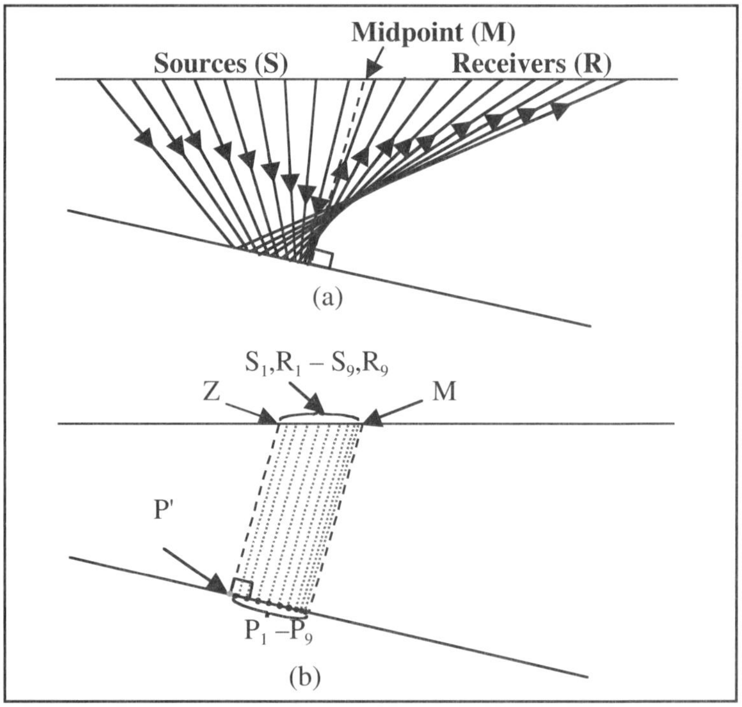

Fig. 6–198 *CMP Traces before NMO Correction and Stack – Dipping Reflector and CMP Traces before NMO Correction and Stack*

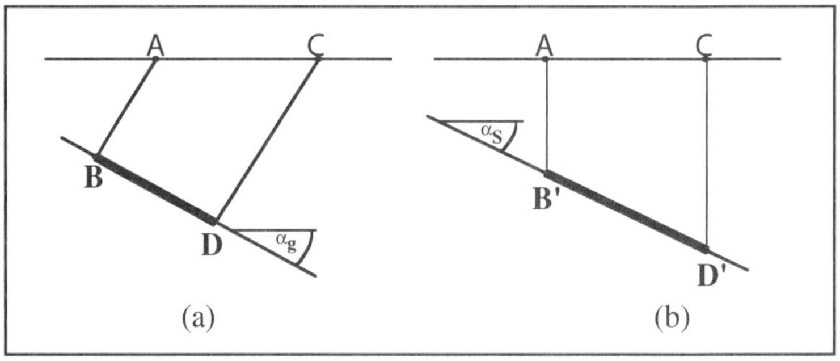

Fig. 6–199 *Dipping Reflector in True Position and as Seen on CMP Stack Section*

In Figure 6–199a, a reflector is shown dipping at an angle α_g. Zero-offset rays are shown from surface positions *A* and *C*. The rays are perpendicular to the reflector at points *B* and *D*. Figure 6–199b shows how a CMP depth section distorts the reflector segment *BD*. The section places reflection points at *B*' and *D*' instead of at *B* and *D*. Note that *AB*'=*AB* and *CD*'=*CD*. However, $B'D' > BD$ and $\alpha_S < \alpha_g$. So, a dipping reflection segment on a CMP stack is shown at a less steep dip and a greater spatial extent than it really is.

Synclines, anticlines, faults, and other structures with sharp edges are also shown distorted on stack sections. Figure 6–200 shows how the bow tie effect occurs on CMP stack sections.

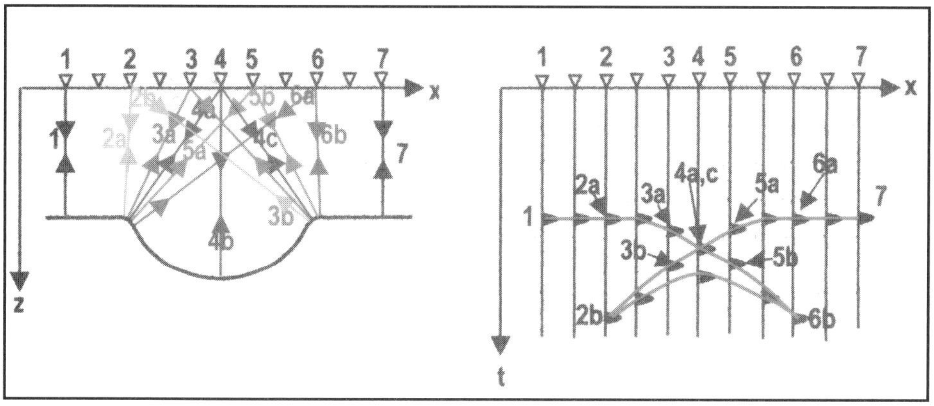

Fig. 6–200 *A Buried Focus or Syncline in Its True Perspective and a Bow Tie*

On the left of Figure 6–200 is a syncline or buried focus. Seven numbered surface positions and zero-offset ray paths from them are shown. There only vertical ray paths at positions 1 and 7. There are two rays at position 2 (2a and 2b) that intersect the curved surface at right angles. At position 3, there are three rays (4a, 4b, and 4c) that intersect the curved surface at right angles. At position 5, there are also two rays (5a and 5b) that intersect the curved surface at right angles.

The right side of Figure 6–200 plots the all the reflections described above as seen on a CMP stack section. On the CMP stack, the times for each two-way ray path are measured, but every time is plotted on the trace on which it was recorded. As a consequence, two reflections from the same reflector are shown on the traces corresponding to receivers between surface positions 3 and 6. Two of the three reflections recorded at position 4 have the same reflection time. The result is the bow tie shape on the section.

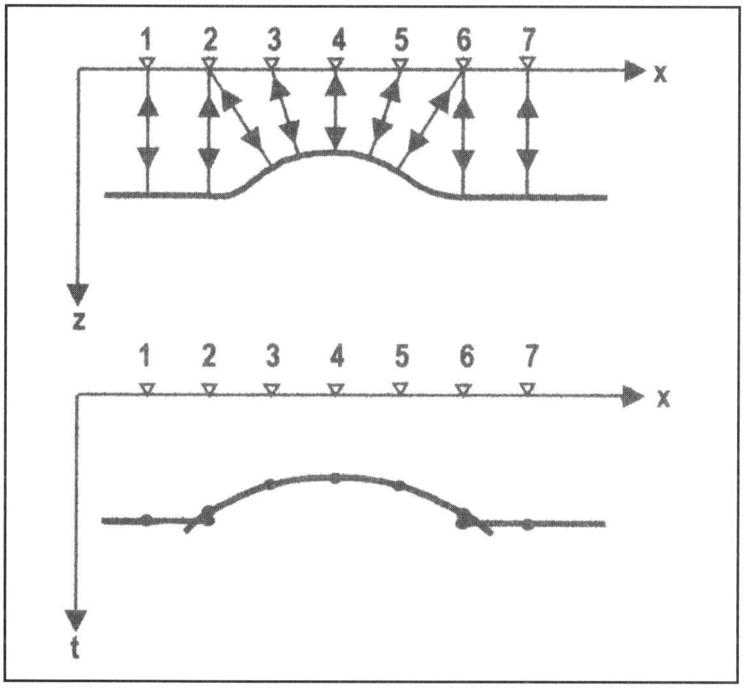

Fig. 6–201 *Anticlinal Reflector and Its Appearance on CMP Stack*

Figure 6–201a shows an anticlinal reflector in its true position and shape. Figure 6–201b shows how the anticline appears on a CMP stack section. Note that the anticline is broader and less steep on the stack section. Also note that on the stack section the anticline extends beyond its true position and crosses through the flat reflections on each side.

Figure 6–202a shows a geologic target consisting of a reef atop a flat, faulted reflector with the up-thrown block on the right of the reef. Figure 6–202b represents the appearance of the geology of Figure 6–202a on a CMP stack section. The end points of every discontinuous reflector becomes a source point and reradiates incident energy in all directions, producing what is known as diffractions on the stack.

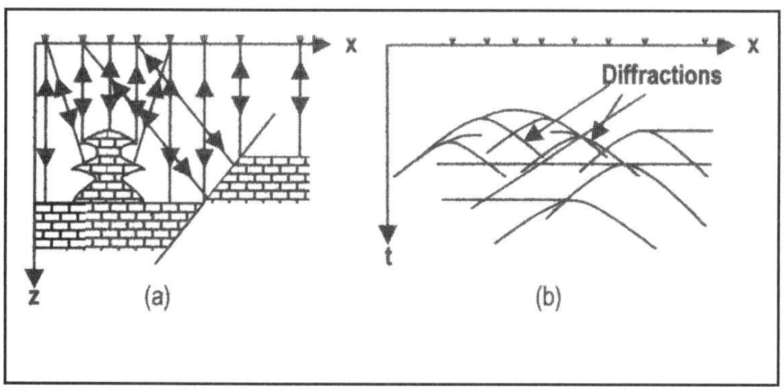

Fig. 6–202 *Geology and CMP Stack*

In summary, CMP stack sections provide a distorted view of structure other than flat reflectors in that:

- dips of reflector segments are decreased and spatial extents are increased
- synclinal features known as buried foci become bow ties
- anticlines are broadened and crossing reflections may be produced on the limbs
- truncated reflectors and sharp edges of structures are extended beyond their true terminations by diffractions

A solution to the problem of distortion on CMP stack sections is required if a correct image of subsurface geology is to be obtained.

The Solution. Figure 6–203a is the same as Figure 194b, a dipping reflector segment as shown on a CMP depth section. Figure 6–197b shows a simple graphical method of migration to its true geologic position. On a CMP time section, the true reflection time is known but not the true reflection point. On a CMP depth section, the true depth is known but not the true reflection point.

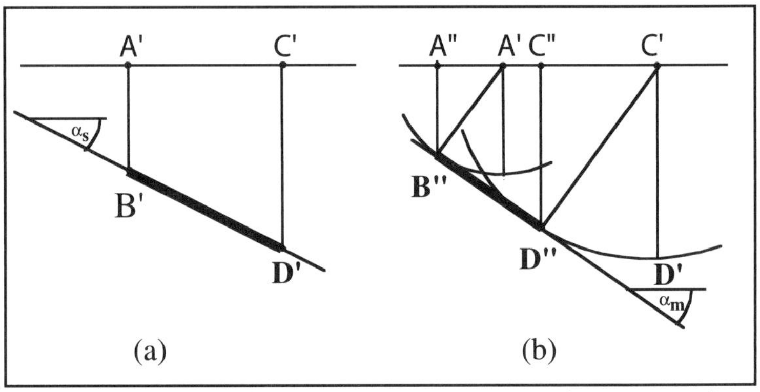

Fig. 6–203 *Dipping Reflector as Seen on CMP Stack Section and After Migration*

That being the case, an arc of radius *AB* defines all the possible reflection points from *A*, and an arc of radius *CD* defines all the possible reflection points from *C*. In addition, it is known that the rays from *A* and *D* to the reflector are perpendicular to the reflector. From geometry, it is known that a line drawn from the center of an arc is perpendicular to the tangent to the arc. In Figure 6–203b *B"D"* is tangent to the arcs about both *A* and *C*. Also, *AB"* and *CD"* are perpendicular to *B"D"*. Thus, *B"D"* is the migrated position of the reflection segment, is of the correct length and dips at the correct angle α_g.

Figure 6–204a is the same as Figure 6–202b, the distorted CMP stack. Figure 6–204b is the migration of the stack. The reef is not completely imaged, but the limbs can be seen and the diffractions have been collapsed so that reflector terminations can be seen.

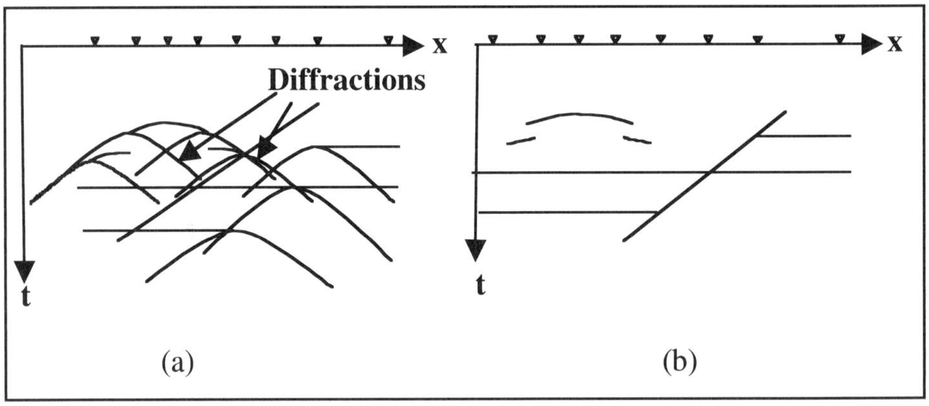

Fig. 6–204 *Stack and Migration*

From the examples presented, it can be seen that migration collapses diffractions, moves dipping events laterally in the up-dip direction and upward in time, and makes dips steeper. The amount of movement depends on dip, velocity, and time of the event. Migration also lowers frequency content. (An explanation for this is provided later.) No matter how good the migration algorithm, the results are only as good as the stack.

The term *imaging* is frequently used to describe a range of processes that are used to extract a picture or image of the subsurface from seismic data.

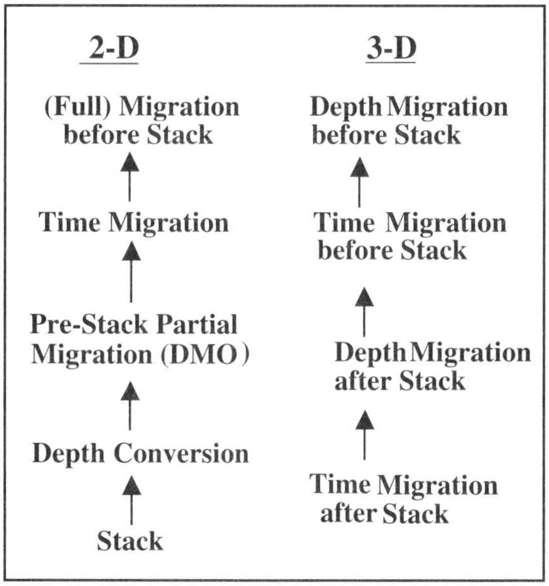

Fig. 6–205 *Imaging Hierarchy*

Figure 6–205 shows a hierarchy of imaging techniques, ranging from fastest and cheapest to slowest and most expensive. As the imaging problem becomes more complex, slower and more expensive migration techniques must be used to solve the imaging problem.

A CMP time section can provide a meaningful subsurface image of relatively simple geology. Many interpreters have learned not only to identify diffractions but also make real use of them in geologic interpretations of seismic data.

Conversion from a time section to a depth section improves the image by producing more realistic thicknesses between reflectors—high velocity produces apparent small thicknesses in time but depth conversion reverses this.

Pre-stack partial migration, or DMO, gives better velocity information and improves the stack section. Go to full pre-stack migration if DMO plus post-stack migration is not good enough.

CMP stack and time migration make the zero-offset assumption. That is, they assume that the CMP Location is coincidental with the zero-offset position. As has been shown, this is true only for zero dip. However, if dip is small, then this is a reasonable assumption. Time migration algorithms do not handle lateral variation in velocity. Depth migration handles lateral velocity variation, but an accurate velocity model is a must.

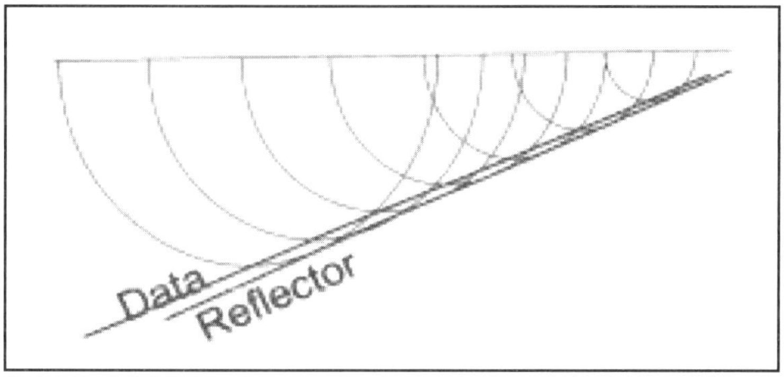

Fig. 6–206 *Circular Migration*

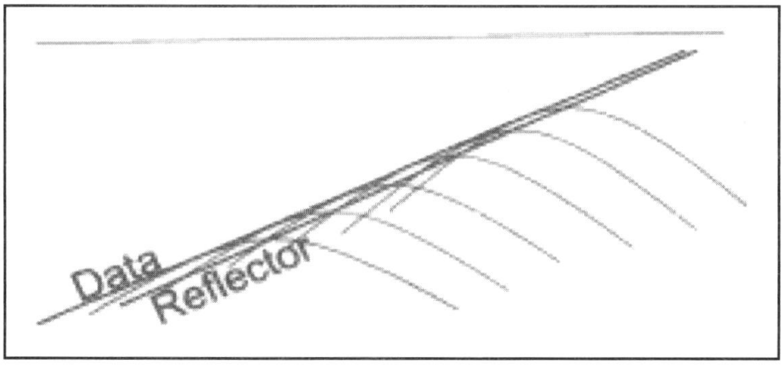

Fig. 6–207 *Hyperbolic Migration*

There are two basic approaches to migrating 3-D data. The first approach is to do 2-D migration in-line, then re-sort into cross-line sections and do another 2-D migration. That is, two orthogonal migrations are done on the data. The other approach is true 3-D migration—one-step migration.

The 2-D migration is not adequate to image a complex structure but the 2-D depth migration before stack process involves migrating every field trace individually. The 3-D time migration before stack is now affordable. Of course, 3-D depth migration before stack is more costly to run, and much more time and resources are required to develop the velocity model.

Migration algorithms. In the days of analog recording and early digital recording, migration was done graphically, much like the migration of Figure 6–206. Data were plotted on cross sections, and arcs were drawn using trace-zero time as the center. Lines were drawn tangent to the arcs to generate the migrated horizon. This is not a promising approach to migration of digital data using signal processing on digital computers.

Figure 6–207 shows another approach. Sets of hyperbolas are drawn with their apexes at the reflection points. Lines drawn tangent to the hyperbolas produce the same migrated reflector as circular migration.

Why use hyperbolas? Diffractions appear on seismic sections as hyperbolas. The assumption that all events on seismic sections are produced by diffractions lead to the migration algorithm called Kirchhoff migration. As discussed previously, Kirchhoff migration is done on the T-X domain.

The sketch of Figure 6–208 suggests another approach. This shows a pond into which a tree has fallen. The tree trunk appears to be bent at a different angle beneath the water. In reality, the tree is not bent but only appears to be because of refraction of light in the water. If some water is released from the pond, more of the tree can be seen and its true direction and extent determined. If the water level continues to be dropped, eventually all of the submerged part of the tree will be seen in its true position and extent.

Fig. 6–208 *Distortion Caused by Refraction of Light*

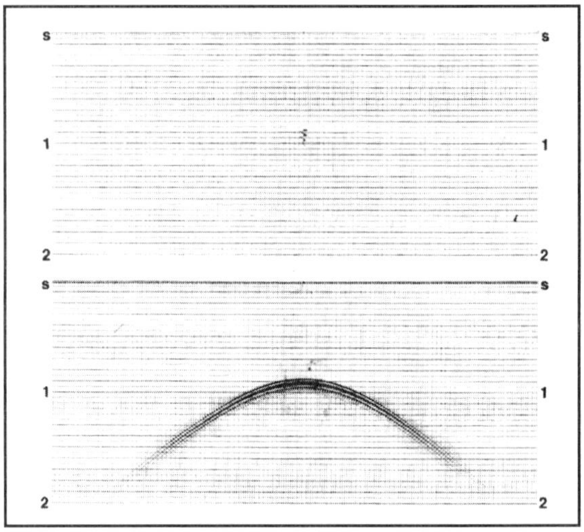

Fig. 6–209 *Representation of a Point Aperture and the Diffraction Hyperbola Produced from It*

A seismic section gives a distorted view of subsurface geology. If the time, or depth, at the top of the section is stripped off, then the distortion is reduced. Eventually the distorted structure is, effectively, at the surface and can be seen undistorted. This approach is called *downward continuation migration*. There are a variety of algorithms used to accomplish downward continuation migration, operating in various domains.

Kirchhoff migration. The top part of Figure 6–209 represents a point aperture by a band-limited impulse. The bottom shows the recording of the diffraction hyperbola produced from it. Perfect migration of the diffraction collapses it into the impulse. The top part of Figure 6–210 represents a set of closely spaced band-limited impulse. The bottom part of Figure 6–210 shows the results of all of the diffraction hyperbolas produced. All diffraction patterns cancel except where there are no other diffraction patterns to destructively interfere at abrupt ends. Superposition of the hyperbolas produces the seismic signature of the actual reflection interface.

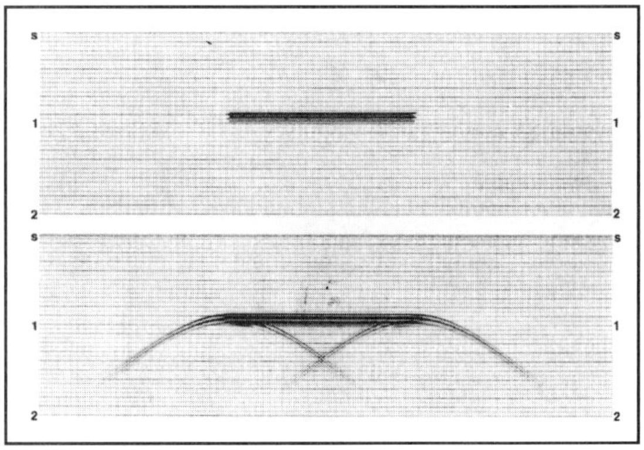

Fig. 6–210 *Set of Closely-Spaced Point Apertures and Resultant of Diffraction Hyperbolas Produced from Them*

To develop the migration solution, look first at the simple problem and then apply the concept to more complex problems. With one point, a simple hyperbolic response is obtained. With a multitude of points, invoke Huygens' principle to show that reflectors can be represented by a series of closely spaced points. That is, reflectors are made of many points that act as a Huygen's secondary sources. The zero-offset section is the superposition of the hyperbolic travel-time response. Whenever there is a discontinuity along the reflector, the diffraction hyperbola appears.

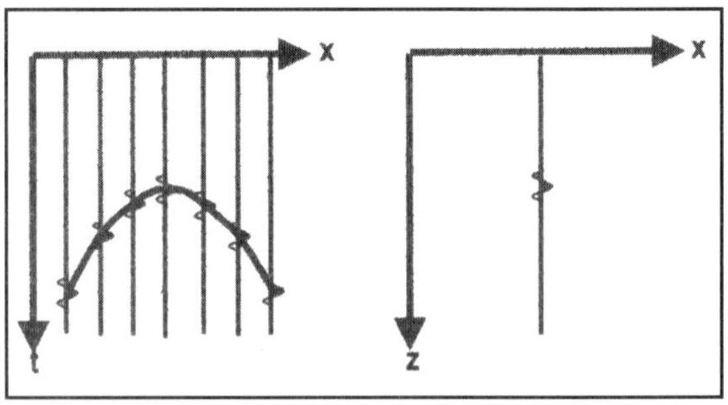

Fig. 6–211 *Summation of Amplitudes along the Hyperbola*

Assume that there is a hyperbolic distribution of amplitude in time and space (Fig. 6–211). The apex of the hyperbola corresponds to the source of the distributed energy. Sum the amplitudes in X and T along the hyperbola. The particular hyperbola depends on the velocity of the medium at the apex of the hyperbola. The depth of the output in z is the product of time multiplied by velocity and divided by two. The x position is that of the apex.

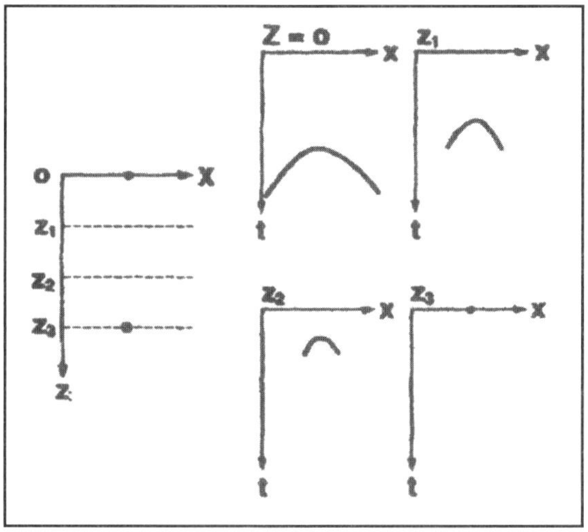

Fig. 6–212 *Downward Continuation*

When 2-D data are summed by gathering along a hyperbola, there is a 45-degree phase shift and amplitudes need to be scaled by the square root of frequency. 3-D summation has a 90° phase shift, and amplitudes need to be scaled by frequency. Diffraction summation procedure for migration that incorporates these three scaling factors is called Kirchhoff migration.

Exploding reflector model (ERM). In the exploding reflector model, the positions of source and reflection points are exchanged and all sources are initiated at time zero. Energy travels out from sources at the same time, and source strengths of are proportional to reflection coefficients. New wave fronts are, in effect, created by the constructive interference of the spherical waves radiating out from the sources. This continues all the way to the receivers at the surface. The ERM uses one-way travel-time—a one-way solution to a two-way problem. The ERM can be made to match the zero-offset model by using half the actual velocity values in it.

Zero offset migration theories are used for migration although the stack section is not really zero offset. However, a close relationship can be drawn between the stack and the exploding reflector model. The idea of representing a reflection as a series of point diffractors is, in effect, the exploding reflector model. The pattern of interfering diffraction curves can be recorded at the surface for the closely spaced secondary sources along the reflector. The individual diffraction patterns for each secondary source point can be considered to be present, but they have been cancelled out because the points are very close together.

No extension of the ERM exists to data recorded at nonzero offsets (pre-stack data). Also, the ERM does not work when there are strong lateral velocity variations, which is why post-stack time migration breaks down in the presence of large lateral velocity variations.

Downward continuation. The left side of Figure 6–212 shows three depth levels—z_1, z_2, and z_3—with a point discontinuity at the depth z_3. The four drawings on the right represent, clockwise from the top left, the actual recording with receivers at z=0, the recording with receivers at $z=z_1$, the recording with receivers at $z=z_2$, and the recording with receivers at $z-z_3$. The diffraction hyperbola occurs at progressively earlier times and with smaller extent, culminating in a single impulse with the receivers at $z=z_3$.

Kirchhoff migration collapses diffractions by summing the amplitudes and placing the sum at apex of the hyperbola. Downward continuation migration maps the hyperbola recorded at one depth to another as if it were recorded at the new depth. It continues mapping the hyperbola downward to new depths in increments of depth $z = V\Delta T$. The hyperbola collapses to a point when the apex is reached. This is the concept of downward continuation.

Downward continuation migration is a very important method that enables simple time migration or depth migration with complex velocity distributions. Downward continuation migration starts from the top of the seismic section $(x, z = 0, t)$, computes the entire volume (x, z, t), and extracts the structure $(x, z, t$=0). Note that the output migration is in depth, the most accurate representation of the subsurface. This is depth migration.

Some methods of downward continuation output portions of the depth slices (x, z=0, t) into a time section. This may form part of the output for a time migration. There are many methods of computing the different layers in a data volume. The volume may also be computed from bottom to top, referred to as reverse time migration.

Comparison of time and depth downward continuation. Vertical movement of data is up toward t = 0 with increased depth. One depth sample per trace is output with each depth layer in depth migration. A band of t time samples per trace is output with each depth layer in time migration.

The downward increment is defined by a depth increment δz in depth migration or time increment and δt in time migration. The depth increment δz must be small enough to prevent aliasing. The time increment δt depends on the velocity model, the algorithm, and the required accuracy. The top of the time-migrated portion of the time section is always at a constant time defined by a multiple of δt. Time migration downward increments are defined by the t step and the velocity.

The top of the depth-migrated time section at t=0 is defined at a constant depth level that may not correspond to a constant time level on the original time section. Data above the tth time level remain unchanged by the migration process and may be used to store the migrated data. A separate storage area is required for the migrated depth output.

The main difference in the downward layers of time and depth migration is the small vertical shift that is applied to the depth migration. This vertical shift (sometimes referred to as the thin lens term) is responsible for applying small relative corrections to data in areas with laterally varying velocities. The accumulation of these small relative corrections adjusts the fit on diffractions to correctly position the apex at the reflection point.

Some algorithms do not apply the whole vertical time shift but only the small relative corrections due to the lateral velocity change. These algorithms have the advantage of the speed of time migrations and the correct lateral positioning of depth migration.

There are three approaches to downward continuation migration—finite difference (T-X domain), Stolt (*F-K* domain), and Gazdag's phase shift (*F-K* domain). Each has its place in migration of seismic data.

Finite difference migration. Finite difference migration extrapolates input data (CMP stack sections) using finite increments of z and predicts what the data would look like at the new depth. There is an output of the wavefield at $t = 0$ for each depth. Prediction and output stop when data are moved back to the reflector.

In Figure 6–213, the input record is at z=0, 0=t= t_{max}, and $x_{minx} = x = x_{max}$. Downward continuation migration, in effect, strips off small depth increments $\delta z = V\delta t$, and predicts the change in the waveform produced by the change in depth. In Figure 6–213, after the depth step from $z = 0$ to $z = z_1$, the time slice δt is migrated. After the depth step from $z = z_1$ to $z = z_2$, the next time slice δt is appended to the first migrated strip. As each depth (or time) increment is stripped off it is appended to the previous strips to produce the migrated section.

Note that as the depth increases, the bottom of the time section remains at the bottom. The top of the migrated part of the time section is decreased by t time increments with each downward layer. The output from each downward layer is defined by the time band t that is at the tth time level.

The size of the t step depends upon the accuracy required for the velocity model and the dip limitations required by the algorithm. A 15° algorithm may require a t step of 50 ms to achieve 15° of migration, while a 65° algorithm may require a step around 4 ms to achieve 65° of dip of migration.

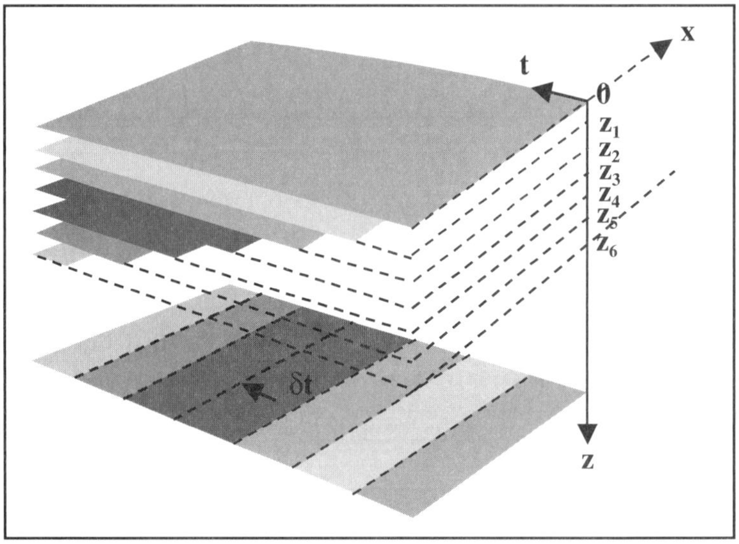

Fig. 6–213 *Downward Continuation Migration with Depth Slices at Different Layers*

The number of time increments to achieve the depth sampling depends on velocity and dip. Velocities used at each depth layer should be the velocities at that depth. Depth slices at different layers should be small enough to prevent aliasing on the output depth migrated trace. Typical processes run close to the Nyquist rate for the migration and then use accurate interpolation methods (sin x/x) to achieve finer sampling and prevent aliasing. The downward propagation of energy from one depth level to the next may be accomplished by a number of algorithms. Usually these algorithms are based on solutions to the one-way wave equation. An example would be the phase shift method.

The velocity for each downward layer is defined at the time level of each t step and may vary for each trace. A complex velocity model may require, a small t step to accurately follow the velocity changes. Conversely, velocity resolution finer than the t step is not required—do not define a velocity model to a 10 ms resolution when the step is 50 ms.

Typical time migration algorithms are usually referred to as 15° or 35° finite difference algorithms. The small angle approximation (±15°) is a parabolic fit to the hyperbolic moveout curve. It is assumed that velocity varies vertically but not horizontally. However, in practice, a smooth lateral velocity change is allowed.

In chapter 3, it was noted that the wave equation is used to describe the wave field. The wave equation is repeated here.

$$\frac{\partial^2 u}{\partial x^2} + \frac{\partial^2 u}{\partial y^2} + \frac{\partial^2 u}{\partial z^2} = \frac{1}{V^2}\frac{\partial^2 u}{\partial t^2}$$

The wave equation as written here is a partial differential equation and is applied to continuous or analog data. Seismic data are sampled, not continuous, and seismic data processing is done on digital computers, so instead of the wave equation above, the scalar wave equation is used. The scalar wave equation uses finite differences instead of differentials and instead of derivatives, such as $\frac{dt}{dx}$, it uses $\frac{\Delta t}{\Delta x}$.

Finite difference migration uses solutions of the scalar wave equation to predict what the amplitudes will be for the new time zero after each depth step. Complications arise because the continuous wave equation is a partial differential equation that contains the second derivatives of the wave field with respect to depth, time, and spatial axes. Setting up the computer algorithm is beyond the scope of this book.

There are two basic finite difference algorithms—*explicit* and *implicit*. The explicit algorithm uses the input data just as they are while the implict employs a kind of feedback. The explicit algorithm can be more correct than the implicit but the explicit can become unstable, i.e., predictions head off in the wrong direction.

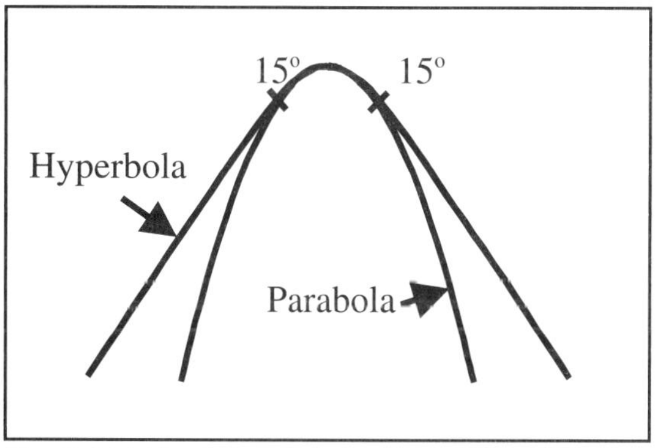

Fig. 6–214 *Fit of Parabola to Hyperbola—Low Order Explicit*

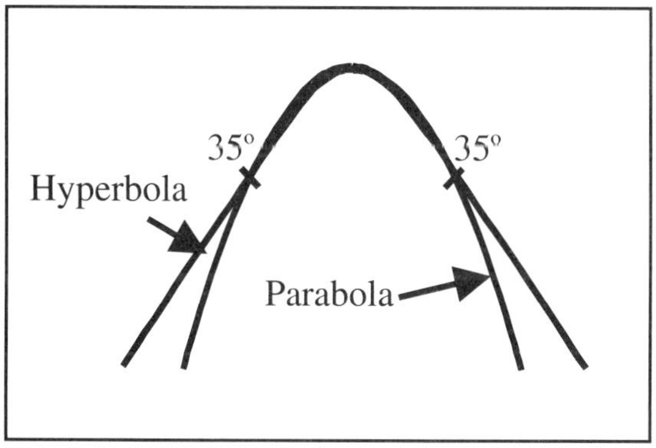

Fig. 6–215 *Fit of Parabola to Hyperbola—High Order Explicit*

There are two types of explicit algorithms—low order and high order. The low order explicit finite difference algorithm has a dip limit of 15°. A parabolic fit is made to the diffraction hyperbolic moveout curve (Fig. 6–214). There is a small error out to a dip of 15°. A parabola is used instead of a hyperbola to avoid recalculations for each sample.

The high order explicit finite difference algorithm has a dip limit of 35°. A parabola is, again, fit to the hyperbolic moveout curve of diffractions, but in this case the parabola is moved upward relative to the hyperbola and a least square error fit is made to the hyperbola (Fig. 6–215). Flat data are not migrated quite correctly, but this is of little concern.

The implicit algorithm uses a technique in which the predicted values are averaged with prior values. The implicit finite difference algorithm has a dip limit of 30°. Finite difference migration allows a vertical velocity variation but no horizontal velocity variation. In practice, however, a smooth lateral velocity variation is permitted.

Stability (the problem of wave amplitudes growing from one extrapolation step to another) of the finite difference operator is an issue for explicit operators. Implicit operators produce stable results because of averaging present and future values. Differential equations used in finite difference migration algorithms have matrix coefficients. Implicit schemes require inversion of this matrix. However, no inversion is needed with explicit schemes, since future values can be written explicitly in terms of only past values.

Stolt *F-K* migration. Figure 6–216 is a flow chart for Stolt constant velocity migration (one velocity for all record times and all traces). This is the first form of Stolt migration introduced and what is usually implied by the term Stolt migration. While the assumption of a constant velocity is quite unrealistic, the algorithm does have applications and is very fast compared to other migration algorithms.

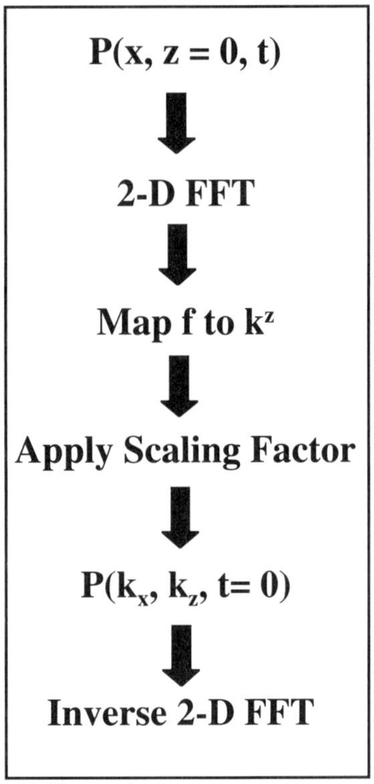

Fig. 6–216 *Constant Velocity Stolt Migration Flow Chart*

The principle of Stolt migration is quite simple.

- Transform from the T-X domain to the *F-K* domain with a direct 2-D FFT.
- Convert frequency to vertical wave number $k_z = \frac{f}{V}$.
- Apply a scaling factor.
- Transform from the *F-K* domain back to the T-X domain with an inverse 2-D FFT.

The Stolt algorithm was later modified to allow some vertical velocity variation by introducing the *stretch-factor W* (Fig. 6–217). It is called *generalized* Stolt when *W* is not equal to 1. The extended Stolt converts time to a pseudo depth by stretching vertical time, migrates in depth, and then converts back to original time axis. Stolt migration with the stretch factor $W = 1$, corresponds to the exact constant velocity depth section and full 90° migration. A stretch factor of $W = 0.5$ causes a semicircle to be truncated to a similar pattern as the finite difference and reduces the dip limit to 45° of geologic dip. If the *W* factor is not equal to 1, then Stolt migration is no longer fast. Stolt migration is super-fast only for constant velocity, which is not a very realistic picture of earth.

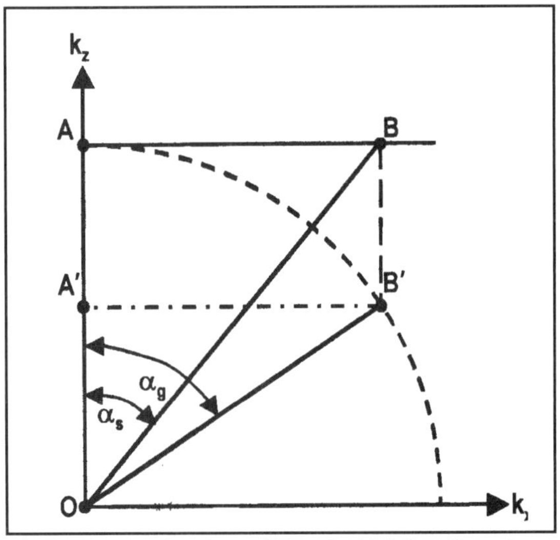

Fig. 6–217 *Reduction of Bandwidth in Migration*

Loss of bandwidth in migration. Previously, it was stated that one effect of migration is to reduce bandwidth. Stolt migration provides an excellent way to demonstrate that. Figure 6–217 shows a circular arc indicating equal values of *k*. The radius of this arc is *OA*. Since $k_z = \frac{f}{V}$, the horizontal line *AB* represents bandwidth. Line *OB* represents a reflection segment on the input CMP stack with an apparent dip angle α_S. A vertical line from *B* intersects the arc of constant *k* at *B*' and the line *OB*' represents the migrated reflection segment. Note that *OB*' < *OB* and $\alpha_g > \alpha_S$, as should be the case. Line *A'B'* is parallel to *AB*, and *A*' is the new k_z value. Since *OA* > *OA*, the bandwidth after migration has been reduced by loss of the higher frequency components. In fact, as the dip increases, the amount of high frequency loss also increases.

Figure 6–218 is an example of Stolt migration of a multi-dip model. The model consists of reflections dipping at angles of 0° through 45° in 5° increments. The zero-offset section (left) shows the expected distortions. The Stolt migration successfully migrates all dips. However, although all have the same bandwidth before migration, the high dips have less high frequency after migration. In the *F-K* plane, a fan has opened and sagged. The higher dips also are aliased. The example substantiates statements made above.

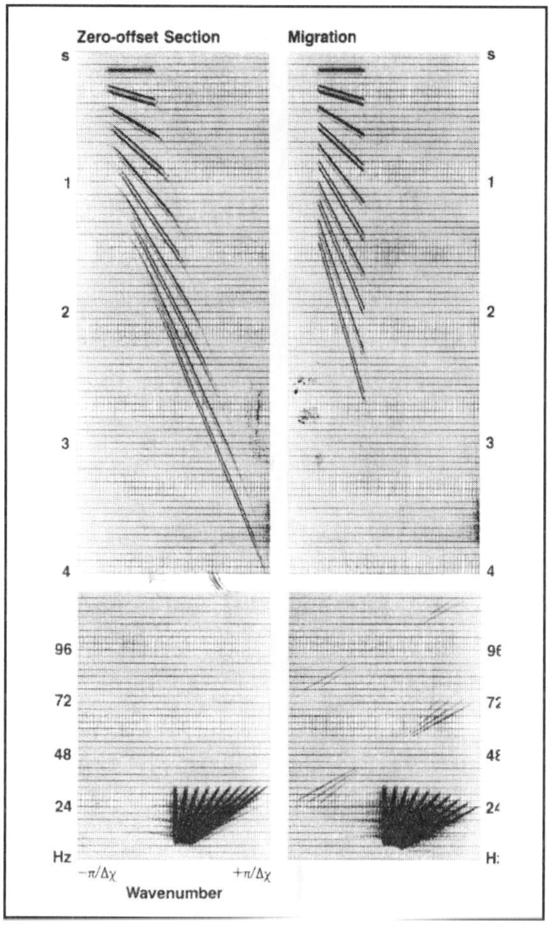

Fig. 6 218 *Stolt Migration Example—Multi-Dip Model*

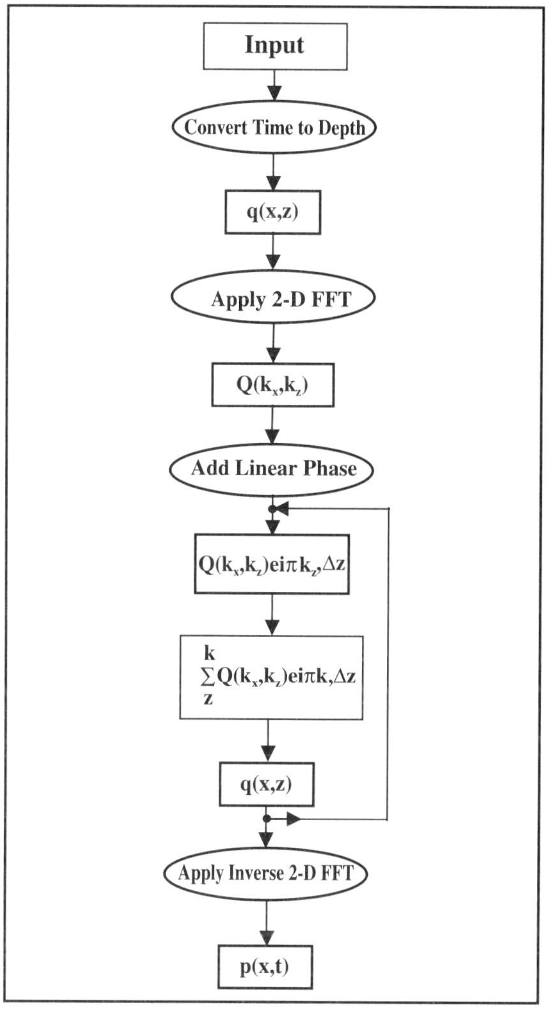

Fig. 6–219 *Gazdag's Phase Shift Migration Flow Chart*

Gazdag's phase shift migration. Gazdag's phase shift migration involves the following steps.

1. Input zero-offset data (stack).
2. Convert time to depth.
3. Perform 2-D forward FFT.
4. Apply linear phase shift ($\pi k_z \Delta z$) to effect depth step (downward continuation).
5. Sum over all vertical wave numbers (k_z).
6. Repeat steps 4 and 5 until all depth steps are completed.
7. Perform 2-D inverse FFT.
8. Convert depth to time, if desired, as output at this stage is depth migration.

Figure 6–219 is a flow chart of phase-shift migration.

Phase shift migration is accurate and migrates to 90° and better. Velocity variation with depth is allowed, and it is also not dispersive. Phase shift migration is the standard against which all other algorithms are compared. It is, however, very expensive.

Notes on Kirchhoff and downward continuation migration. Both Kirchhoff and downward continuation migration use the scalar wave equation (Fig. 6–220). The Kirchhoff approach is essentially an integral solution of the scalar wave equation. Integration is replaced by summation for sampled data. Amplitude corrections are also applied.

The downward continuation approach will solve the differential form of the scalar wave equation. A difference equation replaces the differential equation for sampled data. Kirchhoff and downward continuation migration make different assumptions, and, while neither uses all terms in the scalar wave equation, the terms discarded or deemed negligible are different.

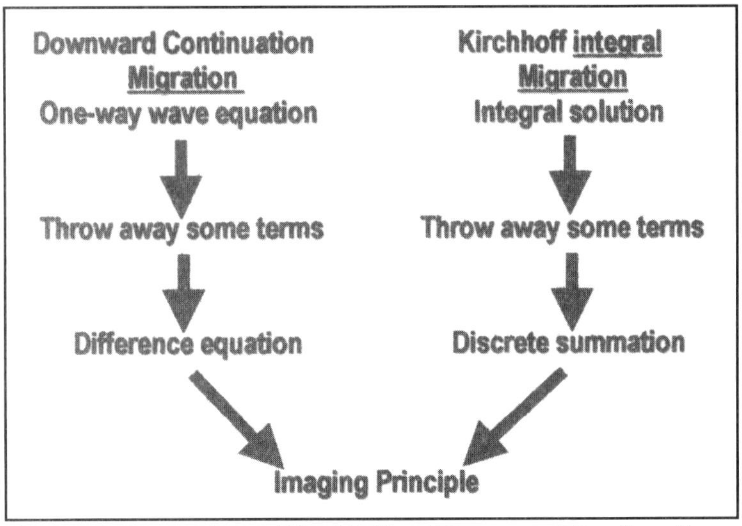

Fig. 6–220 *Comparison of Kirchhoff and Downward Continuation Migration Approaches*

Migration limitations. Migration can be limited by phase, bandwidth, and dip. There is also the problem that seismic data are always spatially limited. That is, data are recorded for only finite lengths of time and over a restricted areal extent. Each of these factors place limits on how well migration performs.

Figure 6–221 illustrates migration of a diffraction. On the far left, the wavelet is mixed phase and has a bandwidth of 12 to 50 Hz, just more than two octaves. It looks over-migrated. The other four have zero-phase wavelets but differ considerably in bandwidth. The best migration is the zero-phase wavelet with a bandwidth of 9 to 110 Hz (3.6 octaves). The zero-phase wavelet with a bandwidth of 9 to 30 Hz (1.7 octaves) is next best. The zero-phase wavelet with a bandwidth of 42 to 110 Hz (1.4 octaves) does not migrate the high dip limbs of the diffraction as well and shows some dispersion. The zero-phase wavelet with a bandwidth of 42 to 68 Hz (0.7 octaves) does a poor job of migration

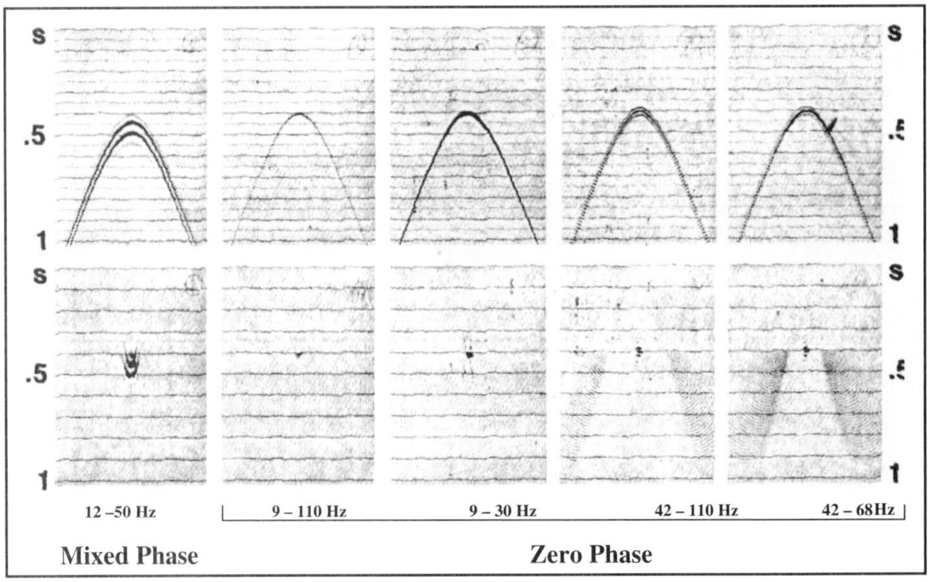

Fig. 6-221 *Effect of Phase and bandwidth on Migration of Diffraction*

It can be seen from this example that migration works best with zero-phase wavelets and a wide bandwidth. Therefore, it is preferable not to filter before migration, but, if there is too much noise, filtering before migration may be required.

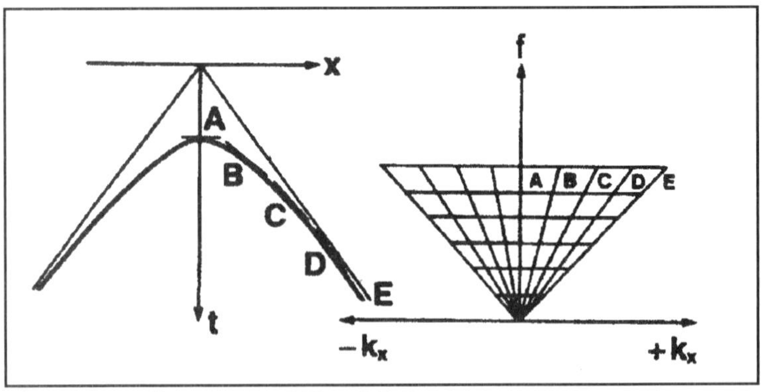

Fig. 6-222 *Diffraction in T-X and F-K*

As shown in Figure 6–222, the diffraction hyperbola in the T-X domain becomes a fan in the *F-K* domain. Steep dips in T-X domain map into

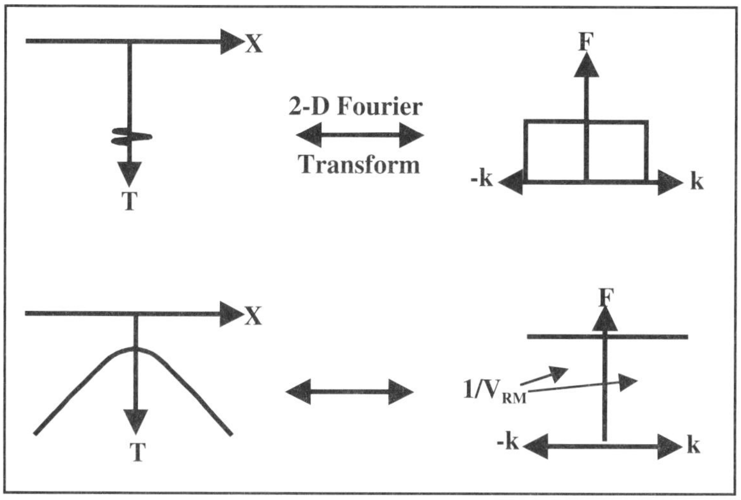

Fig. 6–223 *A Spike in the T-X and F-K Domains*

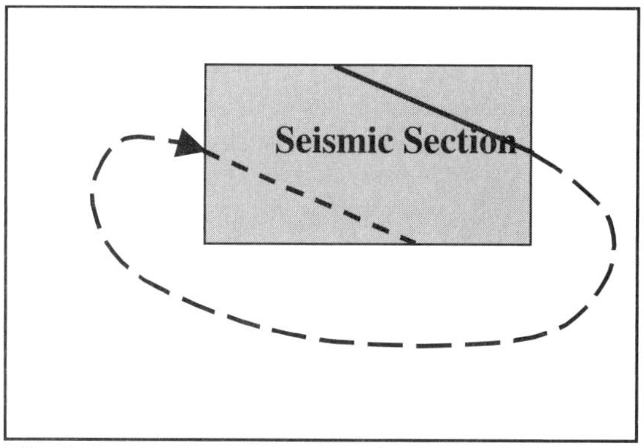

Fig. 6–224 *Data Wrap Around*

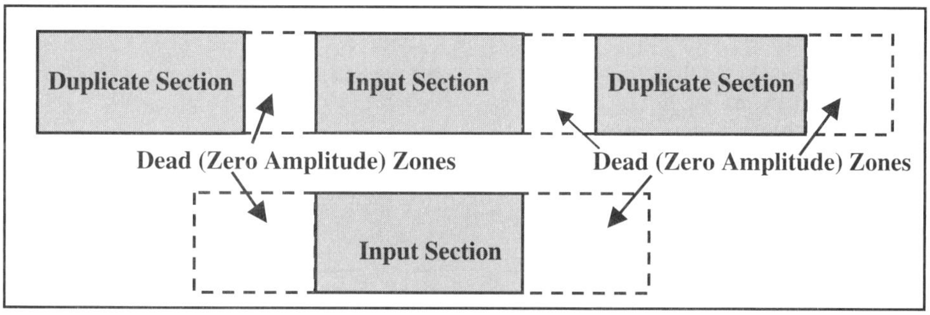

Fig. 6–225 *Preventing Wrap-around Effects*

those parts of the fan near $k = 0$. (Region *A* of the diffraction in the T-X domain.) As dips decrease, they map into regions farther from $k = 0$—regions *B*, *C*, and *D*. Region *E*, in Figure 6–222, corresponds to the asymptote, which has a dip of $1/V_{rms}$. In the *F-K* domain, the fan of dips opens as dips get steeper. Why not a "spike when a diffraction is migrated and vice versa? In order to get a diffraction curve to migrate to a point source, it must have dip-dependent frequency content.

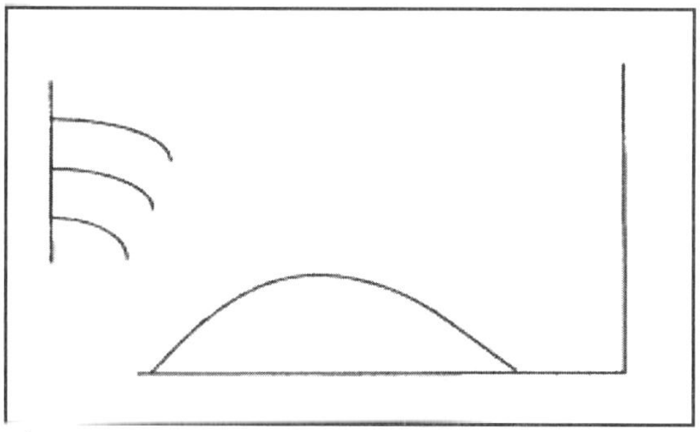

Fig. 6–226 *Truncated Diffractions*

The top of Figure 6–223 shows the ideal spike in the T-X and *F-K* domains. In the center, it is seen that the spike translates into a point and gives rise to a diffraction—the familiar hyperbola in T-X and the fan in *F-K*. After migration (at the bottom of Fig. 6–223), a frequency-limited spike is obtained in the T-X domain and a fan with dip-dependent frequency content in the *F-K* domain.

Seismic data are limited in time and distance, but processes assume sections continue to all time and space. Since they do not, processed data wrap around at the edges of the section. See Figure 6–224. Data must be made to look as if they continued beyond the section boundaries. A method to accomplish this is shown in Figure 6–225. The first method places duplicates of the input section on each side with dead zones—zero amplitude traces—in between. The dead zones catch data that migrates off the profile. In the second method, the section is extended on each side by adding dead zones that act as absorbing boundaries to catch and suppress data that migrates out of the profile. Zero amplitudes are often added to the bottom of sections as well to avoid wrap around effects caused by terminations at the bottom.

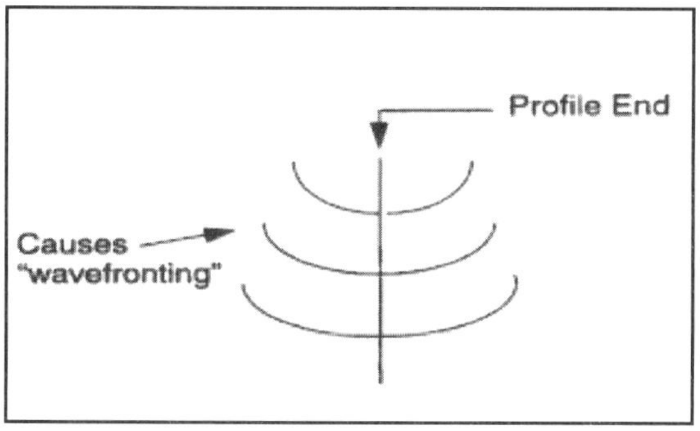

Fig. 6–227 *Migration Wave-Fronting*

When diffractions are cut off, either at the edge or bottom of the section (Fig. 6–226), diffractions will not be completely migrated. In addition, attempts to migrate incomplete diffractions on the section edges can produce wave-fronting because the migration response bounces off the edges (Fig. 6–227).

Migration requires the following:

- full diffraction patterns
- amplitudes as correct as possible
- widest bandwidth possible

The more critical the migration, the more important it is to have these three things

Spatial sampling considerations for migration. In the discussion of 2-D Fourier transforms, it was noted that spatial aliasing occurs when $k_{data} > k_N$, where k_N is the Nyquist wave number. Since $V = f\lambda = \frac{f}{k}$, it follows that $f = kV$. Dip D can be expressed as the reciprocal of velocity $V = 1/D$. Substituting for velocity gives $f = k/D$ or $k = fD$. Thus, $fD > k_N$, is also a criterion for spatial aliasing.

Figures 6–228 and 6–229 show the migration of the multi-dip model using the Stolt frequency wave number, finite difference, and Kirchhoff algorithms. In Figure 6–228, the signal bandwidth is 6 to 36 Hz. No aliasing is seen for any dip. In finite difference, migration data will bounce off the edge of the profile and there is anisotropic dispersion for dips greater than 35°. To get Kirchhoff to start in a model, a spike must be inserted at the first sample on each trace, which accounts for the noisy appearance and the large amplitude a t zero dip in *F-K*. However, in Figure 6–229, with a bandwidth of 6 to 90 Hz, aliasing occurs for dips greater than 20°.

The conclusion. At high dips and high frequencies, there is spatial aliasing. Higher dips lose higher frequencies because those frequencies are aliased to lower, wrong frequencies.

Figure 6–230 shows the effect of spatial sampling on migration of field data, using Kirchhoff, finite difference, and *F-K* (Stolt) methods. The top of the figure shows the input CMP stacks with trace intervals of 41 ft (12.5 m), 82 ft (25 m), and 164 ft (50 m). The next three rows show Kirchhoff, finite fifference, and frequency wave number migrations of the three stacks. As the trace interval increases, higher dips are not migrated in all cases. The smaller spatial sampling produces better migration because of spatial aliasing. It should be remembered that migration does not produce the aliasing. That results from the spatial sampling and signal bandwidth. High dips are not properly migrated because they have been aliased into lower dips!

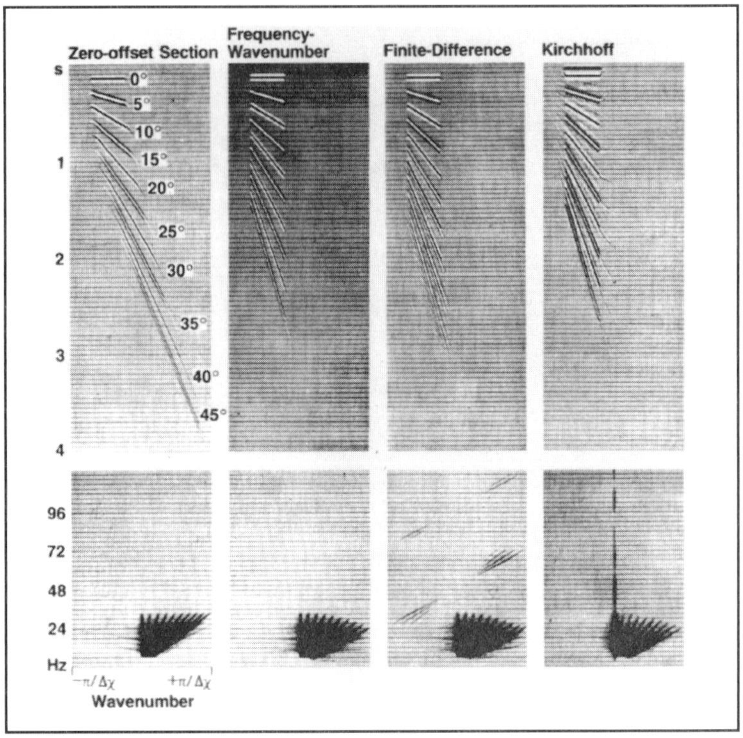

Fig. 6–228 *Migration of Multi-dip Model with 6 to 36 Hz Bandwidth*

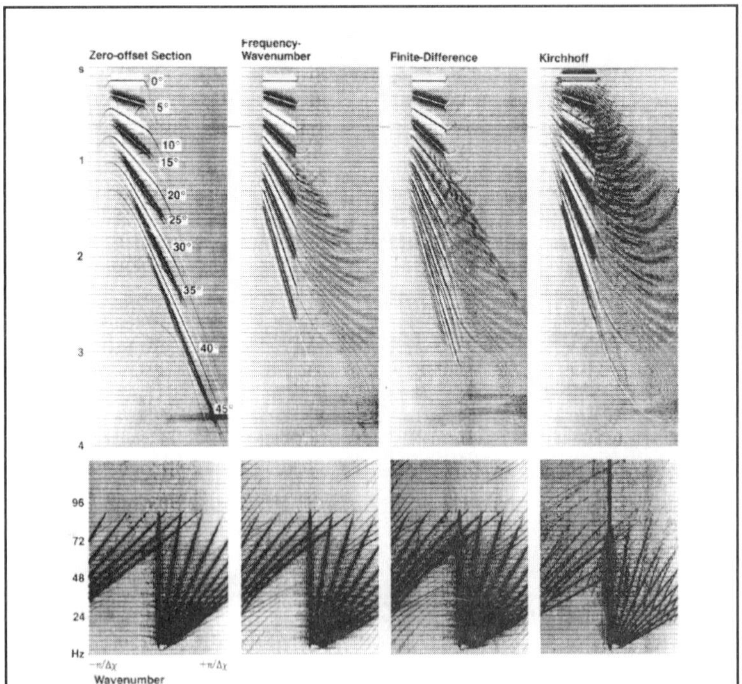

Fig. 6–229 *Migration of Multi-dip Model with 6 to 90 Hz Bandwidth*

Figure 6–231 shows the effect of spatial sampling on the unmigrated and migrated traces using a wavelet band limited to 5 to 80 Hz. The top shows the CMP stack equivalent for the multi-dip model with a trace interval of 48 m (left) and its migration (right). The bottom right shows that reducing bandwidth to 5 to 16 Hz eliminates aliasing. This low, narrow bandwidth is not high resolution. A wide bandwidth is needed for high resolution. The bottom right shows the migration after 4:1 interpolation effectively reduces

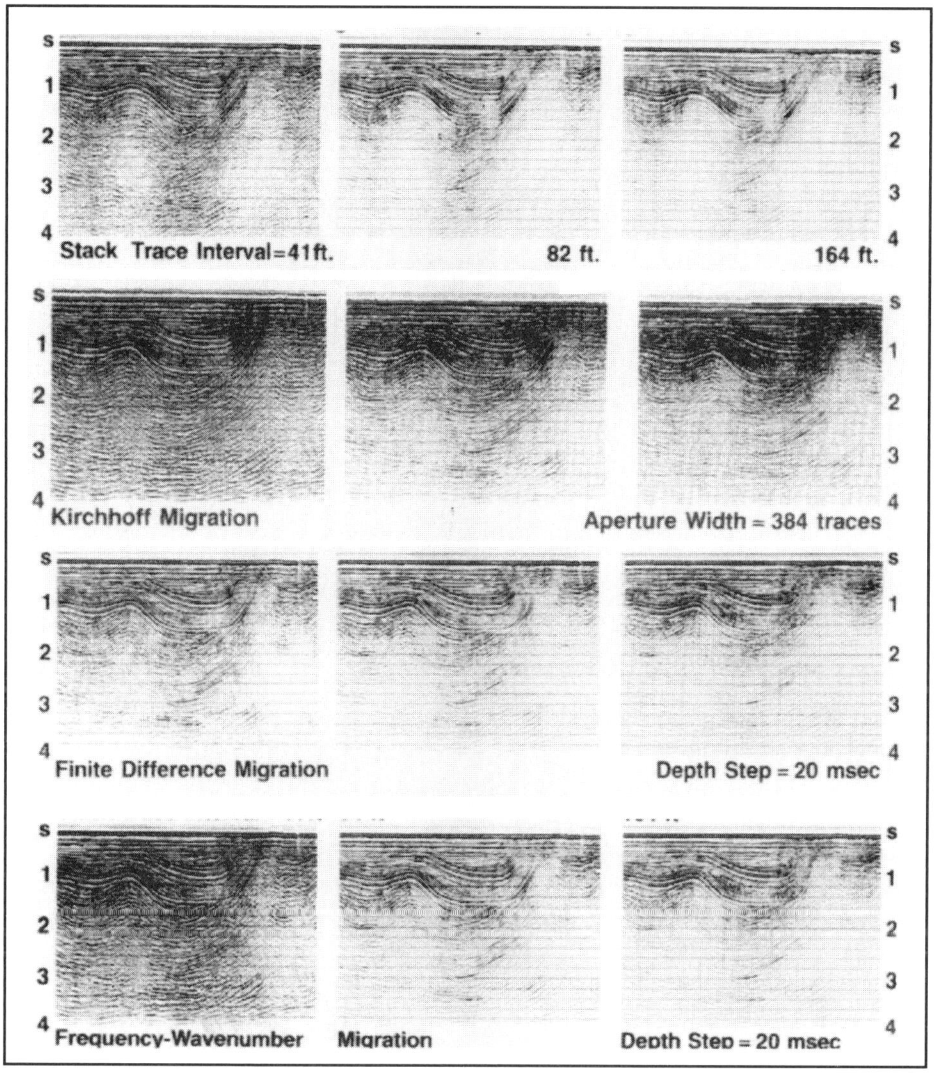

Fig. 6–230 *Effect of Spatial Sampling on Migration*

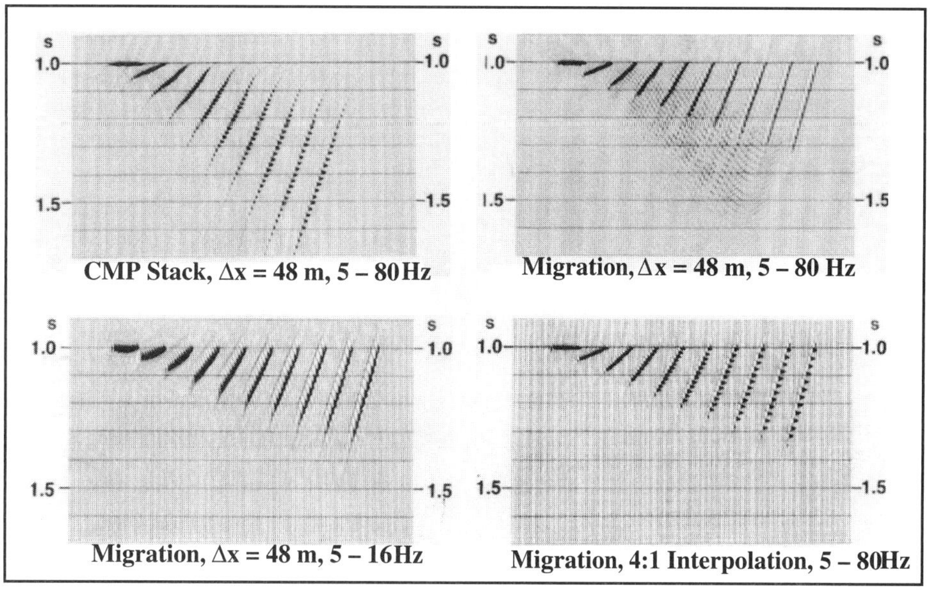

Fig. 6–231 *Spatial Aliasing and Migration—The Role of Interpolation*

the spatial trace interval to 12 m. By making the trace interval smaller, the Nyquist wave number is increased and aliasing at the higher dips and frequencies is eliminated. Trace interpolation may be the answer.

Lower frequencies migrate properly but the higher, aliased frequencies do not. To avoid aliasing effects in migration requires a larger k_N. That requires a smaller Δx (smaller trace interval). Trace interpolation (*tau-p*) before migration may resolve the problem. But this is only a patch. The smaller Δx the better, but smaller Δx increases acquisition costs. Trace interpolation does not work perfectly but is often a good compromise between cost and quality. Trace interpolation (*tau-p*) before migration may resolve the problem.

There is a question as to whether, after interpolation and migration, the interpolated traces should be discarded or kept? There is no real technical answer. Keep or discard is simply a matter of personal or company preferences.

Interpolation results in loss of higher frequencies, as does migration. Since interpolation will cause a drop in the higher frequency part of bandwidth, the question must be asked as to whether interpolation can be done and still satisfy resolution requirements?

Migration comparisons. Migration algorithms differ with regard to:

- computational efficiency (speed)
- accuracy of steep dip migration
- acceptance of significant vertical velocity variations, a more realistic model

Note that time migrations do not allow sharp lateral velocity variations but smoothed lateral changes can be used. Technically, none of the algorithms handle lateral velocity variation.

Some properties are common to all migration algorithms.

- A velocity that is too fast will cause data to over-migrate, while a velocity that is too slow will cause data to under-migrate.
- Data will migrate more if velocity is increased and will migrate less if velocity is decreased.
- 3-D problems are not solved by 2-D migration.
- Multiples will appear as over-migrated events.
- Sufficient trace sampling is required to avoid aliasing.
- Input data must be free of anomalous amplitudes.
- Frequency content of dipping events is reduced proportional to the cosine of the dip angle.

In general, random noise is not attenuated.

Things to compare in migration include the following:

- response to velocity
- constant
- varies with time
- varies in space and time
- the amount of algorithm induced noise
- response to seismic noise
- geologic dip limitations
- run time and costs
- 2-D/3-D applicability, availability
- response to multi-valued problem
- required input velocity field

The following provides information for comparing the different algorithms.

Kirchhoff migration, T-X domain. Figure 6–232 shows the Kirchhoff impulse response on the zero-offset section (top) and the desired migration. It can be seen that Kirchhoff gives a full 90° migration, but, at steep dips, the amplitudes are wrong.

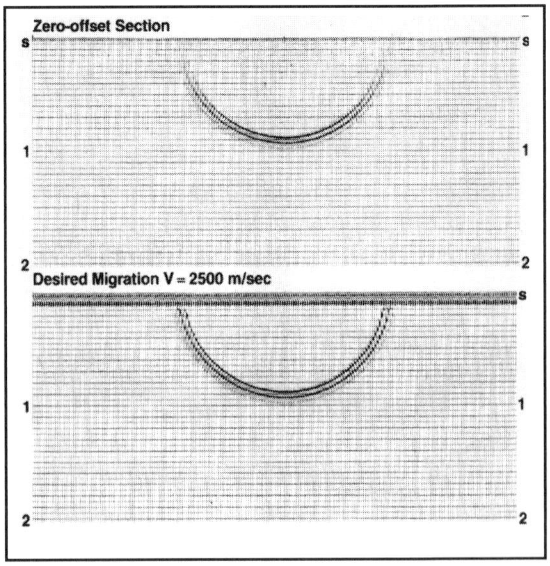

Fig. 6–232 *Kirchhoff Impulse Response*

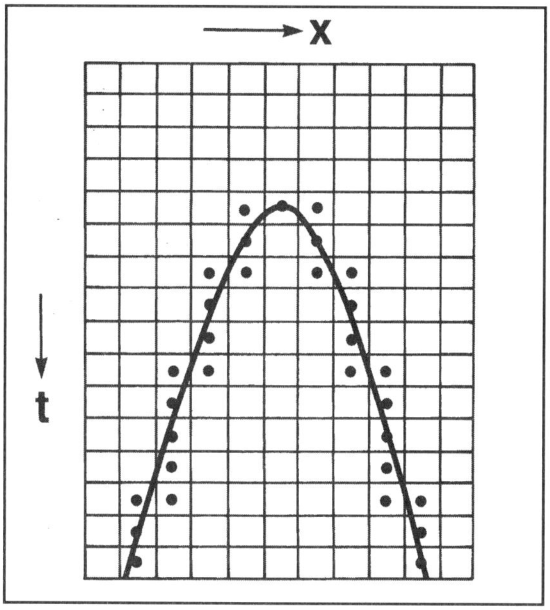

Fig. 6–233 *Chatter on the Limbs of the Kirchhoff Impulse Response*

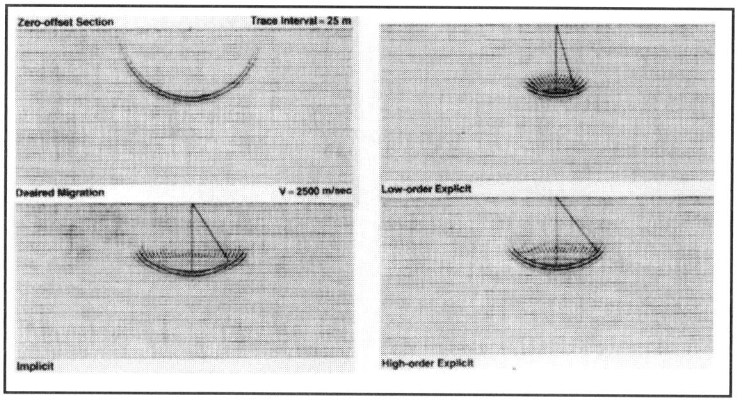

Fig. 6–234 *Finite Difference Impulse Response*

Figure 6–233 shows the Kirchhoff impulse response and problems presented by it. Dip aliasing occurs because of inadequate sampling at steep dip. Chatter on the limbs of the impulse response is caused by aliasing of the diffraction by the operator. Samples are lost—not included in the hyperbolic summation. The result is that amplitudes are not collapsed correctly at the operator apex.

Kirchhoff migration is correct for a constant velocity and very good for a time-variant velocity but poor for time- and space-variant velocity. It can induce noise depending on parameter selection (aperture width, maximum dip). It generates wave-fronts for localized noise and can remove coherent noise by maximum dip selection. Dip is limited by aperture selected.

Kirchhoff migration has a slow run time that is aperture-width dependent (larger aperture, slower running). It handles multi-valued events—conflicting dip or apparently two velocities at same time). It requires migrated V_{RMS} velocity. Tilted, non-symmetric hyperbolas (caused by lateral velocity variation) will not properly migrate to a point.

Finite difference migration, T-X domain. There are two types of finite difference operators—explicit and implicit. Further, there two kinds of explicit operators—low-order and high-order. Thus, Figure 6–234 shows three impulse responses, not just one.

The approximate dip limits for finite difference migrations are

- 30° (implicit, bottom left of Fig. 6–234)
- 15° (low-order explicit, top right of Fig. 6–234)
- 35° (high-order explicit, bottom right of Fig. 6–234)

In finite difference migration noise results from using a parabola instead of a hyperbola. Different portions of the hyperbola represent different dips, with the apex = 0° and limbs up to 90°. The low-order explicit operators fit a parabola to a hyperbola with no shift in apexes. Noise is caused by anisotropic dispersion due to the 15° approximation to the one-way wove equation. Anisotropic dispersion has different properties at various angles. The high-order explicit shifts the parabola to increase dip range, but this puts errors at zero dip (apex of hyperbola). There is less noise in the implicit and high-order explicit because more of the hyperbola is properly migrated.

Finite difference migration is correct for constant velocity. It has a dip-dependent error for time-variant velocity and has a fair response for a velocity that is both time and space variant. The algorithms are dispersive—transferring high dip data to induced noise. This may, however, be cosmetically beneficial. The dispersive algorithm reduces wave fronting from local anomalies. Noise is generally scattered. 15° and 35° algorithms are common, and a 65° algorithm is also available.

Finite difference migration has an acceptable run time, depending on dip recursion. Both 2-D and 3-D algorithms are available. It handles the multi-valued problem and requires a migrated velocity model. It is the best migration algorithm for time- and space-variant velocity.

Stolt migration, *F-K* domain. *F-K* Stolt migration is an *exact* solution when using only one velocity. It can then handle 90° of dip. Stolt requires use of the W-factor when velocity varies with time. It becomes a 45° solution when W = 0.5, and the processing is much slower. See Figure 6–235.

Stolt migration is correct for constant velocity and very good for time-variant velocity and moderate time- and space-variant velocity. There is no induced noise unless dip-limiting options are chosen. It generates wave fronts from localized noise and can remove coherent noise trains. It has a dip range out to 90°.

Stolt migration has acceptable run times with both 2-D and 3-D applicability. It is unable to handle multi-valued problem and uses unmigrated stretch and migrated stretch velocities.

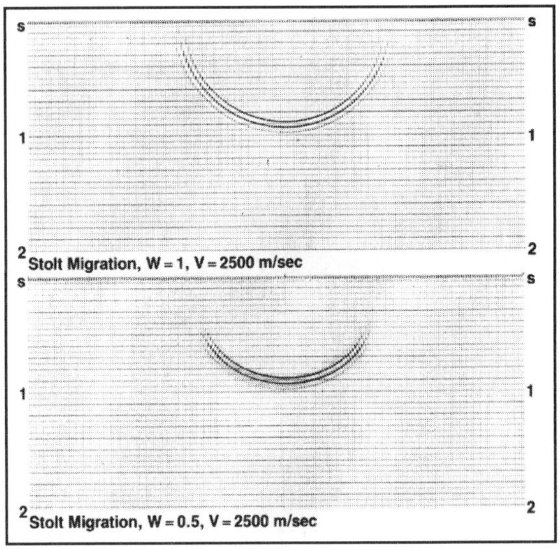

Fig. 6–235 *Stolt Impulse Response When W = 1 and 0.5*

Migration input parameters. The following discusses parameter values for all of the migration algorithms and how to select specific values to solve specific imaging problems. Two synthetic models—the multi-dip model previously used and a three point-aperture model—and a CMP stack section showing complex geology are used to demonstrate parameter testing. Not all models are used for all algorithms and all parameters.

Kirchhoff migration input parameters. Kirchhoff migration parameters are

- aperture width
- maximum dip parameter
- velocity field

Aperture width is a critical parameter. It limits width of diffraction pattern used in summation by specifying the number traces sampled for amplitude on the hyperbolas. It restricts the amount of migration but not sufficient to give total control. The maximum dip parameter is also needed.

The maximum dip parameter does not refer to geological dip but to apparent time dip in diffractions. It can be used as a dip filter to remove time dip greater than the specified value.

A good velocity field must be supplied. This can be derived from but is not identical to the stacking velocity. Generally, a lower velocity is needed for migration.

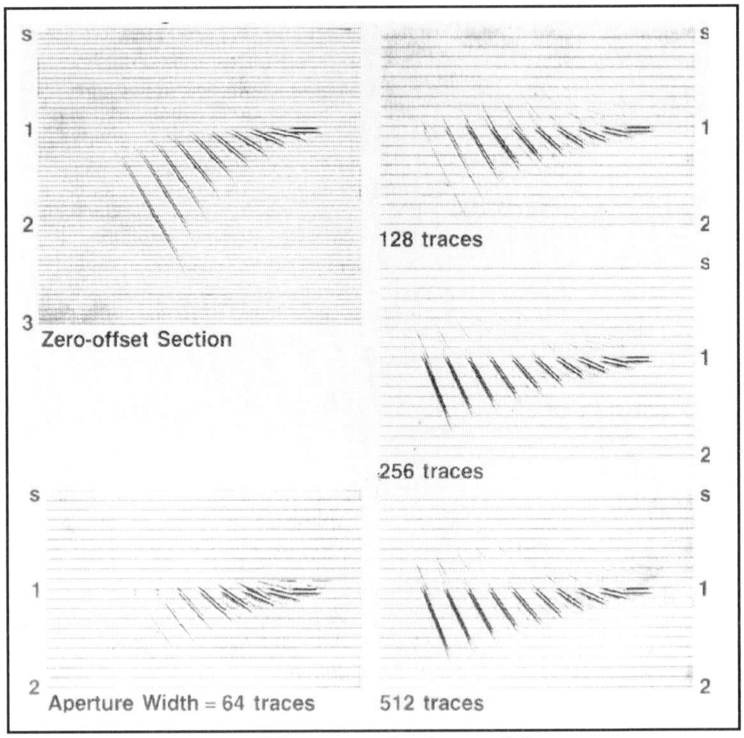

Fig. 6–236 *Aperture Width Test, Multi-dip Model*

Choice of parameter values may depend on shallow versus deep data. For example, an aperture width of 64 traces may be adequate for shallow data. It would see most of the dip, but filter out steeper dip for deeper data.

Aperture width. Figure 6–236 shows Kirchhoff migration of the multi-dip model using values of 64, 128, 256, and 512 traces for the aperture width. The 64-trace and 128-trace apertures do not migrate the highest dips. The 256–trace and 512-trace apertures do migrate all dips. Since the cost of the process is directly proportional to aperture width, use the smallest aperture width consistent with steepest dip to migrate—256 traces in this example.

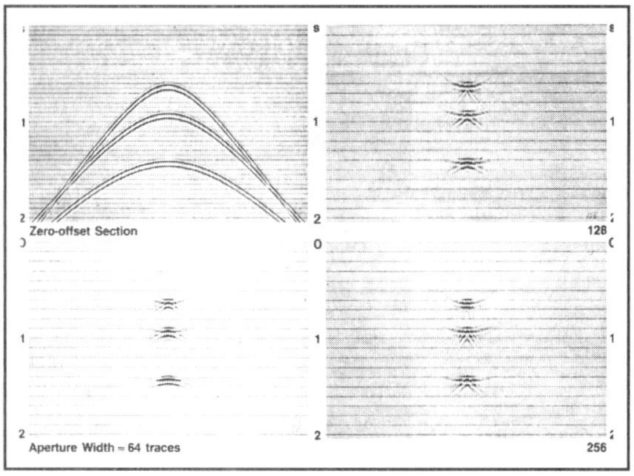

Fig. 6–237 *Aperture Width Test, Diffraction Migration*

Figure 6–237 shows Kirchhoff migration of three diffractions using apertures of 64, 128, and 256 traces. The migration of diffractions is incomplete for all apertures tried. The collapse achieved by the 128-trace aperture appears to be about the same as the 256–trace aperture and is the choice here.

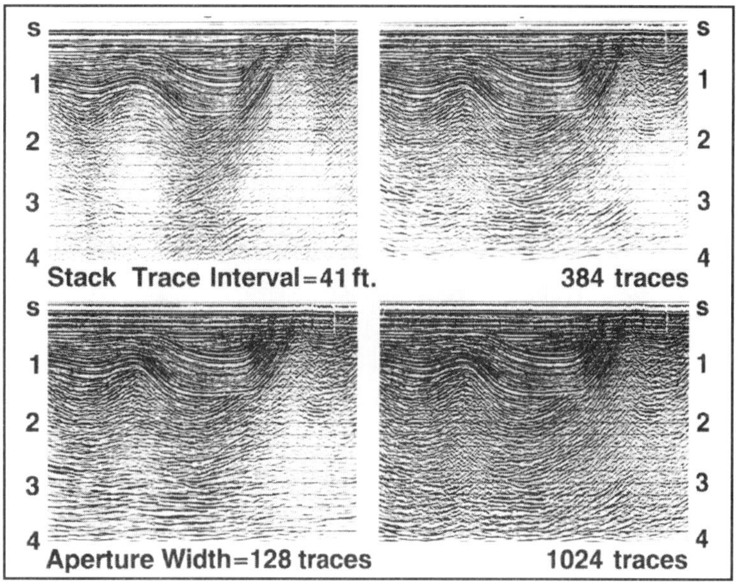

Fig. 6–238 *Aperture Width Test, Field Data*

Kirchhoff migration does not really go out to 70° and thus does not sum in all energy to the correct place. Also, the data are band-limited, spread length is finite, and there is a finite amount of data. Where these conditions exist, the result may be incomplete diffraction patterns and, hence, incomplete collapse of diffractions.

Figure 6–238 shows a CMP stack of field data and Kirchhoff migration using aperture widths of 128, 284, and 1024 traces. The 128-trace aperture is inadequate to handle steepest dips (see far right at about 0.5 to 1 second). The 384-trace aperture handles these dips adequately, as does the1024-trace aperture. The choice is 384 traces in this example.

Some additional comments on aperture width may be helpful.

- Diffractions are truncated if the aperture is too narrow.
- There is also such a thing as an aperture that is too wide. It reaches down and sums in noise.
- Horizontal migration noise is not noticed on real data because ambient noise masks it.
- Noise spikes at the bottom of the section cause wave fronting. The spike causes the impulse response of migration to be imprinted onto the late-time data.
- Kirchhoff migration tends to organize noise as a function of aperture width (Fig. 6–239).
- A narrow aperture gives more linearized noise.
- If the aperture width is used to limit dips, then more linearized noise results.

If the aperture is too wide, then Kirchhoff migration will sum along the hyperbola beyond the point at which the amplitude of the hyperbola leg is less significant than the noise being added in.

Maximum dip. Maximum dip means not just the geologic dip of the data but also the apparent dip seen on the diffraction patterns and fault planes. Maximum dip limit can be specified as ms/trace or degrees. In theory, Kirchhoff can migrate up to 90°, but the practical limit is 35 to 45° because of operator aliasing.

Figure 6–240 shows Kirchhoff migration of the multi-dip model using values of 4, 8, 12, and 24 ms/trace. Note that dips are not migrated when less than 12 ms/trace is used for maximum dip. No improvement in migration is provided by the use of 24 ms/trace, so the choice is 12 ms/trace.

Note that the aperture width used in all cases of this example is 256 traces. In parameter testing, only one parameter is varied at the same time.

Figure 6–241 shows the Kirchhoff migration of field data using values of 4, 8, 16, and 32 ms/trace for maximum dip. The aperture width is 384 traces in each test.

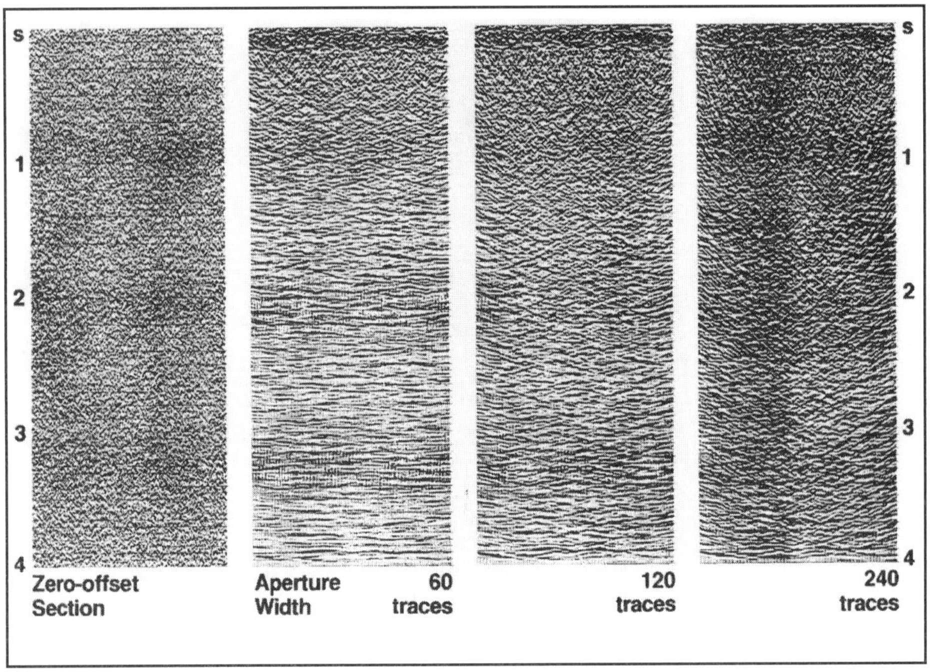

Fig. 6–239 *Aperture Width and Random Noise*

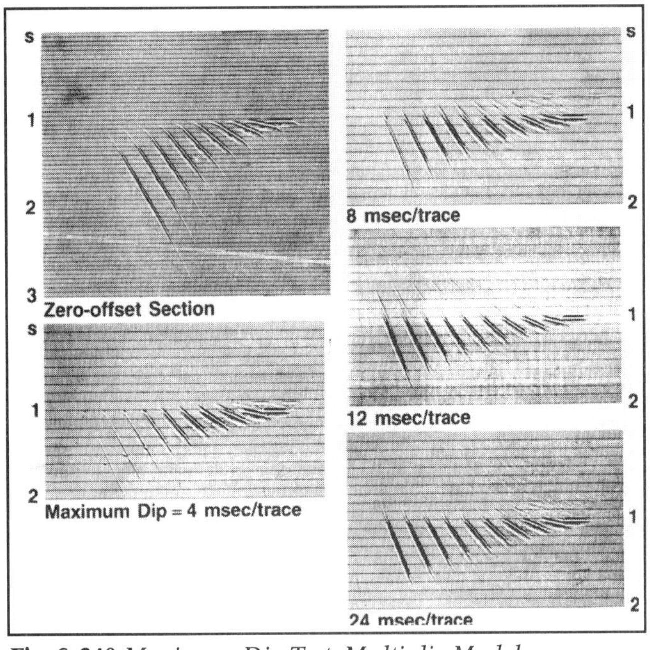

Fig. 6–240 *Maximum Dip Test, Multi-dip Model*

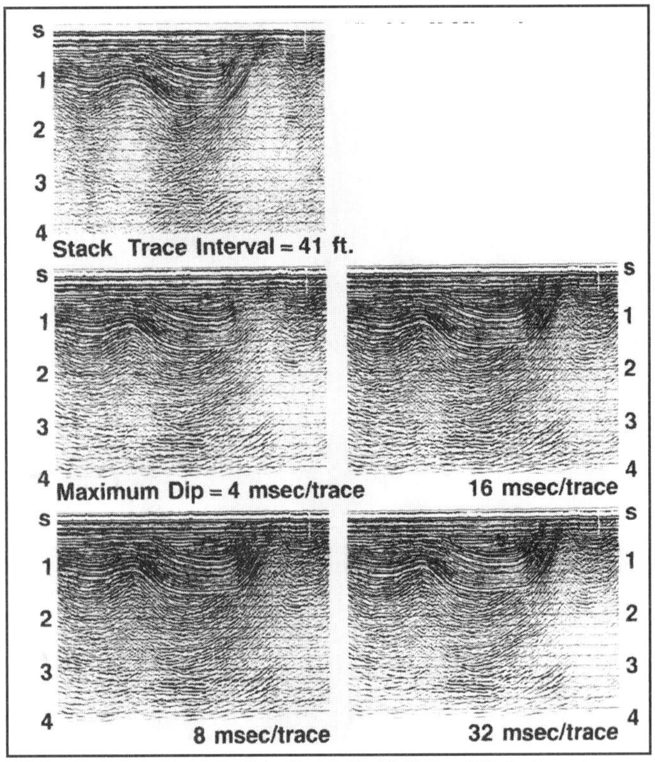

Fig. 6–241 *Maximum Dip Test, Field Data With Aperture Width of 384 Traces*

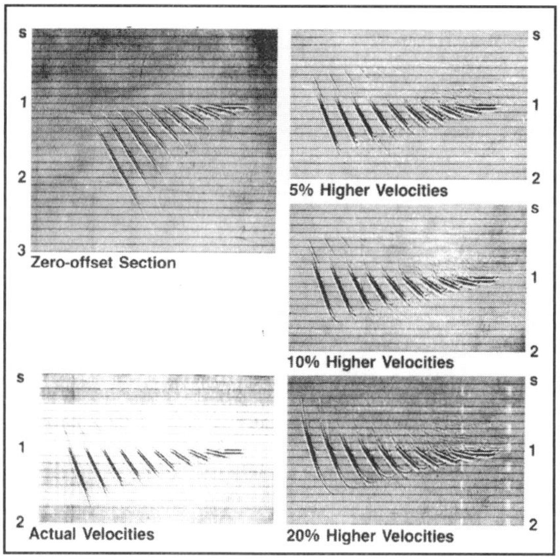

Fig. 6–242 *Effect of Velocity Errors on Kirchhoff Migration*

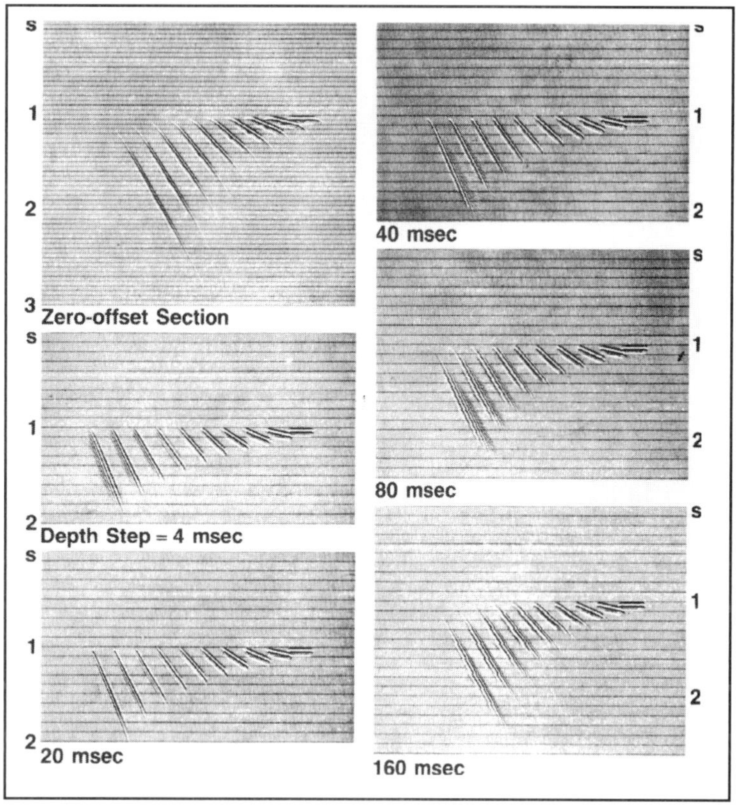

Fig. 6–243 *Depth Step Test, Implicit Finite Difference With a Sample Period of 4 ms*

Velocity errors. Figure 6–242 shows the effects of using velocities that are too high. For low dip, the velocity errors have only a small effect. Remember, too high a velocity moves data too far—over-migrates it—especially steep dip. Anomalies at steep dip are extensions in data caused by operator hyperbola aliasing in Kirchhoff migration.

According to Claerbout in *Imaging the Earth's Interior,* 70% of diffraction energy is within the first Fresnel zone. Migration is focusing/collapsing data within the first Fresnel zone, so diffraction energy within first Fresnel zone must be adequately sampled. The dip of diffraction energy at edge of the Fresnel zone is approximately 15°. Therefore, for adequate spatial sampling, maximum dip should always considered to be not less than 15°.

Using the wrong parameter value for aperture width or maximum dip will cause steep dips to disappear. Small aperture widths generate linearized noise.

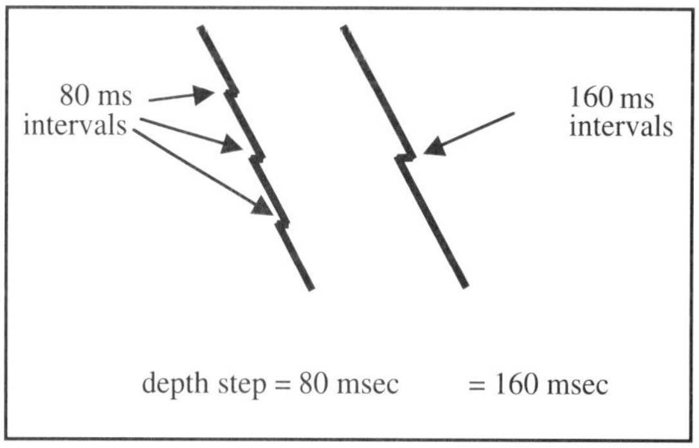

Fig. 6–244 *The Zig-Zag Effect*

Finite difference input parameters. The critical parameter in finite difference migration is depth step, which is specified in ms. The smallest possible depth step is the processing sample period, but using this value maximizes run time and cost. A larger depth step reduces run time but increases prediction error.

Implicit finite difference input parameters. A depth step test for the implicit finite difference algorithm, using the multi-dip model, is shown in Figure 6–243. The sample period is 4 ms. Results, using depth steps of 4, 20, 40, 80, and 160 ms are shown. The 4 ms depth step gives the most accurate migration but shows the dispersive effect of finite difference migration most clearly. Depth steps of 20 and 40 ms under-migrate the 30 to 45° dips but are almost equal, so the choice here is 20 ms.

At 80 and 160 ms, depth steps discontinuities can be seen at each step. This is caused by the increase in prediction error as the time over which the prediction is made increases. This is often called the zig-zag effect. See Figure 6–244.

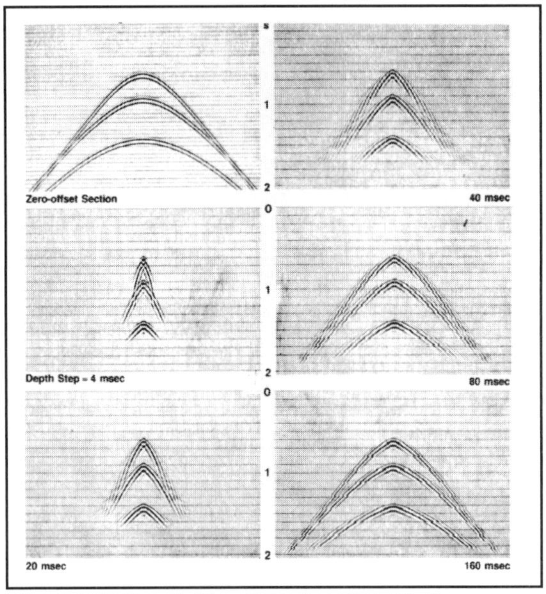

Fig. 6–245 *Depth Step Test, Implicit Finite Difference Migration of Diffractions*

Both explicit algorithms give rise to dispersion because as dip increases, migration error increases and data are spread out over an area. The implicit finite difference algorithm also gives rise to dispersion.

Figure 6–245 shows the implicit finite difference migration of three diffractions. Note the under-migration at 20 and 40 ms depth steps and zig-zag patterns for 80 and 160 ms depth steps.

Figure 6–246 shows a CMP stack section and, below it, a sketch highlighting salient features seen on the stack. Figure 6–247 shows the implicit finite difference migration of the CMP stack section of Figure 6–246 using a depth step of 40 ms. A sketch pointing out significant aspects of the migration is shown below that section of the figure. Note the noise at the bottom of the section. This is caused by diffraction truncation.

Figure 6–248 shows the implicit Finite difference migration of the CMP stack section of Figure 6–246 using a depth step of 20 ms. A sketch pointing out significant aspects of the migration is shown below the section. Migration noise at bottom of section was expected. Two to three more seconds of data were needed to fully migrate diffractions since diffractions were cut off at bottom of the section.

Why not just add more record length? The problem with including extra record length for diffractions is that it reaches down into a portion of the section where the signal-to-noise ratio is low. Diffraction prediction in the lower section will tend to produce more noise than signal. When there are noise spikes in the lower section, they will produce noise smiles after migration. The best processing strategy is to taper amplitudes to zero at the bottom of the time section.

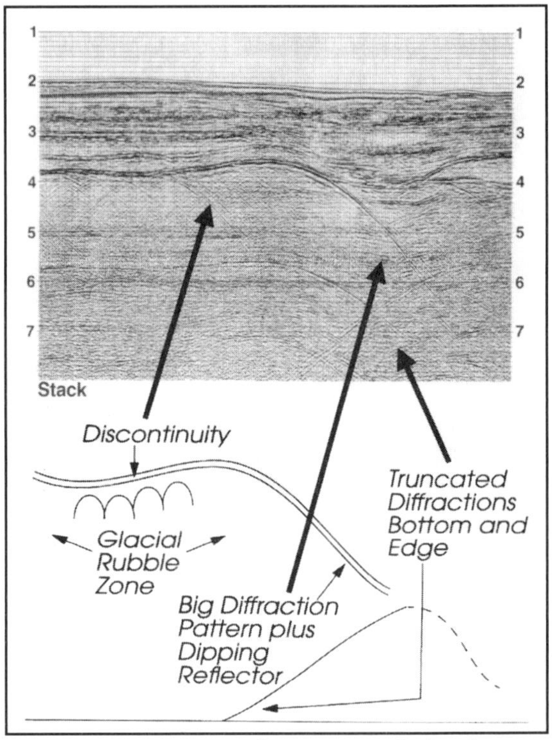

Fig. 6–246 *A CMP Stack Section and Sketch Highlighting Salient Features of the Stack*

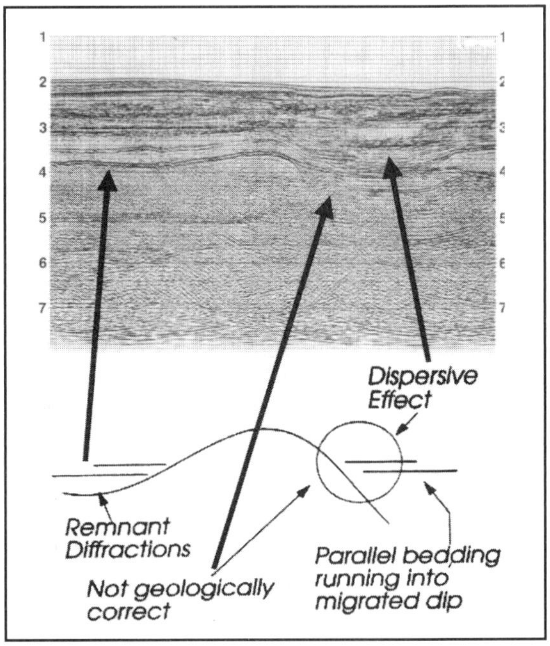

Fig. 6–247 *Implicit Finite Difference Migration of the CMP Stack Section of Figure 6–246 Using a Depth Step of 40 ms and a Sketch Pointing Out Significant Aspects of the Migration*

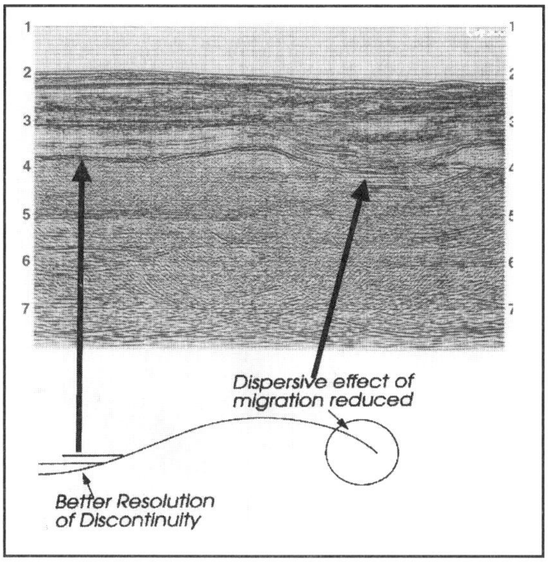

Fig. 6–248 *Implicit Finite Difference Migration of the CMP Stack Section of Figure 6–246 Using A Depth Step Of 20 ms and a Sketch Pointing Out Significant Aspects of the Migration*

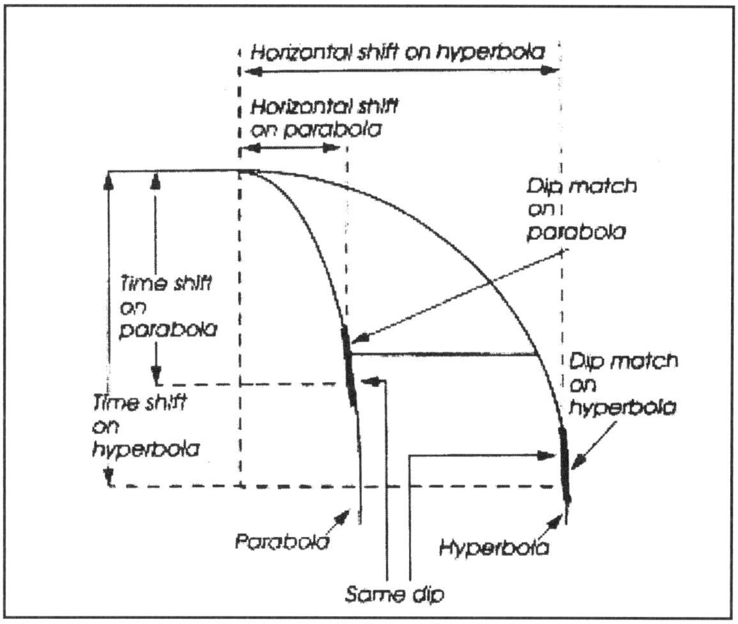

Fig. 6–249 *Comparison of Parabolic and Hyperbolic Time Shifts*

Finite difference migration tends to under-migrate steep dip. It does not shift data enough in distance and time. This is because it uses a parabola instead of a hyperbola. As a result the predicted dip is less than the actual dip. See Figure 6–249.

Finite difference migration under-migrates because of the parabolic approximation. This is comparable to using a velocity that is too slow. If velocities are too high, then Kirchhoff over-migrates. Use of finite difference combined with velocities that are too high tends to balance out the under-migration. Because of this, some people thought that finite difference must be better, but this is only because a velocity error corrected an algorithm defect.

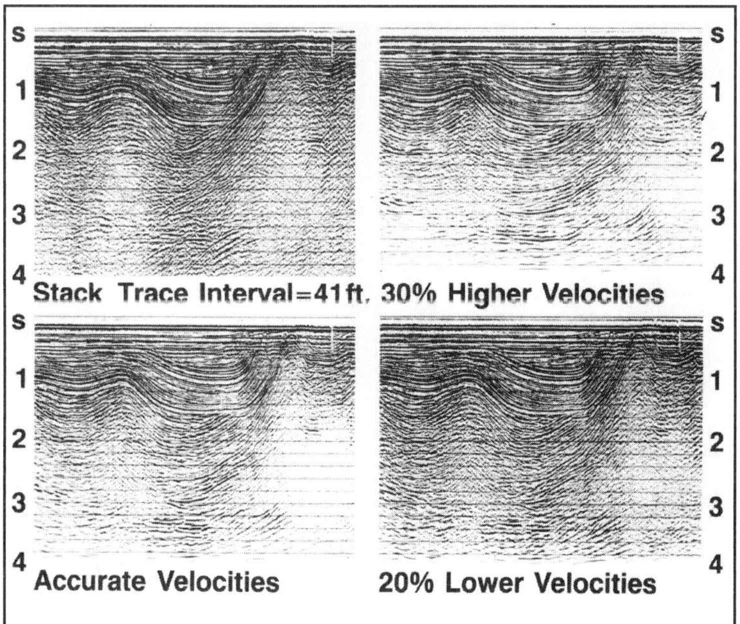

Fig. 6–250 *Velocity Test, Implicit Finite Difference Migration, Depth Step of 20 ms*

Implicit finite difference migration velocity errors. Figure 6–250 shows the implicit finite difference migration of the same CMP stack as in Figure 6–241. Three velocities are used to migrate the data—the correct migration velocity, a velocity 30% higher, and a velocity 20% lower. It is obvious that steeper dips look better at faster velocities. However some lower dips are over-migrated at the higher velocity.

Figure 6–251 uses the multi-dip model to compare migration by implicit, low-order explicit, and high-order explicit finite difference. The depth step is 20 ms in each case. This was the value decided upon in Figure 6–241, which used the same model to test depth step for the implicit algorithm. The high-order explicit appears to migrate the higher dips of the model better than the implicit, but lower dips are a bit over-migrated, and there is a slight indication of zig-zag.

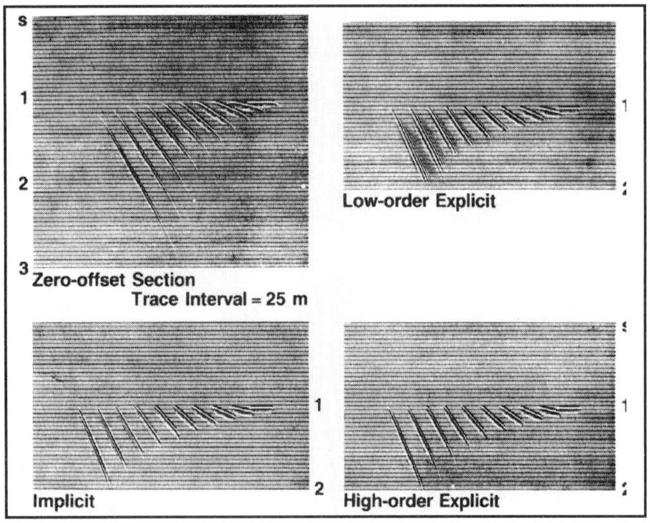

Fig. 6–251 *Comparison of Finite Difference Algorithms, Multi-dip Model*

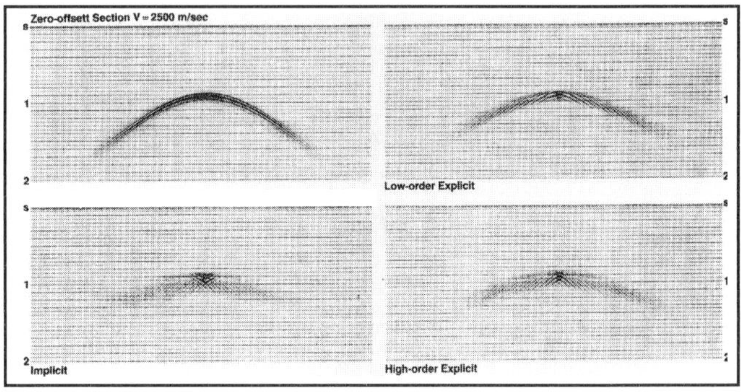

Fig. 6–252 *Comparison of Finite Difference Algorithms, Three-point Aperture Model*

Figure 6–252 compares the three finite difference algorithms using the three-point aperture model. The implicit collapses the diffractions the best but has a lot of dispersion. The low-order explicit collapses the diffractions the least. The high-order explicit is intermediate with regard to diffraction collapse (under-migrates) and dispersion.

The recommended value for depth step in finite difference migration is from one-half to one times the dominant signal period. For steeper dips, use a smaller step in that range.

***F-K* migration parameters.** There are two migration algorithms that operate, nominally, in the *F-K* domain—Stolt and Gazdag's phase-shift migrations. Both algorithms require input of the depth step parameter and a velocity field. Stolt migration also requires the stretch factor W. The original Stolt migration was a constant velocity migration but is called generalized Stolt when a value other than one is used. The generalized Stolt converts time to a pseudo depth by stretching vertical time, migrates in depth, and then converts back to original time axis.

Stolt migration and the stretch factor *W*. Theoretically $0 < W < 2$. $W = 1.0$ is the constant velocity Stolt. However, $W = 0.5$ is the recommended starting value. The effect of W is illustrated by Figure 6–253, which uses a synthetic example consisting of three horizons with large synclines on each, a fault cutting through the right flanks of two shallower synclines, and a small anticline on the middle horizon. The top left of Figure 6–253 is a constant velocity, zero-offset, section. Bow ties that are characteristic of synclines, diffractions, and a distorted anticline can be easily seen.

The other five sections in Figure 6–253 are Stolt migrations of the zero-offset section. The best migration is for $W = 1$. The migrations for $W = 0$ and 0.5 are under-migrated while those for $W = 1.5$ and 1.95 are over-migrated. Since the model is constant velocity, this is not surprising.

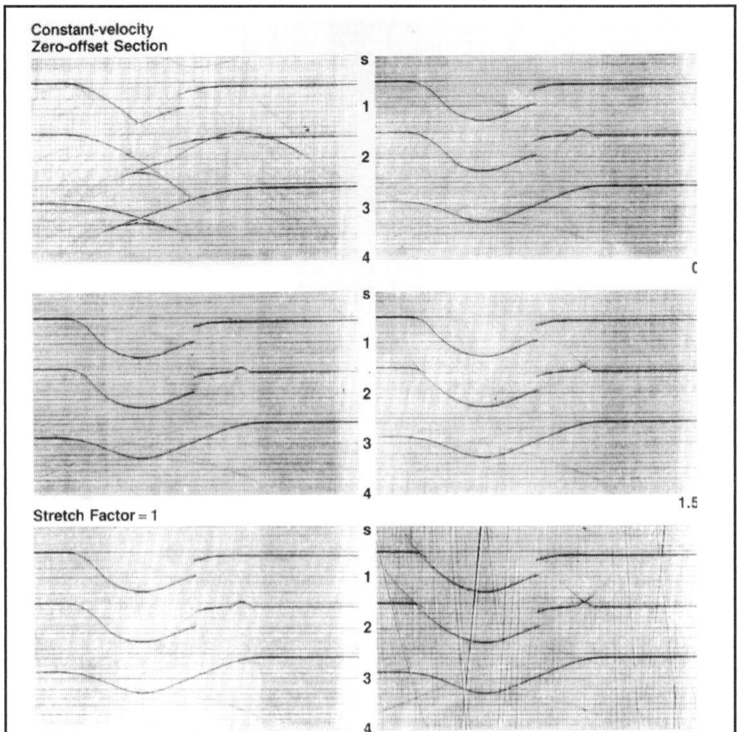

Fig. 6–253 *Parameter Test for Stretch Factor W in Stolt Migration*

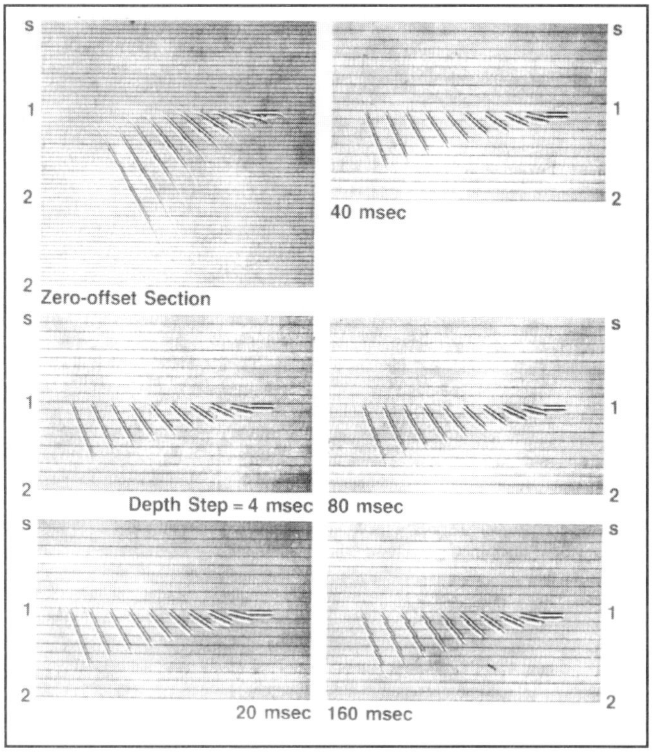

Fig. 6–254 *Depth Step Test, Phase Shift Migration, Multi-dip Model*

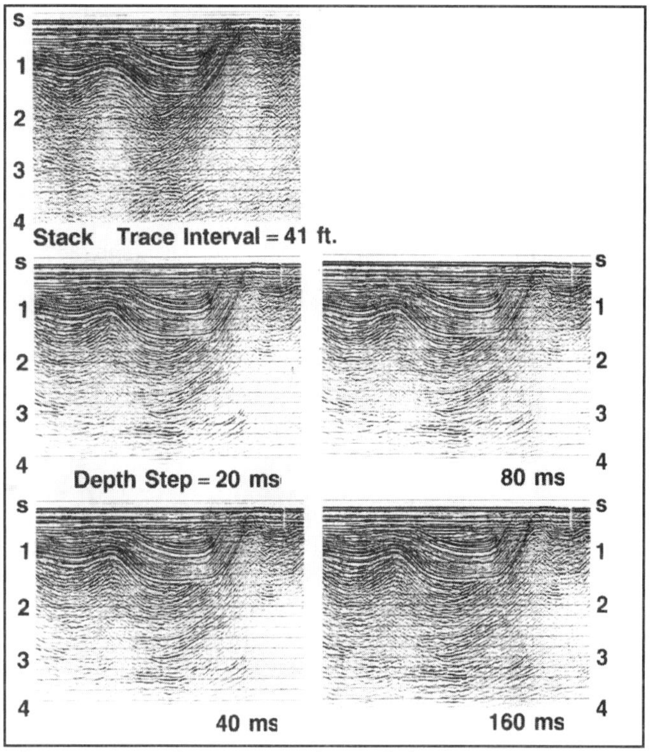

Fig. 6–255 *Depth Step Test, Phase Shift Migration, Field Data*

Gazdag's phase-shift migration input parameters

Depth step. Gazdag's phase-shift migration uses linear phase shift in the frequency-wave number domain. In the K_z domain, a linear change in phase is equivalent to a linear change in depth. As in finite difference and Stolt, depth step is input in ms.

Figure 6–254 shows a test of the depth step parameter for phase-shift migration using the now-familiar multi-dip model. Values of 4, 20, 40, 80, and 160 ms were used. The 4 ms depth step is the most accurate. However, because of the band-limited nature of the wavelets, the 20 ms depth step looks as good as the 4 ms. In synthetic models, which lack random noise, noise generated by the migration process can readily be seen. Figure 6–255 uses field data for a depth-step test. The 20 ms step appears to be nearly as good as the 4 ms step. Field data has noise in the data that tends to cancel or hide migration noise.

Phase-shift migration and velocity errors. Figure 6–256 uses the multi-dip model to illustrate the effect velocity errors have on phase-shift migration. Migrations using the actual velocity and velocities that are 5%, 10%, and 20% higher are shown. Unlike the finite difference migration, faster velocities do not improve the appearance of the output but simply over-migrate the data. The 40 ms depth step appears to perform about as well as the 20 ms depth step.

Steep dips are affected in similar fashion to Kirchhoff migration. However, phase-shift migration, unlike Kirchhoff, does not have edge effects.

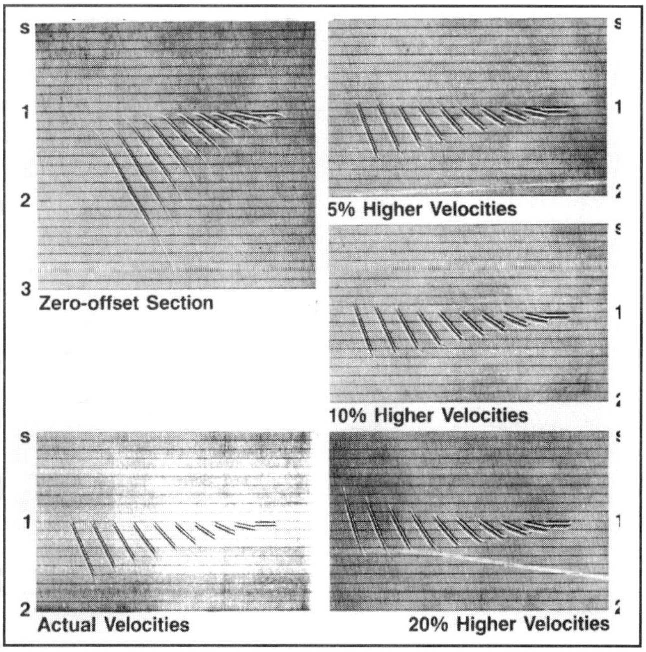

Fig. 6–256 *Effect of Velocity Errors on Phase-shift Migration*

Migration in stages. Things desired in migration are processing speed, accuracy for steep dip, and the capability to handle vertical velocity gradients. Splitting migration into two stages can often extend the accuracy and flexibility of time migration while also lowering the cost of migration.

For example, the constant velocity Stolt, which is fast and accurate out to 90°, can be used as the first stage. Finite difference can be used as the second, variable velocity stage. This procedure requires modified velocities—a combination of reduced velocities in the first (Stolt) stage and residual velocities in the second (finite difference) stage.

It is desirable to have a migration flow that combines different types of migrations. Such combinations may take advantage of each type's benefits while avoiding some of their drawbacks. *Cascading migration* means that the migration process is repeated several times instead of being performed only once.

In cascaded Kirchhoff migration, the aperture width is reduced during each stage. As a result, it treats ray bending caused by vertical velocity variations more accurately. In cascaded Stolt migration, velocity variation in each stage is reduced. This allows accurate Stolt migration in the presence of significant vertical velocity variation.

Cascaded (multi-stage) applications of finite difference migration extend accuracy of the finite difference algorithm beyond its inherent 15° limit. By cascading a number of 15° migrations, the finite difference algorithm is tricked into thinking that the maximum dip is less than 15°. This is accomplished by reducing the velocity used in each stage of migration.

The velocity used in each stage of cascaded migration must conform to Equation 6.73.

$$V^2 = \sum_{i=1}^{N} V_i^2 \tag{6.73}$$

where

V = velocity used in single stage of migration

V_i = effective velocity of the ith stage

Migration in stages introduces *perceived dip*. Perceived dip is the dip migration thinks it is trying to migrate, based on the velocity used for that stage. The lower the velocity is, the smaller the perceived dip. Making the perceived dip smaller improves the effectiveness of migration. For example, the low-order explicit finite difference migration, which is relatively fast and inexpensive, is very effective for dips less than 15°. Another factor is that velocity errors cause large dips to mis-migrate. By effectively reducing dips in each stage by using slower velocities, the effect of velocity errors is also reduced.

The perceived dip θ_i to be migrated is

$$\sin\theta_i = V_i \frac{dt}{dx} \tag{6.74}$$

where

V_i = effective velocity of the ith stage

$\frac{dt}{dx}$ = dip on input section

The dip dt/dx is fixed in data. The velocities for each stage V_i can be made small so that the dip perceived by the migration is small.

Cascaded Finite Difference. Finite difference is good for small amounts of dip, and it allows some lateral velocity variation, at least better than other migrations. Finite difference can be used if the perceived angle is limited to 5° of dip. Do this by cascading finite difference in enough steps of 5° of dip in each stage to build up to dips greater than the 15° limit. As long as Equation 6.73 is satisfied, cascaded finite difference can be done.

This was the original idea for cascaded migration, but it did not work as well in practice as anticipated. The concept was then applied to Stolt *F-K* migration. Use Stolt with time-varying velocity, input the actual velocity, and do a staged migration.

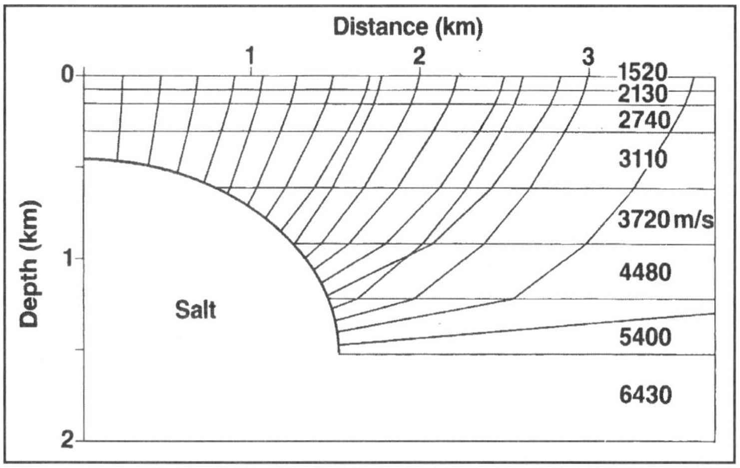

Fig. 6–257 *Salt Model Used to Demonstrate Extended Stolt Migration*

Conventional Stolt with $W = 0.5$ stretch breaks down at 45°. Cascaded Stolt (or extended Stolt) provides migration of dips in the range of 80 to 85° for $W = 0.9$. It handles time-variant velocity but cannot handle lateral velocity variations well. One additional change is necessary for Stolt to handle lateral velocity variations.

In 3-D processing, a velocity field defines a velocity for every CMP position and time. One new velocity field, consisting of the slowest velocity found at each time sample for the entire survey, can be built. When this slowest velocity function $V_z(t)$ is put into the extended Stolt, it gives an 85 to 90° solution with good noise characteristics. One velocity function is best for extended Stolt.

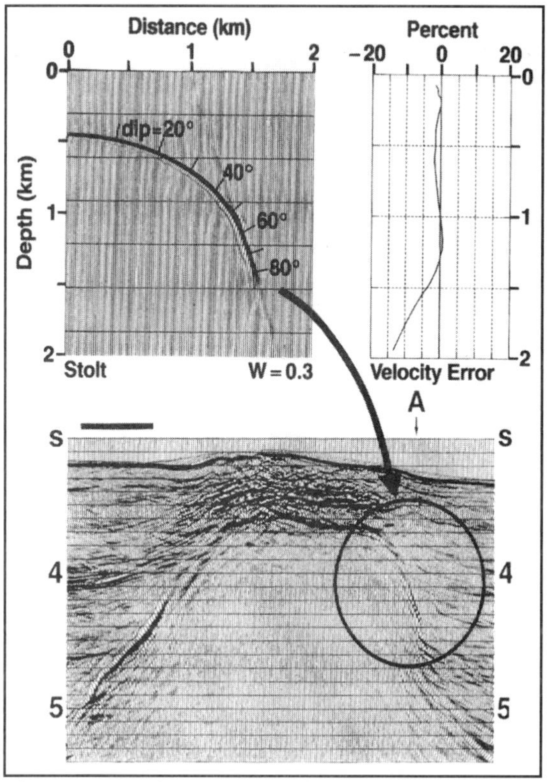

Fig. 6–258 *Stolt Migration of Salt Model, Field Data W = 0.3, and Velocity Error*

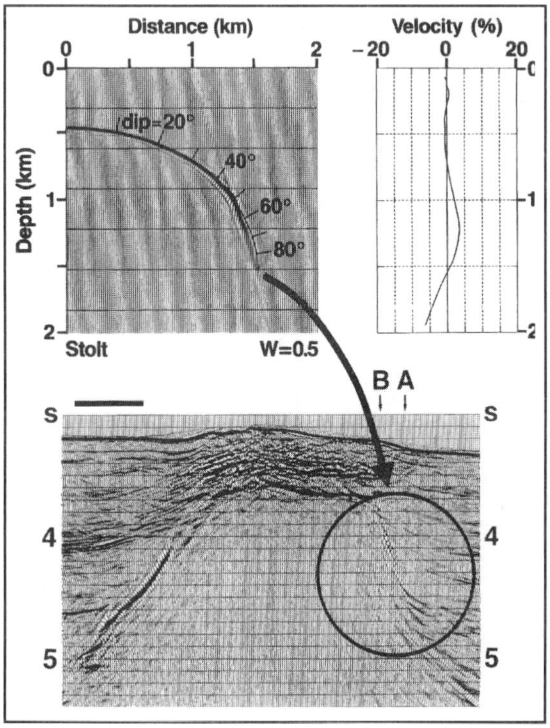

Fig. 6–259 *Stolt Migration of Salt Model, Field Data W = 0.5, and Velocity Error*

This first stage does most of the migration of the data, but they are under-migrated, because the slowest velocities were used. An additional migration stage is required. The velocity field required for this is the difference between the slowest velocity field and all other velocities. This difference or *residual velocity* for the data volume is both time- and space-variant, whereas the slowest velocity field used in the first stage is only time-variant. The residual velocity field is input into a second stage that uses finite difference migration, which can handle lateral velocity variations quite well. Because the residual velocities are relatively small, finite difference sees dips that are less than 15°. This procedure minimizes the disadvantages of Stolt and finite difference while taking advantage of their individual benefits.

Figure 6–257 is a model of the top right flank of a salt dome used for tests of Stolt migration. The model shows interval velocities and some zero-offset ray paths. The salt face is modeled as a spherical surface so that dips from 0 to 90° are represented. Figures 6–258, 6–259, 6–260, and 6–261 show the migrated model, plots of velocity error, and migrations of the section upon which it is based.

Figure 6–258 shows conventional one-pass Stolt using $W = 0.3$. The migration is good out to about 50°. This is clearly seen on the model. The circled part of the salt mass on the section corresponds to the steepest dip on the model. It is seen to be under-migrated.

Figure 6–259 shows conventional one-pass Stolt using $W = 0.5$. It breaks down at 45°, as is obvious on the model migration. The steepest dip of the salt mass on the section (circled) is again under-migrated.

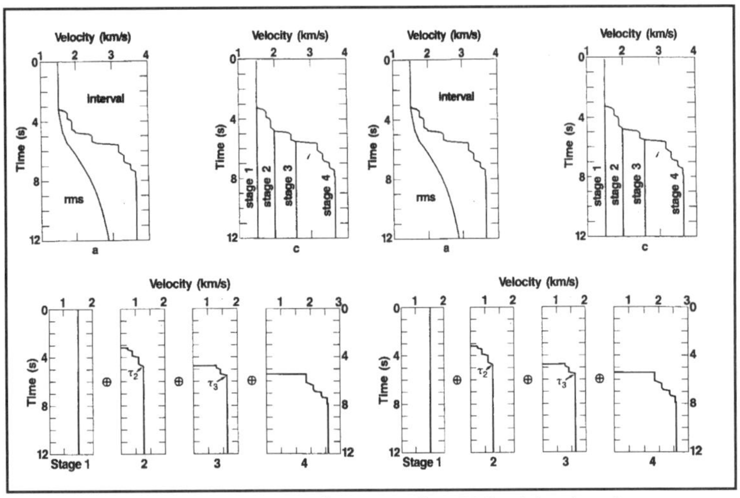

Fig. 6–260 *Velocity Data Used to Develop Velocities for the Four-stage Stolt Migration of Figure 6–261*

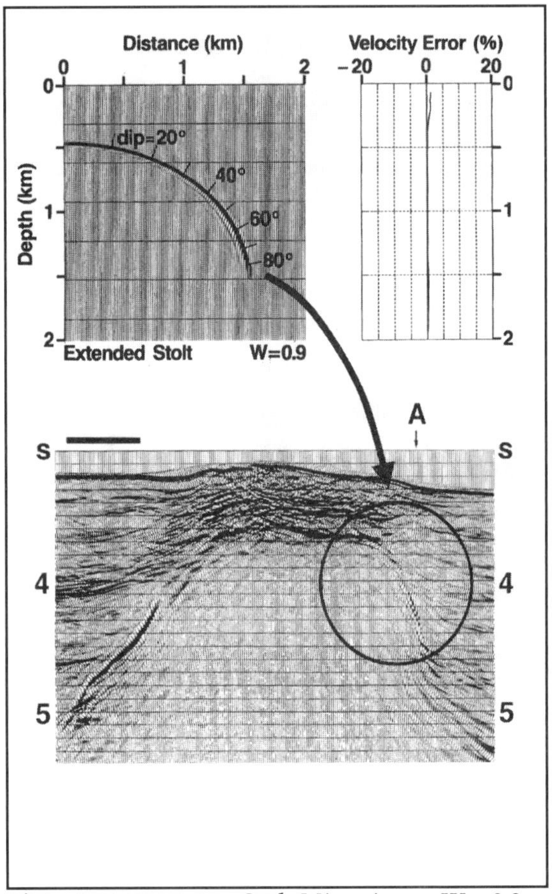

Fig. 6–261 *Four-stage Stolt Migration at W = 0.9 of Model and Field Data*

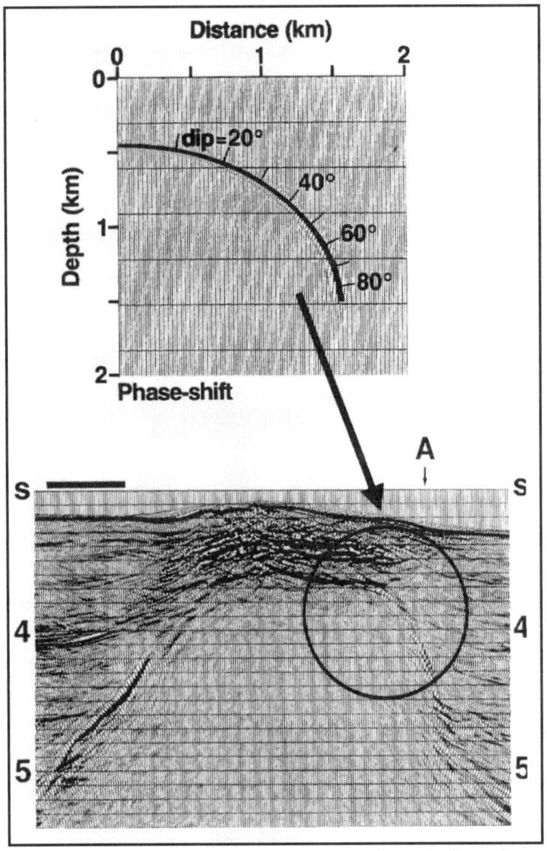

Fig. 6–262 *Phase-shift Migration of Model and Field Data*

Figure 6–260 shows velocities used in a four-stage Stolt migration ($W = 0.9$) of the section. The top left shows rms velocities and derived interval velocities. The top right shows the division of interval velocities into those corresponding to velocities used in stages 1 through 4. The bottom of the figure shows these interval velocities separately. Note that stages 2, 3, and 4 do not migrate all of the section. Stage 1 migrates to time T_1 (about 3.2 s). Stage 2 begins at time T_1 and completes the migration down to time T_2 (about 4.8 s). Stage 3 begins at time T_2 and completes the migration down to time T_3 (about 4.8 s). Stage 4 begins at time T_3 and completes the migration down to the bottom of the section. Note that the actual velocities used in these stages must conform to Equation 6.73—the sum of the squares of the velocities for the four stages must equal the square of the velocity used for one-pass migration.

Figure 6–261 shows the four-stage Stolt migration of the model and the section. The model migration shows that the extended Stolt for $W = 0.9$ migrates to about 85°. The steep dips in the circled part of the field data migration are better migrated than by either of the single-stage migrations. Note also, on the top left, the improvement in velocity error.

Figure 6–262 shows a phase-shift migration of the model and the section. It does not appear to be a significant improvement over the four-stage Stolt migration although its cost is considerably greater.

Post-stack migration summary. Techniques based on summation plus imaging at $t = 0$ include semicircle superposition, diffraction summation, and Kirchhoff integral solution. Kirchhoff is the only one of these normally implemented.

Techniques based on downward continuation include finite difference, Stolt, and phase-shift. Table 6–4 summarizes characteristics, strengths, weaknesses, and other elements for all the preceding algorithms.

Table 6–4 *Post-Stack Migration Summary*

Algorithm	**Kirchhoff**		**Finite Difference**		**Phase Shift**	**Stolt**		
1-Pass	Cascaded	1-Pass	Cascaded	1-Pass Only	1-Pass	Cascaded		1-Pass
Domain	x, t, z	x, t, z	x, t, z	x, t, z	k, f, z	kx, f, kz		kx, f, kz
Speed	Slow	Slow	Good	Fair	Slow	Fast *		Good **
Maximum Dip	75°	90°	15° - 35°	75°	90°	90°	45°	90°
Handles V(z) or V(t)	Fair	Good	Good	Good	N/A	Fair	Good	Fair
Handles V(x)	Fair	Fair	Fair	Fair	N/A	Fair	Fair	Fair
Comments	Newer algorithms have much better noise characterristics	Ray bending compen–sation may speed up 1-pass 3-D	Limited steep-dip accuracy	Accuracy Limited below large interval velocity	Forces constant lateral velocity	Not a practical solution except for water bottom reflection or as first pass of cascaded migration		Accurate with V(z), sensative to V(x)
3-D Comments	Slow	Good						
	1-Pass	1-Pass	2-Pass	2-Pass	2-Pass	2-Pass		1-Pass

* Constant Velocity
**Time- or Depth-Variant Velocity

Time migration is applicable to the time-variant velocity situation. Depth migration is usually needed for the time- and space-variant velocity situation.

Table 6–5 shows processing time for each algorithm relative to constant velocity Stolt, the fastest algorithm. From the table, it is seen that Kirchhoff requires five times as much run time as constant velocity Stolt, and finite difference requires 2.5 times as much run time as constant velocity Stolt. With $W = 0.5$, Stolt run time increases by a factor of 10.

Table 6–5 *Relative Processing Speeds of Migration Algorithms*

Algorithm	Relative Speed
Stolt (W =1)	1
Kirchhoff	5
Finite difference	2.5
Stolt (W = 0.5)	10

Kirchhoff is comparable to Stolt except the amplitudes for dips greater than 80° are reduced. Improvements have been made in generalized Stolt migration by compensating for known errors by parallel processing in multi-CPU systems.

Pre-stack migration. Pre-stack time migration is Kirchhoff migration. As in post-stack migration, all events on the record are assumed to be diffractions. However, since the data are not at zero-offset, the exploding reflector concept cannot be used. Straight rays are assumed, which means a constant velocity between the surface and the point scatterer.

Figure 6–262 illustrates the situation for a trace on a shot record. A source at *S* is at a distance *x* from a receiver at *R*. A point scatterer is at a depth *z*, vertically below a position a distance *d* from the receiver. The point *P*' is the projection of point *P* on the surface. From the geometry of Figure 6–263, the time t_x from *S* to *P* to *R* is given by:

$$t_x = \frac{\overline{SP} + \overline{PR}}{V} \tag{6.75}$$

Applying the Pythagorean theorem to triangles *SPP*' and *PP'R*, gives:

$$\overline{SP} = \sqrt{z^2 + (x-d)^2} \quad \text{and} \quad \overline{PR} = \sqrt{z^2 + d^2}$$

and

Substituting the above into Equation 6.75 yields Equation 6.76.

$$t_x = \frac{\sqrt{z^2 + (x-d)^2} + \sqrt{z^2 + d^2}}{V} \tag{6.76}$$

where

t_x = time from source *S* to point scatterer *P* to receiver *R*

Z = depth to the point scatterer

X = horizontal distance from source *S* to receiver *R*

d = horizontal distance from point scatterer to receiver *R*

V = velocity

Substituting $z = \frac{Vt_0}{2}$ in Equation 6.76, where t_0 is the two-way time from *P* to *P*', gives:

$$t_x = \frac{\sqrt{\left(\frac{Vt_0}{2}\right)^2 + (x-d)^2} + \sqrt{\left(\frac{Vt_0}{2}\right)^2 + d^2}}{V} = \sqrt{\frac{t_0^2}{4} + \left(\frac{x-d}{V}\right)^2} + \sqrt{\frac{t_0^2}{4} + \left(\frac{d}{V}\right)^2} \tag{6.77a}$$

The quantities inside the radicals of Equation 6.77a are similar to the familiar NMO Equation. In fact, for the special case of $d = x/2$, Equation 6.77a) becomes:

$$t_x = \sqrt{\frac{t_0^2}{4} + \left(\frac{x}{2V}\right)^2} + \sqrt{\frac{t_0^2}{4} + \left(\frac{x}{2V}\right)^2} = \sqrt{t_0^2 + \left(\frac{x}{V}\right)^2} \tag{6.77b}$$

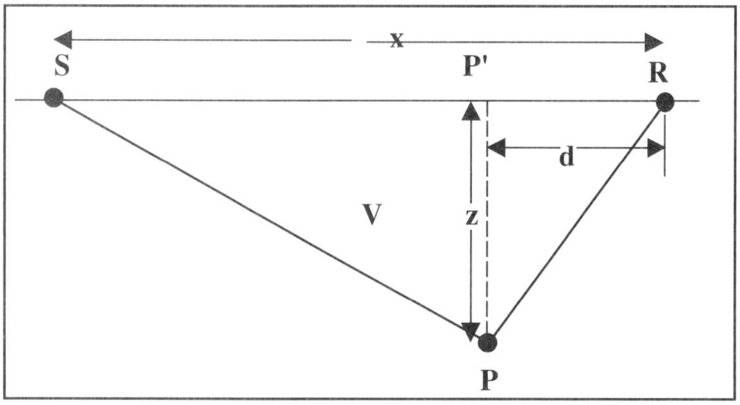

Fig. 6–263 *Point Scatterer Geometry*

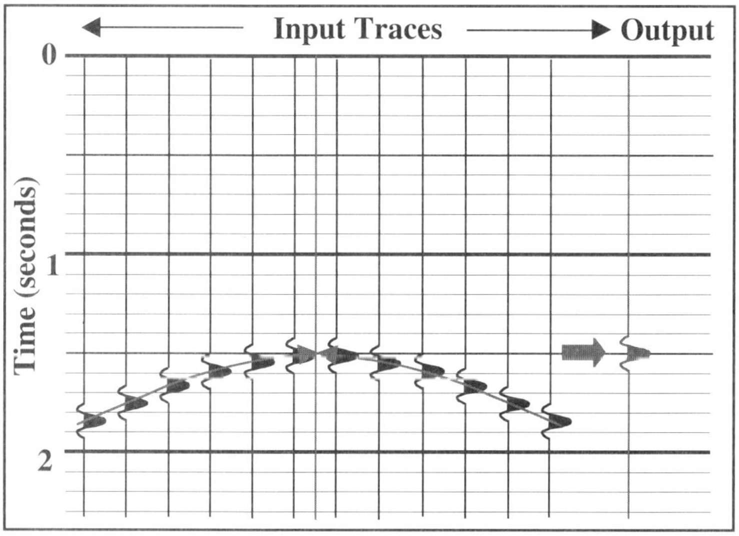

Fig. 6–264 *Kirchhoff Pre-stack Time Migration*

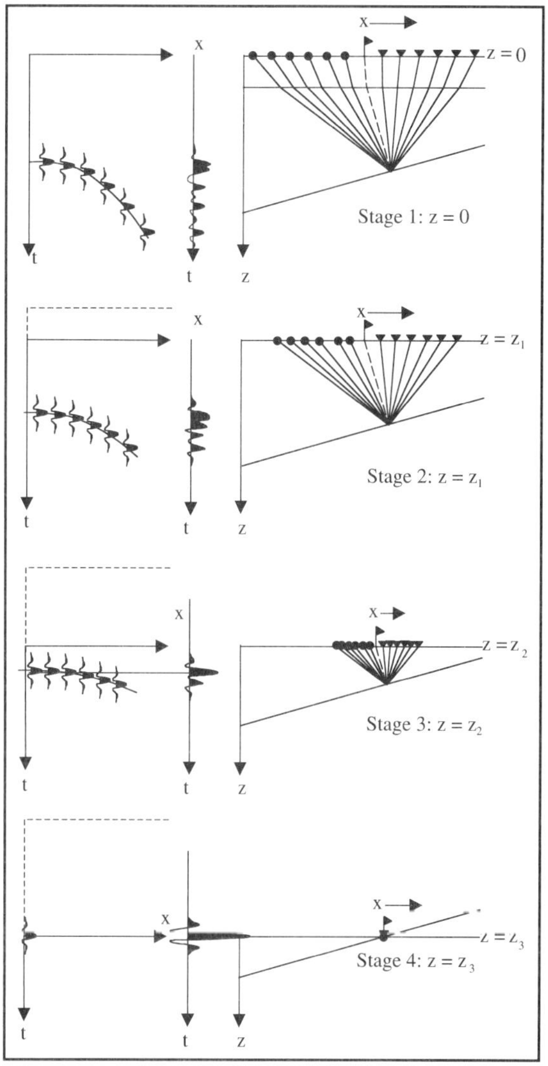

Fig. 6–265 *Conceptual View of Pre-stack Depth Migration*

This is the NMO equation. Hence Equation 6.77 must be a hyperbola.

Data are mapped to the CMP domain and then summed along hyperbolas defined by Equation 6.76 into a single output trace. NMO corrections are not applied because the summation is along, essentially, an NMO curve. See Figure 6–264.

Basically, pre-stack depth migration is a ray-tracing operation. After DMO, all input CMPs have the same zero-offset point and, hence, a common reflection point at all reflecting horizons. Using a downward continuation approach at each depth step, the individual ray paths of a CMP set get closer together and converge to a single event at the reflecting horizon. Interval velocities are required for the process. These can be derived from the stacking velocity using Equation 6.29. Velocities must not be dip-dependent, so this is another reason that DMO should be run before pre-stack migration.

Figure 6–265 provides a conceptual view of this approach. The model has a flat horizontal layer overlaying a dipping layer. For the sake of drawing simplicity, a six-fold CMP is used. Four stages of downward continuation, pre-stack migr-ation are shown. The first stage, the top of Figure 6–265, is the initial condition $z = 0$. The model and ray paths are shown at the right. On the left is the dipping reflection event as seen on the input CMP record. In the center is the stack of the event.. Energy is well distributed in time and space.

The second state is after depth steps have moved the datum to z_2, the bottom of layer 1 in the model. The situation is as if the sources and receivers had moved along the ray paths to this new datum. The wave equation is used to predict the amplitude of the event at the new datum. Note the reduced moveout on the event, which results in a more concentrated stack (center).

In the third stage, the datum has been moved to depth z_2, which is intermediate between the two interfaces in the model. Convergence to the reflection point continues. In addition to sources and receivers moving to the new datum, they also get closer together. The change in depth and spacing results in further reduction in moveout and trace spacing. As a result the stack (center) is even more concentrated.

The fourth stage shows the ideal convergence to the reflecting point. All sources and receivers merge at the same spatial point. All reflection events move to zero offset. The stack is simply six times the input wavelet amplitude. The event has been successfully migrated!

The preceding is rather over-simplified. In practice it may be necessary to do pre-stack migrations in several iterations. If the velocity field used for migration is in error, then the depth at which the reflection events focus will be in error and an out-of-focus migration results. The focusing equation from Yilmaz and Chambers (1984), relates migration velocity and depth as shown below:

$$Z_r V_r = Z_f V_m \tag{6.77}$$

where

Z_r = true depth

V_r = true velocity

Z_f = focusing depth, and

V_m = migration velocity

This relationship and others derived from it provide a means of updating velocities used in succeeding iterations. The more complex the velocity field, in general, the greater the number of iterations.

Figure 6–266 provides a simple illustration of focusing and depth errors. The center column of Figure 6–265 illustrates the focusing of a CMP using the correct velocity. The left column shows the depth levels, starting at the surface $z = 0$. The top drawings in each column represent the moveout of the CMP at input. Subsequent rows of drawings show the reduction in moveout and spatial extent of the reflection event. Using the correct velocity field results in focusing on the true or actual depth just below depth level Z_5. If a slower velocity field is used, then the focusing depth is just below depth level Z_6, but that is too deep if a faster velocity field is used. The focusing depth is just below depth level Z_4, but that is too shallow.

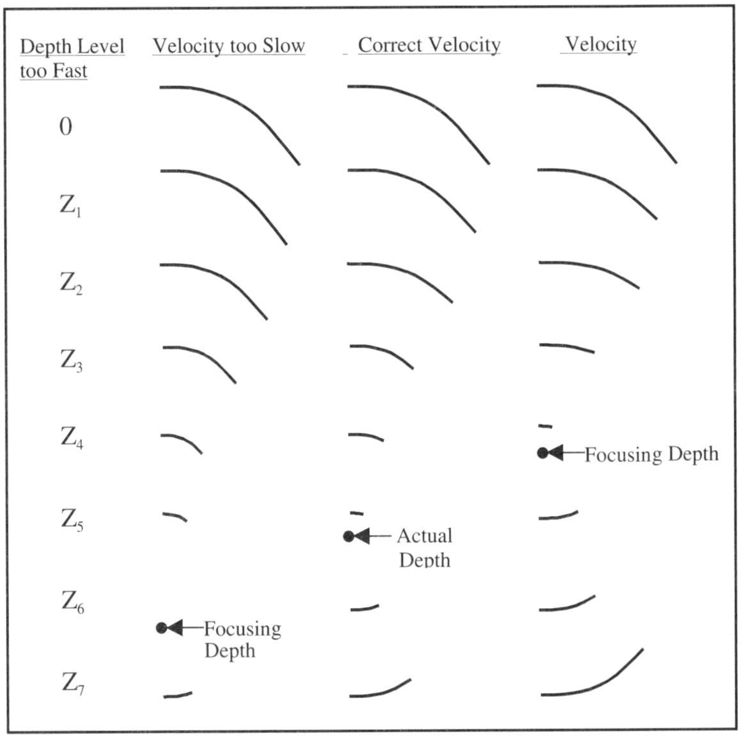

Fig. 6–266 *Pre-stack Depth Migration—Velocity and Focusing*

Another approach to pre-stack depth migration uses up- and down-going waves. The 45°, one-way equation in the F-X domain is used to extrapolate data. A thin lens term is added to allow for lateral velocity changes (Claerbout 1985). After each downward continuation step, the downward propagating source wavelet is cross-correlated with the upward propagating receiver wavelet. A reflection exists where times of the down-going and up-going waves are equal. See Figure 6–267.

Figure 6–268 is an overthrust model and Figure 6–269 is the result of processing synthetic data that used the exact model velocities. A conventional processing sequence that included Kirchhoff DMO, conventional CMP stack, and a steep-dip, post-stack depth migration was performed. Figure 6–269 has the appearance of an over-migrated section.

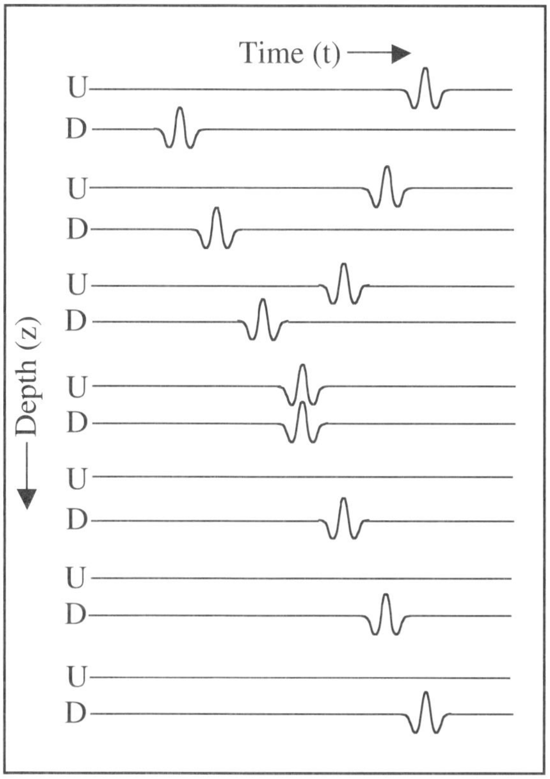

Fig. 6–267 *Up- and Down-going Waves*

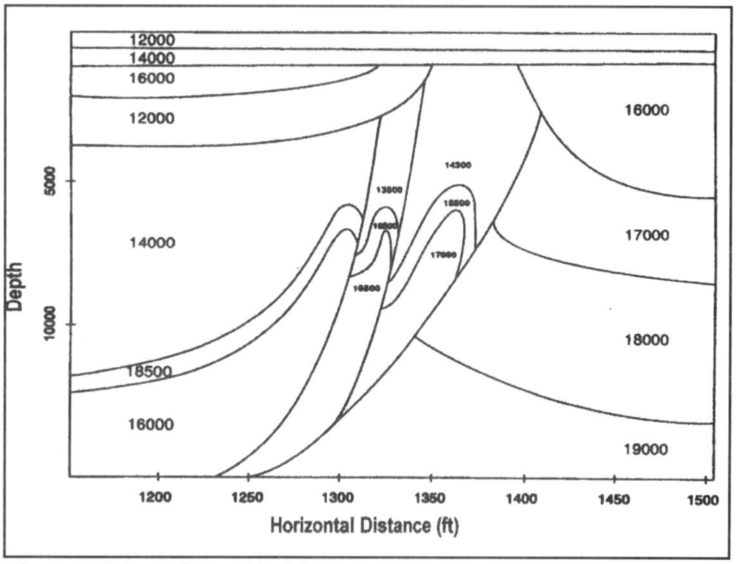

Fig. 6–268 *Overthrust Model*

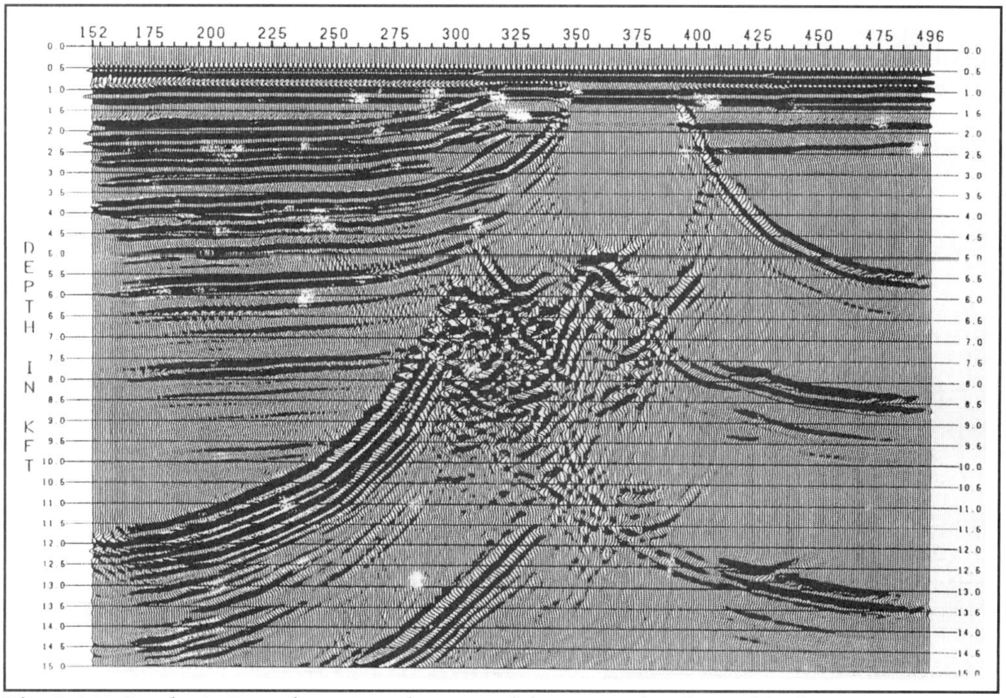

Fig. 6–269 *Synthetic Data from Overthrust Model Processed with Kirchhoff DMO, CMP Stack, and Steep-dip, Post-stack Migration*

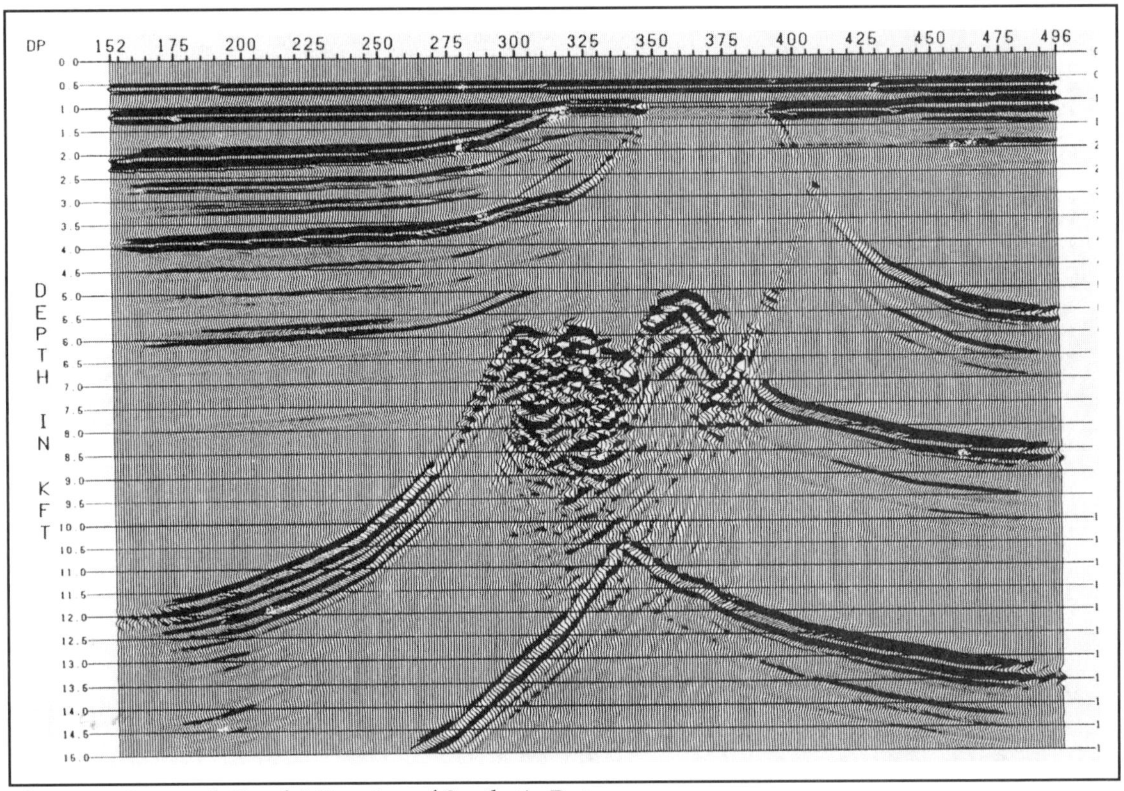

Fig. 6–270 *Pre-stack Depth Migration of Synthetic Data*

Figure 6–270 is a pre-stack depth migration of the same synthetic data used for Figure 6–269. Crests and troughs in the fold belt are much more clearly defined, and some of the fault planes can now be seen. Since a 45° extrapolator was used, the ability to image steeply dipping fault planes was limited. However, the deep event between CMPs 250 and 496 and depths between 10,500 ends 15,000 ft are quite well defined. In Figure 6–269, only the limbs of this event could be seen but even these were not as well defined.

Shot-geophone sinking is a method of pre-stack depth migration that uses Claerbout's double square-root equation to extrapolate, in shot-geophone coordinates. Claerbout's equation is

$$\frac{\partial U}{\partial z} = \left(\left[\left(\frac{i2\pi f}{v(g)} \right)^2 - \frac{\partial^2}{\partial g^2} \right]^{1/2} + \left[\left(\frac{i2\pi f}{v(s)} \right)^2 - \frac{\partial^2}{\partial s^2} \right]^{1/2} \right) U \tag{6.78}$$

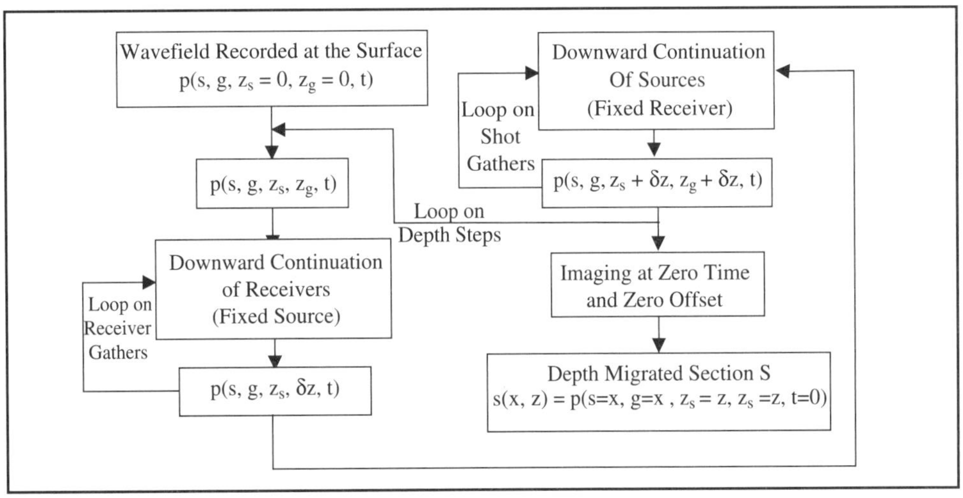

Fig. 6–271 *Shot-geophone Sinking Depth Migration Flow Chart*

Extrapolation, using Equation 6.78 for extrapolation, alternates between shot gathered data and receiver gathered data. Figure 6–271 is a flow chart for the process. In Figure 6–270, *s* and *g* are source and geophone coordinates, respectively.

Table 6–6 provides some final remarks on pre-stack migration.

Table 6–6 Pre-Stack Time and Depth Migration Types

Type	Time	Depth
Full-Time Migration Before Stack	Output is a migrated section. No intermediate unmigrated stack. This is not what the interpreter wants; he must have an unmigrated stack section. It is desirable, in case of conflicting dips, to apply SPM, a simplified method of pre-stack migration	Needed when there is a strong change in the velocity gradient that stacking velocity cannot solve
3-D Migration	Needed when PSPM fails, and when the stack contains cross dips	This is the ultimate goal of everyone, if computer time allows, and if 3-D subsurface velocities are known accurately

Band-pass Filtering

Band-pass filtering is usually the final process applied in a processing sequence. It is done mostly for cosmetic reasons. That is, by filtering out the more noise-dominated frequency components our displays will look cleaner and better.

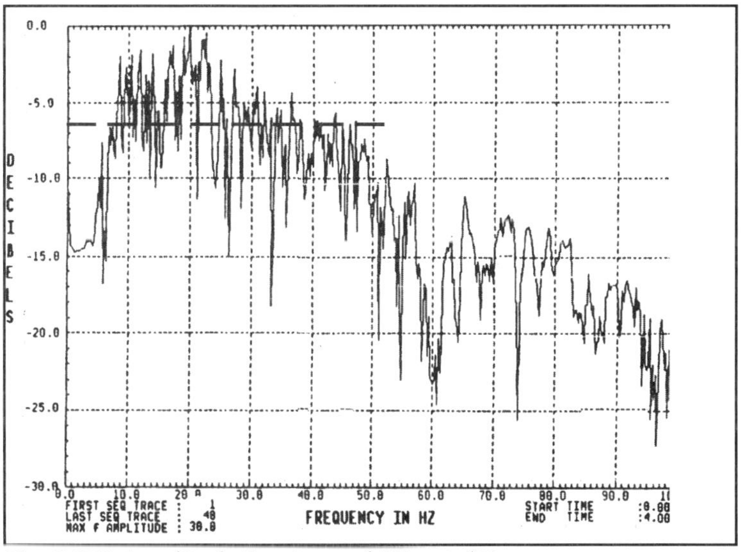

Fig. 6–272 *Amplitude Spectrum of Migrated Trace*

Figure 6–272 represents the amplitude spectrum of a typical trace in a migration output. The dashed horizontal line is at the –6 dB point. Data with amplitudes higher than this level have unity and larger signal-to-noise ratios. So, based on this evidence, and pass band desired is from about 18 Hz to about 38 Hz.

The problem with this approach is that it ignores the time-variant nature of signal-to-noise ratios. A better approach is to run *filter scans*. This is done by applying a range of band-pass filters to selected sets of traces from the migrated sections. Note that these are digital filters. They are specified by four corner frequencies (f_1, f_2, f_3, and f_4) and a filter length. Figure 6–273 shows a pass-band filter response and defines the corner frequencies. The response is 0 at f_1 and f_4 and 1 at f_2 and f_3. From the vertical symmetry of the response, the frequencies at which the response is 1/2 are $\frac{f_1+f_2}{2}$ and $\frac{f_3+f_4}{2}$. These two frequencies determine the filter pass band.

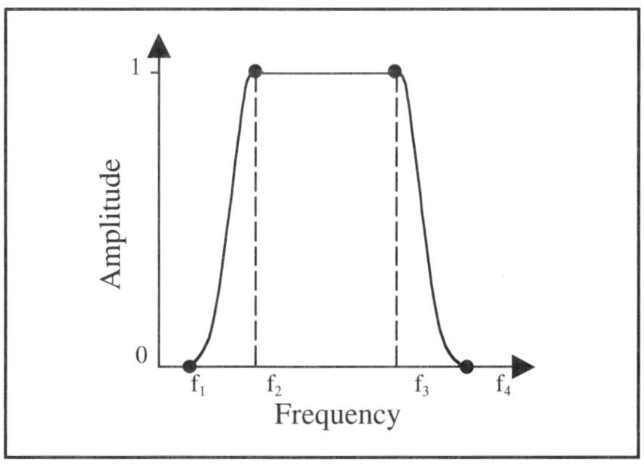

Fig. 6–273 *Band-pass Filter Response and Corner Frequencies*

There are a variety of ways in which filter scans are done. One approach is shown in Figure 6–274. The far left of Figure 6–274 shows the unfiltered traces. The numbers at the top of each filtered set are the corner frequencies. It follows that the pass bands are, from left to right, 6–24 Hz, 9–30 Hz, 12–40 Hz, 15–50 Hz, 21–60 Hz, and 27–90 Hz. Note that in the filter scan example, the pass bands overlap one another and all pass bands are greater than one octave. Some people use very narrow pass bands that may make interpretation of the filter scans difficult.

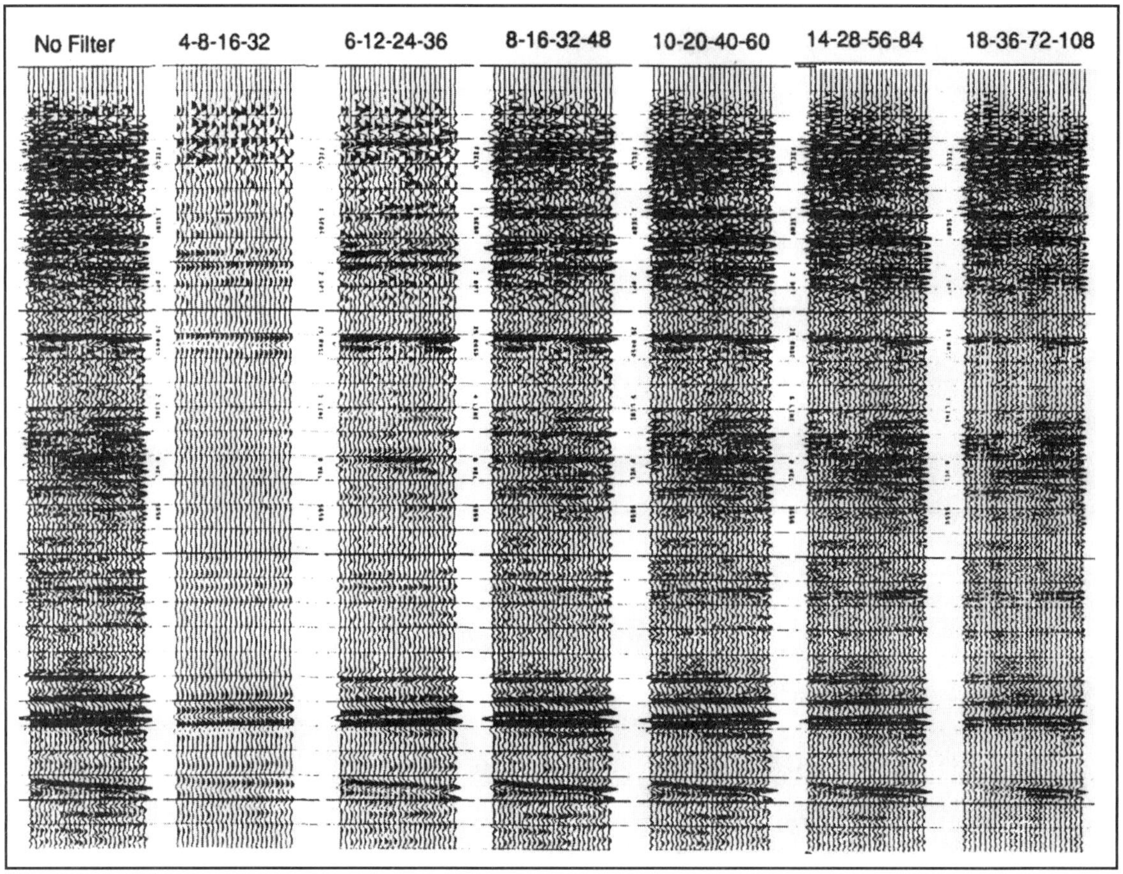

Fig. 6–274 *Filter Scan*

Figure 6–275 is an interpretation of the scan of Figure 6–274. For the first second, signal amplitudes are weak in the first three panels but relatively strong in the last three. Thus corner frequencies of 10–20–72–108 (15–90 Hz) are selected. Between 2 and 3 seconds, signal is strong in the last four panels, so the corner frequencies are 8–16–72–108 (12–90 Hz). For later times, signal is stronger than noise in the second through the fifth panels. Hence, the corner frequencies are 6–12–56–84 (9–70 Hz).

A slightly different but superior, in the authors' opinion, approach is to use overlapping two-octave filters in the scan. Applying this to the data of Figures 6–274 and 6–275 might give the following filter pass bands: 6–24 Hz, 12– 48 Hz, 24–96 Hz, and 48–192 Hz. Filter parameters are selected by inspecting filter panels to determine which have good signal-to-noise ratios over different time intervals.

Another approach is to start with a wide pass band and decrease it alternately from the low end and the high end. Applying this approach to the data of Figures 6–274 and 6–275 might give the following filter pass bands: 6–96 Hz, 12–96 Hz, 24–96 Hz, 6–48 Hz, and 12–48 Hz. The low cut-off frequency is chosen based on the panel where the signal-to-noise ratio first improved to acceptable levels. The high cut-off frequency is selected based on the panel in which high frequency noise becomes dominant. As with other approaches, filter parameters are chosen for different time intervals.

Time-variant filters. TVF are applied in a manner similar to time variant deconvolution, also seen in Figure 6–94. Using the filters selected in Figure 6–275, TVF might be implemented as follows:

- Filter 1: full-on (100%) from start of data to time of 800 ms, decrease linearly to 0 at 1200 ms
- Filter 2: increase linearly from 0 at time of 800 ms to full-on at 1200 ms, decrease linearly to 0 at 2200 ms
- Filter 3: increase linearly from 0 at time of 1800 ms to full-on at 2200 ms to end of data

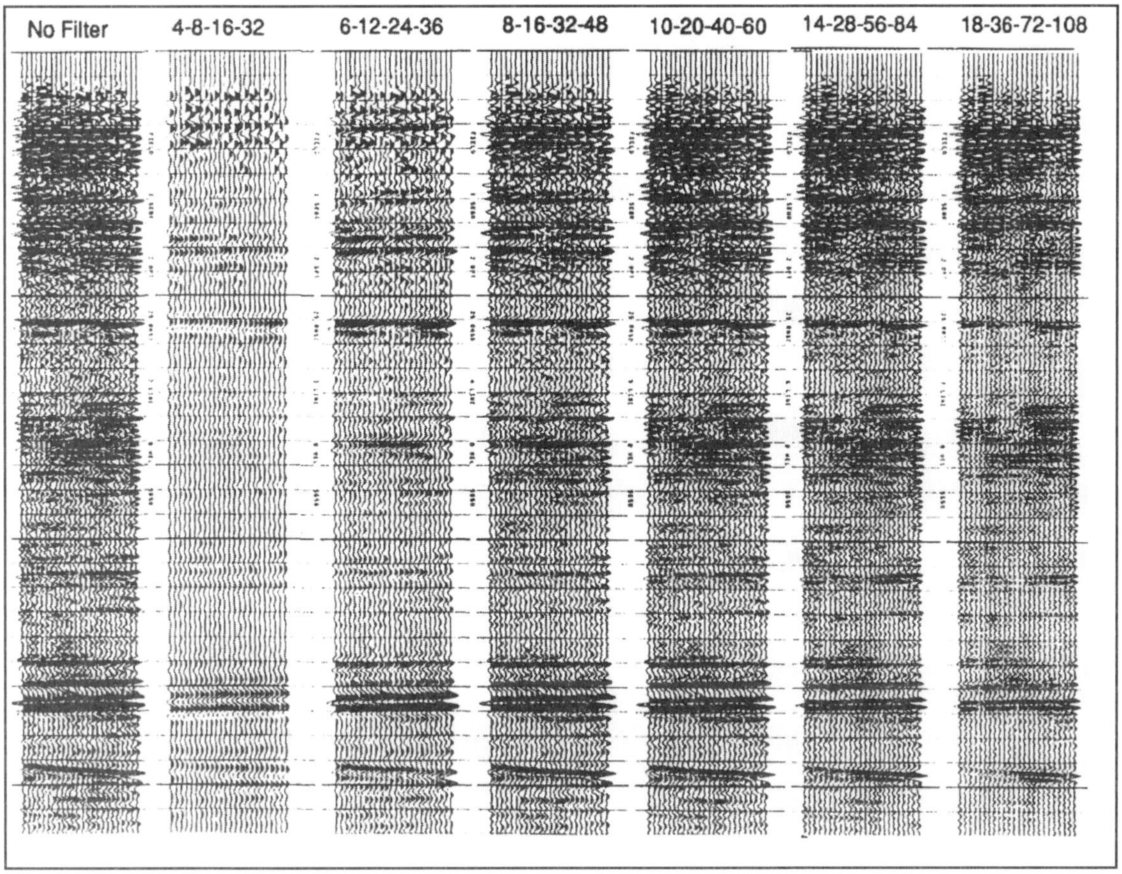

Fig. 6–275 *Filter Scan Interpretation*

Filter scans should be run at various locations in the data volume because noise is usually both time- and space-variant. Filter parameters at locations between those where filter scans were run (control points) are determined by spatial interpolation and extrapolation. Figure 6–276 represents a seismic section with four filter control points. Filter application times are linearly interpolated between control points and linearly extrapolated to section edges from control points one and four.

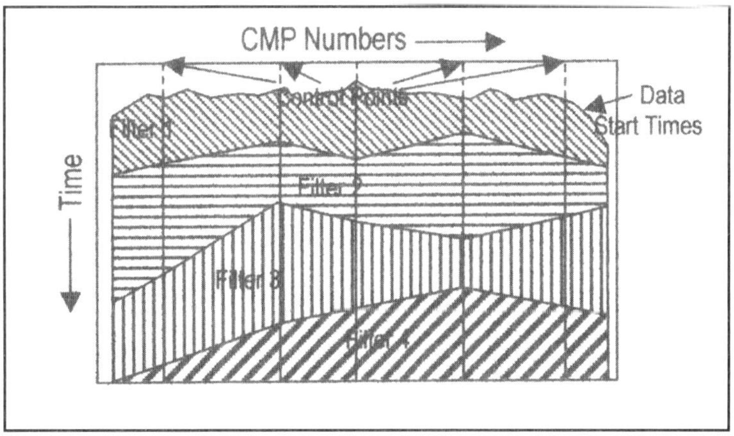

Fig. 6–276 *Interpolation and Extrapolation of Band-pass Filter Application Times*

Figures 6–13 through 6–15 shows that displays are made after major processes are applied to evaluate results and determine, or verify, parameter values. Band-pass filtering is usually applied before these displays are made. However, data passed on to the next process should not be filtered.

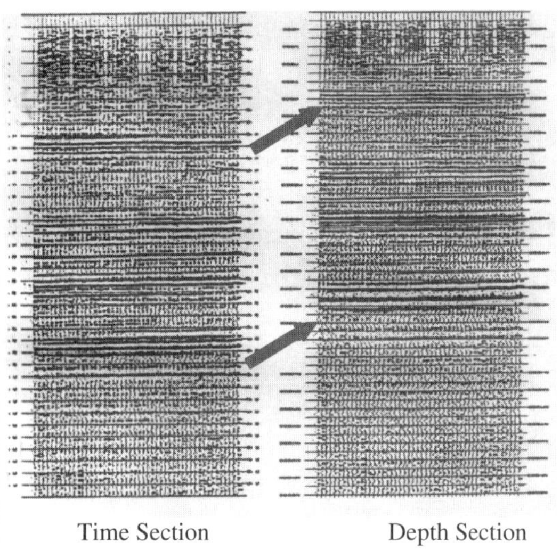

Fig. 6–277 *Time-to-Depth Conversion*

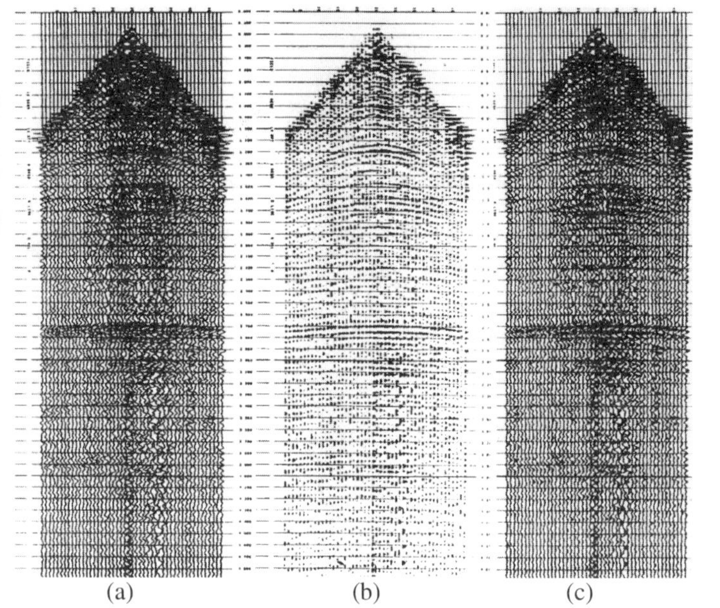

Fig. 6–278 *Trace Display Modes*

Time-to-depth Conversion

If the CMP stack is the final process or if time migration is done, it is necessary to convert from time to depth for the final displays. This conversion requires *average velocity* not stacking velocity. Average velocity can be derived from stacking velocity by first calculating interval velocities using Equation 6.29 and then substituting in Equation 6.20.

Time-to-depth conversion is really part of the total imaging process. In parts of sections corresponding to low interval velocities, events on depth sections appear closer together than on time sections, whereas parts corresponding to high interval velocities appear farther apart on depth sections.

Figure 6–277 compares sections before and after time-to-depth conversion.

Display

Trace displays are usually done in one of three modes.

- Wiggle trace (WT). Amplitudes are shown by deflections of lines from zero positions. A downward deflection usually indicates a positive voltage.
- Variable area (VA). Only filled in peaks are shown, so the larger the peaks the greater the blackened area.
- Wiggle trace/variable density (WT/VA). Combination of wiggle trace and variable area.

In addition to specifying the mode of trace display, it is also necessary to specify display gain in dB (Fig. 6–279), horizontal scale that is usually in traces per inch (Fig. 6–280), and the vertical scale that is usually in inches per second (Fig. 6–281). If VA or WT/VA display modes are selected, bias must be specified. *Bias*, essentially, indicates the level from which the peaks are to be filled.

There are many display options and color schemes available for display of seismic traces and attributes. The top part of Figure 6–282 demonstrates the application of color to the showing wavelet phase. The top left is a conventional wiggle trace/variable area trace display, the center is a dual polarity (peaks in blue, troughs in red), and the top right demonstrates the use of gradational color for the interpretive assessment of how close to zero phase the wavelets are. The bottom of Figure 6–282 demonstrates the use of instantaneous amplitude to obscure the effect of phase distortion.

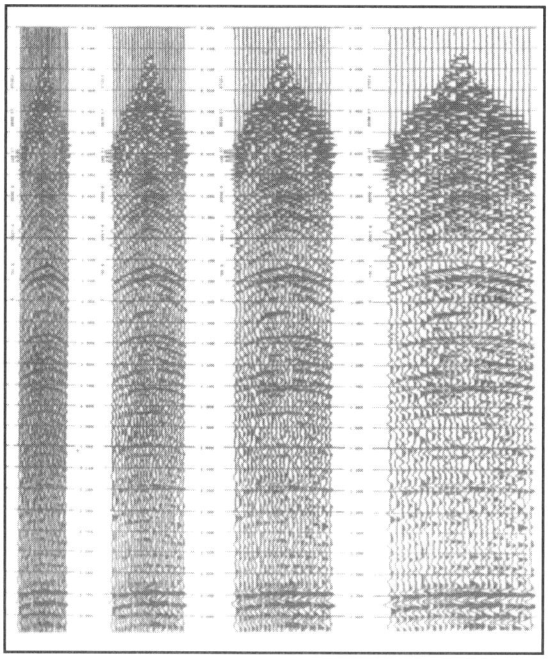

Fig. 6–279 *Display Gain*

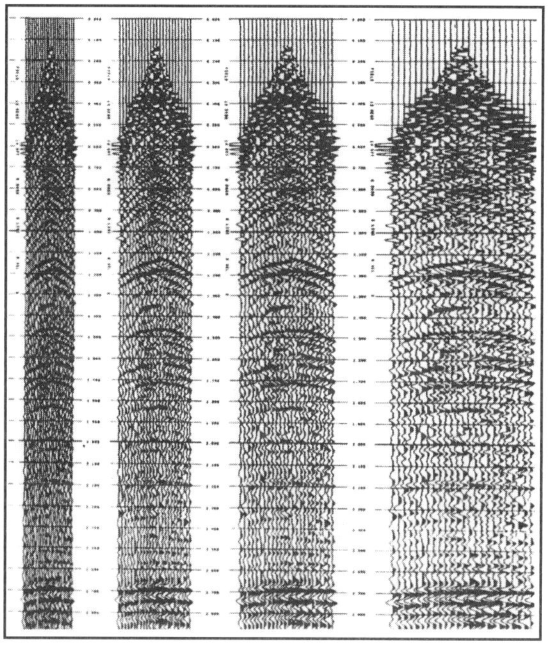

Fig. 6–280 *Variations in Horizontal Scale*

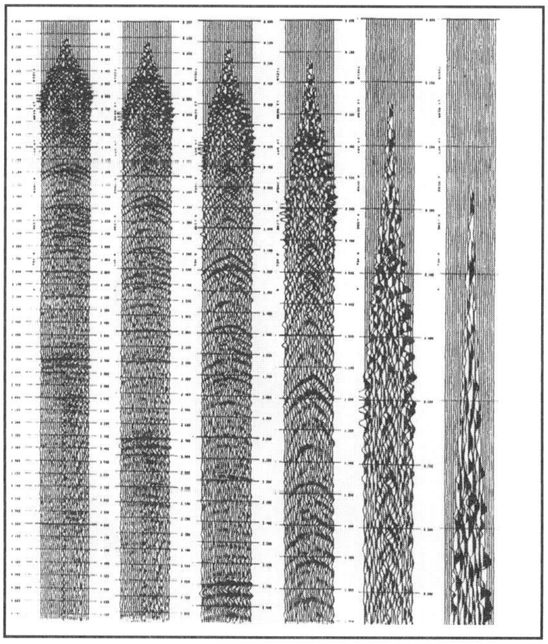

Fig. 6–281 *Variations in Vertical Scale*

There are many more options for displaying 3-D data than 2-D. Consider the lines shown in Figure 6–283, for example. Data wcrc collcctcd along 108 lines of 12.5 km, over a distance (note the scale exaggeration). The in-line direction is east. For 2-D data, the only display option for profiles (seismic sections) is to display lines in directions they were shot. However, 3-D profiles can be displayed as in-line (lines designated by numbers—1, 4, 2, and 3 in Fig. 6–282), cross-line (red lines *A, B, D,* and *C*), or even diagonal lines (blue line).

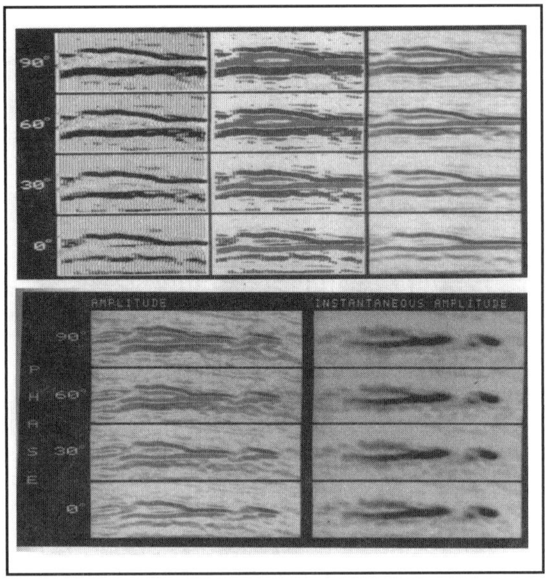

Fig. 6–282 *Color Display Options*

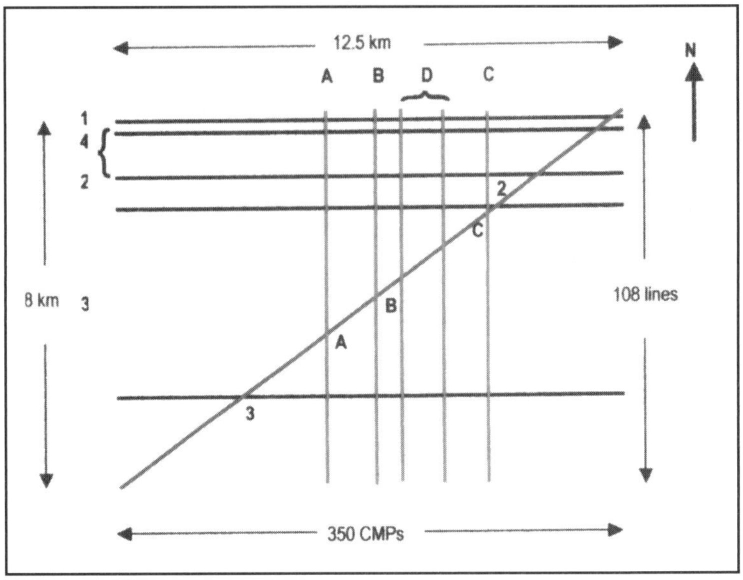

Fig. 6–283 *Line Orientations*

Figures 6–284 through 6–286 show some of the in-line and cross-line profiles of lines shown in Figure 6–282 as well as the diagonal line. One can follow changes in structure and get a feel for the 3-D structural elements. Points at which the diagonal line crosses the lines corresponding to the in-lines and cross-lines marked in Figure 6–283 are indicated at the top of the profile.

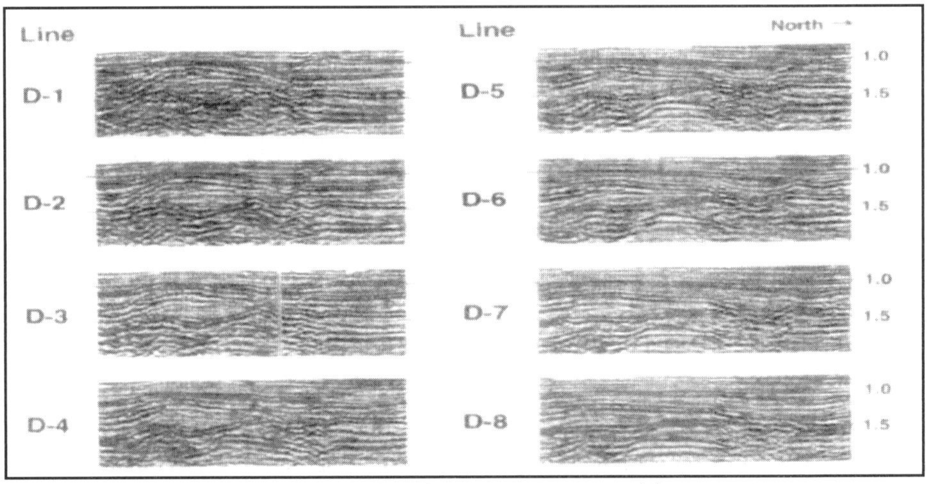

Fig. 6–284 *Migrated Cross-line Profiles for Lines Designated by D in Figure 6–283*

Another option available in display of 3-D data is the plotting of horizontal sections, called *time slices* if the vertical scale is time or *depth slices* if the vertical scale is depth. They portray amplitudes over the entire data volume for a particular two-way time. Horizontal sections have proven to be excellent tools in portraying many target types that are not easily seen on vertical sections. They have other applications, as shown in Figure 6–287.

Figure 6–287 illustrates the use of time slices in constructing time structure maps. Shown is a series of time slices that are 4 ms apart, starting at 2632 ms and ending at 2656 ms. Remember, early time means shallower depth. Amplitudes are shown in dual polarity (red for peaks, black for troughs).

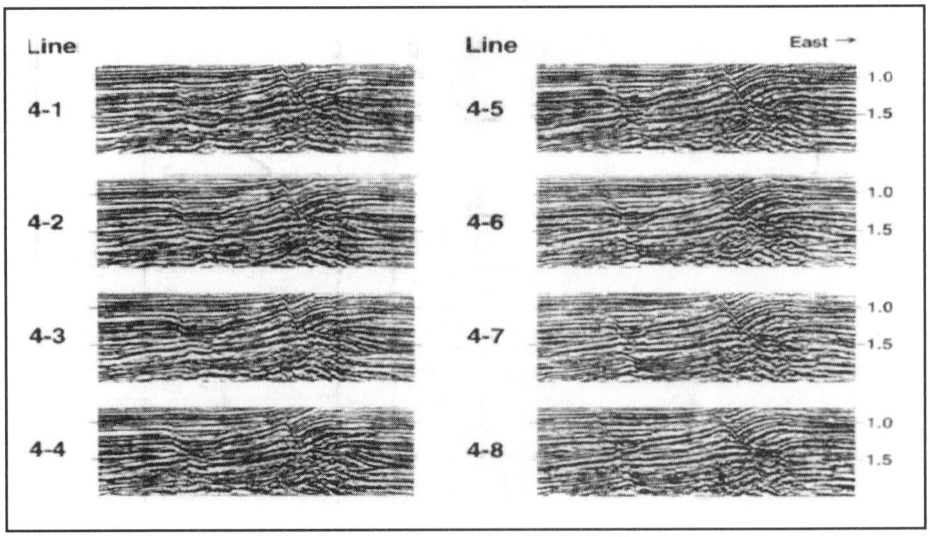

Fig. 6–285 *Migrated In-line Profiles for Lines Designated by D in Figure 6–283*

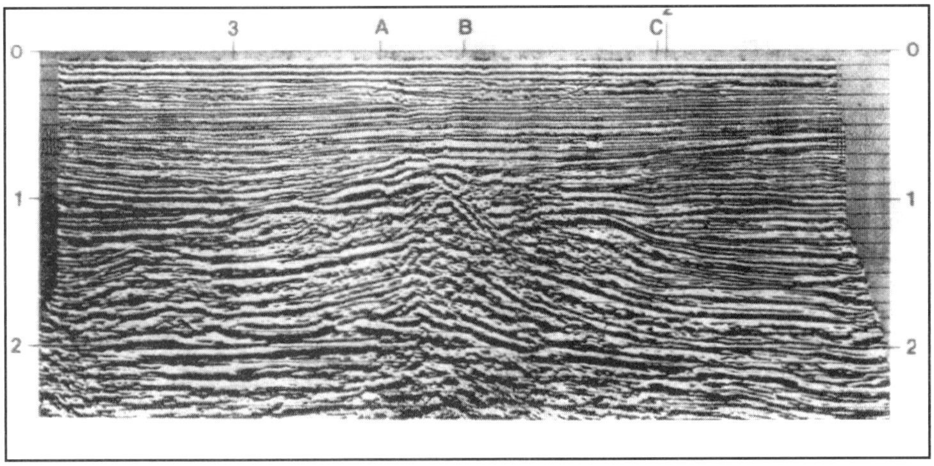

Fig. 6–286 *3-D Migrated Profile for Diagonal Line in Figure 6–283*

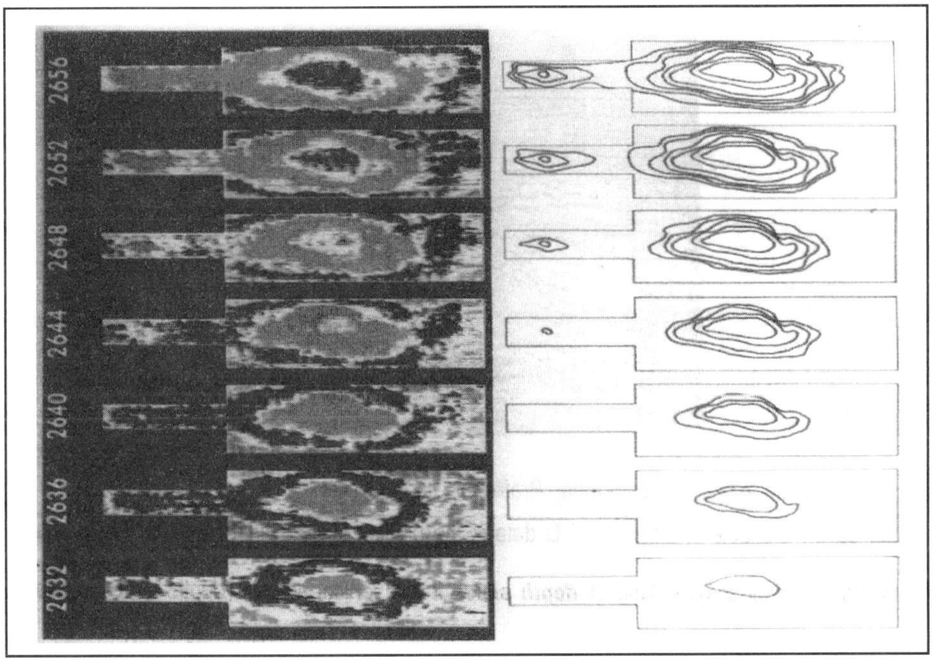

Fig. 6–287 *Constructing Structure Maps from Time Slices*

Fig. 6–288 *Time Slice Through a Salt Dome*

The outline of the red peak of the 2632 ms slice is drawn first. This contour represents a line of equal time called an *isochron*. The outline of the red peak of the 2636 ms slice is drawn around the first outline. The process continues with slices every 4 ms. The 2644 ms slice has the start of a deeper, secondary peak. After all contours are drawn, it is obvious that the structure is a sort of double anticline. If other factors are favorable, the two sets of closed contours could be good drilling targets.

Figure 6–288 is a time slice through a salt dome. The white area near the center is salt. No reflections are generated within the salt body, but the margins of the salt are usually strong reflectors. Many faults are usually present on the flanks of salt domes, resulting in a complex structure.

The examples shown here explore only a few off the many options available in the display of seismic data. Color can be, and is, used extensively. Seismic traces and trace amplitudes are displayed along with many other seismic attributes.

Seismic Data Processing Summary

Seismic field data are not usually amenable to data interpretation. Signal is frequently obscured by noise, including multiple reflections. Data are also geometrically distorted and do not require sufficient resolution (vertical and horizontal). The seismic data processor's function is to correct these deficiencies; improve signal-to-noise ratios, produce correct geometry in the data, and provide resolution required to solve the geological problem.

Input to seismic data processing includes field records on magnetic tape, survey/navigation data on magnetic tape, reports, maps, and more on a variety of media. Output consists of seismic (cross) sections, horizontal sections (3-D), and various special displays.

There are two basic kinds of processes. *Geometrical corrections* remove effects that cause distortions in location, shape, and size of geologic structures. Processes that perform geometrical corrections include normal move out corrections move signals from times at non-zero offsets to zero-offset times. *Static corrections* replace the actual surface with a reference surface and remove the effects of near surface variations on reflection times.

CMP stack produces the view of subsurface that results from having coincident sources and receivers at the reference surface. DMO moves events up-dip to their true zero-offset position.

Signal enhancement processes include the following:

- amplitude recovery corrects amplitudes for geometrical spreading and inelastic attenuation
- velocity filtering eliminates or significantly attenuates linear noise on seismic records
- deconvolution produces shorter duration signal wavelets
- demultiple attenuates reverberations and multiples
- CMP stack attenuates out-of-phase events, including ambient noise
- frequency filtering attenuates frequency components where signal-to-noise ratio is poor
- migration collapses or substantially attenuates diffractions

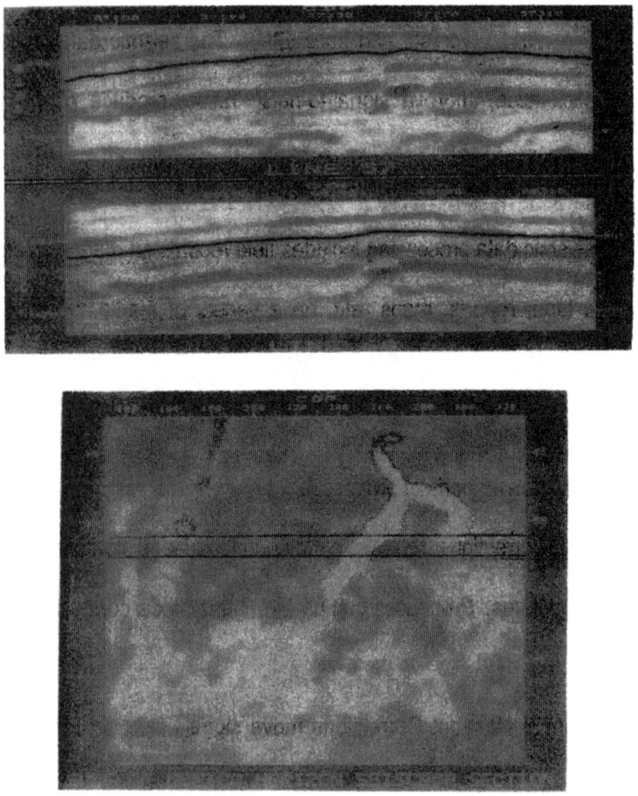

Fig. 6–289 *Migrated In-Line Profiles for Lines Designated by D in Figure 6–283*

Workshop

1. What is the most significant factor in deciding between straight stack and diversity stack of vibrator data?

2. When is minimum-phase correlation of vibrator data preferable to zero-phase correlation?

3. What is the purpose of displaying first breaks after linear move out corrections are applied to land data?

4. How is geometry (relative positions of sources and receivers) obtained in ocean bottom cable processing?

5. How are datum statics calculated if weathering depths and velocities are not known?

6. What is meant by short wavelength and long wavelength statics? What techniques of residual statics analysis work best on wavelength of statics?

7. Which of the following amplitude compensation methods are considered deterministic and can, thus, be applied to data on which amplitude versus offset analyses are to be run?

    a. geometrical spreading correction

    b. programmed gain control

    c. RMS automatic gain control

    d. instantaneous AGC

    e. surface consistent amplitude compensation

8. What types of noises are velocity or *F-K* filtering used to attenuate?

9. In an IVP session, what happens to an event on the center CMP gather when the velocity pick is moved to a much slower velocity? What happens when the pick is moved to a much faster velocity?

10. Does the autocorrelogram in Figure 6–290 indicate the presence of multiples?

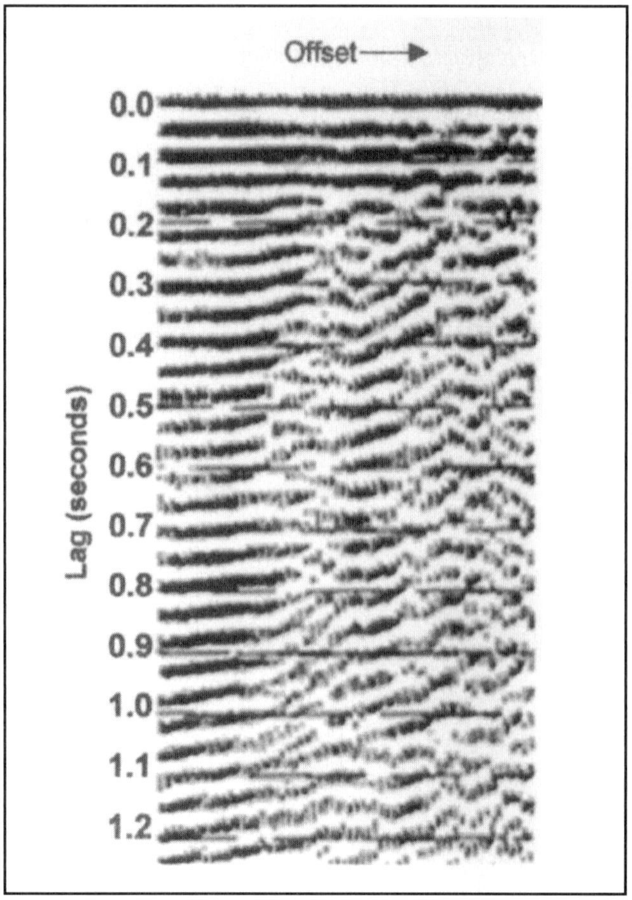

Fig. 6–290

11. If predictive deconvolution is applied to the data, what prediction distance (gap) would you recommend based on the auto-correlogram seen of Figure 6-290?

12. In Table 6–7, write in yes or no in each column to indicate whether the listed process do or do not have the effect on data.

Table 6–7 *Processes and Effect*

Process	Whitens Spectrum	Outputs Zero Phase	Attenuates Multiples
Spiking Deconvolution			
Predictive Deconvolution			
Model-Based Wavelet Processing			

13. What are the significant parameters in reflection residual statics, and how do their values affect results?

14. Datum statics are calculated separately for sources and receivers. What method(s) of residual statics analysis provide separate residual statics for sources and receivers?

15. What causes NMO stretch?

16. The mute applied to a CMP gather results in the following schedule of fold versus record time. How much random signal-to-noise improvement, in dB, is obtained in CMP stack at these times?

Time (ms)	Fold	S/N Improvement
250	6	
500	10	
1000	18	
1500	26	
>1750	30	

17. What are the *two* primary actions of migration?

18. Why is band-pass filtering applied?

19. Draw a sketch to illustrate the difference between wiggle trace and variable area display of seismic traces.

20. How does time-to-depth conversion affect subsurface imaging?

21. Select the optimum value for maximum correlation shift from those shown in Figure 6–291, based on the effect of residual statics on the stack.

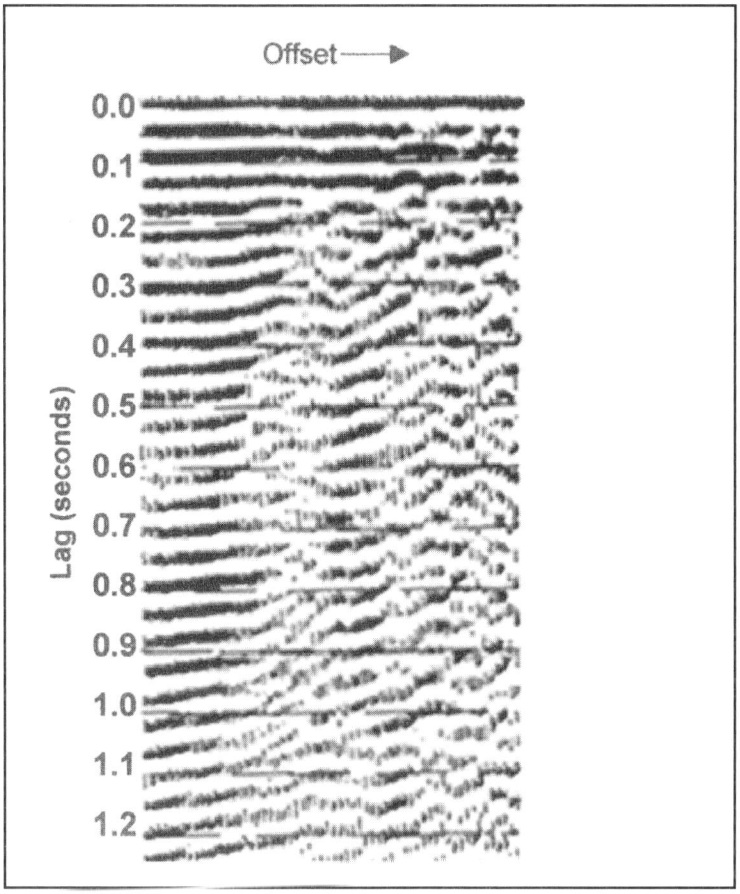

Fig. 6–291

22. Identify the multiple reflections on the records in Figure 6–292 and determine the periods of the multiples.

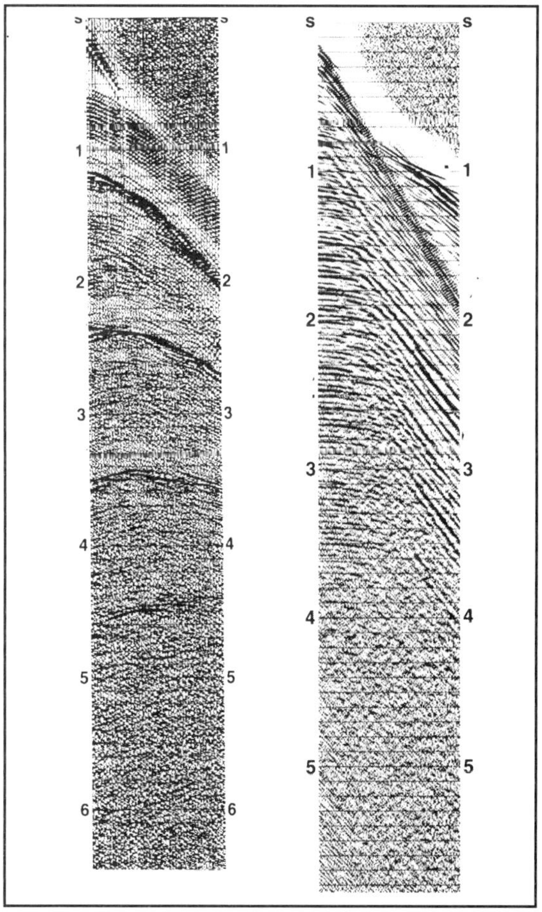

Fig. 6–292

23. Shown in Figure 6–293 are (a) a CMP stack, (b) the migrated stack, and (c) a sketch of the following: a prominent diffraction (D), a dipping event (B) before migration, and a dipping event (A) after migration.

    What are the observations that can be made on the migrated section about the following:

    - the dip angle of reflector *A* in the migrated section compared to the stack section
    - the length of the reflector *A*
    - the direction that reflector *A* moves due to migration

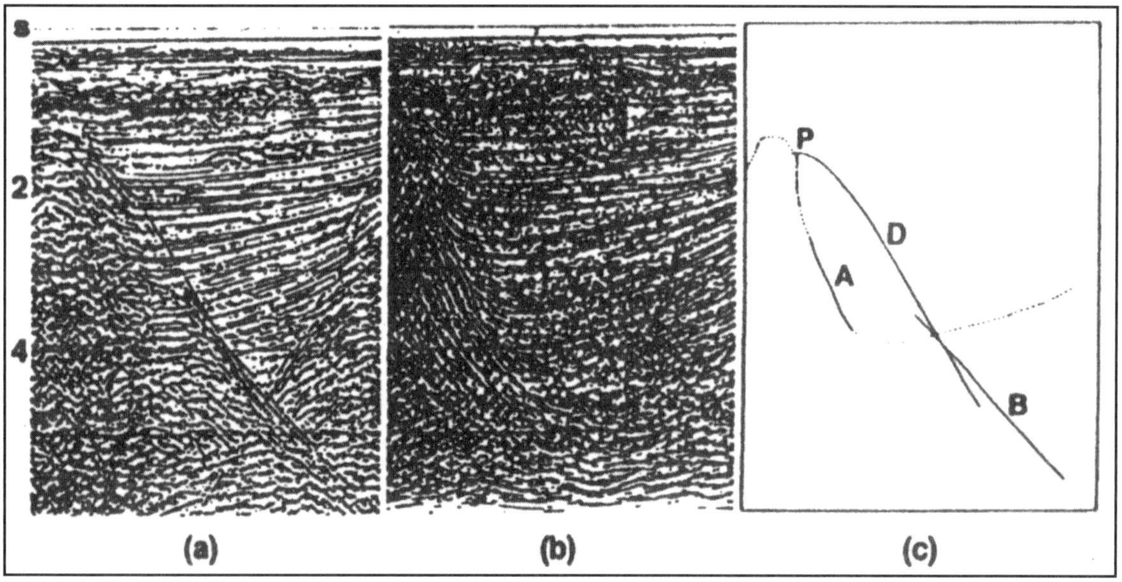

Fig. 6–293

24. Shown in Figure 6–294 is a CMP stack and four migrations of it. Which migration uses the optimum migration velocity? Are the other panels over- or under-migrated?

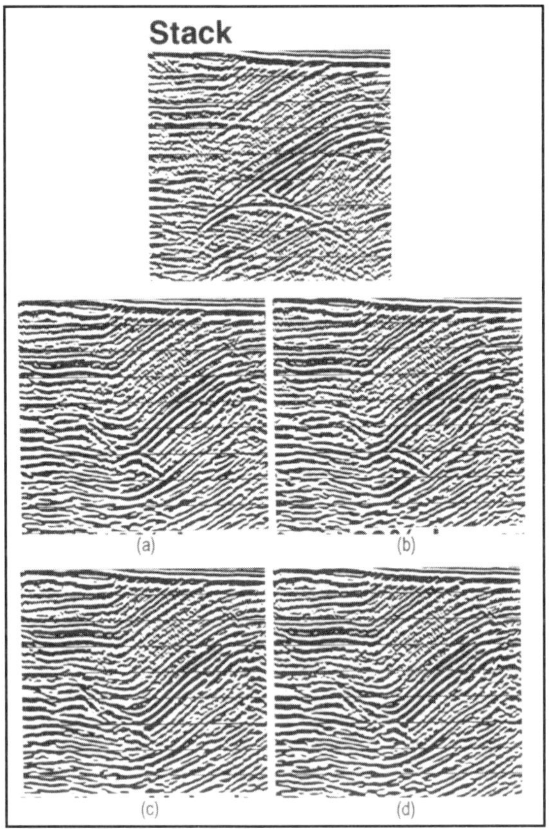

Fig. 6–294

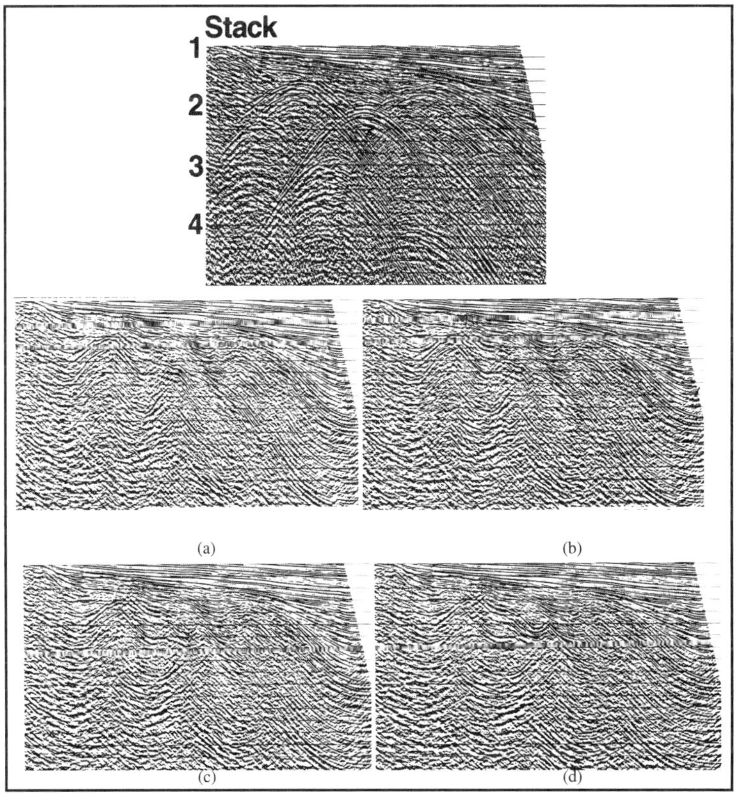

Fig. 6–295

25. Shown in Figure 6-295 is a CMP stack and four migrations of it. Which migration uses the optimum migration velocity? Are the other panels over- or under-migrated?

26. In Figure 6–296 are five migration panels. Which uses the optimum depth step?

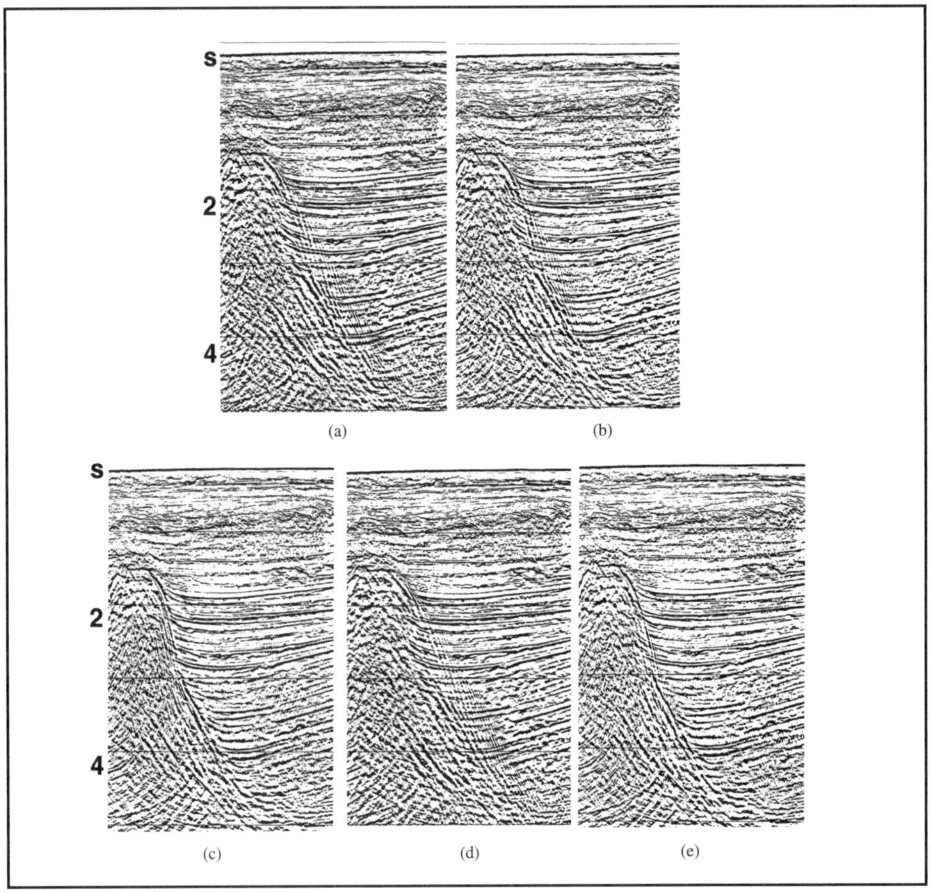

Fig. 6–296

Are the other panels over- or under-migrated?

27. Following are four stacks of the same data but with different trace spacing and the corresponding phase shift migration panels (Figs. 6–297 *a* and *b*). What trace spacing is required to prevent spatial aliasing?

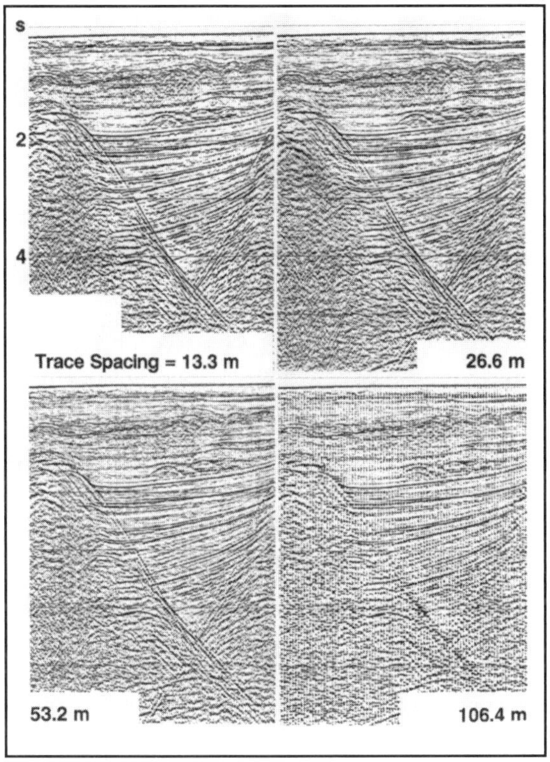

Fig. 6–297a

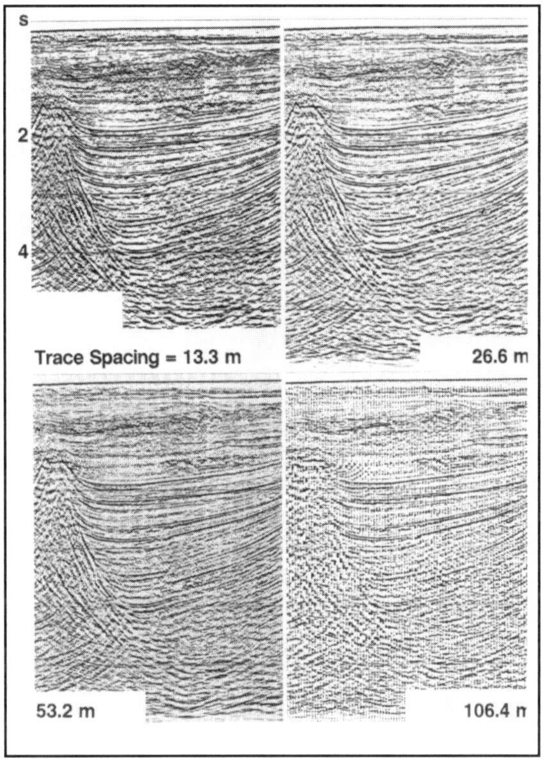

Fig. 6–297b

7 Seismic Reflection Data Interpretation

Introduction

A seismic trace is regarded as the superposition of reflections from many subsurface interfaces where the acoustic impedance (product of velocity and density) changes. This superposition concept describes a seismic wavelet being reflected back at each change of acoustic impedance, with the sign and amplitude of the reflection being proportional to the sign and fractional change of the acoustic impedance. The seismic trace is represented simply by the mathematical operation of convolving (filtering) the wavelet with the Earth's reflectivity. It is a log of acoustic impedance as a function of two-way travel time of the seismic energy.

In conventional seismic interpretation, it is assumed that the acoustic impedance changes are parallel to the bedding planes so that mapping the arrival times of reflections gives the structural picture. Any discontinuity of the reflections is assumed to indicate faulting or some other structural or stratigraphic feature.

For many decades, seismic reflection interpretation involved the arrival times of the markers with little attention given to the amplitude, frequency, and wave-shape variations. Seismic reflection was used mainly in the interpretation of the structural aspects of the hydrocarbon traps with little emphasis on the stratigraphic aspects. Most of the stratigraphic traps found were serendipitous—discovered during the search for something else.

Stratigraphic interpretation of seismic data requires good data quality and involves some art as well as science. The effects of structural complications may overwhelm the effects of stratigraphic variations in a good record area. However, in many areas, seismic stratigraphy can add important geological information and enhance the understanding of the depositional environments, which may help in the understanding the origin, accumulation, and trapping mechanisms of the hydrocarbon deposits.

The seismic traces are trying to tell us the details about the subsurface, but their voices cannot be heard because they speak in a very high-noise environment. With better techniques in field data acquisition and great advancements in data processing, we can have better quality data. This makes it more feasible to extract stratigraphy from the seismic record and makes us believe in the geological significance of the seismic signature of stratigraphic sequence analysis.

Without understanding the basics, it is difficult for one to comprehend the physical meaning of the seismic signature and to relate it to stratigraphy—first in regional aspect, second in local aspect, and finally in individual reflection characteristics. Understanding the basics will help us to identify lithology, depositional environments, and direct hydrocarbon indicators.

The reliability of this information is dependent on the quality of the seismic data. Good-quality data depends on proper field acquisition, careful data processing flow, and a high degree of quality control in order to preserve the frequencies and amplitudes of the reflectors.

In the previous chapters of this text, we discussed data processing techniques necessary to generate a geologically sound seismic section that reveals reliable and meaningful stratigraphic information. Migration was discussed to give a better understanding of the proper imaging of the subsurface. Lateral and vertical resolution of seismic data was considered for better understanding of the limitations of seismic data. At this point, we are prepared to review stratigraphy, stratigraphic traps, depositional environments, facies, geometry of traps, and seismic signature.

Part of this text is devoted to the discussion of forward and inverse modeling approaches and their applications to stratigraphic sequence analysis. The last part of the text is devoted to modern geophysical applications and techniques such as *vertical seismic profiling* (VSP), *amplitude versus offset* (AVO), high-resolution seismic data, shear-wave applications in seismic stratigraphy, and 4-D technology.

Finally, we discuss current research and future developments in borehole measurements, advancements in subsurface imaging, and their role in enhancing the stratigraphic interpretation of seismic data.

Modeling

Introduction. Models are representations of some physical situation used to explain observations or to confirm hypotheses. Physical models are constructed to scale and experiments performed on them, often in an attempt to duplicate observed data. Mathematical models are more often used, usually in conjunction with computers. In seismic exploration, such models most often are used to test or develop interpretations of seismic data.

Geologists have used various models, including plate tectonics models of earth move-ment, to test theory. Paleontological studies are used to model the reservoir conditions of an area. Surface geologic features and available subsurface geology are used to the model subsurface targets of interest.

Geophysicists also use models to represent the subsurface. Resistivity logs, density logs, and interval transit time logs obtained from borehole measurements provide data for models that are most useful for interpolating between drilled wells. There are various methods of checking these models. The most common of these devices are magnetic anomaly mapping, gravity anomaly mapping, and the seismic refraction and reflection mapping schemes. The seismic reflection method is the most detailed of these since it has better resolution and fewer ambiguities than the others. Unfortunately, seismic reflection exploration does have its distortion mechanisms. The following paragraphs investigate seismic reflection exploration measurements, inferences made or calculated from the measurements, and how these may distort the image of the modeled subsurface geology.

Seismic exploration measures the Earth's response to mechanical energy injected into the ground or water. This energy may be impulsive (explosives or airguns) or vibratory (the swept frequency signal of the Vibroseis system). The direct measurements are amplitude as a function of time.

At each interface between subsurface materials differing in velocity and or density, some energy is reflected and some transmitted or refracted. The amplitude of the reflected pressure wave at a given interface, relative to the incident amplitude, is described by the reflection coefficient for that interface. The reflection coefficient depends on the contrast between velocity and density plus the angle of incidence. The Zoeppritz equations are a relatively complex set of four equations that can be used to determine amplitudes of all wave types produced by an incident P- or S-wave. In most seismic reflection applications, a simpler expression (actually valid only for incidence normal to the interface but applicable for small angles usually encountered) can be used. This is:

$$R_v = \frac{\rho_2 v_2 - \rho_1 v_1}{\rho_2 v_2 - \rho_1 v_1} \tag{7.1}$$

where

ρ = density

v = interval velocity

The subscripts denote the Earth layers involved in the reflection process. Units for density and velocity are not important as long as the same units are used in layers 1 and 2.

The response that seismic reflection exploration seeks to measure is the two-way reflection time to interfaces of interest and the reflection coefficients of those interfaces.

Equation 7.1 combined with derived average velocities allows modeling depth of burial, formation thickness, lateral extents, and, in some cases, porosity and hydrocarbon presence. However, it is also necessary to recognize and compensate for distortions from the true subsurface model that the seismic method produces.

Seismic distortions. There are three categories of seismic distortions as follows:

1. errors in position with events displaced laterally and vertically
2. apparent bedding changes caused by reflection response differences
3. noise or unwanted information that interferes with and masks desired information

Historically, the last item—noise—has received the most attention in seismic acquisition and processing. Improvement of the signal-to-noise ratio is a paramount goal in both acquisition and processing. Methods used for this purpose include

1. source/geophone arrays
2. offset distance selection
3. vertical and horizontal stacking
4. frequency filtering
5. velocity filtering
6. deconvolution

Figure 7–1 shows a typical exploration sequence. A prospect is identified by geologists and a seismic program is initiated. The interpreter sees the distorted view of the subsurface on the cross-sections produced by the seismic effort. He must test the validity of the geologists' initial model, which requires study of and compensation for distortion mechanisms. If this is not done, chances of finding oil or gas are poor. All distortions should be examined to determine methods to eliminate them.

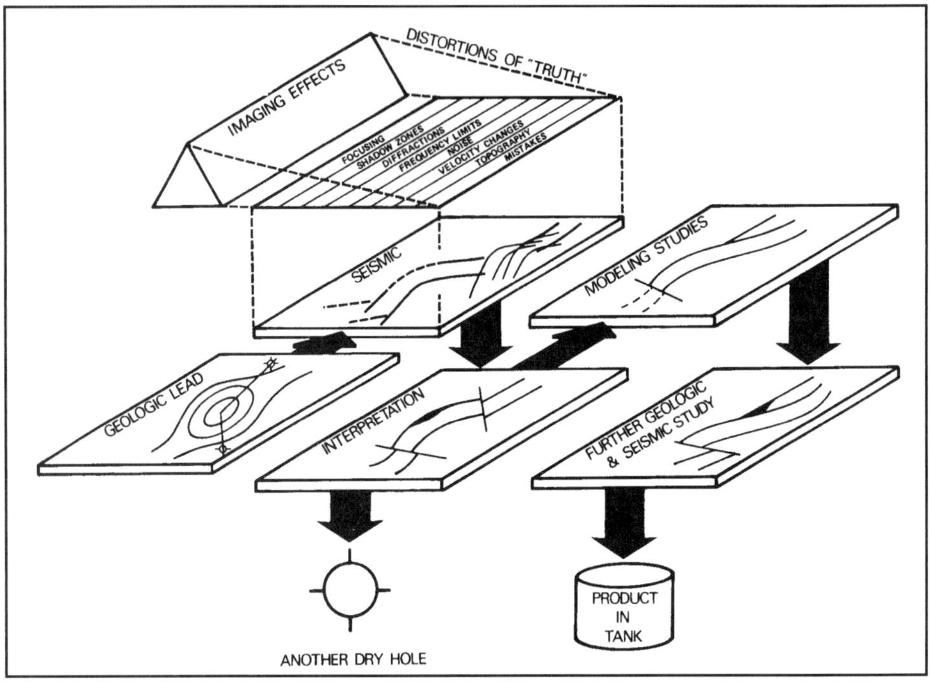

Fig. 7–1 *Typical Exploration Sequence*

Only flat, parallel bedding is correctly represented on CMP stack sections. Irregular and dipping beds are distorted. For example, an anticline (Fig. 7–2) appears larger than its true dimensions because the energy fans out of from the convex surface reflecting it. Figure 7–2 also illustrates the concept of focusing from a syncline or low structure.

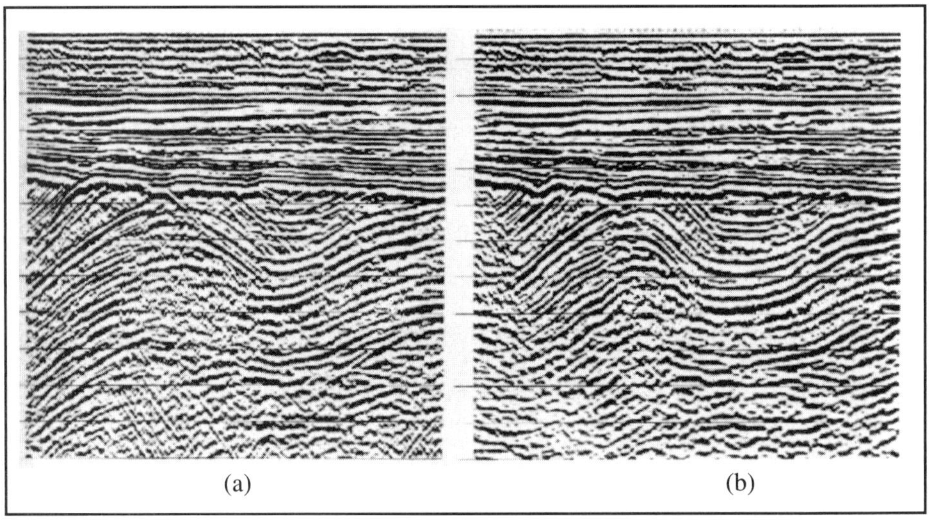

Fig. 7–2 *Focusing in Anticlines and Synclines*

Diffraction occurs when there are discontinuities in the seismic events such as faults, folds, and changes in lithology. In the case of a deep-seated syncline, another phenomenon is observed—two intersecting events with an apparent anticline beneath them. In fact it resembles a bow tie. This phenomenon is explained in the chapter 3 and is referred to as Fermat's principle.

Figure 7–3 shows curved reflecting surfaces (synclines and anticlines).

a) In Figure 7–3a, the zero offset stack shows three synclines and one anticline. Note the bow tie effect shows as the two deep-seated synclines. Diffractions are observed from the anticline D-E. In Figure 7–3b, the process of migration removes the distortion in the stacked section.

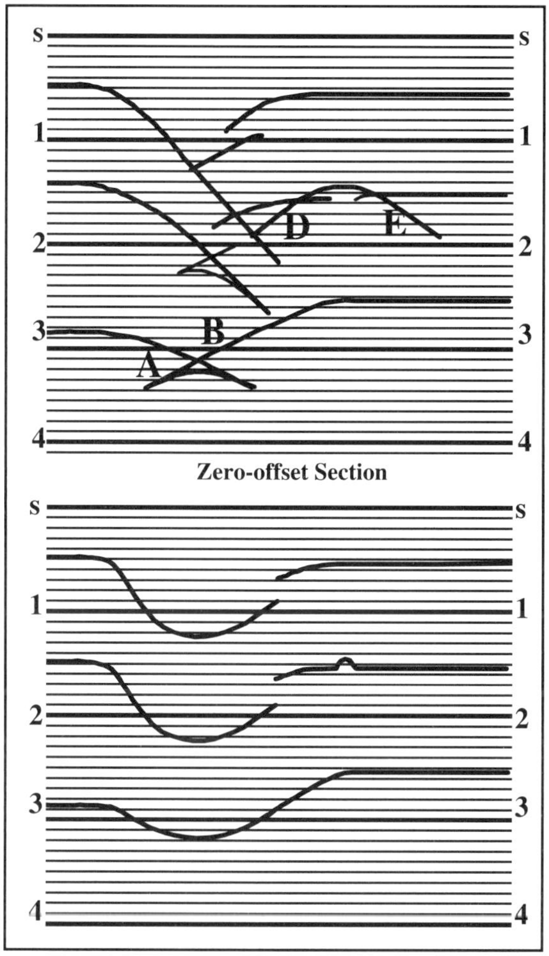

Fig. 7–3 *Bow Tie Effect of Buried Focus*

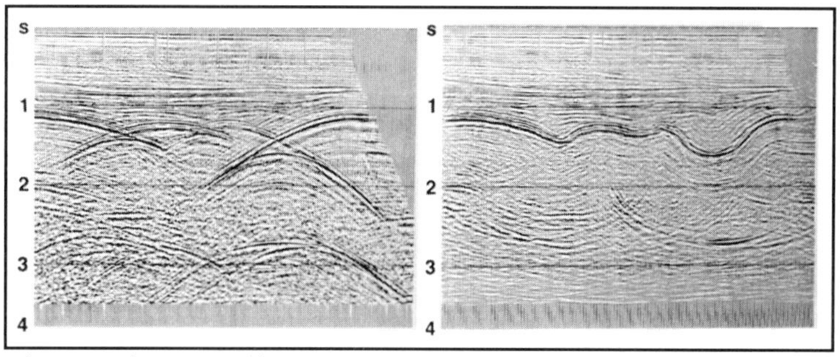

Fig. 7–4 *The Zero-offset Stack Shows the Focusing of the Narrow, Deep-seated Syncline and the Migrated Stack Shows the Bow Tie Untied*

The bow ties were untied and reveal the true subsurface structure as synclines. The anticline is reduced in size of the after the diffraction is collapsed by the migration process. Figure 7–4 is another example of the bow tie effect and its resolution by migration.

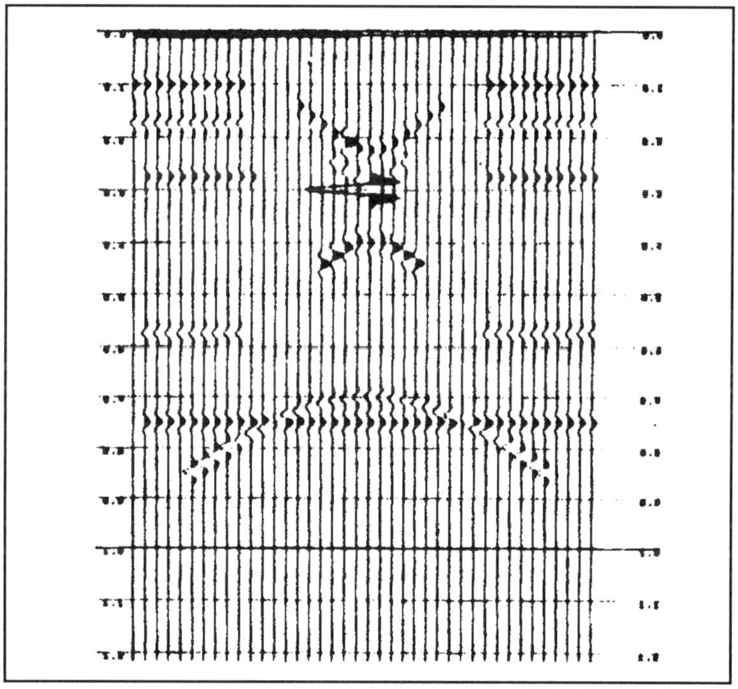

Fig. 7–5 *Shadow Zones*

In Figure 7–5, there are areas with no energy returning (dead areas). These are commonly referred to as *shadow zones*. Shadow zones are common near faults and other discontinuous areas in the subsurface. Here, energy has been focused on other receivers and does not show reflectors in their true positions.

Diffractions occur at discontinuities in the subsurface. Faults and velocity discontinuities, as seen at bright spot terminations, are examples.

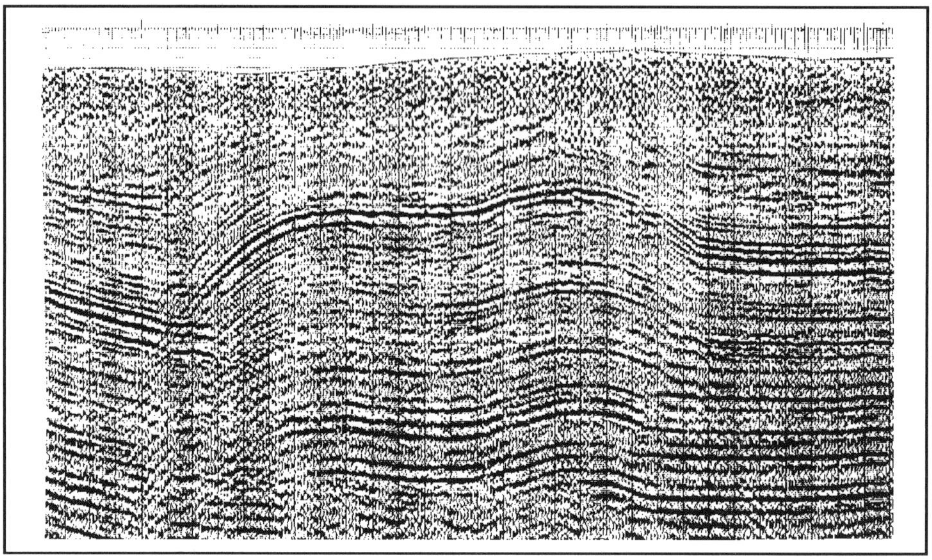

Fig. 7–6 *Zero-Offset Section of Horst Block*

In Figure 7–6, a CMP stack section for a seismic line across a large horst block, diffractions mask the fault plane on the left of the section. Figure 7–7, a normal incidence ray path model of this feature without diffractions, shows the fault zone more clearly. A shadow zone in the fault vicinity can also be seen.

The seismic reflection method uses only the lower end of the acoustic spectrum. This is because the earth attenuates high frequencies much faster than low frequencies. Under normal petroleum exploration conditions, the usable frequency range rarely exceeds 80 Hz. Often, in any one area, closer limits are imposed. Since ground roll and wind noise may cover the lower and higher frequencies in the 5–80 Hz range, the signal frequency band is more often on the order of 20–60 Hz.

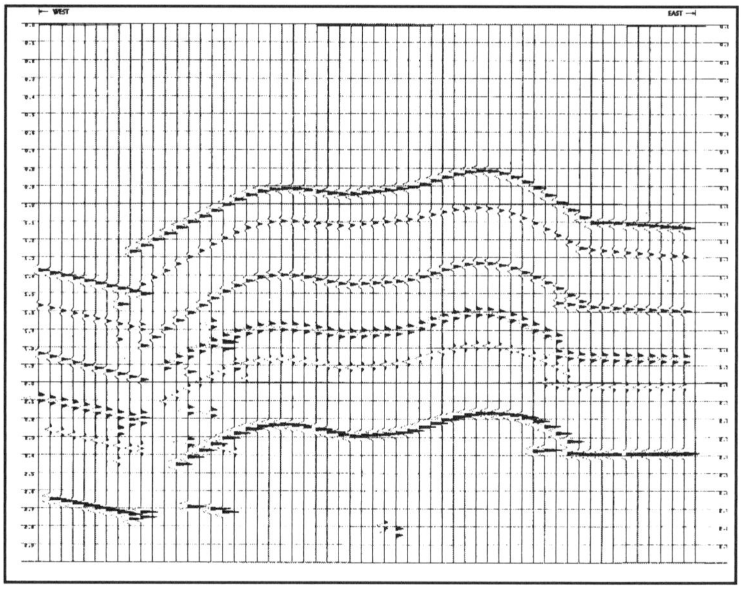

Fig. 7–7 *Normal Incidence Ray Path Model of Horst Block*

The limits on bandwidth carry with them limits on resolution. When layering thickness becomes thin with respect to wavelength—about 100 ft—the wavelets are longer than the time between them, and they interfere with one another. This results in a wavelet complex with amplitude and phase dependent on the convolution of the layering with the band-pass filtering. At very thin layering, the reflections become so close together that the interference becomes so severe that no reflection is present or one strong reflection is present as in Figure 7–8.

With the most advanced software programs and hardware equipments, the acquisition and processing of seismic data is carried out in an optimum manner and residual noise will mask the subsurface to some extent. Examples of noise are

- random noise
- residual coherent noise
- out of plane reflections
- multiples
- ghosts

This interfering energy can cause apparent reflection changes in time (pseudo faults), frequency (pseudo stratification change), amplitude (apparent change in reflection coefficient), or phase (structure or stratigraphic anomaly).

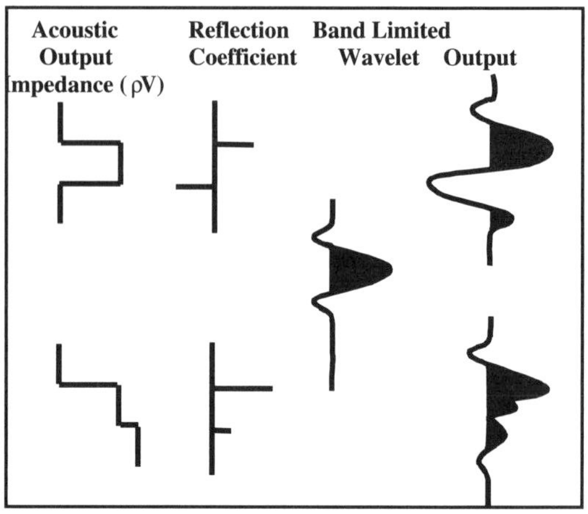

Fig. 7–8 *Thin Bed Response*

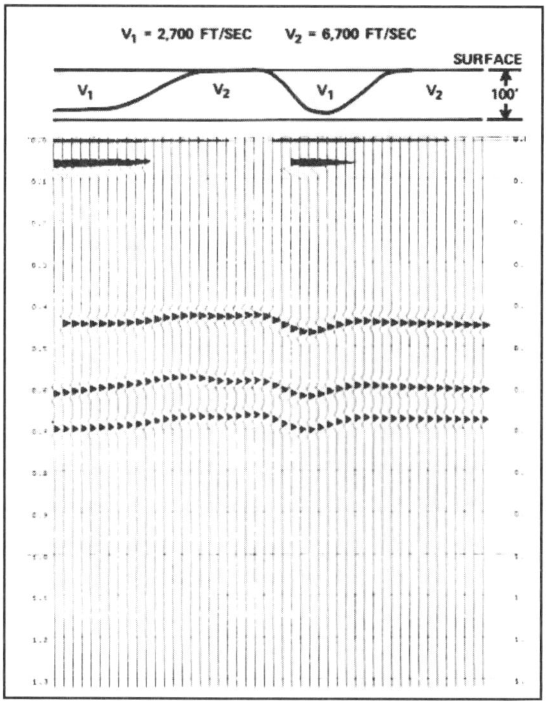

Fig. 7–9 *Distortion in the Seismic Data because Of Lateral Near Surface Velocity Variation*

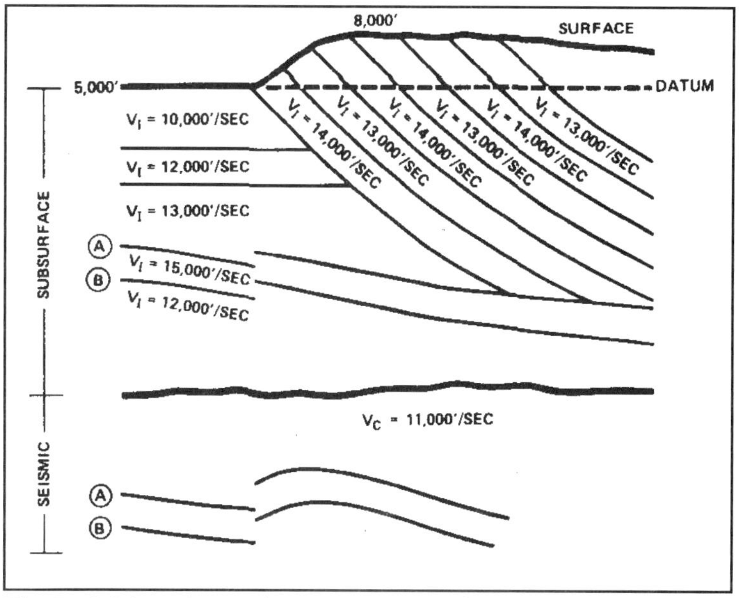

Fig. 7–10 *Distortion in the Deep Structure because of False Turnover against the Fault Plane*

Variations in the surface and near surface are frequent sources of seismic distortion. Surface material changes cause source and receiver response variations that affect reflection quality. Variations in thickness and velocity of the near surface, laterally and vertically, cause event times for traces to indicate false structure. Figure 7–9 shows the effects of a varying near surface velocity the subsurface. Note, in this model, the deep reflector is a flat calibration horizon and shows the distortion directly. Figure 7–10 is a case showing how elevation and near surface velocity causes an apparent turnover into a fault when it really doesn't exist.

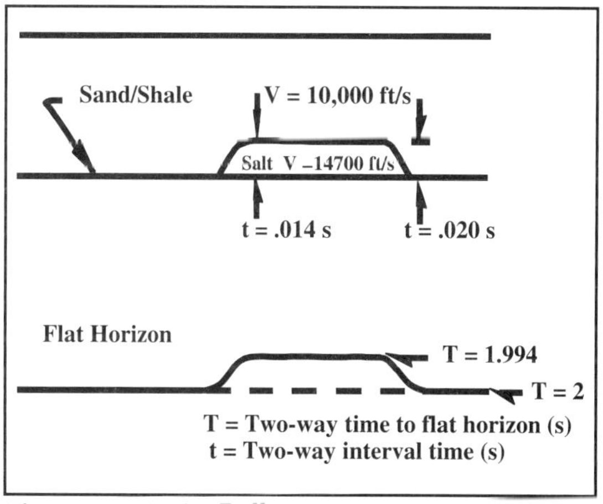

Fig. 7–11 *Velocity Pull-up*

Automatic statics programs and other static correction methods attempt to remove these time shifts before stacking data. Gradual variations in a near-surface layer defy the ability of software programs to eliminate the time variations they cause. Some areas are so complex that the correction velocity to datum needs to vary as the line progresses.

Without changes in velocity variations in the subsurface, there would be no reflections. Velocity variation can, however, cause distortions that distort the view of the subsurface. Some examples of distortions caused by velocity variations are

1. false structure caused by velocity pull-up from shallower beds (Fig. 7–11)
2. apparent thickness changes (Figs. 7–12 and 7–13)
3. pseudo-faults from overlaying anomalies (Figs. 7–14, 7–15, and 7–16)
4. over-pressured shale zones (Figs. 7–17 and 7–18)

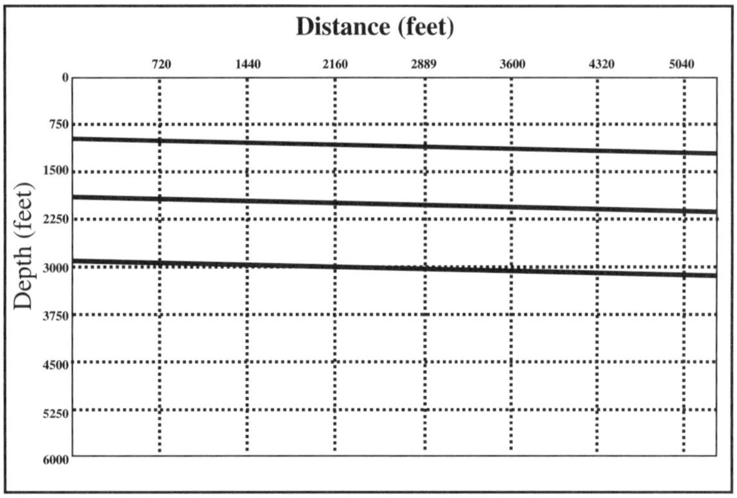

Fig. 7–12 *Subsurface Section—Basinward Thinning*

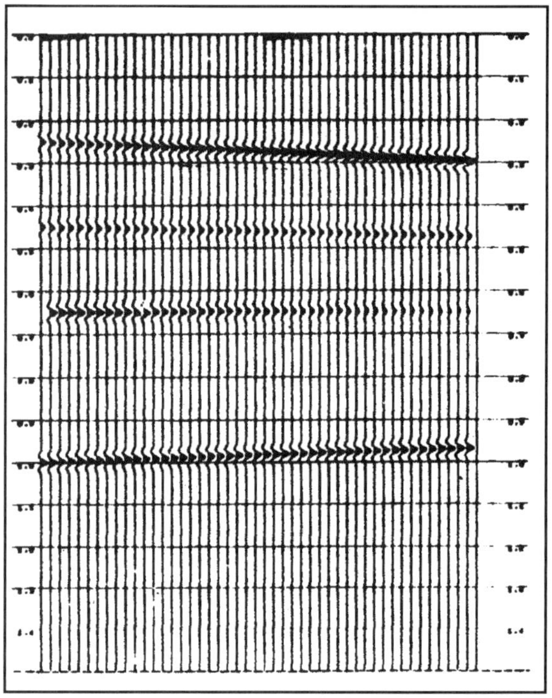

Fig. 7–13 *Seismic Model—Basinward Thinning*

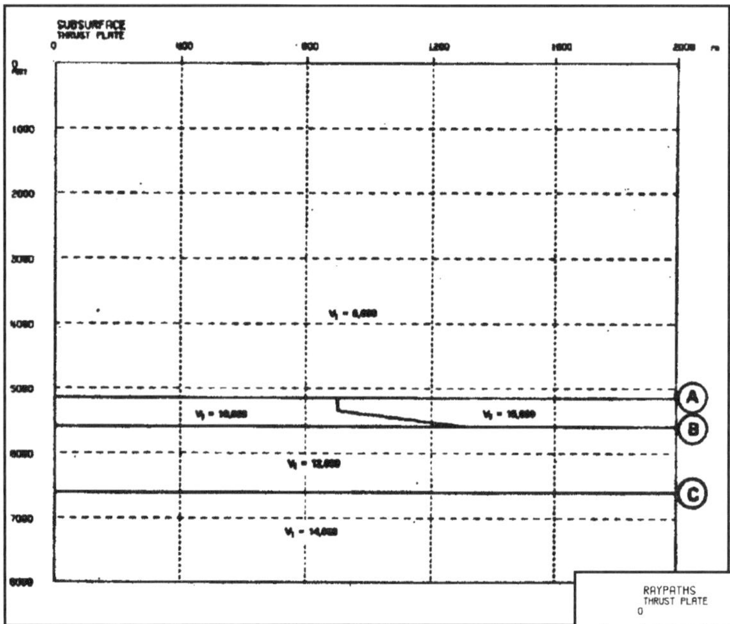

Fig. 7–14 *Subsurface Pseudo Fault Model*

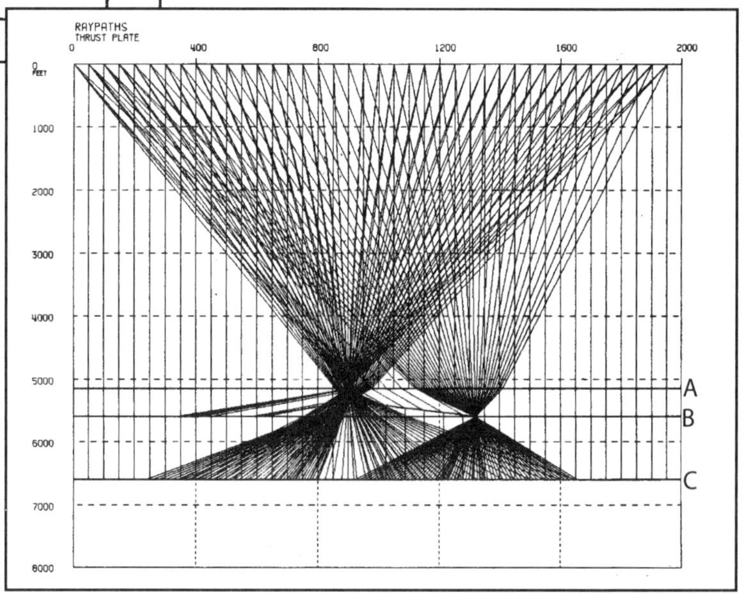

Fig. 7–15 *Ray Tracing For the Subsurface Model of Figure 7–14*

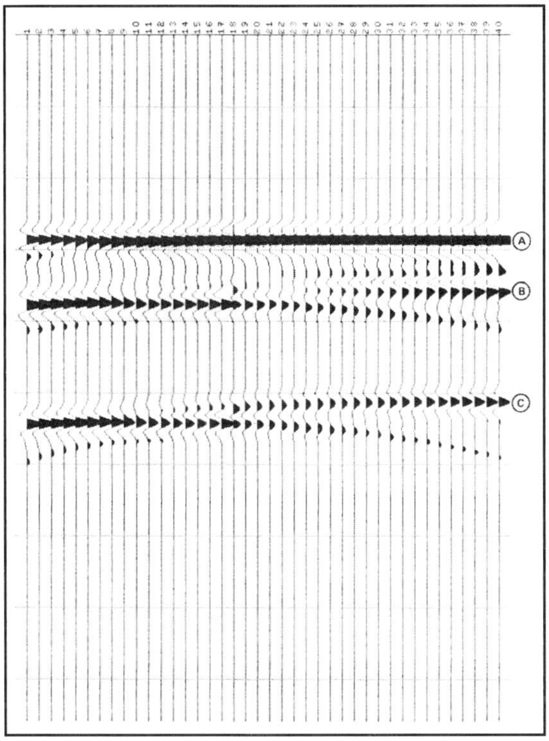

Fig. 7–16 *Seismic Model For the Pseudo Fault*

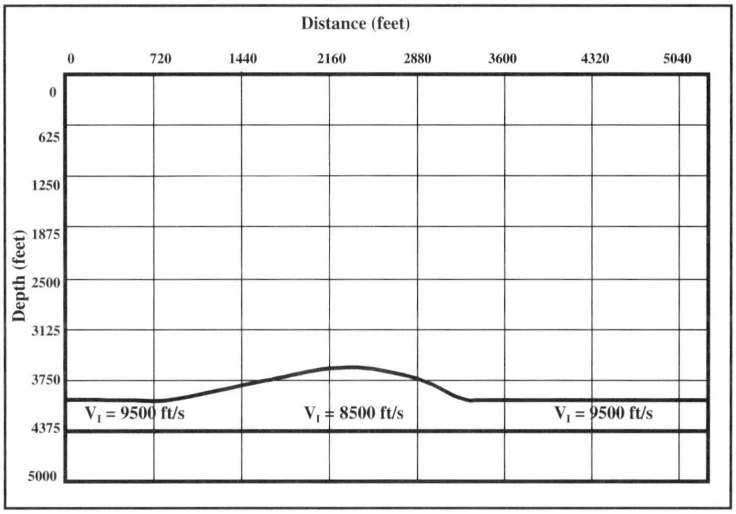

Fig. 7–17 *Over-pressured Shale Model*

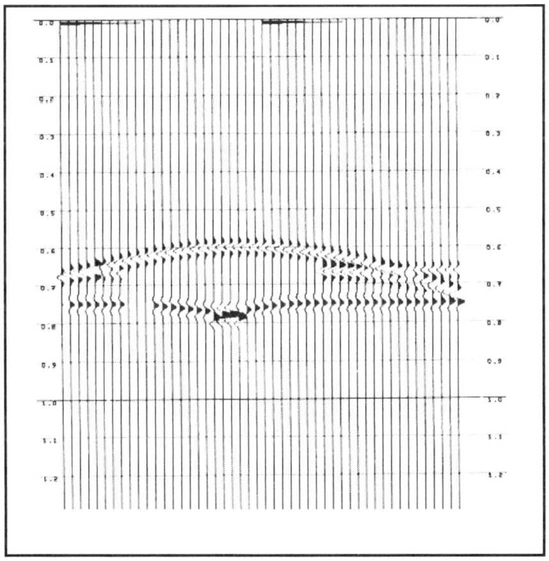

Fig. 7–18 *Seismic Model of Over-pressured Shale*

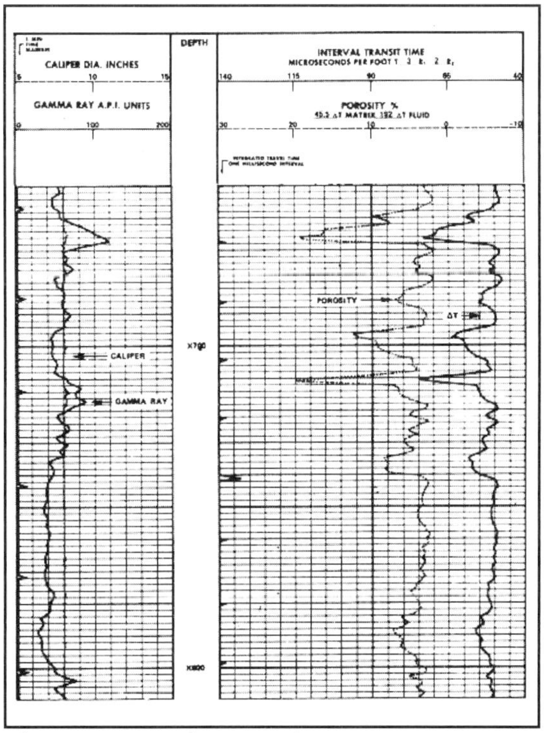

Fig. 7–19 *Interval Transit Time Log*

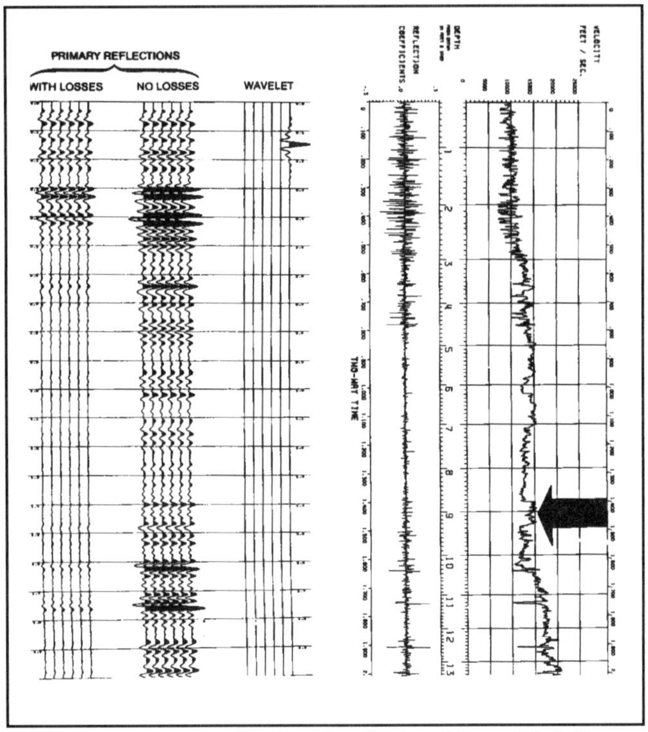

Fig. 7–20 *Primary Reflection Synthetic without Modeling*

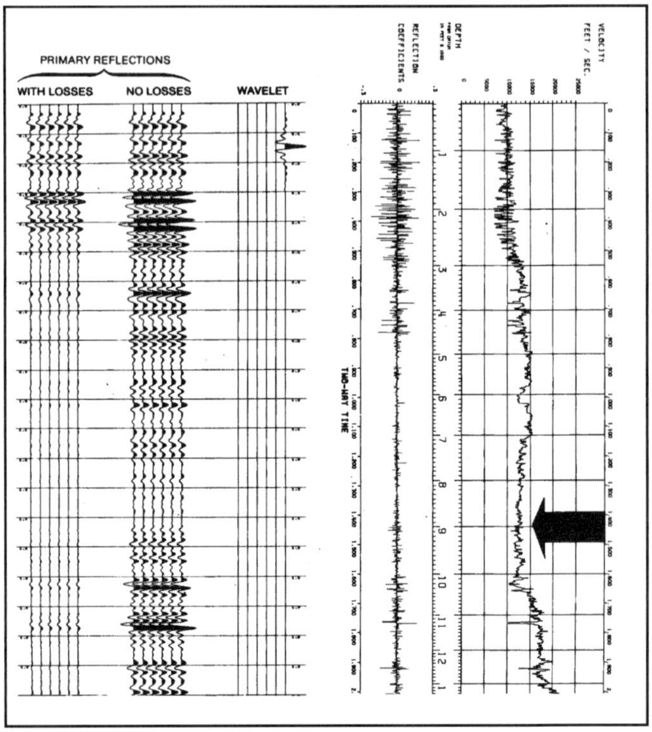

Fig. 7–21 *Primary Reflection Synthetic with Velocity Modified between 8700 and 9350 Ft*

The previous figures are a few of the distortion mechanisms in seismic exploration that must be recognized and the effects of which must be removed to the extent possible. The modeling process is one way of doing this.

Geophysicists perform modeling on an ongoing basis. Some digital processing techniques—velocity spectra, amplitude analysis, post stack velocity analysis, wavelet processing, and migration—help with this modeling. Modeling, programs have been written for direct use in model construction.

Seismic modeling is an important step in upgrading the quality of the play or it may condemn it and save dry hole money

1-D modeling. Synthetic seismograms are one-dimensional models. They are produced from interval transit time logs or continuous velocity logs (Fig. 7– 19).

The method of calculating synthetics is as follows:

1. develop a velocity log (velocity versus depth series)
2. integrate this log to produce a depth versus time curve
3. the log can be combined with a density log, and a reflection coefficient versus time log can be calculated using Equation 7.1

$$R_v = \frac{\rho_2 v_2 - \rho_1 v_1}{\rho_2 v_2 - \rho_1 v_1}$$

4. calculate auxiliary functions as desired
   a. transmission losses $(1 - R_v^2)$
   b. multiples (first order or all multiples)
   c. ghosts
   d. reverberations
5. convolve reflection coefficient logs with a pre-selected wavelet to obtain the synthetic seismogram

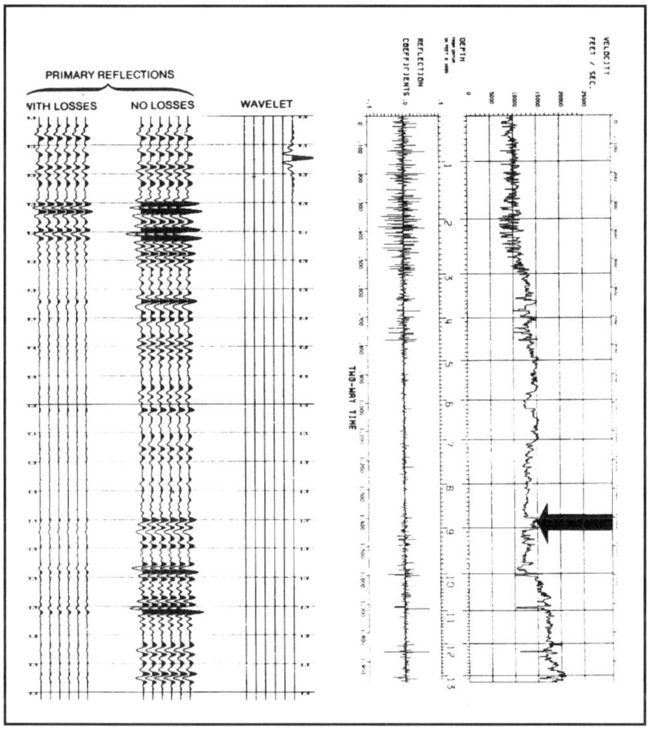

Fig. 7–22 *Primary Reflection Synthetic with Depth Modification at 8700 Ft Bed Thickness Reduced from 430 to 312 Ft*

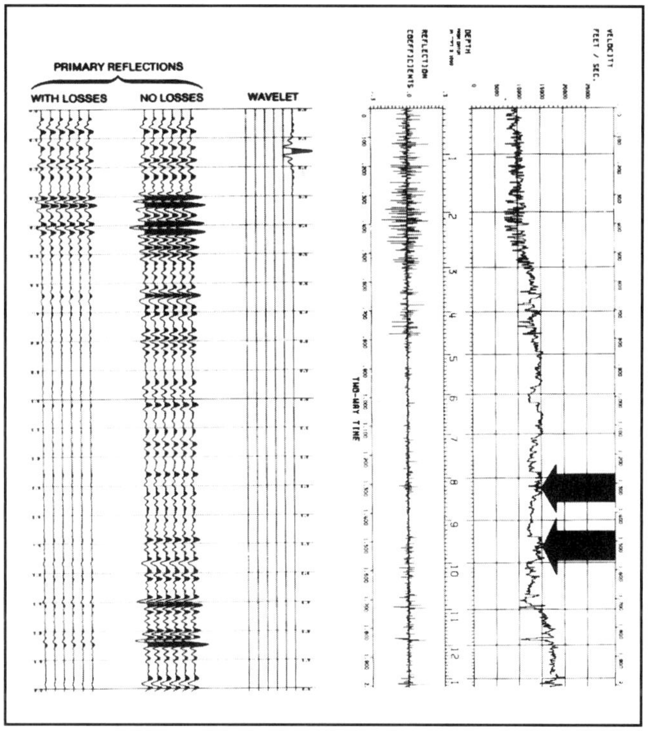

Fig. 7–23 *Primary Reflection Synthetic with Repeat Section to Simulate Thrust Faulting*

This results in the development of a model taken from a well log, which closely resembles a seismic trace taken at the same location.

Geophysicists compare the synthetic to their data and correlate their sections to geologic markers, which are easily seen on the log. A synthetic seismogram is shown in Figure 7–20. Synthetics can also be used to help select acquisition and processing procedures, which will best preserve data most important to the prospect.

Further studies of the subsurface can be done by making assumptions regarding the effect of changes in the geologic section. Changes can be made (modeled) in the synthetic to test the validity of the assumptions. Synthetics may be modeled by the following:

1. changing a bed thickness
2. change bed velocity
3. change bed density
4. merge logs
5. include slabs or ramps of velocity or density in logs

Figures 7–21 through 7–23 demonstrate some of these modeling techniques. The arrow indicates the part of the log that will be modified in these figures.

Further applications of synthetic seismogram uses are shown in Figures 7– 24 and 7–25. Five wells have been collected for geologic and geophysical studies. Figure 7–24 shows the interval velocity logs for the five wells plotted against time. The logs have been time shifted so that horizon *D* is flat, simulating an ancient seabed condition. The high-speed sandstone formation *B-C* is shown as it was deposited. This is a deltaic deposition where most of the material is near shore to the right. Figure 7–25 is the synthetic seismogram traces for the same wells. Notice the amplitude and phase distortion throughout the pinch-out zone of *B-C*.

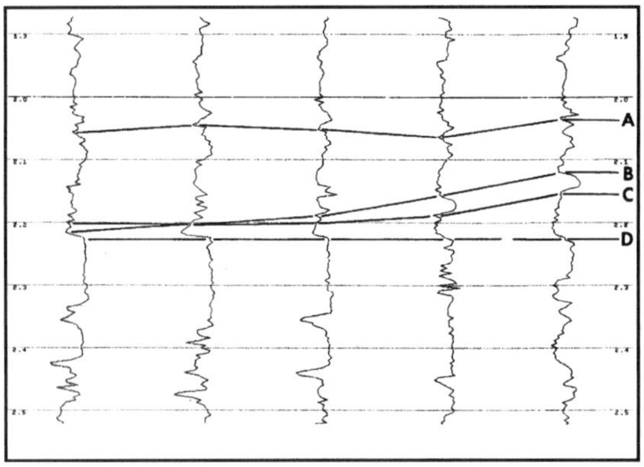

Fig. 7–24 *Model Cross-section Showing Interval Velocity Versus Time*

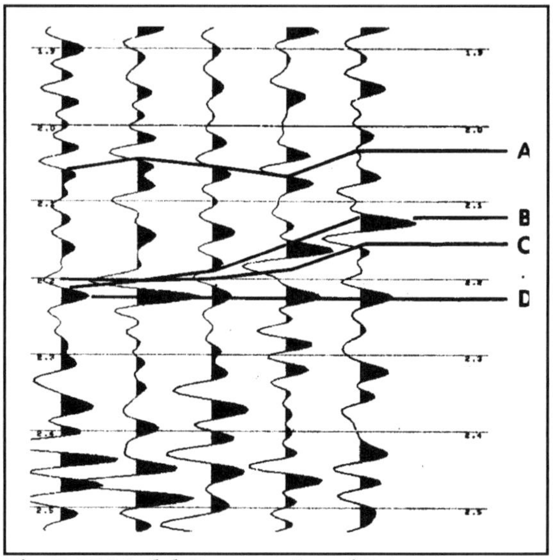

Fig. 7–25 *Model Cross-section of Primary Reflection*

2-D modeling. The one-dimensional modeling techniques can be used to study a few points on a subsurface to resolve detailed changes in a localized area. However, where there are lateral variations between points, focusing, shadow zones, or diffractions, two-dimensional modeling is used. Two-dimensional models range from simple to complex. A list of model types, in increasing order of complexity, is shown in Table 7–1.

Table 7–1 *2-D Models*

Type	Description	Restrictions
Normal incidence straight path	Uses average velocity for times and position	Limited accuracy in focusing diffractions and shadow zones
Vertical path no diffractions	Interpolated one dimensional	No focusing, no shadow zones,
Normal incidence	Snell's law	Poor resolution of tightly folded structure ray tracing (less than a wavelength)
Wave equation	Finite differences, Huygen's principle	Most complete solution, very expensive
CMP ray tracing	Offset rays by Snell's law	Most useful in testing processing parameters, very costly

Interpreters use two-dimensional modeling to test assumed geologic sections against seismic data to confirm or reject the assumptions. The complexity of the modeling scheme depends on the complexity of the subsurface. This, in turn, determines the cost of the modeling. Normal incidence ray tracing is the most widely used modeling technique because it gives the desired accuracy for most problems at a reasonable cost.

Steps in ray-trace modeling include the following:

- make a scaled model of the assumed subsurface geology that includes bedding geometry, interval velocities, and densities if available (Fig. 7–26)
- input the model into a computer in the form required by the software being used
- do the ray-tracing calculations (Fig. 7–27)
- calculate vertical incidence reflection and transmission coefficients for every interface in the model
- use a unit impulse to generate a seismic section (Fig. 7–28a)
- select a source wavelet and convolve it with the section just completed (Fig. 7–28b)
- add random noise if this seems appropriate to the problem (Fig. 7–29)

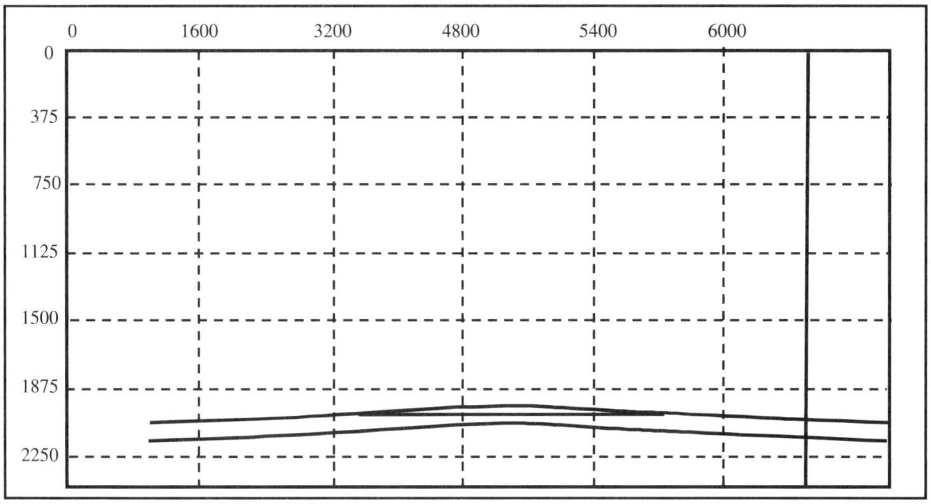

Fig. 7–26 *Subsurface Depth Model*

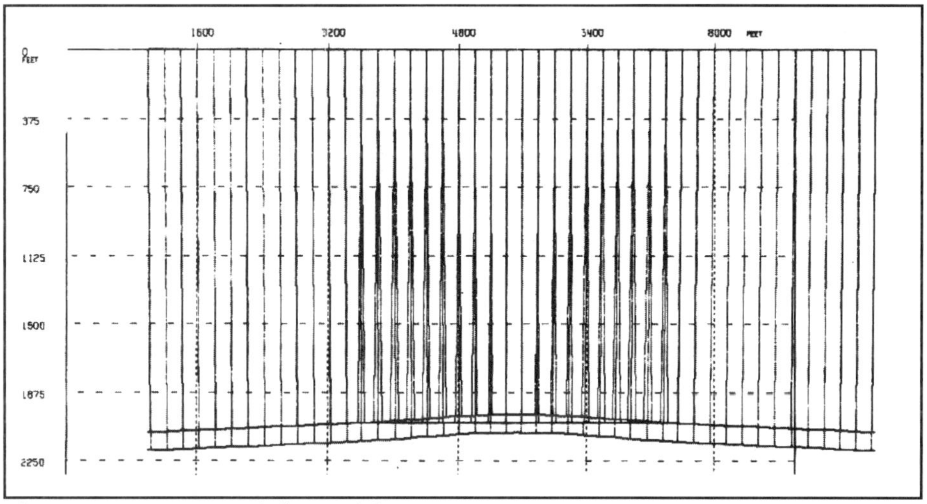

Fig. 7–27 *Ray Tracing of the Model*

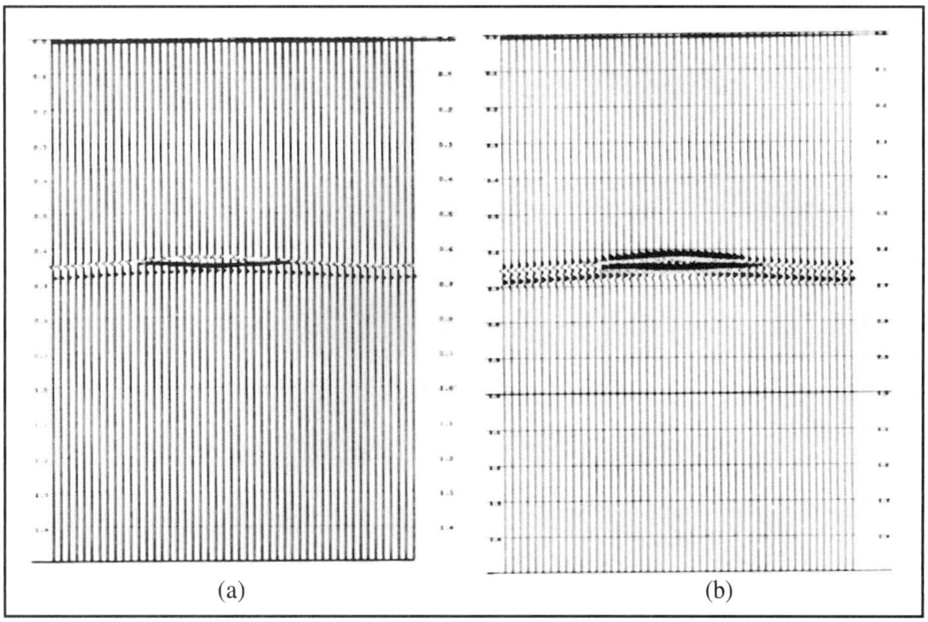

Fig. 7–28 *Spike Seismogram and Wavelet Seismogram from the Model*

The model is then iteratively altered to agree with the seismic section while keeping parameters confined to the local physical limits.

Inverse modeling. This is building a geologic depth model section from seismic time data observed in the field. Ideally, it would approach the function of an interpreter. Inversion is used in a wide application of the exploration, exploitation, and development fields with a great success. For example, in a best-guess velocity model where a first approximation to the depth section is developed then iteratively altered to conform to the physical acquired data.

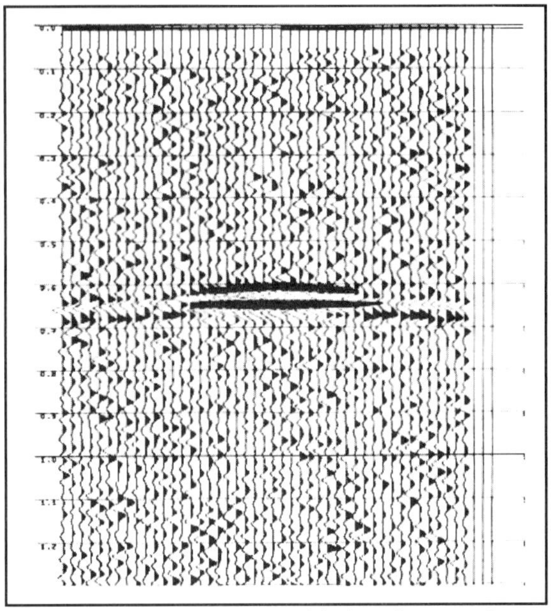

Fig. 7–29 *Random Noise Added to the Wavelet Seismogram*

3-D modeling. Since the subsurface is a three-dimensional solid, two-dimensional assumptions limit the interpreter's analysis. Currently, the seismic industry is using 3-D acquisition and processing, and as it becomes the standard method, modeling will be done in three dimensions also. Some 3-D modeling is now performed, as the computer technology is developed to apply and improve these techniques.

Conclusions. The seismic exploration method attempts to accurately define subsurface geologic structures and stratigraphy. The seismic sections output from seismic data processing can be thought of as seismic or acoustic models of the subsurface. The modeling process uses assumptions regarding the geologic structure and stratigraphy that produced the acoustic model and tests the model against physical constraints and the real data.

There are three advantages in building models.

- Modeling forces a closer look at mechanisms causing seismic distortion.
- Modeling costs much less than relying on the drill to test an interpretation.
- Modeling is the best method available for geophysicists to develop the insights required for data interpretation.

Seismic Inversion and Wavelet Processing

A new technique in signal data processing called *wavelet processing* has been developed to help obtain the better resolution for the detection of thin layers and to provide improved data for stratigraphic interpretation. The objective of wavelet processing is to optimize the shape of the seismic pulse and make it a symmetrical or zero-phase wavelet, which is the simplest form and the one the interpreter desires.

The seismic trace is often described as the convolution of a waveform with the reflectivity series. It can be expressed by the equation:

$$f(t) = c(t) * p(t) + n(t) \qquad (7.2)$$

where

$f(t)$ = the seismic trace

$c(t)$ = random reflection coefficient series

$p(t)$ = pulse waveform

$n(t)$ = additive random noise

$*$ = the convolution notation

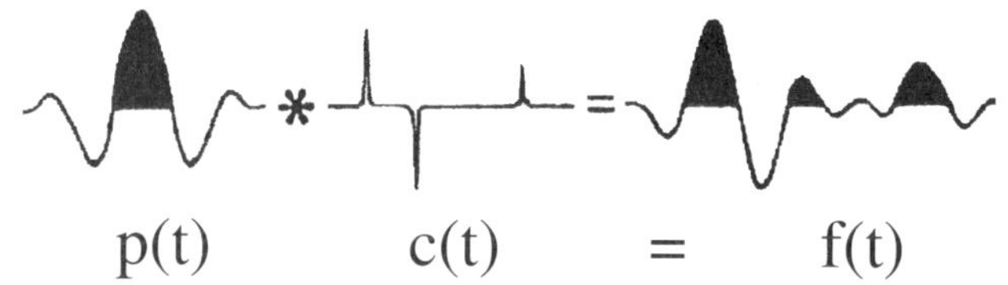

where f(t) = the seismic trace
c(t) = random reflection coefficient series
p(t) = pulse waveform
n(t) = additive random noise
* = convolution notation

Fig. 7–30 *The Seismic Trace*

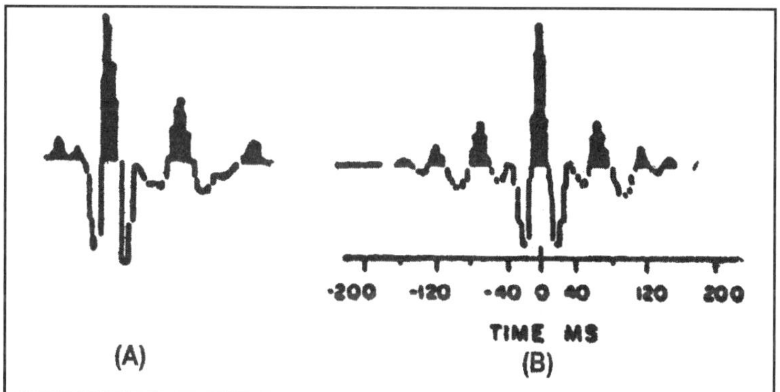

Fig. 7–31 *Direct Measurement of Seismic Waveform*

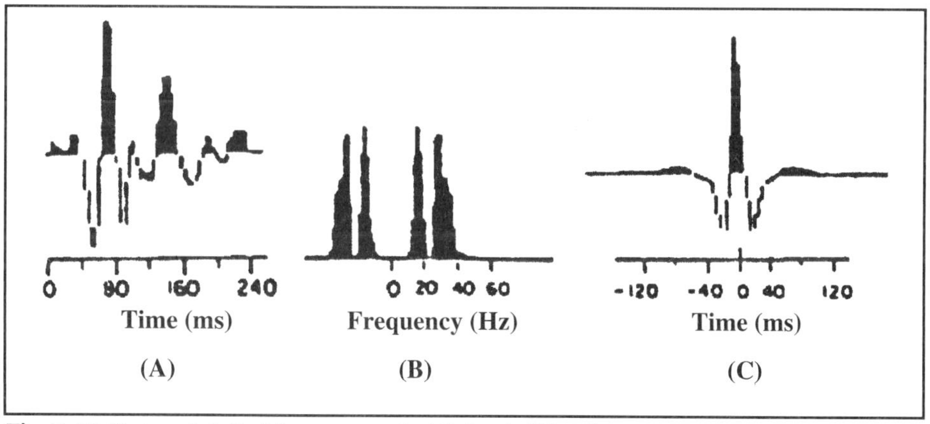

Fig. 7–32 *Deterministic Measurement of Seismic Waveform*

The primary assumption for c(t) is that it is a random and unpredictable series. Figure 7– 30 illustrates Equation 7.2.

Wavelet extraction. Interest in wavelet estimation has generated several methods for extracting the wavelet from the seismic data. Included in these methods are

- direct measurements of seismic waveform
- deterministic wavelet extraction
- statistical estimation of the wavelet

Each of these methods has its merits and its drawbacks.

Direct measurement of noise-free seismic waveforms can be made only in the deep-water environment. Measurements made close to the source are termed *near-field* source signatures, while those made at a distance comparable to that of reflected events are called far-field source signature. The effect of the ghost reflections, source arrays, instrumentation filters, and the cable must be accounted for in order to produce a valid representation of the waveform found on the seismic trace. Figure 7–31a is a measured near-field waveform derived from an airgun signature. Corrections were applied for surface reflection, so this waveform should match the far-field wavelet that propagates along the surface. Wavelet processing with this wavelet will give a zero-phase symmetrical waveform as shown in Figure 7– 31b.

A basic assumption in deterministic wavelet extraction is that in a noise-free, multiple-free seismic trace, the trace may be regarded as a convolution of the reflectivity series with the waveform. If the reflectivity series is determined, then it can be removed to reveal the estimate of the basic wavelet. Sources of the reflectivity series are log measurements or from reasonable geological parameters. A common approach is to assume that the water-bottom reflection event, which is free of refraction arrival events, emulates a good approximation to a single reflection event. This method is shown in Figure 7–32.

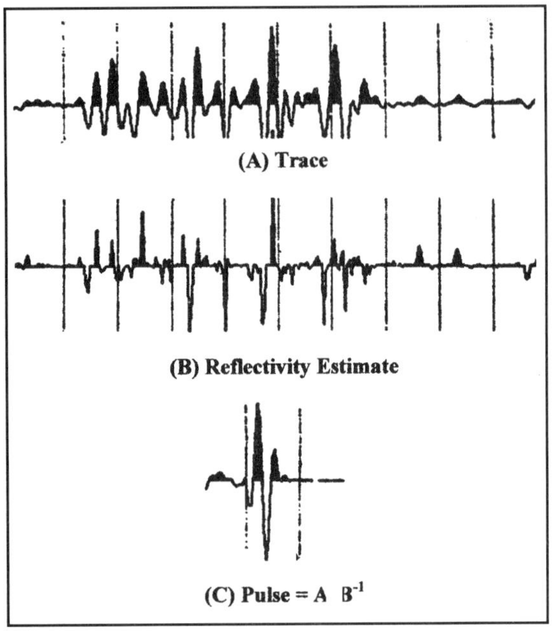

Fig. 7–33 *Statistical Estimation of Wavelet*

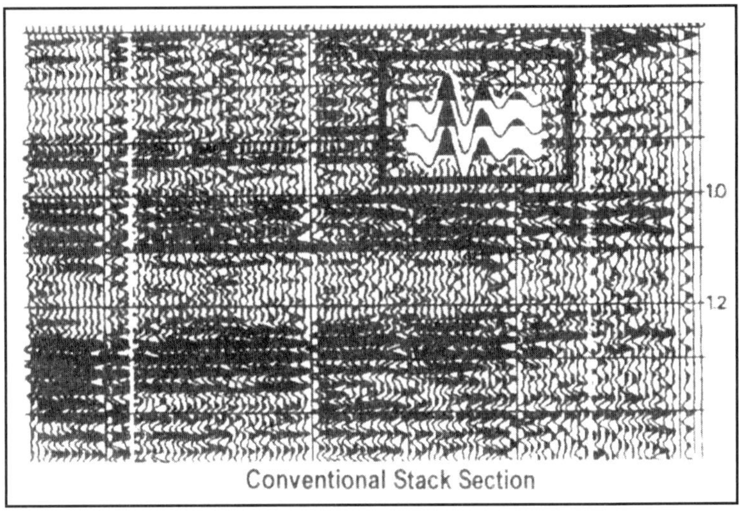

Fig. 7–34 *Statistical Method of Wavelet Processing with 12 Traces*

If the two previous approaches fail in estimating the waveform, then a statistical approach can be applied. One method in this category is *spiking deconvolution* to estimate the reflection coefficient. This approach must reflect the assumption of the earth model (Fig. 7–33).

Other methods of waveform estimation include *homomorphic filters* and *structural deconvolution*. Discussion of these methods is beyond the scope of this book.

Estimates of signal waveforms allow the design of filters that optimize the amplitude and phase spectrum within the recorded bandwidth.

Equation 7.3 provides a means to design a unique operator for each seismic trace.

$$h(t) = \frac{\hat{p}(t)}{P(t)} \tag{7.3}$$

where

$h(t)$ = shaping filter

$p(t)$ = estimated pulse

$P(t)$ = autocorrelation of $p(t)$

^ Denotes time reversal

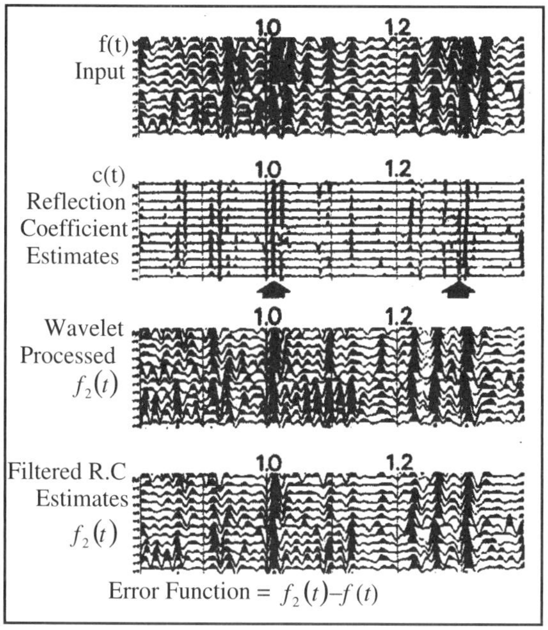

Fig. 7–35 *Statistical Method of Wavelet Processing*

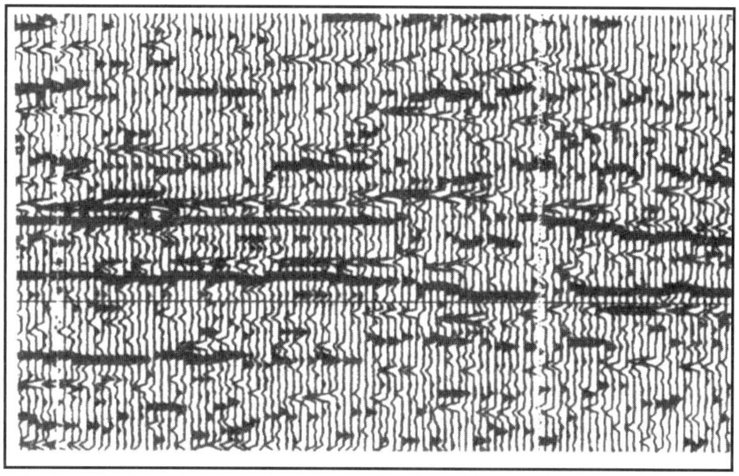

Fig. 7–36 *Estimated Reflection Coefficients*

Division by the autocorrelation levels the frequency spectrum and then corrects the phase spectrum to zero phases. This design is more effective than the standard deconvolution filters. No assumption of the phase is needed, and band limiting of the method suppresses high-frequency noise. Pioneers of this approach include Robinson (1963) and Treitel and Peacock (1969). Stone, et al. (1977) proposed that method for extracting the wavelet from the seismic trace using reflection coefficient estimates. The procedure for estimating, evaluating, and correcting the seismic reflection coefficient from seismic data is summarized as follows:

1. estimate the reflection coefficients by applying a multi-channel, predictive error, adaptive operator (Burg 1975)
2. use the estimates to extract and apply a wavelet that shapes the trace wavelet to zero phase
3. filter the estimates with a zero-phase band pass filter and compare to the wavelet-processed data
4. compute the difference between the filter spike estimates and the wavelet-processed data to create an error function
5. analyze the error function for correction to the estimates

Addition of the corrections to the estimates gives the seismic reflection coefficients. Figures 7–34 and 7–35 illustrate these steps.

Wavelet processing and stratigraphic interpretation. Wavelet processing can solve many stratigraphic interpretation problems. These include removing instrumental filters and establishing correct polarity and arrival time. As a result:

- a symmetrical waveform is assured and resolution of closely spaced layers is optimal for the bandwidth
- well log synthetics can be easily related to the seismic data, as both have a symmetrical waveform
- lines recorded with different sources, methods, and instruments can be made to tie
- look-alike anomalies can be located by lateral correlation on the wavelet section displays

Estimated reflection coefficients can be displayed by using a large plotting scale for a close look and to investigate a particular zone of interest See Figure 7–36.

A popular way to display reflection coefficient estimates is to integrate them:

$$V_{i+1} = \frac{V_i(1 + R_i)}{(1 - R_i)} \tag{7.4}$$

where

V_{i+1} = the ith interval velocity, in unit distance per second

R_i = reflection coefficient

This transforms the reflection coefficient estimates to interval velocities. Such integration often reveals more unnoticed anomalies.

Advantages of using the acoustic impedance log display include

- the interval velocity displays are easier to correlate with well log data
- they are handy and clearly show lateral variations
- zones of exceptional porosity are often detectable

Such a display, devised by Lindeth (1977) from strongly spiked deconvolution of seismic data was called a *seislog.*

Wavelet processing techniques. Figure 7–37 illustrates the sequence of wavelet processing techniques using the statistical method of estimating the reflection coefficient series from the seismic data. Figure 7–37a represents a final stack section, normally of mixed phase. It is easier for the interpreter to work with a simple wavelet shape. Through wavelet processing, the section will be converted to a zero-phase stack. The part of the section inside the rectangle is used to observe effect of the processing sequence.

Figure 7–37b shows the raw reflection coefficient estimates, using the Burg algorithm, which is a predictive, adaptive solution of structural deconvolution. Figure 7–37c is the wavelet-processed, zero phase stack. Figure 7–37d is the refined estimates. Figure 7–37e shows the filtered reflection coefficient with zero-phase wavelet. Note the similarity of *c* and *e*, which gives confidence that this statistical approach is a reliable method to correct the phase of the seismic data to zero-phase pulse.

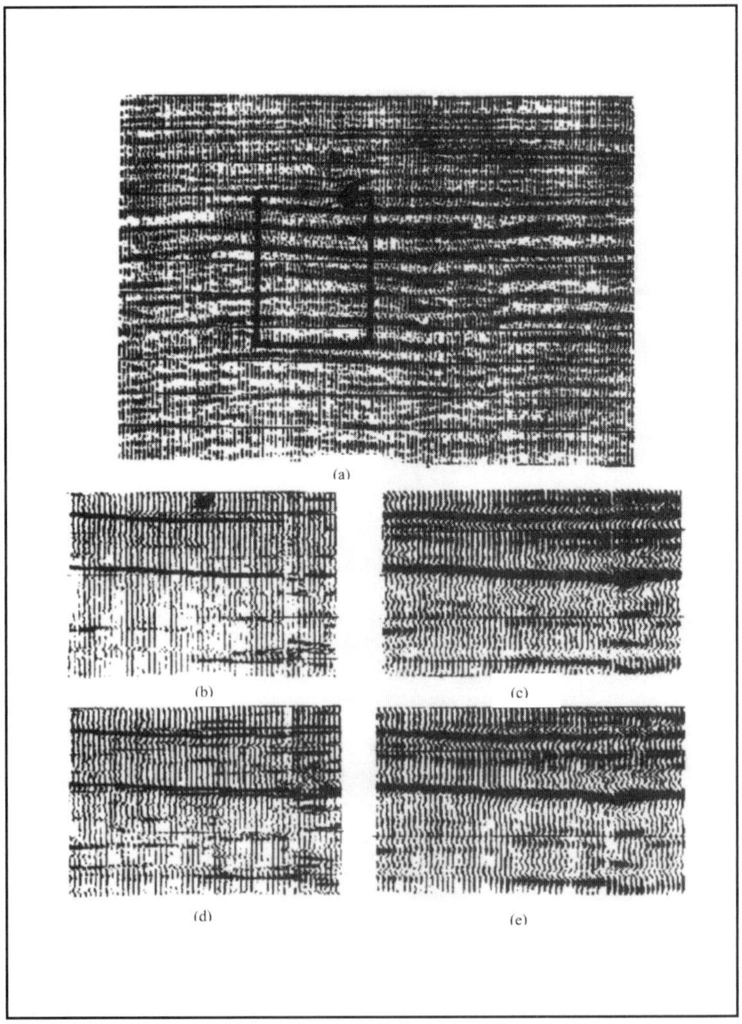

Fig. 7–37 *Sequence of Wavelet Processing Technique*

Figure 7–38a is a seismic section on which arrows indicate two geologic markers. The upper marker is a discrete—well-separated—reflector that is present throughout but not very well defined. The lower one is a limestone layer 85 feet thick. Encased in shale, the top of the limestone should give a strong, positive response followed by a strong negative. There is no indication of this signature in the section. Figure 7–38b is the estimated reflection coefficient. The discrete reflector is clearly separated. Top and bottom of the limestone layer are crisply defined. Interpretation of thin layers can be

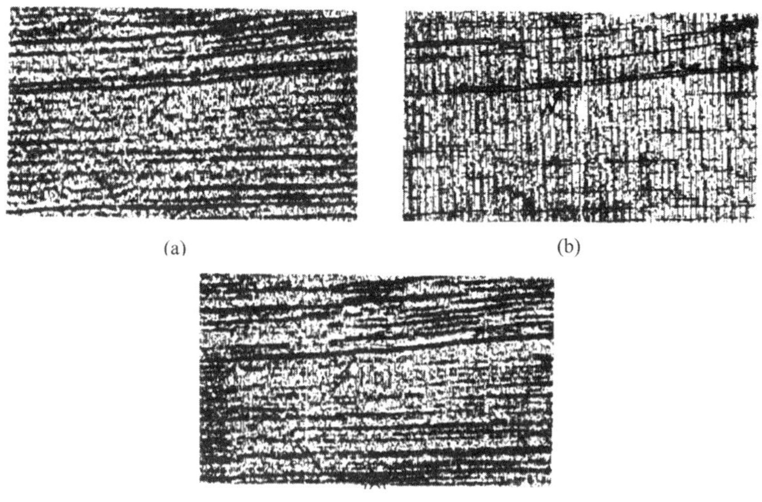

Fig. 7–38 *Enhanced Interpretation from Wavelet Processing*

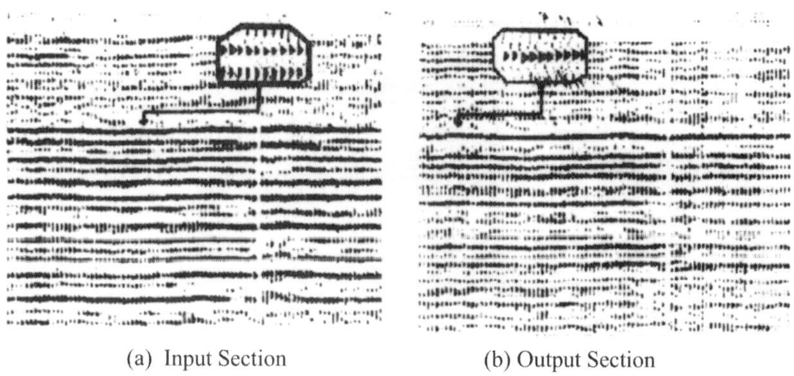

Fig. 7–39 *Wavelet Processing and Better Interpretation*

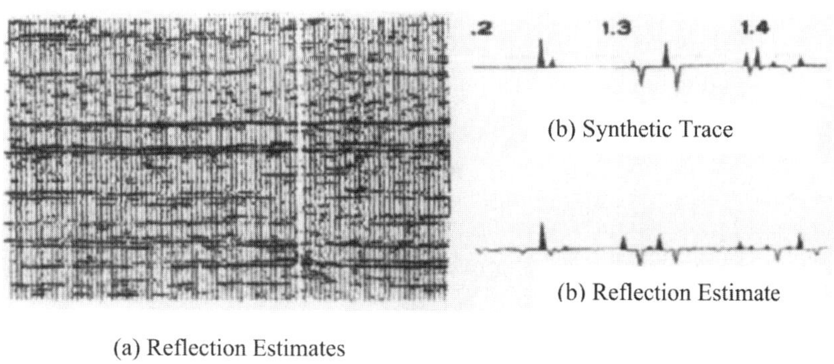

Fig. 7–40 *Reflection Estimates*

attempted on an expanded-scale display of the estimated reflection coefficient. In Figure 7–38c the estimated reflection coefficient was used to derive the basic wavelet, and the wavelet was applied to correct the phase to zero-phase. The limestone is indicated by the expected narrow-band response of a positive followed by a negative.

Figure 7–39 is another illustration of enhanced interpretation through wavelet processing of seismic data. Figure 7–39a shows the top of a chalk marker with overlying shale. The response is not clear; a strong positive reflection coefficient would be expected. Figure 7–39b is the wavelet-processed section. The top and bottom of the chalk are clearly defined, and surface effects resulting in low frequency zone are removed.

Figure 7–40a shows the reflection coefficient estimates. The sharp definition of the reflectors is clearly seen. Figure 7–33b is a synthetic trace and Figure 7–26c is the corresponding reflection estimate. Comparing the reflectivity series derived from well logs and reflection coefficient estimates from seismic data shows a good match, which supports the validity of this approach. Also, note the detection of the transition zone going into the chalk.

Figure 7–41 compares the wavelet-processed portion of a stacked section with a synthetic seismogram that was generated by convolving the reflectivity series from the well log data with a zero-phase wavelet. A very good match is obtained. The blank portion of the synthetic seismogram is due the lack of data below total depth of the well. Note the good match above this point.

Low-frequency data can obscure small faults and make their delineation difficult. Even rather sizeable faults can be less obvious. In Figure 7–42a, the arrow at 1 marks a layer that appears to be continuous, although there is a possible amplitude anomaly just below the arrow at 2. The apparent time sag on the deeper event reinforces the initial interpretation.

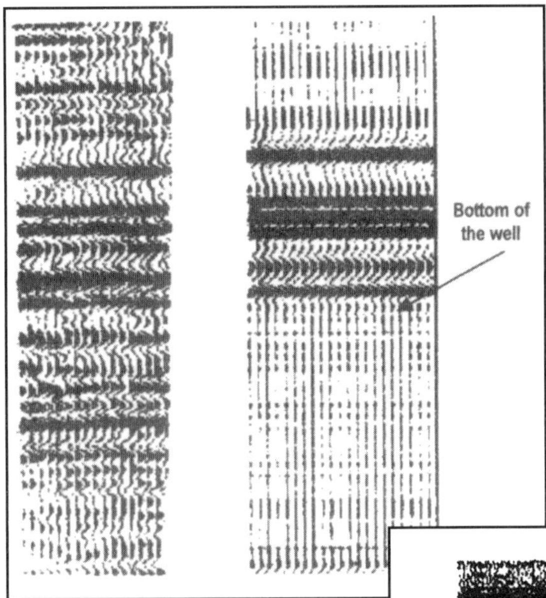

Fig. 7–41 *Synthetic Seismograms and Wavelet Processing*

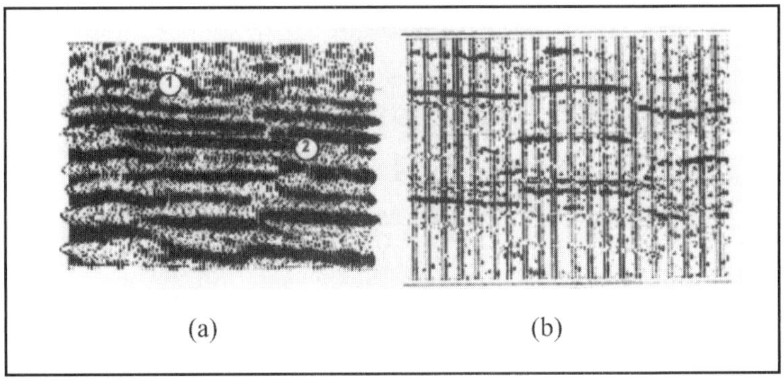

Fig. 7–42 *Fault Detection from Estimated Reflection Coefficient*

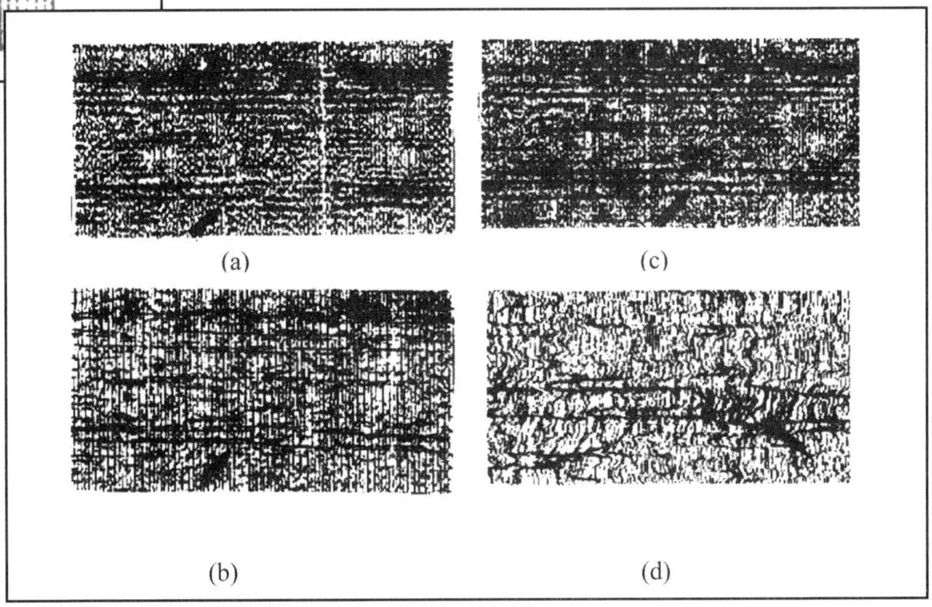

Fig. 7–43 *Stratigraphic Application: Reef Showing (a) Final Stacked Section, (b) Estimated Reflection Coefficient, (c) Wavelet Processed Stack, and (d) Interval Velocity Log*

Figure 7–42b gives a very different impression from the reflectivity estimates of Figure 7–42a. The simplification of the response reveals discontinuities and amplitude contrast. This increased discrimination shows that the structure is a graben fault.

Stratigraphic application. An objective of wavelet processing is to improve the ability to interpret subtle stratigraphic targets. Small reefs are among the most difficult of these targets. Reef signatures are dim, rather than bright, and the structural evidence is very slight. Figure 7–43a is an example of a section over such a target. An arrow indicates the reef-bearing reflection group. Apparent thickening and a reflection discontinuity detect the reef.

Figure 7–43b shows the reflection coefficient estimates. There is a considerable improvement in the lateral character consistency and resolution. The expanded scale in Figure 7–44 shows several reflection layers, which can help in better interpretation of the reef boundaries. Figure 7–43c shows a wavelet-processed stack. Note that the correlation on each side of the reef is more definite as the section becomes zero-phase with simple wavelet shape. Figure 7–43d is the acoustic impedance log. It shows a relative decrease of the velocity (deflection to the left) that is an indication of porosity. This is a qualitative analysis, and can be used to locate similar anomalies.

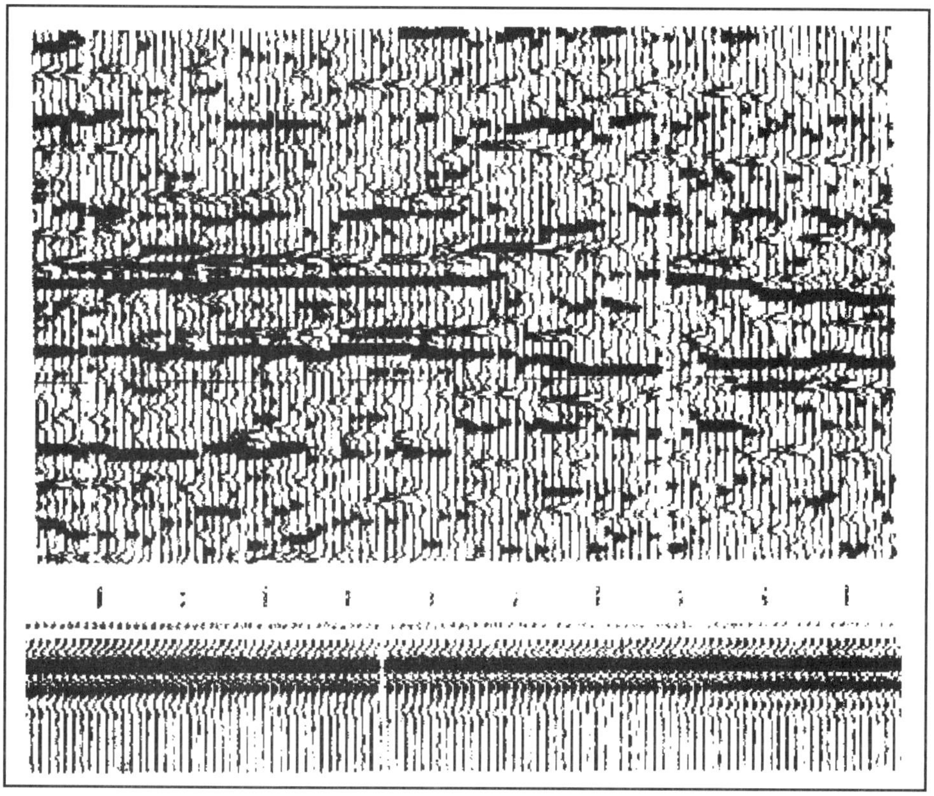

Fig. 7–44 *Stratigraphic Application Using Wavelet Processing*

The top part of Figure 7–44 is an expanded-scale version of the reflection coefficient of Figure 7–43c and the bottom part is a set of extracted wavelets from every trace on the seismic section. The inverse will be applied to make the stack traces zero phase. Figure 7–45a is an expanded-scale version of the acoustic impedance log of Figure 7–43d. The heavy line is a synthetic log, derived from the integration of the sonic and density logs and inserted in the seismically derived acoustic impedance log. A decent match is evident.

Reflection coefficients can be converted to an interval velocity log. If density is incorporated with interval velocity from logs, then an acoustic impedance log can be obtained as shown below. The heavy wiggle is the synthetic seismogram. There is a good match with the seismic acoustic impedance log. Interval velocity or acoustic impedance decrease around the reef is observed. This is a qualitative measure, and can be used to locate similar anomalies along the seismic profile.

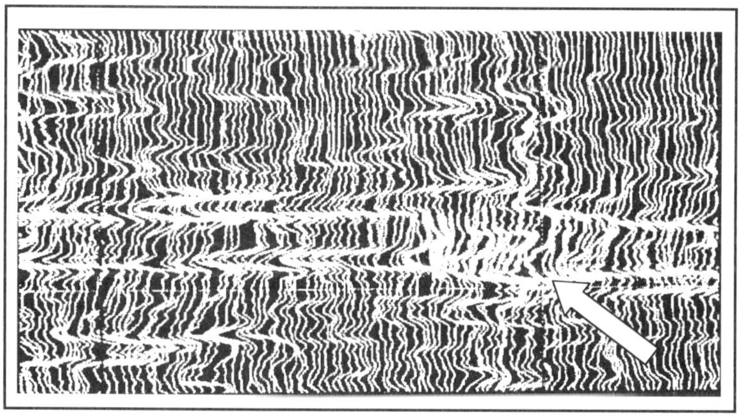

Fig. 7–45 *Interval Velocities on the Reef Example*

Summary. Estimating the reflectivity series provides information needed to optimize the basic wavelet in the seismic data.

There are several methods for wavelet estimations. A technique suited to seismic data is predictive and adaptive deconvolution, known as the *Burg algorithm*. It provides an estimate of the reflectivity series from the seismic trace. An estimated wavelet will be applied to yield a zero-phase seismic section. Some of the benefits that may result from successful wavelet estimation are

- better correlation with well logs, improved resolution of thin layers
- delineation of small fault patterns
- removal of near-surface effects on subtle stratigraphic traps

Seismic Stratigraphy

Introduction. This section introduces the newer sequence-stratigraphic concepts and procedures. These appeared in articles and monograms published, for the most part, over the past five to 10 years. These ideas and approaches build on earlier seismic stratigraphic and depositional systems concepts to introduce some very powerful methods of depositional sequence analysis using seismic, outcrop, and well data. Those receiving most benefit from this part of the text are interpreters with experience in depositional systems and seismic-stratigraphic interpretation.

Seismic Stratigraphic Analysis of Depositional Systems expands seismic facies interpretations and integrates the seismic-stratigraphic concepts more closely with lithogenetic stratigraphic concepts and inferences regarding depositional processes, environments, lithofacies, and, hence, tracts of depositional systems which compose depositional sequences and potential hydrocarbon reservoirs, traps, seals, and source.

Cyclic Sequences, Depositional Systems Tracts, and Their Application to Hydrocarbon Exploration and Development, published in 1991, attempts to summarize the status of the field at the end of the decade, following the burst of publications beginning about mid-decade, mostly by present and former Exxon geoscientists. This focuses on the relationships of systems-tracts to global *eustatic* and relative cycles of sea level, cyclic mechanics, and the significance of sea level in predicting the lithology and hydrocarbon potential of depositional systems.

Seismic-stratigraphic procedures used on a worldwide basis for the past 30 years are primarily based on recognition of unconformities and various *lapout* surfaces that bound relatively concordant groups of rocks called *depositional sequences. Concordant stratal* surfaces (and resulting seismic reflections) are considered to be essentially *isochronous paleodepositional* (and minor hiatal) surfaces. This view is an aid to interpreting depositional processes and the resulting facies. *External geometry,* reflection configurations, reflection continuity, and amplitude are critical seismic parameters for recognizing, mapping, and interpreting these *seismic facies.*

Depositional systems analysis involves the delineation, interpretation, mapping, and integration of component lithofacies commonly recognized in the subsurface by distinctive vertical successions of geophysical log patterns and, where resolvable, by seismic facies interpretations. Interpretation of seismic facies and geophysical log motifs depend on understanding the depositional processes, environments, and the spatial distribution of lithofacies within the spectrum of depositional systems. *Bathymetric interpretations* from microfossils provide supporting documentation of facies interpretations.

Sequence-stratigraphic interpretation depends upon delineation of unconformities and lapout surfaces, which permit recognition of depositional sequences that have been deposited during one cycle of relative sea level. Component systems tracts composed of one or more contemporaneous depositional systems are defined by the position of relative sea level during their deposition. Each cyclic depositional sequence is ideally composed of three systems tracts—transgressive, highstand, and lowstand or shelf-margin tracts. The application of sequence-stratigraphic concepts permits depositional systems tracts to be constrained by regional unconformities or lapout surface, both of which may be recognized and mapped using seismic profiles.

Sequence analysis, therefore, may result in correlation, mapping, and relative sea level interpretation of systems tracts using seismic profiles and verified, calibrated, and/or supplemented by well and lithologic logs, outcrop analysis, and fossils information. The variation in the sea level may be used to understand depositional processes and environments and to predict and map the sedimentary facies and stratigraphic relationships within each system's tract. Such information is a key to predicting the potential reservoir, source beds, and seals.

Before the more recent 1980s concepts of sequence stratigraphy are introduced, a brief summary of basic concepts and ideas of the 1970s will be presented in order to provide a brief review as well as an historical overview of the development of the newer ideas that upgrade searching for stratigraphic traps that may constitute hydrocarbon plays within a basin.

Although the earlier 1970s ideas were valid basic concepts (Mitchum 1977), researchers have continued to improve and elaborate on the fundamental concepts of cyclic sequences. Accordingly, the newest ideas are a stage in the continuing development of a better understanding of sequence stratigraphy.

Acknowledgments. For more than 30 years, former and present geologists and geophysicists of Exxon Production Research Company have provided leadership in the application of seismic stratigraphy to hydrocarbon exploration and development. The publication of much of their research results has provided the industry with most of the ideas, concepts, and procedures now available. Their ideas and concepts will be the main source of information used to teach or write about modern seismic sequence stratigraphy. Most illustrations and material are applied throughout this part of the text, and

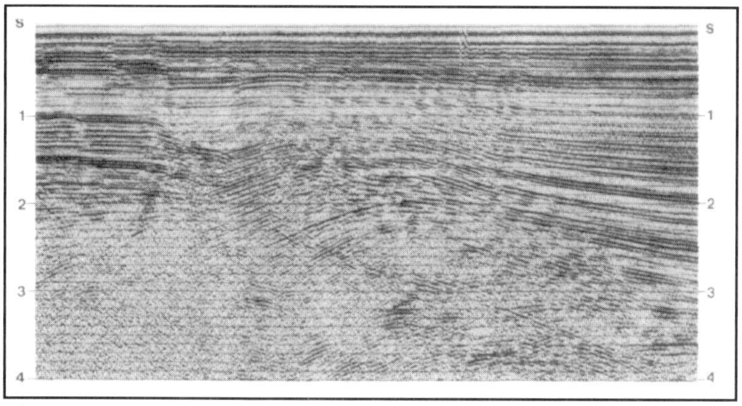

Fig. 7–46 *Seismic Section from the Baltimore Canyon Trough*

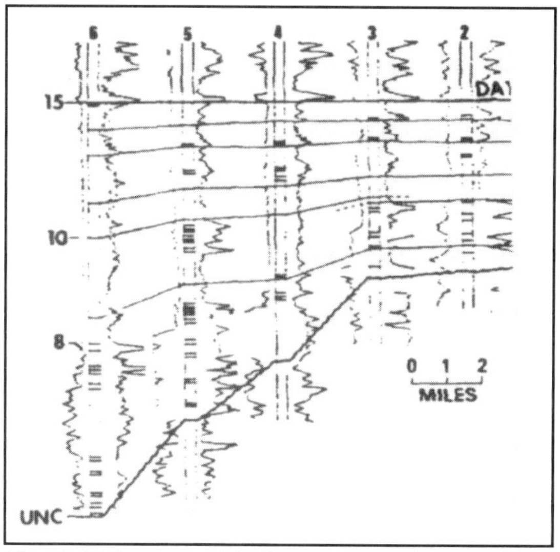

Fig. 7–47 *Stratigraphic Cross-section from Well Log Data Showing Stratal Surfaces*

conclusions reached regarding this field of stratigraphy derive from their publications. Throughout this section, every effort is made to acknowledge fully and accurately this principal source of information about seismic/sequence stratigraphy. Text citations and figures references are included for the authors of the 1977 *AAPG Memoir* 26 and/or the *1988 SEPM Memoir 42*. These two volumes contain most of the published contributions from Exxon research, although other published sources are listed where appropriate.

Review of 1970s concepts. Most of the information on seismic and sequence stratigraphy published prior to the 1980s remains valid and applicable today. Some ideas and concepts have, however, been refined by further research, application, and by critical evaluation of the earlier ideas by the geologist and geophysicists. In this section, the original basic ideas will be briefly reviewed and evaluated in terms of application in modern sequence stratigraphy.

Significance of reflections. As described in chapter 3, seismic reflections occur at geologic interfaces where there are seismic impedance contrasts. Reflection coefficients may be positive or negative, depending upon whether there is a downward increase or decrease in impedance, respectively. A well-processed seismic section can reveal valuable stratigraphic succession within the basin as illustrated in Figure 7–46.

Three principal types of reflectors generate reflections at surfaces within a basin where there is sufficient vertical change in acoustic impedance.

1. stratal surfaces
2. unconformable (discontinuity) surfaces, either erosional or non -depositional
3. diachronous surfaces such as diagenetic, fluid, gas hydrate, and other post-depositional geological interfaces that may cut across stratal surfaces

Multiples, diffraction, and sideswipe reflections may exist on the seismic sections if data processing fails to attenuate them—multiples, sideswipes. and diffractions. Unconformities and stratal surfaces, which can be recognized in outcrop and on well logs, may generate the reflections that provide the information required to carry out sequence-stratigraphic analysis using seismic data (Mitchum et al., 1977a, and Vail et al., 1977d).

Seismic reflections may be generated along stratal surfaces because of strong vertical acoustic impedance contrasts between superposed strata (Figs. 7–47 and 7–48). Stratal reflections are very important in seismic and sequence stratigraphic interpretation (Fig. 7–49). Stratal or bedding surfaces that are relict depositional surfaces separate the products of one minor depositional event from a subsequent event. Superposed strata may be of the same lithology, but commonly overlaying beds have different compositions and accordingly, different acoustic impedance values. Lateral impedance changes are much more gradual than vertical changes, stratal reflections may show considerable continuity.

Some beds, such as marine shale, have relatively widespread distribution and, therefore, may generate well-defined, reference stratal surfaces and, perhaps, good seismic reflections. Stratal surfaces are relict depositional surfaces that represent short term hiatuses in deposition. The surfaces are *chronostratigraphic* (isochronous) because they are, essentially, depositional

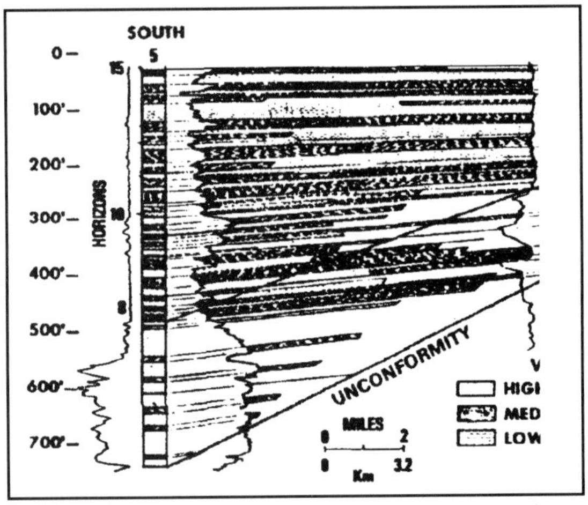

Fig. 7–48 *Seismic Velocity For Tertiary Example*

surfaces or beds that are deposited during the same time interval. If one or more stratal surfaces constructively generate a resolvable reflection that approximates a time line throughout its areal distribution, then it constitutes an isochronous or chronostratigraphic surface.

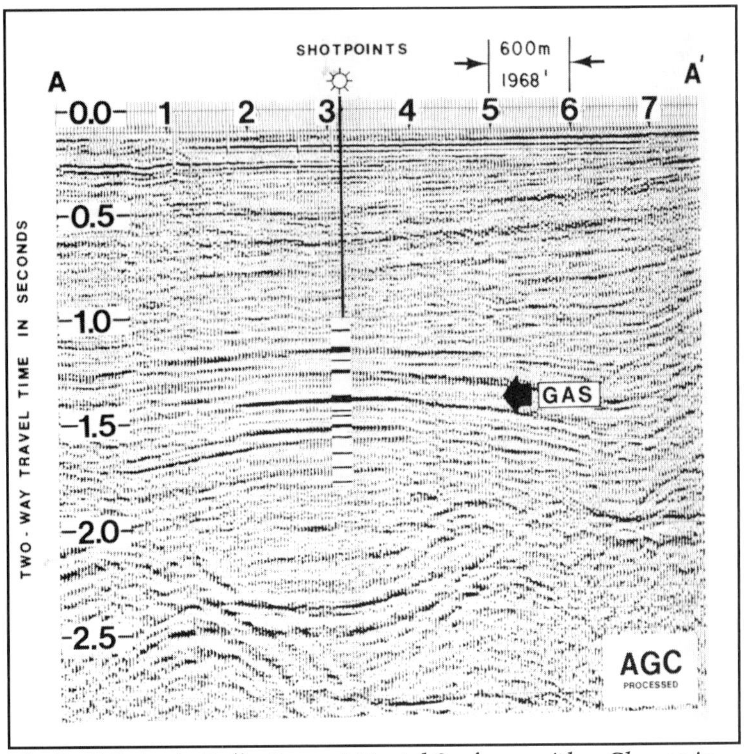

Fig. 7–49 *Seismic Reflections at Stratal Surfaces with a Change in the Acoustic Impedance*

Stratal surfaces and their resulting reflections may pass through lateral facies (Figs. 7–47 and 7–48) changes with perhaps a change in wavelet character. They do not follow Diachronous facies and lithostratigraphic boundaries. The thinnest chronostratigraphic unit that may be resolved seismically will range from about 1/4 to 1/8 of the wavelength of the seismic pulse. This varies with frequency and depth from about 25 to 300 feet in thickness for most hydrocarbon prospects. Using well logs, however, it is possible to recognize stratal surfaces and chronostratigraphic units with much greater precision.

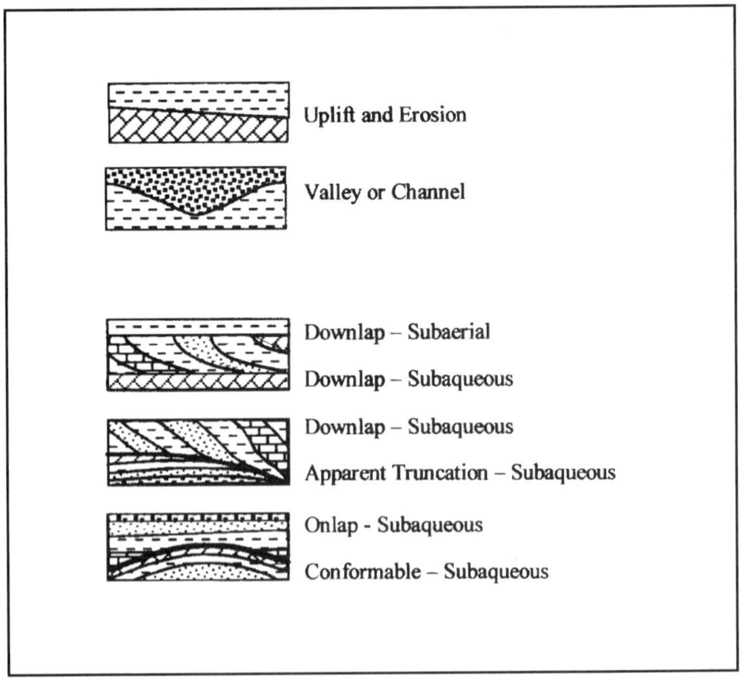

Fig. 7–50 *Discontinuity Surfaces Caused by Erosion and Depositional Hiatus Unconformities*

To realize the maximum advantages in applying sequence-stratigraphic concepts in exploration and development, it is important to integrate seismic and well-log interpretations. The use of chronostratigraphic stratal surfaces, called *flooding surfaces*, to establish local (well log) or regional (seismic) correlations of parasequences and parasequence sets, respectively, will be discussed later in this chapter.

Unconformable surfaces. Surfaces of erosion or non-deposition form unconformities representing time gaps of short or long duration. Unconformities are not chronostratigraphic surfaces because they do not bound equal time intervals, but they have chronostratigraphic significance because rocks above the unconformity are everywhere younger that those below the surface. Where a sufficient acoustic impedance contrast exists between strata above and below the discontinuity, a reflection will generally define the unconformity (Figs. 7–49 and 7–50).

Angular unconformities occur where rocks below the surface have been tilted at an angle and eroded prior to deposition of the younger strata. *Disconformities* define time gaps where strata above and below the surface are parallel.

Whereas angular unconformities can typically be recognized by the truncated surfaces of older strata, disconformities must be verified by paleontological data or by tracing the surface into an angular unconformity. Dating an unconformity requires knowledge of the youngest strata above and the oldest strata below the surface to define the minimum time gap or hiatus. Ideally, this is the point where the unconformity becomes conformable. Age dates are most commonly provided by paleontological interpretation and calibrated with appropriate radiometric dates. Angular unconformities and disconformities may result from subaerial or submarine erosion, although the former is more prone to lengthy periods of uplift and subaerial erosion.

A *nondepositional hiatus* is a surface, that has had little, if any, deposition during short to moderate intervals of time. Periods of very slow deposition typically occur in deep marine or lacustrine basins or on drowned shelves during rising water level and regional transgression. Dilute suspension or in situ *authigenic* and *biogenic* processes produce sediments that are typically composed of very fine-grained siliciclastic particles and rich microorganic matter. These deposits are called *hemipelagic.*

Hemipelagic facies typify sediment-starved episodes. They are characteristic of maximum flooding of the basin margins, eliminating any significant input of sediments. Vail et al. (1984) called hemipelagic facies *marine condensed section*—an important element in sequence interpretation that will be discussed later.

Any type of erosional unconformity may be enhanced on seismic profiles where overlying strata lapout onto the unconformable surface by onlap and downlap, important processes that will be reviewed in this chapter. Where an angular unconformity or a toplapped non-depositional surface was downlapped or onlapped by younger strata, the geometry of the discontinuity may be very distinct on seismic profiles.

Lateral change in lithofacies may occur above and/or below the stratal interface. These lateral changes in lithofacies cause changes in wavelet character, such as amplitude. Various seismic modeling procedures incorporate this concept. Stratal reflections may also disappear when the strata generating the reflections gradually thin below seismic resolution, producing false terminations called *internal convergence* (Fig. 7–51).

A number of geologic and hydrogeologic processes may produce secondary surfaces or boundaries within the stratigraphic succession of a basin. Those with sufficient acoustic impedance variations generate reflections. These surfaces or boundaries with impedance contrast generally cut across the stratal geometries. They are, therefore, diachronous. In some cases, however, there may be some primary stratigraphic control of the secondary feature. Fluid contacts such as gas/water interfaces exhibit strong impedance contrasts and produce the classic bright or flat spot reflections.

Gas hydrates that freeze as pressures are reduced during upward migration of biogenic gas toward the surface produce reflections that generally conform to the sea bottom or land surface. This is similar to the effect of permafrost in arctic regions.

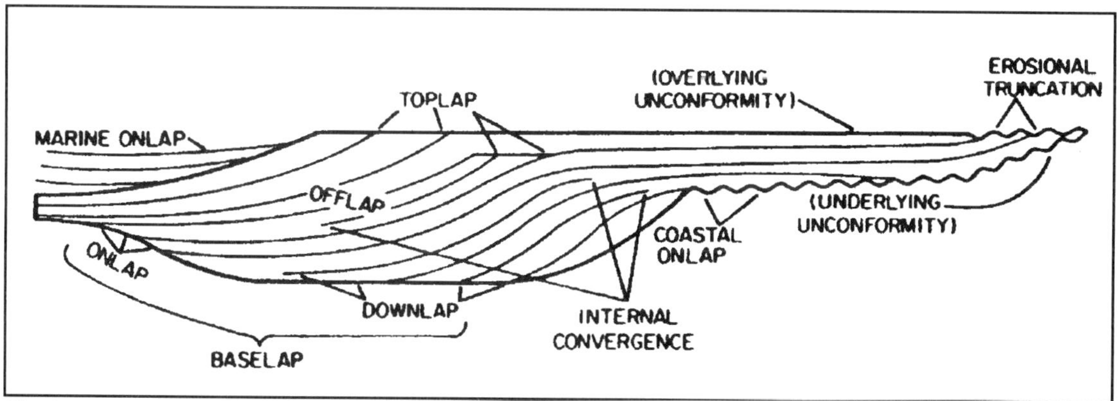

Fig. 7–51 *Terminology Proposed for Reflection Terminations*

Diffractions may be generated at:

- fault surfaces
- igneous intrusions
- relict weathering surfaces
- dissolution fronts, and
- diagenetic alteration fronts

Diffractions are often seen at reflection terminations and may appear to continue reflections beyond their termination points. Diffractions provide useful geologic information and may be useful in basin interpretations.

Primary reflections may disappear laterally when the acoustic impedance contrast that generated them decreases. Such lateral changes are generally accompanied by a gradual decrease in reflection amplitude. Lateral changes in the reflectively of an unconformable surface are caused by lithologic changes of strata above and/or below the surface. When the reflectively generated by stratal surfaces diminishes and disappears, it is normally an indication that a lateral change in lithologic stratal or discontinuity. Unconformable reflections may abruptly disappear laterally by fault termination whether or not the fault surface exhibits sufficient impedance contrast to generate a reflection.

As mentioned previously, stratal or unconformable reflections also may be abruptly truncated by an erosional unconformity. Stratal reflections may abruptly terminate against erosional or nondepositional unconformities. Such terminations are called lapout terminations. They document depositional pinch-out of the stratal units or thinning beyond seismic resolution (Mitchum et al. 1977b). Lapouts occur either above or below major discontinuities in the stratigraphic succession.

Reflection termination terminology. Exxon explorationists have proposed a number of terms to describe reflector terminations. These are described in the following paragraphs.

Stratal lapouts against older, discontinuities are called *base-lap* because they occur at the base of a conformable succession of strata. Two types of base-lap may occur—downlap or onlap. *Downlap* is the generally basinward and downward terminations of one or more reflections in response to pinch-out (or thinning below resolution) at the toes of progradational strata. *Onlap* is the lateral termination of reflections in response to the lateral pinch-out (or thinning below resolution) of aggradational marine, coastal, or even nonmarine strata against an inclined unconformable or non-depositional surface.

Apparent truncation is a term introduced by Vail (1987) to describe a false erosional surface. These may be produced by apparent truncation of reflections in response to the basinward pinch-out (or thinning below minimum seismic resolution) of retrogradational or landward-shifting parasequences, discussed later in this chapter.

Stratal reflections may also terminate upward against a younger nondepositional surface in response to depositional pinch-out (or thinning below seismic resolution) at the top of oblique progradational strata called toplap, this type of stratal termination is commonly caused by the fluvial bypass of sediments across a deltaic or coastal plain.

True toplap is difficult to verify on seismic profiles, because the terminations may be enhanced or caused entirely by subsequent erosion. It can be confirmed if recognizable deltaic/coastal plain sediments overlie the toplap surface. In this case, the toplap boundary is not a sequence boundary but an internal nondepositional hiatus, which is an apparent horizontal, commonly continuous reflection that is actually a series of aligned diachronous reflections. These reflections are generated in response to the acoustic impedance change between delta plain mudstones and diac. The toplap surface, which is defined by the upward termination of several progradational reflections, may also be marked by a diachronous protruded (clinothem) delta front sandstones.

Seismic facies. Early recognition by various workers that distinctive packages of reflections on well processed seismic profiles represented acoustic responses to lithofacies led Exxon geoscientists to introduce and apply the term seismic facies.

They defined seismic facies as an areally definable, three-dimensional unit composed of seismic reflections whose elements, such as reflection configuration, amplitude, continuity, frequency, and interval velocity, differ from the elements of adjacent units. They noted that, "A seismic facies unit is interpreted to express certain lithology, stratification, and depositional features of the deposits that generate the reflections in the unit." They furthermore noted that a number of seismic reflection parameters

Seismic Facies Parameters	Geologic Interpretation
Reflection configuration	• Bedding patterns • Depositional processes • Erosion and paleotopography • Fluid contacts
Reflection continuity	• Bedding continuity • Depositional processes
Reflection amplitude	• Velocity-density contrast • Bed spacing • Fluid Content
Reflection frequency	• Bed thickness • Fluid content
Interval velocity	• Estimation of lithology • Estimation of porosity • Fluid content
External form & areal association environment. of seismic facies units	• Gross depositional • Sediment source • Geologic setting

Fig. 7–52 *Seismic Reflection Parameters Used in Seismic Stratigraphy and Their Geologic Significance*

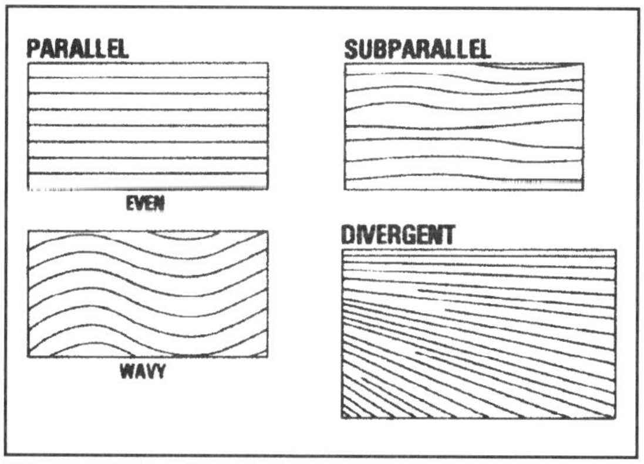

Fig. 7–53 *Parallel, Subparallel, and Divergent Seismic Reflection Configuration*

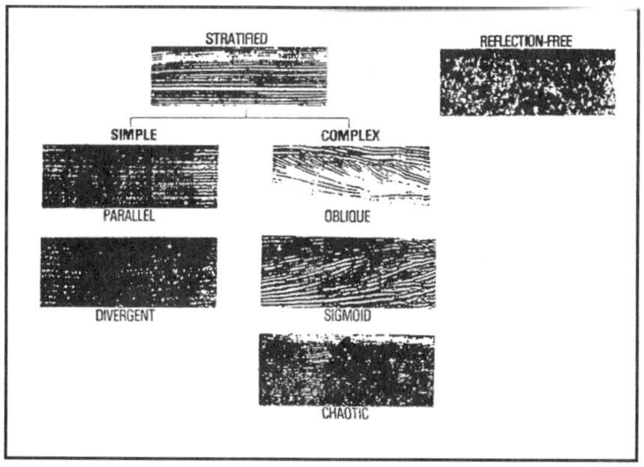

Fig. 7–54 *Stratified Simple and Complex Facies*

were useful in the interpretation of seismic facies units (Fig. 7–52):

1. reflection configuration
2. reflection continuity
3. reflection amplitude and frequency
4. interval velocity
5. external reflection geometry

Types of boundary relationships, whether transitional or terminational, are also important interpretation factors. Seismic facies are, therefore, generated in response to generally conformable strata whose stratal reflections define relict depositional (and essentially isochronous) surfaces, permitting their interpretation in terms of depositional processes limited, of course, by the scale of seismic resolution.

Seismic facies interpretation relies heavily on the interpreter's knowledge of geologic processes, depositional models, and the nature and spatial arrangement of component facies that constitute depositional systems. (horizontal and vertical). At the scale of the seismic resolution, this type of interpretation is a powerful tool in basin analysis if its application is compatible with the concepts of lithogenetic stratigraphy.

At the scale of most seismic data, seismic facies constitute component elements of depositional systems. Seismic facies can be delineated and mapped using a great variety of innovative techniques to display the various critical parameters such as lapout types and directions, time thickness, amplitude variations, and reflection configurations (Mitchum and Vail 1977c).

A variety of general seismic classes or types originally defined on the basis of reflections configuration facies by Mitchum and others (1977b) are useful in describing and communicating information. These include parallel/divergent, progradational, chaotic/reflection-free, mound/ drape, and fill types (Figs. 7–53 through 7–58). These seismic facies types, however, do not represent unique lithofacies, but their application permits the interpreter to focus on those the options available to explain the seismic parameters.

In many respects the process of achieving a unique seismic facies interpretation is a process of elimination. Conclusions are not always possible. Nevertheless, the methods focus on factors that require further analysis using other techniques.

The proper interpretation of seismic facies (Fig. 7–59) depends upon the geological experience of the interpreter, constrained by appropriate concepts of lithogenetic stratigraphy. Its application constitutes a significant core of seismic and sequence analysis, especially the more sophisticated analysis of the internal facies composition of systems tracts (Figs. 7–60 and 7–61).

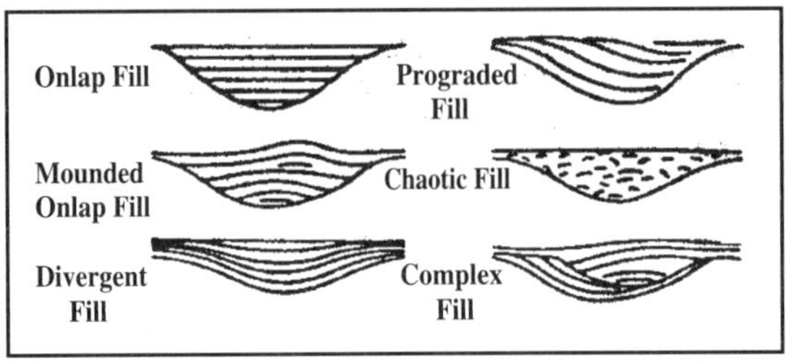

Fig. 7–55 *Fill Facies Units*

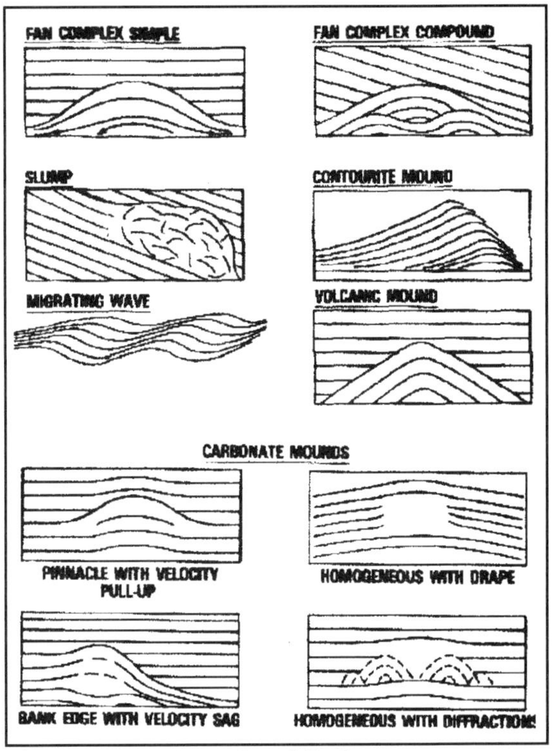

Fig. 7–56 *Seismic Facies of Carbonate*

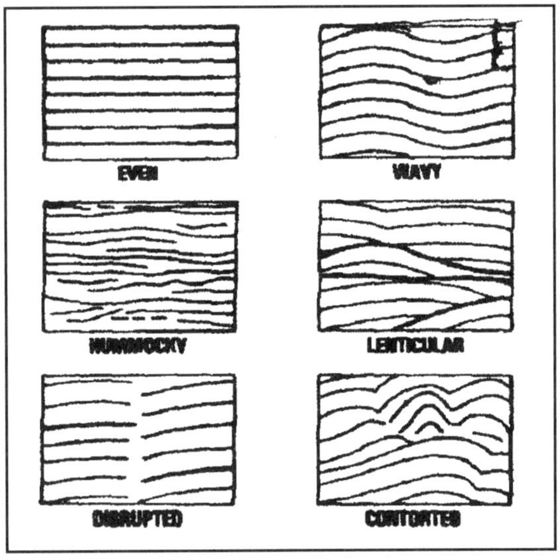

Fig. 7–57 *Modified Seismic Reflections*

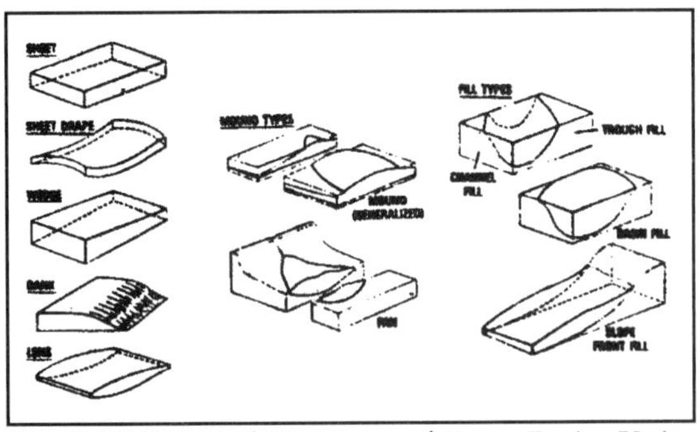

Fig. 7–58 *External Geometry of Some Facies Units*

Reflection Terminations at Sequence Boundaries Units	Reflection Configurations within Sequences	External Forms of Sequences and Seismic Facies
Lapout	Principal Hiatal Configuration	Sheet
Baselap	Parallel	Wedge
Onlap	Subparallel	Bank
Downlap	Divergent	Lens
Toplap	Prograding Clinoform	Mound
Truncation	Sigmoid	Fill
Erosional	Oblique	
Structural	Complex Sigmoid Oblique	
Concordance	Shingled	
(No Termination)	Hummocky Clinoform	
Chaotic		
	Reflection-Free	
	Modifying Terms	
	Even Hummocky	
	Wavy Lenticular	
	Regular Disrupted	
	Irregular Contorted	
	Uniform	
	Variable	

Fig. 7–59 *Geologic Interpretation of Seismic Facies Parameters*

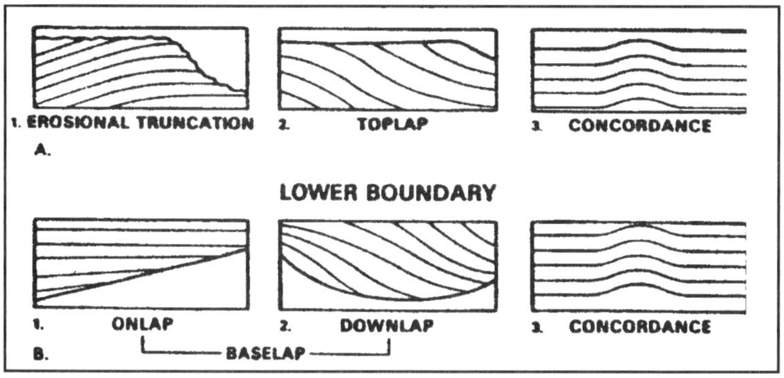

Fig. 7–60 *Stratal Terminations at Upper and Lower Boundaries*

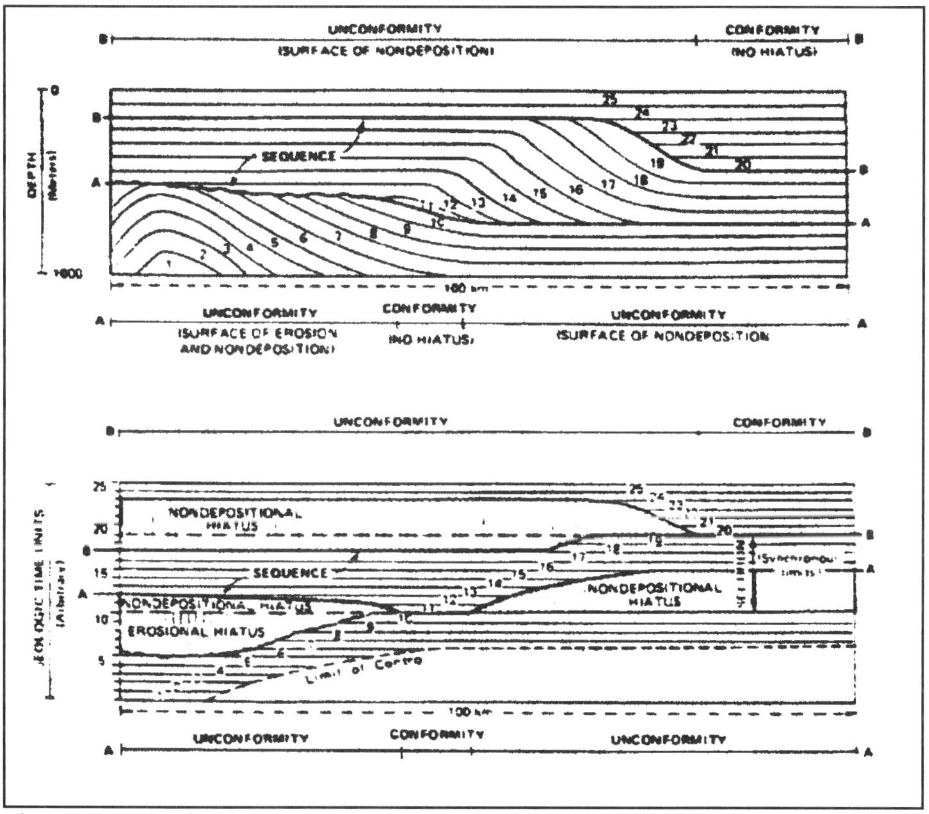

Fig. 7–61 *Generalized Stratigraphic Section on a Sequence and Generalized Chronostratigraphic Section of the Sequence*

Relative sea-level changes. A relative change of sea level (Vail et al. 1977a) was defined as, "An apparent rise or fall of sea level with respect to the land surface." Movement of sea level, the land surface, or both may produce the relative change, which can be local, regional, or global. They further defined a cycle of relative change of sea level as, "An interval of time during which a relative rise and fall of sea level takes place."

In 1977 the cycle was described as consisting of *gradual relative rise*, a period of still stand and a rapid relative fall of sea level. The total gradual rise was inferred to consist of several smaller scale, rapid rises and still stands called paracycles. The *paracycles* are smaller-scale units that may not detected on seismic profiles, but can be delineated on well logs. As noted above, in 1977, Vail and others did not restrict sequence boundaries to erosional unconformities produced by relative falls in sea level but also considered major parasequences (later called parasequence sets or systems tracts) to constitute sequences, typically marked by basal downlap or onlap.

Based on the 1977 sequence definition, one relative cycle of sea level could result in deposition of one or more systems tracts or parasequence sets. According to the Exxon model, several relative cycles exhibit successive rises to higher sea level positions, followed by one or more major relative falls to a lower position. This higher cycle is called a *supercycle*.

Vail and others (1977b) noted that some cycles may be local or perhaps of regional extent, but they stated that most regional cycles are eventually determined be to global in extent and that, "simultaneous relative changes in three or more widely spaced regions around the globe are interpreted as global changes of sea level."

Further, Vail and others (1977b), wrote that indicators of relative sea level changes can be grouped into three basic lines of evidence:

1. coastal onlap indicates a relative rise
2. coastal toplap indicates relative still stand
3. downward shift of coastal onlap indicates a relative (rapid) fall of sea level

They clearly noted that coastal onlap was a better indicator of relative sea level changes than marine onlap or more ambiguous methods such as shoreline transgressions and regressions and deepening and shallowing of water depths. The latter two depend on rate of sediment supply and rate of relative sea-level change (later called rate of accommodation).

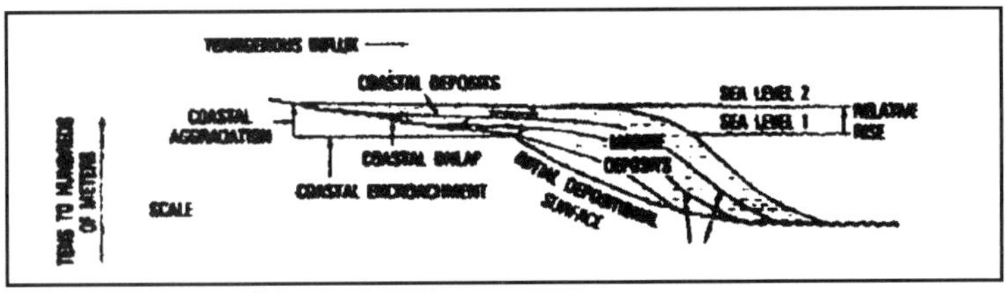

Fig. 7–62 *Coastal Onlap*

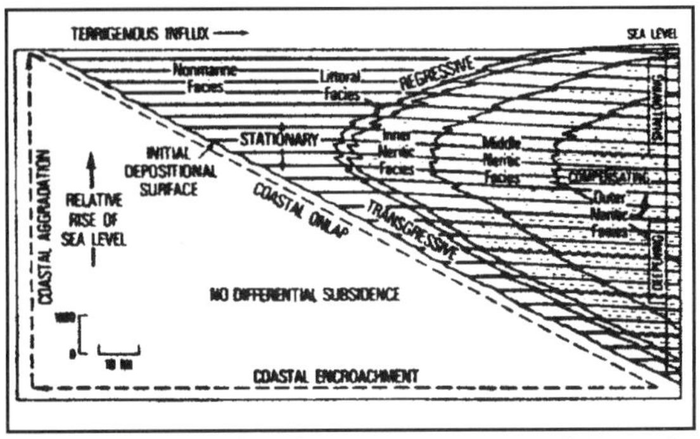

Fig. 7–63 *Coastal Onlap as a Function of Eustatic Level Change, Subsidence, and Sediment Supply*

The following criteria were proposed to identify and interpret relative sea-level changes by variations in coastal onlap.

Coastal onlap, as previously noted, is the progressive landward termination of coastal (paralic) lithofacies (Fig. 7–62).

Vail and others (1977b) noted that coastal onlap may be caused by one of the following:

1. rising sea level when the basin subsides, remains stationary, or rises at a slower rate than sea level
2. sea level remains stationary and the basin subsides
3. sea level falls, but subsidence occurs at a greater rate

Beaches or shoreline facies may onlap ravinement surfaces eroded into subaerial facies, but deltaic or lower alluvial plain facies more commonly onlap the coastal plain unconformity. A schematic model (Fig. 7–62) illustrates this concept. An important point made by Vail and others is that relative sea level rise does not necessarily produces a shoreline transgression, and that shoreline shift during a relative sea-level (coastal onlap) rise is a function of eustatic sea-level changes, subsidence, and sediment supply (Fig. 7–63).

Three scenarios are possible during coastal onlap (Fig. 7–64)—shoreline transgression, regression, or stationary. These three patterns have recognized in ancient basins.

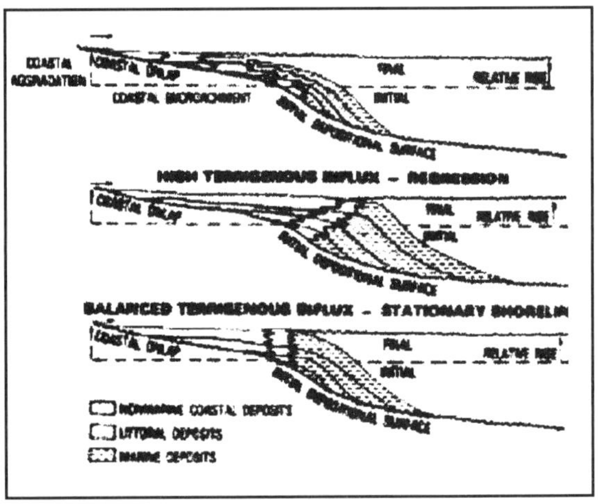

Fig. 7–64 *Three Scenarios during Coastal Onlap*

Coastal toplap (Fig. 7–65) is indicative of a still stand of relative sea level (Vail and others 1977). It indicates that basin subsidence and eustatic sea level remained relatively stationary during an episode of coastal progradation. Since toplap, by definition, results from *oblique progradation*. One can assume that progradation was rapid and probably under the influence of a high rate of sediment supply typified by most deltaic deposition.

Toplap occurs because deltaic or other prograding coastal systems cannot aggrade more than a few meters above sea level unless relative rise of sea level occurs to accommodate aggradation. Therefore, under a stable relative sea level, there is fluvial bypass of the coastal or deltaic plain and a rapid basinward shift of near shore depositional environments.

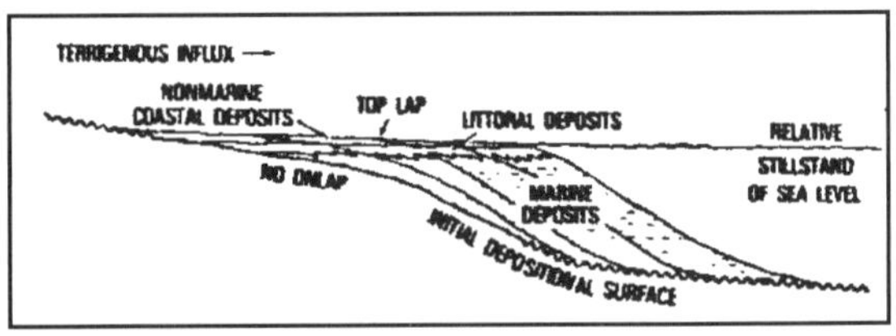

Fig. 7–65 *Coastal Toplap Indicates a Still Stand of Relative Sea Level*

Rate of progradation is controlled by depth of water and rates of subsidence (tectonic and compactional), sediment supply, and eustatic changes. A schematic diagram illustrates the development of toplap (Fig. 7–65).

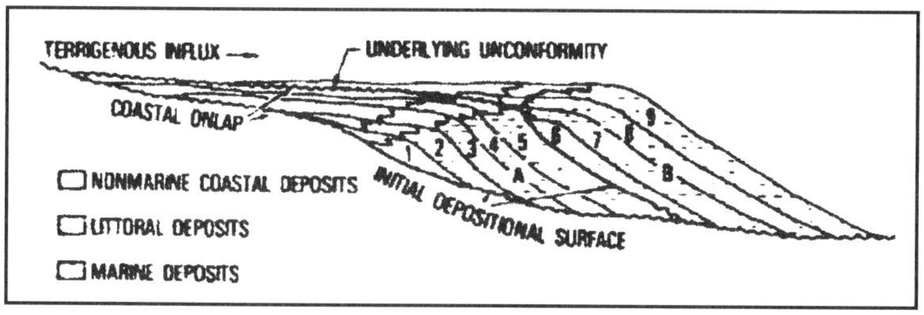

Fig. 7–66 *Downward Shift in Coastal Onlap*

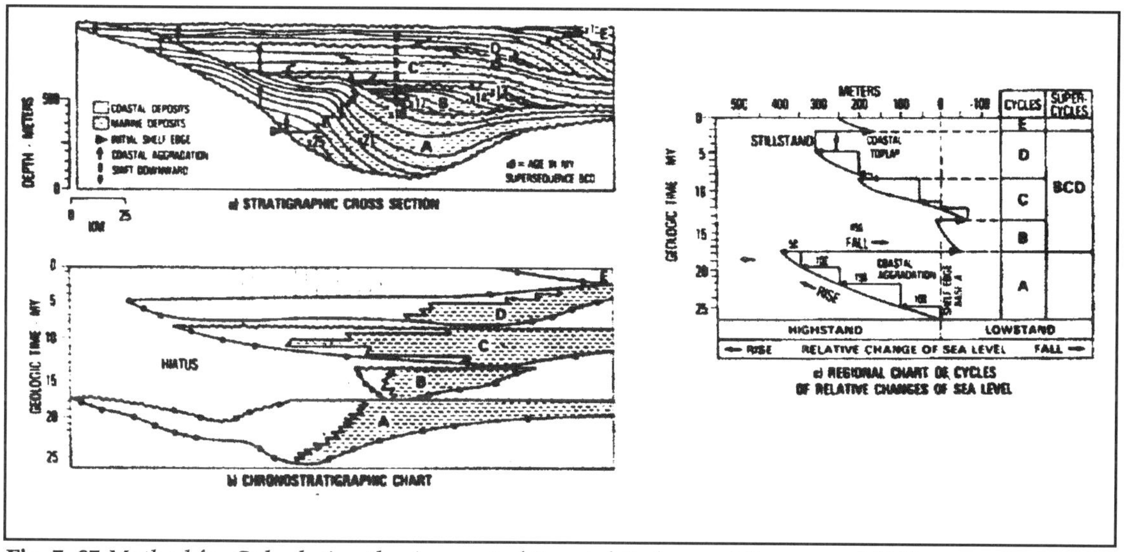

Fig. 7–67 *Method for Calculating the Amount of Coastal Onlap and Downward Shift as a Measurement of Relative Fall of Sea Level*

Downward shift in coastal onlap (Fig. 7–66) indicates a relative (and rapid) fall in sea level. This can occur if:

1. eustatic sea level falls while the basin floor is uplifted, remains stationary, or subsides at a rate lower than eustatic sea level fall
2. sea level remains stationary while the basin floor is uplifted
3. sea level rises but the basin floor is uplifted at a greater rate

Vail and others illustrated downward shift in coastal onlap with a seismic line from the San Joaquin Basin (Fig. 7–66). Well data documented that the Santa Margarita is a coastal deposit rather than a marine onlap deposit. They also illustrate examples from the North Sea Basin. In 1977, the Exxon geoscientists believed that the magnitude of the coastal onlap and downward shift, as measured on the seismic profile (correcting for compaction), could be used to calculate the magnitude of the relative rise and fall of sea level.

The late phases of coastal lap are now recognized to be generally *alluvial*, resulting from subaerial accommodation. Therefore the last onlap cannot be inferred to represent maximum relative sea level (Posamentier and Vail 1988). Posamentier and Vail presented their method for calculating the amount of coastal onlap and downward shift as a measurement of relative fall of sea level (Fig. 7–67).

A conceptual model illustrating their conclusions regarding the deposition of cyclic sequences integrates the application of coastal onlap, coastal toplap, and downward shift of coastal onlap to infer the sea-level cycle (Fig. 7–68). This example represents conditions at the mouth of a river when relative sea level fell. The presented examples are from the San Joaquin basin and from the North Sea (Figs. 7–69 and 7–70). Not shown are conditions that may exist elsewhere along basin margins where there are no rivers to entrench.

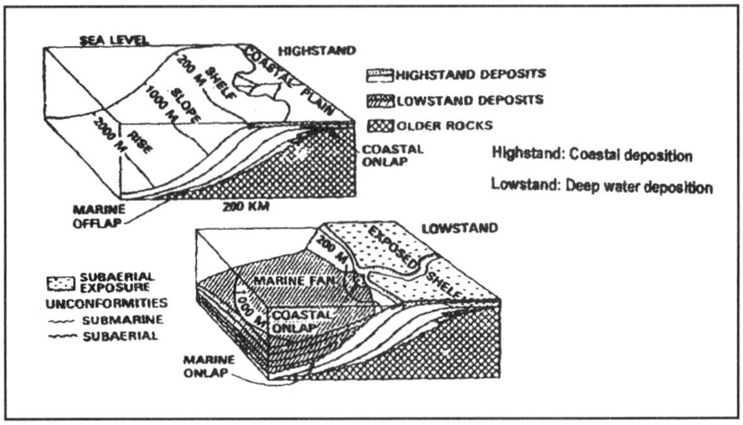

Fig. 7–68 *Continental Margin Clinoform Analysis*

Brown and Fisher (1980) disagreed with this model, suggesting that much of the erosion of submarine canyons that they had observed in Brazil and the Gulf of Mexico Basin occurred during relative rise and highstand of sea level. These differences of interpretation were, in part, the result of the latter's experience along basin-margins where fluvial entrenchment was uncommon and did not resemble the models presented by Exxon geoscientists. Brown later recognized the Exxon model, but still believes that significant canyon erosion may occur in non-fluvial areas during rise and highstand of relative sea level. Regional variations in the depositional model remain to be clarified and documented adequately.

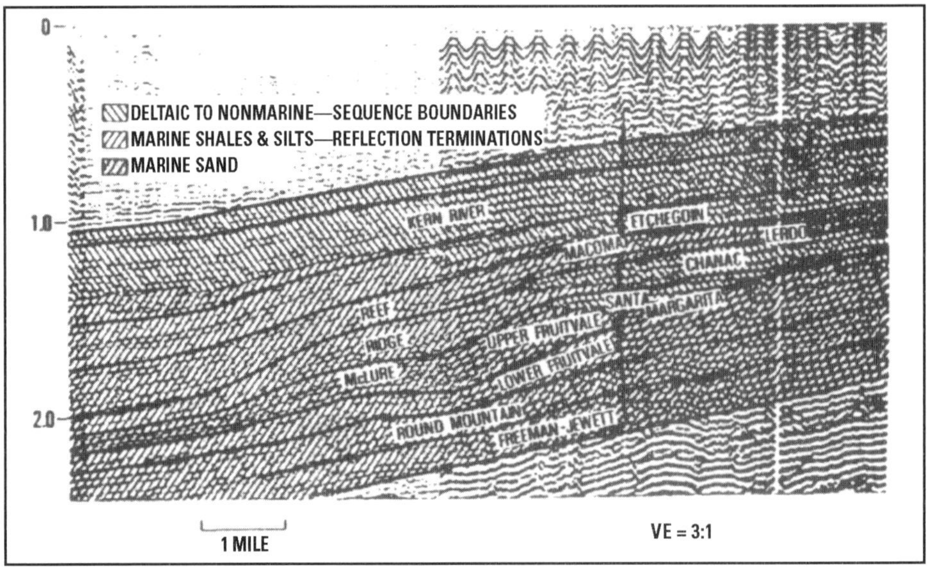

Fig. 7–69 *Downward Shift in Coastal Onlap in San Joaquin, California*

Global cycles of relative sea level. Vail and others, by inferring that the coastal onlap curve defined relative sea level variations (Fig. 7–71), proposed using modal averages of these curves from various basins to construct a global relative sea level chart. They assumed that if they recognized similar coastal onlap patterns in three or more basins, then the curves probably indicated a global distribution of the relative cycles (Fig. 7–72). Of course such global correlations depend upon very accurate paleontologic correlations and upon relatively similar tectonic activity.

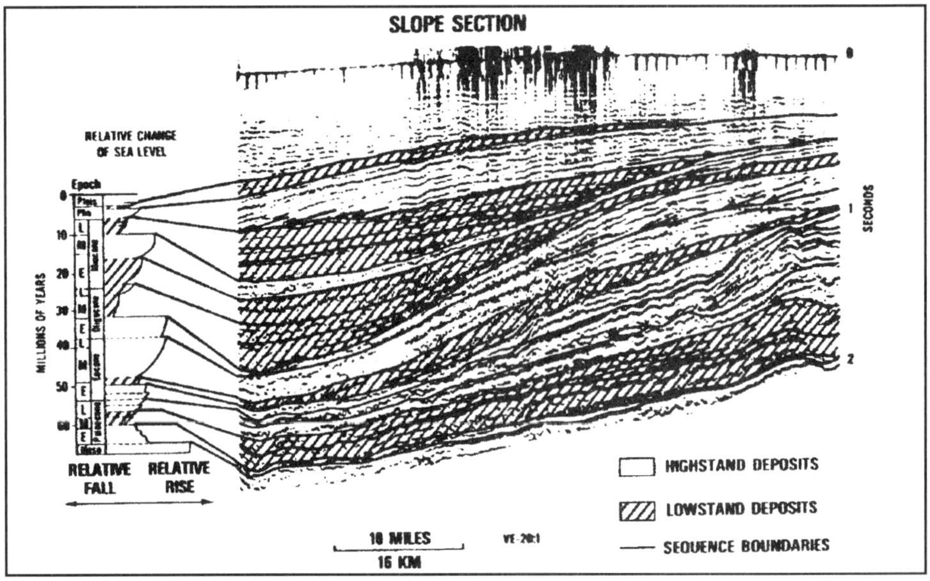

Fig. 7–70 *Progradational and Marine Onlap Cycles, Tertiary, North Sea*

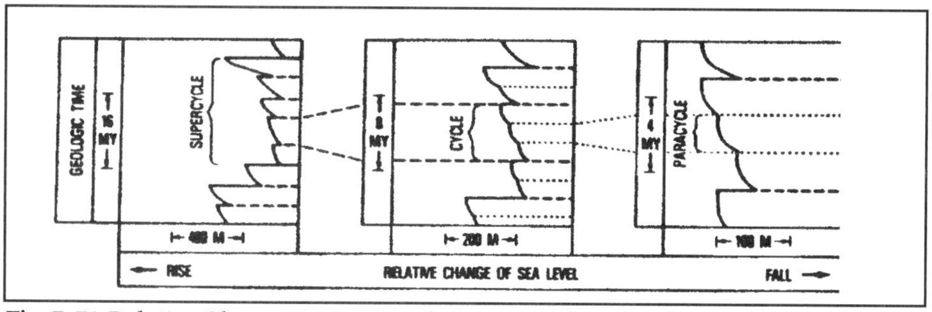

Fig. 7–71 *Relative Changes in Sea Level, Concepts of Paracycles, Cycles, and Super Cycles*

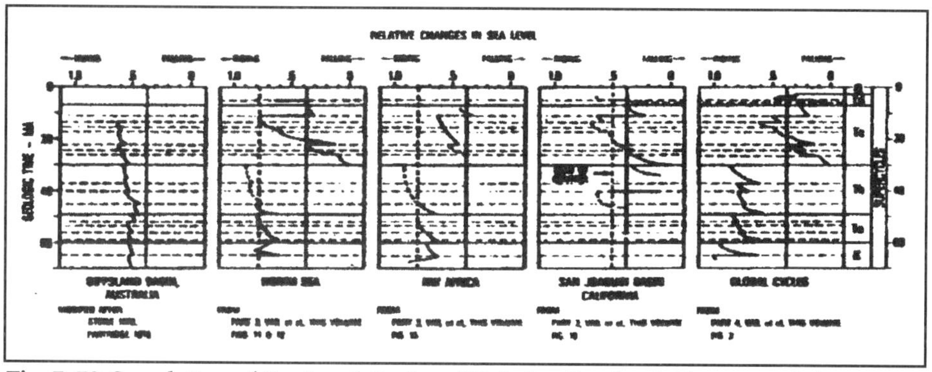

Fig. 7–72 *Correlation of Regional Cycles of Relative Sea-level Change and Averaging to Construct Global Cycles*

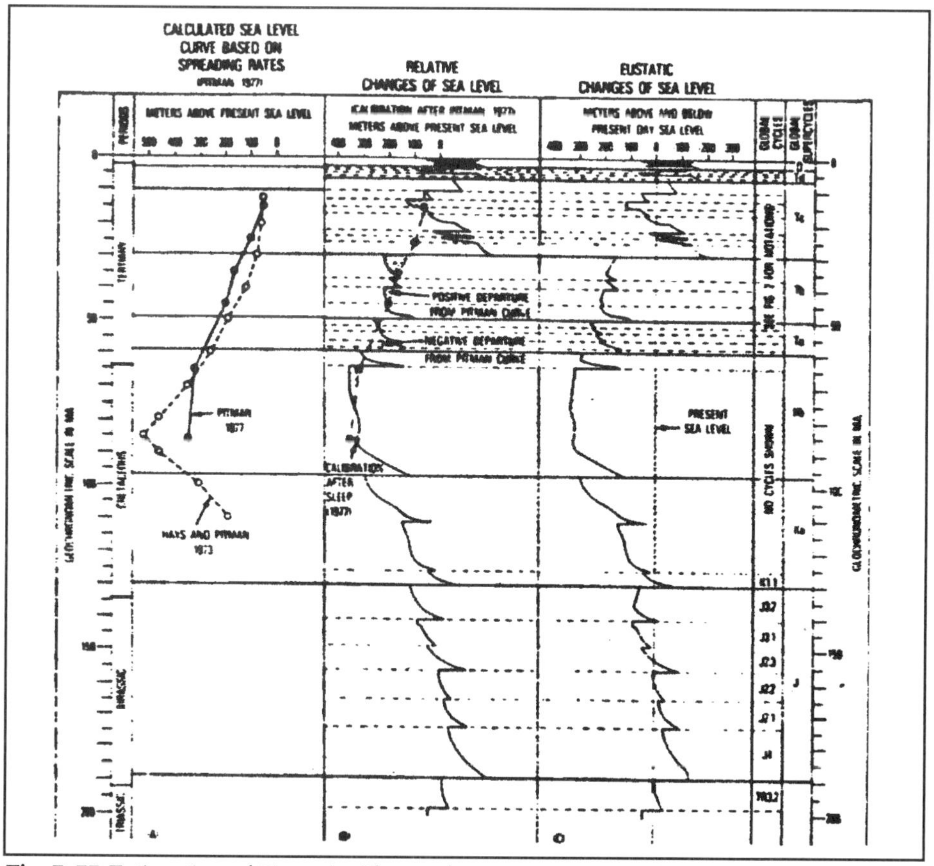

Fig. 7–73 *Estimation of Eustatic Change from Jurassic to Holocene*

Correlation and integration of coastal onlap curves were used to construct their proposed global cycle charts showing relative changes of global sea level (Fig. 7–73). These relative cycles were inferred to correlate directly with inferred eustatic sea level cycles. The relative coastal onlap cycles (called third cycles) generally conformed to and helped define second-order cycles bounded by major unconformities.

The second- and third-order cycles were then adjusted to Phanerozoic first-order cycles derived generally from ocean volume calculations (Pitman 1978) and from stratigraphic evidence of major global highstands and lowstands.

Vail and others have indicated that the first-order cycles span 200 million years, the second-order cycles are of 10–80 million years in duration, and third-order cycles range from 1–10 million with an average of about 3 million years (Figs. 7–74 and 7–75). These numbers have changed during the decade of further research, and second-order cycles are assumed to be of shorter duration. Many more third-order cycles have been described that range from 0.8 to 3.5 million years and average about one million years. Higher order cycles and paracycles exist, but except in basins with unusually slow rates of subsidence, the shorter, lower magnitude cycles normally will not be resolvable on seismic data. They may be recognized in outcrops or on well logs.

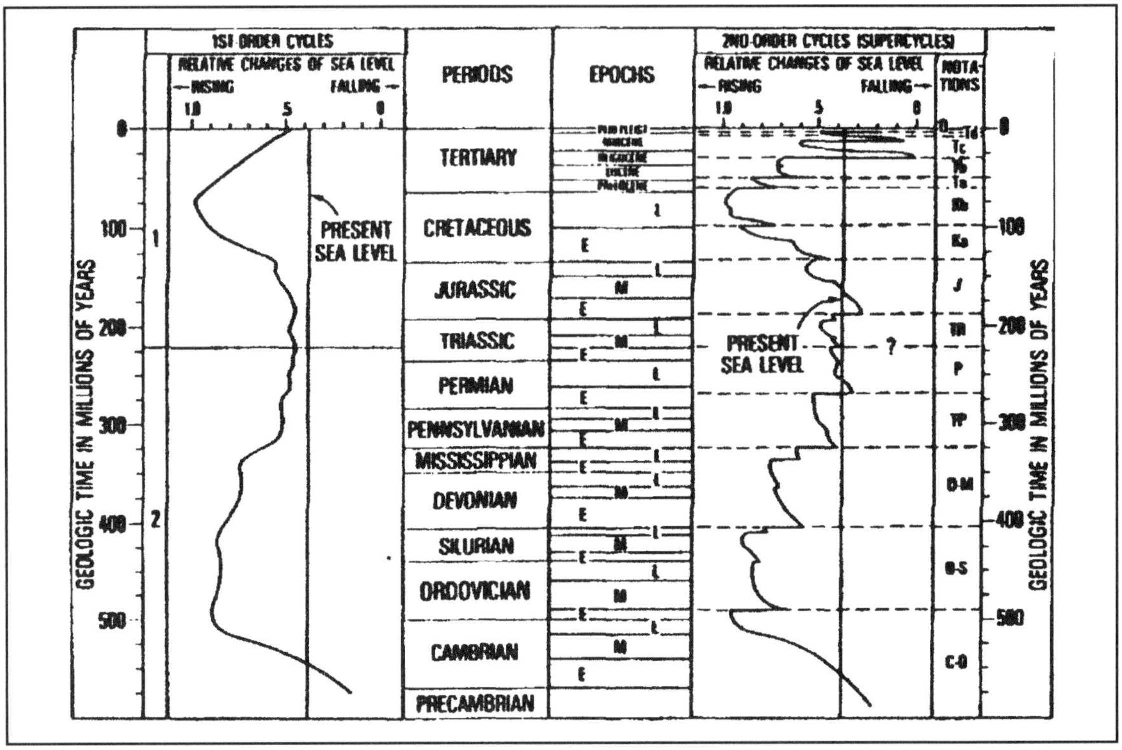

Fig. 7–74 *First- and Second-order Global Cycles of Relative Sea-level Change*

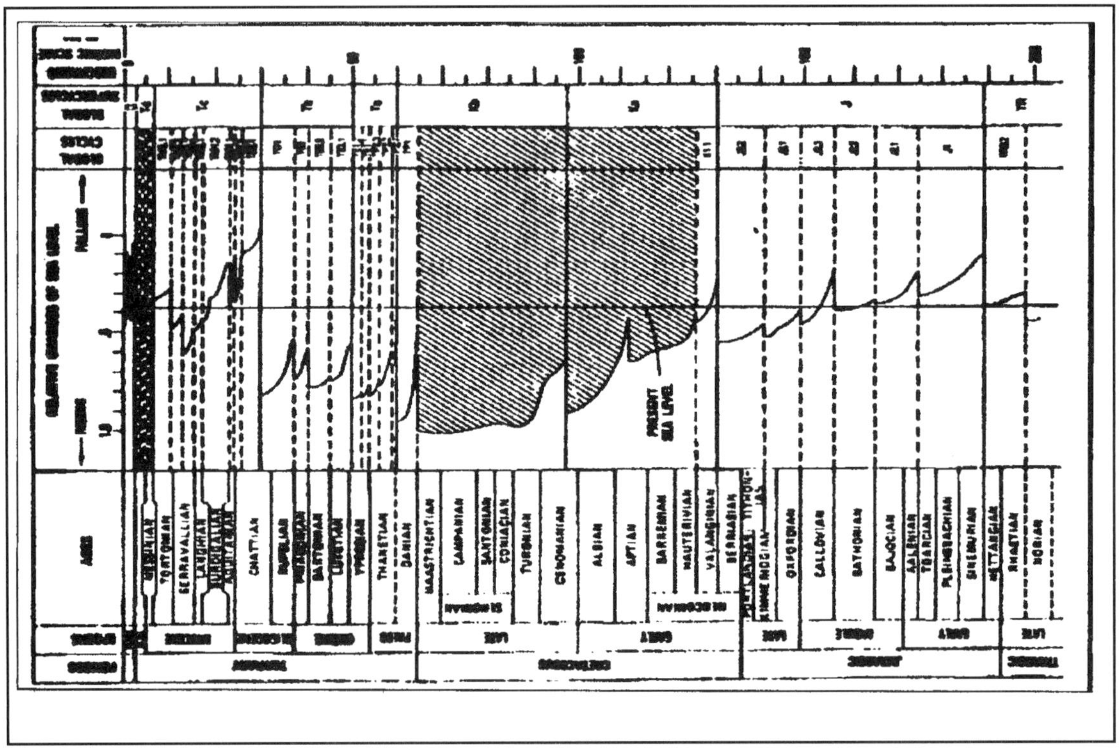

Fig. 7–75 *Global Cycle of Sea-level Changes, Jurassic to Tertiary*

In the late 1970s and early 1980s, Exxon geoscientists recognized that coastal onlap did not track the actual position of relative sea level during late phases of rise, and, consequently, they renamed these curves *relative changes of coastal onlap curves*. The curves, however, indicate the relative direction and to some extent the approximate magnitude and coastal encroachment of relative sea level. The sharp downward shift (saw-tooth) deflection in the coastal onlap curves does not indicate a sharp fall in eustatic sea level, but rather it is a point on the curve when maximum rate of falling eustatic sea level and rate of subsidence generate a rapid fall of relative sea level, which is discussed later in the text.

The 1970s definition of depositional sequences has been changed to restrict its boundaries to erosional unconformities or equivalent concordant surfaces produced by one cycle of eustatic sea level. This modification has in no way affected its application in basin analysis, but has strengthened its use by recognizing its linkage to relative sea level variations.

The concept that coastal onlap marked the level of relative sea level was changed because the late-rise, fluvial onlap component of the curve does not reflect the true position of relative sea level. Criteria for recognizing changes and directions of relative sea level are still valid. Modeling of the interplay of subsidence and eustatic sea level (Posamentier and others1988 and Jervey 1988) has recently shown a more realistic relationship between the two variables.

Global correlations of relative cycles remain controversial, limited by debates on paleontologic and radiometric precision. Arguments continue on precise dating but more and more evidence indicates that the cycles exist.

The new cyclic sequence stratigraphy. The key to strategic use of sequence stratigraphy lies in the interpreter's ability to recognize and trace unconformable truncation surfaces and non-depositional lapout surfaces in order to delineate areally restricted lowstand tracts from more extensive transgressive, highstand, and to some extent, shelf-margin system tracts.

Interpreting the nature of the systems tracts requires an understanding of lithogenetic stratigraphy, depositional systems, and how to relate these from reflection configurations called seismic facies. Geologists and geophysicists with proper skills can learn the process quickly, but some time is required to independently carry out a total sequence analysis.

Vail (1987) recommended a seven-step interpretation procedure for seismic sequence/stratigraphy analysis:

1. seismic sequence analysis
2. well log sequence analysis
3. synthetic well-seismic ties
4. seismic facies analysis
5. interpretation of depositional environment and lithofacies
6. forward seismic modeling
7. final interpretation

It is the purpose of this part of the text to introduce and evaluate the newer and more strategic aspects of sequence stratigraphy and to prepare the reader to begin to apply sequence stratigraphic analysis. The ultimate controls of global cyclic changes in sea level that have been documented by sequences and systems tracts may appear academic for most petroleum explorationists. Recognizing sequences and systems tracts is, however, a critical and highly challenging exercise in hydrocarbon prospecting. This application of cyclic sequence concepts to find and map hydrocarbon plays in a prospective basin may not necessarily depend upon a global system. The global significance of the cycles, however, provides a powerful tool when exploring in frontier basins.

Basic concepts and principles. Vail and Todd (1981) proposed that a depositional sequence should be redefined as a relatively conformable succession of strata bounded by unconformities resulting from eustatic sea-level cycles. This and several other obscure papers (Vail and Mitchum 1979, Vail and Hardenbol 1979, and Vail et al. 1990) initiated sequence redefinition that was further developed by Vail and others (1984) in a paper that contributed significant new ideas to the field of sequence stratigraphy.

Several papers (Haq et al. 1987, van Wagner et al. 1987, and Vail 1987) documented the global importance of Mesozoic and Cenozoic cycles, defined the growing list of new sequence terminology, and summarized the cyclic sequence model and its application to hydrocarbon exploration, respectively.

Finally, the major contributions by present and former Exxon geoscientists to "SEPM Special Paper 42" provide an in-depth look at the new concepts. These include quantitative modeling to explain cyclic siliciclastic deposition (Jervey 1998), a conceptual framework to explain eustatic control of siliciclastic deposition (Posamentier and others 1988), conceptual models of eustatic control of sequence and systems tract deposition (Posamentier and Vail 1988), concepts of carbonate sequence stratigraphy (Sarg 1988), and condensed sections as keys to age determination and correlation of sequences (Loutit and others. 1998). The relationships among the principal factors controlling relative sea-level cycles, inferred mechanics and processes of cyclic erosion and deposition, the stratigraphy and composition of systems tracts, and the applications of sequence stratigraphy to basin analysis developed by the previously listed authors is presented in the following pages.

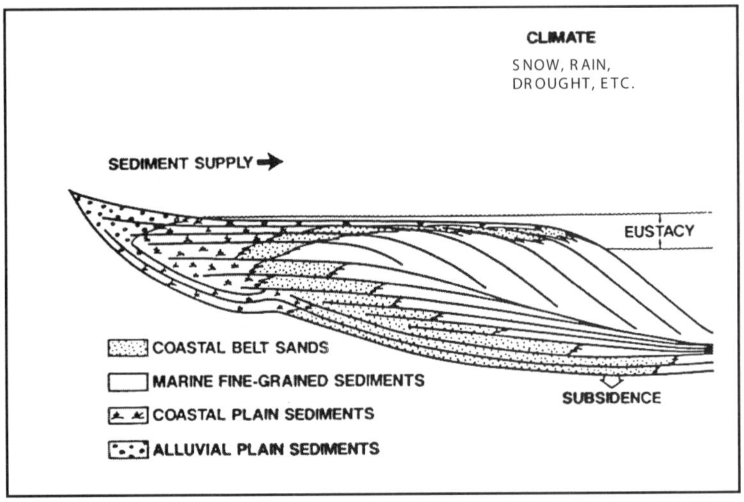

Fig. 7–76 *Control of Sedimentation and Depositional System*

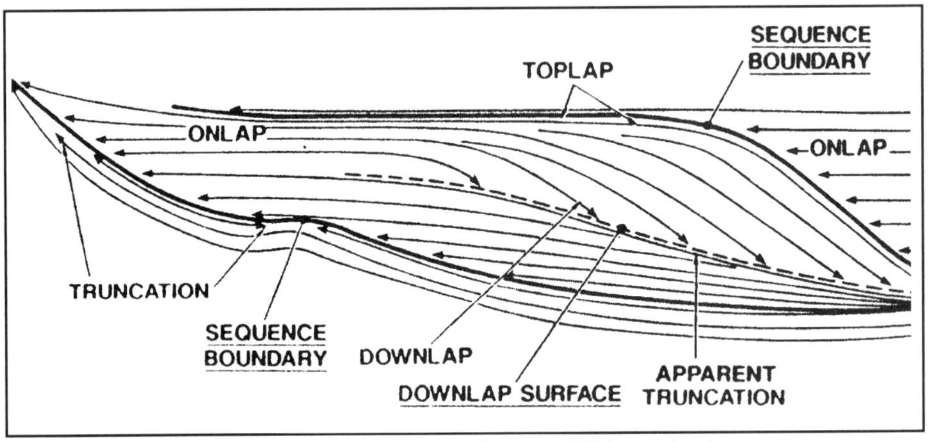

Fig. 7–77 *Reflection Terminations Patterns and Types of Discontinuity*

Basic to understanding the new sequence stratigraphy is the recognition that four critical factors (Fig. 7-76) are important in determining the type, geometry, and composition of depositional sequences—eustacy, basin subsidence, sediment supply, and climate (Posamentier and others 1988). Eustacy, which observations indicate follows a sinusoidal curve, is by definition global and is measured from the center of the earth. Subsidence is the sinking of the basin floor by tectonics—sediment loading, extension, and thermal cooling—and it is the principal factor responsible for creating accommodation space for deposition.

Sediment supply involves the rate and volume of extrabasinal siliciclastic sediments contributed to the basin, principally via fluvial systems and the rate and volume of in situ production of biogenic, carbonate, and evaporite sediments in the basin. Climate strongly, but not entirely, determines whether the sequences will be principally siliciclastic, carbonate/evaporite, or mixed composition.

Reflection termination patterns and types of discontinuities that define cyclic sequences are shown in Figure 7–77.

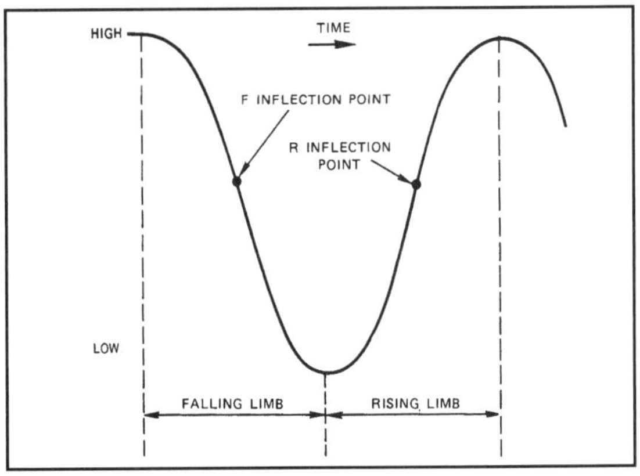

Fig. 7–78 *Reflection Termination Patterns Types of Discontinuities That Define Cyclic Sequences*

Relative sea level is controlled by the interplay of basin subsidence and direction, and magnitude of eustatic sea level changes (Figs. 7–78 and 7–79) is measured relative to an initial underlying depositional surface and provides the principal control of stratal geometries and distribution of lithofacies. The magnitude of relative sea level at any point is the algebraic sum of the two factors and the rate of relative sea level change equals the rate of eustatic change minus the rate of subsidence. The space added or subtracted by changes in relative sea level—eustacy -+ subsidence—is the accommodation space (Fig. 7–80) available for deposition of sediments, and the rate of space added or subtracted and any point is equal to the rate of relative sea level change (Jervey, 1988).

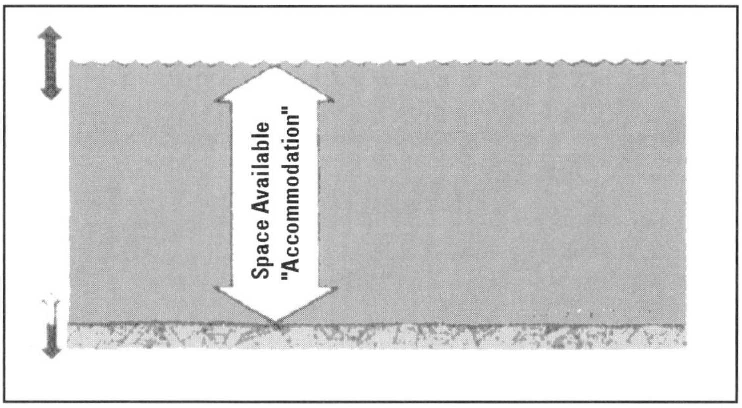

Fig. 7–79 *Accommodation Envelope as a Function of Eustacy and Subsidence*

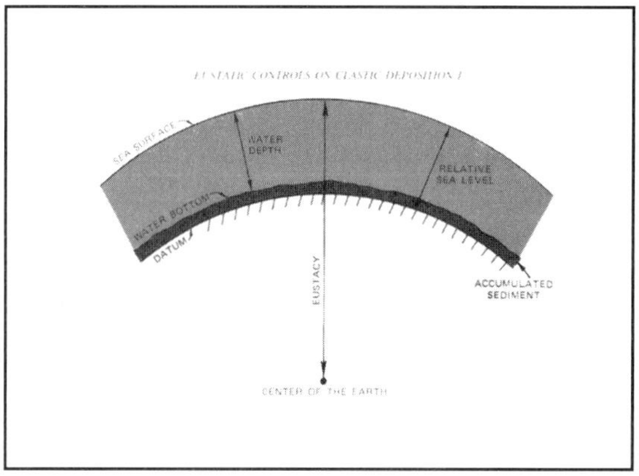

Fig. 7–80 *Eustacy, Relative Sea Level, Water Depth as a Function of Sea Surface, Water Bottom and Datum Position*

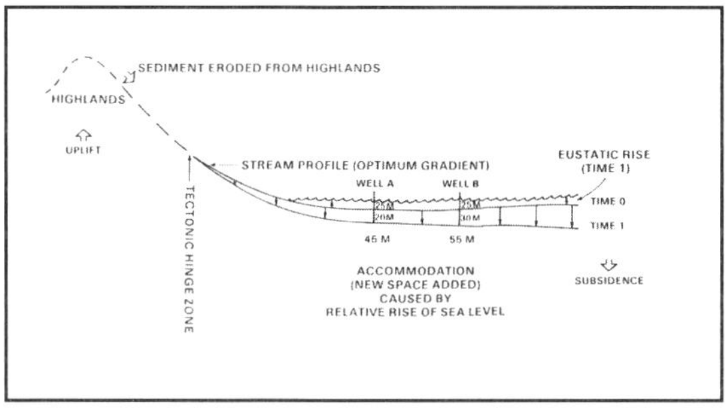

Fig. 7–81 *Accommodation as a Function of Eustacy and Subsidence*

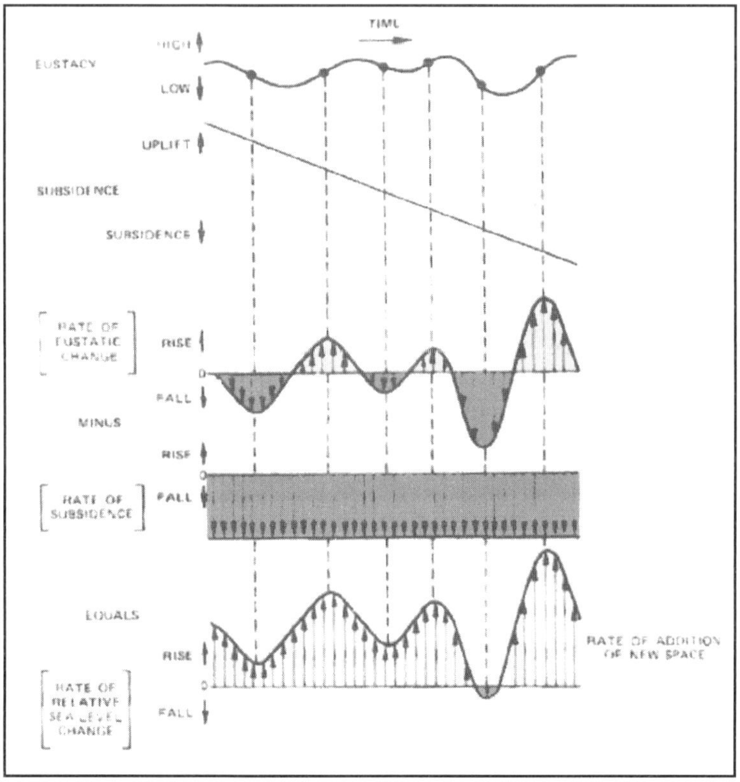

Fig. 7–82 *Relative Sea Level as a Function of Eustacy and Subsidence*

Maximum rates of adding accommodation space occur when eustatic sea level fall passes its *R-inflection point,* the point of maximum rate of eustatic sea-level rise. Minimum accommodation rates coincide with the *F-inflection point,* the point of maximum rate of eustatic sea-level fall (Figs. 7–81 and 7–82).

The interplay of rates of accommodation space added or subtracted and rate of sediment supply determines water depths and shoreline positions. Accommodation space may be filled by two mechanisms—aggradation or up building producing onlap and progradation or lateral/outbuilding producing downlap or perhaps toplap. Consequently, the interplay of factors controlling accommodation strongly influences the geometry of depositional sequences (Jervey 1988).

In considering the interrelationships of eustatic, subsidence, relative sea level, accommodation, and sediment supply, it is important to clearly distinguish between actual values (magnitudes and volumes) and second derivative rates of change. Within a subsiding passive-margin basin, eustatic sea-level cycles and subsidence interact to control direction, magnitude, and rates of relative sea level change. These factors change rates of accommodation space added or subtracted and shift areas of erosion and deposition. In general, subsidence rates along a margin to center-basin profile will remain constant, but the amount of subsidence will increase toward the center of the basin from its marginal tectonic hinge line, the point of no tectonic movement separating basin subsidence from marginal uplift (Fig. 7–83).

Sediment supply rate will certainly vary with time, but any variations will result principally in changing water depths and coastline positions. Modeling of the effects that a eustatic cycle may impose on relative sea level, erosion, and deposition is important in understanding cyclic sequences and systems tracts (Jervey 1988).

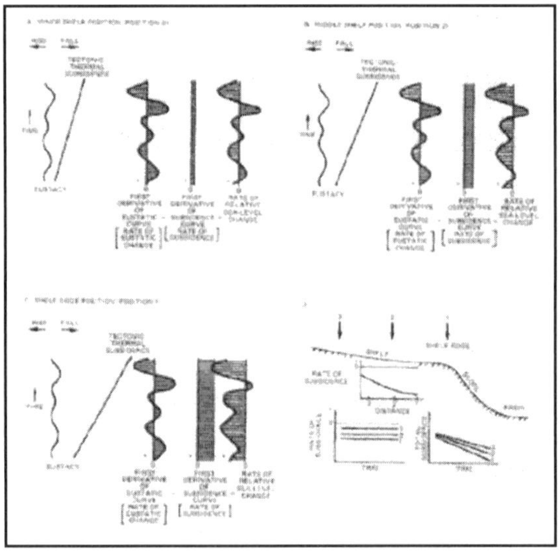

Fig. 7–83 *Response of Relative Sea Level to Differential Tectonic Thermal Subsidence*

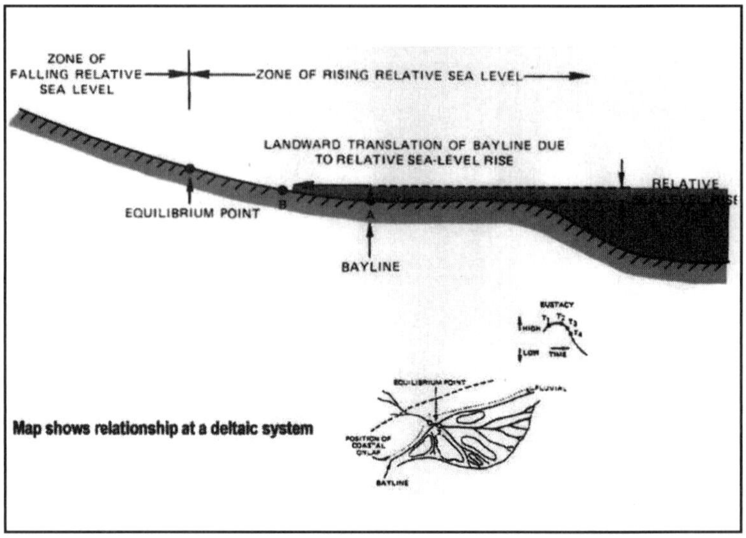

Fig. 7–84 *Effect of Relative Sea Level Rise on Coastline Position*

A relative sea-level cycle, by definition, begins and ends with erosion signaled by a downward shift of the coastal onlap curve. Late in the highstand phase of deposition (Fig. 7–84), the basinward shifting equilibrium point reaches the bayline. The basin shifting equilibrium point is where rates of subsidence and eustatic fall are equal and separate areas of erosion from deposition. The *bayline* is a point located slightly above mean sea level where alluvial profiles are adjusted. The equilibrium point and bayline shift basinward together, producing subaerial accommodation space for alluvial aggradation and coastal/ alluvial onlap.

Erosion. There are two types of erosion—Type 1 and Type 2.

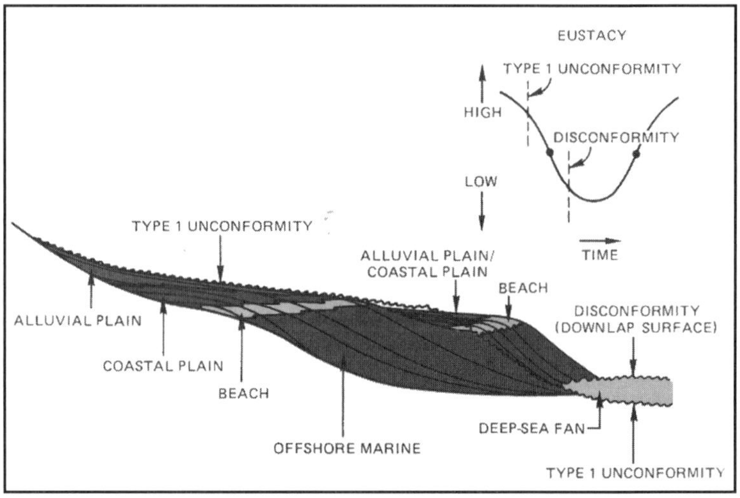

Fig. 7–85 *Type 1 Unconformity*

Type 1 erosion is initiated when falling eustatic sea level approaches and passes the maximum rate or F-inflection point on the falling limb of the eustatic curve. This initiates a rapid basinward shift of the equilibrium point and bayline beyond the pre-existing highstand depositional shoreline break (+30 ft), initiating a Type 1 sequence and lowland systems tract.

The F-inflection point on the eustatic curve that marks the maximum rate of eustatic sea level fall and generates a downward shift that is global and probably recognizable in most basins, worldwide. Bayline and equilibrium point move together across the low gradient shelf until a bathymetric break, such as a relict shelf edge, causes the position of the falling bayline (approximate sea level) to lag behind the equilibrium point. The equilibrium point may move far beyond the relict shelf break (Fig. 7–85). If the bayline shifts basinward of the relict shelf break, extensive erosion of the shelf, intensive valley incision, submarine-canyon erosion, and deposition of basin floor fans will occur (Fig. 7–86).

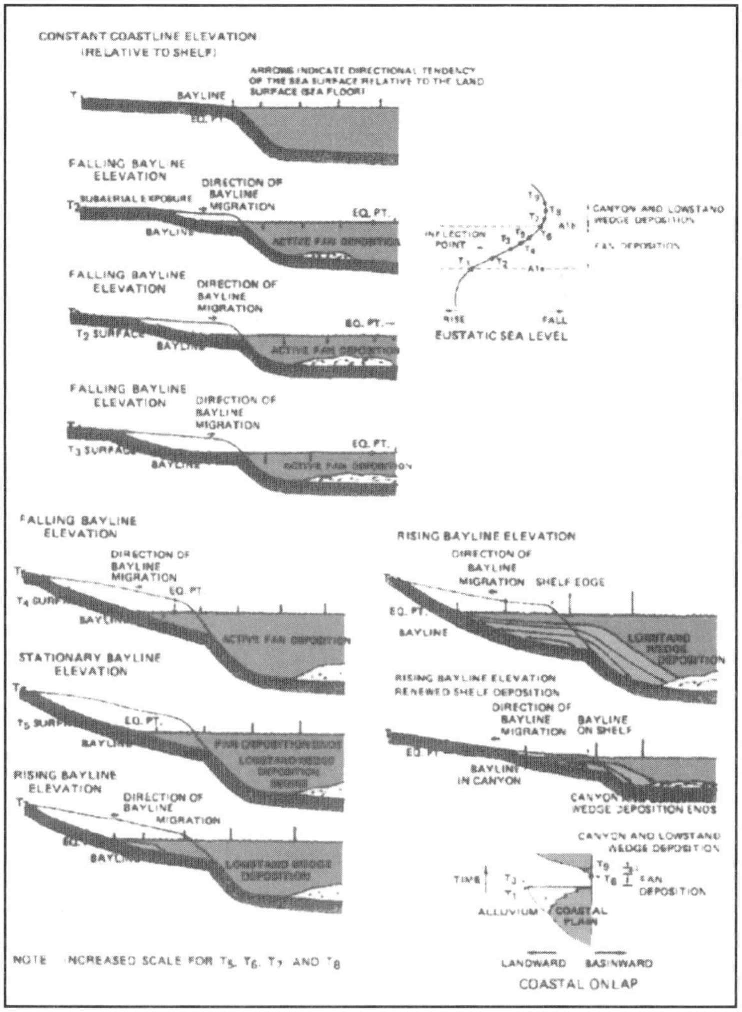

Fig. 7–86 *Response of Sedimentation on an Interval of Rapid Eustatic Fall*

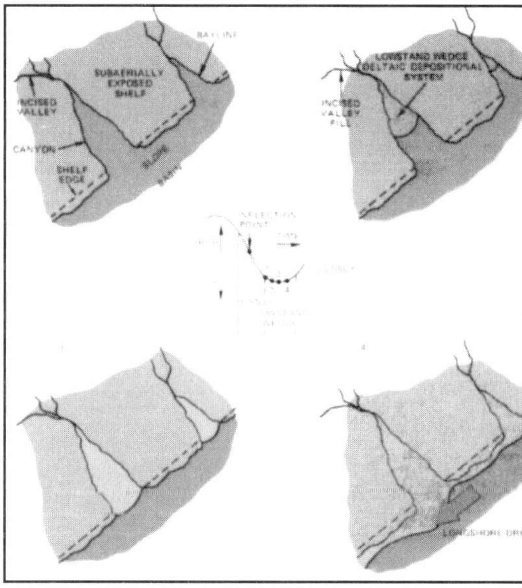

Fig. 7–87 *Distribution of Lowstand Wedge Deposits along The Outer Shelf/Upper Slope*

When eustatic fall reaches its F-inflection point, the equilibrium point has reached its most basinward position and it reverses and shifts landward, at which time levee-slope fans may be deposited. Falling sea level may not always occur in a simple cycle, but may involve short- and long-period falls (Fig. 7–87), probably caused by combined third- and higher-order cycles. The bayline position shifts landward but lags behind the shifting equilibrium point. This shift gradually increases accommodation space permitting the early lowstand progradational wedge to increase aggradation (Fig. 7–88).

Relative rise of sea level (decelerating fall and early rise of eustatic sea level) drowns the wedge, fills lower parts of the incised valleys, and initiates deposition of the *transgressive systems tract* (Fig. 7–89).

Type 2 erosion begins when slowly falling eustatic sea level—*late highstand deposition*—reaches its F-inflection point and most seaward position and the equilibrium point and bayline reverse direction and shift together landward without shifting below the depositional shoreline break (Fig. 7–90). This reversal initiates the deposition of a Type 2 sequence and a shelf-margin systems tract, and it also terminates alluvial aggradation because rising sea level disrupts the alluvial profile (Fig. 7–91).

The bayline (approximate sea level) accelerates its rise as it lags behind but continues to follow the equilibrium point in a landward direction. The equilibrium point moves landward because of the decreasing rate of fall, and the eventual rise of eustatic sea level leads to the flooding of the shelf-margin tract and initiation of transgression. When relative sea level rise increases under the influence of combined subsidence and a diminishing rate of eustatic fall, deposition of the transgressive systems tract begins, strongly influenced by the sediment supply (Fig. 7–90).

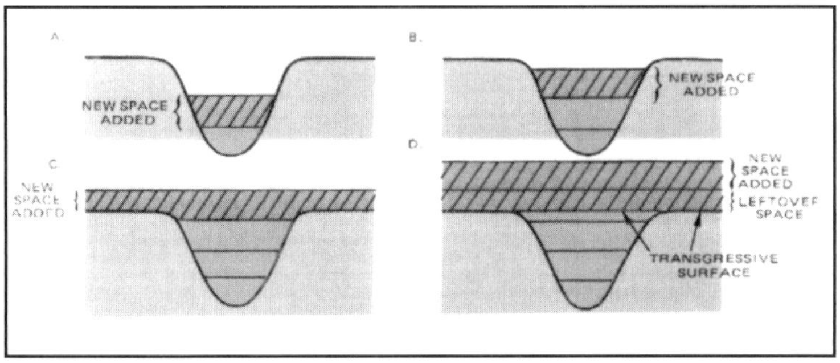

Fig. 7–88 *New Space Added during an Interval of Constant Rate of Relative Sea Level Rise Following Type 1 Unconformity*

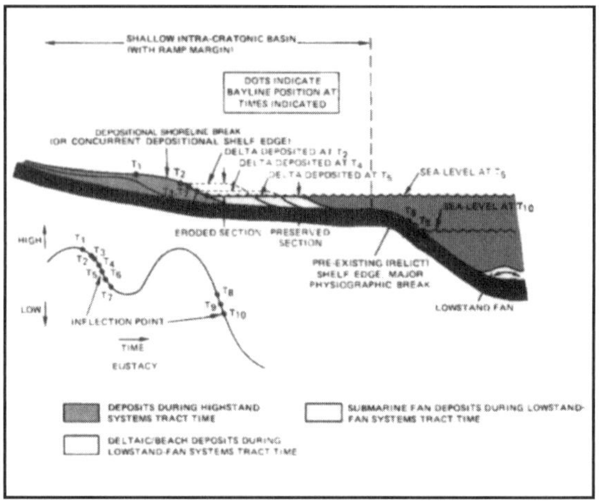

Fig. 7–89 *Bayline Position and Lowstand Deposits during Rapid Short- and Long-Period Eustatic Fall*

Because of early slow rate of relative rise, thick retrogradational sedimentation may occur, but as eustatic rise begins, relative rise accelerates, the bayline moves landward, and the depositional units begin to thin. The equilibrium point retreats landward, passing the tectonic hinge-line at lowest eustatic sea level and reaching its most landward position shortly after the eustatic R-inflection point is reached. Landward of the hinge-line, the equilibrium point is the point where rate of eustatic rise equals the rate of tectonic uplift.

Transgressive deposition generally continues until eustatic sea level rise approximately reaches its R-inflection point on the rising limb of the eustatic curve, and maximum flooding occurs, generally coincident with maximum shoreline transgression and the onset of major marine-condensed deposition (Fig. 7–90).

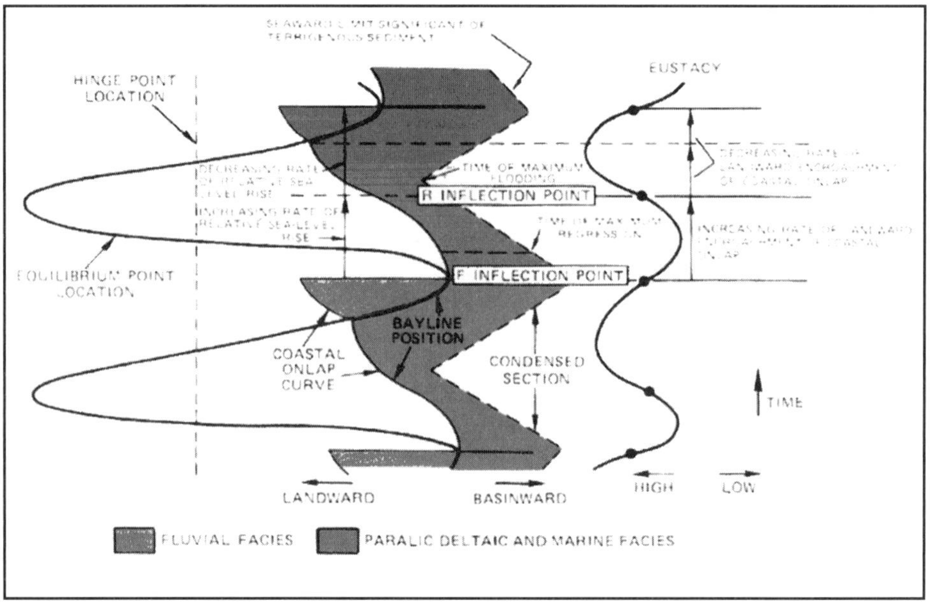

Fig. 7–90 *Elements of Coastal Onlap Curve*

The R-inflection point, which marks the maximum rate of rising eustatic and relative sea levels, generates a flooding event that may be global, essentially contemporaneous, and probably recognizable worldwide. Rising but decelerating relative sea level continues until the basinward shifting equilibrium point reaches the bayline and together they move basinward to initiate *highstand deposition.*

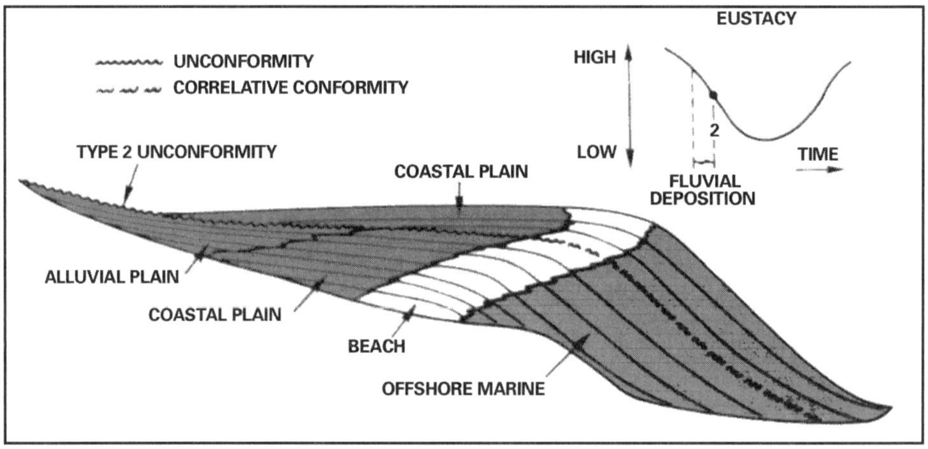

Fig. 7–91 *Type 2 Unconformity*

Under the influence of deposition of highstand systems, tracts begins when the equilibrium point shifts basinward diminishing rates of eustatic sea level rise—it passes the tectonic

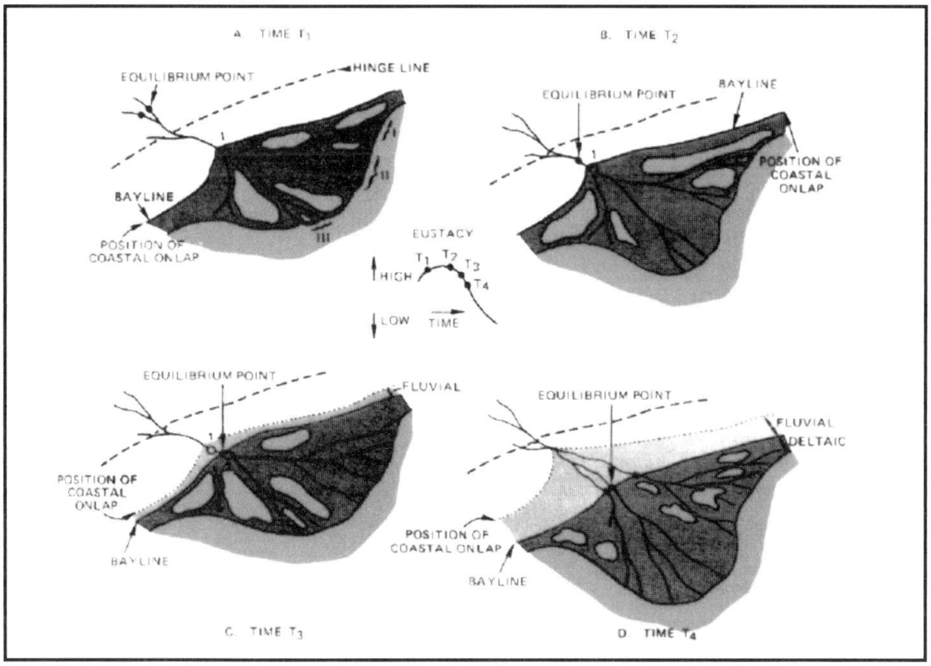

Fig. 7–92 *Effect of Equilibrium Point Migration on Fluvial Deposition in Prograding Environment*

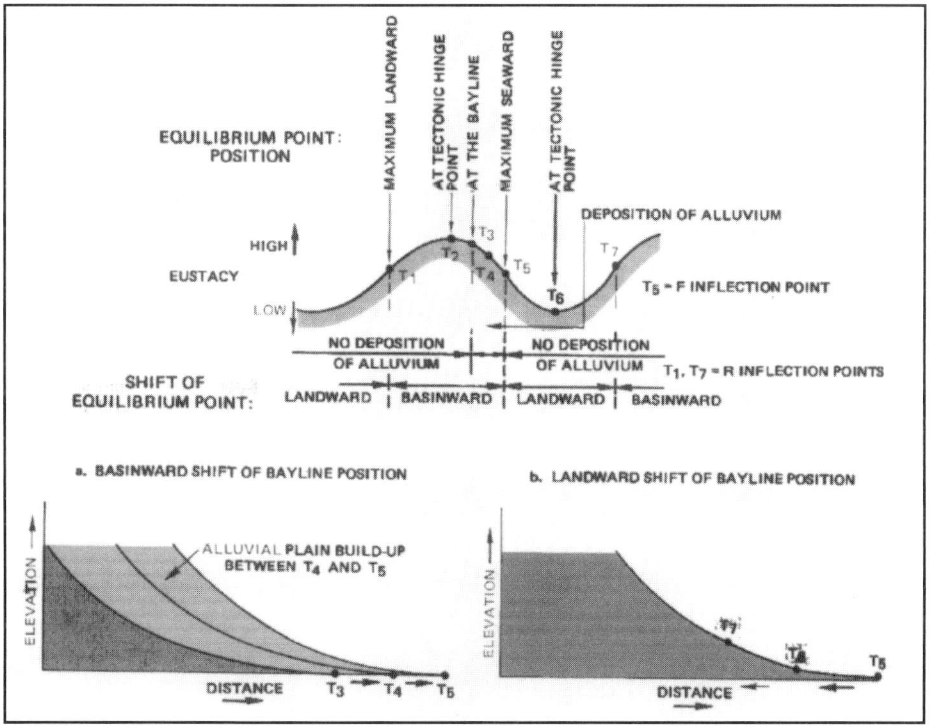

Fig. 7–93 *Effect of Shifting Equilibrium Point on Fluvial Deposition*

hinge-line at highest eustatic sea level—and reaches the still landward shifting, and relatively rising, bayline (Fig. 7–90). At that point, both bayline and equilibrium point together move slowly basinward under slowly falling relative sea level (Fig. 7–92). The seaward shifting bayline permits adjustment of alluvial profiles, creating subaerial accommodation space for extensive fluvial deposition (Fig. 7–93).

After the early phase of aggradational/progradational (sigmoidal) highstand deposition, the diminishing rates of relative rise and initial, slow fall of relative sea level—caused by accelerating rates of early eustatic fall—result in a late phase of increasingly progradational (oblique) deposition (Fig. 7–94).

As noted previously, highstand deposition is terminated when the F-inflection point is approached, initiating either Type 1 or 2 erosion and deposition. Figure 7–95 summarizes the relationships between eustatic sea level and phases of erosion and systems-tract deposition.

Cycles and paracycles. Changes of relative sea level produce within the depositional sequence certain unique stratigraphic relationships, bounding surfaces, internal stratal configurations and geometric patterns, and predictable lithofacies distributions that can be analyzed and interpreted in outcrop, with well logs, and on seismic profiles. A cycle of relative sea level by definition begins and ends with a relative fall at the eustatic F-inflection point, but almost all of the sequence is deposited during rising relative sea level. Rising sea level provides the space required to accommodate the sediments (van Wagoner and others 1987). Relative sea level rises in small, periodic steps followed by still stands called paracycles (Fig. 7–96).

Accelerating rates of relative sea level rise eventually lead to small-scale marine flooding surfaces, but as the rate of rise diminishes, progradational deposition is initiated and continues during late rise and still stand until rising sea level again floods the area. The paracycles probably occur because the failing limbs of minor higher-order eustatic cycles briefly offset relative sea level rise, which is produced by the combined effects of rising, lower order eustatic cycles, and subsidence. The progradational stratigraphic unit (carbonate or siliciclastic) deposited during a paracycle is a shoaling-upward succession of facies called a parasequence.

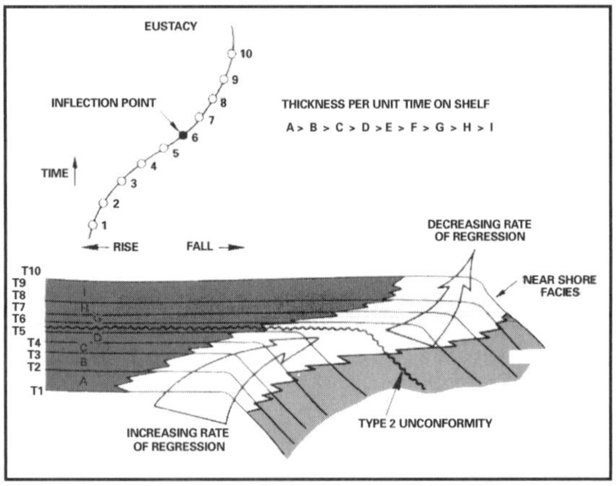

Fig. 7–94 *Response of the Topset Bed Thickness to Eustatic Fall*

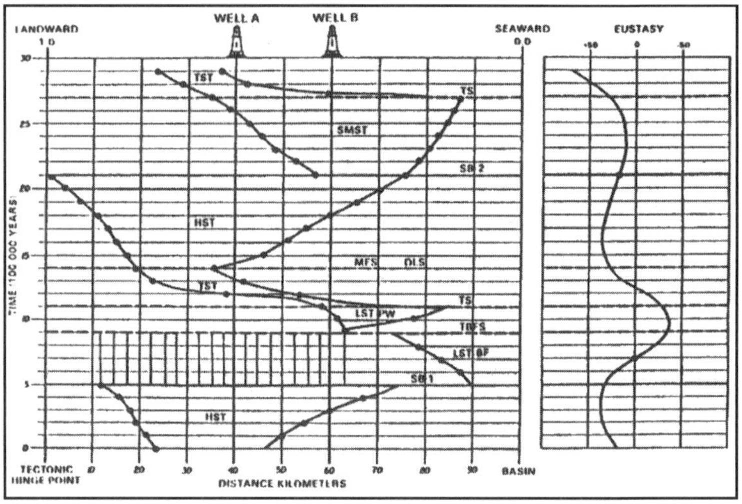

Fig. 7–95 *Relationship between Eustatic Sea Level and Phases of Erosion and System Track Deposition*

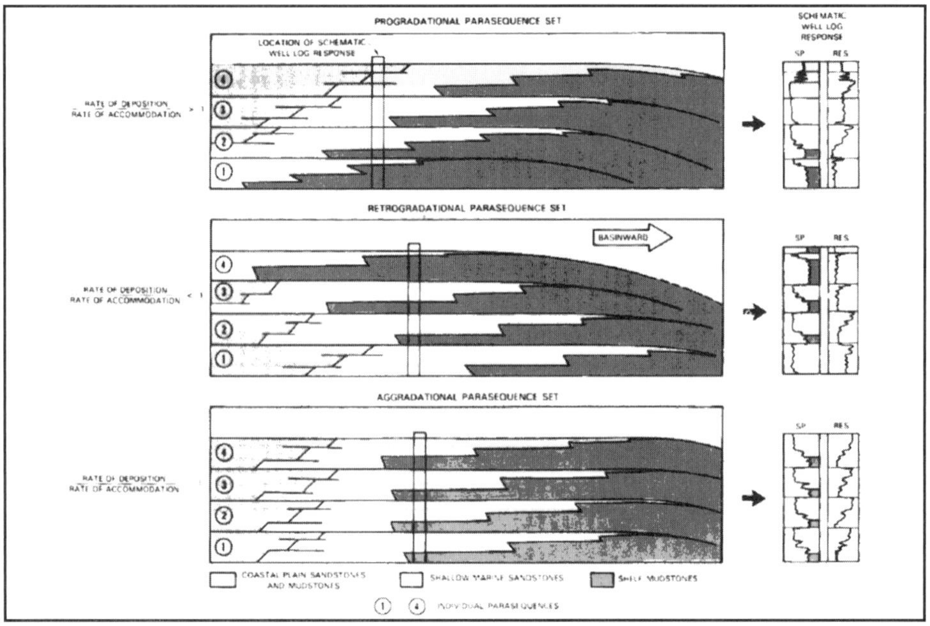

Fig. 7–96 *Types of Parasequence Sets*

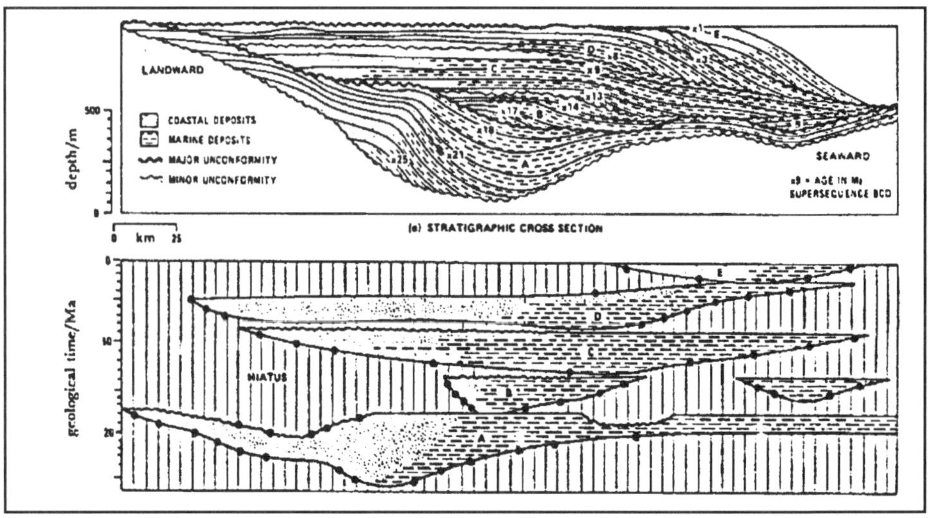

Fig. 7–97 *Recognizing and Dating Unconformities*

A series of stacked parasequences is called a parasequence set, and the various stacking patterns of parasequence sets can be used to demonstrate whether the long-term rate of sediment supply (sed) exceeded, equaled, or was less than the rate of accommodation A.

For example:

Sed > A — progradational patterns or basinward stepping patterns

Sed = A — aggradational patterns or vertical stacking patterns

Sed < A — retrogradational patterns or landward retreating patterns

Each depositional system tract, therefore, may be characterized to a considerable degree by the nature of the stacking patterns of its parasequence sets.

The redefined depositional sequence. Vail and Todd (1981) and Vail and others (1984) developed the redefined sequence model through studies of Jurassic rocks in the North Sea. These are bounded by unconformities that were eroded as a result of falling eustatic sea level and may have been locally enhanced by positive tectonic movements. It was noted that two types of erosional unconformities bound depositional sequences called Type1 or Type 2 sequences. These unconformities, which formed when falling eustatic sea level approached and passed its F-inflection point, can be recognized by extensive erosional surfaces and downward shifts of coastal onlap (Fig. 7–97).

Downward shift follows the basinward shift of the equilibrium point and subaerial exposure and erosion occur landward of the point at which the rate of eustatic fall equals the rate of subsidence. Recognition and dating of unconformities require careful interpretation of reflection terminations and adequate paleontologic control (Fig. 7–97).

Type 1 surfaces are subaerial and submarine unconformities or their concordant equivalent surfaces that developed when eustatic sea level rapidly fell faster (>10 cm/1000 years) than subsidence at the depositional coastline break (Fig. 7–98). Hence, the equilibrium point moved rapidly basinward, commonly beyond the relict shelf edge and the bayline lagged behind but dropped below the relect shelf edge. Exposure of the shelf (*paleosols*), incision of fluvial valleys, and erosion of submarine canyons provided sediment to basin-floor submarine fans.

Slope-front fills are composed of levee-channel slope fans and subsequent aggradational lowstand deltaic and coastal systems (or evaporite/carbonate systems). They coastally onlapped the Type 1 surface and filled the canyons and incised valleys during the subsequent relative sea-level rise. Type 1 erosion, in relatively stable, subsiding basins (Fig. 7–99a) is typified by well-developed entrenched valleys and submarine canyons near the relict shelf edge, but erosional evidence diminishes landward.

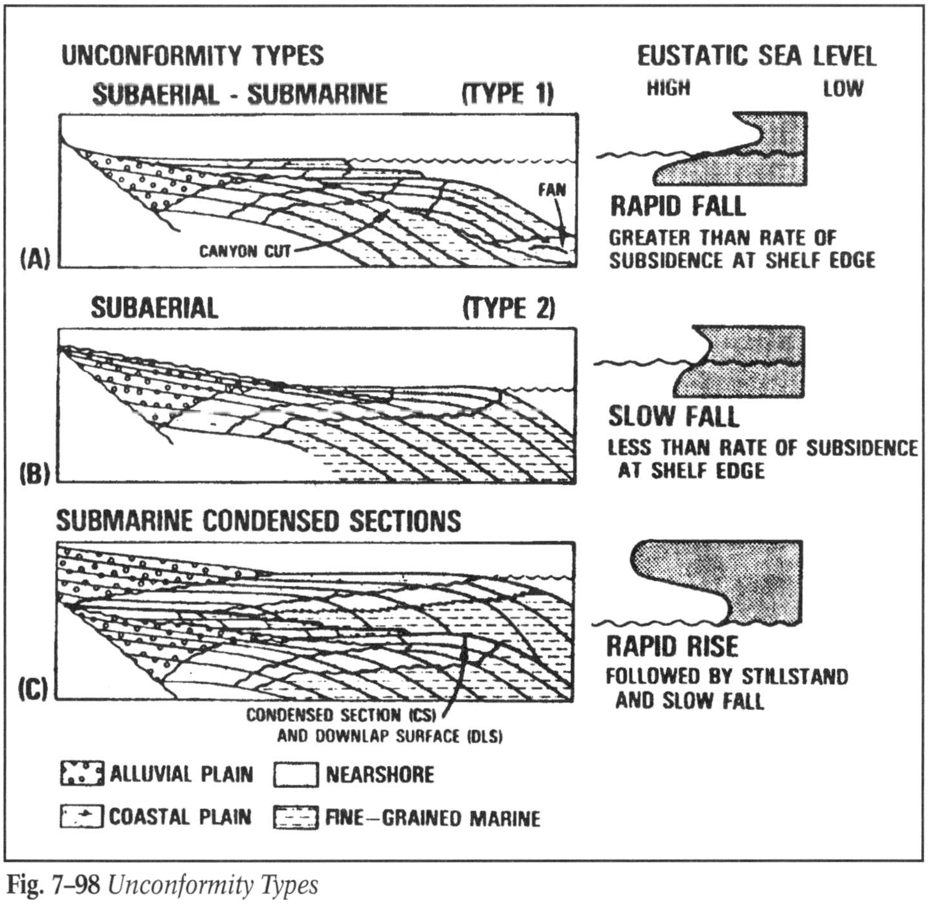

Fig. 7–98 *Unconformity Types*

Type 1 erosion of a basin during tectonic uplift and folding produces maximum truncational erosion in the landward direction, and it tends to diminish toward the shelf edge (Fig. 7–99b). On gently sloping ramps with less gradient than the fluvial profiles, relative sea level falls basinward of the depositional shore-line break, but Type 1 unconformities are less erosive and there is limited submarine erosion and deposition.

Type 2 surfaces are principally subaerial exposure (paleosols) unconformities or their concordant equivalent surfaces. They developed when eustatic sea level slowly fell, causing the equilibrium point and bayline to move basinward together to a point landward of the shelf edge (Fig. 7–100b) where *rates of eustatic fall and subsidence were equal.* Limited deep-water sedimentation occurred, and progradational and aggradational shelf-margin systems coastally onlapped the Type 2 surface during subsequent relative rise of sea level. Type 2 unconformities display limited fluvial incision, and they are principally regional subaerial exposure surfaces.

Downlap surfaces (DLS) are a third important type of discontinuity within the cyclic depositional sequence. It occurs at the top of a starved biogenic/authigenic sediment blanket or hard ground—called the *marine condensed section* (MCS). They formed when eustatic rise approached its rising R-inflection point or maximum rate of rise (Fig. 7–98c), producing the most landward shift of the equilibrium point. The position of rising relative sea level is generally coincident with maximum paleowater depths, marine flooding, and diminishing sediment supply (<1 cm/1000 years).

The marine-condensed sections are downlapped by the prograding clinothem of subsequent highstand deltaic and coastal (or carbonate) systems that normally display maximum landward and basinward distribution of depositional systems within the sequence. Marine-condensed deposition, which blankets transgressive (or retrogressive) depositional systems during rising relative sea level, generally correlates with maximum transgression.

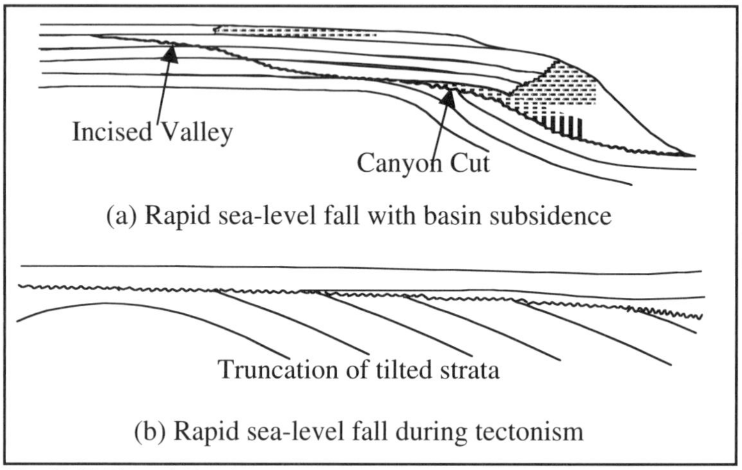

Fig. 7–99 *Type 1 Erosion*

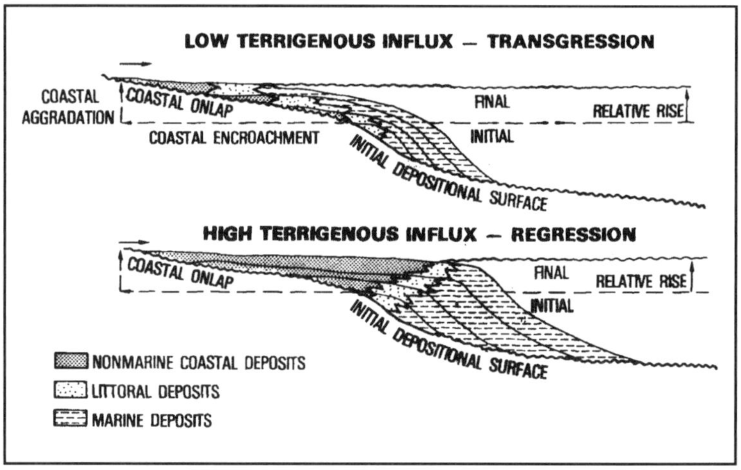

Fig. 7–100 *Relation between Transgression or Regression and Eustatic Sea Level*

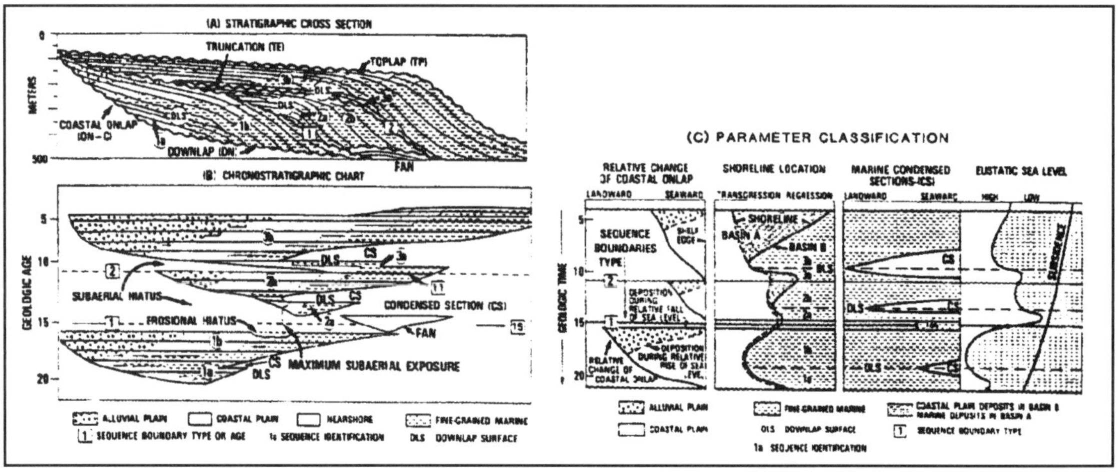

Fig. 7–101 *A Deposition Sequence in Depth and Time and Its Relation to Marine Condensed Section, Coastal Onlap, Shoreline, and Eustatic Sea Level*

Coastal onlap or depositional onlap of coastal systems occurs almost continuously during deposition of a sequence, in response to a landward shifting equilibrium point and rising relative sea level or bayline, which provides accommodation space for the sediments (Fig. 7–101).

Downward shifts of coastal onlap mark very short time gaps in the deposition of a sequence (Fig. 7–100a and b). Coastal onlap curves provide a more precise documentation of relative sea-level changes than do shoreline transgressions and regressions. The latter are controlled by rates of both sea-level change (accommodation) and sediment supply (Fig. 7–101c). For example, with an adequate sediment supply, highstand shoreline regressions (exhibiting stacks of shoaling-upward progradational successions) typically begin while relative sea level is stiff and slowly rising, but eustatic sea level is falling at a rate less than subsidence (Fig. 7–101a).

During relative rise as the equilibrium point and bayline move together basinward, alluvial onlap occurs before it is terminated by the abrupt downward shift in coastal onlap in a Type 1 sequence. Consequently, with different rates of sediment supply and subsidence, regressions may vary among basins.

On the other hand, shoreline transgressions—exhibiting stacks of shoaling, upward, retrogradational successions—though not always synchronous, tend to exhibit better correlation among basins because they are controlled by the combined effect of subsidence and rising eustatic sea level (Fig. 7–101b). Transgressions generally tend to be more rapid than regressions and occur near the base (maximum) and top, and maximum transgression occurs near the middle of typical sequences. These interrelationships are illustrated in Figure 7–102.

Three principal discontinuities (Type 1, Type 2, and MCS) mark variable time gaps, but they each have chonostratigraphic significance because they separate rocks of different ages (Fig. 7–101a and b). Sequences, therefore, are inferred to represent global chonostratigraphic genetic successions of strata if their bounding unconformities correlate globally (Fig. 7–102). The unconformities are most precisely dated at the minimum time gap where they become conformable. The saw-tooth curve indicating a downward shift of coastal onlap (Fig. 7–102) may display global synchronicity, especially in passive-margin basins exhibiting simple diminishing thermal subsidence rates, because the downward shift occurs at the F-inflection point on the eustatic curve.

The MCS, which occurs near the middle of most sequences, is composed of rich pelagic paleofaunas and floras that provide some of the most precise age dates within the sequence. The MCSs are approximately contemporaneous globally but some temporal variation may occur from basin to basin because of possible variations in rates of deposition and subsidence.

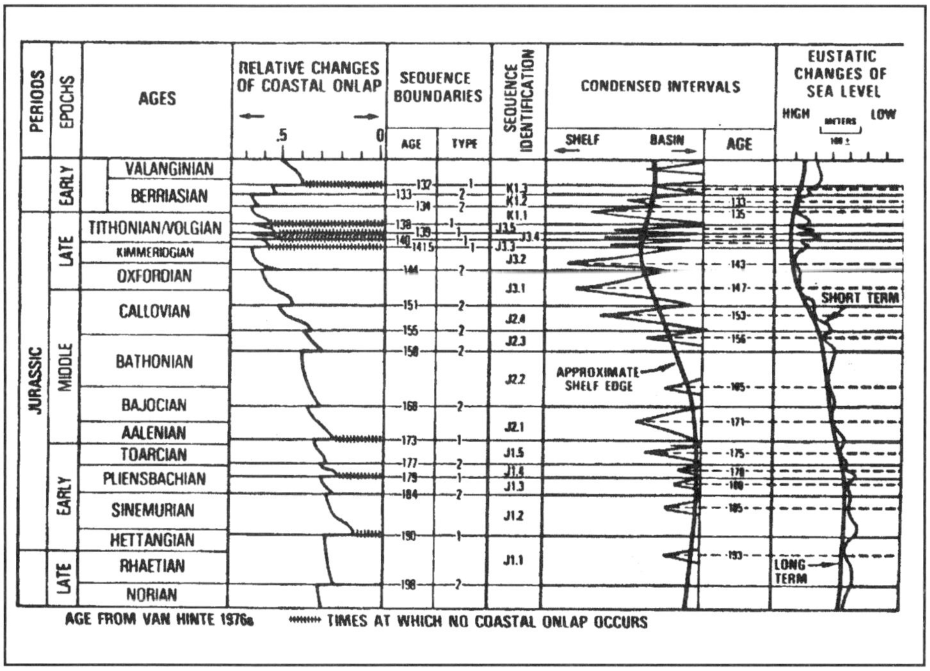

Fig. 7–102 *Relationship of Sequence to Relative Changes of Coastal Onlap, Types and Ages of Unconformities, Condensed Intervals/Ages, and Inferred Eustatic Sea-level Changes*

In their 1984 redefinition of the depositional sequence, Vail and others proposed that cyclic sequences, which are bounded by Type 1 or Type 2 unconformities, are typically composed of three subsequences deposited at specific times during a cycle of relative sea level. They noted, further, that within the unconformity bounded sequences these subsequences are separated by marine-condensed sections, lesser discontinuities, and/or lapout surfaces (Fig. 7–102).

In a series of 1987 papers, Exxon geoscientists named these subsequences *depositional systems tracts.* This term was introduced by Brown and Fisher (1977) who defined it as a lithogenetic unit composed of one or more contemporaneous depositional systems.

The redefinition by Exxon researchers defined four systems tracts:

- lowstand
- shelf margin
- transgressive (retrogradational)
- high-stand

Ideally, the lowstand or shelf margin combine with transgressive and highstand tracts to compose a depositional sequence. The tracts were named by Vail (1987) and van Wagner and others (1987) for their respective sea-level position when the tract was deposited. Posamentier and others (1988) and Posamentier and Vail (1988) provide an in-depth review of systems tracts.

A critical aspect of sequence-stratigraphic analysis involves the recognition, mapping, and interpretation of depositional systems tracts and/or their component elements. They are important because recognition of the type of systems tract permits greater precision in interpreting the depositional systems and their component lithofacies, the key to identifying potential reservoirs, seals, source, and stratigraphic traps. In fact, some petroleum geologists estimate that 85% of the world's hydrocarbons are trapped in reservoirs either deposited or enhanced during lowstands of relative sea level (Fig. 7–103).

Systems Tract	Reserves	
	Gas BBDE	Oil BBD
Lowland Delta	28	217
Valley/Canyon Fill	12	97
Basin Floor Fan	13	96
Highstand Deposits	6	48
Transgressive Deposits	3	24

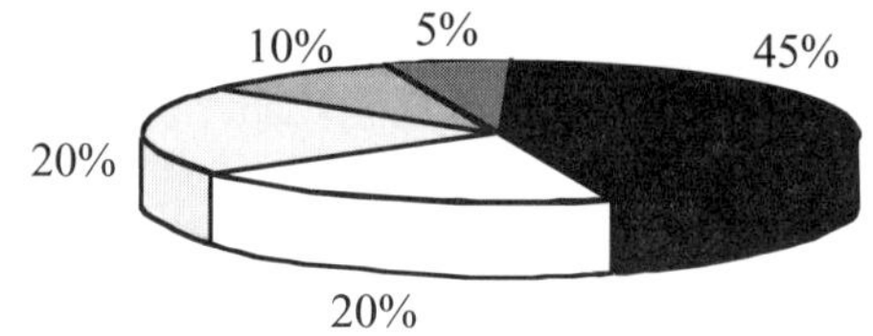

Fig. 7–103 *Estimation of Worldwide Hydrocarbon Reserve in Clastic Depositional Sequences*

These figures are overly optimistic about the distribution of hydrocarbons related to lowstands of sea level. It is probably true, however, that lowstand sand reservoirs and carbonate reservoirs that experienced porosity/permeability enhancement during lowstands of sea level constitute the majority of the world's reservoirs. For these reasons, explorationists and exploitationist should be knowledgeable about systems tracts.

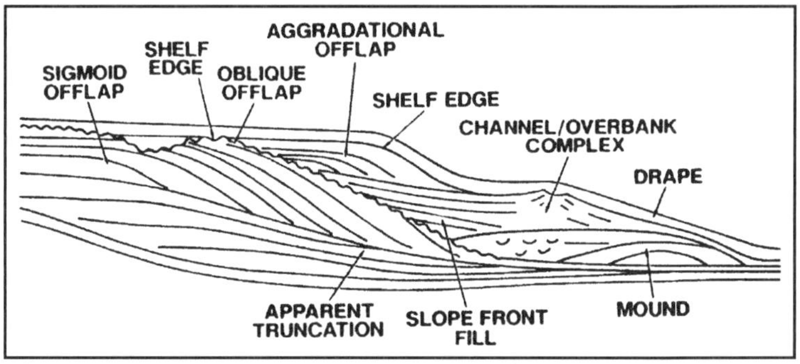

Fig. 7–104 *Diagrammatic Seismic Section Showing Common Stratal Geometries and Terminations*

After this discussion of the four systems tracts in respect to sea-level control, depositional processes, and resulting systems, the hydrocarbon potential of each will be evaluated.

Beginning with falling relative sea level, the chronology of systems tracts within the depositional sequence is, therefore,

1. low-stand or shelf-margin tract
2. transgressive (retrogradational) tract
3. highstand tract

From the stratal configurations and geometries, stratal terminations, and types of discontinuities and surfaces, interpretation of a cyclic scenario is normally possible (Fig. 7–104). By using such a predictive model. One can test the interpretive criteria on the seismic profile (Fig. 7–105). Color codes for surfaces, discontinuities, and systems tracts are useful in delineating, correlating, and eventually mapping the sequence and its systems tracts.

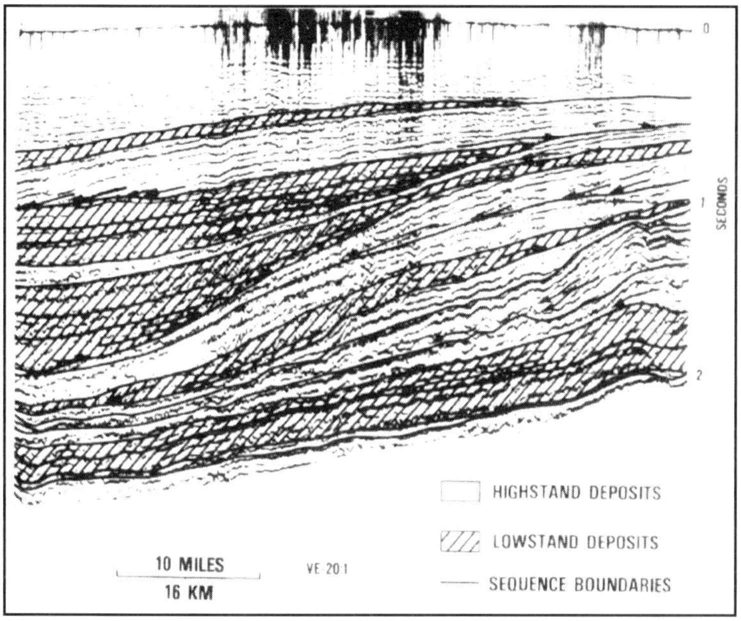

Fig. 7–105 *Seismic Section Showing System Tracts and Other Elements of Depositional Sequences*

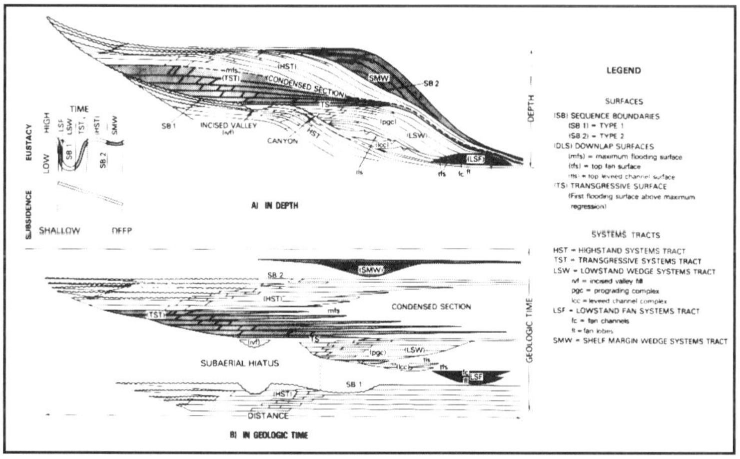

Fig. 7–106 *An Idealized Siliciclastic Depositional Sequence Showing Depositional System Tracts and Their Bounding Surface*

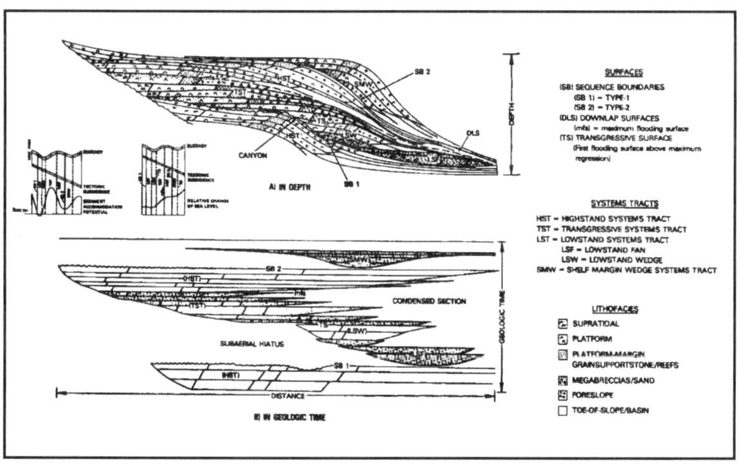

Fig. 7–107 *Schematic Diagram of Carbonate Lithofacies Distribution in a Sequence*

Depositional systems tracts may be siliciclastic, carbonate /evaporite, or of mixed composition. Depositional systems tracts may be recognized and interpreted on the basis of stratal geometries and terminations, as well as their relationship to erosional and non-depositional discontinuities (Figs. 7–106 and 7–107). These characteristics are generally less than ideally exhibited by a sequence, but with sufficient seismic profiles, it is normally possible to recognize the diagnostic criteria needed to delineate the individual systems tracts.

Highstand systems tracts (HST). These systems tracts are deposited during progradation *and downlap* onto the top of subjacent marine-condensed sections (Figs. 7–106 through 7–109) when rising relative sea level slows, permitting an adequate sediment supply to initiate regressive deposition (Vail 1987). During early highstand, aggradational (sigmoidal) parasequence sets are more typically deposited because the landward-shifting equilibrium point and bayline produce moderate rates of relative sea level rise and, hence, new accommodation space.

During later highstand when eustatic sea level begins to fall slowly (but subsidence continues), the equilibrium point shifts basinward, but the bayline continues to shift landward. Relative sea level rise diminishes, decreasing accommodation space and resulting in the deposition of strongly progradational (oblique) parasequence sets. When the basinward-shifting equilibrium point reaches the bayline and they shift together basinward, subaerial accommodation space is added and aggradation of coastal onlapping alluvial systems is initiated in siliciclastic basins. Coastal/alluvial onlap and oblique progradation continue until eustatic fall reaches its F-inflection point. Then the equilibrium point and bayline shift basinward, resulting in erosion of a Type 1 subaerial and submarine unconformity (rapid shift) or shift slowly landward to initiate Type 2.

When carbonate or siliciclastic highstand tracts prograde into deep water basinward of relict shelf edges, they may exhibit gravity failure and/or growth faulting.

Siliciclastic highstand tracts (Fig. 7–110) are typically composed of fluvial/deltaic systems separated along strike by various types of inter-deltaic systems (Posamentier and Vail 1988). They commonly downlap onto well-developed marine-condensed sections and are terminated by either Type 1 or 2 erosion.

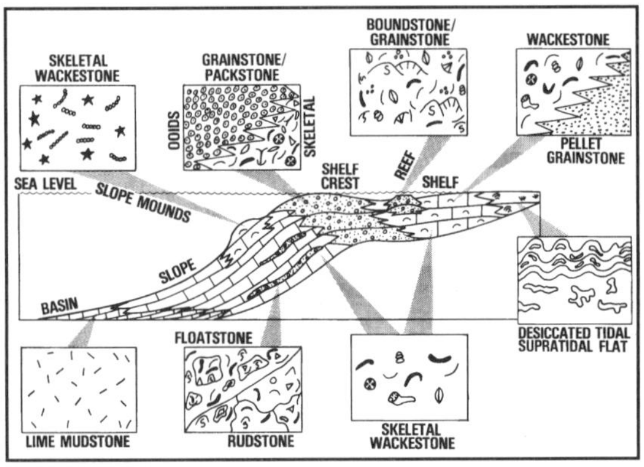

Fig. 7–108 *Carbonate Facies Belts with Representative Textural Types*

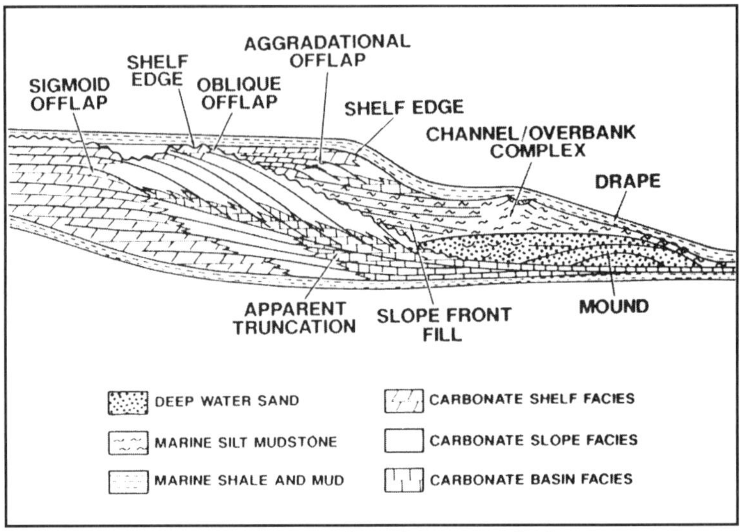

Fig. 7–109 *Diagrammatic Mixed Carbonate and Clastic Sequence Showing Lithologies and Sequence-stratigraphic Elements*

During early highstand, thick aggradational fluvial and deltaic plain facies—coal, mudstones, splays, and stacked point bars—as well as stacked deltaic and coastal sandstones and mudstones may be accommodated to produce sigmoidal geometries. Sedimentation is commonly balanced with moderate rates of relative sea-level rise. As rates of relative rise diminish, accommodation space also decreases resulting in accelerated progradation and deposition of oblique geometries.

When highstand coastlines prograde across submerged shelves to the relict shelf edge, progradational rates generally decrease because of the increased volume of basinal space to fill these shelf-edge deltas, and coastal systems contribute some sediment to the slope and basin by the suspended load but principally by oversteepening, slumping, and dispersed gravity flows. Hummocky to chaotic reflection configurations typify highstand deltaic progradation.

Major highstand systems may mobilize subjacent salt and slope muds into mud or salt ridges and diapers, and they may also initiate listic growth faulting.

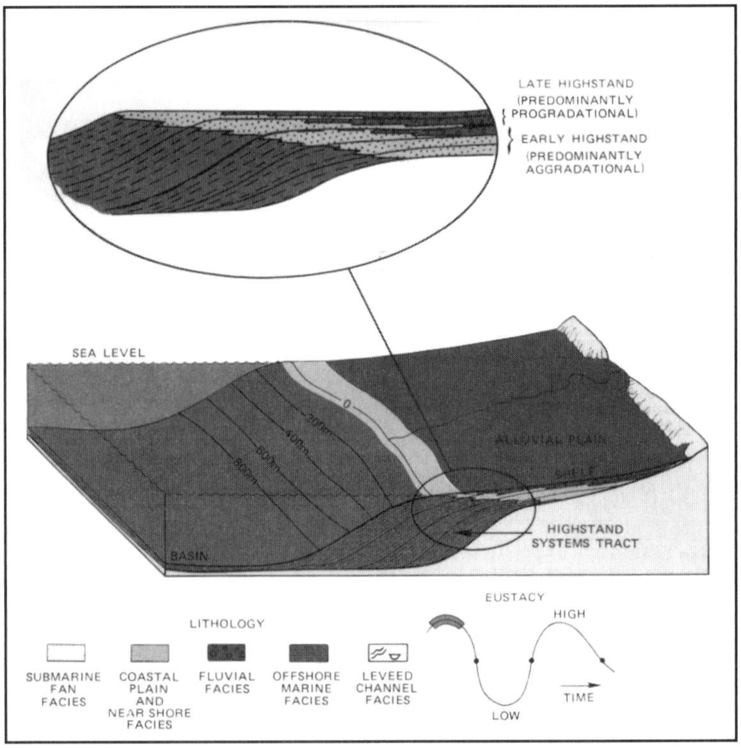

Fig. 7–110 *Highstand Systems Tract*

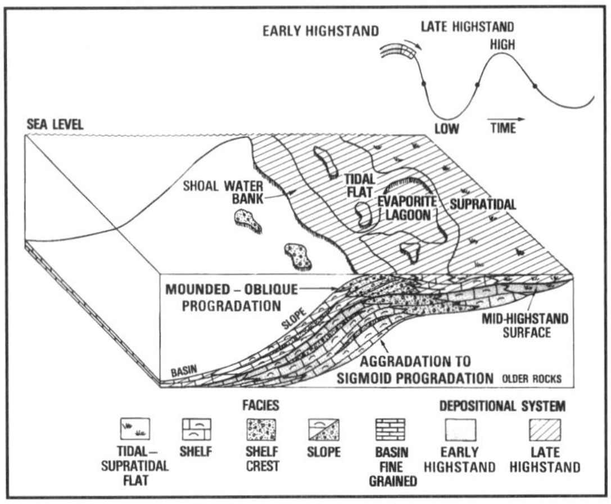

Fig. 7–111 *Carbonate Highstand Deposition*

Carbonate highstand tracts (Fig. 7–111) may exhibit widespread early aggradational (sigmoidal) and later mounded to progradational (oblique) deposition on platforms, ramps, or banks (Sarg 1988). The systems tracts are terminated by downlap onto marine-condensed sections (or hardgound/burrowed surfaces), and they are either Type 1 or 2 erosion and exposure.

During early moderate rates of relative sea level rise, increased rates of added accommodation space result in deeper, more restricted marine conditions on the shelves with lower oxygen content, lower temperature, poor nutrients, and/or high salinity. The result is *catch-up*

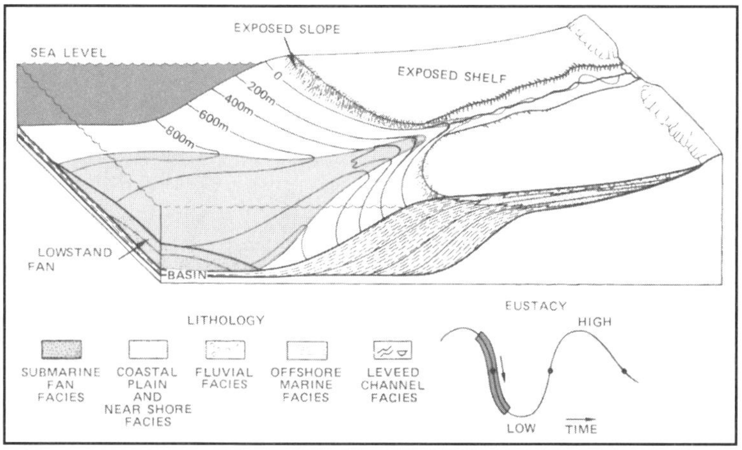

Fig. 7–112 *Lowstand System Tract Basin Floor Fan and Siliciclastic Lowstand Systems Tract during Basin Floor Fan Deposition*

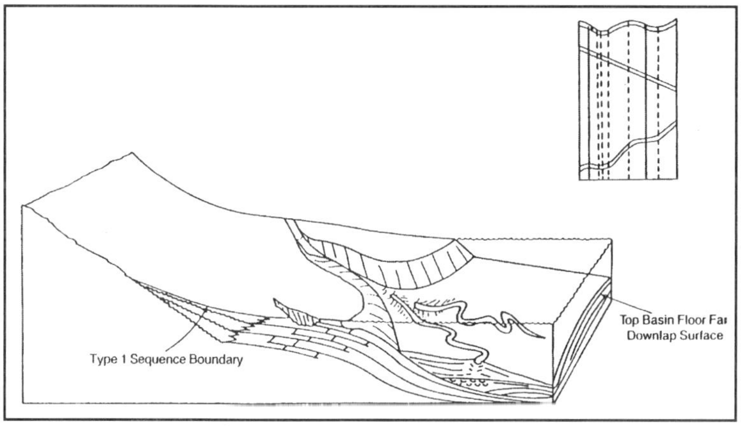

Fig. 7–113 *Lowstand Systems Tract Slope Fan and Siliciclastic Lowstand Systems Tract during Slope Fan Deposition*

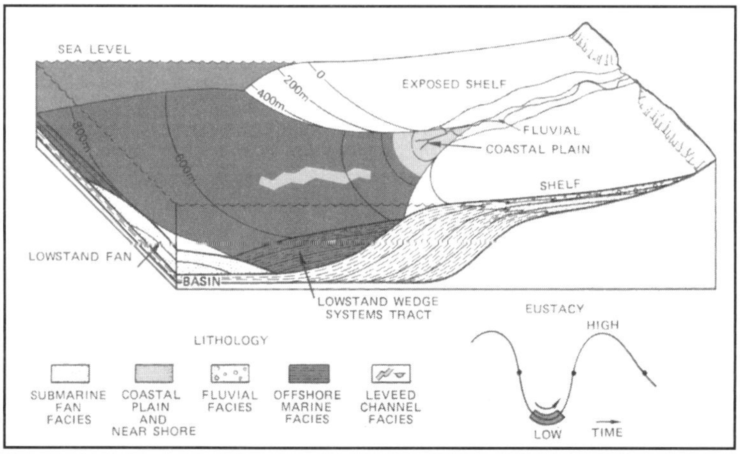

Fig. 7–114 *Lowstand Systems Tract Prograding Wedge and Siliciclastic Lowstand Systems Tract during Prograding Wedge Deposition*

carbonate deposition (Kendall and Schlager 1981) characterized by slower sediment production and depositional rates, deeper water and lower energy micritic textures, and significant early submarine cementation during extended exposure to marine waters.

Later, as relative sea level rise diminishes, there is generally increased circulation, nutrient production. oxygenation, and higher temperatures on the shelves, leading to *keep-up carbonate deposition*. This phase is characterized by rapid sediment production and deposition, shallow environments, grain-rich/mud-poor sediments, and limited marine cementation.

In a siliciclastic basin where a sharp shelf/slope break occurs, a siliciclastic *lowstand system tract* (LST) is ideally composed of four depositional elements:

1. basin floor submarine fan
2. levee-slope fan
3. prograding deltaic/coastal or carbonate wedge
4. incised valley-fill

The basin-floor fan (bf) (Fig. 7–112) is deposited during rapidly falling relative sea level when incised or entrenched river systems contribute significant volumes of coarse clastics to the lowstand coastline. These are redeposited on the basin floor by slumping and gravity flow. As the equilibrium point moves toward and eventually landward of the bayline, relative sea level fall slows and begins to rise, a prograding and onlapping/downlapping levee slope fan (sf) (Fig. 7–113) may be deposited basinward of the lowstand river mouth.

With an adequate sediment supply during these early slow paracyclic rises, a series of progradational parasequence sets are deposited at or near the mouth of the lowstand river to produce a downlapping and coastal-onlapping prograding deltaic/coastal wedge (pw) (Fig. 7–114).

As sea level rise continues to increase, the prograding wedge begins to evolve from progradational to aggradational parasequence sets, which pinch out landward onto the Type 1 surface because of increasing accommodation space. During early sea level rise, fluvial and estuarine sediments may fill incised valleys (iv).

On a gently sloping ramp (Fig. 7–115b) where no sharp shelf/slope break exists and accommodation rates are low, a thin wedge and siliciclastic lowstand tract may be deposited which consists of perhaps too thin coastal onlapping and deltaic or coastal ramp wedges that may exhibit progradational parasequence sets. Little if any submarine erosion and gravity-flow deposition occurs during such a scenario, but incised valley fill (ivf) may exist up-dip of the lowstand coastline if the ramp slope is steeper than the fluvial profile.

Along a growth-faulted shelf/slope break (Fig. 7–115c), siliciclastic lowstand basin- floor fan and slope fan (sf) sediments may be deposited on the downthrown side of the fault and very broad incised valley fill may occur landward on the upthrown side of the fault. Lowstand prograding wedges within the faulted sub-basins are difficult to document, perhaps because contemporaneous fault subsidence, large volumes of sediments, and high accommodation rates precluded well developed progradation but promoted deposition of aggradational parasequence sets.

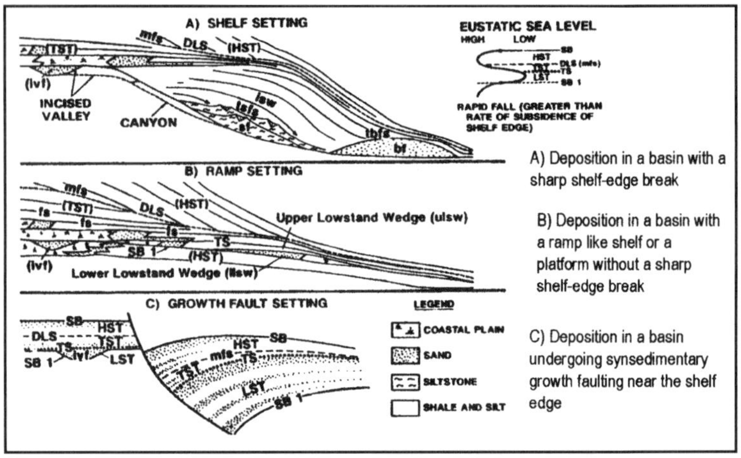

Fig. 7–115 *Different Settings for Lowstand Tract Deposition*

Siliciclastic lowstand tracts may be well developed in cratonic basins where tectonic stability enhances the eustatic cycles, such as glacio-eustatic carboniferous cycles (Fig. 7–116). A Type 1 (lowstand) fall of relative sea level in a *carbonate basin* exposes the previously deposited highstand carbonate tract to meteoric waters, resulting in lenses of fresh water that move across the shelf with falling sea level (Sarg 1988). Dissolution and various processes of meteoric diagenesis result that may enhance the porosity and permeability of highstand facies.

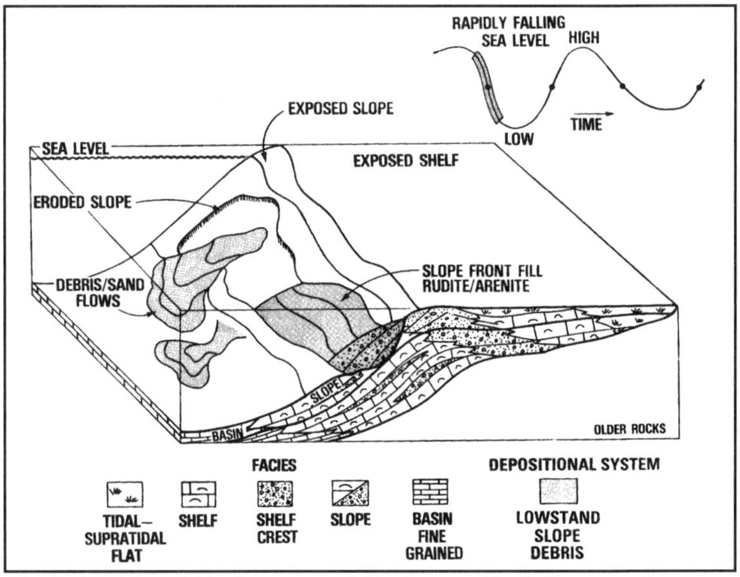

Fig. 7–116 *Type 1 Carbonate and Carbonate Early Lowstand Systems Tract*

Along the outer margins of carbonate platforms, slope-front erosion and deposition produce seaward erosion, canyon cutting, and gravity-transported debris flows that are deposited in slope-front basins and along the toe of the slope. When the equilibrium point approaches and moves landward of the bayline, relative sea level begins to slowly rise and platform margin or ramp-margin lowstand carbonate wedges typically develop on the upper slope. The wedges onlap the exposed shelf and their geometries indicate how sedimentation accommodates the new space added by rising sea level. For example, in restricted basins, biogenic productivity is lower and, therefore, finer grained (catch-up) sedimentation (Kendall and Schlager 1981) does not keep up with new space added and the wedges exhibit sigmoidal (aggradational parasequence sets) geometries.

In open basins. the biogenic productivity is normally higher and, therefore, higher energy and coarser grained (keep-up) deposition efficiently exceeds new space added by relative sea level rise. Consequently, these lowstand wedges typically exhibit mounded and oblique progradational parasequence geometries. If the basin experiences normal marine circulation, then the oblique progradational slopes will be gentler than the steeper, restricted-basin sigmoidal margins. If a basin is highly restricted and the climate is arid, lowstand onlapping evaporite wedges will be deposited as dominant aggradational parasequence sets within the systems tract.

In cratonic basins with mixed siliciclastic and carbonate sedimentation, lowstand system tracts grade along the basin margin from siliciclastic at entrenched fluvial systems to carbonate tracts where siliciclastic sediments were absent (Figs. 7–117 and 7–118). During subsequent accelerated rise of relative sea level, a transgressive ravinement (TS) or first flooding surface is

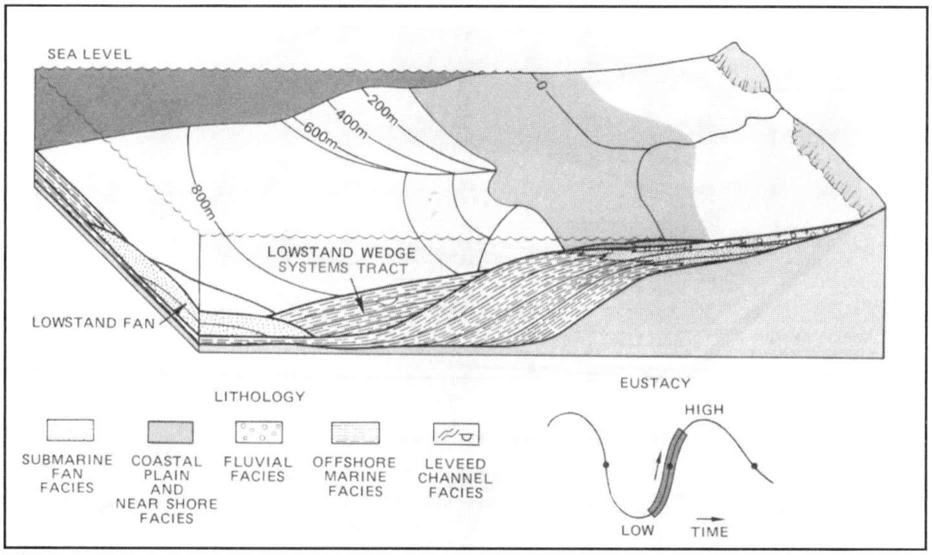

Fig. 7–117 *Transgressive Systems Tract and Siliciclastic Transgressive System Tracts*

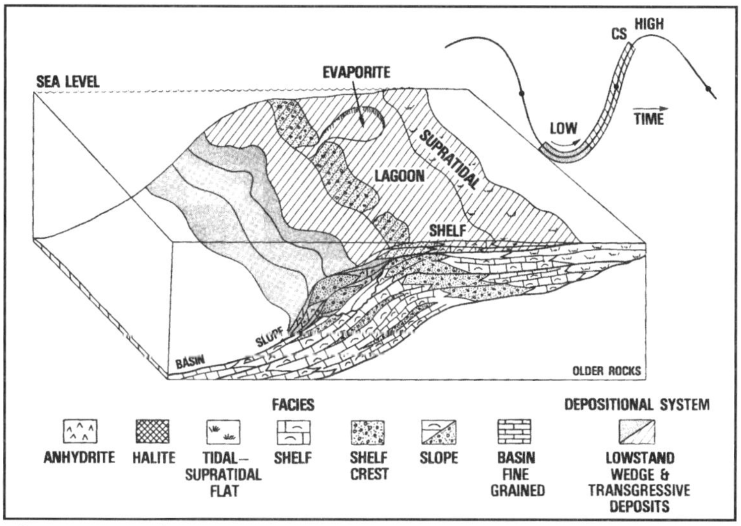

Fig. 7–118 *Carbonate Late Lowstand and Transgressive System Tracts*

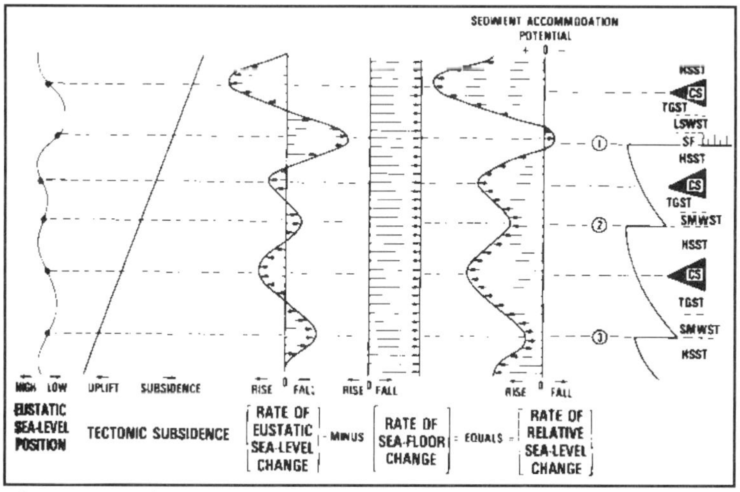

Fig. 7–119 *Sediments Accommodations Potential and Its Relationship to the Marine-Condensed Sediments*

commonly eroded into the top of lowstand wedges. These are overlain by the transgressive systems tract and superposed marine-condensed section. The transgressive surface coalesces up-dip with the subjacent type unconformity.

Transgressive or retrogradational systems tracts (TST) are tracts that coastally onlap transgressive or first marine flooding surfaces on top of lowstand or shelf-margin tracts (Vail 1987) (Fig. 7–119). These are typically local, diachronous ravinement surfaces produced by marine erosion. Up-dip these onlapping surfaces coalesce landward and erosionally enhance the subjacent Type 1 or 2 surfaces. Retrogradational deposition occurs in response to a generally diminishing fluvial sediment input under the influence of accelerating but periodic paracyclic rises and still stands of relative sea level produced by combined subsidence and rising eustatic sea level.

As the equilibrium point moves more rapidly landward, rates of relative sea level rise increases. This adds new accommodation space. The transgressive tracts are composed of progradational parasequences that comprise progressively upward thinning and accelerating landward-shifting retrogradational parasequence sets deposited under periodically deepening water, retreating shorelines, and diminishing sediment supply. The tracts may be composed of siliciclastic or carbonate/evaporite depositional systems.

Siliciclastic tracts depend on a significant sediment supply to offset the paracyclic rises of relative sea level that add new accommodation space (Posamentier and Vail 1988). With adequate fluvial sediment input, fluvial/deltaic and associated coastal systems will prograde the coastlines before being periodically flooded. With limited fluvial sediment supply or along an inter-deltaic/nondeltaic coastline, however, this system tract may be very thin, consisting of extensive coal, drowned and intensively eroded coastal shoreface sand bodies, and extensively reworked transgressive sand sheets.

Carbonate tracts (Fig. 7–119) form under appropriate climatic conditions in the absence of siliciclastic sediment supply (Sarg, 1988). With adequate marine circulation, progradational, keep-up, shallow-water, shoaling- upward, retrogradational, parasequence sets will be deposited

(Sarg, 1988). If marine circulation is restricted, however, carbonate systems may exhibit catch-up deposition composed of lower energy carbonate mudstones that evolve into keep-up deposition near the top of the tract only when relative sea level rise slows sufficiently to permit shallow-water deposition.

Under excessively arid climates, this systems tract may be composed of evaporite systems that exhibit onlapping and periodically progradational parasequences. In regions with climatic conditions conducive to carbonate and/or evaporite deposition, but which were supplied with excessive siliciclastic sediments, the transgressive tract may be composed of carbonate and evaporation systems, even though lowstand and highstand tracts are dominantly siliciclastic. As the paleo-shelf becomes flooded, this lowstand bank is drowned, and a retrogradational transgressive system is deposited over the shelf. Offshore areas become starved, and a condensed section is deposited. The subsequent highstand systems tract downlaps over these underlying systems tracts.

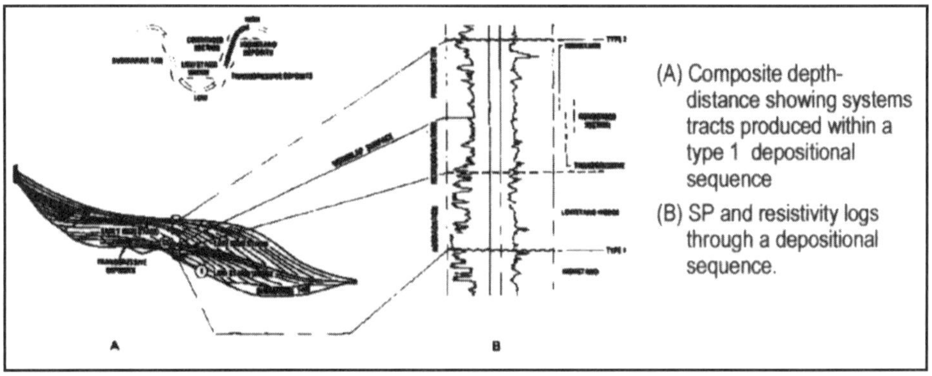

Fig. 7–120 *The Stratigraphic Relationship of Marine–condensed Sections to Others Depositional Systems Tracts*

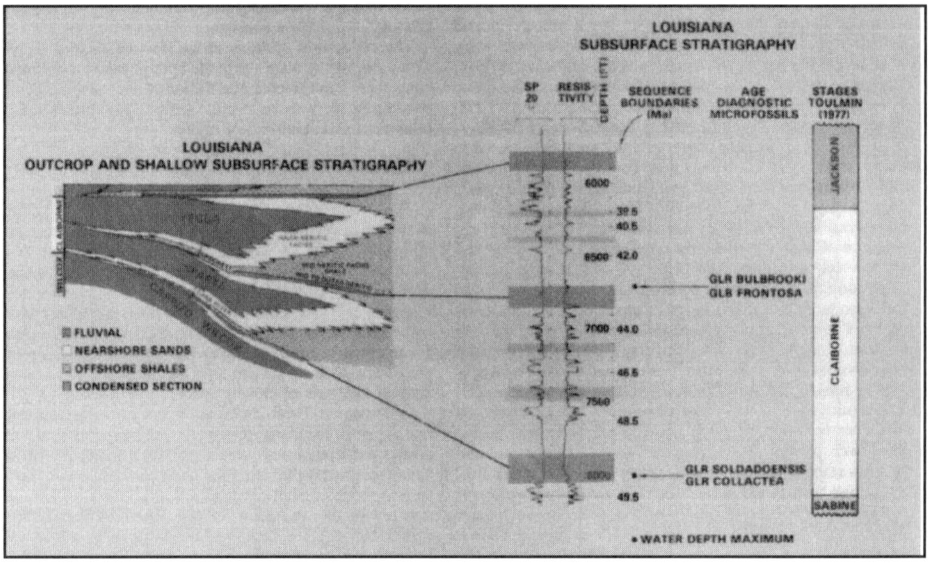

Fig. 7–121 *Marine–condensed Sections and Their Relationship with the Stratigraphic Succession*

Marine-condensed sections. These widespread but thin hemipelagic deposits (Figs. 7–120 through 7–123) signal sediment starvation (Loutit et al. 1988). When rising relative sea level approaches its maximum rate under the influence of accelerating eustatic rise (approaching the R-inflection point) and subsidence, the transgressive tracts are progressively encroached by an extensive blanket of hemipelagic sediments that extends onto the shelves from deep in the basin. Marine-condensed sections mark significance hiatuses in the sequences when the equilibrium point approaches its landward position precluding significant sedimentation.

Siliciclastic transgressive tracts are normally overlain by black phosphatic, glauconitic high-gamma/low resistivity shale comprising the marine-condensed sections, but carbonate tracts (Sarg 1988) may typically be overlain by black, phosphatic, micritic high-gamma limestone and shale or many times by hardground and/or burrowed surfaces.

The marine-condensed sections contain rich paleofaunas and floras, which may grade upward with increasing water depths from shallow-water benthonic assemblages to deep-water pelagic assemblages, dominated under appropriate climates and during geologic times by radiolaria, planktonic foraminifers, and nanofossils. The condensed section, therefore, records changes in water depths associated with late rise and early fall of relative sea level (Fig. 7–121).

The condensed section is recognized more easily in regions shown as a shaly interval between a fining or deepening upward (retrogradational) stacking pattern and a shoaling or coarsening upward (progradational) unit, as in Figure 7–122. The section occurs at or before the time of maximum water depth within each depositional sequence. It fines or deepens upward (retrogradational) in a stacking pattern and a shoaling or coarsening upward (progradational) unit. Downlap surfaces occur at

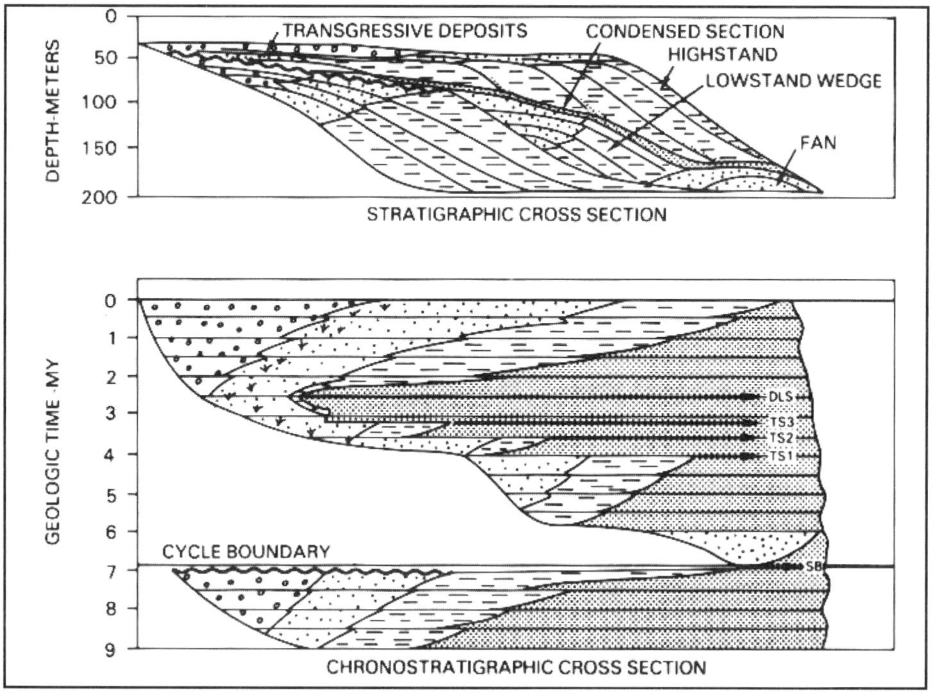

Fig. 7–122 *Depth and Time Sections Showing the Marine–condensed Sections within the Sequence Frame Work*

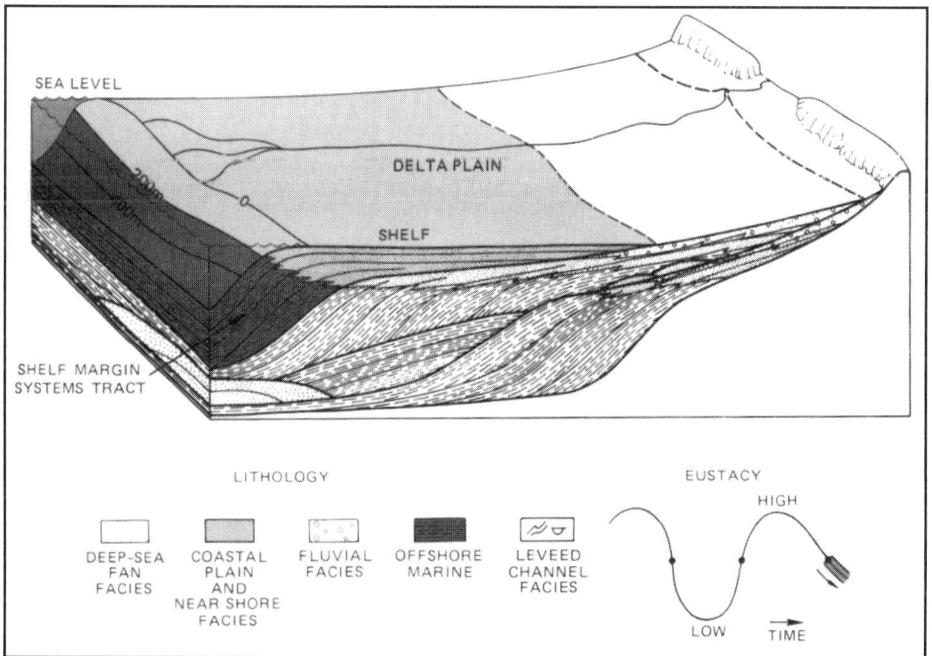

Fig. 7–123 *Siliciclastic Lithofacies and Siliciclastic Shelf-margin Systems Tract*

a point in the section where the parasequence stacking pattern changes from retrogradation to progradation. Downlap surface occurs below downlapping clinoforms of prograding highstand systems tract.

The condensed sections provide the means for excellent paleontological dating and correlation of the highly fossiliferous deep-basinal successions with less fossiliferous marginal marine and intercalated continental facies (Fig. 7–123). Because the clinothem of downlapping retrogradational parasequences terminate progressively landward, they appear to be truncated beneath the marine-condensed section, an effect called apparent truncation. Within the condensed section (near its middle) is the maximum marine flooding surface, and at the top of the condensed section is the downlap surface on which the progradational strata of the subsequent highstand tract terminate.

The marine-condensed section and its contained maximum flooding surface are widespread and distinctive and may be recognized from paleontology, composition, well-log patterns, and seismic reflectively. Low acoustic impedance typically results in marine-condensed sections occurring within troughs between very high amplitude peak reflections. This doublet reflection may be traced landward from basin centers well across shelves separating highstand from subjacent transgressive tracts.

In up-dip areas where transgressive facies are thin, the condensed section or its up-dip equivalent may rest upon or be separated by only a few meters from the underlying Type 1 or 2 unconformities (Fig. 7–123).

In deep basin areas, the condensed sections may be stacked, separated by only meters of distal highstand/lowstand shale. Consequently, the marine condensed section may be recognized even where the bounding unconformities cannot, making it a critical element of the depositional sequence. Because of these attributes, the top of the marine-condensed section (called the downlap surface, but not precisely the maximum flooding surface) has been used by many for boundaries of lithogenetic units

(Fisher and McGowen, 1967; Brown, 1967; Brown and Fisher, 1977; and Galloway, 1988). Using this surface for sequence boundaries, neither distinguishes between the significant differences exhibited by regressive highstand and lowstand tracts. This results in assuming that sea level only rises and never falls.

Shelf-margin systems tracts (SMST) are tracts that are deposited upon Type 2 unconformities or equivalent concordant surfaces when the equilibrium point and bayline move basinward together (Figs. 7–124 through 7–126). However, relative sea level does not fall below the depositional shoreline and break to an equilibrium point landward of the shelf or ramp edge (Vail, 1987). Because of the slow fall, limited erosion and no significant deep-water sedimentation occur, but some meteoric diagenesis may develop in the landward parts of the subaerially exposed shelf or platform. When the equilibrium point and bayline move together landward, the shelf-margin systems tracts coastally onlap the unconformity and the accelerating rise of relative sea level results in early deposition of progradational parasequence sets changing upward into well developed aggradational parasequence sets. This results in sigmoidal geometries.

Type 2 tracts have much greater distribution than the more areally restricted lowstand tracts. Thick shelf-margin wedges may fail and move into the basin by growth faulting or gravity creep.

Siliciclastic tracts (Fig. 7–124) are commonly composed of progradational and highly aggradational fluvial/deltaic and coastal systems, commonly displaying well-developed coal measures (Posamentier and Vail, 1988). If an appropriate climate exists, the siliciclastic tract may grade away from deltaic areas into carbonate or evaporate tracts.

Carbonate tracts (Fig. 7–124) typically consist of platform/bank margin wedges deposited initially at or slightly below the preexisting highstand platform margin (Sarg 1988). In well-circulated basins, biogenic production typically maintains shallow, higher energy deposition, while in restricted basins and deeper water deposition lags and is finer-grained.

Depending upon climate and restriction, evaporite deposition may occur within the shelf-margin carbonate wedge systems. With sufficient basinal restriction and arid climates, shelf-margin onlapping evaporite wedges (Fig. 7–125) exhibiting well-developed aggradational parasequence sets may

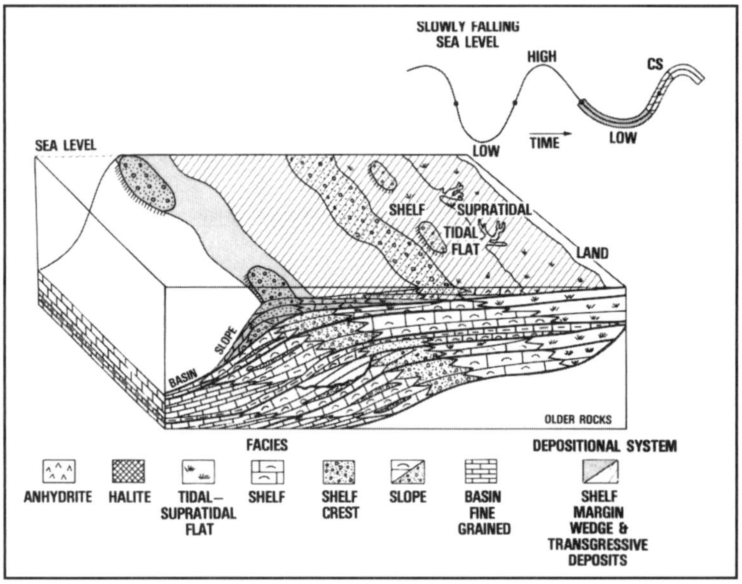

Fig. 7–124 *Type 2 Carbonate Sequence Diagram Showing a Slow Fall of the Sea Level Interpreted as a Type 2 Sequence*

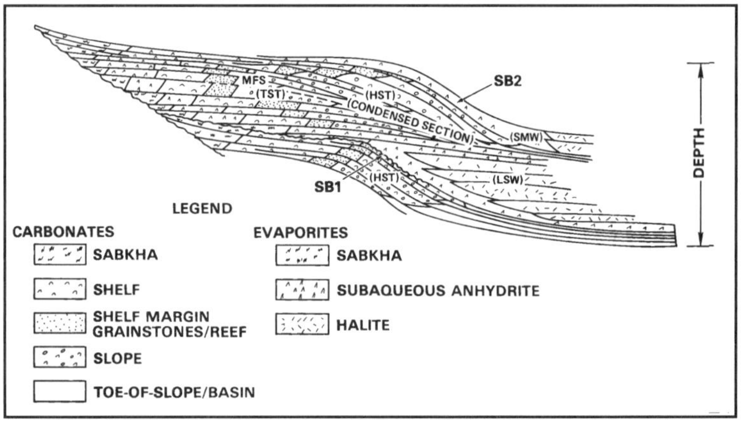

Fig. 7–125 *Sequence Stratigraphy Depositional Model Showing Carbonate and Evaporite Lithofacies, Distribution of Carbonate and Evaporite Lithofacies within the Deposition Sequence Framework*

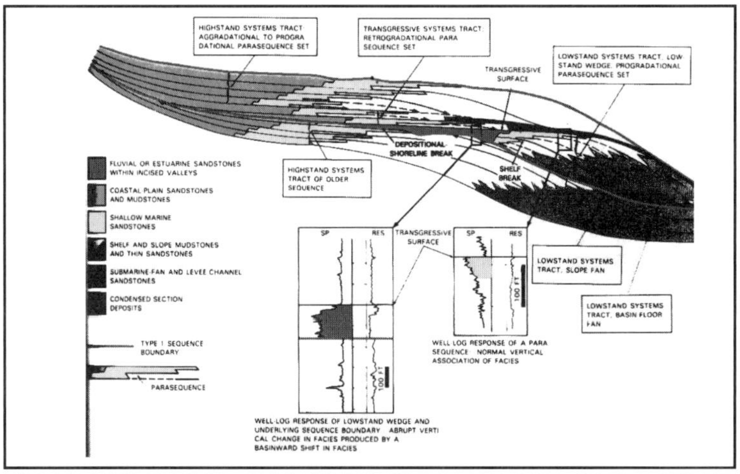

Fig. 7–126 *Stratal Pattern in Type 1 Sequence—Stratal Pattern in Type 1 Sequence Deposited in the Basin with Shelf Break*

dominate the systems tract. Shelf-margin tracts are overlain by transgressive ravinement surfaces and superposed transgressive tracts and marine-condensed sections. The transgressive or first flooding surface coalesces up-dip with the subjacent Type 2 surface.

Exxon's idealized siliciclastic systems-tract model (Fig. 7–123) applies where a major highstand fluvial/deltaic system was entrenched during falling relative sea contributing sediments to lowstand depositional systems. Variable sediment supply, different accommodation rates, and fluvial versus long-shore sediment transport may significantly modify this model (Loutit et al., 1988). System tracts at sites of minor coastal-plain streams or along inter-deltaic and nondeltaic margins are not well developed in Exxon's scenario. Siliciclastic systems may grade laterally into carbonate and evaporite tracts. Also, an increase in subsidence rates along a basin margin can cause Type1 sequences to grade laterally into Type 2 tracts.

A worldwide survey documents a spectrum of systems tracts along ancient margins. In the absence of rivers, basin-floor sediments are supplied locally by headward erosion or cannibalism into relict highstand shelf and/or strandline systems by lowstand contributory valleys and submarine canyons. Submarine erosion continues during subsequent rise and highstand of sea level, and some basinal sediments may be introduced through canyons from active retrogradational and highstand long-shore systems.

Headwardly eroded valleys and canyons were rarely filled during subsequent transgression and highstand, leading to long-term, multiple erosion of the world's major ancient canyons. The nature and distribution of highstand system tracts strongly influences the type and distribution of subsequent lowstand tracts and, therefore, may help guide deep-water exploration along ancient basin margins.

Similarly, variations within carbonate and evaporite basins also exert significant influence on the lateral distribution of systems tracts. These system tracts are perhaps more responsive to tectonic variations than are siliciclastic systems. Tectonic movements may modify marine circulation, imposing important changes in the nature of depositional systems. Submarine canyons eroded into carbonate platforms are typically controlled faults or negative structural axes.

Pleistocene/Holocene models. The concepts of cyclic erosion and deposition can be observed within marine Pleistocene and late Tertiary depositional sequences throughout the world. Unfilled estuaries and incised valleys along depositional coastal plains document our present relative sea-level position. Starved shelves, which compose about 70% of the extant continental margins of the world (Emery, 1968), clearly indicate that during this sea-level cycle, transgressive deposition is essentially over and marine-condensed deposition is extensive (Loutit et al., 1988).

Cores and seismic data on our shelves and coastal zone define the filled incised valleys and proximal submarine canyons that were entrenched into previous highstand systems. These had protruded across the shelf just before sea level fell below the continental shelf edge (Frazier, 1974). Seismic profiles document the occurrence of lowstand wedges below relict shelf edges. The approach of another highstand episode is being signaled by the filling of bays and estuaries and the initial progradation of a few highstand deltas onto shelves. Soon, geologically speaking, sea level will again fall to initiate another cycle.

The global impact of late Cenozoic relative sea-level cycles, driven by combined glacio-eustatic cycles and tectonics, provides an excellent model for the study of depositional sequences, systems tracts, and unconformities. Glacio-eustatic impact on various types of passive-margin and active-margin basins should eventually provide the evidence needed to understand complex ancient assemblages. Consequently, considerable oceanographic, geophysical, sedimentologic, and biologic research is now focused on refining the concepts of sequential stratigraphy and evaluating the inferred global chronology proposed by Exxon scientists.

Several international and national petroleum companies as well as international study groups are actively involved in applying sequence-stratigraphic ideas and methods in their exploration programs. With the onset of increasing offshore stratigraphic exploration, these methods are being evaluated to determine whether they improve the search for subtle traps.

Sequential stratigraphy can be reasonably well understood and applied without a final solution to the questions involving global synchronicity and the mechanics of eustatic sea-level cycles. Although depositional sequences are recognized throughout the Phanerozoic stratigraphic succession, the linkage of each sequence with glacial episodes remains to be documented. Vail believes that, eventually, glaciations may be documented throughout the geologic record. Of course, factors other than climate, such as abrupt plate-tectonic adjustments, may have affected eustatic sea-level cycles.

A model that adequately explains causes and the synchronicity of sea-level cycles observed worldwide will not satisfy everyone for a long time (*Science*, 1988). Nevertheless, the evidence grows that cyclic sequences produced by periodic variations in relative (and probably eustatic) sea level pervade the global stratigraphic record.

Others, however, continue to argue that lithogenetic sequences do not require cyclic sea level and, hence, if pre-Pleistocene lowstand systems tracts do exist, they are not necessarily common (Galloway, 1989). Some argue that lowstand tracts are actually deep-water, highstand systems and that erosion observed on the shelves and slopes occurs during highstand. Evidence from many lowstand tracts clearly documents that coastlines fell below shelf edges. Petroleum explorationists should not permit debates about global eustatic sea levels to preclude the application of defensible sequence stratigraphic concepts and procedures that have the potential to improve hydrocarbon-play identification and prediction.

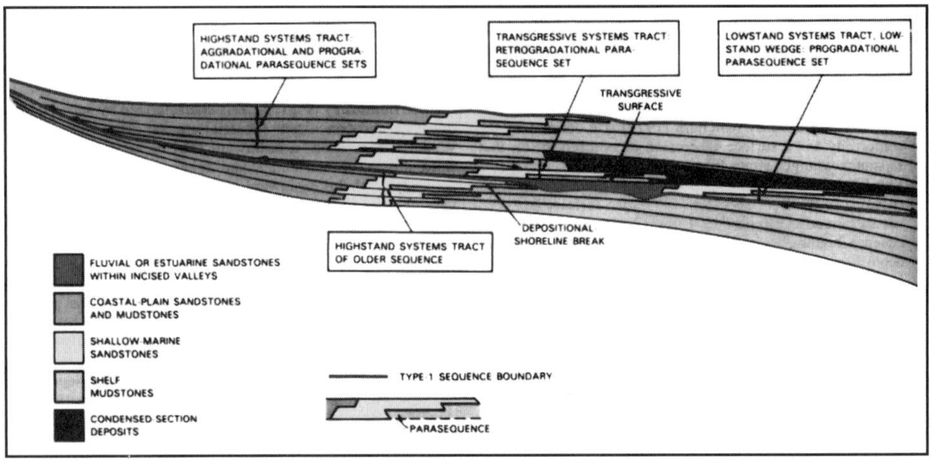

Fig. 7–127 *Stratal Pattern in Type 1 Sequence Deposited in a Basin with Ramp Margin*

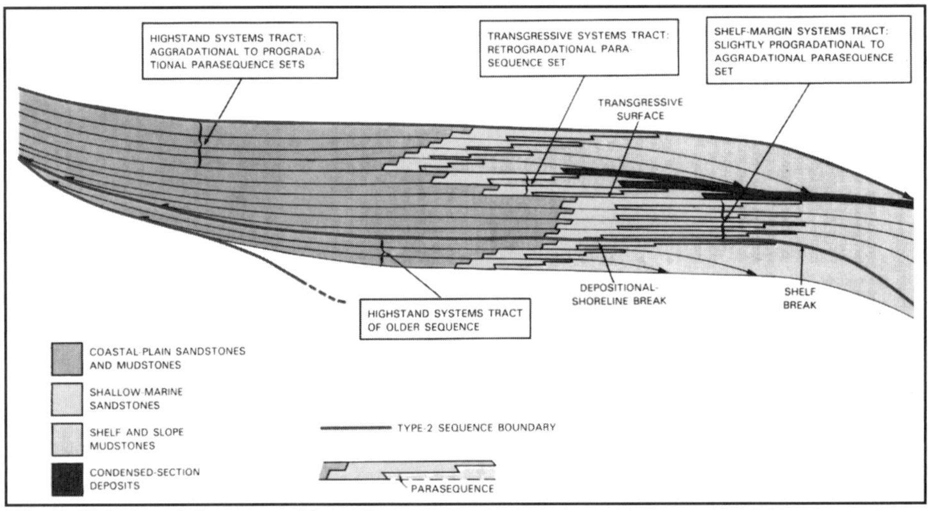

Fig. 7–128 *Type 2 Sequence Boundary*

As important and potentially useful as an acceptable global eustatic system may be, it unfortunately focuses criticism on the entire cyclic sequence concept. Undoubtedly, the debate about tectonics vs. eustacy will continue, and correctly so, but relative sea-level changes have occurred and their recognition offer new directions in stratigraphic research and economic application. The divisive issue about global eustacy will be settled when sufficient data are available to permit researcher to reach a consensus.

In summary, depositional sequences are initiated by rapid relative falls of sea level below the depositional shoreline break (Type 1 sequence) or slow falls that remain above depositional shoreline break (Type 2 sequence). Type 1 sequences may fall to a lowstand position below relict shelf edges (Fig. 7–126) or to a lowstand position on a ramplike shelf/platform (Fig. 7–127). Each depositional sequence is composed ideally of three systems tracts, each of which is characterized in part by distinctive parasequence stacking patterns (van Wagoner and others, 1987).

Type 1 sequences are composed of the following:

1. low-stand system tracts, which typically exhibit early oblique and later sigmoidal progradational stacking patterns
2. transgressive system tracts, which display early aggradational to later strongly retrogradational stacking patterns
3. highstand system tracts that exhibit early sigmoidal and later oblique progradational stacking patterns

Type 2 sequences (Fig. 7–128) are similar except that they contain a basal shelf-margin system tract that displays early progradational and later aggradational stacking patterns. Siliciclastic, carbonate/evaporite or mixed depositional systems may compose the system tracts, but the stratal stacking patterns help to identify the relative sea-level change, sediment supply variation, and accommodation rates which occurred during deposition.

Depositional system tracts and hydrocarbon plays. The term *hydrocarbon play* is defined as an association of geologically similar reservoirs exhibiting the same source, seal(s), and trap characteristics (White, 1980). Plays are delineated primarily according to depositional setting, origin of reservoirs, or less commonly their relationship to regional erosional surfaces or diagenetic facies. The obvious corollary of this definition, written by an Exxon development specialist, is that for the explorationist, the concept also provides a highly genetic logical approach to exploration especially in relatively unexplored basins. Its exploration application is based on determining the fundamental association of elements—reservoir, trap, seal(s) and source—that can collectively constitute a hydrocarbon entrapment and then devising methods for recognizing the association or play. In most basins, the types of data permit some combination of depositional systems, seismic-stratigraphic, and sequence stratigraphic analyses to search for these associations. The concept permits extrapolation of the critical association of play elements once a depositional systems/sequence framework has been constructed. Consequently, extrapolation of plays based on the occurrence of the key play elements provides the basis for establishing a play trend, sometimes called a *play fairway.*

The fairway is a geologically delineated area where the three-dimensional play elements are inferred to occur. They constitute prospective areas inaccessible until after further drilling and analysis are factors that may affect the quality and quantity of the plays—limiting diagenesis, reservoir volumes/quality, adequate seal(s), and migration pathways. If these factors are favorable, the plays may provide an idea of the prospectively of the basin.

Here the hydrocarbon play potential of the various system tracts and their component depositional elements will be considered in terms of reservoir, trap, seal(s), and source by integrating sequence and depositional systems concepts. Reservoirs are component facies within depositional systems that have primary and/or secondary-enhanced porosity and permeability resulting from respective depositional processes and/or diagenetic processes influenced by relative sea-level cycles.

Traps to be described and discussed are those in which:

1. primary stratigraphic pinch-outs related to depositional processes—lateral facies changes, lapouts on to discontinuities, and vertical flooding surfaces—and/or porosity/permeability caused by diagenetic barriers—dissolution, karstification, and cementation—related to subaerial exposure or deep-marine diagenesis, both influenced by relative sea-level cycles
2. non-tectonic structural closures—growth faults, compactional drape—resulting from syndepositional
3. early post-depositional processes—top-seals include flooding surfaces; transgressive, marine-condensed, and downlapping shale; toplapping subaerial mudstones; and diagenetic barriers

Base seal is generally provided by impervious discontinuity surfaces or subjacent shales. Source beds are principally marine-condensed sections, distal highstand mudstones, and delta plain organic (peat/coal) facies that underwent appropriate burial/thermal/temporal maturity and have available pathways for vertical or lateral migration to the reservoir. Principal information on hydrocarbon occurrences within the various systems tracts come from Vail and Sangree (1988), Sangree and others (1988), and Vail and Sangree (1988).

Lowstand system tracts probably contain the greatest potential for stratigraphic hydrocarbon plays because of their several distinctive depositional elements associated with Type 1 erosion, localized lowstand deposition, common juxtaposition to seals and source beds, deep burial, and, in some basins, the occurrence of contemporaneous growth faults near shelf breaks.

At least five potential siliciclastic reservoirs (Vail, 1987) may exist within the tract:

1. basin-floor fans
2. slope leveed-channel fills
3. prograding turbidite wedges
4. prograding/aggrading coastal sands
5. incised valley fills (Fig. 7–129)

Two potential reservoirs may exist in carbonate successions:

1. allochthonous debris-fan wedges
2. autochthonous wedges and secondary porosity and permeability generated in preexisting highstand facies duringsubaerial exposure

Basin-floor fans (Figs. 7–129 through 7–133) are deposited during falling relative sea level. They are supplied by entrenched rivers whose mouths have shifted below the preexisting shelf edge to erode submarine canyons and funnel gravity flows onto the basin floor. Fans may be simple detached lobes, or they may be composed of multiple lobes comprising extensive stratigraphic units. The higher gradient streams carry coarser bed loads making the fans potentially excellent reservoir quality. Fans may exhibit considerable lateral communication among complexly arranged lobes, but some lobes may be isolated.

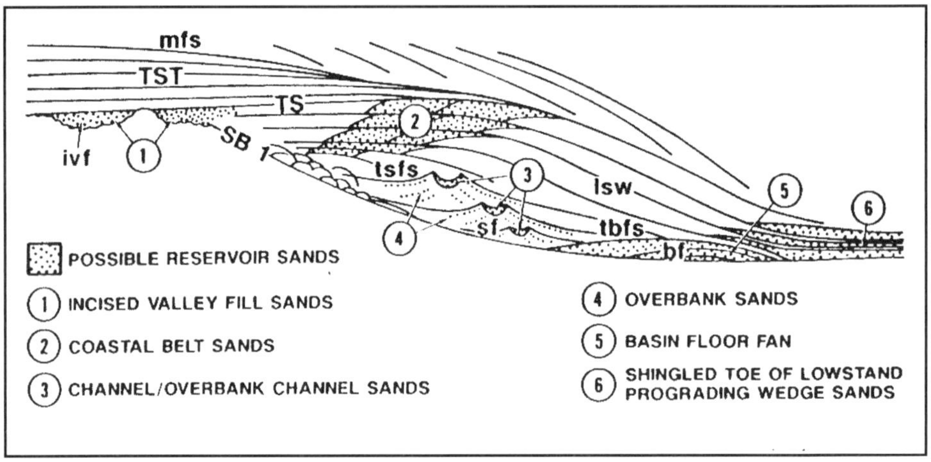

Fig. 7–129 *Possible Reservoir Quality Sand in Siliciclastic Sequence*

Traps are typically lateral stratigraphic sand pinch-outs, but some tectonic or compactional closure commonly enhances closure. Basinward pinch-outs of detached fans onto basin-floor highs provide ideal stratigraphic conditions. Where growth faulting and/or salt mobilization occurred, the isolated fans may possess rollover or fault closure within the downthrown blocks or within slope sub-basins, respectively (Fig. 7–132). Seals are normally excellent, composed of overlying distal prograding wedge muds and/or marine-condensed pelagic shales and distal highstand mudstones.

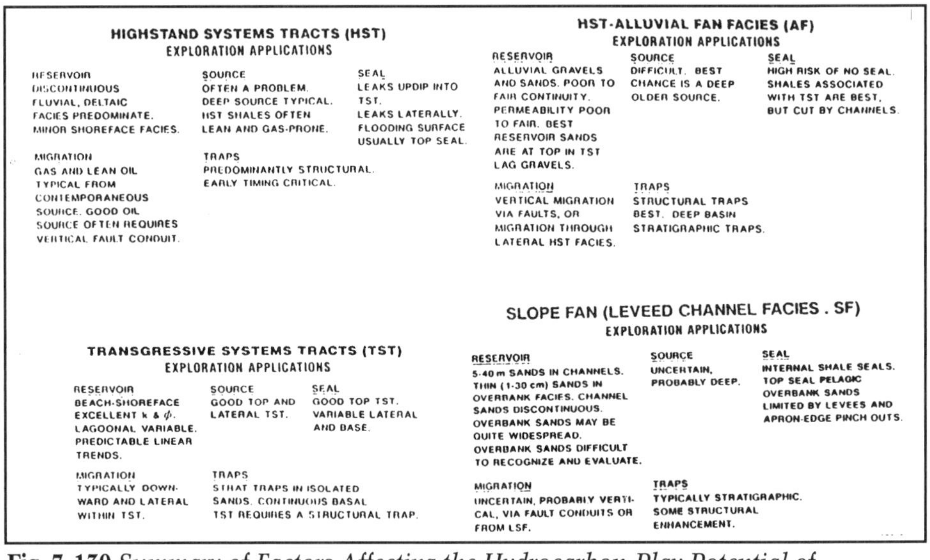

Fig. 7–130 *Summary of Factors Affecting the Hydrocarbon-Play Potential of Siliciclastic Deposition Systems Tracts*

System tracts in the Gulf Coast Basin. The fans lie upon Type 1 surfaces. (Fig. 7–133) Overlying slope fans may preclude good top seals by leaking upward into lowstand shallow coastal and fluvial systems. *Source rocks* are normally underlying or lateral marine-condensed sections, although distal lowstand wedge and highstand shales may also be sources. *Migration* may be vertical from underlying condensed sections, but some lateral migration into the fans is possible. In some carbonate basins, some siliciclastic submarine fans may be derived from headward erosion of canyons into relict siliciclastic tidal-flat facies. Stratigraphic traps, source, and seals are provided by deep-water marls and micritic mudstones.

Levee-channel fills (Figs. 7–130 and 7–131) are deposited as relative sea level fall diminishes and channel mouths supply sufficient sediment to permit the establishment of a depositional slope. Slumping from the incipient deltas and from canyon walls is

PROGRADING COMPLEX (PGC)
EXPLORATION APPLICATIONS

RESERVOIR
VARIABLE: STACKED FLUVIAL, DELTAIC AND SHOREFACE. VARIABLE CONTINUITY.

SOURCE
DEEPER BEDS, OR TST SOURCE AT TOP.

SEAL
GOOD TST TOP SEAL. LATERAL SEAL MAY BE POOR.

MIGRATION
PROBABLY DEPENDS ON FAULT CONDUITS FROM DEEPER SOURCE. POSSIBLE DOWNWARD MIGRATION FROM TST.

TRAPS
TYPICALLY STRUCTURAL. POSSIBLE COMPACTION CLOSURE.

BASIN FLOOR LOWSTAND FAN (BF)
EXPLORATION APPLICATIONS

RESERVOIR
TYPICALLY EXCELLENT k & ϕ. CONTINUITY VARIABLE, OFTEN A PROBLEM IN UPPER CHANNELIZED LOBES.

SOURCE
LEAKAGE FROM DEEPER BEDS. POSSIBLE TOP & LATERAL CONDENSED SECTION (C.S.) SHALES.

SEAL
EXCELLENT, PELAGIC SHALES OF C.S. RISK NO SEAL IF OVERLAIN BY SLOPE FAN.

MIGRATION
VERTICAL FROM DEEPER SOURCE. POSSIBLE DOWNWARD AND LATERAL FROM C.S. SHALES.

TRAPS
TYPICALLY STRATIGRAPHIC.

SUBMARINE CANYON FILL (SCF)
EXPLORATION APPLICATIONS

RESERVOIR
VERY VARIABLE. SUBMARINE CHANNEL SANDS, TURBIDITES. POOR CONTINUITY.

SOURCE
UNCERTAIN. CONTEMPORANEOUS SOURCE IS PROBABLY GAS PRONE.

SEAL
LOCAL SHALE SEALS.

MIGRATION
UNCERTAIN. VERTICAL MIGRATION VIA FAULTS MAY BE BEST.

TRAPS
STRATIGRAPHIC PINCH-OUTS.

INCISED VALLEY FILL (IVF)
EXPLORATION APPLICATIONS

RESERVOIR
BRAIDED STREAM SANDS TYPICAL. GOOD TO FAIR CONTINUITY.

SOURCE
TOP SOURCE FROM TST. POSSIBLE DEEP SOURCES.

SEAL
TST SHALES. POOR LATERAL SEAL.

MIGRATION
DOWNWARD FROM TST. POSSIBLE VERTICAL MIGRATION VIA FAULTS.

TRAPS
TYPICALLY REQUIRES STRUCTURAL CLOSURE OR NOSE.

Fig. 7–131 *Summary of Factors Affecting the Hydrocarbon-play Potential of Siliciclastic Deposition Systems Tracts*

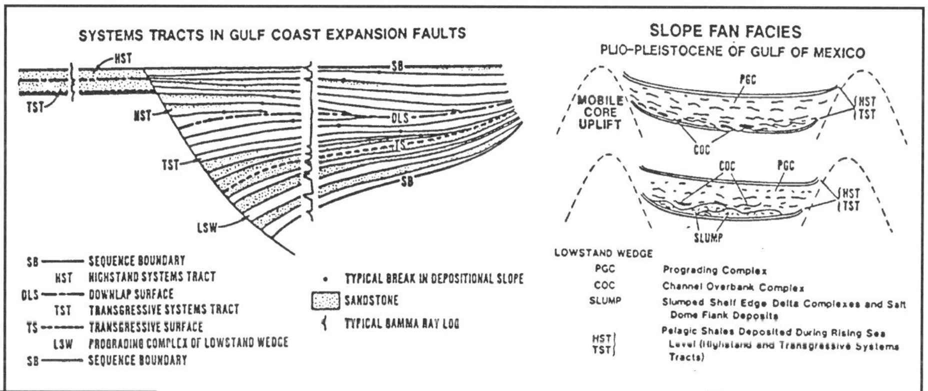

Fig. 7–132 *Systems Tracts in the Gulf Coast Basin*

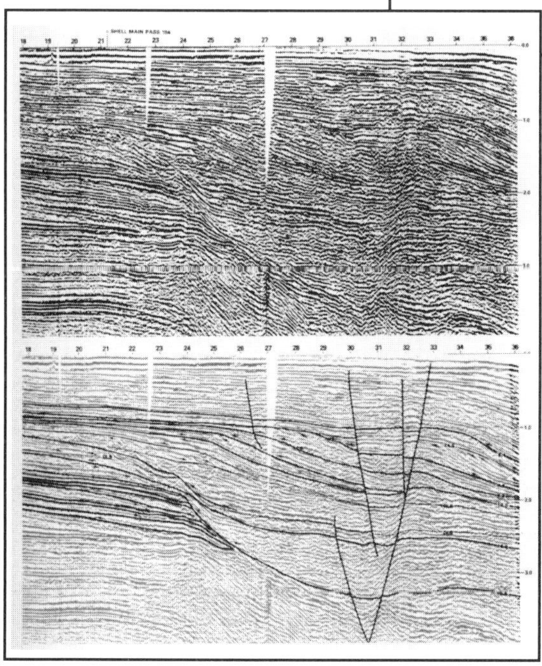

Fig. 7–133 *Systems Tracts within Depositional Sequences Deposited Basinward of USA Gulf Coast Contemporaneous Growth Faults*

common, and such submarine canyon facies are complex and uncertain targets. If a well-developed slope fan exists, then reservoirs, if present, are narrow, straight to meandering, discontinuous channel-fill gravity transported sands (Sangree and others, 1988).

Overbank and levee sheets flank the channels, which collectively compose a complex three-dimensional sand reservoir network that lies within the muddy slope fans. Traps are normally stratigraphic pinch-outs exhibited by channel-fill and levees, but some structural enhancement such as growth faulting is important to preclude up-dip leakage. Seals are mostly the mudstone matrix of the slope fan itself, but overlying marine-condensed sediments or distal prograding wedge mudstones may provide some top seal.

The slope fans rest on the Type I unconformity, typically within submarine canyons. Sources have to be inferred from deep-marine shales and condensed sections below the unconformity or lateral leakage from subjacent basin-floor fans. Migration is, therefore, vertical (along faults if present) or up-dip from basin-floor fans.

In carbonate successions, allochthonous debris flows deposited on the slopes in front of platforms or atolls may contain relict porosity and permeability derived from the eroded highstand shelf edge, and, therefore, constitute reservoirs (Sarg, 1988). Source, seals, and stratigraphic traps typically result from the envelopment by deep-water marls and micritic mudstones.

Prograding siliciclastic wedge complexes may contain fluvial/deltaic/coastal reservoirs, as well as potential prograding or off-lapping turbidite fans. The wedge begins to prograde into the basin below the preexisting shelf edge when relative sea level begins to rise slowly. Its development depends upon sufficient sediment to offset the rise. Extensive interdeltaic systems are rare.

Reservoirs are stacked barrier bar, delta front, and fluvial sands. The reservoirs display protruded and aggraded geometry and may be restricted to the proximal parts of the eroded shelf notch or canyon. Traps are typically structural, commonly with roll-over or fault closure on the downthrown sides of growth faults. Compaction of the sand depocenter may also provide closure. Up-dip stratigraphic Pinch-out is possible if the sands onlap impervious Type 1 unconformities. Top seals are superposed transgressive shales, but lateral up-dip seal is questionable because of leakage into incised valley-fills. The wedges pinch out laterally along strike into shales or onlap the Type I unconformity along the incised canyon/valley walls. Source may be from overlying transgressive or marine-condensed shales or from below via growth-fault systems. Migration is either upward along faults or downward from overlying shales.

At the base of prograding low-stand clinothems, off-lapping shingled turbidite reservoirs may intercalate with clinothem mudstones. These turbidite sheets are deposited during lowstand wedge progradation from sediments that slump from the shallow coastal/deltaic shorelines.

Reservoirs may be extensive, thin sheets of turbidite sands that pinch out basinward and up-dip.

Traps may be up-dip stratigraphic pinch-outs, although they may be connected with the coastal sands via slope channel-fills. Subsequent structural drape over basin-floor highs enhance entrapment.

Seals are muddy lowstand wedge clinothems and transgressive, marine condensed, or distal highstand mudstones.

Source is from overlying marine condensed and transgressive shales or vertically upward from below the Type 1 surface by vertical migration.

Autochthonous lowstand carbonate wedges (Sarg, 1988) may contain shallow-water, high energy keep-up grainstone and packstone reservoirs including some reefal buildups. Traps may be stratigraphic and possess marine-condensed top seal and source beds. During lowstand, subaerially exposed relict highstand facies typically produce secondary plays.

Incised valley fill (Figs. 7–130 and 7–132) provides an up-dip lowstand play that has been exploited extensively. Reservoirs are stacked braided to bed load fluvial sands and conglomerates deposited within entrenched valleys when relative sea level began to rise. Basal reservoirs exhibit considerable continuity but younger aggrading point bar sands may possess considerable lenticularity.

Traps are commonly subtle structural closures, but stratigraphic trapping is possible if there are up-dip pinch-outs against meandering valley walls or mudstone plugs among the discontinuous fluvial sand bodies. Top seals are estuarine and marine transgressive mudstones, but lateral seal must be provided by impervious valley walls (Type 1 unconformities). Up-dip mud-plug seals are critical for stratigraphic trapping.

Source is from overlying transgressive shales and marine-condensed sections or from similar relict sources beneath the entrenched valley. Long-range up-dip migration may occur from the basin via levee-channels and lowstand wedges. Lagoons and bays, which may form landward of the retrograding coastal systems, can provide up-dip pinch-out into mudstones. Subsequent reactivation of growth faults or later tectonic movement may provide structural traps for these systems.

In basins that experience *synsedimentary tectonics*—growth faults and/or salt mobilization—lowstand tracts may constitute significant plays, but their stratigraphic and areal distribution are strongly influenced by the tectonic activity. Therefore, these plays are more complex and resemble plays in more active tectonic basins such as rift, convergent, and strike-slip basins.

Transgressive system tracts. These system tracts, which are deposited during accelerating rise of relative sea level, may contain both siliciclastic and carbonate plays (Fig. 7–131). If well developed, the tracts display retrogradational geometry and onlap a series of diachronous transgressive surfaces. These, when up-dip of the lowstand or shelf-margin tracts, coincide with Type 1 or 2 unconformities.

In siliciclastic successions, the principal reservoir may include early aggradational and later progradational parasequences composed of fluvial, deltaic, and coastal systems. Ravinement erosion and reworking of early transgressive and relict highstand systems may provide some basal reservoirs composed of drowned and truncated barrier and coastal sands.

If sediment supplies are considerable, bed load streams and their coastal equivalent systems may stack to form major reservoirs. Because of retrogradation, reservoir quality diminishes progressively upward in the tract. Coastal onlap against the Type 1 or 2 unconformities may provide up-dip stratigraphic traps, but the discontinuities are rarely impervious in up-dip areas where they have eroded sandy highstand facies.

Top seals are usually excellent, composed of distal transgressive shales and superposed marine-condensed sections. Base seals depend upon the composition of sediments below the transgressive surface. Reservoirs of this tract may be regionally extensive along strike and, hence, typically require some lateral structural termination. *Source beds* are intercalated distal transgressive shales within the tracts and marine-condensed sections above the tract everywhere except perhaps in very proximal areas. Migration, therefore, is landward or up-dip.

In carbonate transgressive tracts, if marine circulation and other factors favor biogenic production in balance with rising relative sea level, thick retrogradational keep-up grainstones, oolites, and packstones with limited marine diagenesis may characterize the tract (Sarg, 1988). Because of typically slow subsidence rates in such basins, the reservoir quality may also be affected by many higher orders of eustatic cycles and paracycles that can induce short-term exposures and, hence, dissolution that enhances porosity and permeability.

Stratigraphic traps, top seals, and sources are typically *distal* transgressive and composed of marine-condensed micritic limestone and marls.

Highstand and shelf-margin system tracts. These two system tracts exhibit very similar depositional systems. Highstand systems (Fig. 7–133) are deposited during late decelerating relative sea level rise, and shelf-margin systems are deposited during early accelerating relative sea level rise following a slow relative fall of highstanding sea level that did not drop below the depositional shoreline break.

In siliciclastic basins, these progradational systems tracts contain major deltaic and widespread interdeltaic coastal systems that evolve from sigmoidal to oblique geometries (highstand tracts) or toward increasingly aggradational geometries (shelf-margin tracts).

According to Sangree and others, reservoirs in fluvial-dominated systems may be discontinuous point bar (*meanderbelt*), distributary channel fill, and delta front sands. In wave-dominated deltas and along interdeltaic coastlines, extensive barrier sand bodies can exhibit regional distribution. Early highstand and late shelf-margin reservoirs display maximum superposition and thickness.

Traps are typically structural, either tectonic or synsedimentary, and trapping must be early to preclude hydrocarbon discharge through the massive and complex fluvial plumbing systems typifying these tracts. Poor up-dip seals preclude large stratigraphic traps, but some local traps typify individual point bars, deltaic lobes, and local barrier segments.

Top-seals and lateral seals are generally poor, especially within wave-dominated deltaic and sandy interdeltaic systems, but within river-dominated systems, delta plain mudstone and muddy interdeltaic facies may provide adequate top seal and lateral seal, respectively. Flooding surfaces may constitute the only top seal, especially in up-dip areas. Erosion of Type I unconformities into highstand tracts may compromise top-seals and permit leakage upward/up-dip into transgressive facies.

Transgressive tracts and marine-condensed sections provide top-seal and superposed source for shelf-margin systems. Abundant gas-prone terrestrial organic source beds (delta plain and prodelta facies) exist in highstand and shelf-margin systems. Sourcing highstand reservoirs from older transgressive and marine-condensed sections via fault migration may occur, especially along growth-faulted margins.

In *carbonate basins*, highstand and shelf-margin system tracts have excellent primary hydrocarbon-play potential if the biogenic productivity and rising relative sea level are balanced, permitting extensive deposition of keep-up grainstones, oolites, and packstone facies (Sarg, 1988).

Classic reservoirs are thick shelf-edge buildups that may occur under optimum rates of sea level rise. During late highstand, hypersaline dolomitization associated with restricted evaporite deposition and meteoric dissolution associated with early failing sea level may continue during subsequent Type 1 falls of sea level to enhance porosity and permeability.

Traps may be structural, but stratigraphic traps may result from complex diagenesis of shelf and platform-edge facies and buried biogenic buildups. Deep-water micritic catch-up sediments, deep-basinal muds and marls, and marine-condensed sections are typical source beds. The marine-condensed section also may provide a top seal for both highstand and shelf-margin tracts.

Chronostratigraphy of depositional sequences. In 1977, Exxon geoscientists first presented documentation showing their concept that unconformity-bounded depositional sequences were produced by relative sea level cycles could be correlated globally. This interpretation strongly suggested that the relative cycles were the result of eustatic cycles superimposed on tectonic subsidence. They provided examples of global sequence correlation and proposed that glacio-eustatic or other unknown processes were responsible for the global sea level cycles.

Sea level cycles of first, second, and third orders were recognized. *Third-order* cycles were bounded by unconformities documented by downward shifts of coastal onlap recognized at the seismic data resolution. When the ages of unconformities in at least three different basins correlated, the unconformity was considered to be global, and, hence, probably the result of eustatic falls of sea level. As mentioned, the Exxon researchers later realized from model studies that downward shifts apparently coincided with maximum rates of eustatic sea level fall (F-inflection points).

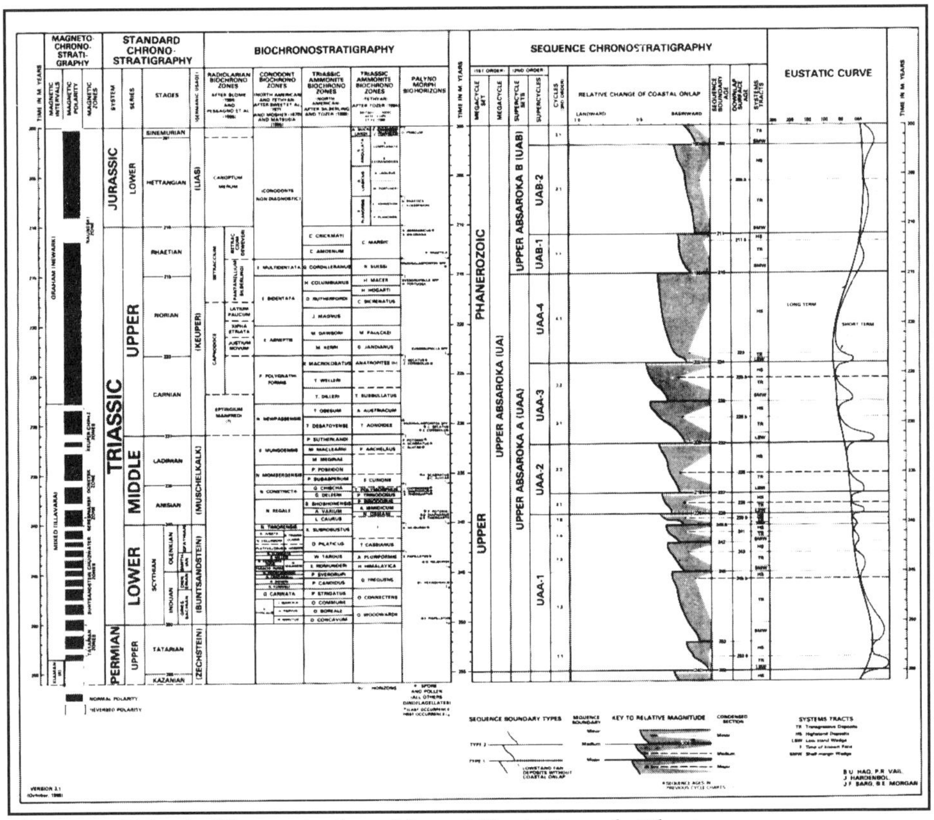

Fig. 7–134 *Triassic Chronostratigraphic and Eustatic-cycle Chart*

Between 1977 and 1987, integrated worldwide outcrop, well log, and seismic-stratigraphic studies further documented Mesozoic and Cenozoic sequences, and their ages were determined from combined biomagnetostratigraphic data. Ages of a variety of magnetic events, based on a broad set of paleontologic information, were further calibrated with available radiometric dates to provide a time framework with which to attempt correlation of principally third-order depositional sequences on a worldwide basis.

Results of this immense integration of information were first published by Haq and others in 1987. A revised version published by Haq and others in the 1988 SEPM *Special Publication 42*, included Triassic, Jurassic, Cretaceous, and Cenozoic cycle charts (Figs. 7–134 through 7–137). Several intensive critiques of the 1987 paper and a response by Haq and others were published in the July 29, 1988 issue of *Science*. Most of the critiques questioned various aspects of the precision of dating, especially the radiometric dates. Results of these reviews, however, did not seriously question the stratigraphic record of cyclic events but only the precision of their calibration with absolute dates.

The paper and charts by Haq and others provide an elegant integration of magnetostratigraphy, biochronostratigraphy, and sequence chronostratigraphy and their correlation with the standard chronostratigraphic systems and present absolute time framework.

Obviously, this is only the beginning of a complex and long-term process, but it provides a system with which to evaluate and analyze basins throughout the world. Some aspects of the system will change, and greater precision will come with time, but the study represents one of the major advancements in stratigraphy.

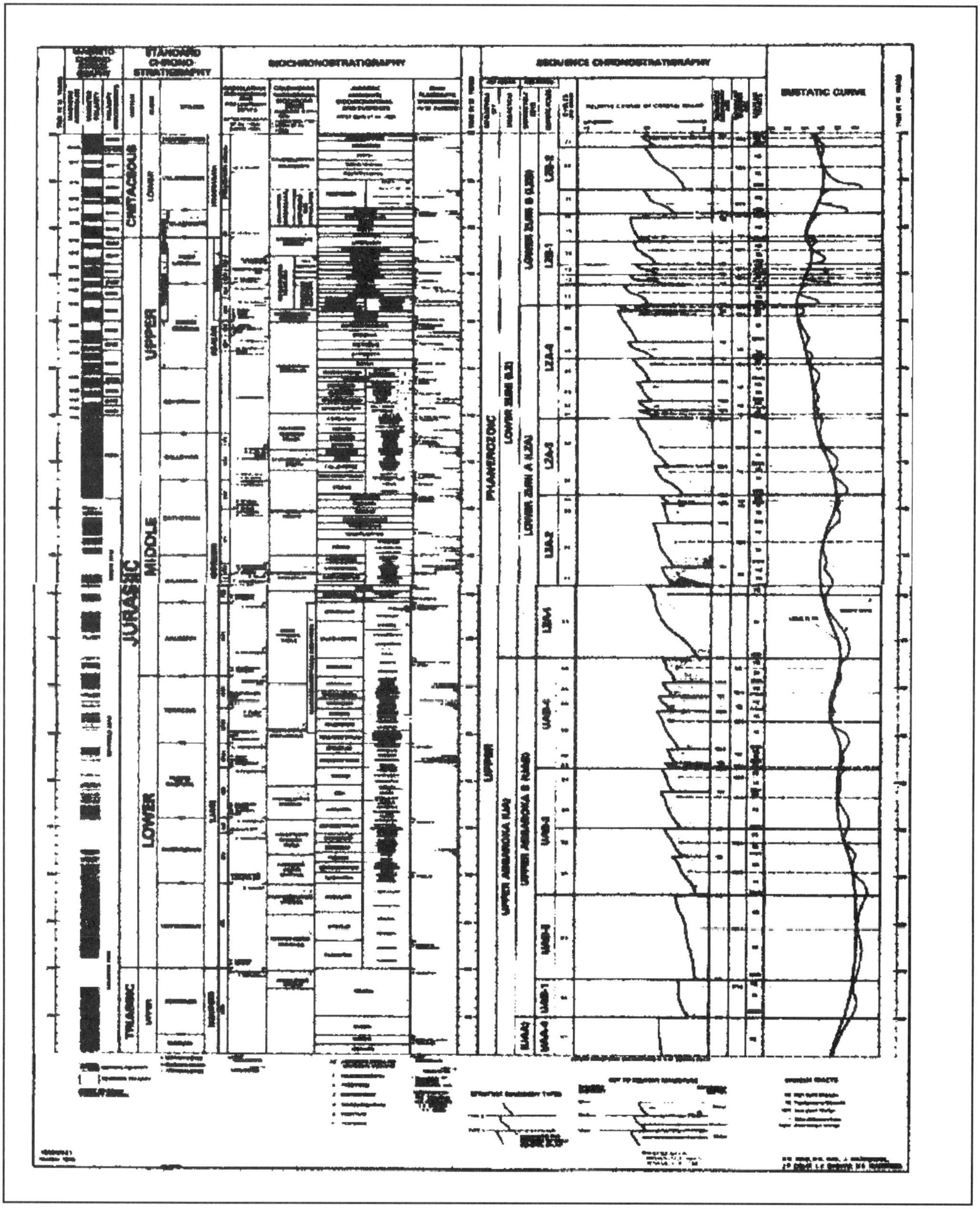

Fig. 7–135 *Jurassic Chronostratigraphic and Eustatic-cycle Chart*

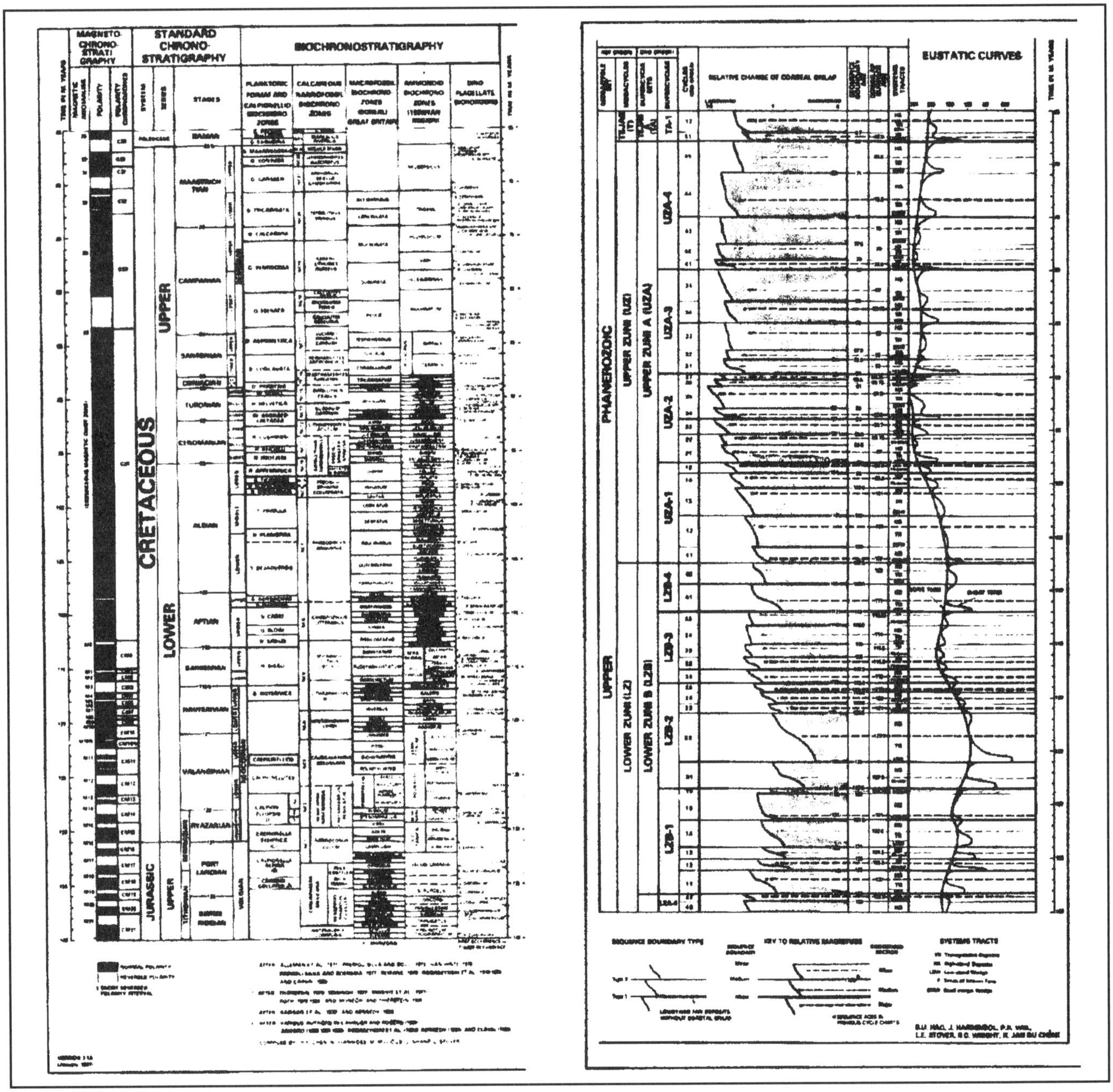

Fig. 7–136 *Cretaceous Chronostratigraphic and Eustatic-cycle Chart*

The sequence chronostratigraphic chart contains a global relative change of coastal onlap curve and best ages of unconformities and marine-condensed sections. System tracts are displayed, and a system of grading sequence boundaries as minor, medium, or major events summarizes the combined efforts of Exxon's staff for more than 20 years. The charts illustrate relative coastal encroachment and magnitudes of eustatic cycles. Anyone seriously interested in sequence stratigraphic analysis, especially in Mesozoic and Cenozoic rocks, should obtain a chart and study it carefully. If one views this global summary with objectivity, it offers the first tangible step toward a global stratigraphic synthesis that will eventually interrelate with global tectonics, paleoclimates, and paleooceanography.

Surface seismic data have been augmented with well log data to reinforce the concept of seismic sequence stratigraphy.

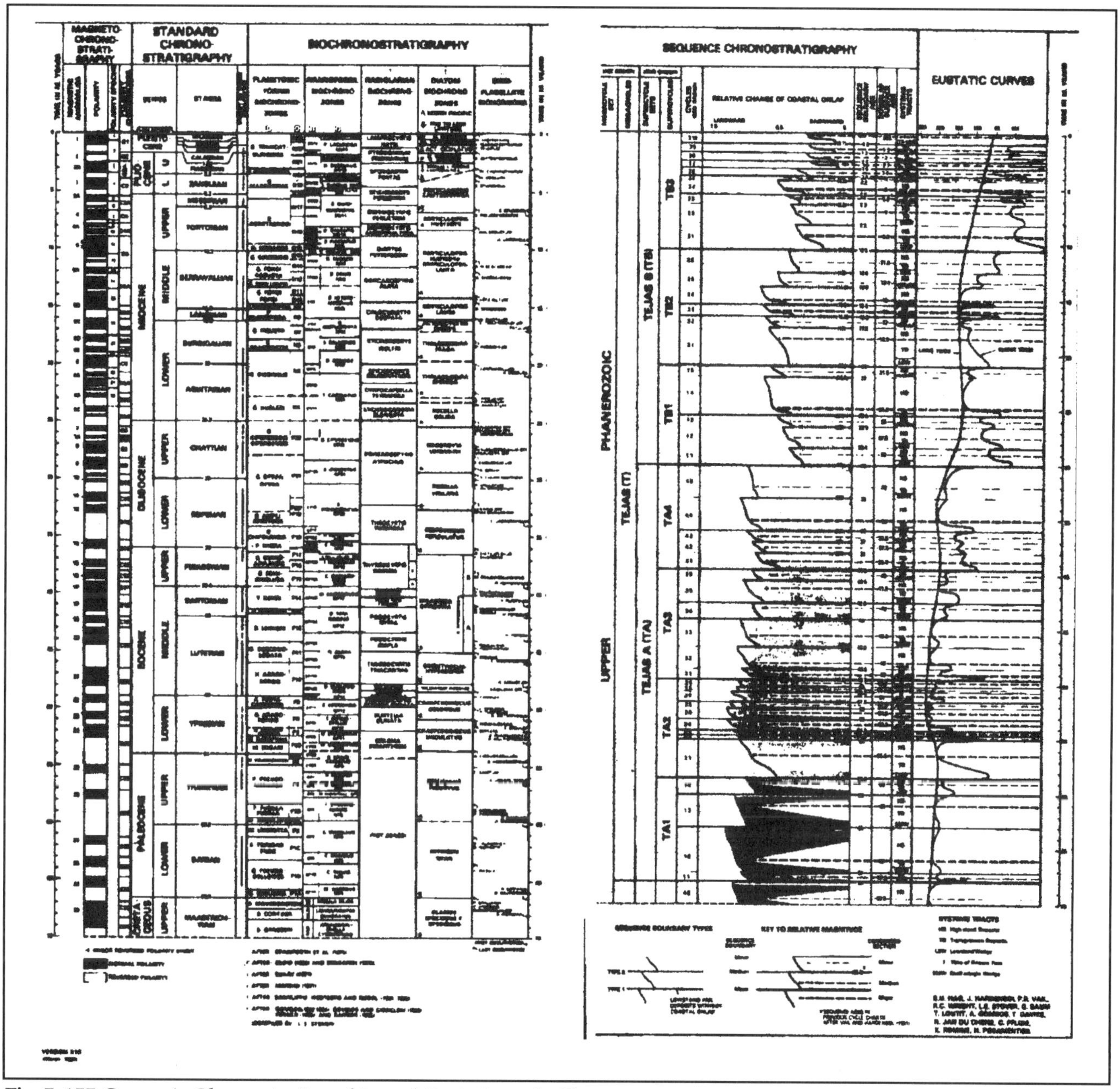

Fig. 7–137 *Cenozoic Chronostratigraphic and Eustatic-cycle Chart*

Modern Techniques and Future Applications of Sequence Stratigraphy

Large hydrocarbon accumulations have been discovered in easy-to-find structural traps. The majority of the stratigraphic traps were discovered as pleasant surprises while searching for the structural traps. However, there is no definite classification of hydrocarbon traps. Three common types are known—structural, stratigraphic, and a combination of the two.

The nature of stratigraphic traps and their depositional environments make it difficult to find and develop them. It is more difficult to conduct a secondary recovery project in a stratigraphic trap because it is subtle in nature. Sand deposits are irregular in pattern and are normally meandering. Carbonate build-ups are localized and their trends are sometimes difficult to recognize.

With the increasing demand for hydrocarbons, it became necessary to develop new tools and techniques in order to be able to image better the fine details in the subsurface. Many new techniques have been developed and implemented with great success; more are under development, and others are still in the research stage.

One fact that needs to be mentioned is that geophysical applications for all aspects of petroleum exploration, exploitation, and development are always moving forward in continuing efforts to optimize the methods we now use as well as research to develop completely new techniques.

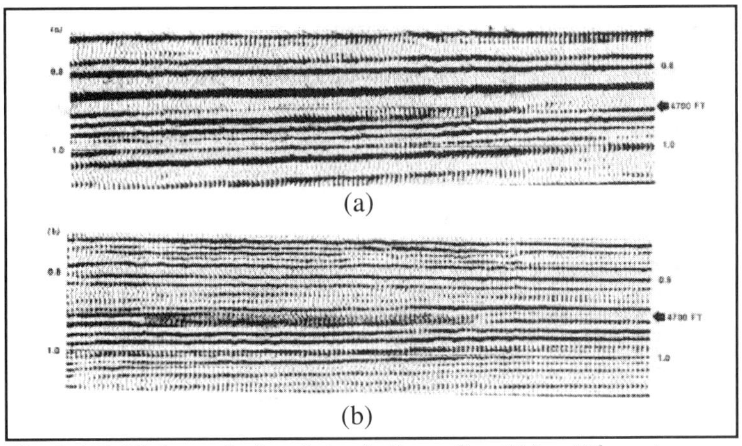

Fig. 7–138 *High-resolution Seismic Recording Using Vibroseis*

Some of the modern exploration techniques are:

1. high-resolution seismic data acquisition
2. vertical seismic profiling
3. amplitude versus offset
4. shear wave and compressional wave ratio
5. 4-D seismic technology

The basic theory and application of each of these techniques are discussed in the following pages.

High-resolution seismic data acquisition

The subtle nature of stratigraphic features makes new developments in data acquisition necessary to permit defining thin beds and limited, localized traps. Recording systems with as many as 10,000 data channels have been used to sample the surface at intervals of 50 feet or less. Sampling intervals of a fraction of a millisecond are used to record high frequencies, which, in the case of a Vibroseis energy source, are in the range of 10 to 180 Hz. In offshore data acquisition, frequencies between 700 and 800 Hz have been observed on the seismic record to a depth of approximately 1000 feet when using a sparker as the energy source.

To illustrate the advantage of using a high-frequency energy source for high-resolution data, Figure 7–138 compares a seismic line recorded with a 90–25 Hz sweep to the same line recorded with a 58–12 Hz sweep. The objective was to identify a sand channel at a depth of 4700 feet.

Figure 7–139 is a seismic section that has been converted from time to depth. The data were acquired with the explosive sources and the geophones in shallow (4 ft) water-filled holes to minimize the near-surface effect. Frequencies in the range of 60–500 Hz were recorded. The section reveals a subsidence fault, whose existence was confirmed by drilling.

A marine line shot with a sparker as the source is shown in Figure 7–140. Frequencies up to 700 Hz can be identified at a reflector depth of 1000 feet. Note the expanded scale of the display and the clearly defined faults in the upper part of the section, above 200 ms.

Vertical seismic profiling (VSP)

In recent years, geologists and engineers have used VSP to obtain valuable information that helped them in drilling and reservoir-engineering problems. Geophysicists, as well, have used VSP in their exploration, exploitation, and development efforts.

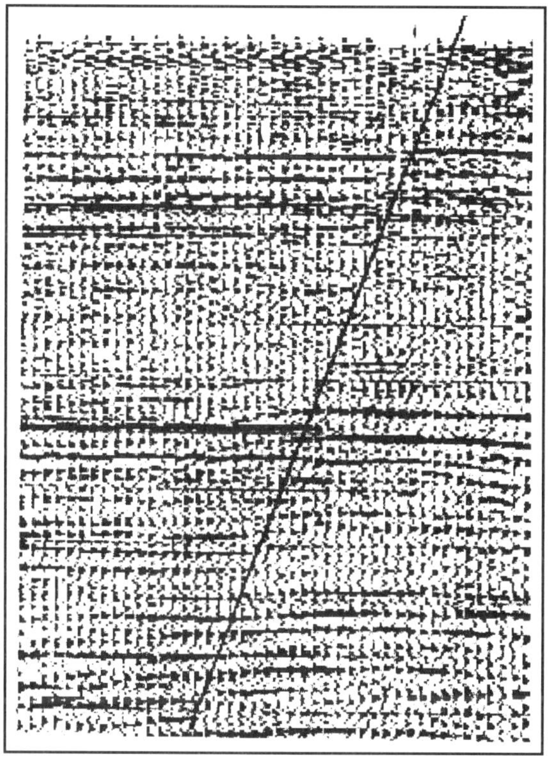

Fig. 7–139 *High-resolution Seismic Section*

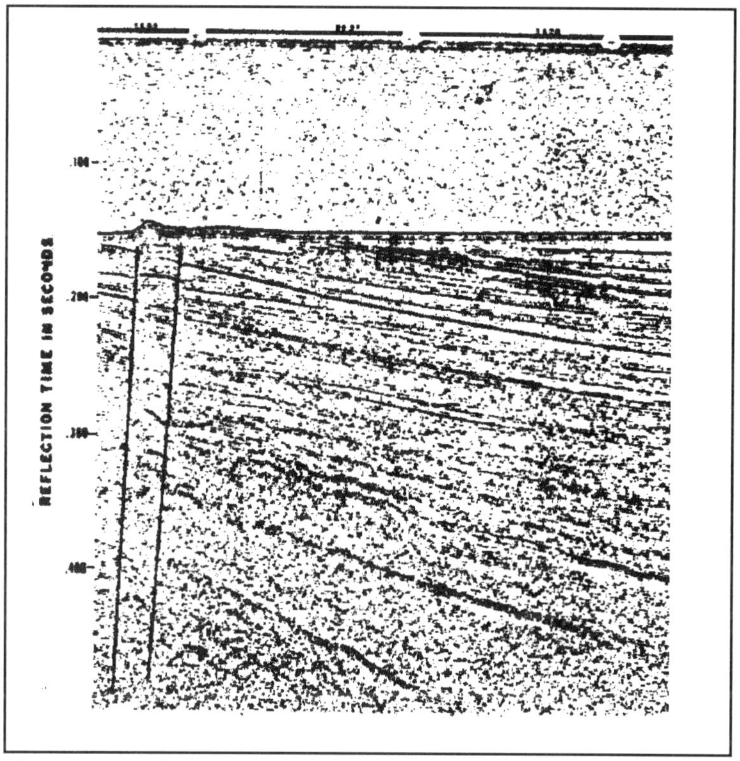

Fig. 7–140 *High-frequency Marine Seismic Section*

Long-offset VSP was implemented to delineate sand channels, and has been used successfully to exploit and develop sands of a meandering nature. Many publications have shown the effectiveness of the VSP method to image a salt dome flank, which surface seismic had failed to define due to the near-surface problem and to the steep dip of the flank that distorts the subsurface picture. Using VSP, the receiver element is located in the subsurface so the surface layers have no effect on data quality.

The concept of vertical seismic profiling. VSP is done by generating a seismic signal at the surface and recording it by geophones placed at various depths in a drilled well. A geophone is located in the subsurface when recording VSP data. It responds to both upgoing and downgoing seismic events, whereas only upgoing events can be recorded by a geophone planted at the surface. Figure 7-141 shows the Vertical Seismic Profiling concept.

The VSP is closely related to a velocity survey, since the source and receiver geometry is the same for both measurements. The major differences between VSP and velocity survey are the distance between geophone recording depths is much less for VSP (15–40 m). Velocity survey shot levels are separated by hundreds of meters. First break times are the critical information needed from down-hole signals in a velocity survey, but first breaks plus upgoing and downgoing events that follow the first breaks must be recorded in VSP surveys to be able to separate the upgoing from the downgoing events.

The basic components required to conduct a VSP survey are

- a borehole
- an energy source
- a down-hole geophone
- a recording system

Design of the equipment will not be discussed here, but there are lists of references in the bibliography that can be used to learn more about the equipment.

VSP Data Acquisition. There must be a borehole before a VSP can be run. Things that should be considered when selecting a hole include

1. borehole deviation
2. casing and cementing
3. borehole diameter
4. borehole obstructions
5. independently collected data on the borehole

Borehole deviation means that the hole is not drilled vertically. It is preferable to run a VSP survey in a *vertical* hole. It costs less to collect the data and the interpretation is easier. The position of the down-hole geophone relative to the energy source is uncertain in a deviated hole. The source may be moved to several different locations during the data acquisition phase, which complicates the situation. Therefore, the vertical depths at which primary and multiple reflections appear are doubtful in deviated holes. Offshore wells are often highly deviated and sound interpretation is sometimes difficult.

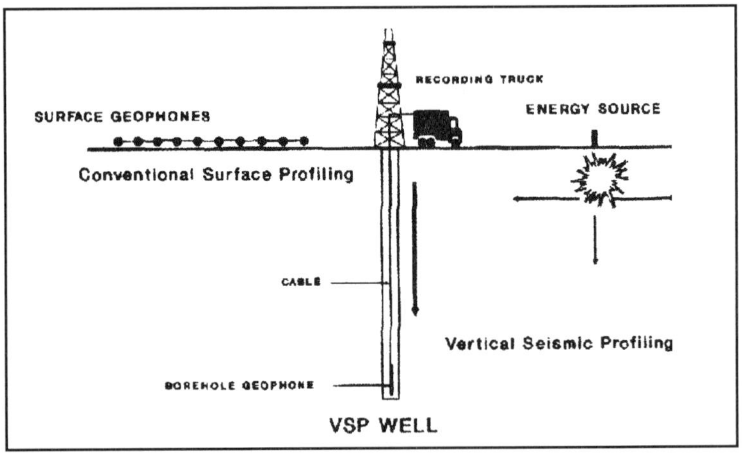

Fig. 7–141 *Vertical Seismic Profiling Concepts*

On the other hand, there are some advantages to recording a VSP in a deviated well, since this type of well allows the subsurface beneath the borehole be imaged laterally with great resolution. If the intent is to identify the depth and one-way time of primary reflectors, a vertical hole is the better choice. VSP can be recorded more quickly and easily, and the results are more accurate in a vertical hole.

The condition of the drilled hole—whether it is cased or not—has great effect on the quality of the data recorded. It is preferable to record the VSP in a cased hole, since the geophone is protected from sloughing and differential pressure sticking problems. The casing should be cemented to provide a medium to transmit seismic energy between the casing and the borehole. Cement has been found to be the best medium.

If the hole is uncased, hole *rugosity* will affect the clamping of the geophone to the formation. Large washouts are most troublesome because the geophone locking arm may be too short to reach the borehole wall, making clamping impossible.

Packers or trace rings that can prevent the borehole geophone from reaching the desired depth intervals may be present in a cased hole. Whether there are obstruc-tions must be determined before starting a VSP survey. Running a cheap, disposable tool of the same diameter as the geophone is one way to check the well for possible obstructions.

Using independent data that specifies the physical properties of the formations around the borehole can enhance reliability of VSP data interpretation. A borehole in which caliper, sonic, density, resistivity, and radioactive logs were recorded, in addition to having cores taken and drill cuttings preserved, is much preferred over a borehole for which no data are available. A cement bond log and measurements of the depths of all casing strings are required to be sure of the nature of the acoustic coupling between the VSP geophone and the formation.

The VSP energy source must produce a highly consistent shot wavelet time after time. Otherwise it will be too difficult to correlate equivalent features of upgoing and downgoing wavelets throughout the vertical section over which data is recorded preferred energy source wavelet. Energy sources used in VSP surveys follow.

- **Dynamite**. Dynamite charges loaded in drilled holes are excellent producers of seismic body wave energy and are widely used as the surface energy source for VSP. However, when shooting 30, 40, 50 or more shots, it becomes very difficult to maintain *repeatable wavelet shape*. Reasonable invariance in shot wavelet shape can be realized by exercising great care in the field.

- **Other impulsive sources.** There are many varieties of seismic sources that can generate seismic energy by applying a vertical impulsive force to the surface. Such sources should be tested to determine whether they are acceptable as VSP energy sources. These surface sources often provide only a limited frequency bandwidth. There is also a tendency to generate severe shallow reverberations. It is recommended that field tests be run at each onshore position to determine if valid information can be recorded.
- **Vibrators.** The variety of vibratory energy sources available makes them attractive as VSP energy sources. Their mobility allows VSP experiments that use many different source locations. The input pilot sweep to the ground can be varied to meet resolution requirements for a particular VSP recording. Vibrators are preferred for VSP surveys because their signals can be designed to suit the bandwidth requirements of an area and its signal waveform is nearly invariant.
- **Air guns.** In offshore vertical seismic profiling, air guns are used almost exclusively. The air gun can be fired at a fixed location near the well head if the well is vertical, allowing many VSP objectives to be met. The air gun is suspended from a work crane, making shooting very simple. Air guns are also attractive as onshore VSP energy sources as well as offshore. They are small and portable, can be fired at intervals of a few seconds, and produce highly repeatable wavelets. However, they must be submerged in water in order to function properly.

The physical appearance and construction of a geophone used for surface recording and the instrument used to record a VSP survey are considerably different. The down-hole geophone must be placed within a casing to protect it from the high pressure and temperatures existing in the deeper parts of the borehole. The mechanical deployment system that anchors the geophone to the borehole wall and electronic amplifying and telemetry circuits are located in the same housing.

There are rigid standards regarding resolution, dynamic range, gain, and recording format that VSP recordings should meet. Both the down-hole geophone data and near-field monitor geophone responses should be recorded with enough resolution (at least 12 bits including sign) in order to capture high-resolution waveforms. Recording the near-field wavelet is important in all marine VSP surveys. This is especially important when performing source-signature deconvolution with an energy source such as an untuned air gun that because of the bubble effect results in a long wavelet.

When the geophone is at shallow depths, more input energy is sometimes required to enable weak late arrivals from deep reflectors to be recorded. In fact, the energy input needed when recording a VSP at shallow depths may be two or three times that required when the geophone is at the bottom of a deep hole. To get the required energy, multiple shots can be taken at the same level to maintain the same wavelet character. Increasing the strength of the input energy often creates a wavelet with a completely different character from the wavelet already recorded. See chapter 5 for more detail.

A VSP survey is recorded as the borehole geophone is raised from the bottom of the hole to the surface in increments of a few meters. As the geophone is lowered into the well, exploratory data should be recorded at intervals of 300 to 500 meters. These measurements allow quality control personnel to select the appropriate recording parameters, such as source energy and the number of shots to be summed to enhance S/N ratio.

Data are recorded at constant intervals as the geophone tool is pulled out of the borehole using an energy source with repeatable wavelet shape. Multiple records are normally recorded at each geophone depth and vertically summed in the data processing stage. The depth increment around the target horizon is normally small (100 feet or less) and becomes progressively coarser as the geophone is pulled out of the hole.

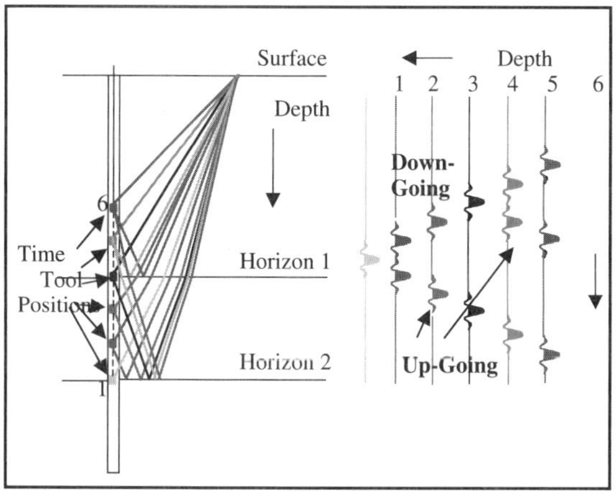

Fig. 7–142 *Upgoing and Downgoing Events*

Processing of VSP data. Unprocessed VSP field data are extremely difficult to interpret. The downgoing events dominate, and it is difficult to do any interpretation on the upgoing events. Also, the data contain both random and coherent noise. Considerable data processing must be done before the maximum benefit from the VSP recorded data can be realized.

Figure 7–142 demonstrates recording of downgoing and upgoing events. On the left is shown a diagram of VSP acquisition showing four successively deeper tool positions and ray paths between a surface source and the tool. The downgoing events travel directly from the source to the tool. Traces representing recordings of the event are on the right. Event waveforms and ray paths are color coded for correlation. Event times increase with depth for downgoing events. Upgoing events travel from source to receiver via reflection from two horizons. Event times decrease with depth for downgoing events.

The center of Figure 7–143 shows raw VSP survey data. The horizontal scale is geophone depth in the hole in units of distance (feet or meters), and the vertical scale is time in seconds. The first strong amplitude events are *direct arrivals* or *downgoing waves*. Their orientation trend from shallow time to the deeper times is from right to left. Following the downgoing waves on the VSP recording are downgoing multiples. *Upgoing waves* have an orientation trend that goes from deep time at the right to shallower time at the left—a mirror image of the downgoing events. The upgoing and downgoing waves must be separated to allow information from each type can be used.

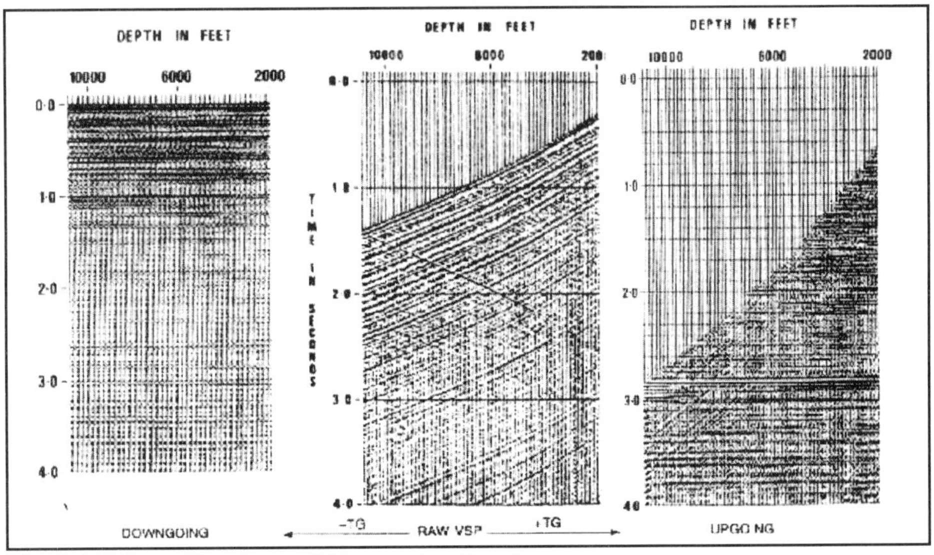

Fig. 7–143 *Raw, Upgoing and Downgoing Events*

To tie the surface seismic data, presented in two-way travel time, to the upgoing waves from the VSP, data must be converted to two-way time. The travel time of the downgoing waves is short a *TG* to make all events have two-way travel time. *TG* is the first-break time for the VSP trace recorded at geophone position *G* in Figure 7–143. Therefore, by adding *TG* to the raw VSP data, the upgoing events are obtained, and by sub-tracting *TG* from the raw data, the downgoing events are obtained. Figure 7-144 shows the use of VSP to define seismic reflections.

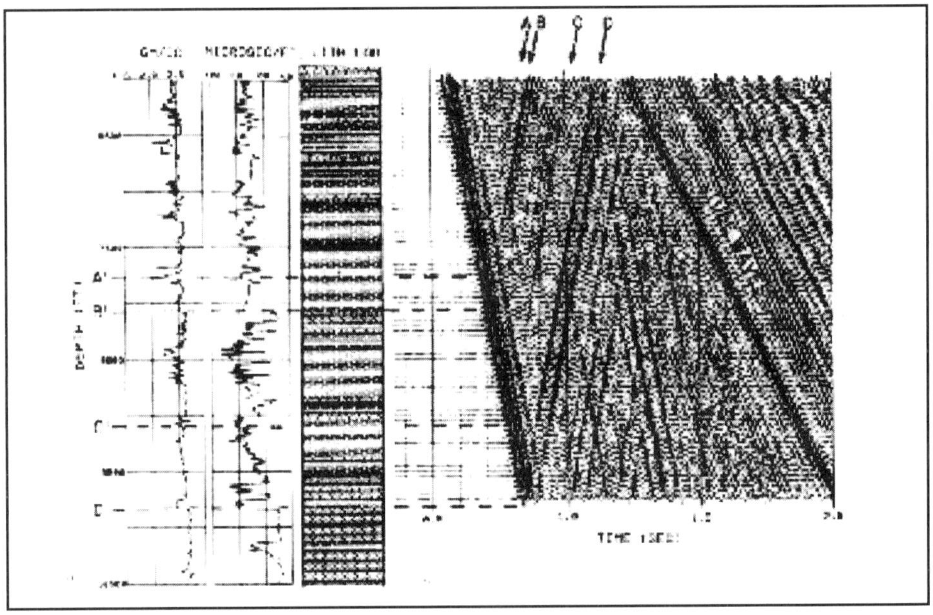

Fig. 7–144 *Identification of Seismic Reflectors*

To complete the separation of the downgoing from the upgoing events, apply a 2-D Fourier transform. As shown in Figure 7–145, the upgoing events fall in the negative half-plane in F-K while the downgoing events fall in the positive half plane. So a filter that rejects all negative dips outputs downgoing events, and one that rejects positive dips outputs upgoing events.

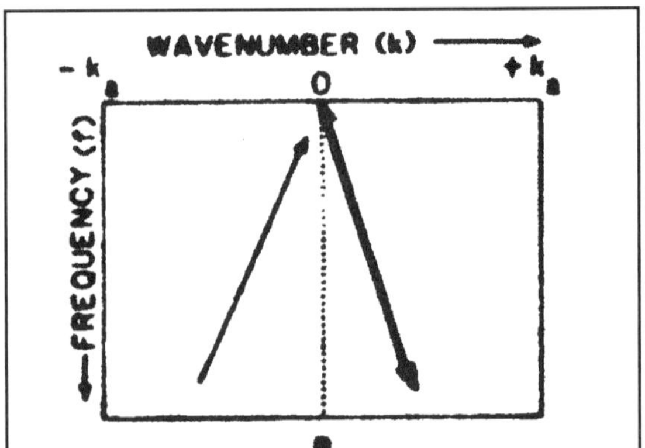

Fig. 7–145 *Separation of Up- and Downgoing Events in F-K Space*

VSP applications. VSP is a very flexible tool with many applications, and, using creative imagination, one can test his ideas using VSP. Some selective applications of VSP are listed here in two main categories. Some applications related to stratigraphic sequence will be discussed.

Exploration applications of VSP data include estimation of reflection coefficients, identification of seismic reflectors, comparison of VSP with synthetic seismograms, determination of Fresnel zone size and VSP horizontal resolution, seismic amplitude studies, determination of subsurface rocks' physical properties, seismic wave attenuation, and thin bed stratigraphy.

VSP has several reservoir engineering and drilling applications as well. Among these are prediction of seismic reflectors' depths, predicting rock conditions ahead of the bit, defining reservoir boundaries, locating faults, monitoring secondary recovery processes, seismic tomography and reservoir description, predicting high-pressure zones ahead of the bit, and detection of man-made fractures.

Exploration applications of VSP include identification of seismic reflectors. A good interpreter will attempt to relate the surface-acquired reflection seismic data to subsurface stratigraphy and depositional facies. When doing such a task, one should keep in mind that good-quality VSP data are capable of defining the depth at which each upgoing primary reflection is created in a stratigraphic section near the borehole. Therefore, by VSP, the correct stratigraphic interpretation of a surface-acquired reflection seismic section can be reached.

Using VSP data with a high signal-to-noise ratio, an interpreter can answer such questions as:

a. Is a reflection generated at a litho-stratigraphic boundary or a chronostratigraphic boundary?

b. Which lithological boundaries can be seen with seismic data and which cannot?

c. How reliably do synthetic seismograms made from well log data identify primary and multiple reflections?

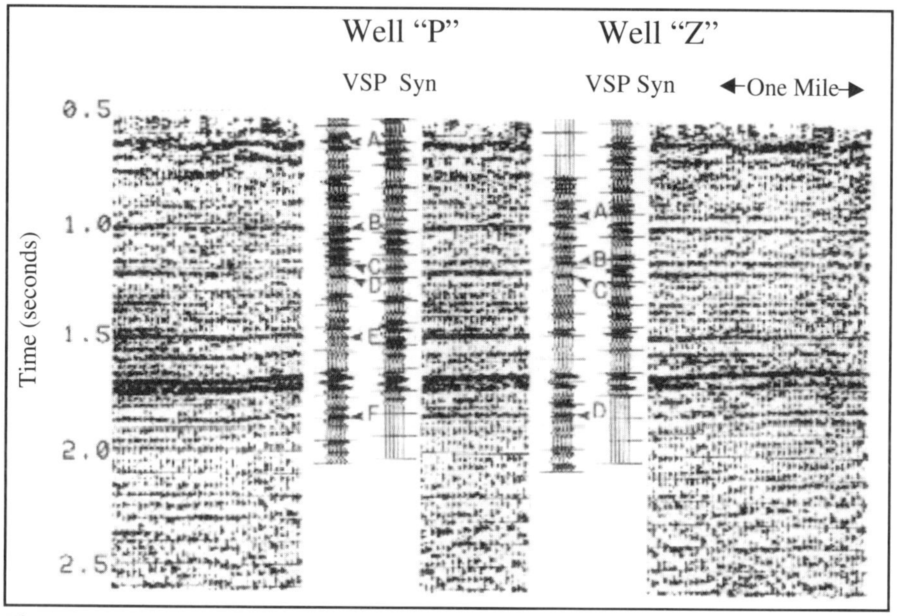

Fig. 7–146 *Comparison of VSP with Synthetic Seismogram*

Figure 7–146 shows VSP data recorded in a well where the stratigraphic and lithological conditions that create seismic reflections can be identified (Hardage 1983). As an example of the reliability with which VSP can often identify primary seismic reflectors, consider the four upgoing reflections that are shown by the lineup of black peaks labeled *A*, *B*, *C*, and *D*. The subsurface depth of the interfaces that generated each reflection can be defined by extrapolating the apexes of the black peaks downward until they intersect the first break loci of the downgoing compressional event. These depths are labeled *A'*, *B'*, *C'*, and *D'*.

Figure 7–146 shows raw field data. No processing has been done other than an AGC function has been applied to equalize all the amplitudes.

The traditional tool established to correlate between the subsurface stratigraphy and surface-measured seismic data is the synthetic seismograph. VSP can be used to identify lithology and subsurface stratigraphy with a high degree of accuracy. In contrast, a synthetic seismogram is only an artificial representation of seismic measurements. Vertical seismic profiling can use the same type of source, the same geophone, and the same instrumentation that was used to record the surface seismic data. Synthetic seismogram calculations can only approximate these aspects of the total seismic recording process.

Figure 7–146 compares of surface seismic data crossing VSP well *P* and well *Z* with a synthetic seismogram. The lettered arrowheads show where the VSP data are a better match to the surface data than the synthetic seismogram data are.

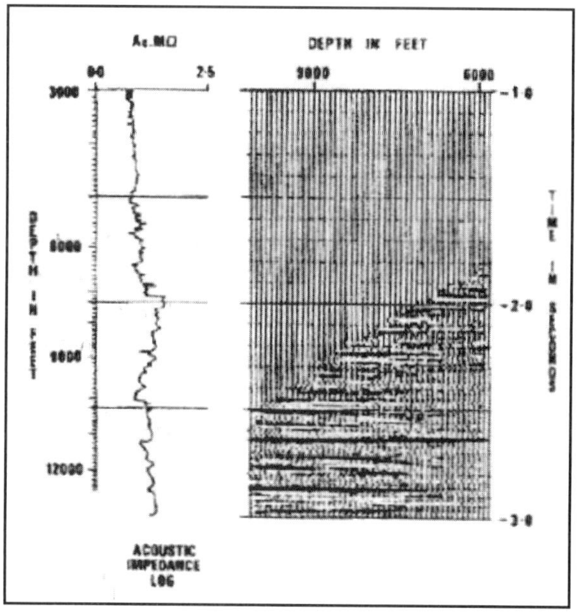

Fig. 7–147 *Predicting Interval Velocity Ahead of the Bit*

One output of VSP data processing is a plot of acoustic impedance versus depth. Since most rock densities are within a narrow range, a single constant value can be assumed. A plot of interval velocity versus depth can be obtained in addition to the velocity value in the interpretation. See Figure 7–147. Interval velocities can be used in engineering applications, as will be demonstrated later. For example, velocity inversions indicate high pressure zones.

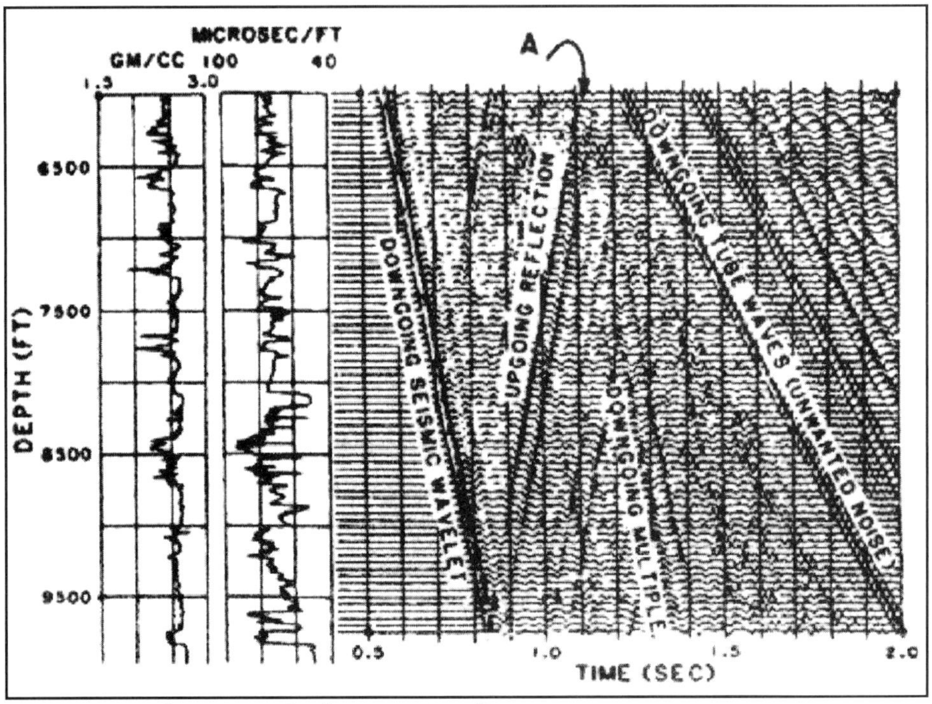

Fig. 7–148 *Predicting Depth of a Seismic Reflector*

Predicting drilling depth to key seismic reflectors is a common activity in oil and gas exploration, and many geophysicists are able to make quite accurate depth estimates from surface determinations of seismic velocities. However, these estimates become more difficult and less accurate when a well is in a wildcat area, where there is little drilling history, or in a poor-quality seismic recording area. Using VSP to predict reflector depth will yield the greatest benefits in those areas where seismic reflection is poor. A factor that works in favor of VSP is that the geophone is located deep in the hole that is usually a seismically quiet environment. Certainly, wells have been drilled and will continue to be drilled in areas where seismic data are of poor quality and where the depths of reflectors are speculative. Figure 7–148 demonstrates this application.

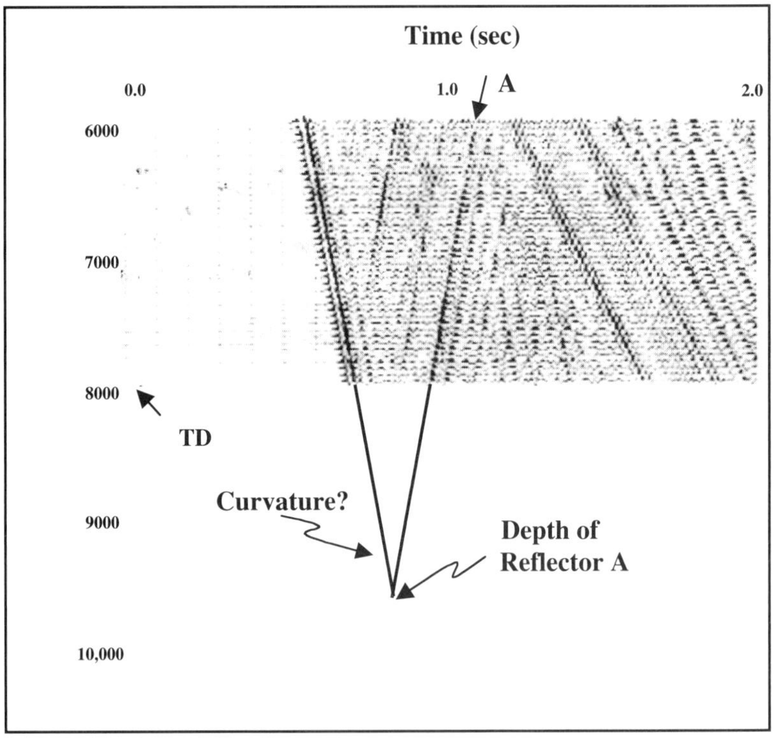

Fig. 7–149 *Looking Ahead of the Bit*

Figure 7–149 illustrates a way that VSP data has been used to predict the distance from the drill bit to a deeper formation. It is assumed that the well was drilled to 8000 feet and that VSP data are recorded from that depth upward far enough so that deep reflection events can be seen and interpreted. This means that data should be recorded from the bottom of the hole to about 2000 feet above bottom with constant depth increment.

The question to be answered is, "How far below the current drilling depth is reflector *A*?" As a first approximation, it can be assumed that the downgoing first arrival wavelet continues below 8000 feet with the same depth curvature as the recorded data intervals from 6000–8000 feet. The intersection of the downgoing extension line and upgoing event *A* will intersect at the depth of reflector *A*.

Recommendations for using multi-offset VSP follow.

- Seismic modeling of the proposed survey should be conducted before the survey to verify that the resolution required to solve the problem at hand can be obtained and to assist in designing survey parameters such as source offset and geophone level increment and to assist the interpreter in understanding the record area.
- Surveys should be conducted in either a completely cased hole or in the open hole before casing is set because poorly cemented casing causes serious degradation of the VSP data quality.
- Multi-offset VSP surveys should be designed to utilize all the possible existing well control to confirm modeled results.
- A near offset VSP should be run with the far offset to establish velocity control to aid in correlating data with the well logs.

VSP is one of the geophysical methods that can be applied in developing a field economically.

Amplitude versus offset analysis (AVO)

In chapter 3 it was shown that amplitudes of reflected seismic signals depend upon the angle of incidence, acoustic impedances for P-waves, acoustic impedances for S-waves, and bulk densities on each side of a reflecting interface. If the acoustic impedances and/or densities do not change laterally, then the relationships among incident and reflected wave amplitudes are described by the Zoeppritz equations and reflected amplitudes vary only with angle of incidence. As shown in Figure 7–150, the angle of incidence increases as offset between source and receiver increases. Thus, amplitudes also vary with offset.

In porous rocks, P-wave velocities depend on fluid content of the pores as well as the rock matrix. For a porous rock:

$$V_P = \varphi V_f + (1 - \varphi)V_m \tag{7.5}$$

where

V_P = effective P-wave velocity of the rock

V_f = P wave velocity of the pore fluid

V_m = P-wave velocity of the rock matrix

φ = porosity of the rock

In general, three fluids may exist in a reservoir rock—oil, gas, and water. The P-wave velocities of oil and water are not markedly different, but the P-wave velocity of natural gas is significantly lower than that of oil and water. Thus, when a P-wave is incident on the gas-filled part of a reservoir rock, the observed amplitude may differ substantially from amplitudes reflected below the gas-water boundary in the reservoir. In other words, amplitude is anomalous. Anomal-ous amplitude variation with offset can be an indicator that hydrocarbons are present. Such amplitude anomalies are most often observed in gas-sands.

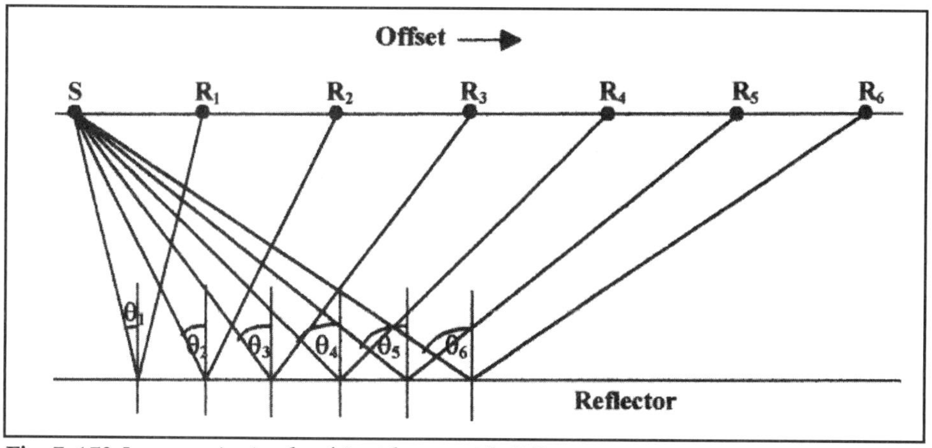

Fig. 7–150 *Increase in Angle of Incidence with Offset*

Common midpoint stack (CMP) tends to conceal amplitude anomalies since each CMP trace is the average of individual trace amplitudes in the common midpoint gather. These traces were recorded over a wide range of offsets. Thus AVO analysis must be performed on pre-stack data if amplitude anomalies are to be observed and analyzed.

The anomalies sought in AVO analysis are changes in the reflection amplitude with the angle of incidence. A *bright spot* is an increase in the relative true amplitude as the angle of incidence increases. Bright spots can occur in gas sand reservoirs. A *dim spot* is a decrease in relative true amplitude as the offset increases. Dim spots are observed in carbonate reservoirs. In other cases, little or no change in the amplitude of the seismic data with lithology and fluid or gas content variations in the reservoir is observed.

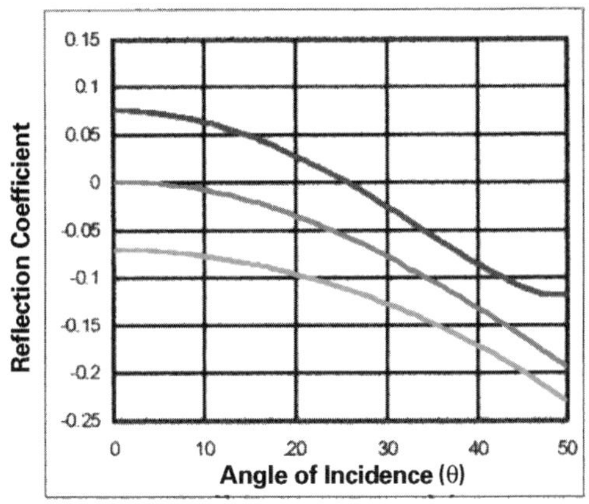

Fig. 7–151 *AVO Classes*

As noted above, P-wave velocity in porous media depends on fluid content of the pores and formation matrix. By contrast, S-wave velocity is independent of pore fluids. Consequently, abrupt changes in the *P* to *S* velocity ratio may be indicative of fluid changes in a porous reservoir rock.

Equation 3.5 of chapter 3 expresses the ratio of *P* to *S* velocities as a function of Poisson's ratio:

$$\frac{V_P}{V_S} = \sqrt{\frac{2(1-\sigma)}{1-2\sigma}}$$

where

σ = Poisson's ratio.

This equation can be solved for Poisson's ratio:

$$\sigma = \frac{-0.5\left(\frac{V_P}{V_S}\right)^2 - 1}{\left(\frac{V_P}{V_S}\right)^2 - 1} \tag{7.6a}$$

Equation 7.6a shows that Poisson's ratio is also affected by pore fluid change and that AVO is a seismic fluid indicator as well as lithology tool.

When dealing with both P- and S-wave velocities across an interface, it is more convenient to use α for P-wave velocity and β for S-wave velocity. Thus, the velocities in layer 1 are α_1 and β_1, while those in layer 2 are α_2 and β_2. With this revised nomenclature, Equation 7.6a becomes:

$$\sigma = \frac{0.5\left(\frac{\alpha}{\beta}\right)^2 - 1}{\left(\frac{\alpha}{\beta}\right)^2 - 1} \tag{7.6b}$$

Rutherford and Williams, 1989, defined three classes of gas-sand AVO anomalies (Fig. 7–151). For compressional waves, class 1 shows a strong reflection coefficient that decreases as the angle of incidence increases. A phase change is observed at larger angles of incidence.

In class 2, reflection coefficient values are about zero for normal incidence and become increasingly negative. The phase change occurs at a smaller angle of incidence compared to class 1. Class 3 demonstrates a negative normal incidence reflection coefficient that becomes more negative as the angles of incidence increase. This is a typical bright spot anomaly (gas sands).

The Zoeppritz equations express reflection coefficient $R(\theta)$ as a function of reflection angle or angle of incidence. Shuey developed a simplified approximation to the Zoeppritz equation:

$$R(\theta) = R_0 + \left[A_0 R_0 + \frac{\Delta\sigma}{(1-\sigma)^2}\right] \sin^2\theta + \frac{\Delta\alpha}{\alpha} \tan^2\theta \sin^2\theta \tag{7.7}$$

where

α is the average P-wave velocity across the interface, $= \frac{1}{2}(\alpha_1 + \alpha_2)$

$\Delta\alpha$ is the contrast in P-wave velocity, $\sigma_2 + \sigma_1)$

σ is the average Poisson's ratio across the interface, $= \frac{1}{2}(\sigma_1 + \sigma_2)$

$\Delta\sigma$ is the contrast in Poisson's ratio, i.e. - $\Delta\sigma = (\sigma_2 + \sigma_1)$

Other versions of Equation (7.6) have been developed, including a later one by Shuey. The one used here is Mallick's 1993 version:

$$R_{pp} \approx R_0 + \left[\frac{1}{2}\frac{\Delta\alpha}{\alpha} - 2\frac{\Delta\mu}{\rho\alpha^2}\right] \sin^2\theta + \frac{1}{2}\frac{\Delta\alpha}{\alpha} \sin^2\theta \tan^2\theta \tag{7.8}$$

In Equation 7.8, $\beta = \frac{1}{2}(\beta_1 + \beta_2)$, $\Delta\beta$ is the. shear velocity contrast. $= \beta_2 - \beta_1$, $\Delta\mu$ is the shear modulus contrast, $\mu_2 - \mu_1$, and relative P-wave reflection amplitude is equated to the P-wave reflection coefficient as a function of angle of incidence—R_{pp}~$R(\theta)$.

Equation (7.8) can be written as:

$$R_{pp} \approx A + B\sin^2\theta + C\tan^2\theta\sin^2\theta \tag{7.9}$$

Values of the constants A, B, and C are used in various ways to analyze amplitude anomalies.

Equation 7.8 expresses relative P-wave amplitude as a function of angle of incidence, not offset. So, the direct analysis is amplitude versus angle (AVA). Angle gathers are used to reformat input data in terms of angle of incidence. Angle gathers are a series of angle traces extracted from NMO-corrected CMP gathers that simulate data recorded at reflection angles rather than fixed offsets. Each trace in an angle gather corresponds to a narrow range of angles of incidence for a particular CMP gather (Todd and Backus, 1985, and Todd 1986).

Two-Term AVO inversion. The two-term AVO inversion ignores the last term of Equation 7.9, producing the slope intercept equation:

$$R_{pp} \approx A + B\sin^2\theta \tag{7.10}$$

For every sample time of the angle gathers, amplitudes of each trace are plotted against $\sin^2\theta$, where θ is the trace angle (Fig. 7–152). The constants A and B are then determined by linear regression. The constant A is called the *AVO Intercept* and B is called the *AVO Gradient.* Figure 7–153 illustrates two-term AVO inversion. Note that the AVO intercept is often symbolized by R_0 and the AVO gradient by G.

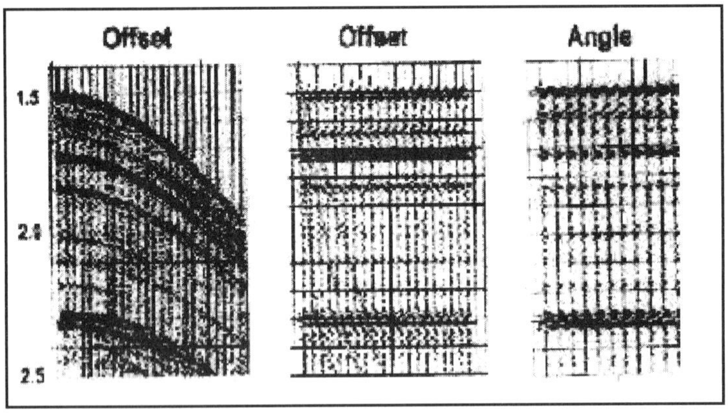

Fig. 7–152 *Angle Gathers*

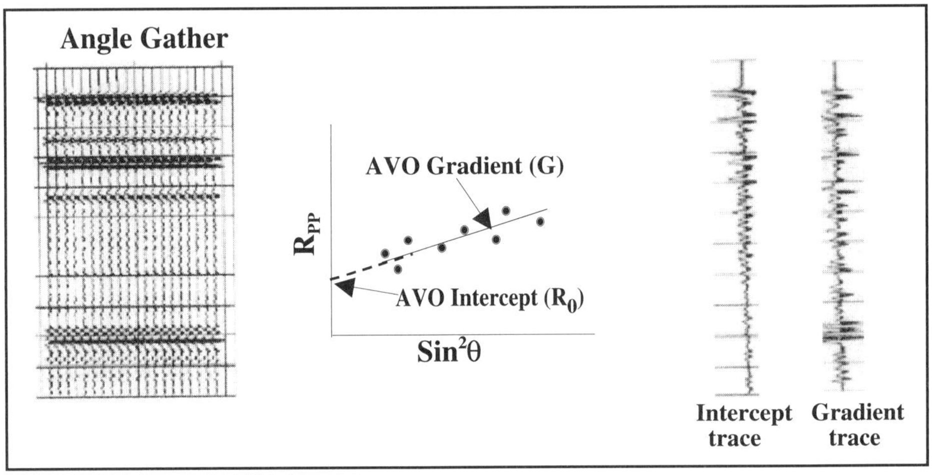

Fig. 7–153 *Two-term AVO Inversion*

Using Mallick's version of the equation:

$$A = R_0 = \frac{\rho_2\alpha_2 - \rho_1\alpha_1}{\rho_2\alpha_2 - \rho_1\alpha_1} \approx .5\left[\frac{\Delta\alpha}{\alpha} + \frac{\Delta\rho}{\rho}\right] \tag{7.11}$$

$$B = \frac{1}{2}\frac{\Delta\alpha}{\alpha} - 2\frac{\Delta\mu}{\rho\alpha^2} \tag{7.12}$$

From chapter 3, Equation 3.3, shear wave velocity β can be calculated from:

$$\beta = \sqrt{\frac{\mu}{\rho}} \tag{7.13}$$

Hence, $\mu = \rho\beta^2$. Applying differential calculus obtains:

$$\Delta\mu = \beta^2\Delta\rho + 2\beta\rho\Delta\beta \tag{7.14}$$

Substituting Equation 7.14 into Equation 7.12 and assuming that $\frac{\alpha}{\beta} \approx 2$ yields:

$$B = \frac{1}{2}\left(\frac{\Delta\alpha}{\alpha} - \frac{\Delta\rho}{\rho}\right) - \frac{\Delta\beta}{\beta} \tag{7.15}$$

Equations 7.11 and 7.15 can be used to produce various elastic parameters, as shown:

Pseudo–S Wave Section:
$$\frac{1}{2}(A - B) = \frac{1}{2}\left(\frac{\Delta\beta}{\beta} + \frac{\Delta\rho}{\rho}\right) \quad (7.16)$$

Poisson's Ratio Contrast:
$$\Delta\sigma = \frac{4}{3}(A + B) \quad (7.17)$$

Figure 7–154 shows portions of sections that plot the AVO intercept, pseudo-S wave, and Poisson's ratio contrast as well as a similar portion of the corresponding CMP stack section.

Three-term AVO inversion. Three-term AVO inversion uses all terms of Equation 7.10.

$$R_{pp} \approx A + B\sin^2\theta + C\tan^2\theta\,\sin^2\theta$$

The constants *A*, *B*, and *C* are determined by three-term linear regression using the amplitudes and reflection angles of the angle gathers in a manner similar to that used in two-term inversion.

As before, adopting Mallick's Equation 7.9. A and B are defined by Equations 7.11 and 7.12. Constant C is given by Equation 7.18.

$$C = \frac{1}{2}\frac{\Delta\alpha}{\alpha} \quad (7.18)$$

$$\frac{\Delta\alpha}{\alpha} = 2C \quad (7.19)$$

$$\frac{\Delta\beta}{\beta} = 2C\text{–}A\text{–}B = \frac{\Delta\alpha}{\alpha} - \frac{1}{2}\left[\frac{\Delta\alpha}{\alpha} + \frac{\Delta\rho}{\rho}\right] - \left[\frac{1}{2}\left(\frac{\Delta\alpha}{\alpha} - \frac{\Delta\rho}{\rho}\right) - \frac{\Delta\beta}{\beta}\right] \quad (7.20)$$

$$\frac{\Delta\rho}{\rho} = 2(A\text{–}C) = \frac{\Delta\alpha}{\alpha} + \frac{\Delta\rho}{\rho} + \frac{\Delta\alpha}{\alpha} \quad (7.21)$$

Figure 7–154 shows portions of a CMP stack section along with corresponding portions of an AVO intercept section, a pseudo-shear wave section, and a Poisson's ratio contrast section. Note that the time scale in all four sections are the same. That is, the derived quantities are plotted at the same time and position as the P-wave CMP stack traces.

Data processing. Factors to consider in setting up a processing sequence for AVO include:

- record quality
- signal-to-noise ratio
- near surface problem (land and OBC data acquisition)
- multiple reflections (offshore seismic data)

Since the aim is to study relative true amplitudes and relate them to variations to lithology and fluid content, processes that do not maintain fidelity of the relative true amplitude must be avoided. Multi-channel operations—velocity filtering and Radon transforms—may improve signal-to-noise ratio but the outputs will not be relative true amplitude. Applying gain functions is often desirable in conventional processing but they change the amplitude relationship.

Weiner deconvolution is statistical in nature, and as TVD employs trace summation, will change the amplitude relationship.

Surface consistent deconvolution, surface consistent static corrections, and surface consistent gain should be applied to AVO data.

Applications of AVO. There are three broad areas of AVO application:

1. geophysical
2. engineering
3. exploration

Geophysical applications include verification of direct hydrocarbon indicators—bright spots, dim spots, and any other amplitude anomaly—performing 2-D analysis versus 1-D analysis, CMP versus stacked trace analysis, and investigation of offset (angle) dependence. Engineering applications include documentation of producing anomalies that can be used to identify look-like anomalies using AVO analysis. Exploration applications include defining and delineating reservoir boundary and prediction of high-pressure gas zones. AVO is a good tool for field development.

AVO summary. The effects of variations in subsurface parameters such as velocity and density plus angles of incidence on P-wave reflection amplitudes have been known for many years. The existence of anomalous amplitudes on CMP stack sections has also been known for a long time. AVO is a reliable way to verify direct hydrocarbon indicators such as bright spots in gas sands and dim spots in carbonate reservoirs as well as related amplitude anomalies. Poisson's ratio can be estimated from AVO, allowing determination of various other elastic properties of rocks.

Position and areal extent reservoirs can be delineated from AVO analysis. The accuracy of this delineation depends upon the adequacy of the data and methods used to process it.

Some problems with AVO analysis have been encountered. Noise interference tends to be a greater problem for data recorded at longer offsets. These data will, of course, have larger angles of incidence, which increases the problem. Careful attention to mutes is needed to avoid false AVO responses caused by NMO stretch on far offset traces.

Analysis of amplitude anomalies through use of approximations to the Zoeppritz equation is not always advisable. Approximations should be tested against the full Zoeppritz equation and used only when shown to be sufficiently accurate.

Amplitude gain control, velocity (F-K) filters, or similar software programs that mix or alter amplitudes should not be used to process data for AVO analysis. Such processes change relative amplitudes, making assumptions of AVO invalid.

Shear waves and seismic stratigraphy. Shear waves or S-waves naturally occur in all seismic surveys. In conventional surveys, acquisition and processing parameters discriminate against S-waves, but when both S- and P-waves (pressure waves) are recorded much more information about subsurface rocks can be obtained (Fig. 7–154).

In the compressional wave reflection method, geophones respond to the vertical movement of the

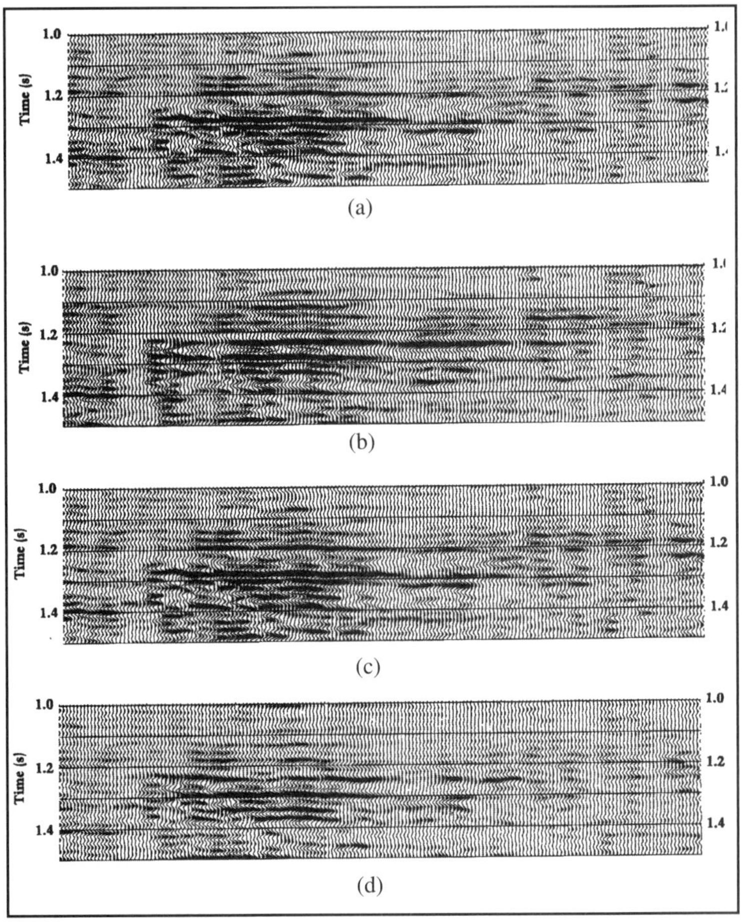

Fig. 7–154 *Portion of a CMP Stack Section Showing a Bright Spot, P-Wave Intercept Section, Pseudo S-Wave Section, and Poisson's Ratio Section*

earth. Depending on the type of shear wave motion being measured, they may be recorded by groups of single, horizontal geophones all pointing in the same direction (SH), by two-component geophones (SV), or by three-component geophones.

Three-component geophones are normally used in orthogonal groups (three units mutually perpendicular) with two in the horizontal plane and the third one vertical. There are three-component units in which all three elements are similar, arranged orthogonally, and oriented in the manner of a tripod with the geophones inclined at 54.7° to the vertical. The combinations of inputs from the three geophones, after gain adjustment, give the vertical and horizontal components of the motion.

Basically, shear waves are recorded with the same arrangement as P-waves. That is, the CMP method of gathering data.

S-wave and P-wave orientation. As shown in Figure 7–155a, P-wave reflection is obtained from a wave whose particle motion is along the ray path (perpendicular to the wavefronts if the material is isotropic). An S-wave striking a reflector at normal incidence is reflected without change. However, in nature and in practice, normal incidence is only a theoretical concept. Due to the dip and anisotropy of the earth formations, a shear wave is reflected at a different orientation than the incident wave. Shear waves have particle motion at right angles to the ray path. Arbitrarily oriented S-waves can be split into SV and SH components. The SV component has a particle motion in the vertical plane (perpendicular to the horizontal reflector) whereas the SH wave has a particle motion in the horizontal plane.

The SH-waves are reflected and refracted without change, provided that the dip of the reflector is aligned with the survey line. Under any other conditions the incident waves are no longer purely SH and SV—both SV and P reflections and refractions are obtained. For practical purposes, this discussion will concentrate on SH waves for the following reasons.

1. It is known that the P-wave and/or SV will be partly converted into one another upon striking any reflecting horizon at any angle other than normal incidence.
2. It is easier to generate SH waves alone by using horizontal energy at right angles to the direction of the surface traverse, thereby minimizing the conversion of wave type at near-horizontal boundaries.

Shear wave generation. Since angles of emergence in seismic exploration tend to be rather small, P-waves have their maximum component of motion in the vertical direction but some P-wave energy is recorded by a radial component geophone. SV-waves are expected to have their maximum component of motion in the radial direction (horizontal, along the line of geophones). SH-wave motion is in the transverse direction only (horizontal, perpendicular to the line). This assumes 2-D type operation. In 3-D recording, the different azimuths will result in recording of some P- and SV wave energy on all three components. Some SH-wave energy will be seen on the radial as well as the transverse component.

While many schemes have been used to generate SH-waves, horizontal vibrators are often used. When SH vibrator trucks are used, they are usually pointed transverse to the direction of wave propagation. The geophones must be arranged so that polarization marks on the cases are all on the same side of the line as it progresses. This is true whatever the direction of the line.

The basic operation of horizontal vibrators is the same as vertical vibrators. The base plate moves from side to side, perpendicular to the line of survey rather than up and down. The bottom of the base plate is usually not flat for vertical vibrators but has pyramids that sink into the ground to improve coupling and transfer of SH motion into the ground.

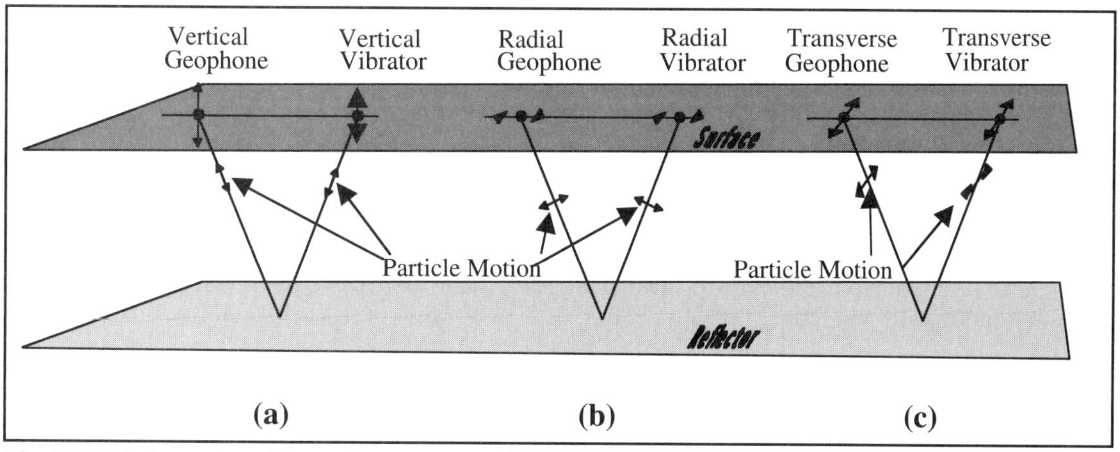

Fig. 7–155 *Orientation of P- and S-wave Particle Motion*

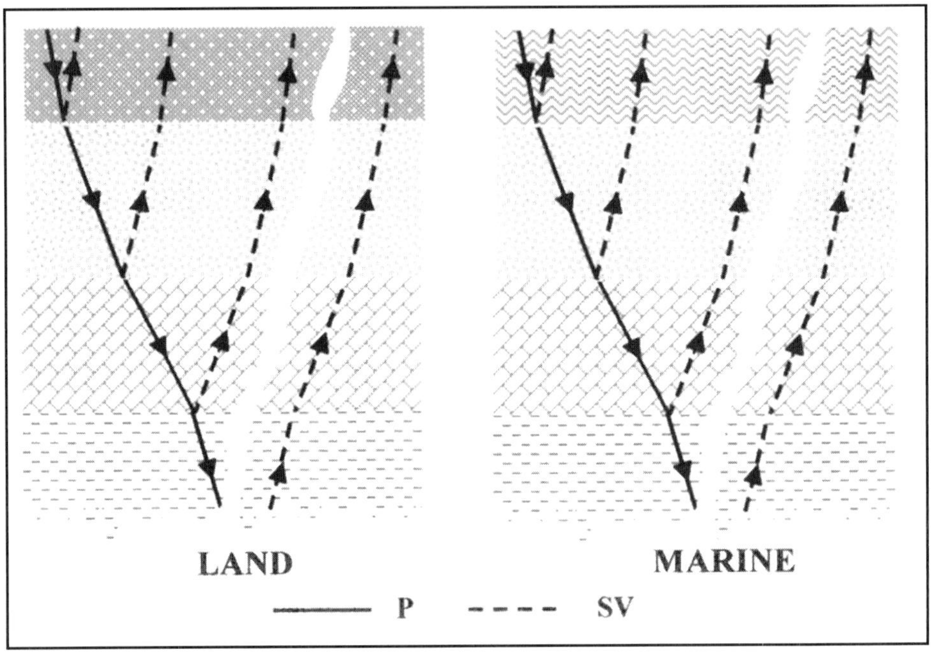

Fig. 7–156 *Mode Conversion of Ray Paths*

The signal put into the ground is not, however, pure SH. A lower-level P-wave signal also results. This is thought to result from the rocking motion of the base plate as it sweeps from side to side. The frequency of this P-wave signal is twice that of the S-wave. If the sweep is from 5 Hz to 30 Hz for SH, then the P-wave signal produced will be from 10 Hz to 60 Hz. P-wave records extracted from horizontal vibrator data are found to be worse than P-wave records generated by vertical vibrators with respect to signal-to-noise ratio, signal bandwidth, and depth of penetration.

Sweeps for horizontal vibrators are generally lower in frequency than for vertical vibrators. This is because S-wave data seem to be richer in lower frequencies than P-wave data. However, S-wave velocities are also lower so S-wave data have about the same or somewhat shorter wavelengths than the P-wave data. Resolution is thus about the same.

Sources used to generate SH-waves can, in general, be used to generate SV--waves by properly aligning the source mechanism with the spread. The simplest and easiest way to get SV-waves is through mode conversion. When P--waves are incident on an interface between solid media at non-vertical angles, reflected and refracted SV-waves are generated. In the case of a P-wave propagating in a fluid, such as water, there is only a refracted SV-wave. This phenomenon is called *mode conversion.*

Figure 7–156 shows ray paths for P-SV conversion. On land, the surface layer is solid so there are reflected P- and SV-waves. The S-wave velocity of the second layer is almost sure to be higher than the P-wave velocity in the surface layer. Hence, the SV ray in layer 2 is bent toward the horizontal. Note the asymmetrical paths. Reflection points are not midway between source and receiver as in P-wave surveys.

In the marine environment, no SV-wave exists in the surface (water) layer. The mode conversion in this case is P-SV-P. While S-waves cannot be recorded directly in marine surveys, consideration of the P-SV-P ray path geometry, and proper extraction techniques in processing allows SV to be used in marine recording. When using SV-waves from mode conversion, spread layouts should be somewhat different. Mode conversion creates SV energy more efficiently at larger angles of incidence and the velocity of Rayleigh waves is about 92% of S-wave velocity. To circumvent this situation, the offset to the first group should be somewhat longer than for P-wave surveys. The shorter wavelengths of S-waves may also make it advisable to use shorter group intervals.

S-wave versus P-wave seismic section. Since the V_S/V_P ratio is approximately 0.5, the S-wave records are longer than P-wave records for a given penetration depth. Accordingly, the correlation process could take twice as long for an S-wave record.

Theoretically, the sample rate should be halved for S-wave records. If a 4 ms sample interval is used for the P-wave, an 8 ms interval should be used for the S-wave. However, the same sampling rate is used for both, and the processing time is normally doubled to get the same subsurface details of the P-wave section.

One source of coherent noise, ground roll, recorded with the P-wave has a counterpart, called, in the SH-wave record. This noise is just as difficult to attenuate as ground roll.

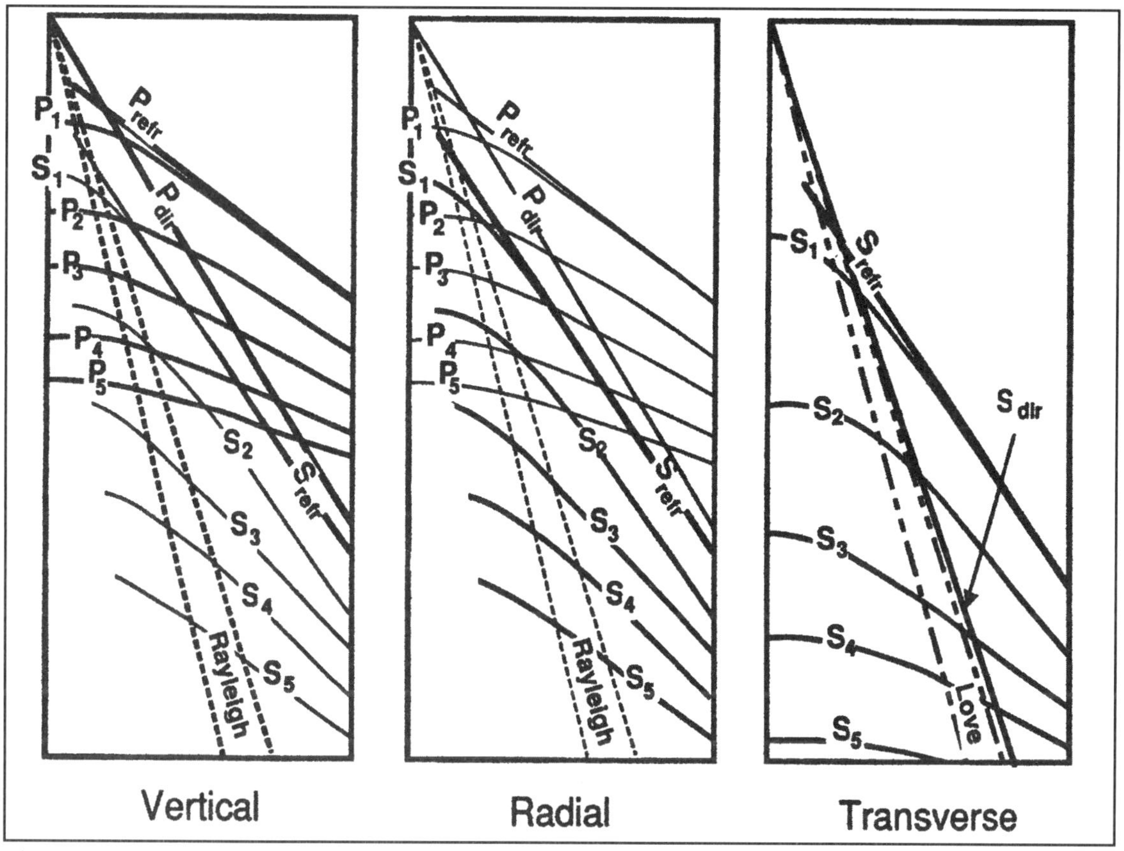

Fig. 7–157 *Schematic Representation of Three-component Records*

The statics problem for shear waves is more complicated than for P-wave generated statics—the static delay is frequency sensitive. Much research is under way to solve these static problems in order to enhance the image of the subsurface using S-wave seismology.

Processing S-wave data. It is necessary to extract and separate the different modes (P, SV and SH) before normal processing begins. Processes applied to P-wave data are, in general, also applied to S-wave data, although there are some differences in the way processes are applied, particularly in the attenuation of coherent noise, static corrections, and trace gathers.

For three-component recording, the preferred receiver orientation is vertical, radial (horizontal in the direction of the spread), and transverse (horizontal, perpendicular to the spread). However, P- and SV-waves are not completely separated even in this orientation. If there is structure with relatively steep dip striking parallel to the spread, the transverse component may not be pure SH. Further, anisotropy may lead to S-wave splitting so that more than one S-wave arrival is observed for the same reflection events.

Figure 7–157 shows, schematically, the situation described above for a surface spread. The vertical and radial component records include P reflections (P1–P5), S reflections (S1–S5), the P direct arrival, P and S refractions from the top of the first consolidated layer, and dispersed Rayleigh waves (ground roll). The radial component record has only SV events, and the transverse component has only SH events. Line thicknesses indicate relative amplitudes. Note also that SV -reflections start at longer offsets as reflection time increases. This is because P--SV conversion is negligible at small angles of incidence.

The obvious differences between P and S NMO curves suggest F-K filtering to separate P and S events as well as to attenuate the linear events. Unfortunately, as shown in Figure 7–158, P and S energy is not adequately separated for this approach to work well. The best approach for this situation is to first apply NMO corrections using S velocities, then transform to the F-K domain and apply a band-pass filter. As shown in Figure 7–159, this still does not completely separate P and S energy. Further, the narrow pass zone will produce a ringy output (T-X domain) record.

The preferred method to separate the P and S events is to apply the Radon transform, as discussed in chapters 3 and 5. Figure 7–160a shows the vertical component record of Figure 7–157 and Figure 7–160b shows its Radon transform. The P reflections are concentrated in an approximately triangular zone (shown shaded) with S-reflections represented by only portions of S1 and S2 in this zone.

The linear events in the T-X domain are transformed into points or very small areas. For refractions, the coordinates of these points are τ = intercept time and $p = \sin i_c / V_r$. Note that i_c = critical angle and V_r = refraction velocity. For direct arrivals, the coordinates are $\tau = 0$ and $p = 1/V_w$ where V_w is the weathering velocity. For Rayleigh waves (ground roll), the coordinates are $\tau = 0$ and $p = 1/V_R$, where V_R is the Rayleigh wave propagation velocity.

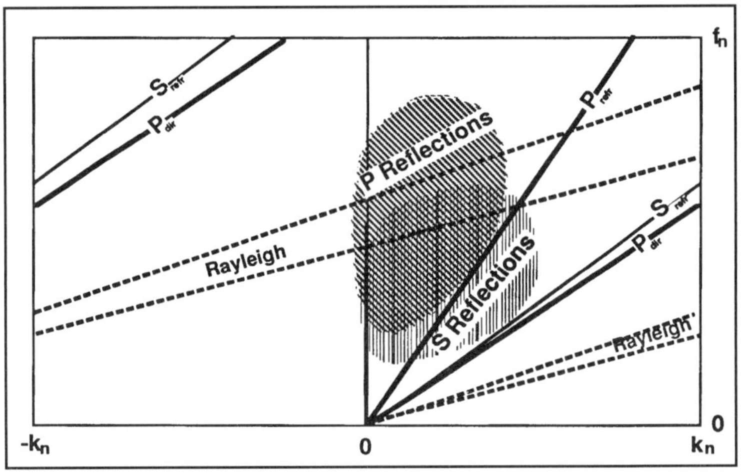

Fig. 7–158 *F-K Domain Representation of Vertical Component Record in Figure 7–157*

Figure 7–160c represents the inverse transform of the shaded part of *b* to the T-X domain. Except for short segments of S_1 and S_2, the *P* data have been isolated from the *S* and the low-velocity linear events. (These S-wave segments should be well attenuated by CMP stack.) A forward Radon transform of the radial component record followed by an inverse transform area of the *S* reflection zone yields similar results for SV data.

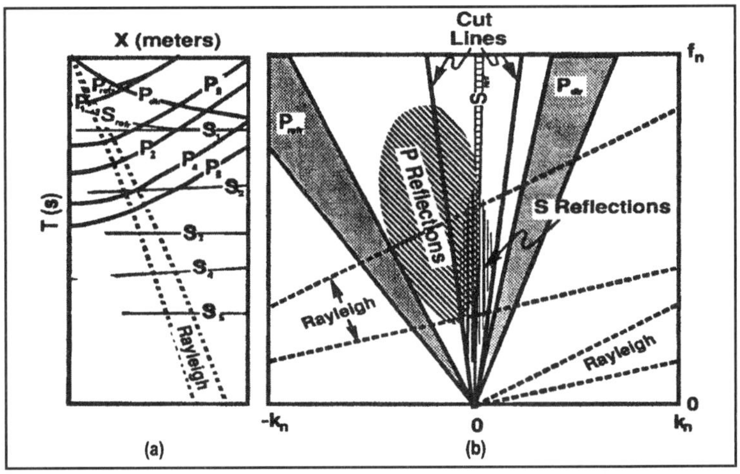

Fig. 7–159 *Vertical Component Record of Figure 7–157 after NMO Corrections Using S-wave Velocities*

A further extraction problem is presented by S-wave splitting, mentioned earlier. This phenomenon, which results from anisotropy, results in the recording fast and slow S-waves. Note that fast and slow do not necessarily correspond exactly to SV and SH. At least one algorithm (Alford, 1986) exists for rotation of the two S-wave arrivals into SV and SH and it is expected that more will be developed to handle other types of anisotropy. The linear events in the T-X domain are transformed into points or small regions.

Static corrections for S-waves range from about 2 to 10 times those for P-waves. This is, in part, because S-wave velocities in the near surface are lower and more variable than P-wave velocities. Another reason is, as shown in Figure 7–161, that the base of the low-velocity layer for P-waves generally corresponds to the bottom of the water table whereas the base of this layer for S-waves corresponds to the geologic boundary separating the

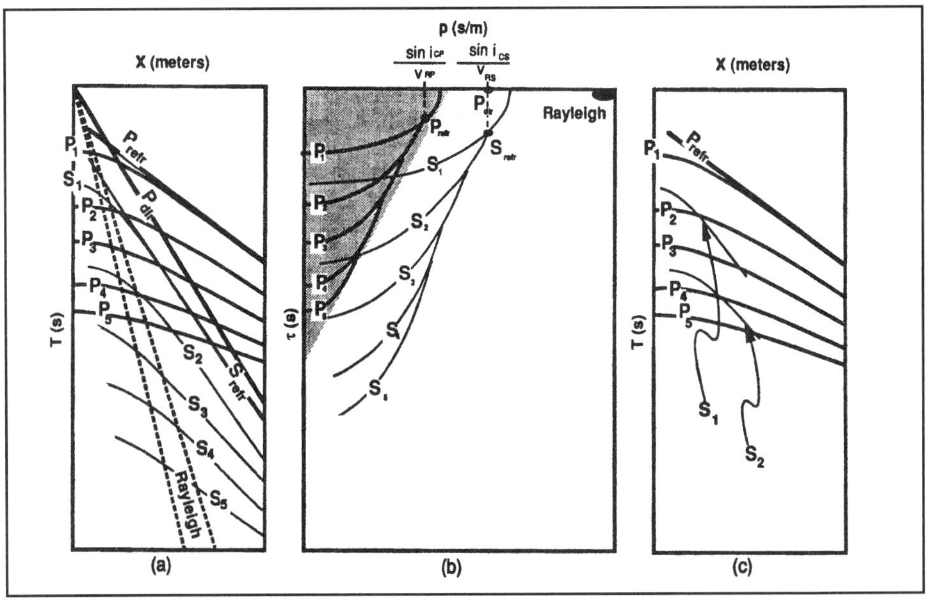

Fig. 7–160 *Vertical Component Record of Figure 7–157*

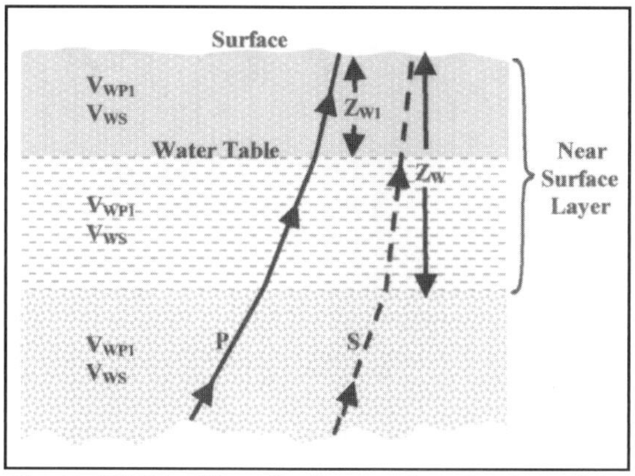

Fig. 7–161 *The Near Surface as Seen by P- and S-waves*

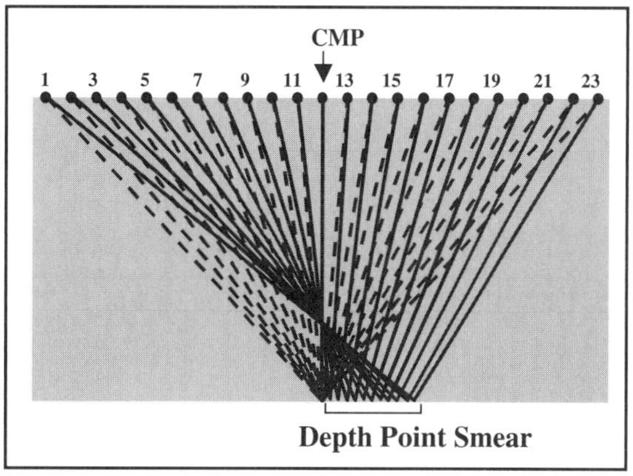

Fig. 7–162 *Comparison of P-P and P-SV Ray Paths*

less consolidated, lower velocity material from the truly consolidated rock below. This follows from the effect of fluids being greater for P-waves than for S-waves.

If in Figure 7–161, the base of the near surface layer is the reference surface, the receiver static corrections for P- and S-waves, RSC_p and RSC_s, respectively, are given by:

$$RSC_P = \frac{Z_{W1}}{Z_{WP1}} + \frac{(Z_W - Z_{W1})}{V_{WP2}} \tag{7.22}$$

and

$$RSC_P = \frac{Z_W}{V_{WS}} \tag{7.23}$$

Further, $Z_{W1} = Z_W/2$, $V_{WP1} = 2V_{WS}$ and $V_{WP2} = 5V_{WS}$, $RSC_p = 0.35\ RSC_s$.

Except for mode-converted (P-SV) data, surface-consistent statics algorithms can be used for S-wave statics as well as for P-wave statics. In the case of P-SV waves, the source end is a P-wave while the receiver end is an S-wave so the static does not depend on surface location alone. Also, as shown below, the pattern of source and receiver numbers on gather records is different for P and S data.

If an S-wave source is used, the resulting SH and/or SV traces can be gathered into Common-MidPoint (CMP) records, each of which is later stacked to produce a CDP stack output. In the case of P-SV data, this procedure will produce depth-point smear in the stack. This is because of the non-symmetrical travel paths that are illustrated in Figure 7–162.

Figure 7–162 shows raypaths for *P-P* (incident P, reflected P) in dashed lines and for *P-SV* (Incident P, reflected SV) in solid lines. The *P-P* ray paths have a common reflection point (CRP) at the CMP. The *P-SV* ray paths do not have a common reflection point. If the same traces are gathered and stacked for *P-SV* as for *P-P*, there will be considerable depth-point smear. That is, any structure on the reflector will cause reflections on the CMP record to be misaligned, giving poor stack response and distorting the actual structure.

Figure 7–163 shows the correct gather for *P-SV* data. The ray paths were drawn with $V_p/V_s = 2$. Thus, for angle of incidence ϕ_i, the angle of reflection, is given by $\phi_i = \sin^{-1}(0.5\sin\phi_i)$ and offset X_i is given by $X_i = Z(\tan\phi_i\ e + \tan\phi_i)$ where Z is depth to the reflector.

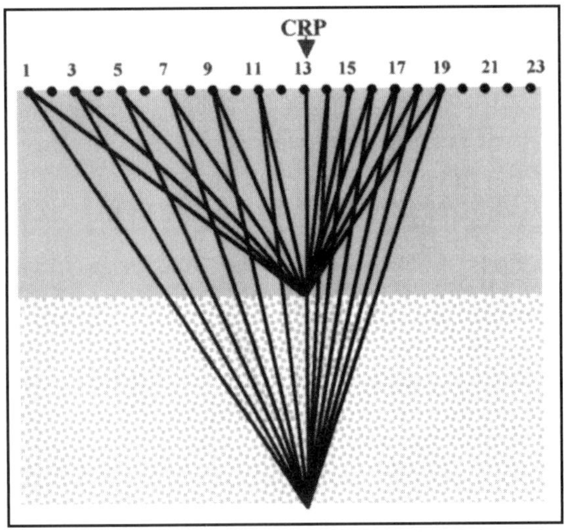

Fig. 7–163 *Ray Paths of CRP Traces for P-SV*

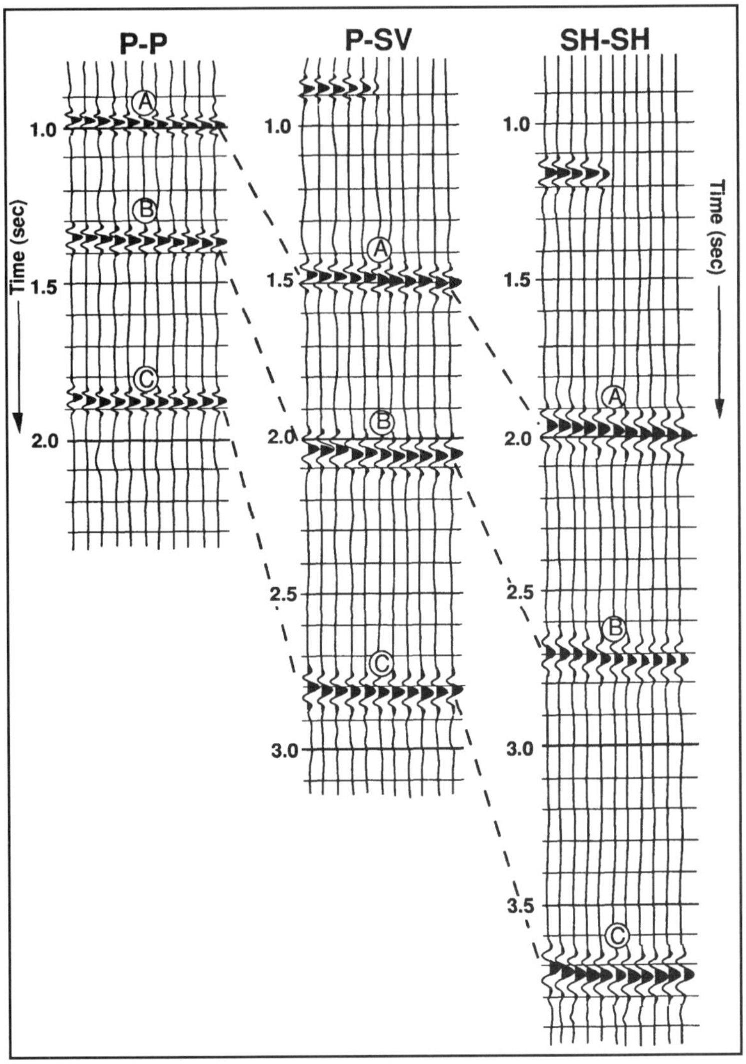

Fig. 7–164 *Representative P-P, P-SV, and SH-SH Traces*

For small angles (0 < 10° or $n/18$ radians), sinθ = tanθ = 0. In such cases, for Figure 7–163, $X = 1.5\ Z\theta_i$.

As shown in Figure 7–163, this condition occurs for the shorter offsets and deeper reflections. For the shallower reflection in Figure 7–163, the reflection points for the short offsets fall very nearly below the point labeled CRP (2/3 of the offset distance measured from the source), but for the longer offsets the reflection points are closer to the receiver. All reflection points fall on the CRP for the deeper reflection.

Table 7–2 presents a further illustration of the differences between trace gathers for *P–P* and *P-SV*.

Table 7–2 *P-P and P-SV CRP Trace Attribute Comparisons*

Source Location	*P-P* Receiver Location	*P-P* Offset/Δx	*P-SV* Receiver Location	*P-SV* Offset/Δx Offset/Δx
1	23	22	19	18
3	21	18	18	15
5	19	14	17	12
7	17	10	16	9
9	15	6	15	6
11	13	2	14	3

Note that Figures 7–162 and 7–164 plus Table 7–2 are used to illustrate general concepts and are not typical of any field or processing parameters for P-SV data.

As shown in Figure 7–164, the same events occur at different times on *P-P*, *P-SV* and *SH-SH* records. Correlation of events on these records is facilitated by applying a scaling factor to the display parameter that controls the time scale on the section display. For example, if the *P-P* section is displayed using $TSCALE_{P\text{-}P} = N$ inches per sec, than the *P-SV* and *SH-SH* sections will be displayed with:

$$\text{TSCALE}_{\text{P-SH}} = \left(\frac{1 + \left(V_P / V_S \right)_{reg}}{2} \right) \text{TSCALE}_{\text{P-P}} \tag{7.24}$$

and

$$\text{TSCALE}_{\text{P-SH}} = \left(V_P / V_S \right)_{reg} \text{TSCALE}_{\text{P-P}} \tag{7.25}$$

where

$(V_p/V_s)_{reg}$ is the gross regional average of the P to S velocity ratio

Figure 7–165 shows the traces of Figure 7–164 after scaling for velocity differences. Figures 7–166 and 7–167 show the same contrasts for real data.

Figures 7–166 is a comparison of P-wave and SH–wave seismic sections from west Texas. Figure 7–167 also compares S-wave and P-wave sections. Note that the S-section has more reflections indicated by *. The comparison is remarkable, in that the correlation at $V_s/V_P = 0.5$ has been obtained for the pre-Tonkwa section. From this point to the surface, correlations are difficult, partly due to the presence of unpaired shear and P-wave reflections at shot point 355.

Enough work has been done to establish the S-wave as a valuable tool for removing uncertainties in the origin of bright spot. If the S-wave can penetrate the weathered layer, then shear wave reflections can be expected. This is a plus, even though there may not be a correlation between P- and S-waves.

Once correlation has been definitely established (by the use of VSPs for P- and S-waves), V_s/V_P can be calculated. Given good continuity in the reflection, V_s/V_P changes can be followed through the surveyed area. Lithology determination using sections recorded by P-and S-waves has not reached a high degree of reliability simply because there is insufficient data on localized lithology, fracture densities, porosities, and pore fluids. It is possible that a simple relationship involving two parameters, such as V_s/V_P and V_P, can be used to discriminate between rock types. Unfortunately, how-ever, the lack of data is hindering development of such relationship.

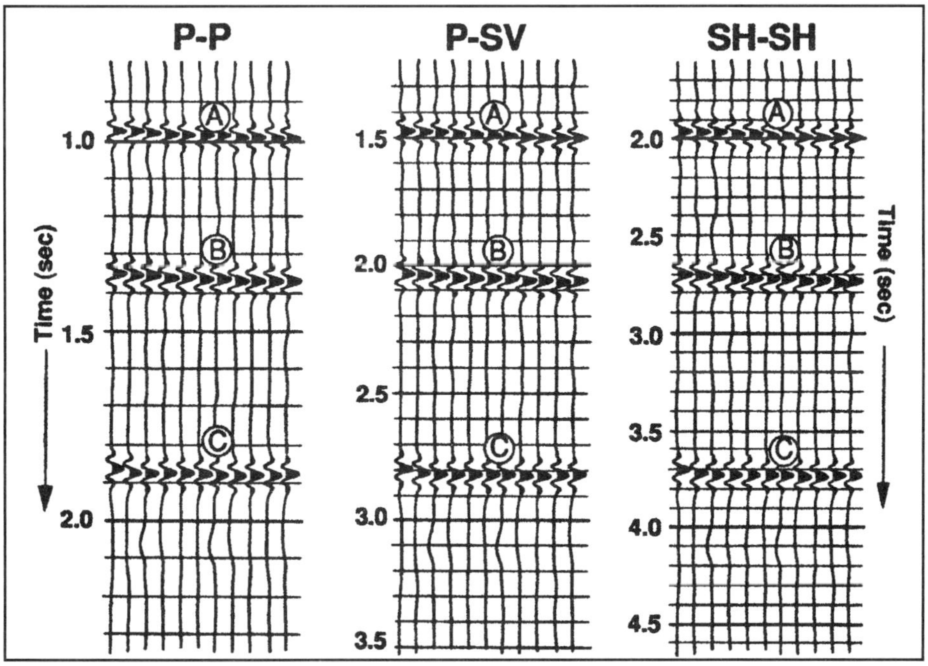

Fig. 7–165 *Traces of Figure 7–164 after Time Scaling to Enhance Event Correlation*

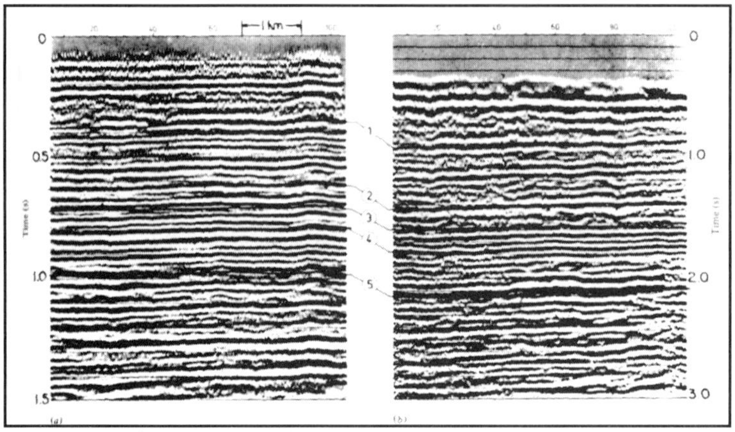

Fig. 7–166 *Shear-wave and P-wave Sections*

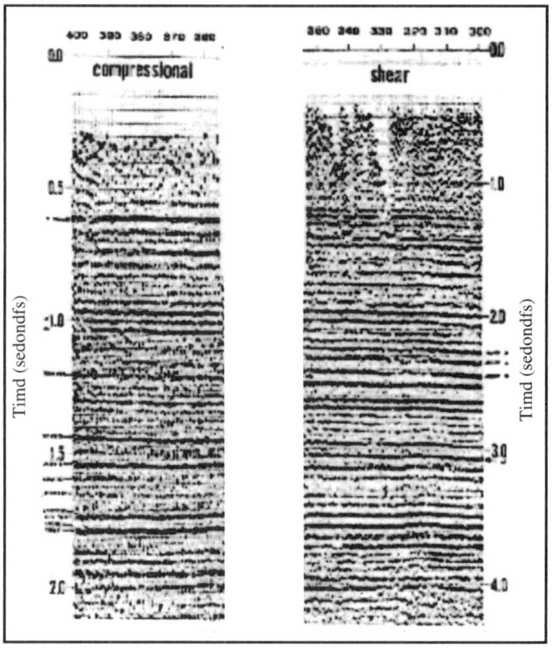

Fig. 7–167 *Unpaired Reflections*

Statistical discrimination, based on multiple variables, is possible to relate rock type to velocity. These variables may include

1. V_s/V_P
2. V_s or V_P separately
3. time of occurrence or depth of burial
4. anisotropy ratio
5. attenuation constant (Q)

Wave velocity in rocks has been measured by at least four different methods. These methods, with their strengths and weaknesses, are summarized in the following list.

- Ultrasonic measurements have been made on specimens cut from cores. The problem with this method is that the measurements are made under unnatural conditions of overburden, pressure, and temperature.
- Velocity logging measures refracted wave traveling in the formations surrounding a drilled hole, where the acoustic conditions are altered. Moreover, the formation and fluid temperatures and pressures are disturbed by the mud filtrate.
- Seismic data measurements use the standard velocity spectrum method of obtaining velocities present. This velocity type analysis has good accuracy for stacking, but the accuracy of determining interval velocities falls rapidly as depth increases. It is difficult to use for interval velocities of individual rock types unless thick, homogeneous formations are present.
- Down-hole measurements or vertical seismic profiling add a new method, in which reflection times for two different wave types and measured depths can be correlated. This method offers the best chance of obtaining values of V_s/V_P and correlating them with lithology.

Much research effort has been devoted to relating seismic velocities to rock properties. However, because of uncertainties and the number of factors involved, the correlations have been contradictory at times. Seismic velocity or velocity ratio can, under limited conditions, be related to rock properties such as lithology, porosity, fracturing, or connate fluid.

The following equations are empirical relationships that were derived for specific geologic provinces.

The Gardener equation:

$$\rho = 229.5V_p^{.25} \tag{7.26}$$

This equation relates the density of the rock to compressional interval velocity. In this equation, ρ is the density in Kg/m^3 and V_P is the compressional interval velocity in m/sec.

Ludwig et al. (1970) found a complex relationship between V_S, V_P, and V_s/V_P. M. Yoshimura et al. (1982) suggested the following relationship, which is valid for shear-wave velocity (VS) in the range of 0.4 to 2.0 km/se:

$$\frac{V_S}{V_P} = 0.576 - \frac{.139}{V_S} \tag{7.27}$$

Equation 7.22 is known as the *Yoshimura equation.*

Clay and soils display values of Poisson's ratio greater than 0.4 and *VS/VP* can reach values as high as 6.0 in Gulf Coast sediments. Thus, the presence of such materials within the pores of clastic sediments greatly affects the relationship between velocities and rock properties. Anderson et al. (1972) and Tixier et al. (1973) derived an empirical relationship between what they call a *Shaliness factor q* and Poisson's ratio, The shaliness factor is computed from the equation:

$$q = \frac{(\varphi_z - \varphi_e)}{\varphi_z} \tag{7.28}$$

where

φ_Z = total pore space between grains that support the overburden (equal to density log porosity)

φ_e = porosity available for water and hydrocarbons (equal to sonic log porosity)

Poisson's ratio σ is computed from q by:

$$\sigma = 0.125q + 0.27 \tag{7.29}$$

where

$q = 0$ (fluids can fill all the pore space)

$\sigma = 0.27$, which corresponds to $V_S/V_P = 0.5613$

If $q = 0.4$, then $\sigma = 0.32$, and $V_S/V_P = 0.5145$

Many studies have been done to relate V_S/V_P and rock type. Pickett (1963), using numerous measurements, demonstrated that the ratio of P-wave slowness (reciprocal of velocity) to S-wave slowness, could be related to rock type. Figure 7–168 illustrates the slowness plot and the relationship between V_S/V_P and different rock types such as limestone, dolomite, and sandstone. The points for carbonates all have V_S/V_P ratios less than 0.56, while sandstone points have ratios greater than 0.56. Limestone exhibits the lowest ratios at about 0.53.

Benzing (1978) made measurements of V_S/V_P in the laboratory by applying a rock pressure of 10,000 psi and a fluid pressure of 4350 psi to a 100% water-saturated rock. Figure 7–169 shows the results for all types of rock. All sandstone points are included within the irregular line.

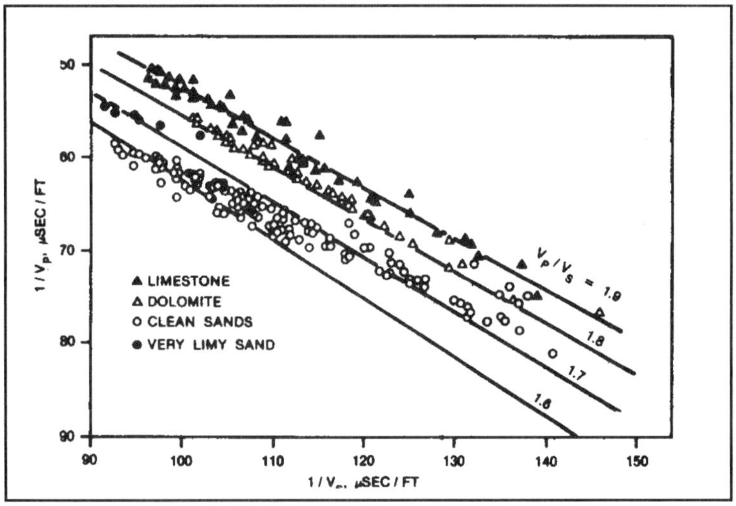

Fig. 7–168 *Rock Velocities Versus Lithology from Well Logs*

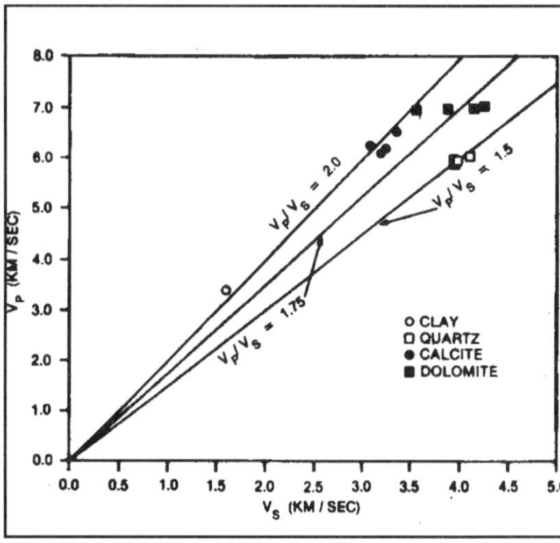

Fig. 7–169 *Rock Velocity Versus Lithology from Laboratory Examples*

At low velocities, $V_S/V_P = 0.50 \pm 0.08$. At higher velocities, quartzite $V_S/V_P = 0.55 \pm 0.05$ and carbonates (average) $V_S/V_P = 0.46 \pm 0.03$. The separation between carbonates and sandstone is not as clear as in Pickett's chart Figure 7–168.

Figure 7–169 is a plot of V_P versus V_S for numerous rock samples measured in the laboratory, with simulated pressure and temperature. The separation between sandstones and carbonates by the almost closed boundary is based on physical description. It would be difficult to identify on basis of a constant V_S/V_P zone.

If a geologic marker can be obtained for S- and P-wave reflections, then V_S/V_P and t_P/t_S can be determined. Well logs can be used to identify lithology, but it is not simple to find corresponding reflections. Using VSP can solve this problem. VSP will show the seismic marker that should correspond to the lithology boundary. As this is one of the most important applications of shear wave seismology, the determination of V_S/V_P in different areas can be related to lithology and other properties, defining shale, salt, and other soft rocks.

The use of modern sonic logs coupled with VSPs is the only way to obtain accurate V_S/V_P values. This method is being used more and more in exploration and development efforts.

There are some reservations concerning the acquisition of SH data.

- There may be some areas in which surface conditions make it difficult to generate deep shear waves.
- Because of rapid decay of S-wave energy, the shear wave records become narrower band with depth, and accordingly lose definition.
- Near-surface coherent noise (Love waves) can be recorded and cause some trouble. Experimental fieldwork may be conducted in an effort to attenuate this noise pattern.
- Present sources of energy can cause surface damage that may limit acquisition permits.

The future of shear waves in reservoir analysis. Shear wave data acquisition and processing are still in the stage of development, and there is much to be learned about the information contained in S-waves. Research is advancing rapidly in this area. It has been reported that shear-wave dispersion can be used as an indicator of the presence of permeability in a reservoir rock.

With the advancement in field data acquisition techniques and the utilization of borehole tools to record the three wave components, it is evident that borehole geophysics will play a vital role in describing the petrophysical properties of the reservoir rock in greater detail.

Shear-wave recording will be helpful in identifying the rock type, change in lithology, and detection of permeability. In time, and with much research, there will be more applications that can be refined using shear-wave velocities.

With the demand for hydrocarbons increasing worldwide and the scarcity of undeveloped structural traps, it is obvious that stratigraphic traps must be the main source for additional oil production. Seismic stratigraphy, or seismic analysis of stratigraphic sequences, can be used effectively to solve various problems related to the irregular and subtle nature of stratigraphic traps.

Well-designed and carefully conducted 2-D or 3-D seismic surveys can obtain high-resolution data using compressional and/or shear wave energy. Through high-resolution data processing using optimum processing parameters and by analyzing the seismic signature variations, one can obtain fine details about unconformities, facies changes, depositional environ-ments, and direct indications of hydrocarbons. By studying the continuity of events, phase changes, and amplitude anomalies such as bright spots for gas sand and dim spots for carbonates, one can greatly increase the level of confidence in a certain play.

For more than two decades borehole seismic receivers have played a vital role in better imaging of the subsurface. VSP has helped in solving a number of complex stratigraphic problems, such as developing meandering channel sand reservoirs, identifying of salt dome flanks, and defining reservoir boundaries.

Much research is under way, seeking to get more measurements in the bore-hole and measurements between wells. This research focuses on a down--hole source that can generate compressional, horizontal shear, and vertical shear energy in one borehole and a three-component receiver that responds to the three types of energy in another borehole. We can expect that, in the near future, the recorded velocities can be related to petrophysical properties away from the boreholes, which will help in better description of the reservoir.

Recorded velocities have been used to identify lithology, rock type, porosity, fluid saturations, and other properties. It has been reported in published articles that it is even possible to relate seismic dispersion to permeability.

Tomography, a technique used extensively in medicine for imaging the human body, will find greater use in imaging formations between wells. This will provide earth scientists with information to permit better calculations of oil in place, to better under-stand reservoir performance and design better recovery methods. All these new techniques in seismology will have a great impact on seismic stratigraphy.

In the future, the success ratio in developing stratigraphic traps will improve. We will gain better understanding of reservoir performance, allowing us to enhance our methods and techniques in secondary and tertiary recovery projects. Imaging of the subsurface will become more accurate and detailed, leading to more economical and efficient exploration, development, and exploitation.

4-D seismic technology

4-D seismic technology is based on 3-D surveys over the same area at different times. The earlier 3-D survey may have been done before or after production has taken place. The later 3-D survey must always be done after production has taken place since the purpose of 4-D is to measure changes in areal and depth distributions of fluid, temperature, and pressure in producing reservoirs with time. The measured seismic differences are combined with production and other data to better define the reservoir and changes that have occurred in it. Geologists and reservoir engineers are usually involved in planning, implementation and interpretation of a 4-D program. This assures better results in optimizing production, improving ultimate hydrocarbon recovery and reducing costs.

4-D technology shows changes in reservoir condition as differences in data attributes in successive 3-D seismic surveys. Analysis of the observed changes allows reservoir engineers to adapt field development to match geologic complexities of reservoirs. The potential benefits are increased reserves—produced faster and at a lower cost.

4-D technology has been applied in both onshore and offshore environments, to clastic and carbonate reservoirs, and a where a variety of production techniques have been used. These include

- steam and fire floods
- miscible solvent flood
- CO_2 injection
- conventional waterfloods
- gas injection

4-D has also been used over natural gas storage reservoirs.

Success in 4-D programs requires that changes in reservoir conditions, such as change in fluid levels, produce measurable changes in seismic attributes such as amplitude. If this is the case, reservoir simulation predictions can be confirmed or modified based on seismic data anomalies. 4-D data may be used to produce more detailed and accurate reservoir models that result in better fluid flow predictions and reservoir performance. Changes in seismic character allow tracking of reservoir fluid movement as a function of time. This information can be used to change production techniques and well patterns that improve ultimate recovery and profit.

Conventional production data provides data from wells. Seismic data can provide production information between existing wells. For example, identification of reserves not drained by existing well completions allows siting of new wells, deepening existing wells, or recompletions. Additional constraints to reservoir modeling are provided through history matching, resulting in more accurate and more detailed models.

The basis of 4-D is the measurement of differences among 3-D seismic surveys, separated in time. How well this is done depends on things such as:

- level of random and coherent noise (including seasonal or environmental variation in noise)
- repeatability of seismic signal in both generation and recording
- accuracy and repeatability of navigation and survey.
- limits of resolution and detection (dependent on frequency content of signal)
- predictions of seismic response changes
- reservoir depth and complexity and lithologic characteristics overlying geologic structures
- existence of man-made obstructions
- variability and complexity of the near-surface

A successful 4-D program requires integration of geophysical and production technologies and organizations. Change in reservoir conditions *must* be such that a change in the seismic reflection response is observed. Since response changes are usually small, extremely high quality seismic data acquisition and processing are required. Techniques and parameters used in acquisition and processing must be chosen to enhance the anticipated fluid movement indicators.

Before conducting a 4-D project, a feasibility study is advised. This should include evaluation of the expected reservoir conditions, modeling, and data review. Also, a pilot study over a small area with good potential for success is a good step before field-wide application of the project.

It is *not required* that the same acquisition equipment and techniques or same data processing flow be used for the subsequent 3-D surveys. Use of the best technology and the flexibility to modify survey geometries and parameters as project objectives and problems are needed to sucessfully implement a 4-D project. Processing and analysis technology should take care of differences in acquisition and processing.

Figure 7–170 is a flowchart for a 4-D project run over a reservoir that has been water flooded.

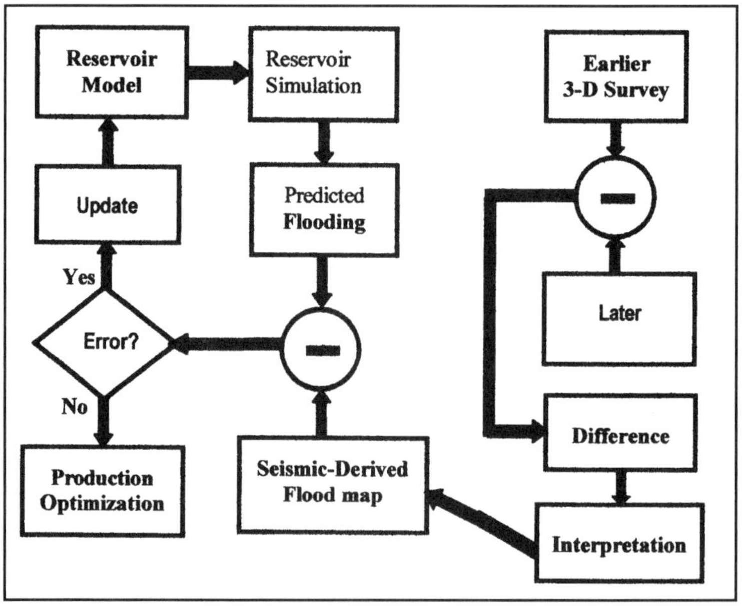

Fig. 7–170 *Flowchart for a 4-D Project*

Two examples, based on papers published in the *Leading Edge*, a publication of the Society of Exploration Geophysicists, are presented. The first example is a *legacy time-lapse* study. That is, a time-lapse study was done using two independent 3-D data sets from the same area. The second was not done with 4-D in mind. This example was reported by a group of WesternGeco employees (Xuri Huang, Robert Will, Mashiur Khan, and Larry Stanley) in the March 2001 issue of *The Leading Edge*.

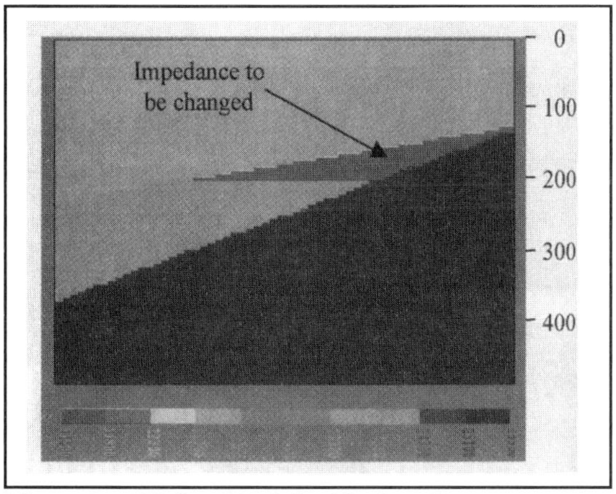

Fig. 7–171 *Wedge Model with Gas Cap*

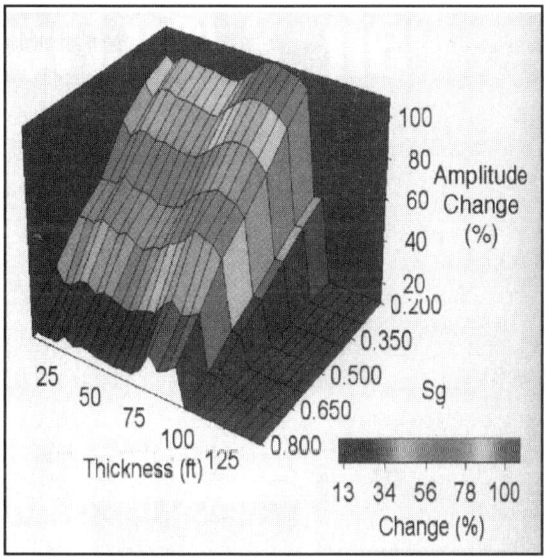

Fig. 7–172 *Relationships among Reservoir Thickness, Gas Saturation, and Amplitude Change*

The second example makes use not only of a legacy 3-D but also a two-phase time-lapse program that produced some surprising results. This was reported Wayne D. Pennington, Horacio Acevedo, Joshua J. Haataja, and Anastasia Minaeva of the Michigan Technological University in the October 2001 issue of *The Leading Edge.*

In the first example, the reservoir selected for study was the shallowest of a series of stacked oil and gas pays in the Gulf of Mexico. This reservoir produces gas from a faulted anticlinal trap at a depth of about 3000 ft that is normally pressured and exhibits a strong water drive. Three wells were drilled but only one was producing gas due to high water production in the others.

The first of the two 3-D seismic surveys was done in 1987 and the second in 1995. Gas production had started before the first 3-D survey. In the time between surveys, approximately 26.6 bcf of gas was produced. Well logs and production data were combined with the legacy 3-D data to provide spatial constraints on estimates of produced and remaining reserves.

Modeling the expected acoustic response of the reservoir to simulate fluid saturation changes resulted in the adoption of a patchy saturation model. This model predicted an acoustic impedance increase of approximately 10% in the swept zone (30% gas replaced by water). The reservoir model used (Fig. 7–171) is a wedge with thickness varying from 0–150 ft and a gas cap. Table 7–3 shows the change in acoustic impedance with change in water saturation in the gas cap. Figure 7–172 shows the relationships among seismic amplitude, wedge thickness, and gas saturation change for a frequency of 60 Hz. The results are based on model runs that assumed a patchy saturation distribution.

Modeling was also done to evaluate the effects of signal frequency changes and saturation on amplitude difference. It was determined that if the data sets have similar frequency content for a given thickness of 70 ft and change of gas saturation of 40%, then the amplitude change will be around 80%. This closely approximated their observations from the 3-D data over this field.

Table 7–3 *Acoustic Impedance Change Caused by Gas Saturation Change*

Gas Saturation (%)	Acoustic Impedance	% Change
88	11330	
80	11740	3.6
70	12213	7.8
60	12768	12.7
50	13138	15.9
40	13655	20.5
30	14095	24.4
20	14444	27.5

The two 3-D data sets were acquired and processed by different contractors. As a result, there were differences in source-receiver offset ranges, spatial sampling density, and 3-D migration algorithms. Little post stack treatment of the data sets was done. Frequency contents for both surveys were shown to be quite similar, with the second being slightly richer in high frequencies.

Sample intervals of 50 m in-line and 60 m cross-line were selected for the analysis. Data were spatially resampled or repositioned to allow meaningful differences between the two surveys to be measured.

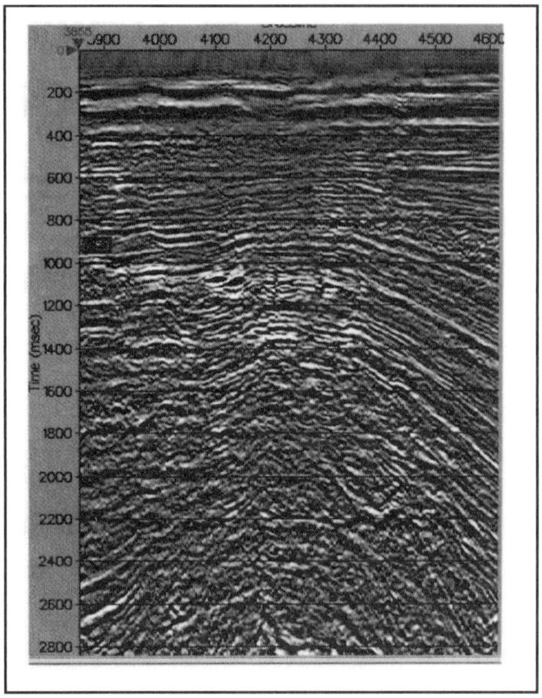

Fig. 7–173 *Raw Difference after Applying Global Equalization with a Single Scaler*

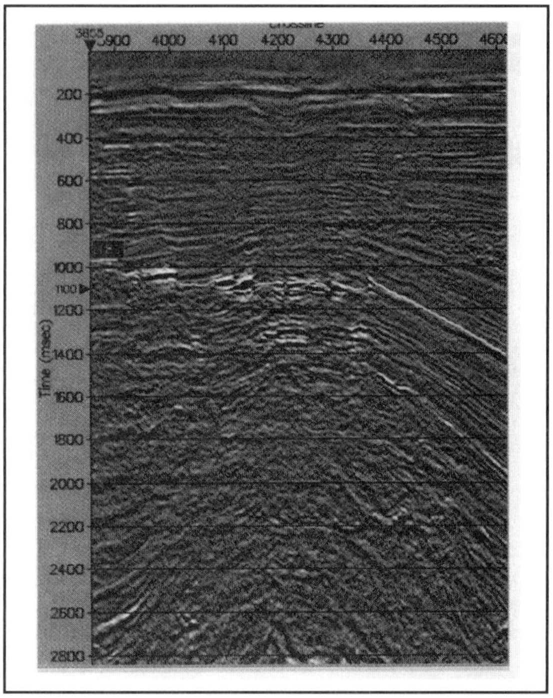

Fig. 7–174 *Difference after Global Phase and Amplitude Match*

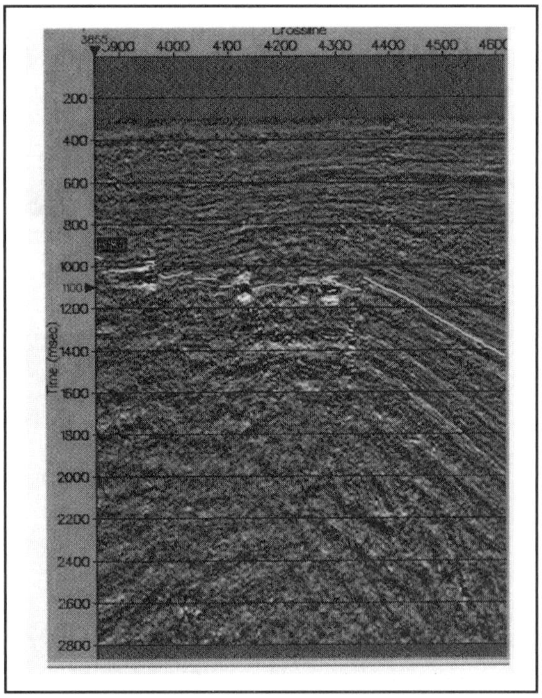

Fig. 7–175 *Difference after Time- and Space-Variant Cross-Equalization*

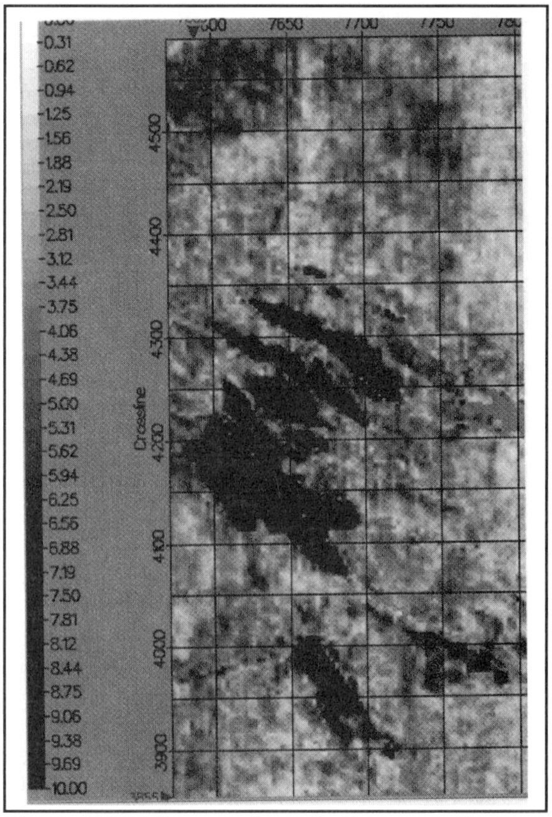

Fig. 7–176 *Difference along the Reservoir Horizon after Global Equalization*

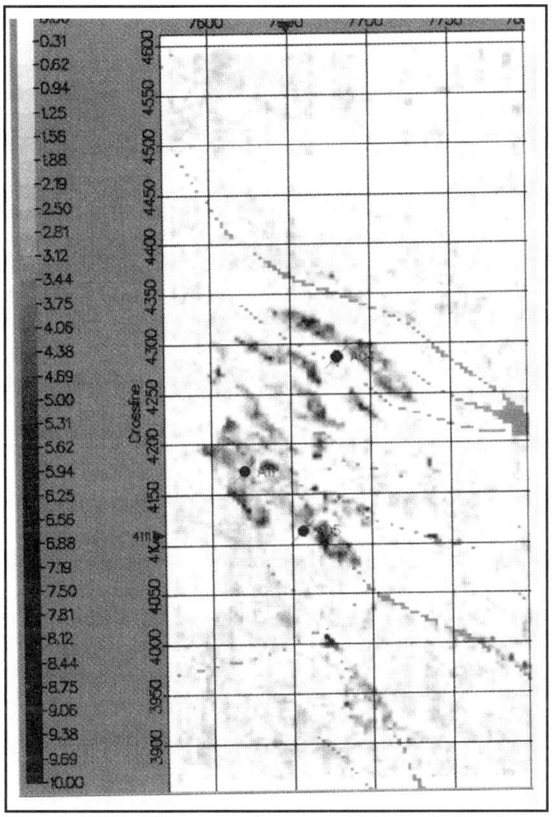

Fig. 7–177 *Difference along the Reservoir Horizon after Local Equalization*

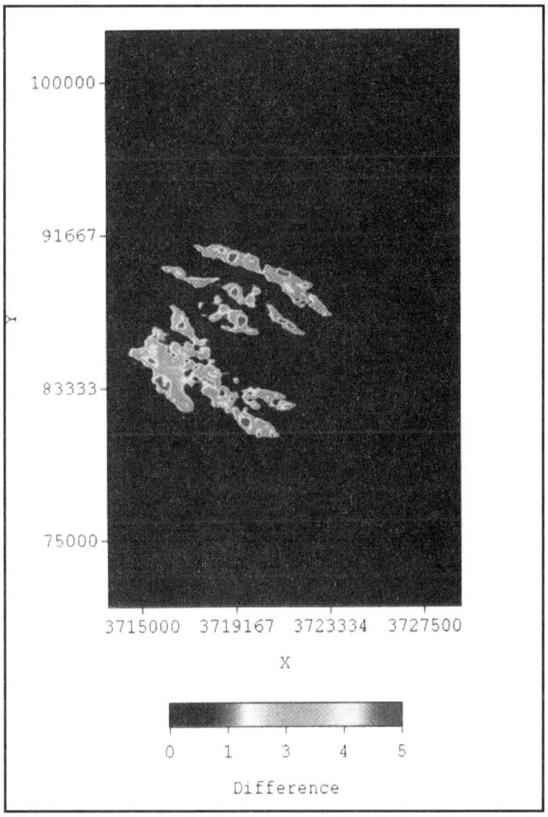

Fig. 7–178 *Seismic Difference after Matching with Cumulative Production*

Steps were also necessary to assure differences in amplitudes were not data-dependent. Data over a non-producing region below the field's original hydrocarbon-water contact was selected for this operation. This assured that amplitude equalization was not compromised by production-related changes. A rather complex method of amplitude equalization was used. This consisted of:

- band-limiting data sets to a common frequency bandwidth.
- scaling with a single global equalization scaler. Figure 7–173 shows the measured differences for an inline section. Significant differences are seen outside the reservoir interval.
- applying a global cross-equalization filter. The amplitude differences between reservoir and non-reservoir zones are significantly improved. Compare Figures 7–174 and 7–173.
- Applying a time- and space-variant cross-equalization. Figure 7–175 shows the difference for the same inline as in Figure 7–173 and 7–174. Additional significant improvement in repeatability was observed after this step.

Another seismic attribute used in this 4-D analysis was the difference in amplitude envelope along the reservoir horizon. Figure 7–176 shows this attribute after application of global equalization, and Figure 7–177 shows the difference along the same horizon after application of local correlation operators to the globally equalized volume. It was found that differences outside the reservoir boundaries were further reduced by the second equalization step.

Estimates of reservoir gas volume were made from production data, as well as from seismic differences. The two estimates required reconciliation for the seismic results to be reliable. A seismic amplitude threshold that best approximated the boundary of the swept area was determined and used to effect the reconciliation. Figure 7–178 shows the seismic difference over the reservoir horizon after matching with cumulative production.

The three wells were used to calibrate the seismic data. This calibration was combined with the results shown in Figure 7–173 to produce the residual gas saturation map of Figure 7–179.

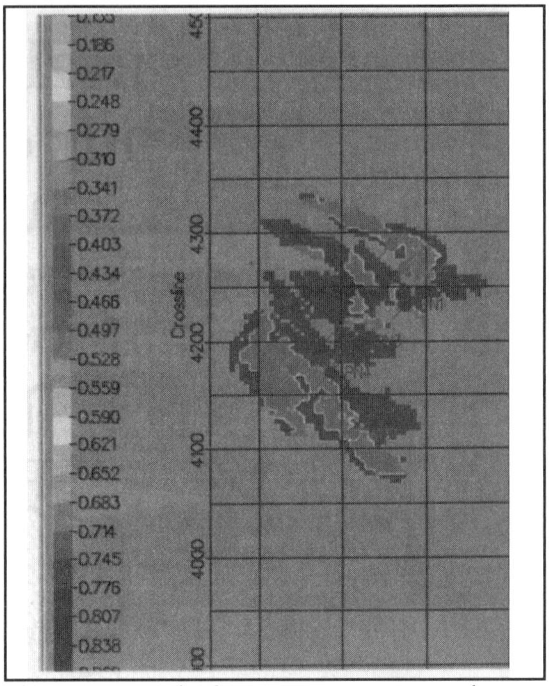

Fig. 7–179 *Residual Gas Saturation Map after Material Balance Matching and Calibration*

The second example was also a study done on reservoirs in the Gulf of Mexico. Data used were from a pre-production 3-D survey and two additional 3-D surveys (Phase I and Phase II). The pre-production 3-D survey was a conventional marine survey using streamers. Phases I and II were OBC surveys using 4-D equipment. (Each OBC group contained three-component geophones plus a hydrophone).

Figure 7–180 is a three-dimensional representation of the sand structure containing the two reservoirs covered in the report. The arrow points north. The box outlining the volume extends from 1.250 ms to 1750 ms in two-way travel time, and is roughly 8000 ft (2500 m) on each side.

The production history of the 4500 ft reservoir is shown in Figure 7–181. Note times of Phases I and II are also indicated in the figure. Pennington (SEG 2000 *Expanded Abstracts*) predicted that the change in the AVO class III reservoir would be more complicated than what conventional wisdom predicted. The latter prediction was that there would be continued brightening of the bright spot at all offsets. Pennington predicted that, after the initial brightening, the near offsets would dim.

Petrophysical predictions from time-lapse 3-D data are based on a petrophysical model. Because of inconclusive log data, the legacy 3-D data volume was inverted for acoustic impedance (Fig 7–182). The model developed was checked by using the acoustic impedance in the water sands to accurately predict acoustic impedance in the oil sands. This provided the rock properties needed to make predictions of seismic response during production.

Figure 7–183 shows the amplitude differences between Phase I and Phase II at the 4500 ft horizon while Figure 7–184 shows the difference in a perspective view. Both the 4500 ft reservoir (*A* in Fig. 7–180) and the Little Neighbor reservoir (*B* in Fig. 7–180) show significant changes. This is surprising since no wells had been drilled into the Little Neighbor reservoir.

Figure 7–185 shows changes in P-wave velocity V_P, Poisson's ratio PR, and acoustic impedance with production time. The upper graph shows V_P and *PR* resulting from fluid calculations only (dotted lines) and with frame stiffening included (solid lines). It can be seen that the frame stiffening effect more than offsets the reduction in velocity caused by fluid substitution and enhances the Poisson's ratio effect. The lower graph in Figure 7–185 shows that changes in density when oil replaces gas lessens the effect of frame stiffening on acoustic impedance. However, after the large initial drop, impedance continues to rise.

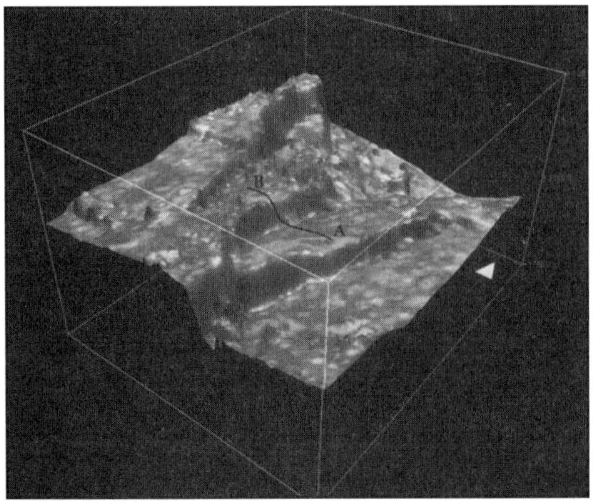

Fig. 7–180 *Perspective View of the Sand Structure Containing the Currently Producing 4500 Ft Reservoir*

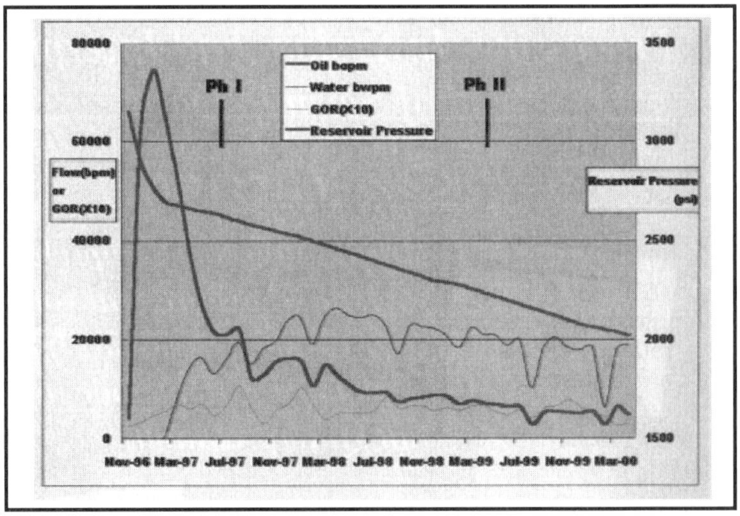

Fig. 7–181 *Smoothed Production History of the 4500 ft Reservoir*

Fig. 7–182 *Inverted Legacy Data Volume Showing Acoustic Impedance 12 ms below the Top of the Tracked 4500 ft Horizon*

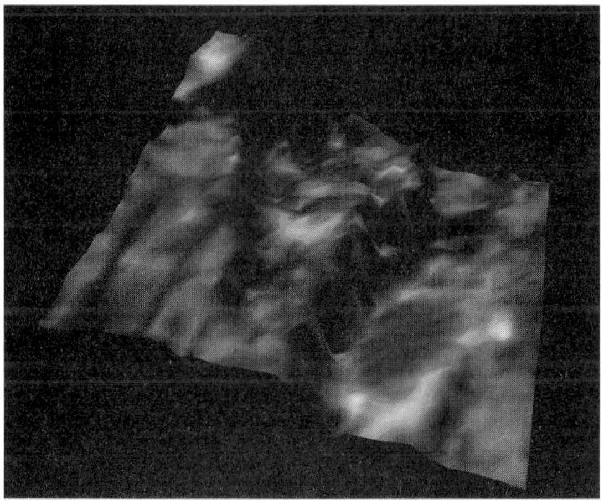

Fig. 7–184 *Changes in P-wave velocity, Poisson's Ratio, and Acoustic Impedance with Time of Production*

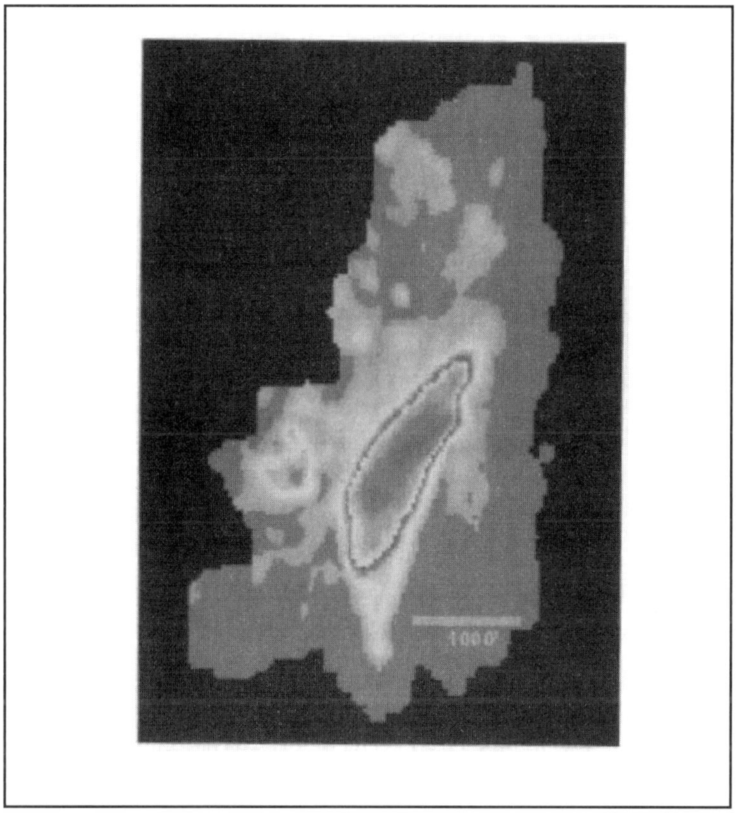

Fig. 7–183 *Time-lapse Difference Mapped on the 4500 ft Reservoir*

The model adopted for fluid substitution and frame stiffening caused by changing gas saturation predicted that the P-wave velocity would initially decrease then significantly increase during production. It also predicted that Poisson's ratio would continuously decrease during production. These changes lead to an AVO effect that includes initial brightening of the bright spot at all offsets followed by a dimming at near offsets and continued brightening at far offsets, as Pennington had predicted.

Investigation of pre-stack behavior of reflections from both the 4500 ft and Little Neighbor reservoirs substantiated the AVO behavior described above and also showed that fluid substitution had occurred in the Little Neighbor reservoir. Figure 7–186 shows partial stack amplitudes for three different offset ranges. The bottom parts of Figure 7–186 show wiggle trace amplitudes (both Phase I and Phase II) and color-coded amplitudes. The top parts of the figure are diagrams based on measurements on the upper peak and lower troughs of each trace shown below each diagram. The data of Figure 7–186 indicate that near-offset trace amplitudes remained essentially unchanged between Phases I and II, but far offset trace amplitudes increased over the same period. This provided at least partial agreement with the model used.

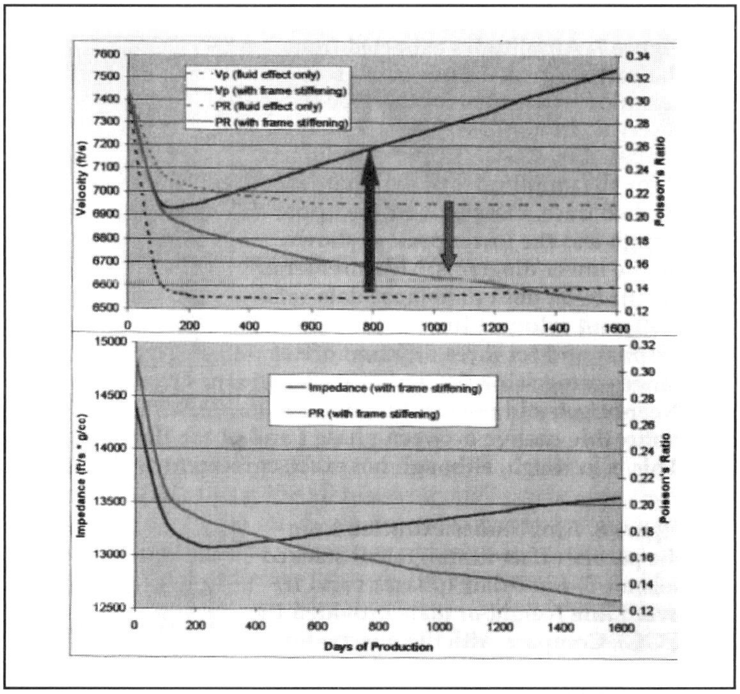

Fig. 7–185 *Amplitudes Extracted from Partial-Offset Stacked P-Wave Data for the 4500 Ft Reservoir from Phases I and II*

Figure 7–187 provides the same information for the Little Neighbor reservoir as in Figure 7–186 for the 4500 ft reservoir. In this case near-offset amplitudes decreased and the far offset amplitudes increased between Phases I and II. This is very good agreement with the model used. This data was provided by CCG (Pennington et al, 2001).

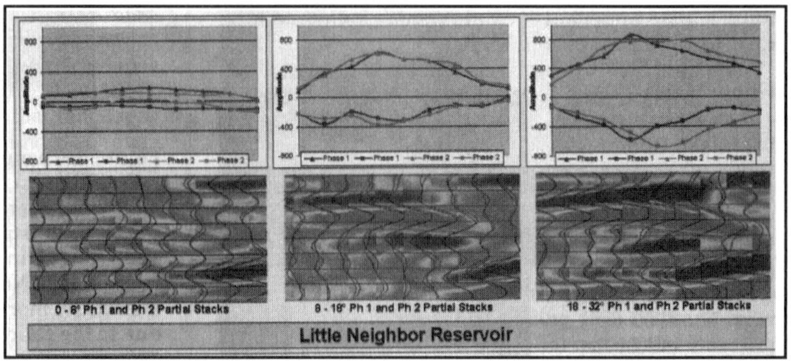

Fig. 7–186 *Amplitudes Extracted from Partial-offset Stacked P-wave Data for the Little Neighbor Reservoir from Phases I and II*

The authors note that the observed behavior contains two surprises for the conventional viewpoint.

1. Amplitudes do not monotonically increase as additional gas is released. In the interval between time-lapse surveys, the near-offset amplitudes eventually decrease after initially increasing and far offset amplitudes increase.

2. Similar changes over the Little Neighbor reservoir indicate it is responding to production in a similar manner as the 4500 ft reservoir. Hence there must be a decline in pressure in this reservoir because of production in the 4500 ft reservoir.

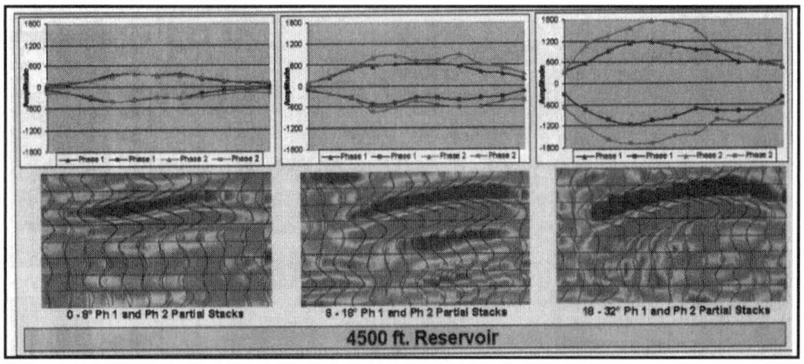

Figure 7–187 *Amplitudes extracted from Partial-Offset (Unmigrated) Stacked P-wave Data for the Litte Neighbor Reservoir from Phases I and II*

The authors' explanation for the behavior In the Little Neighbor reservoir and the connection with production in the 4500 ft reservoir is as follows:

a) production in the 4500 ft reservoir begins—here is, as yet, no sign of water encroachment and OOWC is the original oil-water contact

b) production in the 4500 ft reservoir results in a free gas cap and water encroachment and this causes a pressure decrease that is communicated through the water sands

c) the drop in pressure creates a free gas cap in the Little Neighbor reservoir and downward displacement of the remaining oil—the oil escapes through the spill-point and is, perhaps, lost forever

The authors concluded that this application of 4-D technology yielded some completely unexpected results. The phenomenon of production in one reservoir leading to loss of hydrocarbons in another unproduced reservoir may or may not be widespread in the Gulf of Mexico but without the time-lapse program carried out here, it would not have been recognized in this case.

Summary and conclusions. Time-lapse 3-D or 4-D technology is now one of the more important applications of seismology. With the advent and continued enhancement of this technology, seismic applications to the petroleum industry have matured from a suspect method of locating potential reservoirs to a reliable method of delineating petroleum reservoirs and optimizing production of petroleum from such reservoirs.

Design, application, and interpretation of 4-D technology is a multi-disciplinary activity. To have maximum effectiveness, geologists, reservoir engineers, and exploration seismologists must work in close concert in developing geologic and production models, determining what seismic responses to expect based on the models, and in interpreting the results of the surveys.

Systems are available that use permanently implanted sensors within boreholes and on sea floors. These allow reservoir monitoring at virtually any desired time interval. Continued increase in the sophistication of 4-D technology is likely as is the use of 4-D technology.

Appendix A Refraction Calulations

Calculating layer thickness, dip, and velocity can be very tedious when many layers are involved. Properly programmed computers can, of course, relieve much of this, but the calculations still require a lot of input. The following is presented to show how in-line seismic refraction data are interpreted. Cases presented are

- single horizontal layer
- two horizontal layers
- single dipping layer
- two dipping layers

Single horizontal layer $V_1 > V_0$

Figure A–1 shows the model and notation used in the derivations that follow.

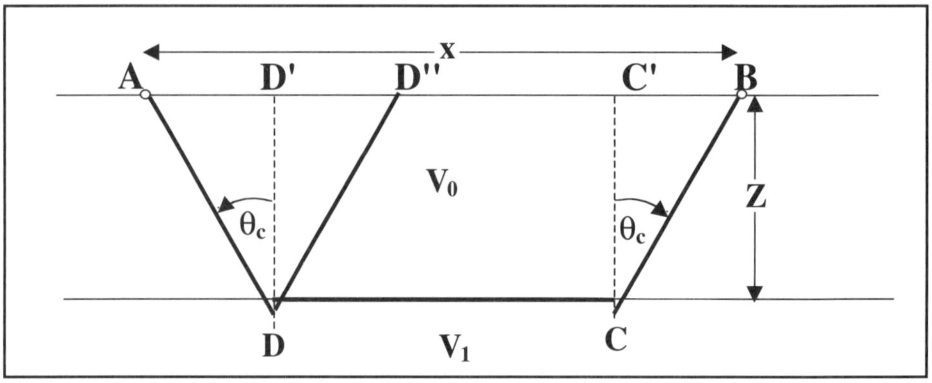

Fig. A–1 *Earth Model for a Single Horizontal Layer*

Assume a source is at *A* and a receiver at *B*. The refraction path is from *A* to *D* to *C* to *B*. Propagation from *A* to *D* and from *C* to *B* is at velocity V_0. From *D* to *C* propagation is at velocity V_1. It follows then that refraction time *T* can calculated from:

$$T = \frac{\overline{AD}}{V_0} + \frac{\overline{CD}}{V_1} + \frac{\overline{BC}}{V_0}$$

From Figure A–1 it can be seen that:

$\overline{AD} = \overline{BC} = \frac{Z}{\cos\theta_c}$ and $\overline{CD} = x - (\overline{AD'} = \overline{C'B}) = x - 2Z\tan\theta_c$

where

Z = thickness of layer 1 = depth to refractor

θ_c = critical angle

Substituting in the original equation gives:

$$T = \frac{Z}{V_0\tan\theta_c} + \frac{x\text{–}2Z\tan\theta_c}{V_1} + \frac{Z}{V_0\cos\theta_c} = \frac{x}{V_1} + 2Z\left[\frac{1}{V_0\cos\theta_c} - \frac{\tan\theta_c}{V_1}\right]$$

Noting that $\tan\theta_c = \dfrac{\sin\theta_c}{\cos\theta_c}$,

$$\text{T} = \frac{x}{V_1} + \frac{2Z}{V_0\cos\theta_c}\left(1 - \frac{V_0\sin\theta_c}{V_1}\right)$$

But $\sin\theta_c = V_0/\,V_1$, so substituting in the previous equation:

$$\text{T} = \frac{x}{V_1} + \frac{2Z}{V_0\cos\theta_c}(1 - \sin^2\theta_c) = \frac{x}{V_1} + \frac{2Z\cos\theta_c}{V_0} = \frac{x}{V_1} + T_0 \qquad \text{(A.1)}$$

where

$$\text{T}_0 = \text{delay time} = \frac{2Z\cos\theta_c}{V_0} \qquad \text{(A.2)}$$

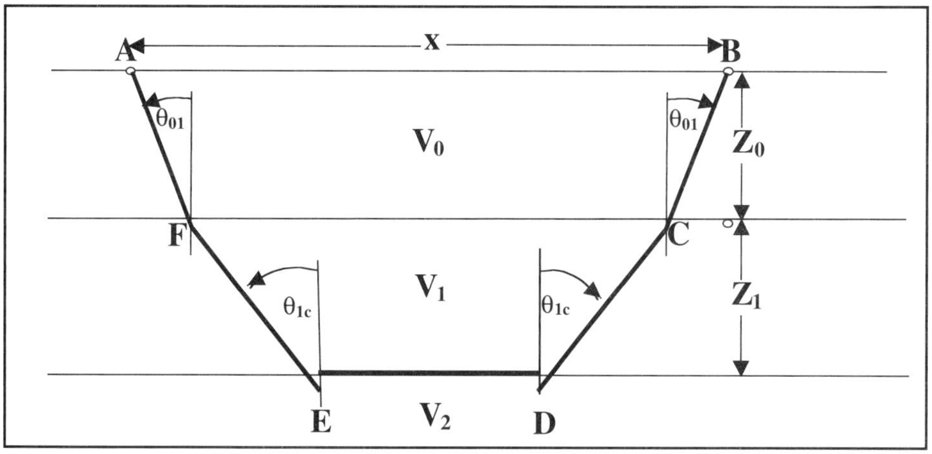

Fig. A-2 *Two Horizontal Layers Earth Model*

It follows that depth Z can be calculated from:

$$Z = \frac{V_0 T_0}{2\cos\theta_c} \quad \text{(A.3)}$$

Another parameter used in refraction is the *critical distance* or the distance at which the reflection at the critical angle occurs. This is also the minimum distance at which the refraction can be observed. From Figure A–1 the critical distance can be calculated as follows:

$$x_c = 2Z\tan\theta_c = 2Z\left(\frac{\sin\theta_c}{\cos\theta_c}\right) = 2Z\left(\frac{\sin\theta_c}{\sqrt{1-\sin^2\theta_c}}\right) = 2Z\left(\frac{\frac{V_0}{V_1}}{\sqrt{1-\frac{V_0^2}{V_1^2}}}\right) = 2Z\left(\frac{V_0}{\sqrt{V_1^2-V_0^2}}\right) \quad \text{(A.4)}$$

Two horizontal layers $V_2 > V_1 > V_0$

Figure A–2 shows the model and notation used in the derivations that follow.

Assume a source is at *A* and a receiver at *B*. The refraction path is from *A* to *D* to *C* to *B*. Propagation from *A* to *D* and from *C* to *B* is at velocity V_0. From *D* to *C*, propagation is at velocity V_1. It follows, then that refraction time T can calculated from:

$$T_2 = \frac{\overline{AF}}{V_0} + \frac{\overline{FF}}{V_1} + \frac{\overline{DE}}{V_2} + \frac{\overline{CD}}{V_1} + \frac{\overline{BC}}{V_0}$$

$$\overline{AD} = \overline{BC} = \frac{Z_0}{\cos\theta_{01}},\ \overline{EF} = \overline{CD} = \frac{Z_1}{\cos\theta_{lc}},\ \overline{DE} = x - 2Z_0\tan\theta_{01} - 2Z_1\tan\theta_{lc}$$

Substituting into the first equation, above:

$$T_2 = \frac{Z_0}{V_0\cos\theta_{01}} + \frac{Z_1}{V_1\cos\theta_{lc}} + \frac{x - 2Z_0\tan\theta_{01} - 2Z_1\tan\theta_{lc}}{V_2} + \frac{Z}{V_1\cos\theta_{lc}} + \frac{Z_0}{V_0\cos\theta_{01}}$$

$$= \frac{x}{V_2} + 2Z_0\left(\frac{1}{V_0\cos\theta_{lc}} - \frac{\tan\theta_{01}}{V_2}\right) + 2Z_1\left(\frac{1}{V_0\cos\theta_{lc}} - \frac{\tan\theta_{lc}}{V_2}\right)$$

$$= \frac{x}{V_2} + \frac{2Z_0}{V_0\cos\theta_{lc}}\left(1 - \frac{V_0\sin\theta_{lc}}{V_2}\right) + \frac{2Z_1}{V_1\cos\theta_{lc}}\left(1 - \frac{V_1\sin\theta_{lc}}{V_2}\right)$$

Since $\frac{\sin\theta_{01}}{V_0} = \frac{\sin\theta_{lc}}{V_1} = \frac{1}{V_2}$

$$T_2 = \frac{x}{V_2} + \frac{2Z_0\cos\theta_{lc}}{V_0} + \frac{2Z_1\cos\theta_{lc}}{V_1} = \frac{x}{V_2} + T_{lc} + T_{02} \tag{A.5}$$

where

$$T_{01} = \text{first delay time} = \frac{2Z_0\cos\theta_{lc}}{V_0} \text{ and} \tag{A.6a}$$

$$T_{02} = \text{second delay time} = \frac{2Z_1\cos\theta_{lc}}{V_1} \tag{A.6b}$$

The critical distance for this case is given by:

$$x_c = 2Z_0\tan\theta_{lc} + 2Z_1\tan\theta_{lc}$$

$$= 2Z_0\left(\frac{V_0}{\sqrt{V_2^2 - V_0^2}}\right) + 2Z_1\left(\frac{V_1}{\sqrt{V_2^2 - V_1^2}}\right) \tag{A.7}$$

N Horizontal Layers $V_0 < V_1 < V_2 < \ldots < V_N$

Using the model of Figure A–3 and extrapolating from the two layer case, the expression for the time of a refraction from the Nth refractor is:

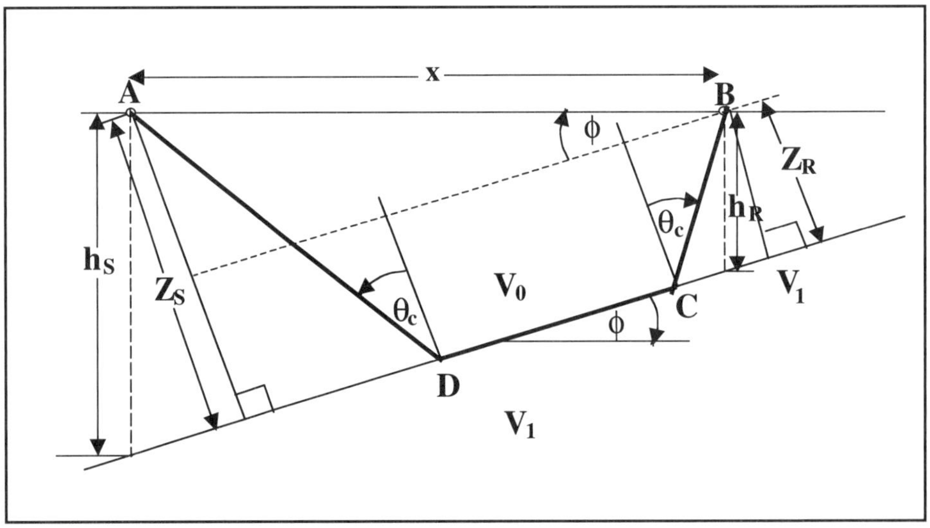

Fig. A–3 *Earth Model for a Single Dipping Layer*

$$T_1 = \frac{x}{V_N} + \sum_{n=1}^{N} T_{0n} \tag{A.8}$$

where

$$T_{0n} = \frac{2Z_{n-1}\cos\theta_{n-1,n}}{V_{n-1}} \tag{A.9}$$

Single dipping layer $V_1 > V_0$

Figure A–3 provides an Earth model and shows annotation used in the following derivations. As shown in the T-X plot of Figure 4–2, when the refractor dips, the refraction velocity in the up-dip direction appears to be different from that in the down-dip direction. From Figure A–3 the refraction time from A to B is:

$$T_1 = \frac{\overline{AD}}{V_0} + \frac{\overline{CD}}{V_1} + \frac{\overline{BC}}{V_0}$$

$$\overline{AD} = \frac{Z_S}{\cos\theta_c},\ \overline{BC} = \frac{Z_R}{\cos\theta_c},\ \text{and}\ \overline{CD} = x\cos\phi - (Z_S - Z_R)\tan\theta_c$$

Substituting this information into the starting equation:

$$T_1 = \frac{Z_S}{V_0\cos\theta_c} + \frac{Z_R}{V_0\cos\theta_c} + \frac{x\cos\phi - (Z_S - Z_R)\tan\theta_c}{V_1}$$

$$= \frac{x\cos\phi}{V_1} + (Z_S + Z_R)\left(\frac{1}{V_0\cos\theta_c} - \frac{(Z_S - Z_R)\tan\theta_c}{V_1}\right) = \frac{x\cos\phi}{V_1} + \frac{(Z_S + Z_R)}{V_0\cos\theta_c}\left(1 - \frac{V_0\sin\theta_c}{V_1}\right)$$

$$= \frac{x\cos\phi}{V_1} + \frac{(Z_S + Z_R)\cos\theta_c}{V_0}$$

Shooting up-dip (from A to B), substitute $Z_R = Z_S - x\sin\phi$.

$$T_{1u} = \frac{x\cos\phi}{V_1} + \frac{(2Z_S - x\cos\phi)\cos\theta_c}{V_0} = x\left(\frac{\cos\phi}{V_1} - \frac{\sin\phi\cos\theta_c}{V_0}\right) + \frac{2Z_S\cos\theta_c}{V_0}$$

$$= x\left(\frac{\sin\theta_c\cos\phi}{V_0} - \frac{\sin\phi\cos\theta_c}{V_0}\right) + \frac{2Z_S\cos\theta_c}{V_0} = \frac{x\sin(\theta_c - \phi)}{V_0} + \frac{2Z_S\cos\theta_c}{V_0}$$

Let $V_{1u} = V_0/\cos(\theta_c - \phi)$ and $Z_S = h_S\cos\phi$. Substituting these into the last equation gives:

$$T_{1u} = \frac{x}{V_{1u}} + \frac{h_s\cos\theta_c\cos\phi}{V_0} = \frac{x}{V_{1u}} + T_{01u} \tag{A.10}$$

Shooting down-dip (from B to A), substitute $Z_S = Z_R + x\sin\phi$.

$$T_{1d} = \frac{x\cos\phi}{V_1} + \frac{(2Z_R + x\sin\phi)\cos\theta_c}{V_0} = x\left(\frac{\cos\phi}{V_1} + \frac{\sin\phi\cos\theta_c}{V_0}\right) + \frac{2Z_S\cos\theta_c}{V_0}$$

$$= x\left(\frac{\sin\theta_c\cos\phi}{V_0} + \frac{\sin\phi\cos\theta_c}{V_0}\right) + \frac{2Z_S\cos\theta_c}{V_0} = \frac{x\sin(\theta_c+\phi)}{V_0} + \frac{2Z_S\cos\theta_c}{V_0}$$

Let $V_{1u} = V_0/\cos(\theta_x+\phi)$ and $Z_R = h_R\cos\phi$. Substituting these into the last equation gives:

$$T_{ld} = \frac{x}{V_{ld}} + \frac{h_s\cos\theta_c\cos\phi}{V_0} = \frac{x}{V_{ld}} + T_{0ld} \tag{A.11}$$

Different *apparent* velocities are obtained when shooting up-dip $[V_{1u} = \dfrac{V_0}{\sin(\theta_c - \phi)}]$

and down-dip $[V_{1d} = \dfrac{V_0}{\sin(\theta_c + \phi)}]$. Solving for $\sin(\theta_c - \phi)$ and $\sin(\theta_c + \phi)$ gives

$$\sin(\theta_c - \phi) = \frac{V_0}{V_{lu}} \text{ and } \sin(\theta_c + \phi) = \frac{V_0}{V_{ld}}.$$

But $\sin(\theta_c - \phi) + \sin(\theta_c - \phi) = \sin\theta_c\cos\phi - \cos\theta_c\sin\phi + \sin\theta_c\cos\phi + \cos\theta_c\sin\phi = 2\sin\theta_c\cos\phi = \dfrac{V_0}{V_{lu}} + \dfrac{V_0}{V_{ld}} = V_0\left(\dfrac{1}{V_{lu}} + \dfrac{1}{V_{ld}}\right)$

The true velocity, V_1 is given by:

$$V_1 = \frac{V_0}{\sin\theta_c} = 2\cos\phi\left(\frac{1}{\frac{1}{V_{lu}} + \frac{1}{V_{ld}}}\right) = 2\cos\phi\left(\frac{V_{lu}V_{ld}}{V_{lu} + V_{ld}}\right) \tag{A.12}$$

Also,

$$\theta_c = \frac{1}{2}[\sin^{-1}(\theta_c - \phi) + \sin^{-1}(\theta_c + \phi)] = \frac{1}{2}\left[\sin^{-1}\left(\frac{V_0}{V_{ld}}\right) + \sin^{-1}\left(\frac{V_0}{V_{ld}}\right)\right] \tag{A.13}$$

and

$$\phi = \frac{1}{2}[\sin^{-1}(\theta_c - \phi) - \sin^{-1}(\theta_c - \phi)] = \frac{1}{2}\left[\sin^{-1}\left(\frac{V_0}{V_{ld}}\right) - \sin^{-1}\left(\frac{V_0}{V_{lu}}\right)\right] \tag{A.14}$$

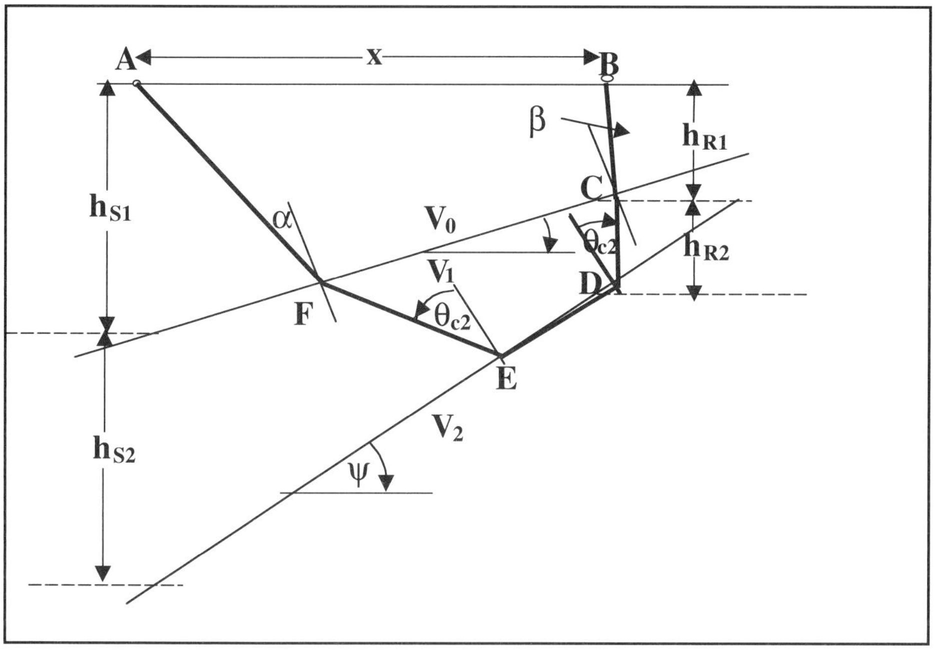

Fig. A–4 *Two Dipping Layers Earth Model*

Two dipping layers $V_2 > V_1 > V_0$

Derivation is similar to the one-layer dipping case. (See Fig. A–4. Up-dip is relative to layer 1.)

$$T_{2u} = \frac{x}{V_{2u}} + \frac{h_{S1}[\cos(\alpha + \phi) + \cos(\beta - \phi)] + 2h_{S2}\sin\theta_{c1}\cos\theta_{c2}\cos\psi}{V_0} + \frac{x}{V_{2u}} + T_{01u} \quad \text{(A.15)}$$

$$T_{2d} = \frac{x}{V_{2d}} + \frac{h_{R1}[\cos(\alpha + \phi) + \cos(\beta - \phi)] + 2h_{R2}\sin\theta_{c1}\cos\theta_{c2}\cos\psi}{V_0} + \frac{x}{V_{2d}} + T_{01d} \quad \text{(A.16)}$$

where $V_{2u} = \dfrac{V_0}{\sin(\beta - \phi)}$ and $V_{2d} = \dfrac{V_0}{\sin(\alpha + \phi)}$ (A.17)

Interpretation of seismic refraction data is done one layer at a time. Single layer equations are used for the first layer, two layer equations for the second, etc.

Appendix B SEG Standard Tape Formats

Introduction

At the beginning of digital recording (early- to mid-1960s), a great variety of tape formats were in use by contractors and clients. The SEG adopted standard formats to eliminate chaos. Listed below are the dates that the formats were adopted and brief descriptions of them.

- 1967 – SEG A (field data, multiplexed, NRZI)
  SEG B (field data, multiplexed, NRZI)
  SEG X (data exchange, demultiplexed, NRZI)
- 1972 – SEG C (field data, multiplexed, NRZI or PE) introduced to accommodate IFP recorders.
- 1975 – SEG Y (demultiplexed, NRZI or PE) introduced as new data exchange format to accommodate computer field equipment and newer processing hardware.
- 1980 – SEG D (multi-purpose, mpx, or demux) introduced to accommodate further advances in data acquisition and processing. Revised in 1994 to accommodate other developments, including 24-bit recording.

Although unlikely, it is possible that someone might want to reprocess very old data. In such a case, knowledge of the SEG A, B, X, and C might be required. For this reason, and for historical purposes, descriptions of those formats are included. SEG Y is still in use by some people. SEG D is the most used field data format, because of its versatility.

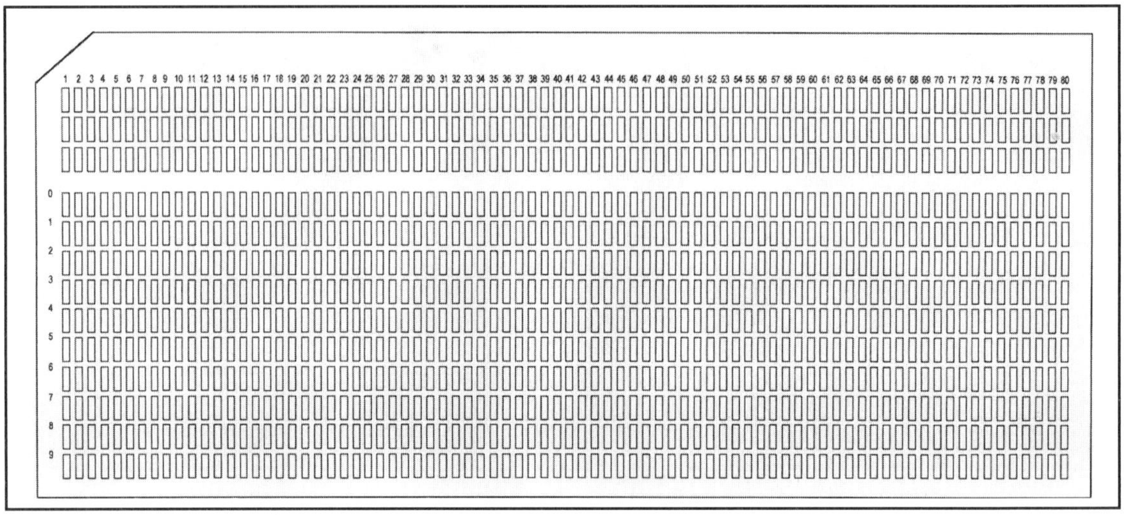

Fig. B–1 *Data Input Card*

Comments and notes

All formats consist of one or more *header blocks* that give information about the way the data were recorded and, in some cases, acquisition parameters. Note that bytes 3 and 4 of the first, or only header, give the format code. Data blocks follow the header(s).

SEG A and B use three types of parity or QC checks. The parity bit P is used to check for errors in the recording of a byte, data recorded across the tracks. The longitudinal parity check (LPC) is a parity check along the tracks. The cyclic redundancy check (CRC) is a diagonal or zig-zag parity check. The value written for each track is 0 for an even number of 1s along that diagonal path or 1 for an odd number of 1s.

Multiplexed data are recorded in *scans*. A scan records samples from all channels in one sample period. Early recording systems had a very limited number of data channels available.

SEG X and SEG Y formats employ *card images*. That is, they mimic the input data cards (popularly known as IBM cards) used to input data in the early days of seismic data processing) These cards had 80 columns and 13 rows of rectangles. Holes were punched in the rectangles so that each column represented one alphanumeric character. See Figure B–1.

The SEG D format is actually a family of formats. There are several standard header blocks that can be used. The real flexibility, however, is in the data block formats. A few of the more popular ones are described here. Note that multiplexed and demultiplexed formats differ by only one bit per sample.

Standard tape formats

Figure B–2 shows track/bit conventions for a 9-track tape. These are rarely, if ever, used today. However, each multiple of nine tracks will have the bit assignments indicated.

Figures B–3 through B–7 show header and data block formats for SEG A and SEG B. Figure B–8 illustrates the SEG X format. Figure B–9 gives the SEG C format. SEG Y is shown by Figures B–10 through B–15.

Tape Track Number	1	2	3	4	5	6	7	8	9
Tape Bit Order	5	7	3	P	2	1	0	6	4
Bit Number	7	6	5	4	3	2	1	0	P
Typical Packed BCD Value	1	2	4	8	1	2	4	8	P
Typical Data Word Value	2^7	2^8	2^9	2^{10}	2^{11}	2^{12}	2^7	S	P
	0	2^0	2^1	2^2	2^3	2^4	2^5	2^6	P

Fig. B–2 *Track/Bit Conventions*

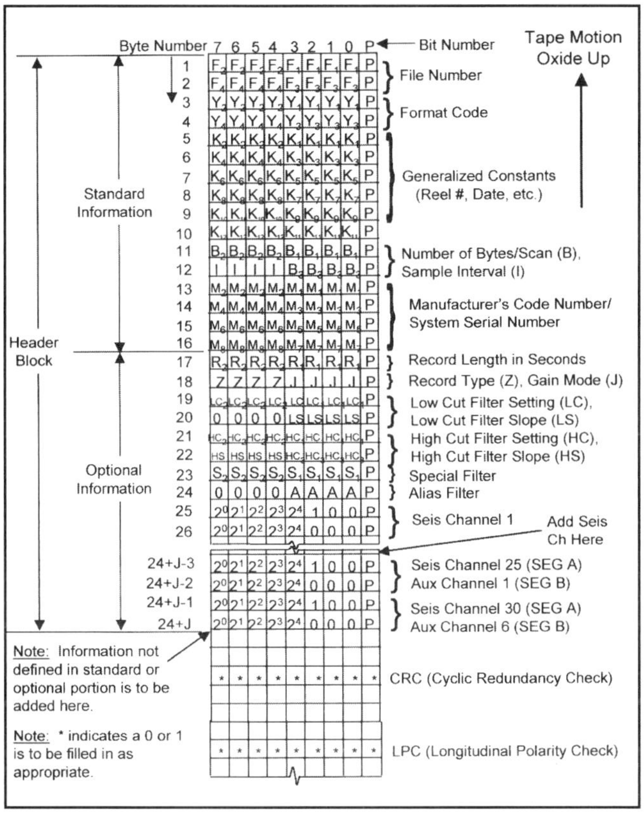

Fig. B–3 *SEG A and B Formats—Header Block*

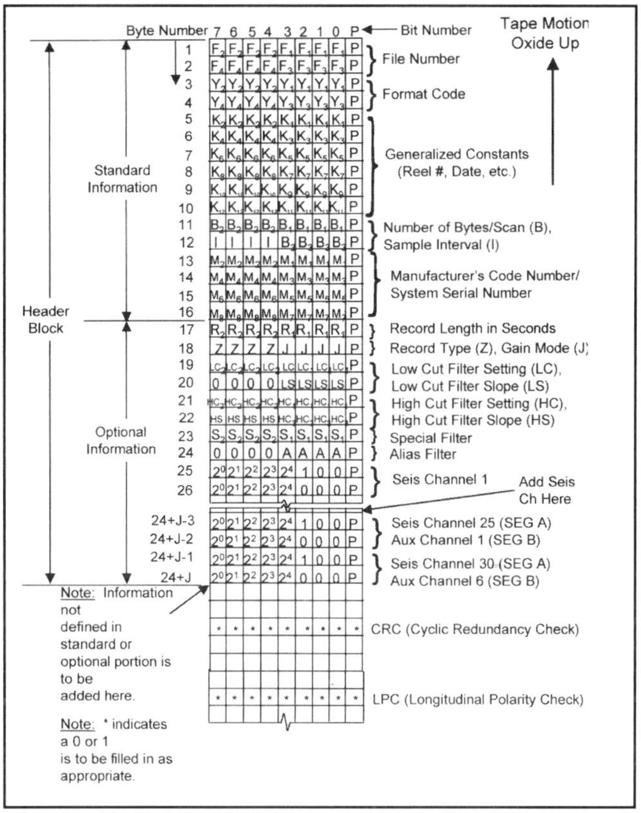

Fig. B–4 *SEG A and B Formats—Header Block (continued)*

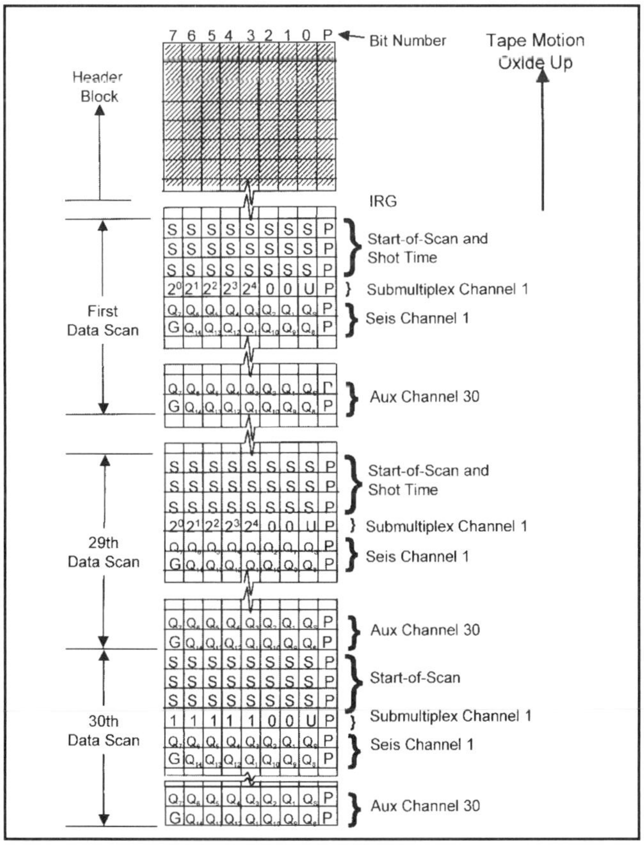

Fig. B–5 *SEG A Format—Data Block*

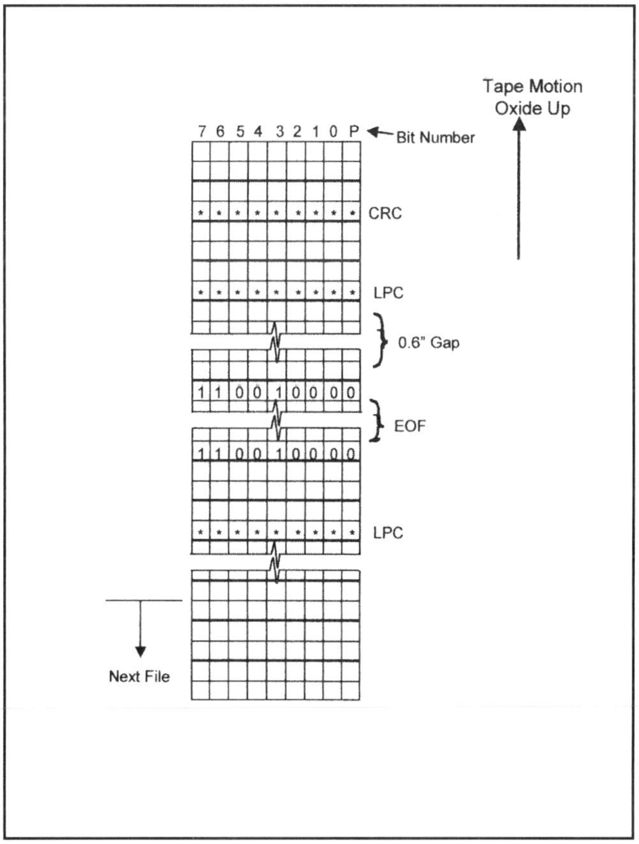

Fig. B–5b *SEG A Format—Data Block*

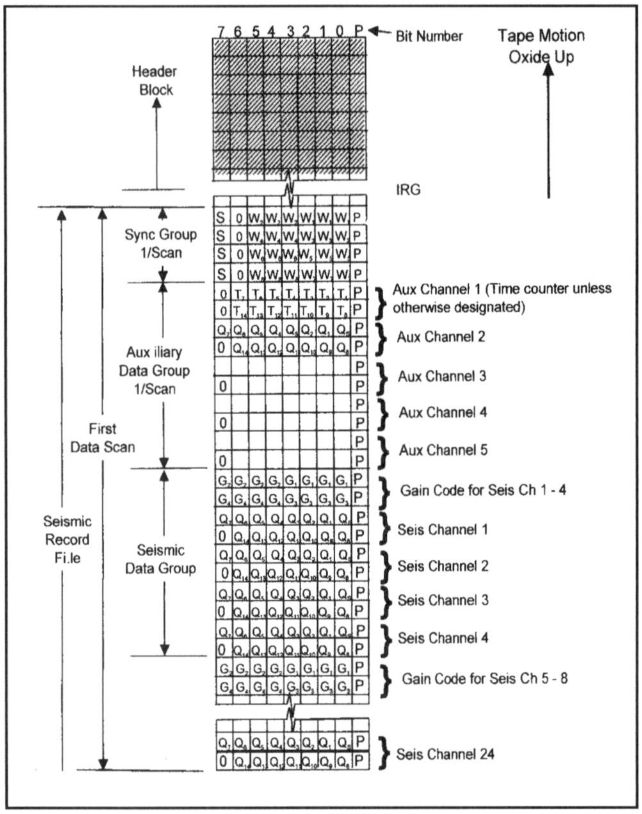

Fig. B–6 *SEG B Format—Data Block*

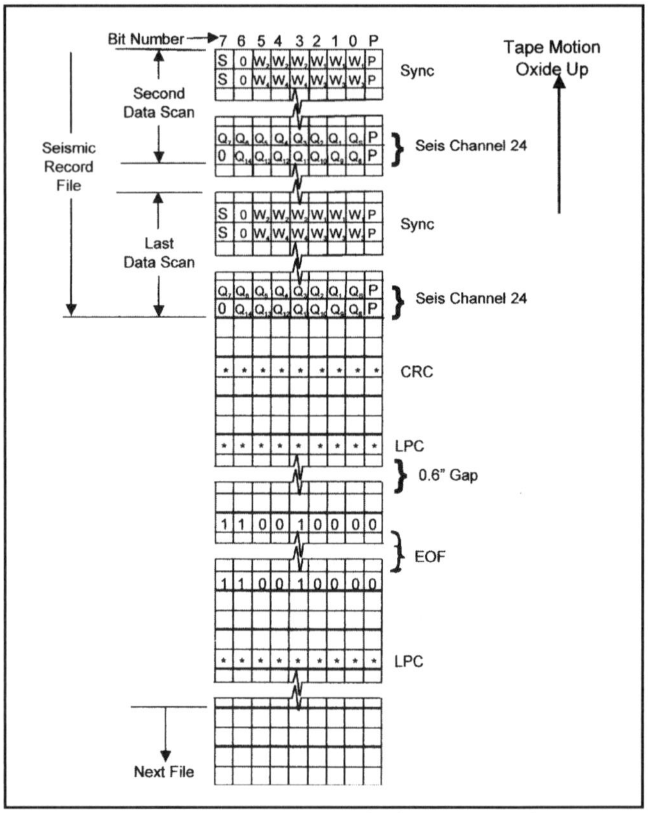

Fig. B–7 *SEG B Format—Data Block (continued)*

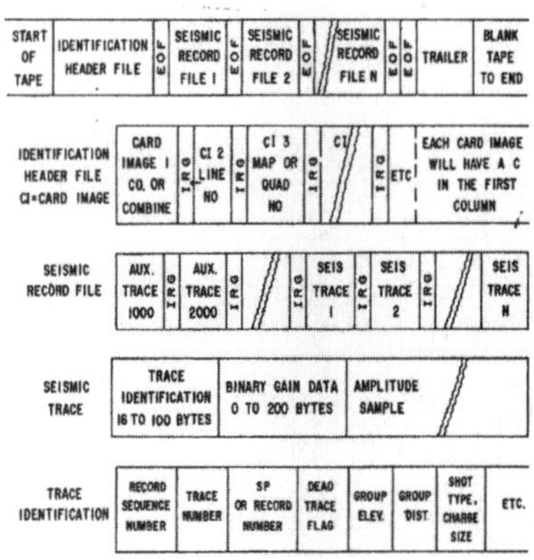

HEADER IDENTIFICATION FILE CONTENTS

A. Identification
  Company or combine name
  Line Number
  Map Number or quad name
B. Data type
  Seismic source type
  Fold
  Original data equivalent
  State of processing
and
  Hardware recording parameters
C. Equipment manufacturer, model number and identification
  Sample rate of data
  Gain mode and rate of change
  Restrictions
  Recording cable parameters
  Specifications for determination of time breaks and water breaks
D. Field recording parameters
  Distance units (ft or m)
  Shothole spacing and configuration
  Group spacing
  Number of groups
  Number of geophones/group
  Geophone spacing and configuration
  Type, frequency, hookup and damping of geophones
E. Data format
  Number of seismic records
  Number of traces/record
  Number of samples/trace
  Tape format: number of tracks
  packing density
  Number of bytes (characters) of identification/trace (16-100)
  Number of bytes (characters) of gain information/trace (0.-200)
  Number of bytes (characters) per amplitude sample
  Negative number system (1's or 2's comp.)
  Number of contributors per inverse gain or amplitude trace
  Location and number of water-break detectors
F Other comments
  Date of start of line

Fig. B–8 *SEG X Format*

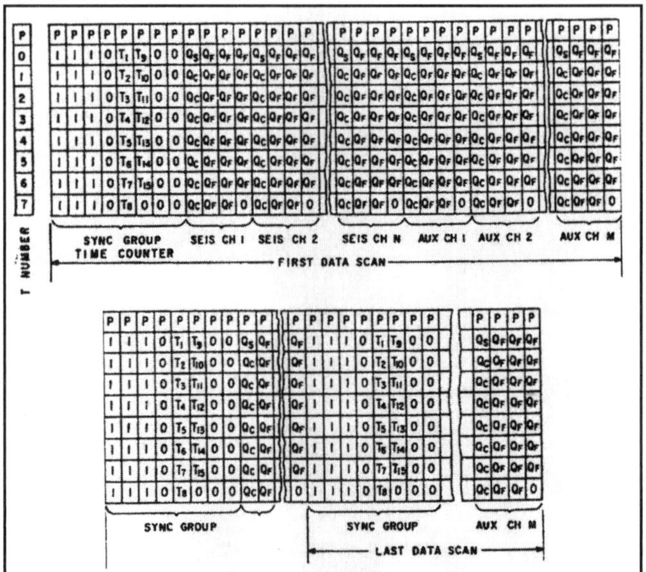

Zero data are recorded in any tape data scan words remaining after all seismic channels and all available auxiliaries have been recorded.

Fig. B–9 *SEG C Format—Data Block*

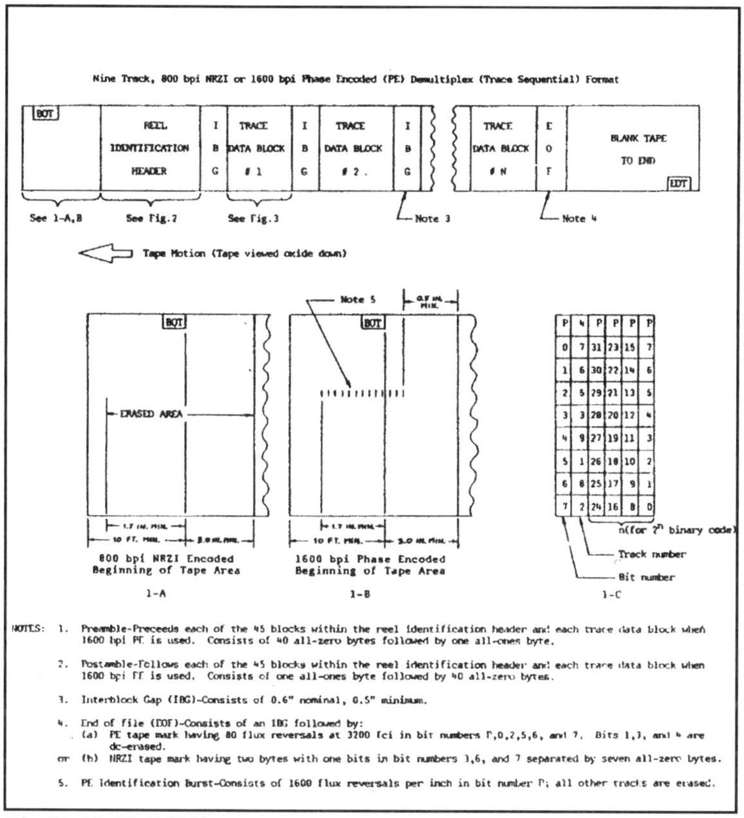

Fig. B–10 *SEG Y Format*

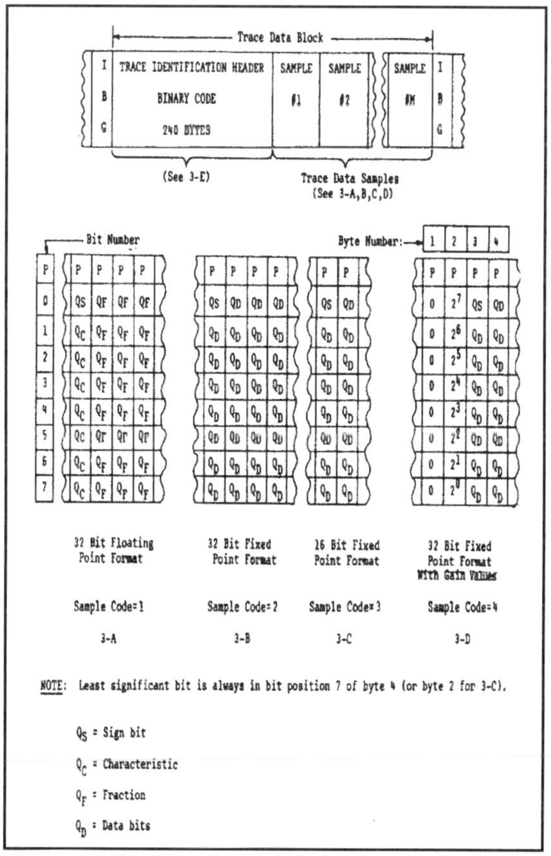

Fig. B–11 *SEG Y Format, Reel Identification Header—EBCDIC Image Block*

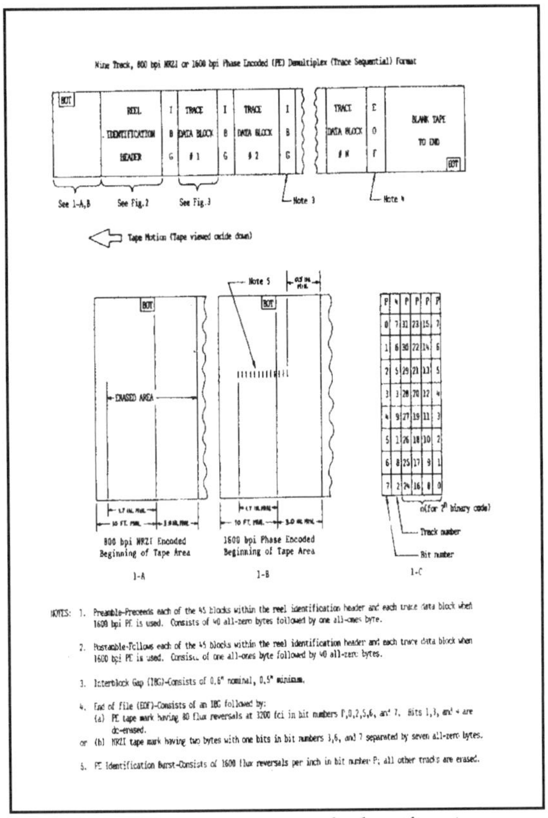

Fig. B–12 *SEG Y Format, Reel Identification Header—Binary Coded Block*

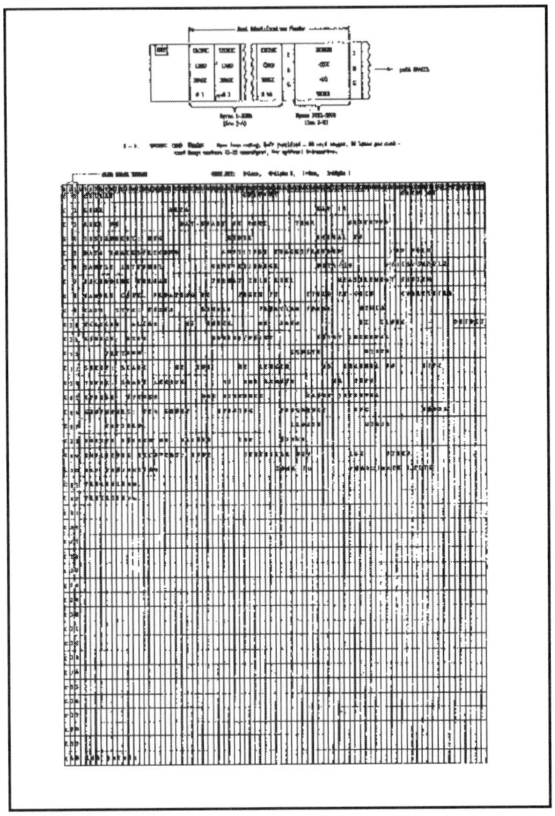

Fig. B–13 *SEG Y Format, Trace Data Block*

BINARY CODE-Right justified

Byte Numbers		Description
3201-3204		Job identification number.
3205-3208	*	Line number (only one line per reel).
3209-3212	*	Reel number.
3213-3214	*	Number of data traces per record (includes dummy and zero traces inserted to fill out the record or common depth point).
3215-3216	*	Number of auxiliary traces per record (includes sweep, timing, gain, sync, and all other nondata traces).
3217-3218	*	Sample interval in μsec (for this reel of data). } Designated in microseconds to accommodate sample intervals less than one millisecond.
3219-3220		Sample interval in μsec (for original field recording). } Designated in microseconds to accommodate sample intervals less than one millisecond.
3221-3222	*	Number of samples per data trace (for this reel of data).
3223-3224		Number of samples per data trace (for original field recording).
3225-3226	*	Data sample format code: 1 = floating point (4 bytes); 2 = fixed point (4 bytes); 3 = fixed point (2 bytes); 4 = fixed point w/gain code (4 bytes). Auxiliary traces use the same number of bytes per sample.
3227-3228	*	CDP fold (expected number of data traces per CDP ensemble).
3229-3230		Trace sorting code: 1 = as recorded (no sorting); 2 = CDP ensemble; 3 = single fold continuous profile; 4 = horizontally stacked
3231-3232		Vertical sum code: 1 = no sum, 2 = two sum, ..., N = N sum (N = 32,767)
3233-3234		Sweep frequency at start.
3235-3236		Sweep frequency at end.
3237-3238		Sweep length (msec).
3239-3240		Sweep type code: 1 = linear; 2 = parabolic; 3 = exponential; 4 = other
3241-3242		Trace number of sweep channel.
3243-3244		Sweep trace taper length in msec at start if tapered (the taper starts at zero time and is effective for this length).
3245-3246		Sweep trace taper length in msec at end (the ending taper starts at sweep length minus the taper length at end).
3247-3248		Taper type: 1 = linear; 2 = cos²; 3 = other
3249-3250		Correlated data traces: 1 = no; 2 = yes
3251-3252		Binary gain recovered: 1 = yes; 2 = no
3253-3254		Amplitude recovery method: 1 = none; 2 = spherical divergence; 3 = AGC; 4 = other
3255-3256		Measurement system: 1 = meters; 2 = feet
3257-3258		Impulse signal Polarity: 1 = Increase in pressure or upward geophone case movement gives negative number on tape. 2 = Increase in pressure or upward geophone case movement gives positive number on tape.
3259-3260		Vibratory polarity code: Seismic signal lags pilot signal by: 1 = 337.5° to 22.5°; 2 = 22.5° to 67.5°; 3 = 67.5° to 112.5°; 4 = 112.5° to 157.5°; 5 = 157.5° to 202.5°; 6 = 202.5° to 247.5°; 7 = 247.5° to 292.5°; 8 = 292.5° to 337.5°
3261-3600		Unassigned--for optional information.

*Strongly recommended that this information always be recorded.

Fig. B–14 *SEG Y Format, Trace Identification Header Written in Binary Code*

Byte Numbers	Description
1 - 4	* Trace sequence number within line--numbers continue to increase if additional reels are required on same line.
5 - 8	Trace sequence number within reel--each reel starts with trace number one.
9 -12	* Original field record number.
13-16	* Trace number within the original field record.
17-20	Energy source point number--used when more than one record occurs at the same effective surface location.
21-24	CDP ensemble number
25-28	Trace number within the CDP ensemble--each ensemble starts with trace number one.
29-30	* Trace identification code: 1 = seismic data 4 = time break 7 = timing 2 = dead 5 = uphole 8 = water break 3 = dummy 6 = sweep 9----. N = optional use (N = 32,767)
31-32	Number of vertically summed traces yielding this trace. (1 is one trace, 2 is two summed traces, etc.)
33-34	Number of horizontally stacked traces yielding this trace. (1 is one trace, 2 is two stacked traces, etc.)
35-36	Data use: 1 = production. 2 = test.
37-40	Distance from source point to receiver group (negative if opposite to direction in which line is shot).
41-44	Receiver group elevation; all elevations above sea level are positive and below sea level are negative.
45-48	Surface elevation at source.
49-52	Source depth below surface (a positive number).
53-56	Datum elevation at receiver group.
57-60	Datum elevation at source.
61-64	Water depth at source.
65-68	Water depth at group.
69-70	Scaler to be applied to all elevations and depths specified in bytes 41-68 to give the real value. Scaler = 1, ±10, ±100, ±1000, or ±10,000. If positive, scaler is used as a multiplier; if negative, scaler is used as a divisor.
71-72	Scaler to be applied to all coordinates specified in bytes 73-88 to give the real value. Scaler = 1, ±10, ±100, ±1000, or ±10,000. If positive, scaler is used as a multiplier; if negative, scaler is used as divisor.
73-76	Source coordinate - X.
77-80	Source coordinate - Y.
81-84	Group coordinate - X.
85-88	Group coordinate - Y. (bytes 73-88: If the coordinate units are in seconds of arc, the X values represent longitude and the Y values latitude. A positive value designates the number of seconds east of Greenwich Meridian or north of the equator and a negative value designates the number of seconds south or west.)
89-90	Coordinate units: 1 = length (meters or feet), 2 = seconds of arc.
91-92	Weathering velocity.
93-94	Subweathering velocity.
95-96	Uphole time at source.
97-98	Uphole time at group.
99-100	Source static correction.
101-102	Group static correction.
103-104	Total static applied. (Zero if no static has been applied.)
105-106	Lag time A. Time in ms. between end of 240-byte trace identification header and time break. Positive if time break occurs after end of header, negative if time break occurs before end of header. Time break is defined as the initiation pulse which may be recorded on an auxiliary trace or as otherwise specified by the recording system.
107-108	Lag Time B. Time in ms. between time break and the initiation time of the energy source. May be positive or negative.
109-110	Delay recording time. Time in ms. between initiation time of energy source and time when recording of data samples begins. (For deep water work if data recording does not start at zero time.)
111-112	Mute time--start.
113-114	Mute time--end.
115-116	* Number of samples in this trace.
117-118	* Sample interval in μs for this trace.
119-120	Gain type of field instruments; 1 = fixed. 2 = binary. 3 = floating point. 4 --- N = optional use.
121-122	Instrument gain constant.
123-124	Instrument early or initial gain (db).
125-126	Correlated: 1 = no. 2 = yes.
127-128	Sweep frequency at start.
129-130	Sweep frequency at end.
131-132	Sweep length in ms.
133-134	Sweep type: 1 = linear. 2 = parabolic. 3 = exponential. 4 = other.
135-136	Sweep trace taper length at start in ms.
137-138	Sweep trace taper length at end in ms.
139-140	Taper type: 1 = linear. 2 = $\cos^2$. 3 = other.
141-142	Alias filter frequency, if used.
143-144	Alias filter slope
145-146	Notch filter frequency, if used.
147-148	Notch filter slope.
149-150	Low cut frequency, if used.
151-152	High cut frequency, if used.
153-154	Low cut slope
155-156	High cut slope
157-158	Year data recorded.
159-160	Day of year.
161-162	Hour of day (24 hour clock)
163-164	Minute of hour.
165-166	Second of minute.
167-168	Time basis code: 1 = local. 2 = GMT. 3 = other.
169-170	Trace weighting factor--defined as 2^{-N} volts for the least significant bit. (N = 0, 1, ..., 32,767.)
171-172	Geophone group number of roll switch position one.
173-174	Geophone group number of trace number one within original field record.
175-176	Geophone group number of last trace within original field record.
177-178	Gap size (total number of groups dropped).
179-180	Overtravel associated with taper at beginning or end of line: 1 = down (or behind). 2 = up (or ahead).
181-240	Unassigned--for optional information.

* Strongly recommended that this information always be recorded.

Fig. B–15 *SEG Y Format, Trace Identification Header Written in Binary Code (continued)*

The versatility of the SEG D format requires a great variety of header blocks and data blocks. Figure B–16 shows the general layout of the SEG D format and the various header and data blocks involved. Figure B–17 illustrates the SEG D general header. Figures B–18 and B–19 illustrate the channel set descriptor, which can be used to describe situations in which recording parameters are different from those in other channels. Figure B–20 shows the start-of-scan and timing word formats. Figure B–21 shows the revised SEG D format demultiplexed trace header. Figure B–22 lists all the SEG data recording formats and their codes. Figures B–23 through B–26 illustrate some of the more popular data recording formats in SEG D.

By the time the SEG D format was introduced, instrument sophistication had reached the point that different recording parameters could be used for some channels rather than only one set for all channels. Thus, the channel set descriptor, a means to identify sets of channels and their different parameter values, was necessary.

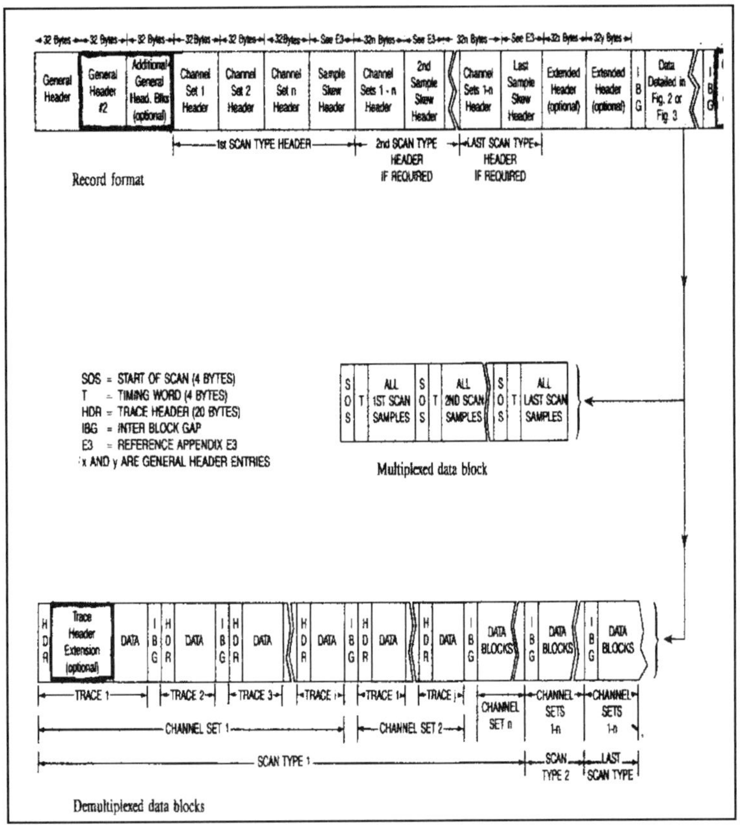

Fig. B–16 *SEG D Format*

BIT NO	P	0	1	2	3	4	5	6	7	
BCD VALUE MSD		8	4	2	1	8	4	2	1	LSD
BINARY VALUE MSB		128	64	32	16	8	4	2	1	LSB
FILE NUMBER		F_1	F_1	F_1	F_1	F_2	F_2	F_2	F_2	1
		F_3	F_3	F_3	F_3	F_4	F_4	F_4	F_4	2
FORMAT CODE		Y_1	Y_1	Y_1	Y_1	Y_2	Y_2	Y_2	Y_2	3
		Y_3	Y_3	Y_3	Y_3	Y_4	Y_4	Y_4	Y_4	4
GENERAL CONSTANTS		K_1	K_1	K_1	K_1	K_2	K_2	K_2	K_2	5
		K_3	K_3	K_3	K_3	K_4	K_4	K_4	K_4	6
		K_5	K_5	K_5	K_5	K_6	K_6	K_6	K_6	7
		K_7	K_7	K_7	K_7	K_8	K_8	K_8	K_8	8
		K_9	K_9	K_9	K_9	K_{10}	K_{10}	K_{10}	K_{10}	9
		K_{11}	K_{11}	K_{11}	K_{11}	K_{12}	K_{12}	K_{12}	K_{12}	10
YEAR		YR_1	YR_1	YR_1	YR_1	YR_2	YR_2	YR_2	YR_2	11
# BLKS IN GEN HDR		GH_3	GH_2	GH_1	GH_0	DY_1	DY_1	DY_1	DY_1	12
DAY (DY)		DY_2	DY_2	DY_2	DY_2	DY_3	DY_3	DY_3	DY_3	13
HOUR		H_1	H_1	H_1	H_1	H_2	H_2	H_2	H_2	14
MINUTE		MI_1	MI_1	MI_1	MI_1	MI_2	MI_2	MI_2	MI_2	15
SECOND		SE_1	SE_1	SE_1	SE_1	SE_2	SE_2	SE_2	SE_2	16
MANUFACTURERS CODE		M_1	M_1	M_1	M_1	M_2	M_2	M_2	M_2	17
		M_3	M_3	M_3	M_3	M_4	M_4	M_4	M_4	18
		M_5	M_5	M_5	M_5	M_6	M_6	M_6	M_6	19
BYTES PER SCAN		B_1	B_1	B_1	B_1	B_2	B_2	B_2	B_2	20
		B_3	B_3	B_3	B_3	B_4	B_4	B_4	B_4	21
		B_5	B_5	B_5	B_5	B_6	B_6	B_6	B_6	22
BASE SCAN INTERVAL		I_3	I_2	I_1	I_0	I_{-1}	I_{-2}	I_{-3}	I_{-4}	23
POLARITY (P) S/B		P	P	P	P	S/B_{x3}	S/B_{x2}	S/B_{x1}	S/B_{x0}	24
SCANS/BLK EXPONENT		S/B_7	S/B_6	S/B_5	S/B_4	S/B_3	S/B_2	S/B_1	S/B_0	25
RECORD TYPE (Z)		Z	Z	Z	Z	R_1	R_1	R_1	R_1	26
RECORD LENGTH (R)		R_2	R_2	R_2	R_2	R_3	R_3	R_3	R_3	27
SCAN TYPES/RECORD		ST/R_1	ST/R_1	ST/R_1	ST/R_1	ST/R_2	ST/R_2	ST/R_2	ST/R_2	28
CHAN SETS/SCAN TYPE		CS_1	CS_1	CS_1	CS_1	CS_2	CS_2	CS_2	CS_2	29
SKEW BLOCKS		SK_1	SK_1	SK_1	SK_1	SK_2	SK_2	SK_2	SK_2	30
EXTENDED HEADER BLK		EC_1	EC_1	EC_1	EC_1	EC_2	EC_2	EC_2	EC_2	31
EXTERNAL HEADER BLK		EX_1	EX_1	EX_1	EX_1	EX_2	EX_2	EX_2	EX_2	32

Fig. B–17 *SEG D Format, General Header*

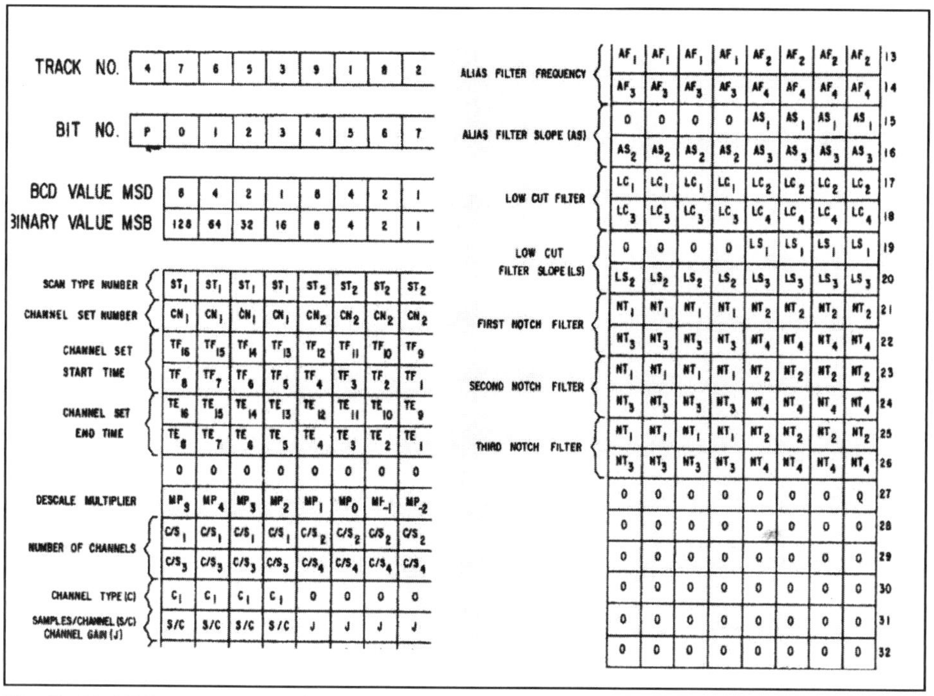

Fig. B–18 *SEG D Format, Channel Set Descriptor*

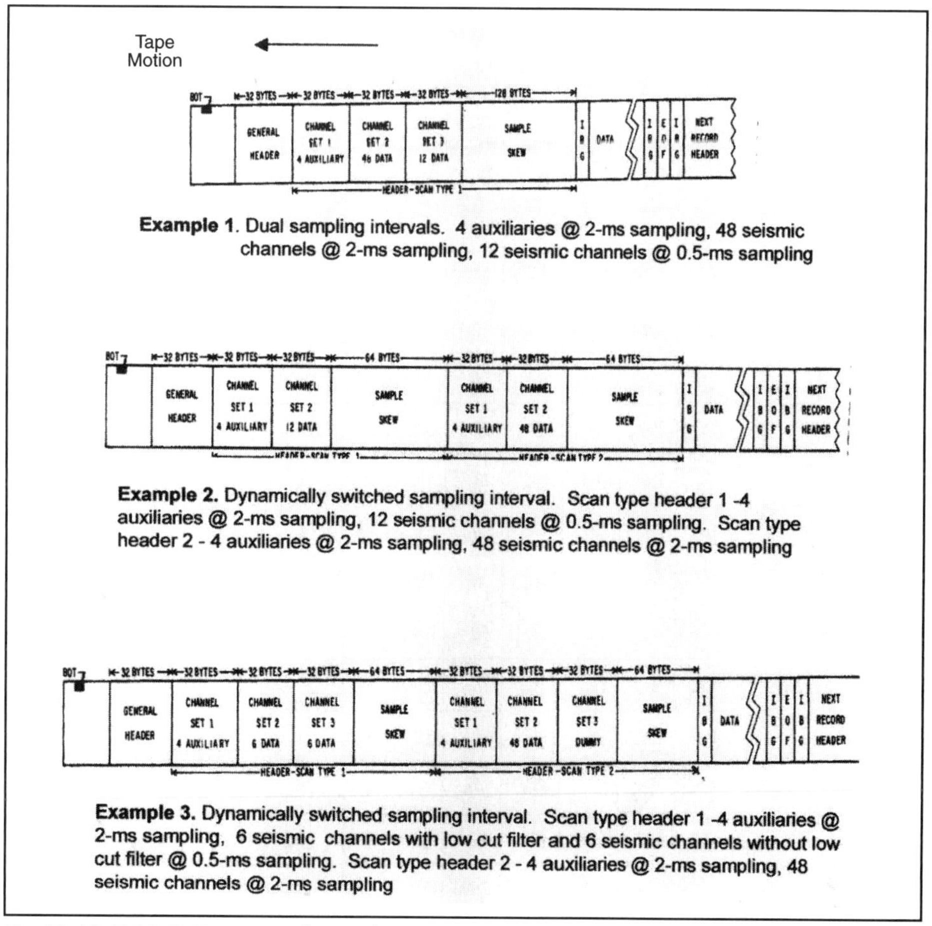

Fig. B–19 *SEG D Format, Channel Set Examples*

TRACK NO.		4	7	6	5	3	9	1	8	2	
BIT NO.		P	0	1	2	3	4	5	6	7	
BCD VALUE MSD			8	4	2	1	8	4	2	1	LSD
BINARY VALUE MSB			128	64	32	16	8	4	2	1	LSB
START OF SCAN			1	1	1	1	1	1	1	1	1
			1	1	1	1	1	1	1	1	2
			1	1	1	1	1	1	1	1	3
			x	TWI	ITB	DP	x	x	0	1	4
TIMING WORD			T_{15}	T_{14}	T_{13}	T_{12}	T_{11}	T_{10}	T_9	T_8	5
			T_7	T_6	T_5	T_4	T_3	T_2	T_1	T_0	6
			T_{-1}	T_{-2}	T_{-3}	T_{-4}	T_{-5}	T_{-6}	T_{-7}	T_{-8}	7
			0	0	0	0	0	0	0	0	8

Fig. B–20 *SEG D Format, Start of Scan and Timing Word*

BIT NO	P	0	1	2	3	4	5	6	7	
FILE NUMBER		F_1	F_1	F_1	F_1	F_2	F_2	F_2	F_2	1
		F_3	F_3	F_3	F_3	F_4	F_4	F_4	F_4	2
SCAN TYPE NUMBER		ST_1	ST_1	ST_1	ST_1	ST_2	ST_2	ST_2	ST_2	3
CHANNEL SET NUMBER		CN_1	CN_1	CN_1	CN_1	CN_2	CN_2	CN_2	CN_2	4
TRACE NUMBER		TN_1	TN_1	TN_1	TN_1	TN_2	TN_2	TN_2	TN_2	5
		TN_3	TN_3	TN_3	TN_3	TN_4	TN_4	TN_4	TN_4	6
FIRST TIMING WORD		T_{15}	T_{14}	T_{13}	T_{12}	T_{11}	T_{10}	T_9	T_8	7
		T_7	T_6	T_5	T_4	T_3	T_2	T_1	T_0	8
		T_{-1}	T_{-2}	T_{-3}	T_{-4}	T_{-5}	T_{-6}	T_{-7}	T_{-8}	9
TRACE HEADER EXTEN.		THE_7	THE_6	THE_5	THE_4	THE_3	THE_2	THE_1	THE_0	10
SAMPLE SKEW		SSK_{-1}	SSK_{-2}	SSK_{-3}	SSK_{-4}	SSK_{-5}	SSK_{-6}	SSK_{-7}	SSK_{-8}	11
TRACE EDIT		TR_7	TR_6	TR_5	TR_4	TR_3	TR_2	TR_1	TR_0	12
TIME BREAK WINDOW		TW_{15}	TW_{14}	TW_{13}	TW_{12}	TW_{11}	TW_{10}	TW_9	TW_8	13
		TW_7	TW_6	TW_5	TW_4	TW_3	TW_2	TW_1	TW_0	14
		TW_{-1}	TW_{-2}	TW_{-3}	TW_{-4}	TW_{-5}	TW_{-6}	TW_{-7}	TW_{-8}	15
EXTENDED CHANNEL SET NUMBER		EN_{15}	EN_{14}	EN_{13}	EN_{12}	EN_{11}	EN_{10}	EN_9	EN_8	16
		EN_7	EN_6	EN_5	EN_4	EN_3	EN_2	EN_1	EN_0	17
EXTENDED FILE NUMBER		EFN_{23}	EFN_{22}	EFN_{21}	EFN_{20}	EFN_{19}	EFN_{18}	EFN_{17}	EFN_{16}	18
		EFN_{15}	EFN_{14}	EFN_{13}	EFN_{12}	EFN_{11}	EFN_{10}	EFN_9	EFN_8	19
		EFN_7	EFN_6	EFN_5	EFN_4	EFN_3	EFN_2	EFN_1	EFN_0	20

Fig. B–21 *SEG D Format, Demultiplexed Trace Header*

Code	Format
0015	20-bit binary multiplexed
0022	8-bit quaternary multiplexed
0024	16-bit quaternary multiplexed
0036	24-bit 2's complement integer multiplexed
0038	32-bit 2's complement integer multiplexed
0042	8-bit hexadecimal multiplexed
0044	16-bit hexadecimal multiplexed
0048	32-bit hexadecimal multiplexed
0058	32-bit IEEE multiplexed
8015	20-bit binary demultiplexed
8022	8-bit quaternary demultiplexed
8024	16-bit quaternary demultiplexed
8036	24-bit 2's complement integer demultiplexed
8038	32-bit 2's complement integer demultiplexed
8042	8-bit hexadecimal demultiplexed
8044	16-bit hexadecimal demultiplexed
8048	32-bit hexadecimal demultiplexed
8058	32-bit IEEE demultiplexed
0000	Illegal, do not use
0200	Illegal, do not use

Fig. B–22 *Format Codes*

Bits

Bytes	0	1	2	3	4	5	6	7	Description
1	C_3	C_2	C_1	C_0	C_3	C_2	C_1	C_0	Exponent for
2	C_3	C_2	C_1	C_0	C_3	C_2	C_1	C_0	Channels 1-4
3	S	Q_{-1}	Q_{-2}	Q_{-3}	Q_{-4}	Q_{-5}	Q_{-6}	Q_{-7}	Channel 1
4	Q_{-8}	Q_{-9}	Q_{-10}	Q_{-11}	Q_{-12}	Q_{-13}	Q_{-14}	0	Mantissa
5	S	Q_{-1}	Q_{-2}	Q_{-3}	Q_{-4}	Q_{-5}	Q_{-6}	Q_{-7}	Channel 2
6	Q_{-8}	Q_{-9}	Q_{-10}	Q_{-11}	Q_{-12}	Q_{-13}	Q_{-14}	0	Mantissa
7	S	Q_{-1}	Q_{-2}	Q_{-3}	Q_{-4}	Q_{-5}	Q_{-6}	Q_{-7}	Channel 3
8	Q_{-8}	Q_{-9}	Q_{-10}	Q_{-11}	Q_{-12}	Q_{-13}	Q_{-14}	0	Mantissa
9	S	Q_{-1}	Q_{-2}	Q_{-3}	Q_{-4}	Q_{-5}	Q_{-6}	Q_{-7}	Channel 4
10	Q_{-8}	Q_{-9}	Q_{-10}	Q_{-11}	Q_{-12}	Q_{-13}	Q_{-14}	0	Mantissa

Value of data word = $(-1)^S$ .QQQQQQQQQQQQQQ x 2^{CCCC}

Fig. B–23 *SEG D Format, Data Recording Method, 2½ Byte Binary Exponent-multiplexed*

Bits

Bytes	0	1	2	3	4	5	6	7	Description
1	C_3	C_2	C_1	C_0	C_3	C_2	C_1	C_0	Exponent for
2	C_3	C_2	C_1	C_0	C_3	C_2	C_1	C_0	Samples 1-4
3	S	Q_{-1}	Q_{-2}	Q_{-3}	Q_{-4}	Q_{-5}	Q_{-6}	Q_{-7}	Sample 1
4	Q_{-8}	Q_{-9}	Q_{-10}	Q_{-11}	Q_{-12}	Q_{-13}	Q_{-14}	Q_{-15}	Mantissa
5	S	Q_{-1}	Q_{-2}	Q_{-3}	Q_{-4}	Q_{-5}	Q_{-6}	Q_{-7}	Sample 2
6	Q_{-8}	Q_{-9}	Q_{-10}	Q_{-11}	Q_{-12}	Q_{-13}	Q_{-14}	Q_{-15}	Mantissa
7	S	Q_{-1}	Q_{-2}	Q_{-3}	Q_{-4}	Q_{-5}	Q_{-6}	Q_{-7}	Sample 3
8	Q_{-8}	Q_{-9}	Q_{-10}	Q_{-11}	Q_{-12}	Q_{-13}	Q_{-14}	Q_{-15}	Mantissa
9	S	Q_{-1}	Q_{-2}	Q_{-3}	Q_{-4}	Q_{-5}	Q_{-6}	Q_{-7}	Sample 4
10	Q_{-8}	Q_{-9}	Q_{-10}	Q_{-11}	Q_{-12}	Q_{-13}	Q_{-14}	Q_{-15}	Mantissa

Fig. B–24 *SEG D Format, Data Recording Method, 2½ Byte Binary Exponent-demultiplexed*

Bits →

Bytes	0	1	2	3	4	5	6	7
1	S	C_6	C_5	C_4	C_3	C_2	C_1	C_0
2	Q_{-1}	Q_{-2}	Q_{-3}	Q_{-4}	Q_{-5}	Q_{-6}	Q_{-7}	Q_{-8}
3	Q_{-9}	Q_{-10}	Q_{-11}	Q_{-12}	Q_{-13}	Q_{-14}	Q_{-15}	Q_{-16}
4	Q_{-17}	Q_{-18}	Q_{-19}	Q_{-20}	Q_{-21}	Q_{-22}	Q_{-23}	0

Value of data word = $(-1)^S$ .QQQQQQQQQQQQQQQQQQQQQQ x $16^{(CCCCCCC-64)}$

Fig. B–25 *SEG D Format, Data Recording Method, Four Byte Hexadecimal Exponent-multiplexed*

Bits →

Bytes	0	1	2	3	4	5	6	7
1	S	C_7	C_6	C_5	C_4	C_3	C_2	C_1
2	C_0	Q_{-1}	Q_{-2}	Q_{-3}	Q_{-4}	Q_{-5}	Q_{-6}	Q_{-7}
3	Q_{-8}	Q_{-9}	Q_{-10}	Q_{-11}	Q_{-12}	Q_{-13}	Q_{-14}	Q_{-15}
4	Q_{-16}	Q_{-17}	Q_{-18}	Q_{-19}	Q_{-20}	Q_{-21}	Q_{-22}	Q_{-23}

Note: Q_{-23} must equal 0 in demultiplexed format

$e_{10} = C_7C_6C_5C_4C_3C_2C_1C_0$

$f_{10} = .Q_{-1}Q_{-2}Q_{-3}Q_{-4}Q_{-5}Q_{-6}Q_{-7}Q_{-8}Q_{-9}Q_{-10}Q_{-11}Q_{-12}Q_{13}Q_{-14}Q_{-15}$
$Q_{-16}Q_{-17}Q_{-18}Q_{-19}Q_{-20}Q_{-21}Q_{-22}Q_{-23}$

y = value represented by the above bytes

If e = 255 and f =0, y is not a number; If e = 255 and f =0,
$y = (1)^{S}*$, i.e.. overflow

If 0 < e < 255, y = = $(-1)^{S}*2^{(e-127)}*1.f$(normalized); If e = 0 and f =0,
$y = (-1)^{S}*2^{(e-127)}*0.f$ (denormalized); If e = 0 and f =0,
$y = (-1)^{S}*0$ (+0)

Fig. B–26 *SEG D Format, Data Recording Method, 32-Bit IEEE Format*

Appendix C Workshop Answers

The following pages give the answers to the odd-numbered exercises in all Workshops.

Geological background workshop answers

1. Prior to the break-up of Pangea, there was a small sea between what is now New Brunswick, Canada and Morocco. Since oil has been found off the coast of New Brunswick, do you think this makes it more or less likely that oil can be found off the Atlantic coast of Morocco?

   Answer: It should make it likely that oil would be found off of Morocco in late Triassic and older strata since these rocks would have been formed in the same type of environment as those off New Brunswick.

3. What processes are involved in the formation of clastic sedimentary rocks?

   Answer: Weathering, erosion, deposition, compaction, and cementation.

5. Shale is formed from clay particles and sandstone is formed from sand. Which type of rock would you expect to form farther from shore in a sea or ocean?

   Answer: Shale, since the clay particles are smaller and lighter than sand.

7. How are the geologic laws used, along with fossil evidence, to determine the geologic time scale?

   Answer: The Law of Superposition, in particular, is used since it states that older rocks are found at deeper depths. Thus, fossils found at the bottom of a rock sequence are older than those found near the top of a sequence. The Law of Uniformitarianism is used to infer the environment in which the fossilized organisms once lived.

9. In North America, the Pennsylvanian and Mississippian are considered to be Periods of the Paleozoic Era. In Europe, the Carboniferous Period covers essentially the same time span, and the Pennsylvanian and Mississippian are not recognized as Periods. What does this imply about the orogeny that separated the Pennsylvanian and Mississippian Periods?

   Answer: The orogeny which demarcates the Pennsylvanian from the Mississippian was not as geographically extensive as those which marked the close of the Triassic and start of the Jurassic, for example.

11. Based on outcrops, a sandstone bed is found to have a strike of N 23° W. When looking along the strike, the bed dips downward to the right. In what direction should the dip be measured?

   Answer: Dip is measured perpendicular to strike. This means, in this case, that the maximum dip is at N 23° W + 90° or S 67° W.

13. Sketch a normal fault that has a heave of 20 m and a throw of 50 m. What is the dip angle?

   Answer: See Figure C–1.

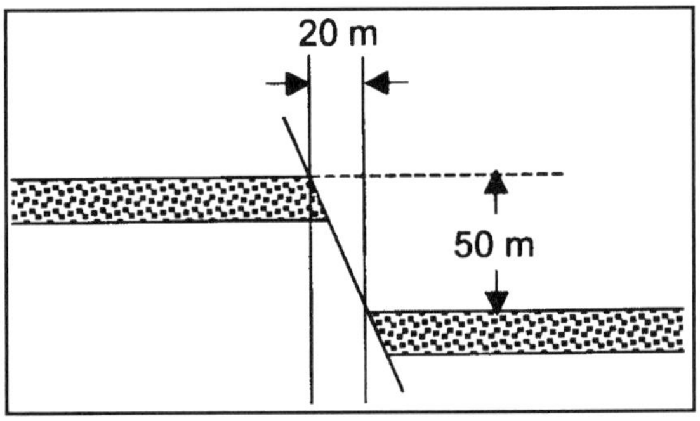

Fig. C–1 *Heave and Throw in a Normal Fault*

15. Why are sulfur and oxygen undesirable elements in crude oil?

   Answer: Sulfur is undesirable because it is a pollutant (forms hydrogen sulfide). Oxygen is undesirable because it means partial oxidation has occurred. Burning of oil is simply rapid oxidation.

17. We know that temperature is a measure of heat content and temperature increases with depth in the earth. Further, heat is actually the amount of molecular motion in a solid or fluid. How might these facts account for the absence of the larger hydrocarbon molecules when the source rocks are below the oil window?

   Answer: Larger and longer molecules are more likely to break apart as temperature so molecular motion increases.

19. Determine the trap type in each illustration.

   Answer: Correct trap types are indicated in sketches numbered Figures C–2a through C–2f.

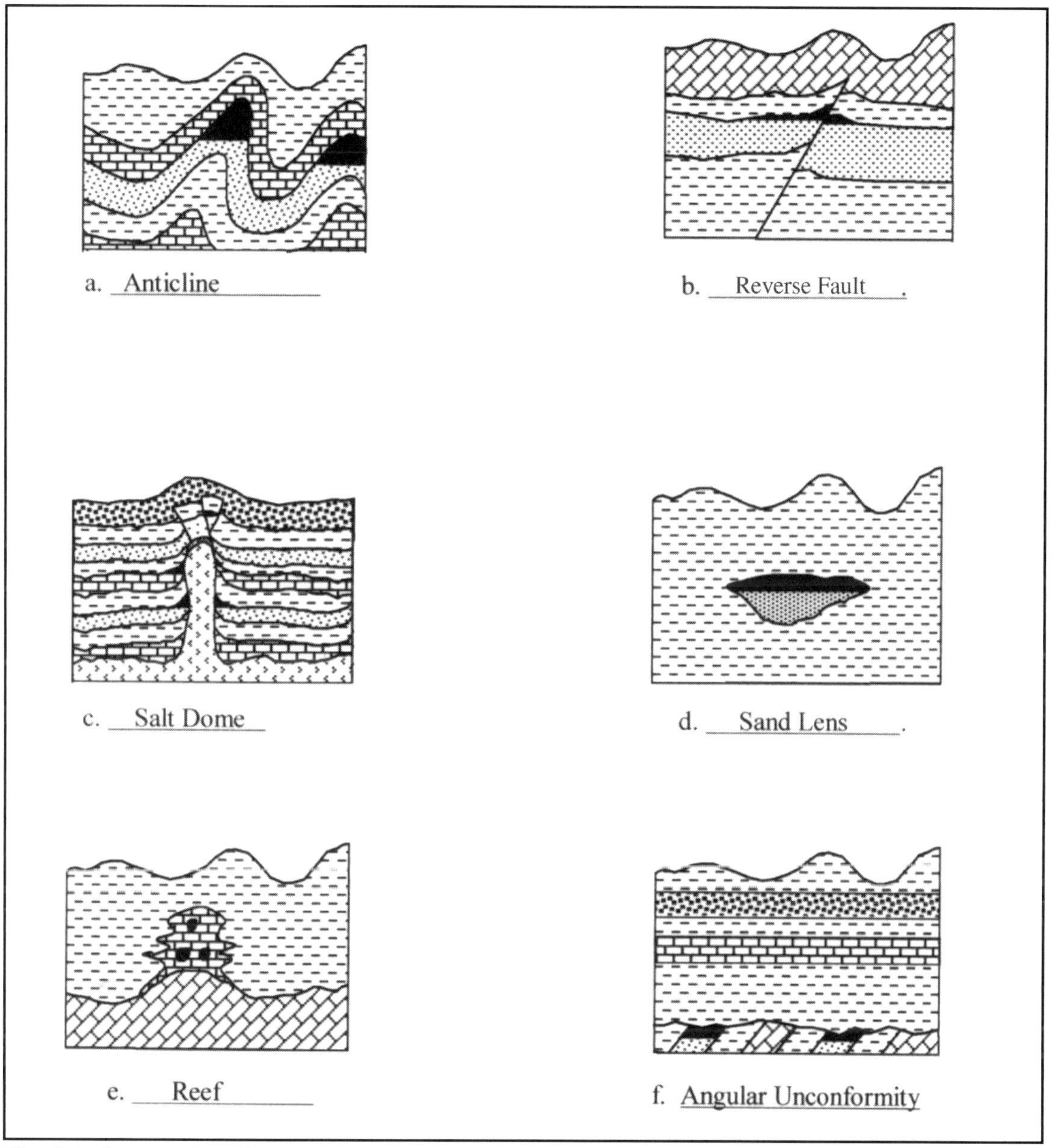

Fig. C–2 *Designation of Trap Types*

Geophysical and mathematical background workshop answers

1. Match the list of properties with the wave types by writing the appropriate letter(s) after the wave type.

   A. Particle motion in the direction of propagation

   B. Particle motion orthogonal to direction of propagation

   C. Particle motion describes retrograde ellipse

   D. Propagates through body of medium

   E. Propagates along surface of medium

   F. Propagates in solids

   G. Propagates in fluids

P-wave: A, D, E, F, and G

S-wave: B, D, E, and F

Rayleigh wave: C, E, and F

Love wave: B, E, and F

3. A P-wave that is propagating in a medium having a velocity of 2000 m/s is incident on a medium having a velocity of 2500 m/s at an angle of 15° from the normal to the interface. Determine the angles at which all resulting P-waves propagate in both media.

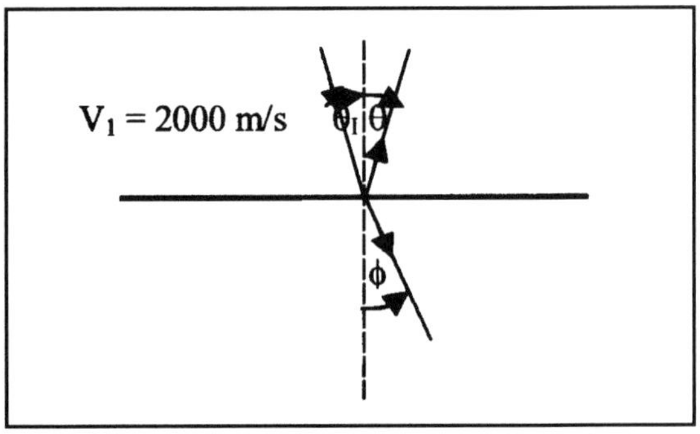

Fig. C–3 *Incident, Reflected, and Refracted Rays*

Law of reflection: $\theta_i = \theta_r so\ \theta_r = 15°$

$$\phi = \sin^{-1}\left(\frac{V_2 \sin\theta_i}{V_1}\right) = \sin^{-1}\left(\frac{2500 \sin 15°}{2000}\right)$$

$$= \sin^{-1}(0.3235) = 18.88°$$

5. Complete the table below by determining the output frequencies for each input frequency and sample period.

Table C–1 *Output Frequencies*

Input Frequency (Hz)	Output Frequency (Hz) 1 ms sampling	Output Frequency (Hz) 2 ms sampling	Output Frequency (Hz) 4 ms sampling	Output Frequency (Hz) 8 ms sampling
60	60	60	60	60
120	120	120	120	5
180	180	180	70	55
240	240	240	10	10
300	300	50	50	50

7. Given wavelet $\alpha = \{5, -2\}$ and wavelet $b = \{-3, 1\}$, calculate the cross-correlations, $\phi ab(t)$ and $\phi ba(t)$.

Answer:

	5	−2			
−3	1			5×1=	5
	−3	1		5×−3+(−2×1)=	−17
		−3	1	5×1=	5
			−3 1	−2×−3=	6

$$\phi ab(\tau) = \{5, -17, 6\},\ \phi ba(\tau) = \phi ab(-\tau) = \{6, -17, 5\}$$

9. Which of the following are minimum phase wavelets? (The first value in each case is at time zero.)

a. 6, −1, −2, Minimum phase

b. 3, 4, −4, Mixed phase

c. 0, 12, −1, −6, Mixed phase

d. −2, 5, −2, Linear phase

e. 28, −27, 5, Minimum phase

11. Which wavelet has the larger bandwidth, A or B?

Answer: B has the larger bandwidth. The larger the bandwidth, the shorter the duration in time.

13. It is desired to have frequencies up to 50 Hz in the target zone. If the velocity and maximum dip are expected to be 3600 m/s and 35°, respectively, what is the largest group interval that will assure having the desired frequency?

$$\Delta x = \frac{V}{3f_{max}\sin\theta} = \frac{3600}{3 \times 50 \times \sin 35°} = 41.8m$$

15. Determine the inverse of wavelet a in exercise 7.

$$W_a^{-1}(z) = \frac{1}{W_a(z)} = \frac{1}{5 - 2z} = \frac{1}{5}\left[1 + \frac{2}{5}z + \left(\frac{2}{5}\right)^2 z^2 + \left(\frac{2}{5}\right)^3 z^3 + \left(\frac{2}{5}\right)^4 z^4 + \ldots\right]$$

Seismic refraction exploration workshop answers

1. Use the T-X plot shown to determine the dip angle j, the critical angle qc, and the true velocity of the second layer V_1.

$$\theta_c = \frac{1}{2}\left[\sin^{-1}\left(\frac{V_0}{V_{1d}}\right) + \sin^{-1}\left(\frac{V_0}{V_{1u}}\right)\right] = \frac{1}{2}\left[\sin^{-1}\left(\frac{2000}{2390}\right) + \sin^{-1}\left(\frac{2000}{4434}\right)\right]$$

$$= \frac{1}{2}(56.806° + 26.812°) = 41.809°$$

$$\varphi = \frac{1}{2}[\sin^{-1}\left(\frac{V_0}{V_{1d}}\right) - \sin^{-1}\left(\frac{V_0}{V_{1u}}\right)] = \frac{1}{2}\ (56.806° + 26.812°) = 14.997°$$

$$V_1 = 2\cos\varphi\left(\frac{V_{1d}V_u}{V_{1d}+V_u}\right) + 2\cos(14.997°)\left(\frac{2390 \times 4344}{2390 \times 4344}\right) = 3000m/s$$

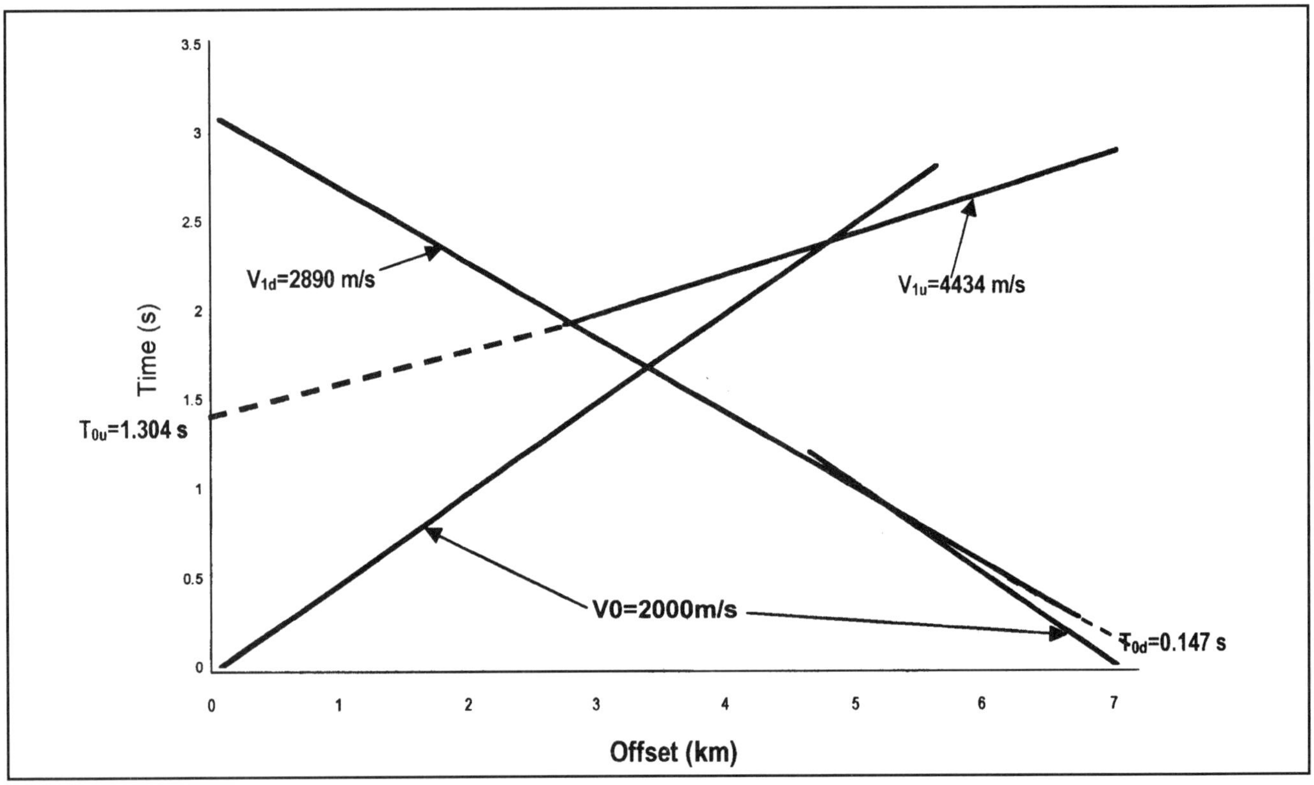

Figure C–4 *T-X Plot with Velocities and Delay Times Labeled*

Seismic reflection data acquisition workshop answers

1. A marine vessel is pulling a 6 km long streamer at a speed of 6 knots along a boat track of N75°E. There is a current of 1.5 knots in the direction S30°W. What will be the angle between the vessel heading and the streamer?

   How many m of cross-line deflection will result at the end of the streamer? (Assume the current acts uniformly on the streamer.)

   Answer:

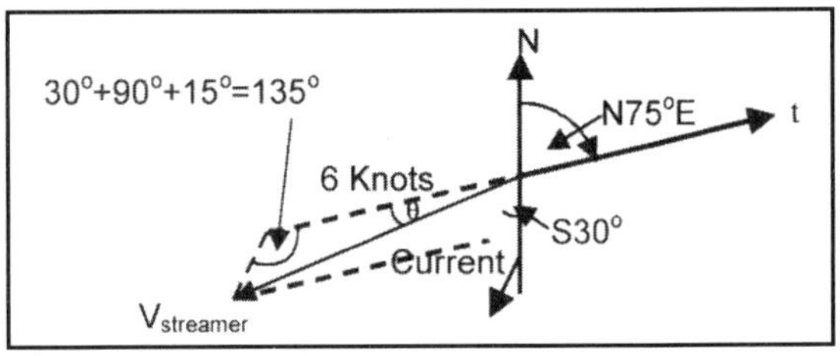

Fig. C–5 *Vector Diagram*

$$V_{streamer} = \sqrt{1.5^2 + 8^2 - 2 \times 1.5 \times 6 \times \cos 135°} = \sqrt{38.25 + 18\sin 45°} = 7.14$$

30°+90°+15°=135°

3. Shown here are sketches representing the far-field signatures and amplitude spectra for a one-pound charge and an eight-pound charge of dynamite. Based on these, sketch the signature and amplitude spectrum for two one-pound charges fired simultaneously in two holes.

Answer: See Figure C–6.

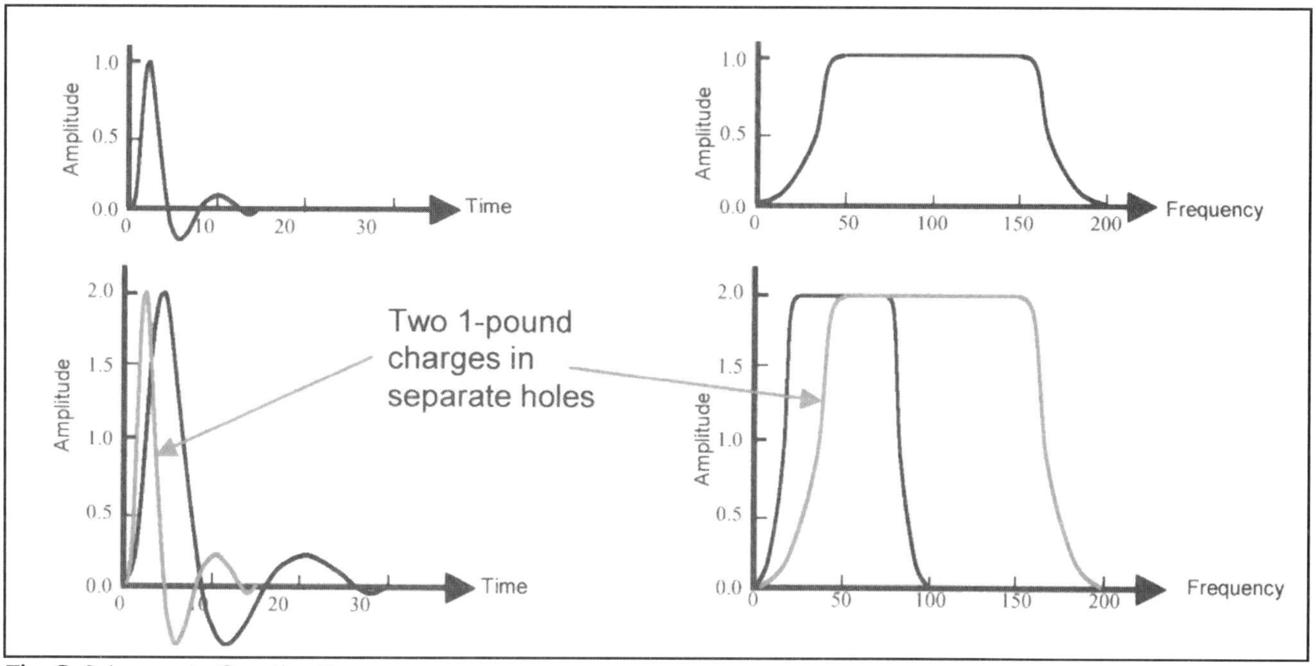

Fig. C–6 *Answer to Question 3*

5. If the airgun arrays are at a depth of 6 m and the streamer is being towed at a depth of 10 m, at what frequencies would the ghost notches be seen? (Assume water velocity is 1500 m/s.)

$$f_{source} = \frac{1500}{2x6} = 125Hz \quad f_{receiver} = \frac{1500}{2x10} = 75Hz$$

7. A 4-element source array and an 18-element receiver array are to be used. The group interval is 110 ft. If source-generated noise is expected, what spacing between elements in the two arrays will maximize noise attenuation?

$$k_{min} = \frac{20}{1800} = 0.01111cpf$$

$$k_{max} = \frac{30}{1140} = 0.02632cpf$$

Determine the largest interval between elements that will give an array length less than group interval and values of $kd < 0.5$.

Answer: Select d_{source} =19 ft and $d_{receiver}$ = 6 ft.

Calculate ranges for kd.

Source:

$$kd_{\max} = \frac{30 \times 19}{1140} = 0.5$$

$$kd_{\min} = \frac{30 \times 19}{1800} = 0.3167$$

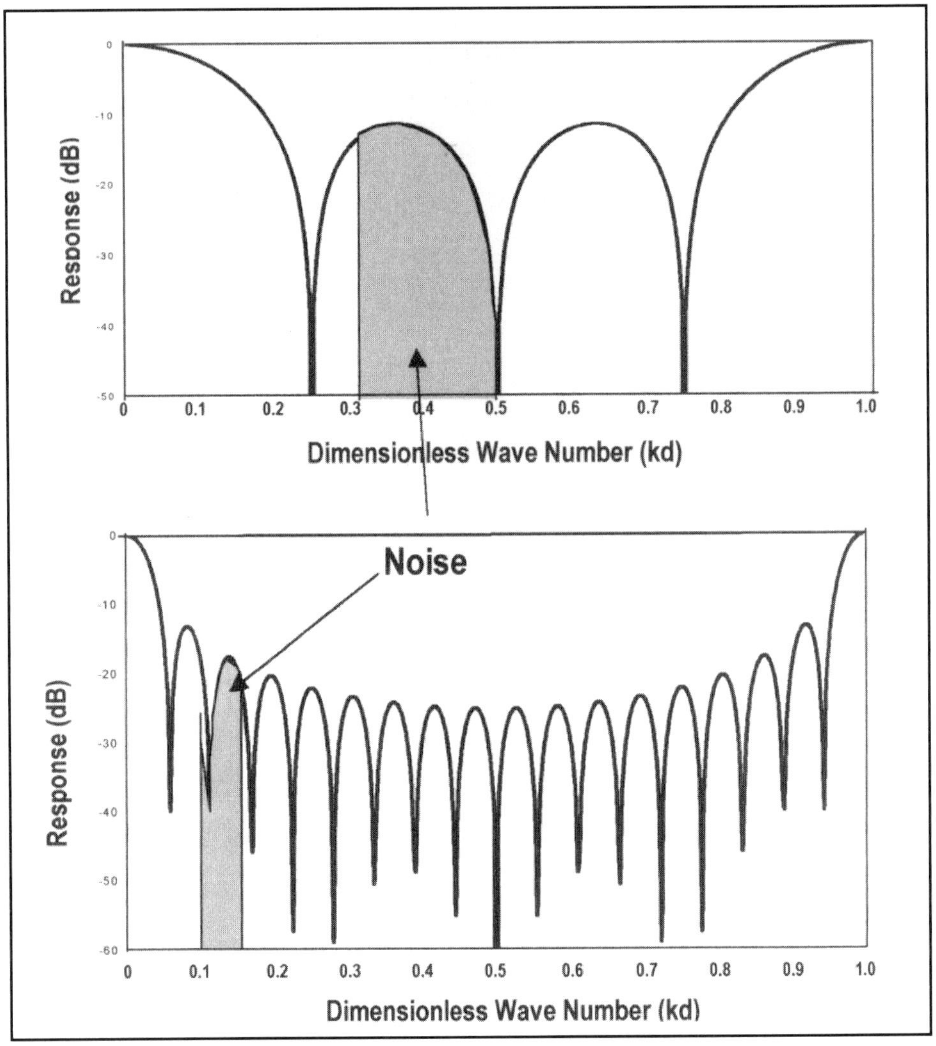

Fig. C-7 *Receiver and Source Array Response and Combined Source and Receiver Response*

Receiver:

$$kd_{max} = \frac{30 \times 6}{1140} = 0.1579$$

$$kd_{min} = \frac{30 \times 6}{1800} = 0.1$$

9. A tape dump yields the following information:

File Number Format Code

First four bytes of General Header: 0371 8048

Bytes 101–104 of first Data Block: 43F8 8000

What amplitude is written in bytes 101–104 of first Data Block?

Format 8048 is 4-Byte Hexadecimal Exponent, so converting bytes 101–104 to binary gives
0100 0011 1111 1000 1000 0000 0000 0000

S Exponent Mantissa

Answer:

Value =0.111110001000000000000000x16$^{(67-64)}$ =111110001000_2 = 379610

Seismic reflection data processing workshop answers

1. What is the most significant factor in deciding between straight stack and diversity stack of vibrator data?

   Answer: Presence of noise bursts usually calls for diversity stack.

3. What is the purpose of displaying first breaks after LMO corrections are applied to land data?

   To determine whether source points are located as observer logs/trace headers indicate and to move source positions to be consistent with first break refraction arrivals.

5. How are datum statics calculated if weathering depths and velocities are not known?

   Answer:

$$SSC = \frac{E_{SS} - E_R}{V_{AVD}} \quad \text{RSC} = \frac{E_{RS} - E_R}{V_{AVD}}$$

where

SSC = source static

RSC = receiver static

E_{SS} = elevation at source

E_{RS} = elevation at receiver

E_R = elevation of datum

V_{AVD} = average velocity from surface to datum

7. Which of the following amplitude compensation methods are considered deterministic and can, thus, be applied to data on which AVO analyses are to be run?

   Answer:

   a. geometric spreading correction

   e. surface consistent amplitude compensation

9. In an IVP session, what happens to an event on the center CMP gather when the velocity pick is moved to a much slower velocity? What happens when the pick is moved to a much faster velocity?

   Answer: When the pick is moved to a slower velocity the event in question, as well as nearby events, becomes over-corrected. When the pick is moved to a faster velocity the event in question, as well as nearby events, becomes under-corrected. Either case causes a deterioration in stack response.

11. If predictive deconvolution is applied to the data, what prediction distance (gap) would you recommend based on the autocorrelogram?

    Answer: Using the second zero-crossing, a gap of about 40 ms would be selected

13. What are the significant parameters in reflection residual statics and how do their values affect results?

    Answer:

    correlation window(s)

    number and length

    maximum correlation shift

    NMO correction

15. What causes NMO stretch?

    Answer: The amount of NMO correction decreases as time increases because both velocity and zero-offset time increase. Thus, the earlier samples in a wavelet are moved farther in time than later samples.

17. What are the two primary actions of migration?

    Answer: Move dipping events in time and space to their true location and collapse difractions.

19. Draw a sketch to illustrate the difference between wiggle trace and variable area display of seismic traces.

    Answer: See Figure C–8.

Fig. C–8 *Wiggle Trace (left) and Variable Area (right) Trace Displays*

21. Select the optimum value for maximum correlation shift from those shown, based on the effect of residual statics on the stack.

Answer: See Figure C–9.

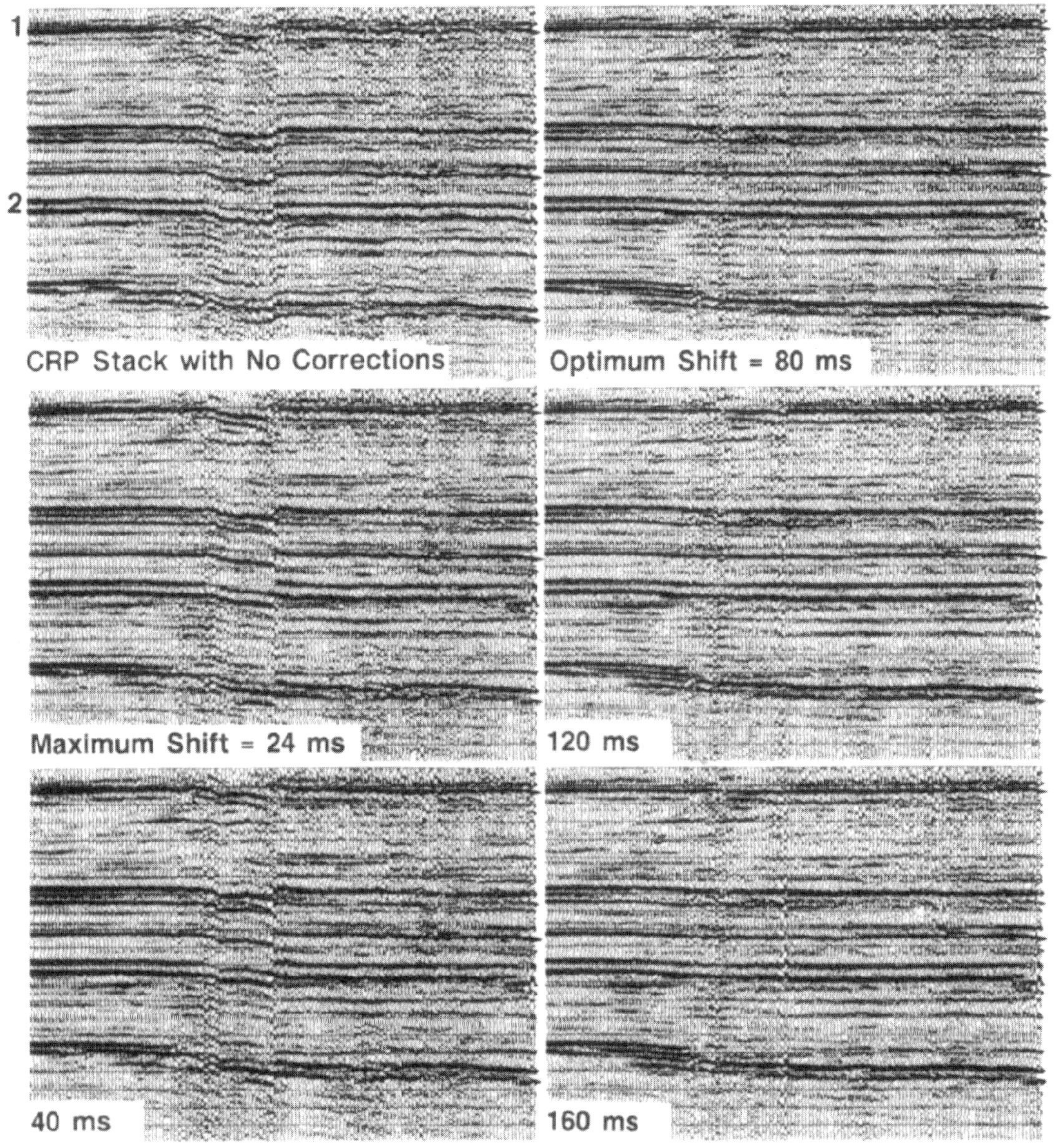

Fig. C–9 *Optimum Value for Maximum Correlation Shift*

23. Shown in Figure C–10 are:

    (a) a CMP stack

    (b) the migrated stack, and

    (c) a sketch of the following—a prominent diffraction *D*, a dipping event *B* before migration, and a dipping event *A* after migration.

Answer:

- the dip angle of reflector *A* is steeper in the migrated section compared to the stack section
- reflector *A* is shorter after migration
- migration moves reflector *A* up and to the left (up-dip)

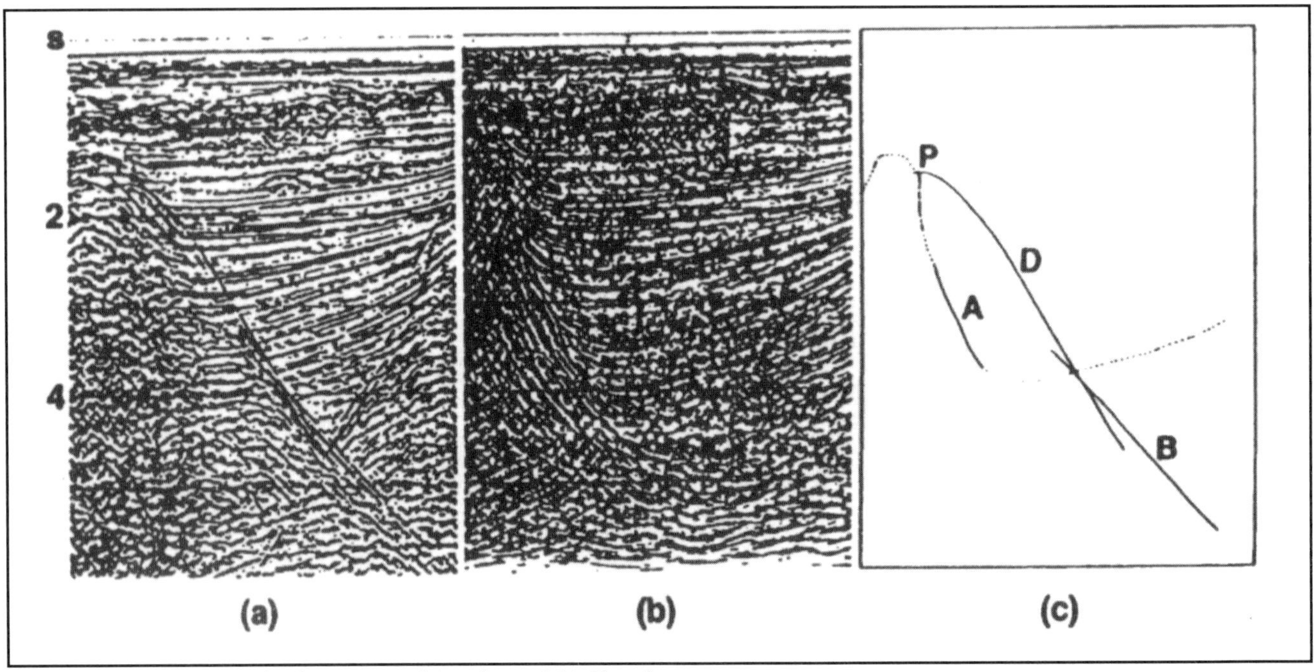

Fig. C–10 *(a) CMP Stack, (b) Migrated Stack, and (c) Sketch*

25. Shown in Figure C–11 is a CMP Stack and four migrations of it. Which migration uses the optimum migration velocity? Are the other panels over- or under-migrated?

Answer: See Figure C–11.

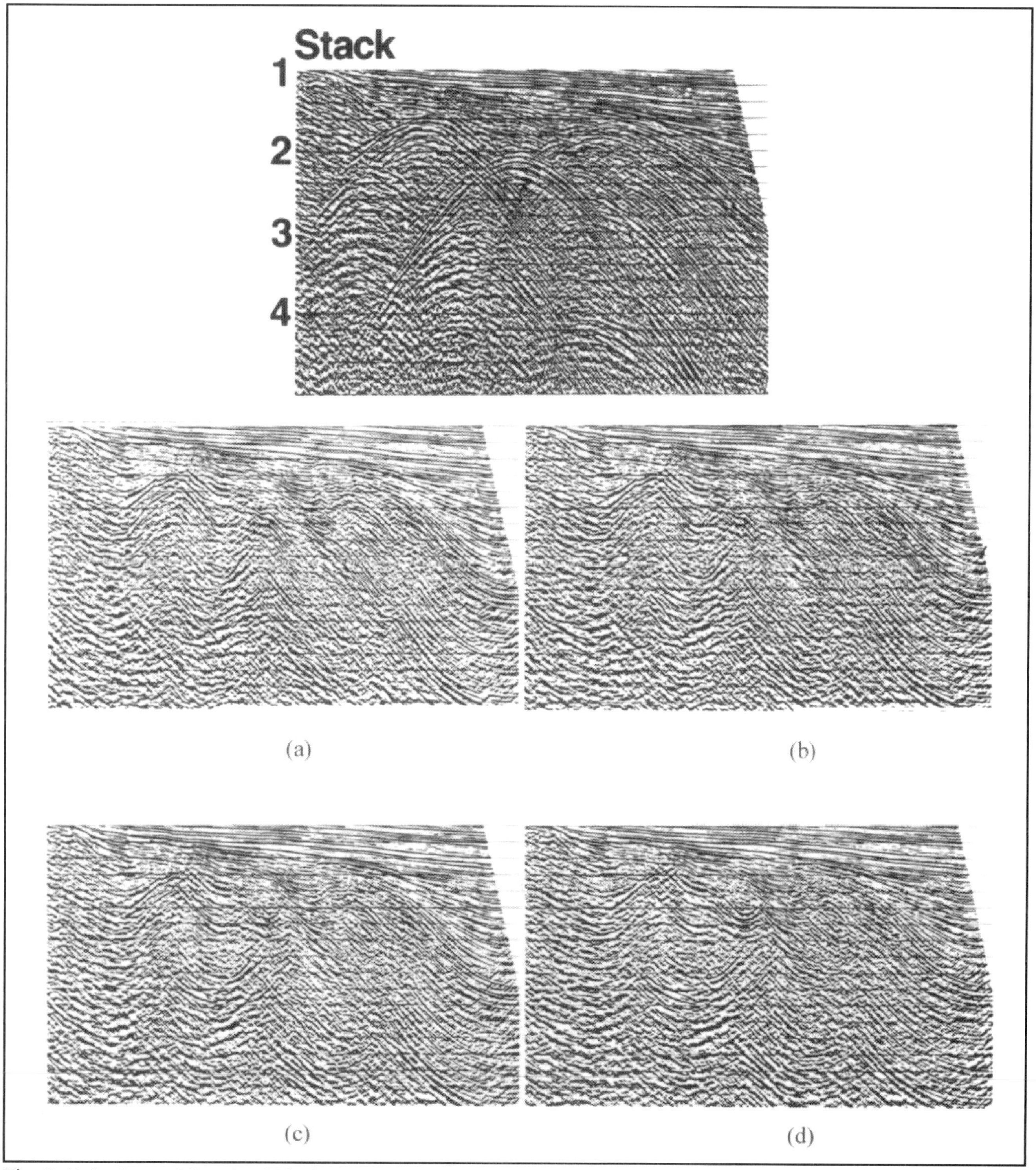

Fig. C–11 *Optimum Migration Velocity*

27. Shown in Figure C–12 are four stacks of the same data but with different trace spacing. Figure C–13 shows the corresponding phase shift migration panels. What trace spacing is required to prevent spatial aliasing?

    Answer: 26.6 m.

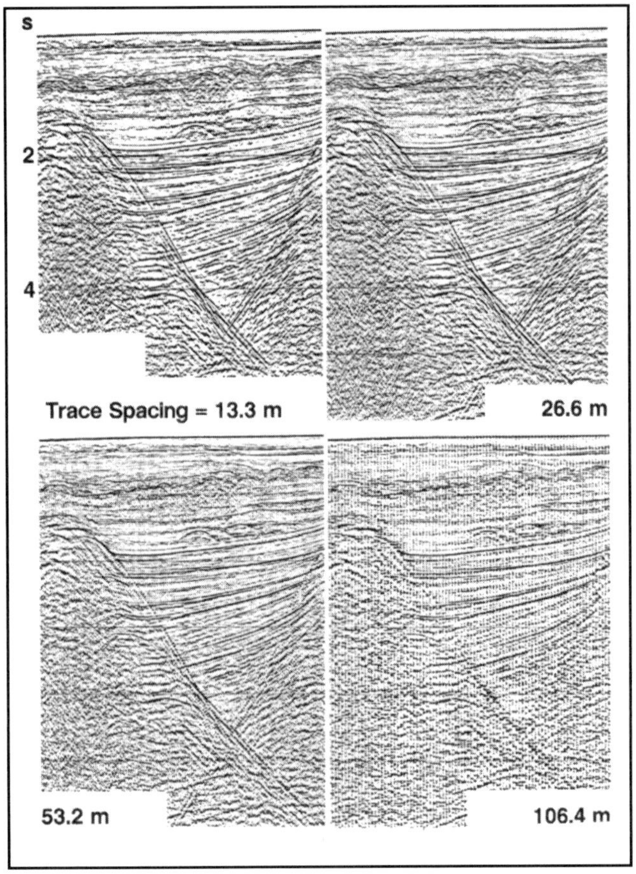

Fig. C–12 *Four CMP Stacks of Same Data with Different Trace Spacings*

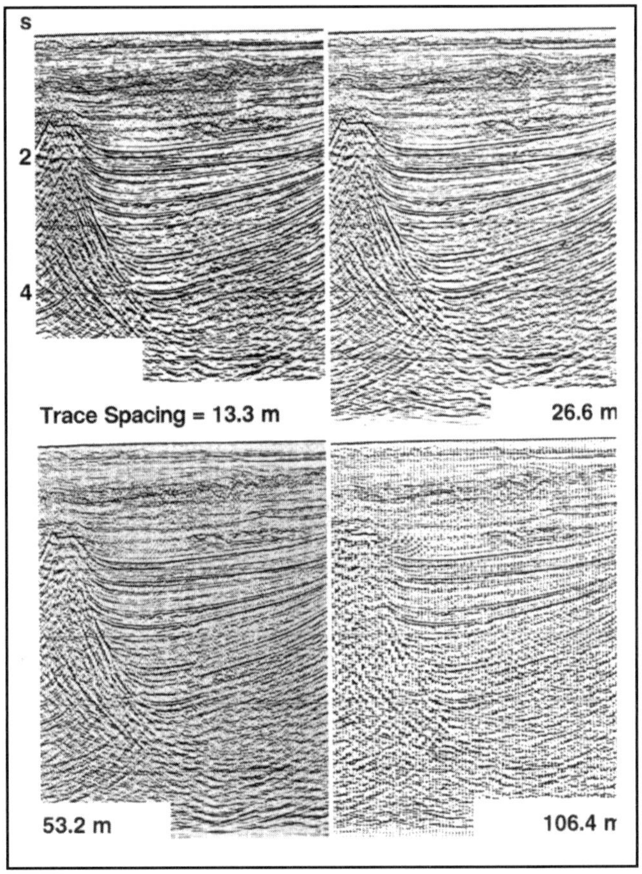

Fig. C–13 *Phase Shift Migrations of the Stacks Shown in Figure C–12*

Appendix D
Color Plates

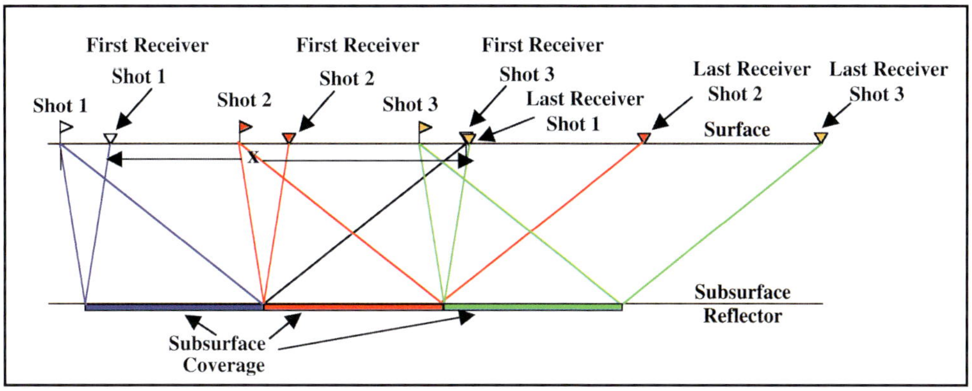

Plate 1–4 *Continuous or Single-Fold Subsurface Coverage*

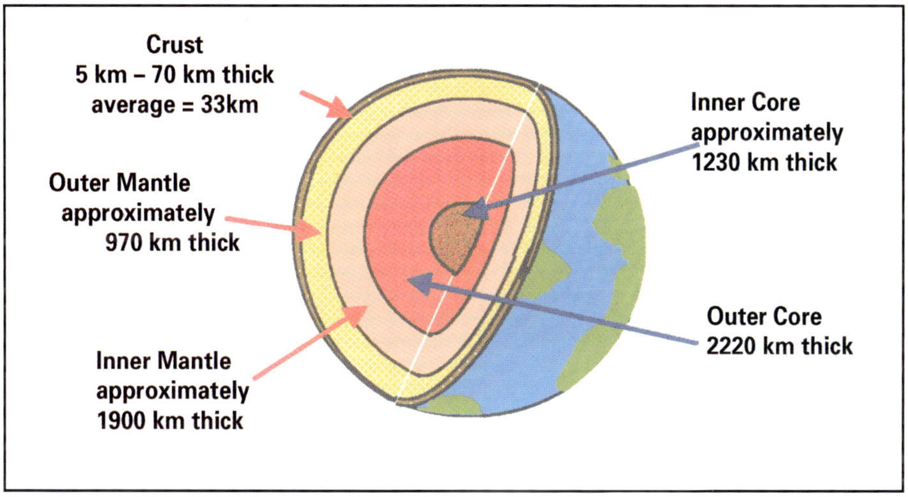

Plate 2–3 *Internal Structure of the Earth*

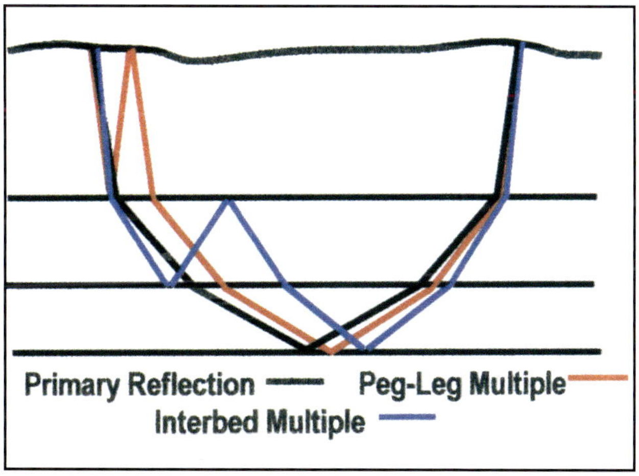

Plate 3–19 *Multiple Reflections*

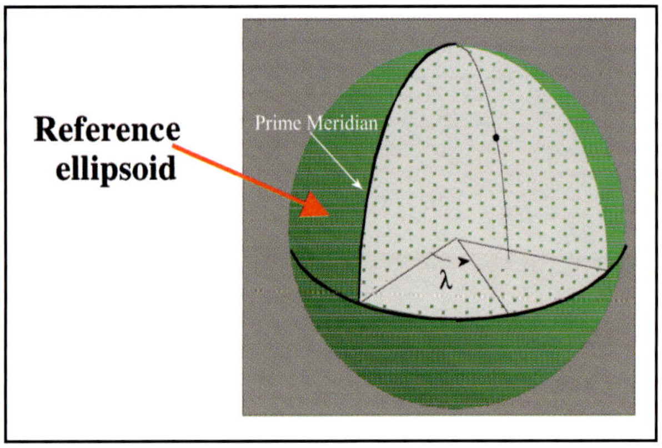

Plate 5–6 *Longitude Measurement*

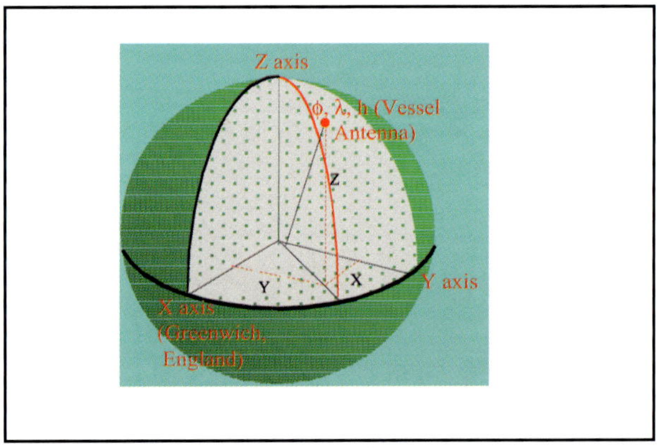

Plate 5–7 *Longitude Measurement*

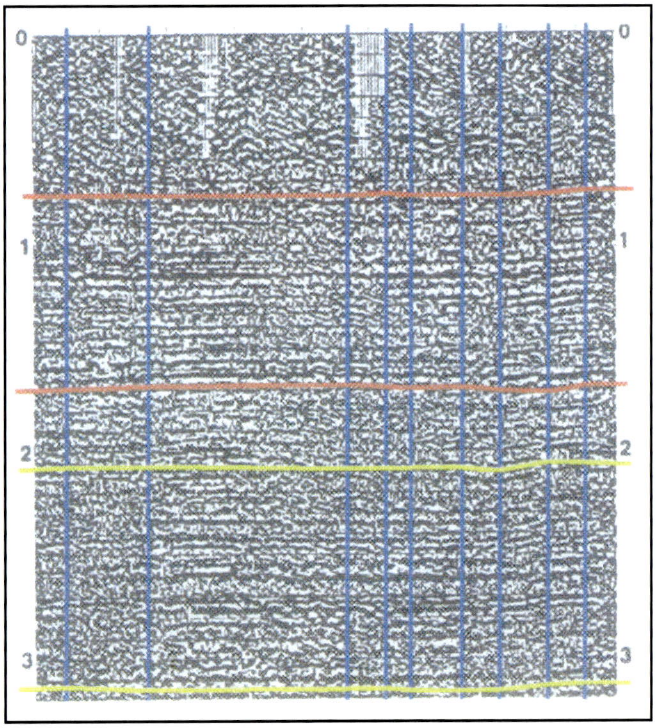

Plate 6–131 *Defining Windows for Residual Statics Analysis*

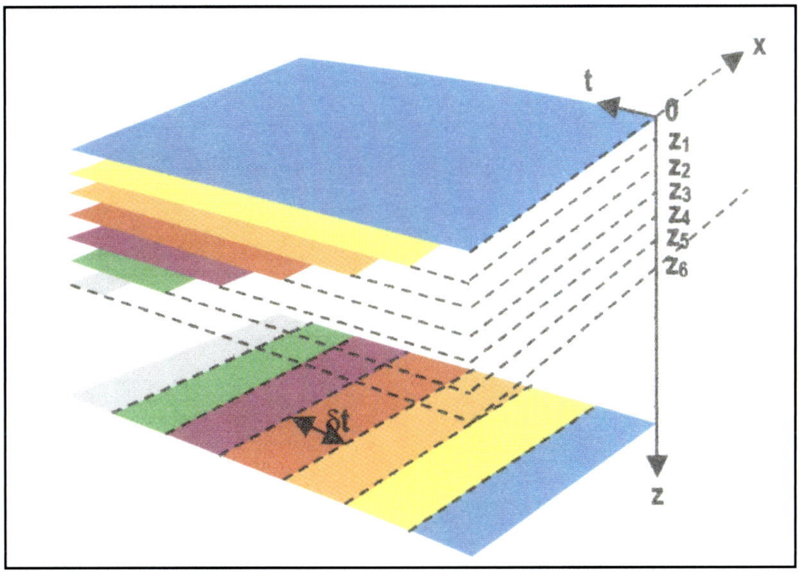

Plate 6–213 *Downward Continuation Time Migration with Depth Slices at Different Layers*

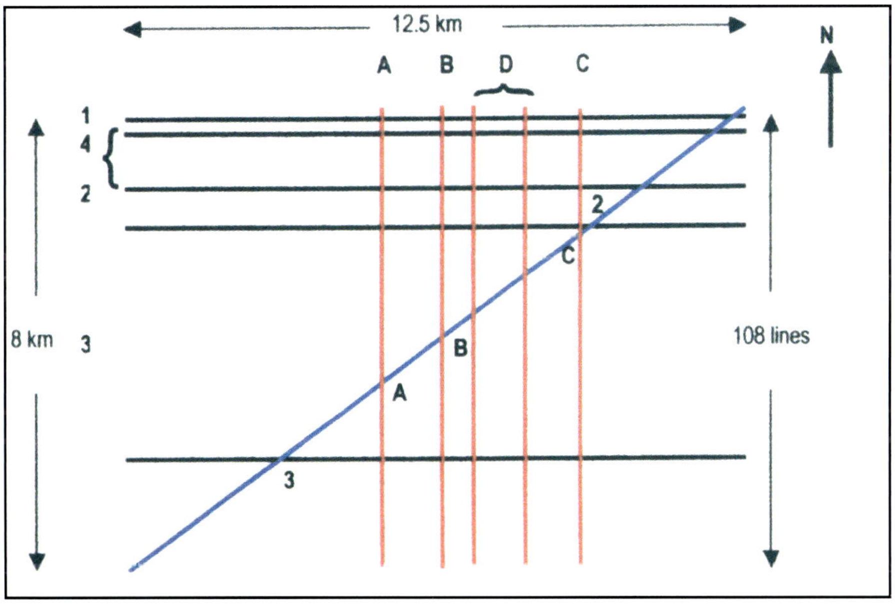

Plate 6–282 *Line Orientations*

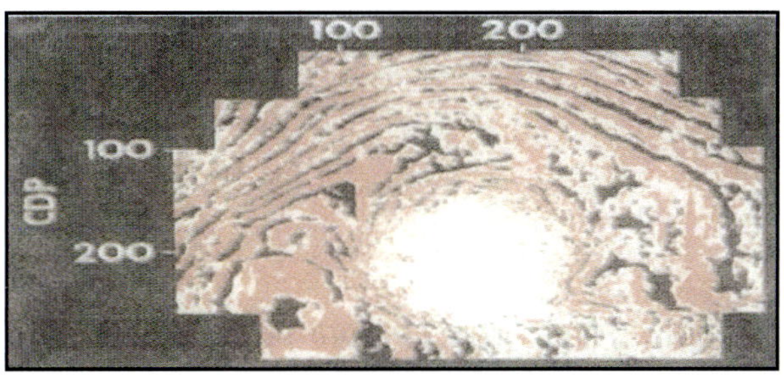

Plate 6–287 *Time Slice through a Salt Dome*

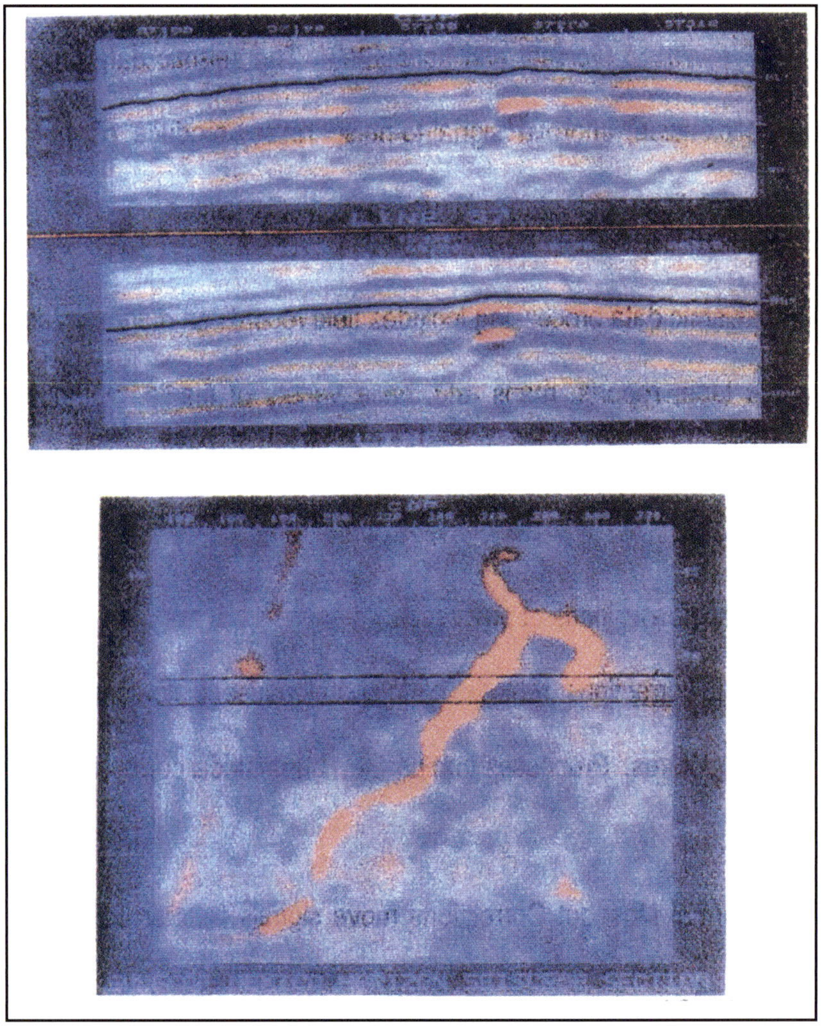

Plate 6–288 *Bright Spots on Vertical and Horizontal Sections*

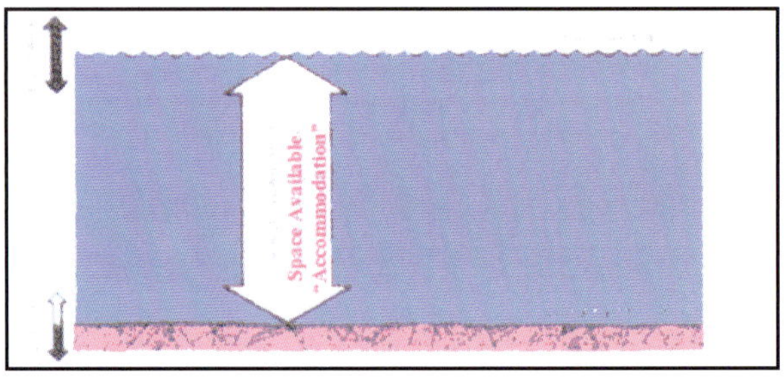

Plate 7–79 *Accommodation Envelope as a Function of Eustacy and Subsidence*

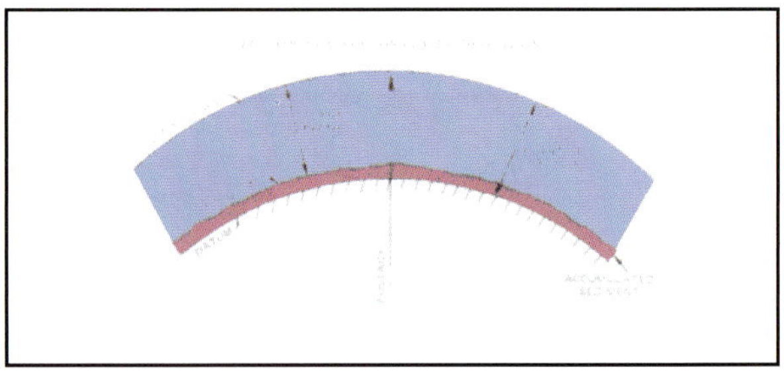

Plate 7–80 *Eustacy, Relative Sea Level, Water Depth as a Function of Sea Surface, Water Bottom and Datum Position*

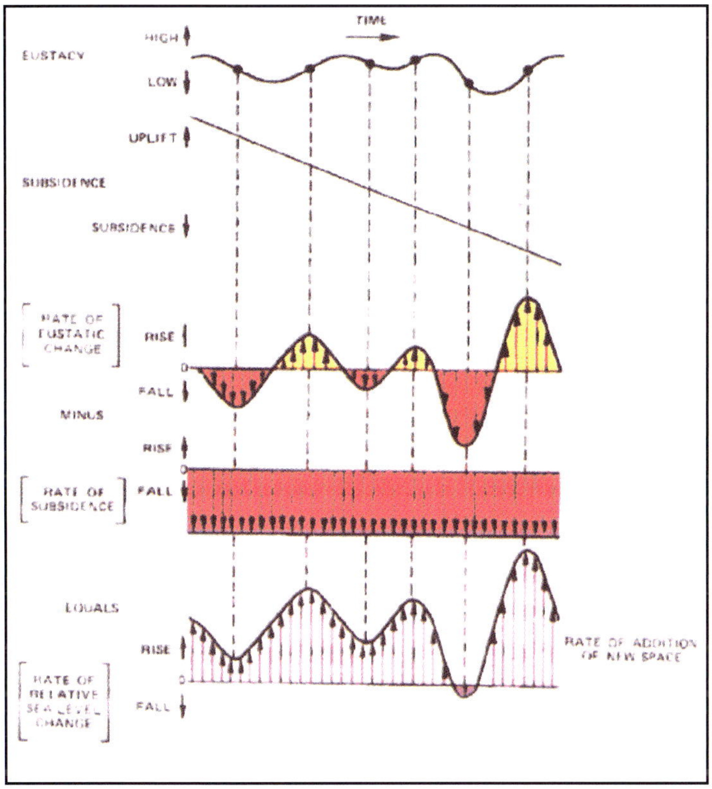

Plate 7–82 *Relative Sea Level as a Function of Eustacy and Subsidence*

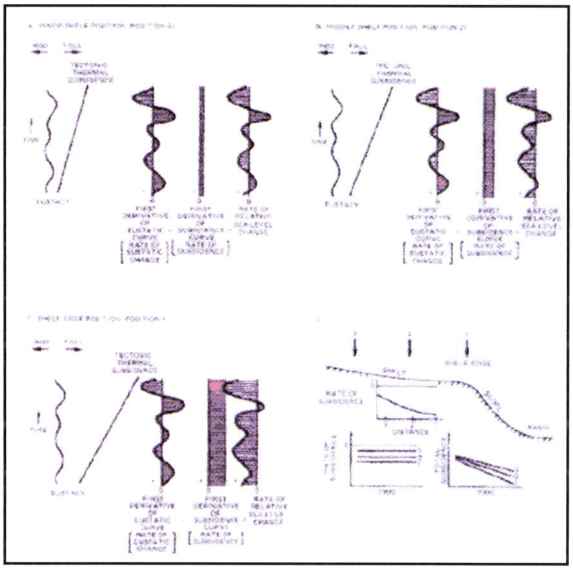

Plate 7–83 *Response of Relative Sea Level to Differential Tectonic Thermal Subsidence*

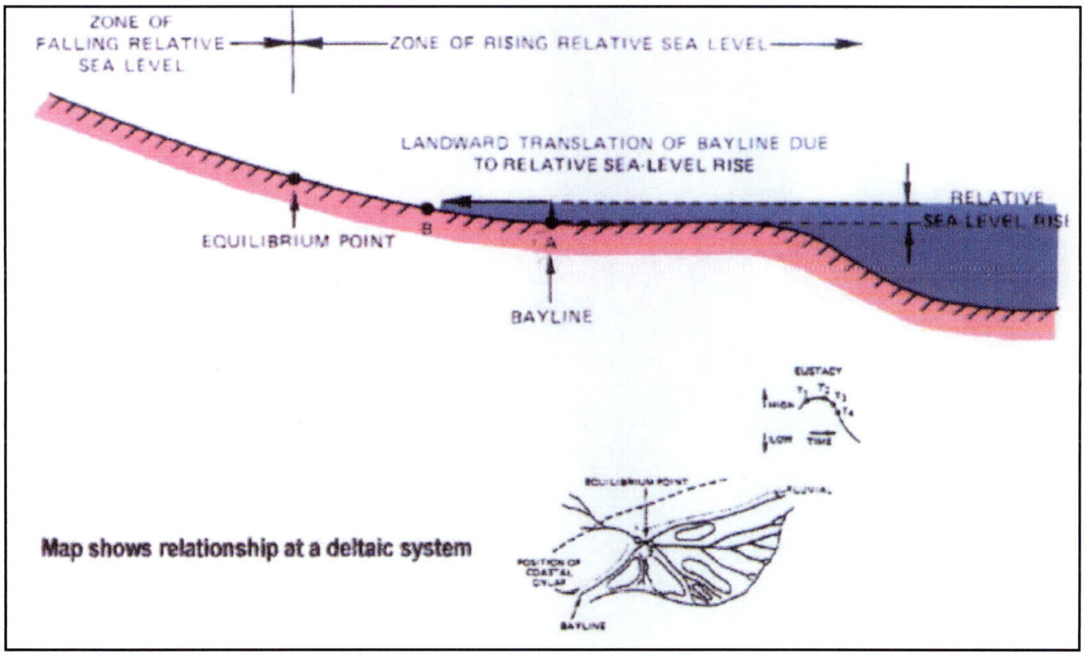

Plate 7–84 *Effect of Relative Sea Level Rise on Coastline Position*

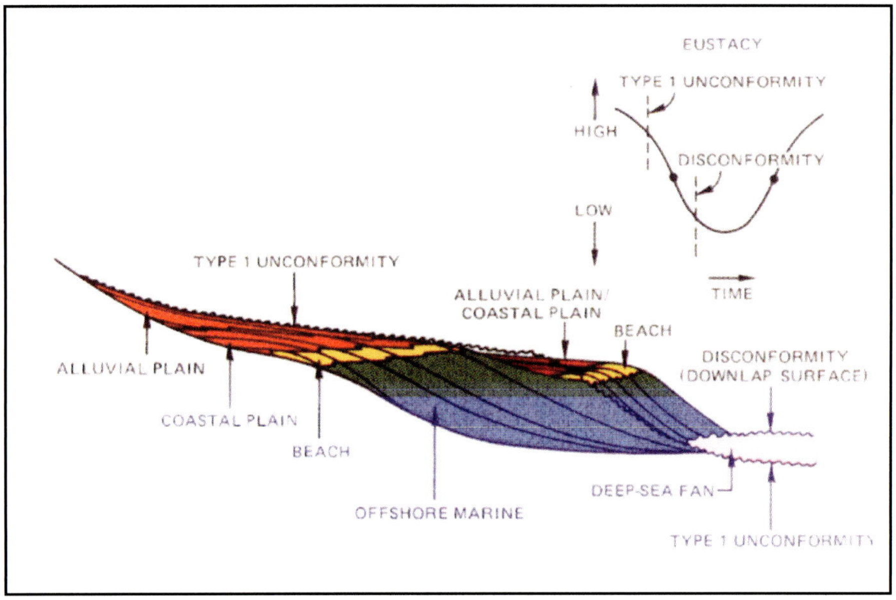

Plate 7–85 *Type 1 Unconformity*

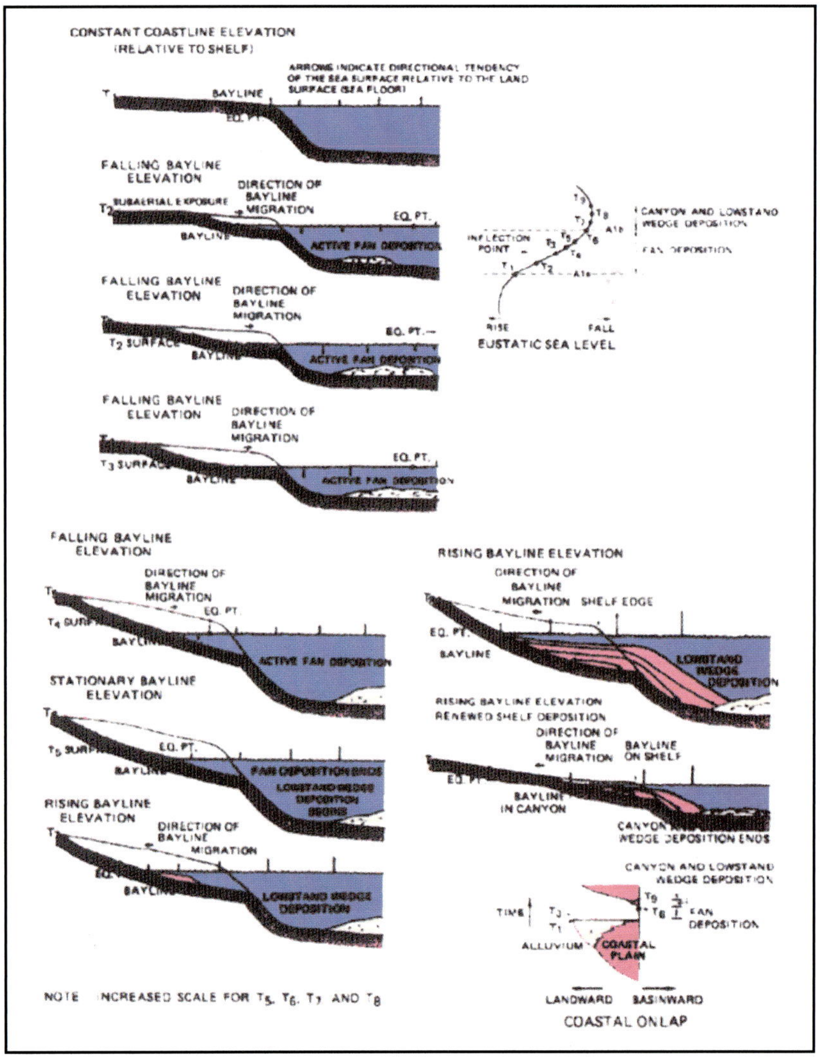

Plate 7–86 *Response of Sedimentation on an Interval of Rapid Eustatic Fall*

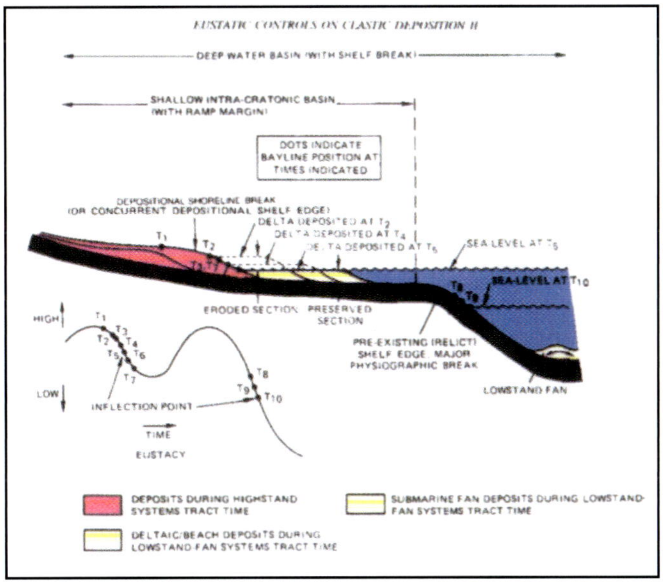

Plate 7–89 *Bayline Position and Lowstand Deposits during Rapid Short- and Long-period Eustatic Fall*

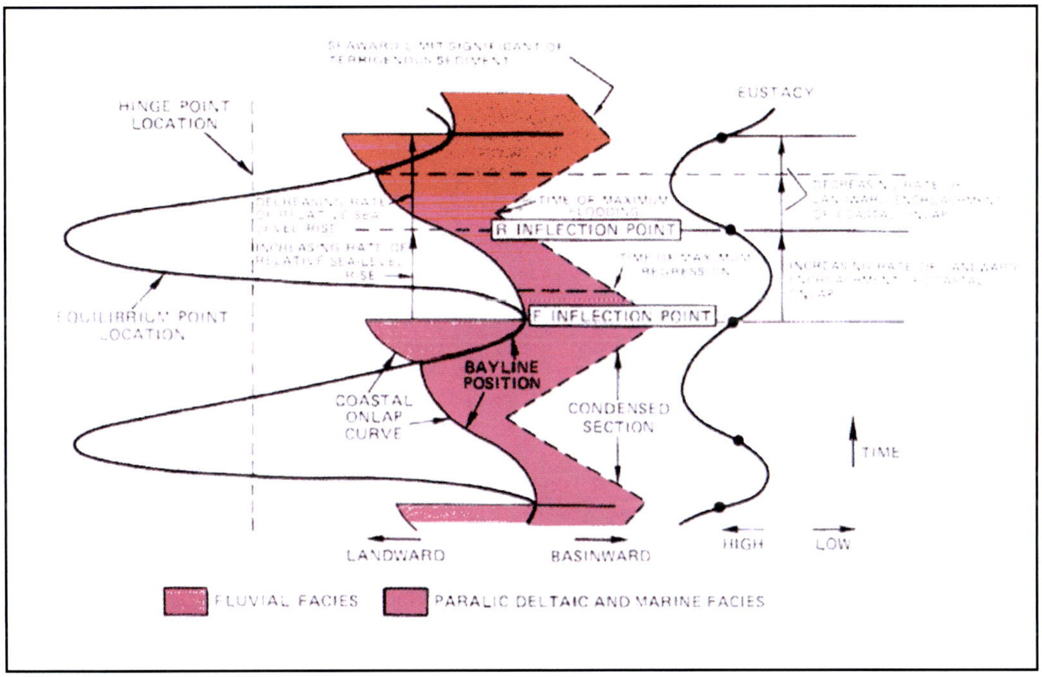

Plate 7–90 *Elements of Coastal Onlap Curve*

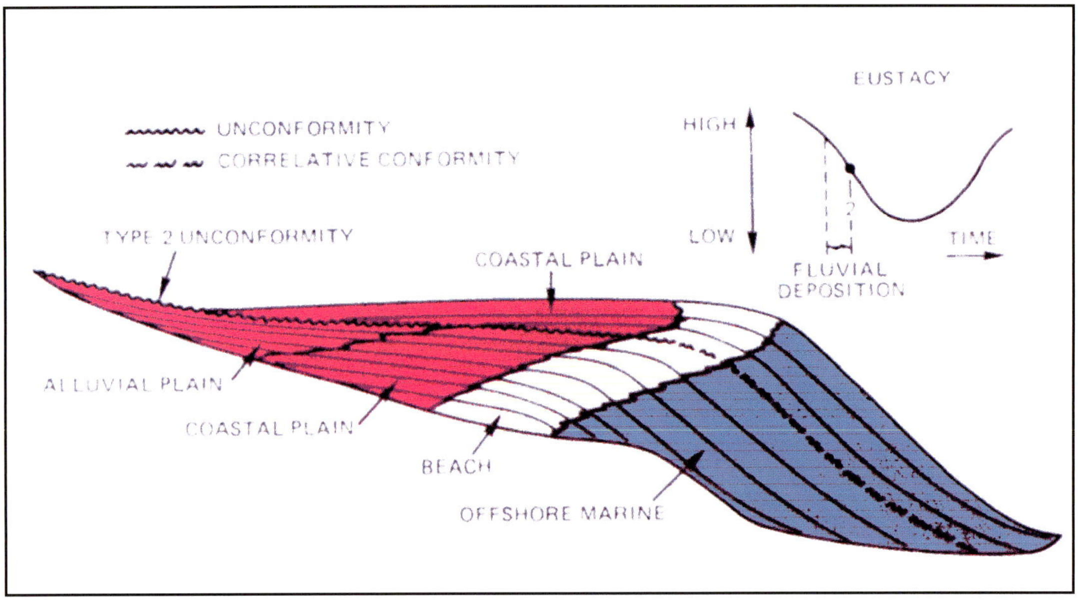

Plate 7–91 *Type 2 Unconformity*

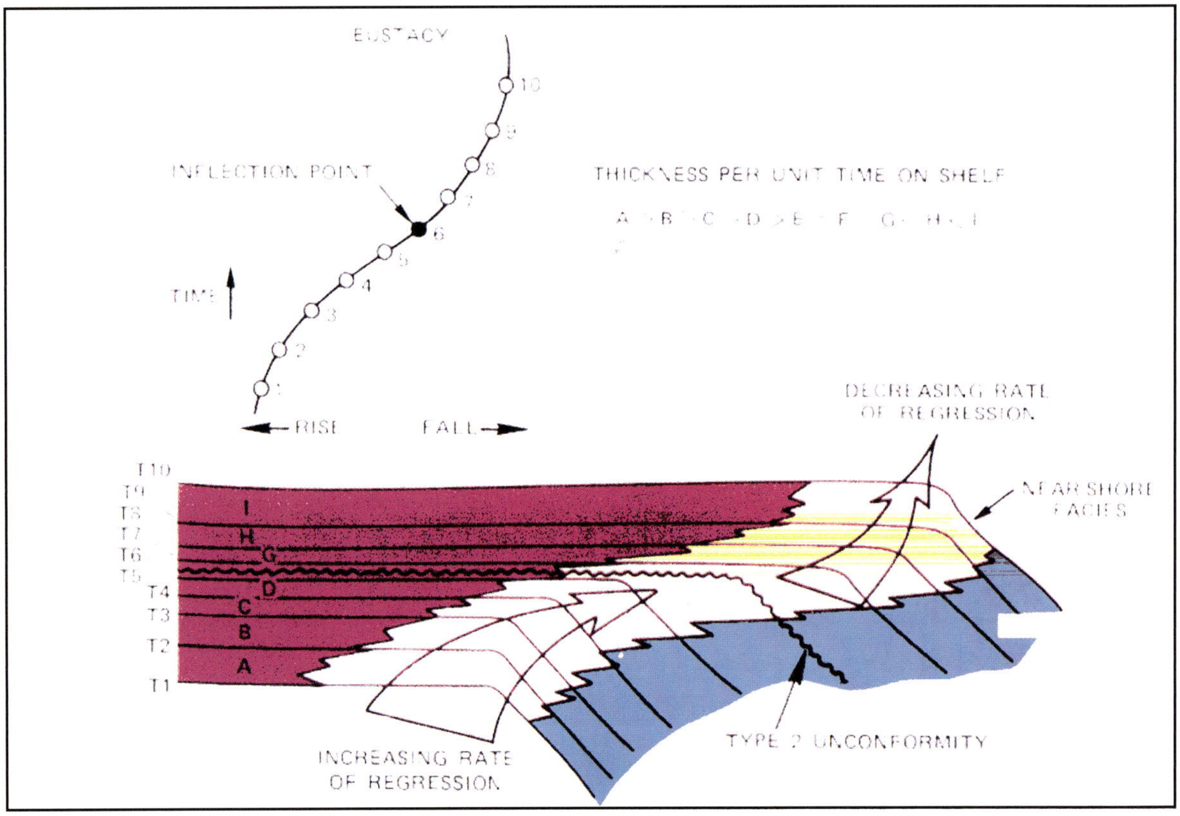

Plate 7–94 *Response of the Topset Bed Thickness to Eustatic Fall*

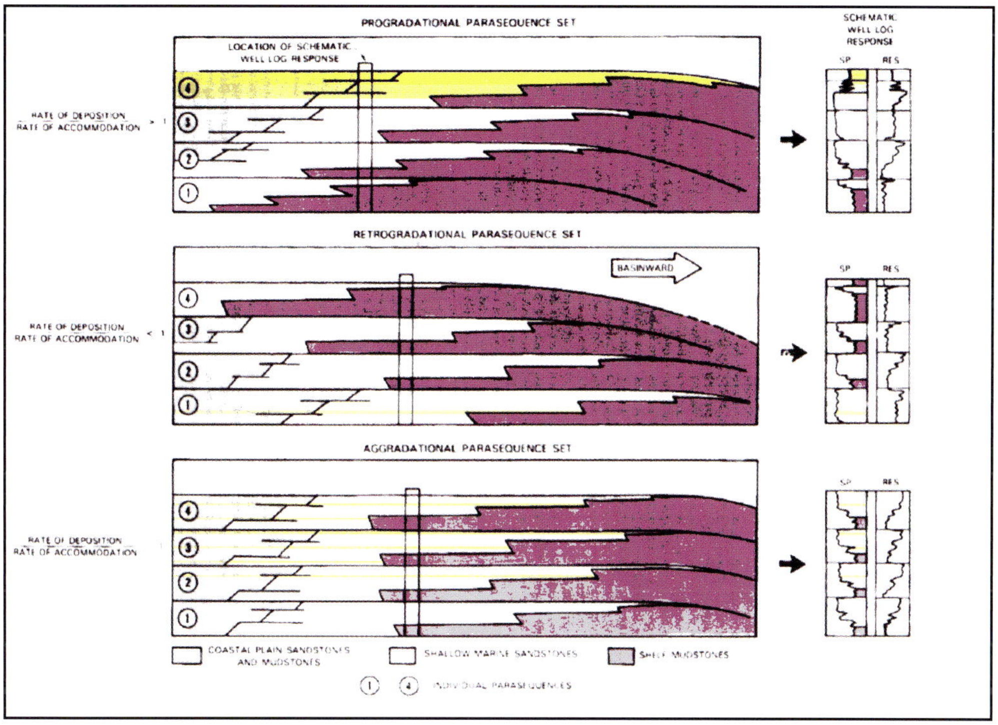

Plate 7–96 *Types of Parasequence Sets*

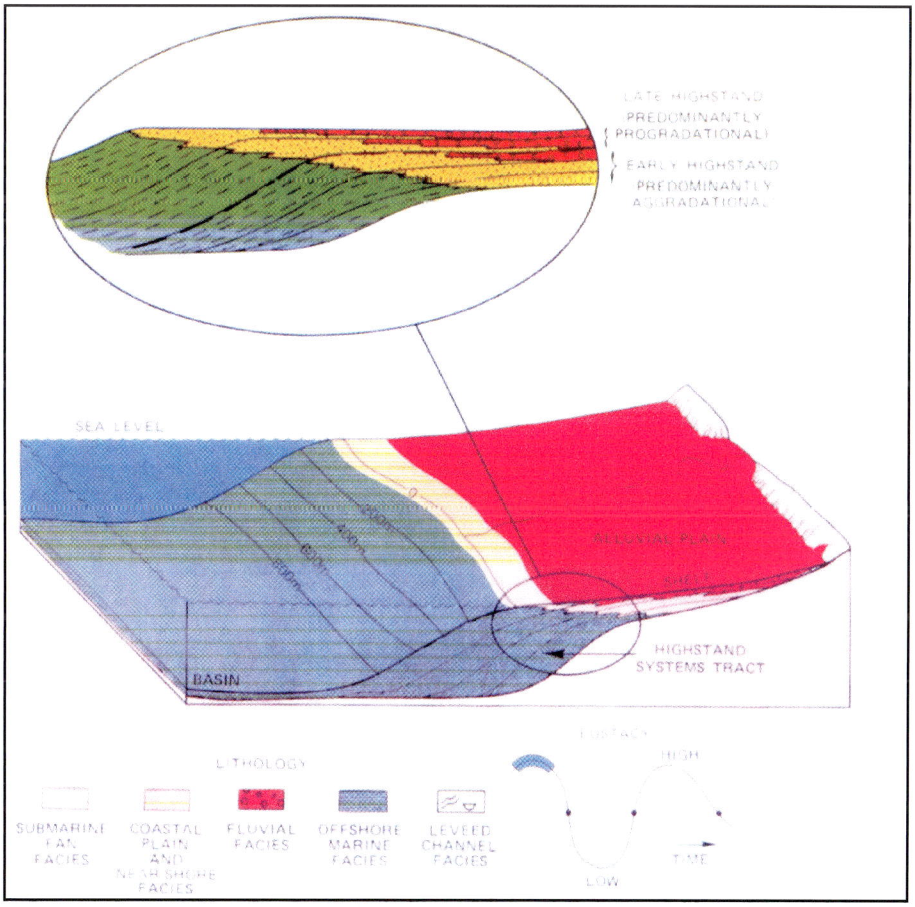

Plate 7–110 *Highstand Systems Tract*

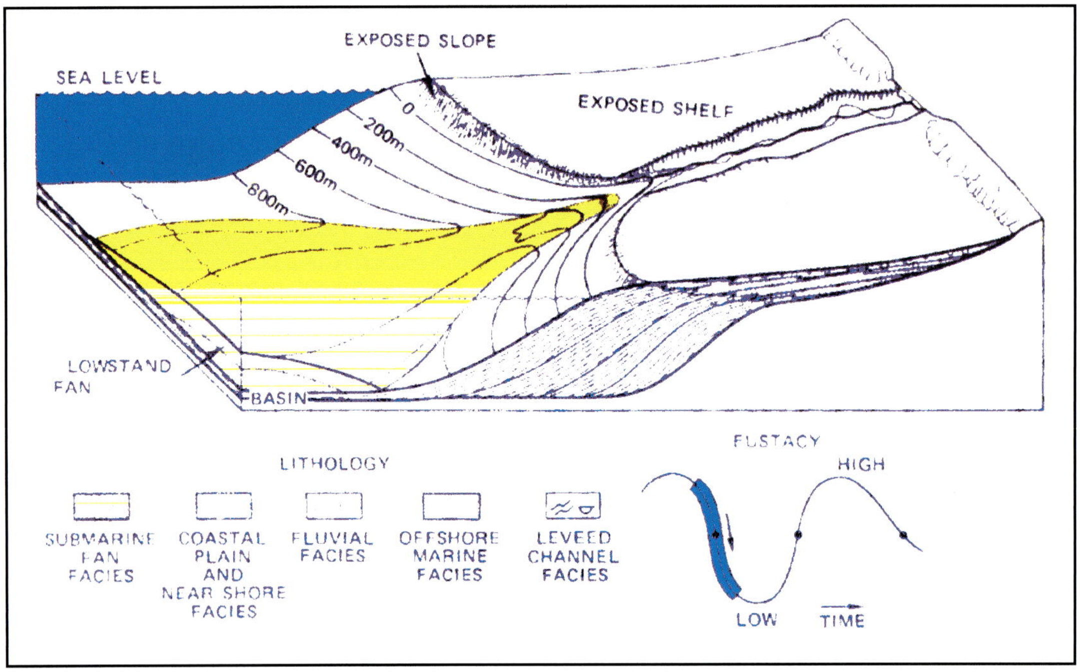

Plate 7–112 *Lowstand System Tract—Lowstand Fan*

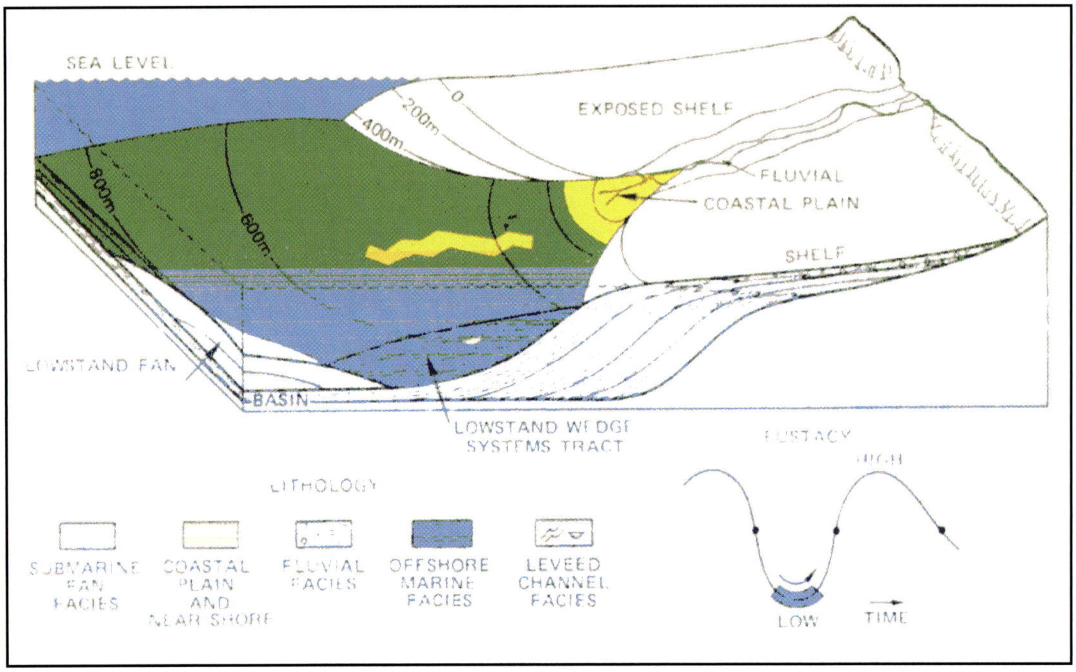

Plate 7–114 *Lowstand System Tract—Lowstand Wedge*

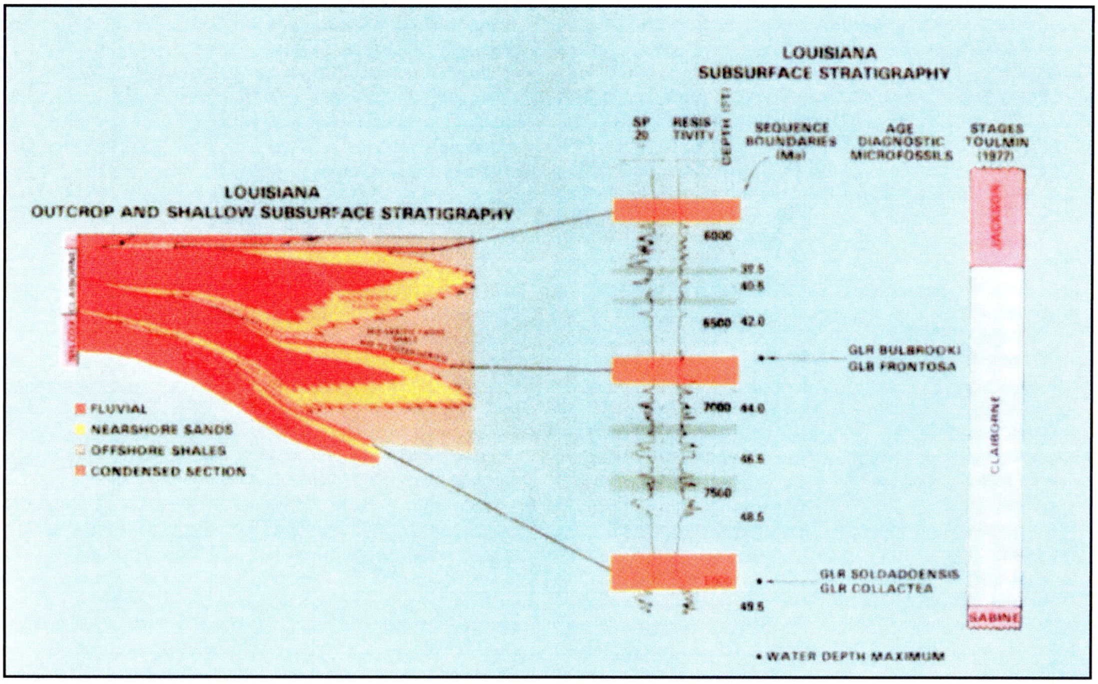

Plate 7–121 *Marine–condensed Sections and Their Relationship with the Stratigraphic Succession*

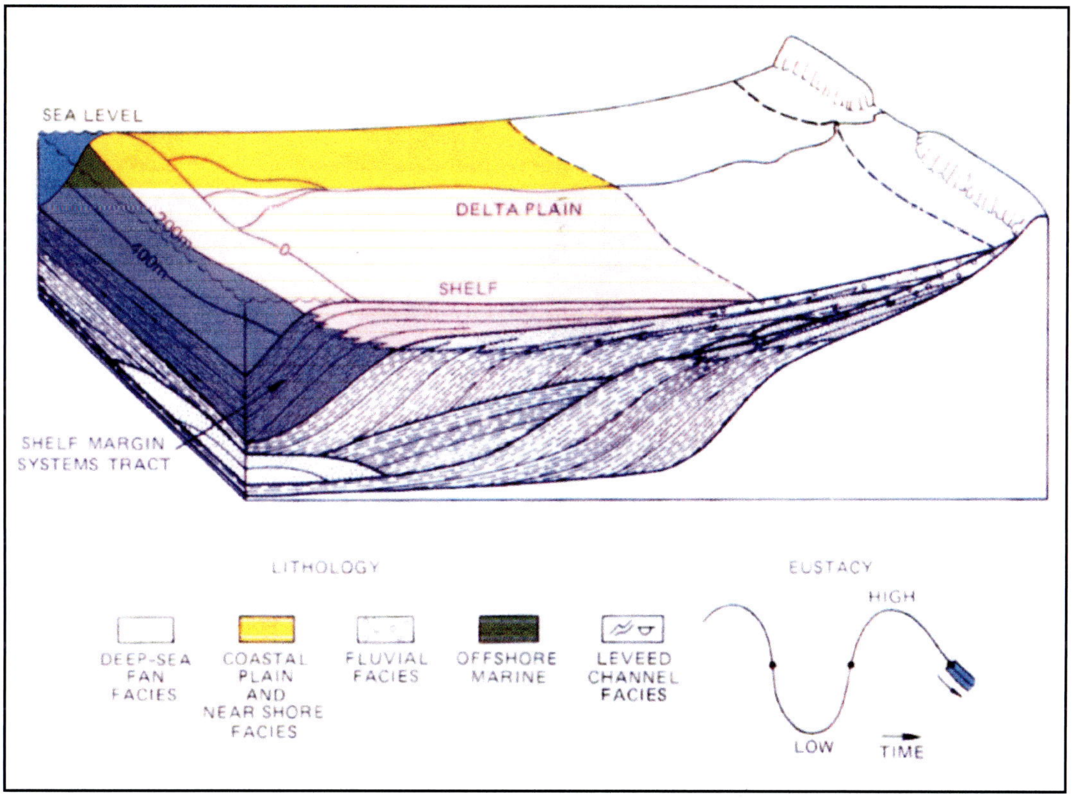

Plate 7–123 *Shelf-margin Systems Tract*

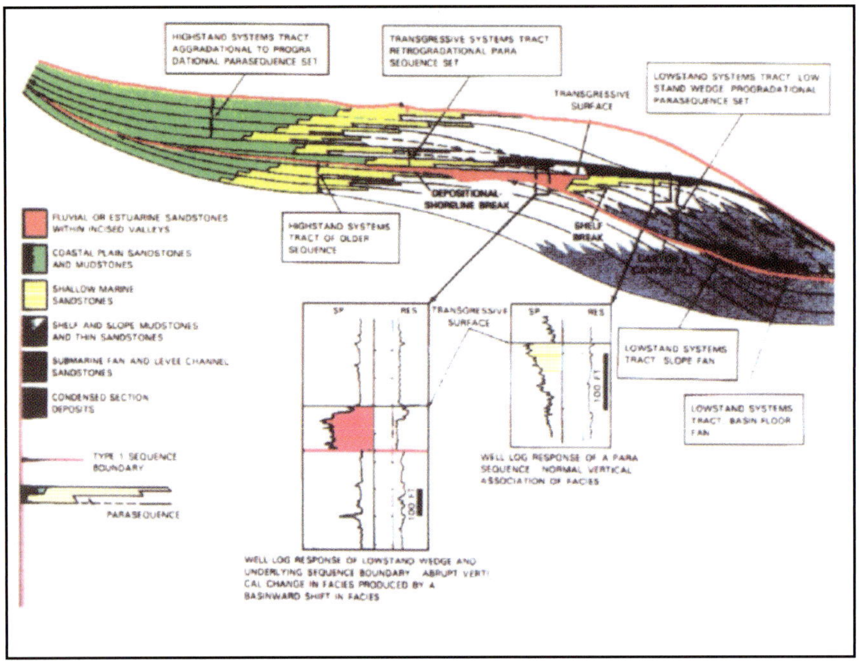

Plate 7–126 *Stratal Pattern in Type 1 Sequence. Stratal Pattern in Type 1 Sequence Deposited in the Basin with Shelf Break.*

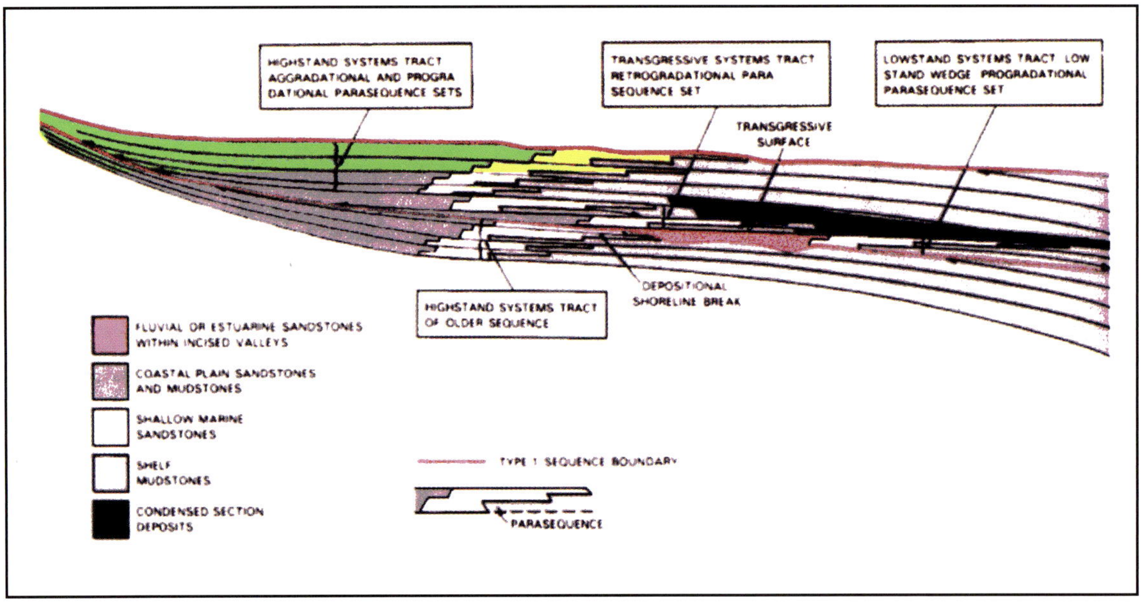

Plate 7–127 *Stratal Pattern in Type 1 Sequence Deposited in a Basin with Ramp Margin*

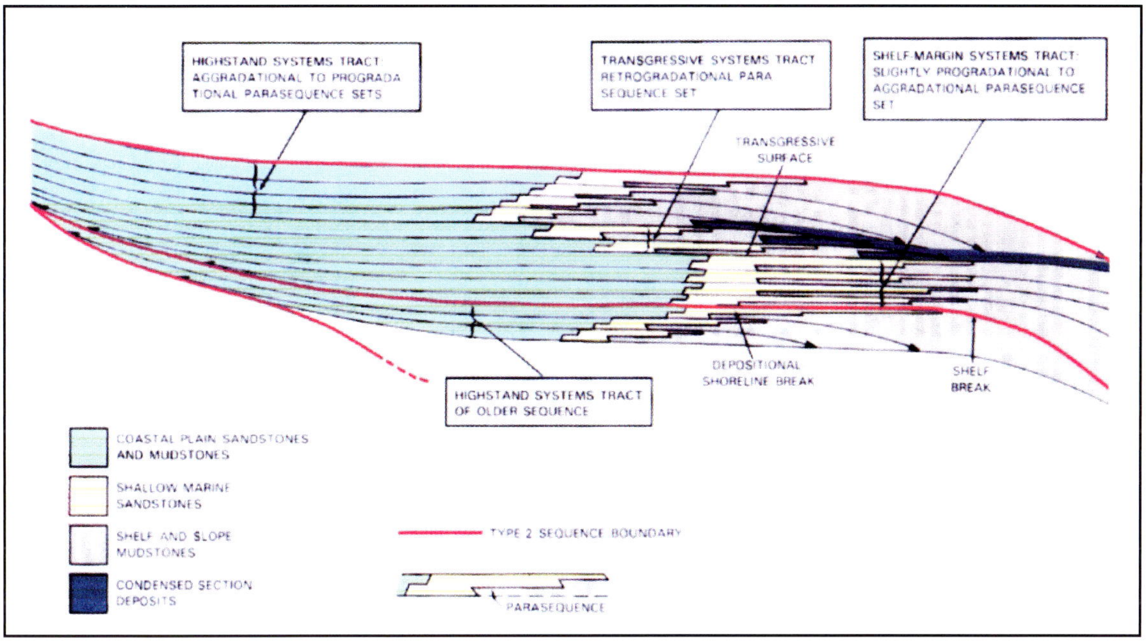

Plate 7–128 *Type 2 Sequence Boundary*

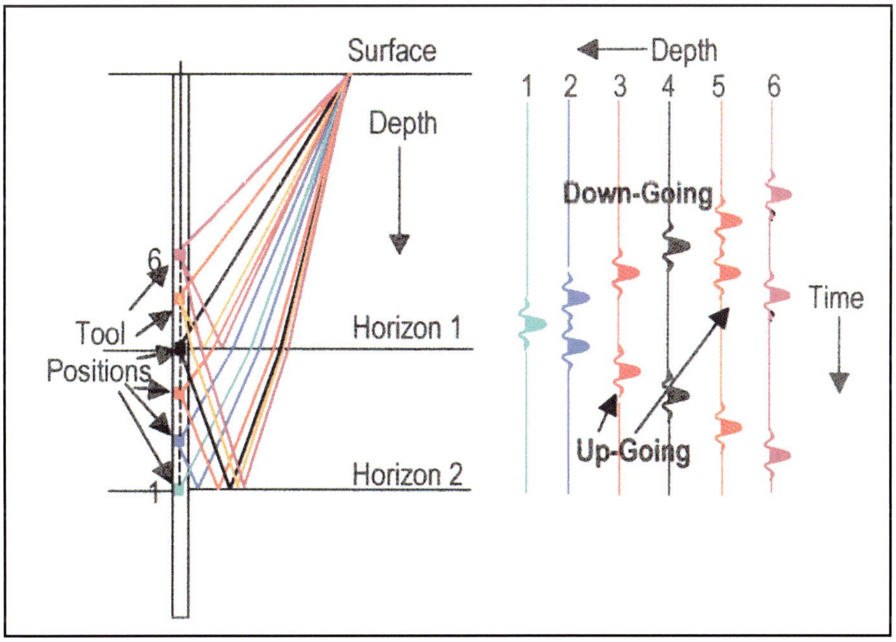

Plate 7–142 *Up- and Down-going Events*

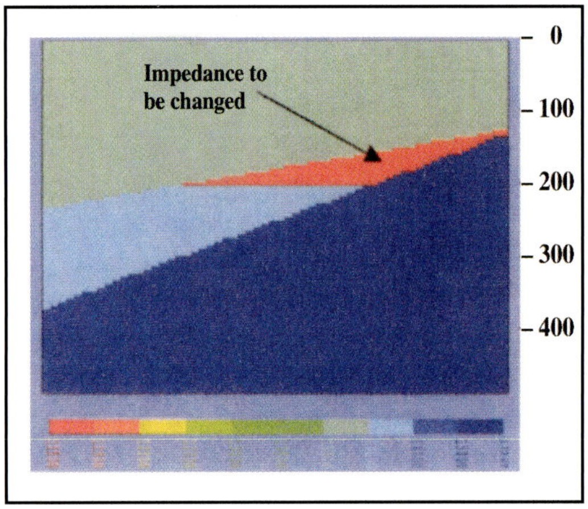

Plate 7–171 *Wedge Model with Gas Cap*

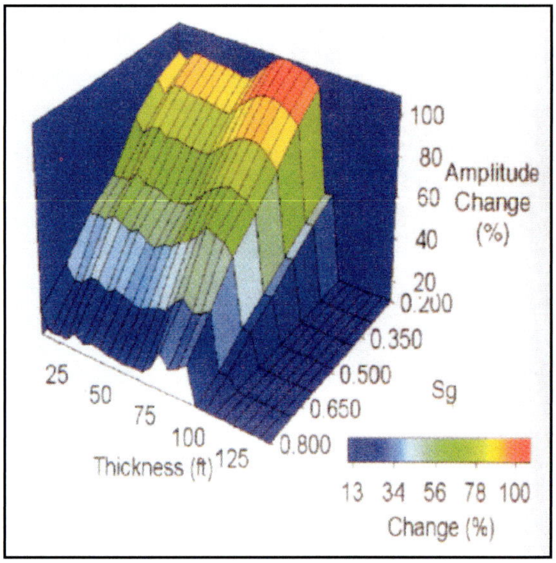

Plate 7–172 *Relationships among Reservoir Thickness, Gas Saturation and Amplitude Change*

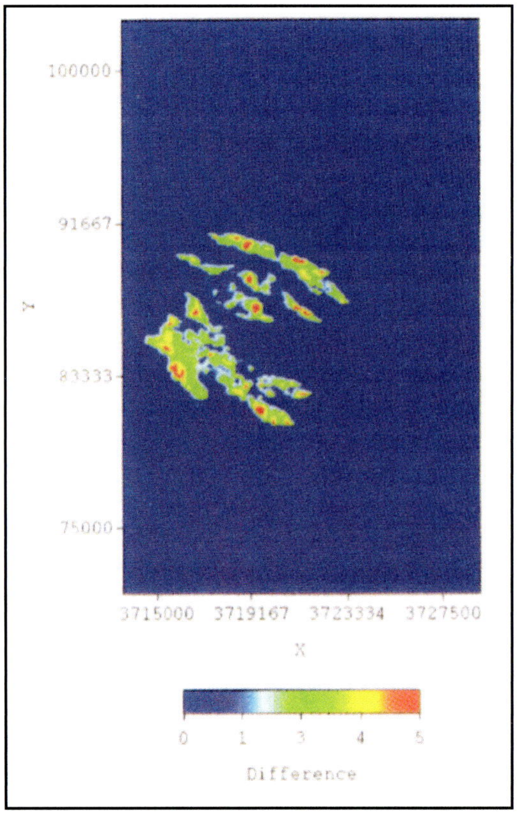

Plate 7–178 *Seismic Difference after Matching with Cumulative Production*

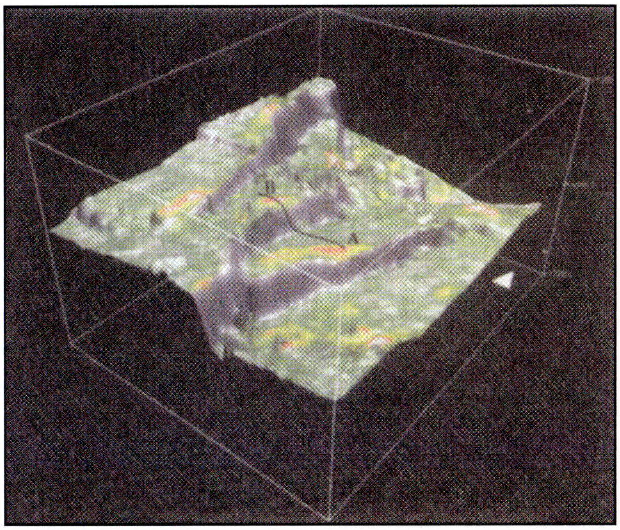

Plate 7–180 *Perspective View of the Sand Structure Containing the Currently Producing 4500-ft Reservoir*

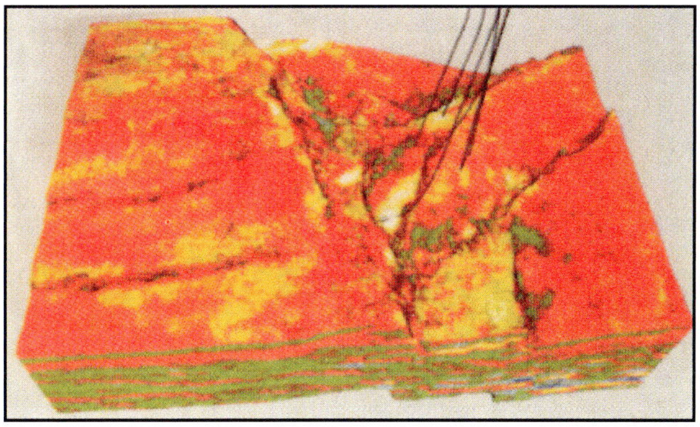

Plate 7–182 *Inverted Legacy Data Volume Showing Acoustic Impedance 12 ms Below the Top of the Tracked 4500-ft Horizon. Green Indicates High Impedances (Shales), Red Indicates Intermediate Impedances (Water Sands), and Yellow Indicates Low Impedances (Oil Sands).*

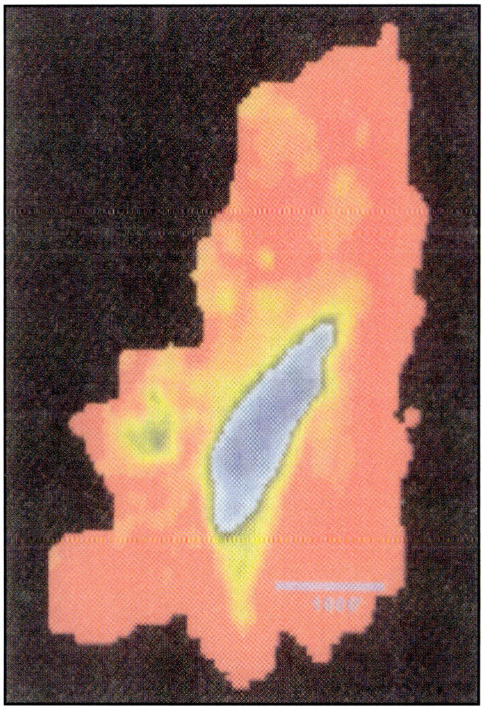

Plate 7–183 *Time-lapse Difference Mapped on the 4500-ft Reservoir. The Blue and Green Colors Indicate Significant Amplitude Changes.*

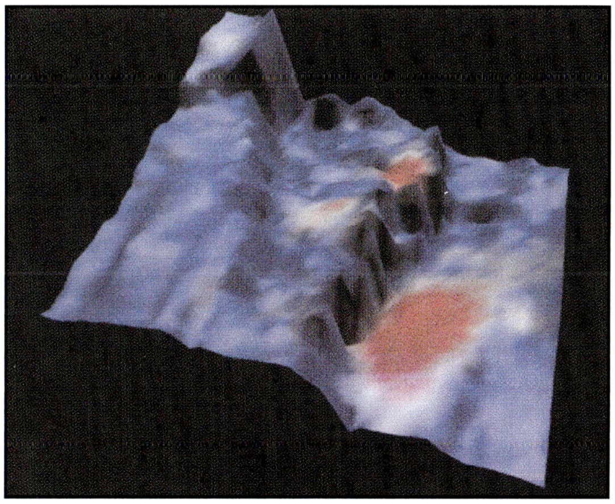

Plate 7–184 *Perspective View Showing Difference (Phase II–Phase I) for Both 4500-ft and Little Neighbor Reservoirs*

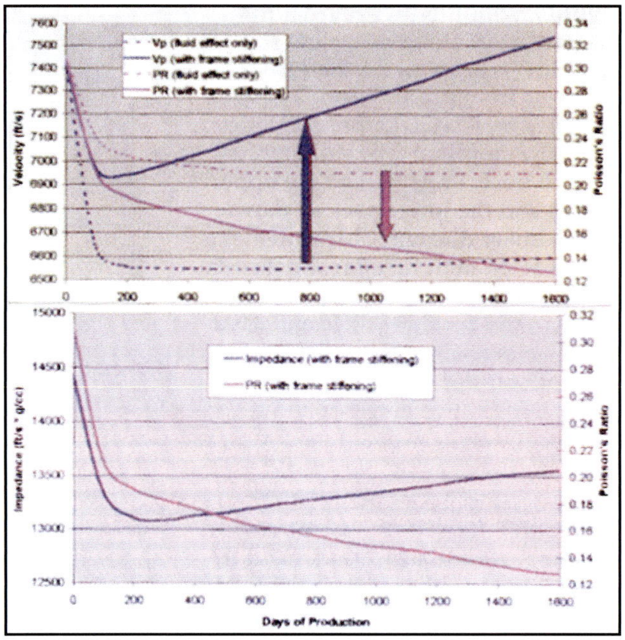

Plate 7–185 *Changes in P-wave Velocity (V_p), Poisson's Ratio (PR), and Acoustic Impedance with Time of Production*

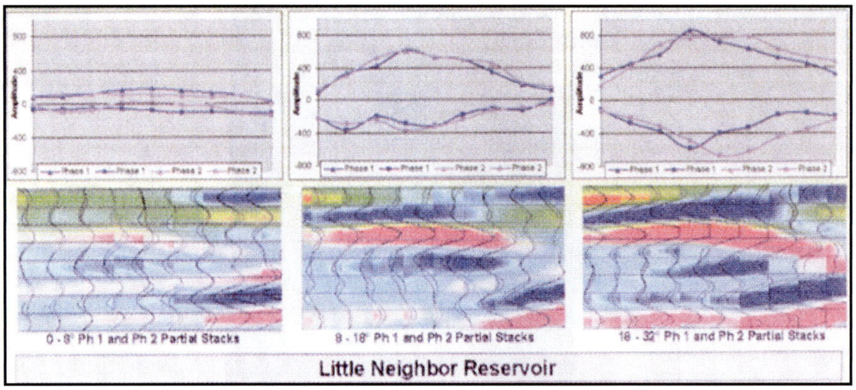

Plate 7–186 *Amplitudes Extracted from Partial-offset (Unmigrated) Stacked P-Wave Data for the 4500-ft Reservoir from Phases I and II.*

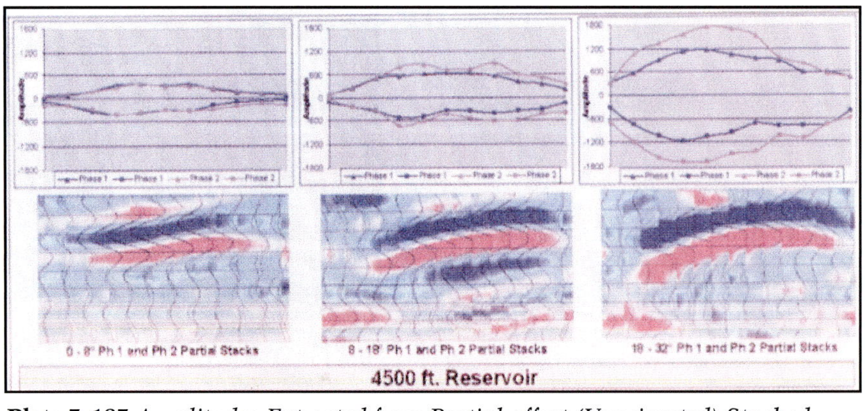

Plate 7–187 *Amplitudes Extracted from Partial-offset (Unmigrated) Stacked P-wave Data for the 4500-ft Reservoir from Phases I and II.*

Bibliography

Al-Sadi, H. N. *Seismic Exploration: Technique and Processing*. Birkhauser Verlag, 1982.

Angeleri, G. P. and Loinger, E. "Amplitude and Phase Distortions Due to Absorption in Seismograms and VSP." Paper presented at 44th annual meeting of EAE, 1982.

Anstey, N. A. "Attacking the Problems of the Synthetic Seismogram." *Geophysical Prospecting*, Vol. 8, 242–260.

Arya, V. K. and H. D. Holden. *A Geophysical Application: Deconvolution of Seismic Data.* Digital Signal Processing. N. Hollywood, CA, Western Periodicals, 1979, 324–338.

Backus, M. M. "The Reflection Seismogram in a Layered Earth." *AAPG Bulletin*, 1983, Vol. 67, 416–417.

Balch, A. H. and M. W. Lee. "Some Considerations on the Use of Downhole Sources in Vertical Seismic Profiles." Paper presented at 35th Annual SEG Midwestern Exploration Meeting, 1982.

Balch, A. H., M. W. Lee, J. J. Miller and R. T. Ryder. "The Use of Vertical Seismic Profiles in Seismic Investigations of the Earth." *Geophysics*, Vol. 47, 906–918.

Baranov, V. "Film Synthétique Avec Réflexions Multiples—Théorie et Calcul Pratique." *Geophysical Prospecting*. Vol. 8, 315–325.

Baranov, V. *Synthetic Seismograms With and Without Multiple Reflections.* AFTP, July 1959.

Barry, K. and T. R. Shugart. "Zero-Phase Seismic Sections." SEG preprint, Tulsa: SEG, 1975.

Barton, D. C. "The Seismic Method of Mapping Geologic Structure." *Geophysical Prospecting* (American Institute of Mining and Materials Engineering), Vol. 1, 572–624.

Berg, O. R. "Seismic Detection and Evaluation of Delta and Turbidite Sequences: Their Application to Exploration for the Subtle Trap." *AAPG Bulletin*, Vol. 66, 1271–1288.

Bernard, H. A., R. J. LeBlanc, and C. F. Major. "Recent and Pleistocene Geology of Southeast Texas." *Geology of the Gulf Coast and Central Texas and Guidebook of Excursions.* Houston Geological Society, 1962, 175–205.

Berryman, J. G. "Long-Wave Elastic Anisotropy in Transversely Isotropic Media." *Geophysics*, Vol. 44, 896–917.

Biot, M. A. "Propagation of Elastic Waves in a Cylindrical Bore Containing a Fluid." *Journal of Applied Physics*, Vol. 23, 997–1005.

Blair, D. P. "Dynamic Modeling of In-Hole Mounts for Seismic Detectors." *Geophysics Journal of the Royal Astronomy Society*, Vol. 69, 803–817.

Bois, P., M. Laporte, M. Laverne, and G. Thomas. "Well-to-Well Seismic Measurements." *Geophysics*, Vol. 37, 471–480.

Brewer, H. L. and J. Holtzscherer. "Results of Subsurface Investigations Using Seismic Detectors and Deep Bore Holes." *Geophysical Prospecting*, Vol. 6, 81–100.

Brown, A. R. "Interpretation of Three-Dimensional Seismic Data." Second Edition. *AAPG Memoir* 42, Tulsa, 1988.

Brown, L. F. Jr. and W. L. Fisher. "Seismic Stratigraphic Interpretation and Petroleum Exploration." AAPG, Continuing Education Course Note Series, No. 16, 1980.

Burg, J. P. *Maximum Entropy Spectrum Analysis.* PhD dissertation, Stanford University, Stanford, CA, 1975.

Busch, D. A. and Link, D. A. *Exploration Methods for Sandstone Reservoirs.* Tulsa: OGCI Publications, 1985.

Butler, D. K. and J. R. Curro. "Crosshole Seismic Testing—Procedures and Pitfalls." *Geophysics*, Vol. 46, 23–29.

Chapman, W. L., G. L. Brown, and D. W. Fair. "The Vibroseis System, a High Frequency Tool." *Geophysics*, Vol. 46, 1657–1666.

Cheng, C. H. and M. N. Toksoz. "Elastic Wave Propagation in a Fluid-Filled Borehole and Synthetic Acoustic Logs." *Geophysics*, Vol. 46, 1042–1053.

Cheng, C. H. and M. N. Toksoz. "Generation, Propagation and Analysis of Tube Waves in a Borehole." Paper P, Trans. SPWLA 23rd Annual Logging Symposium, Vol. I., 1982.

Cheng, C. H. and M. N. Toksoz. "Tube Wave Propagation and Attenuation in a Borehole." Paper presented at Massachusetts Institute of Technology Industrial Liaison Program Symposium, Houston, 1981.

Cherry, J. T. and K. H. Waters. "Shear-wave Recording Using Continuous Signal Methods." *Geophysics*, Vol. 33, 229–239.

Chun, J., D. G. Stone, and C. A. Jacewitz. "Extrapolation and Interpolation of VSP Data." Tulsa: Seismograph Service Companies Report, 1982.

Claerbout, J. F. *Fundamentals of Geophysical Data Processing.* McGraw-Hill, 1976, pp 136, 143, 157–159.

Claerbout, J. F. *Imaging the Earth's Interior,* Blackwell Scientific Publication, 1985.

Collins, F. and C. C. Lee. "Seismic Wave Attenuation Characteristics from Pulse Experiment." *Geophysics*, Vol. 21, 16–40.

Courtier, W. H. and H. L. Mendenhall. "Experiences with Multiple Coverage Seismic Methods." *Geophysics*, Vol. 32, 230–258.

Cramer, P. W., "Reservoir Development Using Offset VSP Techniques in the Denver-Julesburg Basin." *Journal of Petrochemical Technology*, February 1988, 197–205.

Crampin, S. "A Review of Wave Motion in Anisotropic and Cracked Elastic Media." *Wave Motion*, Vol. 3, 343–391.

Crampin, S. "Evaluation of Anisotropy by Shear-wave Splitting." *Geophysics*, Vol. 50, 142–152.

Crawford, J. M., W. E. N. Doty, and M.R. Lee. "Continuous Signal Seismograph." *Geophysics*, Vol. 25, 95–105.

Danborn, S. H. and S. N. Domenico. *Shear-wave Exploration*, Society of Exploration Geophysicists, 1987.

Delaplanche, J., R. F. Hagemann, and P. G. C. Bollard. "An Example of the Use of Synthetic Seismograms." *Geophysics*, Vol. 28, 842–854.

Dalley, R. M. et al. "Dip and Azimuth Displays for 3D Seismic Interpretation." *First Break,* March 1989, 86–95.

Dennison, A. T. "An Introduction to Synthetic Seismogram Techniques." *Geophysical Prospecting,* Vol. 8, 231–241.

DeVoogd, N. and H. Den Rooijen. "Thin Layer Response and Spectral Bandwidth." *Geophysics*, Vol. 48, 12–18.

DiSiena, J. P. and J. E. Gaiser. "Marine Vertical Seismic Profiling." Paper OTC 4541, Offshore Technology Conference, 1983, 245–252.

Dix, C. H. "Seismic Velocities from Surface Measurements." *Geophysics*, Vol. 20, 68–86.

Dobrin, M. B. *Introduction to Geophysical Prospecting*. New York: McGraw-Hill International, 1960.

Domenico, S. N. "Elastic Properties of Unconsolidated Porous Sand Reservoirs." *Geophysics*, Vol. 42, 1339–1369.

Douze, E. J. "Signal and Noise in Deep Wells." *Geophysics*, Vol. 29, 721–732.

Dunkin, J. W. and F. K. Levin. "Isochrons for a Three-dimensional System." *Geophysics*, Vol. 36, 1099–1137.

Durschner, H. "Synthetic Seismograms from Continuous Velocity Logs." *Geophysical Prospecting*, Vol. 6, 272–284.

Edelman, H. A. K. "Shover-shear-wave Generation by Vibration Orthogonal to the Polarization." *Geophysical Prospecting*, Vol. 29, 541–549.

Faust, L. Y. "A Velocity Function Including Lithologic Variation." *Geophysics*, Vol. 18, 271–288.

Fisher, W. L. and L. F. Brown. "Clastic Depositional Systems—A Generic Approach to Facies Analysis: An Annotated Outline and Bibliography." The University of Texas at Austin Bureau of Economic Geology, 1972.

Fontaine, J. M., R. Cussey, J. Lacase, and L. Yapaudjia. "Seismic Interpretation of Carbonate Depositional Environments." *AAPG Bulletin*, 1987, 281–297.

Futterman, W. I. "Dispersive Body Waves." *Journal of Geophysical Research*, Vol. 67, 5279–5291.

Gaizer, J. E. and J. P. DiSiena. "VSP Fundamentals that Improve CDP Data Interpretation." Paper S12.2, 52nd Annual International Meeting of SEG, Technical Program Abstracts, 1982, 154–156.

Galperin, E. I. "Vertical Seismic Profiling." SEG, *Special Publ.* No. 12, Tulsa, 1974.

Gardner, D. H. "Measurement of Relative Ground Motion in Reflection Recording." *Geophysics*, Vol. 3, 40–45.

Gardner, G. H. F., L. W. Gardner, and A. R. Gregory. "Formation Velocity and Density—The Diagnostic Basics for Stratigraphic Traps." *Geophysics*, Vol. 39, 770–780.

Gary, M., R. McAfee, Jr., and C. L. Wolf. *Glossary of Geology,* American Geological Institute, Washington: 1987.

Gassaway, G. S. and H. J. Richgels. "SAMPLE, Seismic Amplitude Measurement for Primary Lithology Estimation," 53rd SEG Meeting, Las Vegas, 1983, 610–613.

Gelfand, V., P. Ng, and K. Larner. "Seismic Lithologic Modeling of Amplitude-versus-offset Data," 56th SEG Meeting, Houston, 1986, 332–334.

Gerritsma, P. H. A. "Time to Depth Conversion in the Presence of Structure." *Geophysics*, Vol. 42, 760–772.

Geyer, R. L. and S. T. Martner. "SH-waves from Explosive Source." *Geophysics*, Vol. 34, 893–905.

Giles, B. F. "Pneumatic Acoustic Energy Source." *Geophysical Prospecting*, Vol. 16, 21–53.

Godfrey, L. M., J. D. Stewart, and F. Schweiger. "Application of Dinoseis in Canada." *Geophysics*, Vol. 33, 65–77.

Goupillaud, P. L. "An Approach to Inverse Filtering of Near-surface Layer Effects from Seismic Records." *Geophysics*, Vol. 26, 754–760.

Greaves, R. J. and T. J. Flup. "Three-dimensional Seismic Monitoring of an Enhanced Oil Recovery Process." *Geophysics*, Vol. 52, 1175–1187.

Griffiths, D. H. and R. F. King. *Applied Geophysics for Engineers and Geologists*. London: Pergamon, 1965.

Haq, B. U., J. Hardenbol, and P. R. Vail. "Chronology of Fluctuating Sea Levels Since the Triassic. *Science*, V. 235, 1156–1166.

Hardage, B. A. "A New Direction in Exploration Seismology is Down." *The Leading Edge*, Vol. 2, No. 6, 49–52.

Hardage, B. A. "An Examination of Tube Wave Noise in Vertical Seismic Profiling Data." *Geophysics*, Vol. 46, 892–903.

Haskey, P., P. Duke, and J. Fargher. "Which 3-D?" 2nd Latin American Congress on Hydrocarbons, Rio de Janeiro, 1988.

Hill, N. R. and I. Lerche "Acoustic Reflections from Undulating Surfaces." *Geophysics*, Vol.51, 2160–2161.

Hilterman, F. J. "Amplitudes of Seismic Waves—A Quick Look." *Geophysics*, Vol. 40, 745–762.

Hilterman, F. J. "Three-Dimensional Seismic Modeling." *Geophysics*, Vol. 35, 1020–1037.

Hindlet, F. "Thin Layer Analysis Using Offset/Amplitude Data." 56th SEG Meeting, Houston, Expanded Abstracts, 1986, 332–334.

Huang, X., R. Will, M. Khan, and L. Stanley. "Integration of Time-lapse Seismic and Production Data in a Gulf of Mexico Gas Field." *The Leading Edge*, Vol. 20, 278-289

Jervey, M. T. *Quantitative Geological Modeling of Siliciclastic Rock Sequences and Their Seismic Expression*. 1988.

Jolly, R. N. "Deep-hole Geophone Study in Garvin County, OK." *Geophysics*, Vol. 18, 662–670.

Jolly, R. N. "Investigation of Shear Waves." *Geophysics*, Vol. 21, 905–938.

Kearney, P. and M. Brooks. *An Introduction to Geophysical Exploration*. Blackwell Scientific Publication. 1984.

Kelly, K. R., R. W. Ward., S. Treitel, and R. M. Alford. "Synthetic Seismograms: A Finite Difference Approach." *Geophysics*, Vol. 41, 2–27.

Kennett, P., R. L. Ireson, and P. J. Conn. "Vertical Seismic Profiles—Their Applications in Exploration Geophysics." *Geophysical Prospecting*, Vol. 28, 676–699.

Kluesner, D. F. "Champion Field: Role of Three-dimensional Seismic in Development of a Complex Giant Oilfield." *AAPG Bulletin*, 1988, 72, 207.

Koefoed, O. and N. DeVoogd. "The Linear Properties of Thin Layers, with an Application to Synthetic Seismograms over Coal Seams." *Geophysics*, Vol. 45, 1254–1268.

Lang, D. G. "Downhole Seismic Combination of Techniques Sees Nearby Features." *Oil and Gas Journal*, Vol. 77, No. 29, 63–66.

Lang, D. G. "Downhole Seismic Technique Expands Borehole Data." *Oil and Gas Journal*, Vol. 77, No. 28, 139–142.

Lash, C. C. "Investigation of Multiple Reflections and Wave Conversions by Means of Vertical Wave Test (Vertical Seismic Profiling) in Southern Mississippi." *Geophysics*, Vol. 47, 977–1000.

Lavergne, M. and C. Willm. "Inversion of Seismograms and Pseudo Velocity Logs." *Geophysics*, Vol. 25, 231–250.

LeBlanc, R. L. "Significant Studies of Modern and Ancient Deltaic Sediments." Broussard, M. L. (Ed), *Deltas, Models for Exploration,* Houston Geol. Soc., 1975, 13–85.

Lee, M. W. and A. H. Balch "Computer Processing of Vertical Seismic Profile Data." *Geophysics*, Vol. 48, 272–287.

Lee, M. W. and A. H. Balch. "Theoretical Seismic Wave Radiation from a Fluid-filled Borehole." *Geophysics*, Vol. 47, 1308–1314.

Levin, F. K., J. F. Bayhi, J. W. Dunkin, J. D. Lea, D. B. Moore, R. K. Warren, and G. H. Webster. "Developments in Exploration Geophysics 1969–1974." *Geophysics*, Vol. 41, 209–218.

Levin, F. K. and R. D. Lynn. "Deep Hole Geophone Studies." *Geophysics*, Vol. 23, 639–664.

Lidz, B. H. (Ed). "Sea-Level Change—An Integrated Approach." SEPM, No. 42, 71–108.

Lindseth, R. O. "The Seislog, A New Formation Logging Method." *AAPG–SEG Pacific Section*, Bakersfield, 1977.

Lindsey, P., N. Neidell, and F. Hilterman. *Interpretive Uses of Seismic Modeling with Emphasis on Stratigraphy and Seismic Resolution*. Tulsa: SEG, 1975.

Loutit, T. S., J. Hardenbol, P. R. Vail, and G. R. Baum. "Condensed Sections, The Key to Age Dating and Correlation of Continental Margin Sequences." Lidz, B. H., (Ed) *Sea-level change—an integrated approach*, SEPM, No. 42, 1988, 71–108.

Marr, J. D. and E. F. Zagst. "Exploration Horizons from New Seismic Concepts of CDP and Digital Processing." *Geophysics*, Vol. 32, 207–224.

Mayne, W. H. "Common Reflection Point Horizontal Stacking Techniques." *Geophysics*, Vol. 27, 927–938.

Mayne, W. H. "Practical Considerations in the Use of Common Reflection Point Technique." *Geophysics*, Vol. 32, 225–229.

Mayne, W. H. and R. G. Quay. "Seismic Signatures of Large Air Guns." *Geophysics*, Vol. 36, 1162–1173.

McCollum, B. and W. W. Larue. "Utilization of Existing Wells in Seismograph Work." Early Geophysical Papers, Vol. 12, 119–127. Also, *AAPG Bulletin*, Vol. 15, 1409–1417.

McQuillin, R., M. Bacon, and W. Barclay. *An Introduction to Seismic Interpretation*. Houston: Gulf Publishing Co., 1984.

Meissner, R. and M. A. Hegazy. "The Ratio of the PP- to SS-reflection Coefficient as a Possible Future Method to Estimate Oil and Gas Reservoirs." *Geophysical Prospecting,* Vol. 29, 533–540.

Mitchum, R. M. Jr. "Seismic Stratigraphic Expression of Submarine Fans." Berg, O. R. and E. G. Woolverton (Eds), *Seismic Stratigraphy II: An Integrated Approach: AAPG Memoir 39*, 1985, 117–138.

Mitchum, R. M. Jr. and P. R. Vail. "Seismic Stratigraphy and Global Changes in Sea Level, Part 7: Seismic Stratigraphic Interpretation Procedure." Payton, C. E. (Ed), *Seismic Stratigraphy—Applications to Hydrocarbon Exploration: AAPG Memoir* 26, 1977, 135–143.

Mitchum, R. M. Jr., P. R. Vail, and J. B. Sangree. "Seismic Stratigraphy and Global Changes in Sea Level, Part 6: Stratigraphic Interpretation of Seismic Reflection Patterns in Depositional Sequence." Payton, C. E. (Ed), *Seismic Stratigraphy—Applications to Hydrocarbon Exploration: AAPG Memoir* 26, 1977, 117–133.

Mitchum, R. M. Jr., P. R. Vail, and S. Thompson, III. "Seismic Stratigraphy and Global Changes in Sea Level, Part 2: The Depositional Sequence as a Basic Unit for Seismic Stratigraphic Analyses." Payton, C. E. (Ed), *Seismic Stratigraphy—Applications to Hydrocarbon Exploration: AAPG Memoir* 26, 1977, 53–62.

Neitzel, E. B. "Seismic Reflection Records Obtained by Dropping a Weight." *Geophysics*, Vol. 23, 58–80.

Nettleton, L. L. *Geophysical Prospecting for Oil*. New York: McGraw-Hill, 1940.

Ostrander, W. J. "Plane Wave Reflection Coefficients for Gas Sands at Non-normal Angles of Incidence." *Geophysics*, Vol. 49, 1637–1648.

Parr, J. O. Jr. and W. H. Mayne. "A New Method of Pattern Shooting." *Geophysics*, Vol. 20, 539–564.

Peacock, R. B. and D. M. Nash, Jr. "Thumping Technique Using Full Spread of Geophones." *Geophysics*, Vol. 27, 952–965.

Pennington, W. D., H. Acevedo, J. L. Haataja, and A. Minaeva. "Seismic Time-lapse Surprise at Teal South, That Little Neighbor Reservoir is Leaking." *The Leading Edge*, Vol. 20, 1172–1175.

Peterson, R. A., W. R. Fillipone, and F. B. Coker. "The Synthesis of Seismograms from Well Log Data." *Geophysics*, Vol. 20, 516–538.

Pittmen, W. C. III. "Relationship Between Eustacy and the Stratigraphic Sequences of Passive Margins." *Geol. Soc. Am. Bulletin*, V. 89, 1389–1403.

Posamentier, H. W., M. T. Jervey, and P. R. Vail. "Eustatic Controls on Clastic Deposition I." Lidz, B. H. (Ed), *Sea-level Changes—An Integrated Approach*: SEPM, No. 42, 125–154.

Posamentier, H. W., M. T. Jervey, and P. R. Vail. "Eustatic Controls on Clastic Deposition II." Lidz, B. H. (Ed), *Sea-level Changes—An Integrated Approach:* SEPM, No. 42, 125–154.

Poulter, T. C. "The Poulter Seismic Method of Geophysical Exploration." *Geophysics*, Vol. 15, 181–207.

Pullin, N., L. Matthews, and K. Hirsche. "Techniques Applied to Obtain Very High Resolution 3-D Seismic Imaging at a Athabasca Tar Sands Thermal Pilot." *The Leading Edge*, Vol. 6, 10–15.

Quarles, M. "Vertical Seismic Profiling—A New Seismic Exploration Technique." Paper presented at the 48th Annual International Meeting of SEG, 1978.

Rascoe, B. "Regional Stratigraphic Analysis of Pennsylvanian and Permian Rock in Western Mid-Continent, Colorado, Kansas, Oklahoma, and Texas." *AAPG Bulletin*, 1962, No. 8, 1345–1370.

Rice, R. B., et al. "Developments in Exploration Geophysics 1975–1980." *Geophysics*, Vol. 46, 1088–1099.

Rich J. L. "Three Critical Environments of Deposition and Criteria for Recognition of Rocks Deposited in Each of Them." *Geol. Soc. Am. Bulletin* (1951), V. 62, 1–20.

Riggs, E. D. "Seismic Wave Types in a Borehole." *Geophysics*, Vol. 20, 53–60.

Ritchie, W. "Role of the 3D Seismic Technique in Improving Oilfield Economics." *Journal of Petrochemical Technology*, July 1986, 777–786.

Robertson, J. D. "Reservoir Management Using 3D Seismic Data." *Journal of Petrochemical Technology*, 1986, 663–667.

Robinson, E. A. *Migration of Geophysical Data*. Boston: IHRDC, 1983.

Robinson, E. A. *Robinson-Treitel Reader*. Tulsa: Seisomograph Service Corp., 1969.

Robinson, E. A. *Seismic Velocity Analysis and the Convolutional Model*. Boston: IHRDC, 1983.

Robinson, E. A. *Structural Properties of Stochastic Processes*. New York: John Wiley and Sons, 1963.

Sangree, J. B., P. R. Vail, and R. M. Sneider. "Evolution of Facies Interpretation of the Shelf-Slope: Application of the New Eustatic Framework in the Gulf of Mexico." Offshore Technology Conference, 1988, 133–144.

Sangree, J. B., D. C. Waylett, D. E. Frazier, G. B. Amergy, and W. J. Fennesey. "Recognition of Continental-Slope Facies, Offshore Texas–Louisiana." Bouma, A. H., G. T. Moore, and J. M. Coleman. (Eds), *Framework, Facies, and Oil-trapping Characteristics of the Upper Continental Margin*: AAPG Studies in Geology, No. 7, 87–116.

Sangree, J. B. and J. M. Widmier. "Interpretation of Depositional Facies from Seismic Data." *Geophysics*, Vol. 44, No. 2, and 131–160.

Sangree, J. B. and J. M. Widmier. "Seismic Stratigraphy and Global Changes in Sea Level, Part 9: Seismic Interpretation of Clastic Depositional Facies." C. E. (Ed), *Seismic Stratigraphy—Applications to Hydrocarbon Exploration: AAPG Memoir* 26, 1977, 165–184.

Sarg, J. F. "Carbonate Sequence Stratigraphy." Lidz, B. H. (Ed) *Sea-level Change—An Integrated Approach*. SEPM, No. 42, 1988, 155–181.

Schoenberger, M. and F. K. Levin, "Reflected and Transmitted Filter Functions for Simple Subsurface Geometries." *Geophysics*, Vol. 41, 1305–1317.

Sengbush, R. L., P. L. Laurence, and F. J. McDonal, "Interpretation of Synthetic Seismograms." *Geophysics*, Vol. 26, 138–157.

Sheriff, R. E. *Encyclopedic Dictionary of Exploration Geophysics*. Society of Exploration Geophysicists, Tulsa: 1961.

Sheriff, R. E. "Factors Affecting Amplitudes—A Review of Physical Principles in Lithology and Direct Detection of Hydrocarbons Using Geophysical Methods." Symposium of the Geophysical Society of Houston, 1973. Also, *Geophysical Prospecting*, Vol. 25, 125–138.

Shuey, R. T. "A Simplification of Zoeppritz Equations." *Geophysics*, Vol. 50, 609–614.

Shultze-Gatterman, R. "Physical Aspects of the Air Pulser as a Seismic Energy Source." *Geophysical Prospecting*, Vol. 20, 155–192.

Sloss, L. L. "Sequences in the Cratonic Interior of North America." *Geological Society of America Bulletin*, Vol. 74, 93–113.

Stephens, R. B. and P. Sheng. "Acoustic Reflections from Complex Strata." *Geophysics*, 1100–1107.

Stewart, R. R., R. M. Turpening, and M. N. Toksoz. "Study of a Subsurface Fracture Zone by Vertical Seismic Profiling." *Geophy. Res. Lett.*, Vol. 9, 1132–1135.

Stone, D. G. *Estimation of Reflection Coefficients from Seismic Data*. Tulsa: SEG, 1977.

Stone, D. G. *Robust Wavelet Estimation by Structural Deconvolution*. Tulsa: SEG, 1973.

Stone, D. G. "VSP—The Missing Link." Paper presented at the VSP short course sponsored by the SE Geophysicists Society in New Orleans, 1981.

Stone, D. G. and J. L. Atherton. *Decomposition of Seismic Trace.* Tulsa: SEG preprint, 1979.

Tatham, R. H. "V_P/V_S and Lithology." *Geophysics*, Vol. 47, 336.

Tatham, R. H. and M. D. McCormack. *Multicomponent Seismology in Petroleum Exploration*, SEG, 1991.

Tatham, R. H. and P. L. Stoffa. "V_P/V_S—A Potential Hydrocarbon Indicator." *Geophysics*, Vol. 41, 837–849.

Treitel, S. "Seismic Wave Propagation in Layered Media in Terms of Communication Theory." *Geophysics*, Vol. 31, 17–32.

Treitel, S. and K. Peacock. "Predictive Deconvolution: Theory and Practice." *Geophysics*, Vol. 34, 155–169.

Trorey, A. W. "Theoretical Seismograms with Frequency and Depth Dependent Absorption." *Geophysics*, Vol. 27, 766–785.

Vail, P. R., J. N. Bubb, W. G. Hatlelid, R. M. Mitchum, J. B. Sangree, S. Thompson III, R. G. Todd, and J. M. Widmier. "Seismic Stratigraphy and Global Changes of Sea Level, Parts 1–11." Payton, C. E. (Ed), *Seismic Stratigraphy—Application to Hydrocarbon Exploration: AAPG Memoir 26*, 1977, 49–212.

Vail, P. R., J. Hardenbol, and R. G. Todd. "Jurassic Unconformities, Chronostratigraphy, and Sea-level Changes from Seismic Stratigraphy and Biostratigraphy." Schlee, J. S. (Ed), *Inter-regional Unconformities and Hydrocarbon Accumulation: AAPG Memoir 36*, 1984, 129–144.

Vail, P. R., R. M. Mitchum, Jr., and S. Thompson III. "Seismic Stratigraphy and Global Changes of Sea Level, Part 3: Relative Changes of Sea Level from Coastal Onlap." Payton, C. E. (Ed), *Seismic Stratigraphy—Application to Hydrocarbon Exploration: AAPG Memoir 26*, 1977, 49–212.

Vail, P. R., R. M. Mitchum, Jr. and S. Thompson III. "Seismic Stratigraphy and Global Changes of Sea Level, Part 4: Global Cycles of Relative Changes of Sea Level." Payton, C. E. (Ed), *Seismic Stratigraphy—Application to Hydrocarbon Exploration: AAPG Memoir 26*, 1977, 83–97.

Vail, P. R. and J. B. Sangree. "Sequence Stratigraphy Interpretation of Seismic, Well, and Outcrop Data Workbook." Short Course, Bureau of Economic Geology, Austin, Texas, April 8–9, 1988.

Vail, P. R. and R. G. Todd. "North Sea Jurassic Unconformities, Chronostratigraphy, and Sea Level Changes from Seismic Stratigraphy." Proceedings of the Petroleum Geology of the Continental Shelf, Northwest Europe, 1981, 216–235.

Vail, P. R., R. G. Todd, and J. B. Sangree. "Seismic Stratigraphy and Global Changes of Sea Level, Part 5: Chronostratigraphic Significance of Reflections." Payton, C. E. (Ed), *Seismic Stratigraphy—Application to Hydrocarbon Exploration: AAPG Memoir 26*, 1977, 99–116.

van Sandt, D. R. and F. K. Levin. "A Study of Cased and Open Holes for Deep Seismic Detection." *Geophysics*, Vol. 28, 8–13.

Van Wagoner, J. C., R. M. Mitchum Jr., H. W. Posamentier and P. R. Vail. "Key Definitions of Seismic Stratigraphy." Bally, A. W. (Ed), *Atlas of Seismic Stratigraphy*, Vol. 1, AAPG, 11–14.

Walton, G. G. "Three-dimensional Seismic Method." *Geophysics*, Vol. 37, 417–430.

Water, K. H. *Reflection Seismology, A Tool for Energy Resource Exploration.* New York: John Wiley and Sons, 1987.

Wayland, R. J. and D. Lee. "Seismic Mapping of EOR Processes." *The Leading Edge*, 1986, 36–40.

White, R. E. and P. N. S. O'Brien. "Estimation of the Primary Seismic Pulse." *Geophysical Prospecting*, Vol. 22, 627–651.

Widess, M. B. "How Thin is a Thin Bed?" *Geophysics*, V. 38, 1976.

Wright, J. "Reflection Coefficients at Pore-fluid Contacts as a Function of Offset." *Geophysics*, Vol. 51, 1858–1860.

Wuenschel, P. C. "Seismogram Synthesis Including Multiples and Transmission Coefficients." *Geophysics*, Vol. 25, 106–219.

Wyatt, K. D. "Synthetic Vertical Seismic Profile." *Geophysics*, Vol. 46, 880–991.

Wyatt, S. B. "The Propagation of Elastic Waves Along a Fluid-filled Annular Region." Master of Science Thesis, University of Tulsa, 1979.

Yilmaz, O. and R. Chambers. "Migration Velocity Analysis by Wave Field Extrapolation." *Geophysics*, Vol. 49, 1664–1674.

Yilmaz, O. *Seismic Data Processing*. SEG, 1987.

Yoshimura, M., S. Fujii, K. Tanaka, and K. Morita. "On the Relationship Between P- and S-wave Velocities in Soft Rock." Paper given at SEG annual international meeting, Dallas, 1982.

Young, G. B. and L. W. Braile. "A Computer Program for the Application of Zoeppritz's Amplitude Equations and Knott's Energy Equations." *Bulletin of the Seismological Society of America,* Vol. 66 (6), 1881–1885.

Yu, G. "Offset-amplitude Variation and Controlled-amplitude Processing." *Geophysics*, Vol. 50, 2697–2708.

Index

A

B

C

D

E

F

G

J

K

L

M

N

O

P

Q

R

T

U

V

W

Y

Z